Specialist Periodical Reports Online

Application for Free Access to Electronic Chapters

Organophosphorus Chemistry
Volume 43

Customers purchasing a print volume are now entitled to free site-wide access to the electronic version of that title. (For the definition of a site, please consult **www.rsc.org/subagree**)

To apply for free access, please complete and return this form.

Contact Name: _____

Job Title: _____

Organisation: _____

Address: _____

_____

_____

_____

Town: _____

Country: _____

Post/Zip Code: _____

E-mail: _____

Telephone: _____

– Side 1 –   Continued over

– Side 2 –

Please give IP addresses of the site overleaf in the format w.x.y.z.
Wildcards and ranges are allowed

e.g.   for Class B 128.128.*.*
        for Class C 192.192.192.*
        for a range 123.456.1-99.*

Please put each entry on a new line:

_____

_____

_____

_____

(please use a continuation page if necessary)

I have read and agree to the terms of the Electronic Information Licence Agreement at **www.rsc.org/subagree**

Signed: _____   Date: _____

☐ Please contact me about access for additional sites outside the site definition in the Electronic Information Licence Agreement

☐ Please arrange username/password access and not IP address control

Please return this form by non-electronic means to the Royal Society of Chemistry at the address below.

(Photocopies or facsimiles are not acceptable.)

Books Department
Royal Society of Chemistry | Thomas Graham House
Science Park | Milton Road | Cambridge | CB4 0WF | UK

T +44(0)1223 420066 | F +44(0)1223 420247
E SPR@rsc.org   Registered charity number 207890

> The Royal Society of Chemistry will store the information you supply on its electronic records in order that information about its activities, products and services may be sent to you by mail, telephone, email or fax.
>
> If you **DO NOT** wish to receive information, please put a tick in the box ☐.

Organophosphorus Chemistry

Volume 43

A Specialist Periodical Report

Organophosphorus Chemistry

Volume 43

A Review of the Literature Published between
January 2012 and August 2013

Editors
D. W. Allen, *Sheffield Hallam University, Sheffield, UK*
J. C. Tebby, *Staffordshire University, Stoke-on-Trent, UK*
D. Loakes, *Laboratory of Molecular Biology, Cambridge, UK*

Authors
P. Bałczewski, *Polish Academy of Sciences, Łódź, Poland and Jan Długosz University in Czestochowa, Poland*
G. Keglevich, *Budapest University of Technology and Economics, Budapest, Hungary*
R. Narukulla, *Argenta, a Galapagos Company, Harlow, UK*
M. Noè, *Universita Ca' Foscari Venezia, Italy*
R. Pajkert, *Jacobs University Bremen gGmbH, Germany*
A. Perosa, *Universita Ca' Foscari Venezia, Italy*
G.-V. Röschenthaler, *Jacobs University Bremen GmbH, Germany*
M. Selva, *Universita Ca' Foscari Venezia, Italy*
J. Skalik, *Polish Academy of Sciences, Łódź, Poland*
R. N. Slinn, *University of Liverpool, UK*
F. F. Stewart, *Idaho National Laboratory, Idaho Falls, ID, USA*
Y.-Z. Xu, *Open University, Milton Keynes, UK*

If you buy this title on standing order, you will be given FREE access to the chapters online. Please contact sales@rsc.org with proof of purchase to arrange access to be set up.

Thank you.

ISBN: 978-1-84973-942-9
ISSN: 0306-0713
DOI: 10.1039/978-1-78262-397-7

A catalogue record for this book is available from the British Library

© The Royal Society of Chemistry 2014

All rights reserved

Apart from fair dealing for the purposes of research for non-commercial purposes or for private study, criticism or review, as permitted under the Copyright, Designs and Patents Act 1988 and the Copyright and Related Rights Regulations 2003, this publication may not be reproduced, stored or transmitted, in any form or by any means, without the prior permission in writing of The Royal Society of Chemistry, or in the case of reproduction in accordance with the terms of licences issued by the Copyright Licensing Agency in the UK, or in accordance with the terms of the licences issued by the appropriate Reproduction Rights Organization outside the UK. Enquiries concerning reproduction outside the terms stated here should be sent to The Royal Society of Chemistry at the address printed on this page.

Published by The Royal Society of Chemistry,
Thomas Graham House, Science Park, Milton Road,
Cambridge CB4 0WF, UK

Registered Charity Number 207890

For further information see our web site at www.rsc.org

Printed in the United Kingdom by CPI Group (UK) Ltd, Croydon, CR0 4YY, UK

Preface

David Allen,[a] David Loakes[b] and John Tebby[c]
DOI: 10.1039/9781782623977-FP005

This volume, No. 43 in the series, covers the literature of organophosphorus chemistry published in the period from January 2012 to January 2013, and continues our efforts in recent years to provide an up to date survey of progress in this topic which continues to generate a vast amount of research. The 19th International Conference on Phosphorus Chemistry was held in Rotterdam, The Netherlands, in 2012 and papers from this event have now appeared in issues 1–3 of volume **188** of *Phosphorus, Sulfur, Silicon*, (2013). We are pleased to announce that coverage of 'phosphonium salts and ylides' has resumed in this Specialist Periodical Review volume with a two-year review (covering 2011 and 2012) provided by a new team of authors led by Professor M. Selva of the University of Venice. Unfortunately, we have again been unable to secure coverage of the P(III) acid derivatives area but hope to make amends in the next volume.

As in recent years, the use of a wide range of tervalent phosphorus ligands in homogeneous catalysis has once again continued to be a major driver in the chemistry of both traditional P–C-bonded phosphines and also that of tervalent phosphorus acid derivatives. Interest has continued to grow in the application of tertiary phosphines as nucleophilic catalysts in the reactions of electrophilic unsaturated systems involved in new synthetic approaches. Interest has also continued in the reactions of sterically-crowded arylphosphine-arylboranes (Frustrated Lewis Pair (FLP) systems) in the activation of small molecules such as dihydrogen and carbon dioxide. Increased interest is also apparent in the chemistry of low coordination number phosphorus species, *e.g.*, phosphenium ions (R_2P:$^+$ and RP:$^{2+}$, and related monophosphorus cationic species) and also phosphinidenes (RP:). In phosphine chalcogenide chemistry, interest in the development of methods for their synthesis, and their applications as new components in opto-electronic devices, has shown considerable growth. The chemistry of phosphonium salts and related ylides has also shown remarkable activity, with particular reference to catalytic applications and, in particular, to the synthesis and applications of phosphonium salts as new types of ionic liquids that display higher thermal and electrochemical stabilities compared to related ammonium salts and also have potential as new solvents in organic synthesis and as stabilisers for nanoparticle systems.

2013 marked the 60th anniversary of the publication of the DNA double helix by Watson and Crick, and our knowledge and control of nucleic

---

[a]*Biomedical Research Centre, Sheffield Hallam University, Sheffield, S1 1WB, UK*
[b]*Medical Research Council, Laboratory of Molecular Biology, Hills Road, Cambridge, CB2 0QH, UK*
[c]*Division of Chemistry, Faculty of Sciences, Staffordshire University, Stoke-on-Trent, ST4 2DE, UK*

systems has undergone radical changes since then. Now we are able to understand far more complex systems, such as the ribosome, and have a much greater range of tools to understand them. This year we welcome a new member to our team, Dr. Yao Xu from the Open University, who contributed the chapter on mononucleotides. As a chemist, the chapter focuses largely on synthesis of mononucleotides and on some of their therapeutic applications. The chapter describes the synthesis of modified building blocks for the synthesis of oligonucleotides, either as phosphoramidites or nucleoside triphosphates, as well as a range of masked phosphates, nucleoside phosphoramidates and phosphonates, used most frequently as therapeutic agents. The synthesis and applications of nucleoside di- and poly-phosphates is also described.

The chapter on modified oligonucleotides is again the largest chapter in this volume, and attempts to review the vast number of modified backbones, sugars and nucleobases that have been incorporated into oligonucleotides, as well as some of their applications. Modified nucleobases continue to be one of the main areas of research for a broad range of applications including therapeutic, in understanding their role in biological systems, and to introduce some novel functionality into oligonucleotides. The section on aptamers and aptazymes is a two year review, and is more cursory but it is an ever increasing field of research, frequently carried out for the development of a diagnostic assay. The range of cargoes attached to oligonucleotides is vast, from fluorophores to modifications that allow the development of nanostructures and nanodevices. The section on X-ray crystallography and NMR is again a two year review, but developments in these techniques have allowed for solving the structures of more and more complex systems. A broad range of other techniques for structure determination is also described, in particular the various techniques used in electron microscopy.

Coverage of pentavalent phosphorus acid compounds reflects the literature concerning phosphoric, phosphonic and phosphinic acids and their derivatives, highlighting some of the most important developments. The interest in the area of synthesis of phosphorus (V) acids and their derivatives is large and their use in the synthesis of a broad range of organic systems has been described. Bulky phosphoric acid derivatives based on BINOL substituted in the 3,3′-positions have been successfully applied as catalysts in various reactions to afford high chemical yields and excellent stereoselectivities. In this chapter, a sub-section concerning the use of chiral phosphoric acids as catalysts in various chemical reactions has been included and reveals a continuing interest in this area. The interest in the area of phosphonic acids and their derivatives is greater than in the previous review period. A number of new transformations have been reported using phosphonic acids, including the total syntheses of biologically-active compounds, such as Tamiflu. In the area of phosphinic acids and their derivatives, a decreasing interest, as in previous review periods, has been observed, both in sections of synthesis, reactions and biological aspects.

There have been significantly fewer studies on the synthesis of novel pentacoordinated compounds although the asymmetric synthesis of

P-chiral pentacoordinated spirophosphoranes has been reviewed. Novel pentacoordinated phosphoranes were generated from phosphine-mediated cycloisomerization of alkynyl hemiketals as well as by the phosphorylation of aromatic hydrazides. More emphasis has been placed on understanding chemical and biochemical mechanisms involving pentacoordinated species. Examples include: semi-pinacol rearrangements of diols, reduction of phosphine oxides to the corresponding phosphines, formation of bis-phosphonium ylides, hydrolysis of phosphate triesters, zinc-catalysed cleavage and isomerization of uridine 3′-alkylphosphates and spontaneous reactivation of sarin-phosphorylated wild-type enzyme of human butyrylcholinesterase. Theoretical studies of the Wittig-reaction, have been based on the dynamic processes observed in the structure of 1,2-oxaphosphethanes. Studies of hexacoordinated compounds included oxidative addition to phosphorus(III) halides and the thermolysis of hexafluorophosphates, compounds bearing a transannular N–P bond as well as betaines and the chemistry of *meso*-triarylcorroles.

The ability to insert functional groups into phosphazenes facilitates numerous chemical attachments, such as drug conjugates and the development of synthetic tissue. Significant advances on immunoadjuvants based on polyphosphazene have been reviewed. In non-biological applications there have been advances in understanding of the controlled decomposition and thermal stability of phosphazenes. Novel elastomeric materials continue to emerge from the use of polyphosphazenes. Further developments of fire-resistant materials include a blend of hexa-(4-nitrophenoxy)cyclotriphosphazene and poly[ethylene terephthalate] (PET). There have been numerous studies of complexes of phosphazenes with metal centres to create new catalysts. The study of cyclomatrix 4,4′-sulfonyldiphenol structures has continued with a variety of new applications including the use of a bifunctional pendant group to create a highly cross-linked network of cyclotriphosphazene rings which condensed can form nanostructures. Computational methods continue to provide insight into the structure and conformation of complex three dimensional systems. For example, the Amsterdam Density Functional package was used to probe adducts of B, Al, Ga, In, Tl with the ring nitrogen atoms of a cyclotriphosphazene.

Most areas of physical methods have continued to expand – in particular theoretical and computational studies. Organophosphorus chemistry is well suited for study by modern aspects of NMR spectroscopy, often in combination with other physical methods. Highlights of other methods include the first detailed vibrational characterization of five analogues of *N*-benzylamino-(boronphenyl)methylphosphonic acids using FT-IR, FT-Raman, and surface-enhanced Raman spectroscopy (SERS), along with DFT/B3LYP theoretical calculations. Also a stable, crystalline, singlet *bis*(imidazolidin-2-iminato)phosphinonitrene has been isolated and a single-crystal XRD study showed that the phosphorus atom is in a planar environment, the P–N bond length being very short at 1.457 Å. Competitive fragmentation processes in mass spectrosopy were used to differentiate stereoisomers and selective deuterated analogues and product- and precursor-ion mass spectra allowed the elucidation of the fragmentation mechanisms.

CONTENTS

Cover

A selection of organophosphorus molecules. Image reproduced by permission of Dr David Loakes.

Preface   v

David Allen, David Loakes and John Tebby

Phosphines and related P–C-bonded compounds   1

D. W. Allen

    1   Introduction   1
    2   Phosphines   1
    3   p_π-Bonded phosphorus compounds   26
    4   Phosphirenes, phospholes and phosphinines   31
    References   34

Phosphine chalcogenides   52

G. Keglevich

    References   80

Phosphonium salts and P-ylides   85

Maurizio Selva, Alvise Perosa and Marco Noè

    1   Introduction   85
    2   Phosphonium salts   85
    3   Phosphonium-based ionic liquids (PILs)   95
    4   P-Ylides (Phosphoranes)   101
    References   108

Nucleotides and oligonucleotides: mononucleotides 117
Yao-Zhong Xu and Raman Narukulla

1	Nucleoside monophosphates	117
2	Nucleoside phosphoramidites and phosphoramidates	126
3	Nucleoside phosphonates	129
4	Dinucleoside phosphates and other nucleotides	133
5	Tri- and poly-phosphates	135
6	Summaries	141
	References	142

Nucleotides and nucleic acids; oligo- and poly-nucleotides 146
David Loakes

1	Introduction	146
2	Aptamers and (deoxy)ribozymes	176
3	Oligonucleotide conjugates	179
4	Nucleic acid structures	193
	Summary	200
	References	200

Quinquevalent phosphorus acids 246
P. Bałczewski and A. Bodzioch

1	Introduction	246
2	Phosphonic acids and their derivatives	284
3	Phosphinic acids and their derivatives	333
	References	340

Pentacoordinated and hexacoordinated compounds 348
Romana Pajkert and Gerd-Volker Röschenthaler

1	Introduction	348
2	Pentacoordinated phosphorus compounds	348
3	Hexacoordinated compounds	361
	References	365

Phosphazenes 366
Frederick F. Stewart

1	Introduction	366
2	Non-ionic phosphazenes for drug delivery	366
3	Phosphazenes for tissue engineering applications	370
4	Phosphazenes as immunoadjuvants	372
5	Thermal stability and decomposition	372
6	Elastomeric polyphosphazenes	374
7	Cyclotriphosphazenes	380

8	Phosphazenes in ring systems	385
9	Inorganic complexes	395
10	Nanomaterials	403
11	Computational methods	408
	Acknowledgement	409
	References	409

Physical methods 413
Robert N. Slinn

1	Introduction	413
2	Theoretical and computational chemistry methods	413
3	Nuclear magnetic resonance spectroscopy	423
4	Electron paramagnetic (spin) resonance spectroscopy	428
5	Vibrational (IR and Raman) spectroscopy	430
6	Electronic spectroscopy	431
7	X-ray diffraction (XRD) structural studies	434
8	Electrochemical methods	436
9	Acidities, basicities and thermochemistry	438
10	Mass spectrometry techniques	438
11	Chromatography and related separation techniques	440
12	Kinetics	442
	References	443

Abbreviations

BAD	Benzamide adenine dinucleotide
cDPG	Cyclodiphospho D-glycerate
CE	Capillary electrophoresis
CK	Creatine kinase
CPE	Controlled potential electrolysis
Cpmp	1-(2-chlorophenyl)-4-methoxylpiperidin-2-yl
CV	Cyclic voltammetry
DETPA	Di(2-ethylhexyl)thiophosphoric acid
DMAD	Dimethylacetylene dicarboxylate
DMF	Dimethylformamide
DMPC	Dimyristoylphosphatidylcholine
DRAMA	Dipolar restoration at the magic angle
DSC	Differential scanning calorimetry
DTA	Differential thermal analysis
ERMS	Energy resolved mass spectrometry
ESI-MS	Electrospray ionization mass spectrometry
EXAFS	Extended X-ray absorption ne structure
FAB	Fast atom bombardment
Fpmp	1-(2-uorophenyl)-4-methoxylpiperidin-2-yl
HPLC	High-performance liquid chromatography
LA-FTICR	Laser ablation Fourier Transform ion cyclotron resonance
MALDI	Matrix assisted laser desorption ionization
MCE	Micellar electrokinetic chromatography
MIKE	Mass-analysed ion kinetic energy
PAH	Polycyclic aromatic hydrocarbons
QDA	Hydroquinone-O,O′-diacetic acid
PMEA	9-[2-(phosphonomethoxy)ethyl] adenine
SATE	S-acyl-2-thioethyl
SIMS	Secondary ion mass spectrometry
SSAT	Spermidine/spermine-N1-acetyltransferase
SSIMS	Static secondary ion mass spectrometry
TAD	Thiazole-4-carboxamide adenine dinucleotide
tBDMS	tert-Butyldimethylsilyl
TFA	Triuoroacetic acid
TGA	Thermogravimetric analysis
TLC	Thin-layer chromatography
TOF	Time of ight
XANES	X-Ray absorption near edge spectroscopy

Phosphines and related P–C-bonded compounds

D. W. Allen

DOI: 10.1039/9781782623977-00001

1 Introduction

This chapter covers the literature published during 2012 relating to the above area, apart from a few papers from 2011 in less accessible journals which came to light in *Chemical Abstracts* in 2012. The number of papers published in 2012 is similar to that in 2011 and again it has been necessary to continue to be selective in the choice of publications cited. Nevertheless, it is hoped that most significant developments have been noted. The year under review has again seen the publication of a considerable number of review articles and many of these are cited in the various sections of this report. Although the use of a wide range of tervalent phosphorus ligands in catalysis continues to be a major driver in the chemistry of traditional P–C-bonded phosphines (and also that of tervalent phosphorus acid derivatives, usually covered in detail elsewhere in this volume), a noteworthy feature of the literature reviewed here is a significant increase in the number of papers reporting studies of the reactivity of phosphines, in particular those involving nucleophilic attack at a carbon atom of an electrophilic substrate. Recent general reviews of phosphine ligand chemistry relevant to the catalysis area have provided coverage of recent developments in the asymmetric synthesis of chiral phosphines, (and other classes of organophosphorus compound),[1] the synthesis of planar chiral phosphines derived from [2,2]paracyclophane,[2] routes to P-stereogenic oligomers, polymers and related cyclic phosphines,[3] and the stoichiometric and catalytic synthesis of alkynylphosphines.[4] Butler has published a 'practioner's perspective' on simple routes to ferrocenyl phosphine ligands.[5]

2 Phosphines

2.1 Preparation

2.1.1 From halogenophosphines and organometallic reagents. This route has continued to be applied widely, with most work involving the use of organolithium reagents. Although only a few reports of Grignard and related organomagnesium-based procedures have appeared, these reagents have found use, in combination with chlorophosphines, in the synthesis of new dibenzo-$7\lambda^3$-phosphanorbornadiene derivatives, *e.g.*, (1), using magnesium anthracene,[6] a range of new air-stable *P,N*-triarylphosphines bearing cyclic secondary amine substituents in the *ortho* position, *e.g.*, (2),[7] new chiral trialkylphosphines based on the

Biomedical Research Centre, Sheffield Hallam University, Sheffield, S1 1WB, UK.
E-mail: d.w.allen@shu.ac.uk

2,5-dialkylphospholane unit, *e.g.*, (**3**),[8] and the *P*-stereogenic benzodiphosphetane (**4**), easily transformable into chiral C_2-symmetric *o*-diphosphines, *e.g.*, (**5**), *via* alkylation with MeOTf, followed by addition of a nucleophile.[9] Buchwald's group has developed both Grignard-[10] and organolithium[11]-based routes to a series of new highly substituted phosphinobiphenyls, *e.g.*, (**6**), (the phosphination steps being catalysed by copper(I) chloride). Organolithium reagent-halogenophosphine routes have been widely employed in the synthesis of a variety of new phosphines, particularly those bearing a heteroaryl substituent either directly bound to phosphorus or as a C-substituent on an arylphosphine. New phosphines of these types include the imidazolylphosphines (**7**)[12] and the C-oxazolinyl-functionalised indolylphosphines (**8**),[13] the phosphinitoaryl-functional imidazolylphosphine (**9**),[14] and new phenothiazolinyldiphenylphosphines.[15] The Arduengo and Streubel groups have jointly reported the synthesis of new 4-phosphanylated 1,3-dialkylimidazole-2-thiones, *e.g.*, (**10**).[16] A range of new arylphosphines (**11**), bearing an *ortho*-(2-benzimidazolyl) substituent, has also been prepared.[17] Also reported are new dihydrophenophosphazines (**12**),[18] phosphines bearing cyclopentadienyl- substituents, *e.g.*, (**13**)[19] and (**14**),[20] a series of (ferrocenylethynyl)phosphines, $R_2P(C\equiv CFc)$,[21] various troticenylmonophosphines and carbaphospha-[2]-troticenophanes, *e.g.*, (**15**),[22] and new chiral camphane-based phosphinoarylcarboxamide,[23] *ortho*-sulfinylaryl[24]- and *ortho*-sulfonatoaryl[25]-phosphine ligands. Organolithium-halogenophosphine routes have also been developed for the synthesis of new atropoisomeric *P,N*-ligands, *e.g.*, (**16**),[26] the proton-sponge-phosphines (**17**), (having potential as reversibly-chargeable ligands for ESI-MS analysis),[27] a series of mono- and di-phosphino-biferrocenes,[28] cyclophosphazenes bearing phosphinoaryloxy substituents,[29] and a range of di- and poly-phosphino-ligands including the 4,5-diphosphinoimidazolium cation (**18**), precursor to a diphosphinocarbene ligand,[30] diphosphino-functional regioisomeric chiral diols, *e.g.*, (**19**),[31] and diphosphino-functional spiroketals (**20**),[32] open-chain triphosphines of the type $R^1Me_2SiCH_2P(PR^2{}_2)_2$, ($R^1$ = Me or Ph; R^2 = Me$_3$Si, Cy or Ph),[33] and the tetraphosphinobiphenyl (**21**).[34]

(1) R = Bu^t, (Me$_3$Si)$_2$N or Pr$^i{}_2$N

(2) X = CH$_2$, O, NH or NMe
    n = 0–2

(3) R = Cy, Bu^t, Pr^i, Bu or Me

(4)

(5) R = Me or Ph

(6) R = Ad, Bu^t or Cy

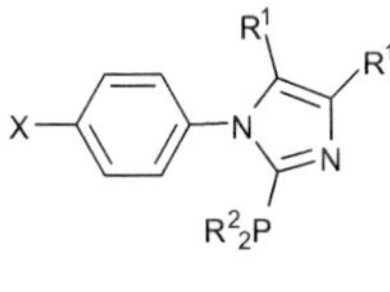

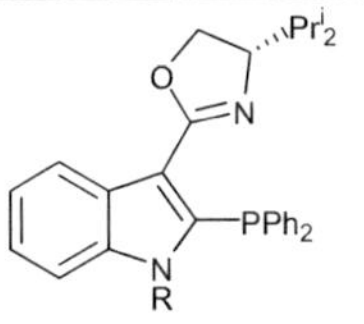

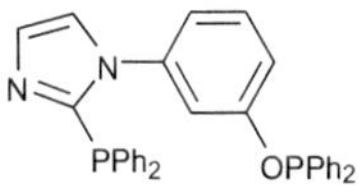

(7) R¹ = H or Me; R² = Ph, Cy or 2-furyl (8) R = Me, MOM, Bn, CONPr$^i_2$, Pr or Ph (9)

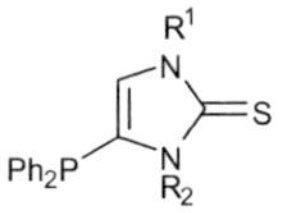

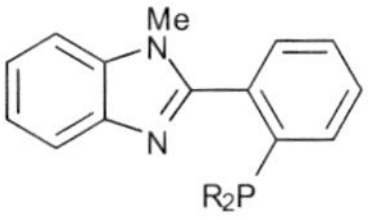

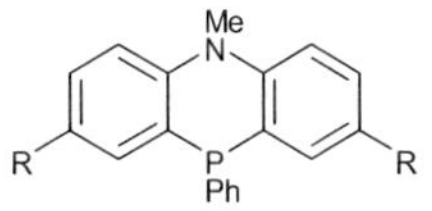

X = H, Br or I

(10) R¹ = Me, Ph, Pr^i, Bu or Bu^t (11) R = Pr^i, Cy or Ph (12) R = Me or Bu^t
R² = Me, Ph or Pr^i

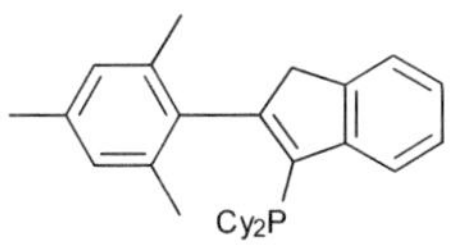

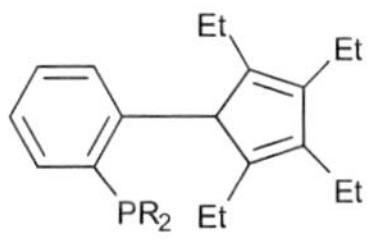

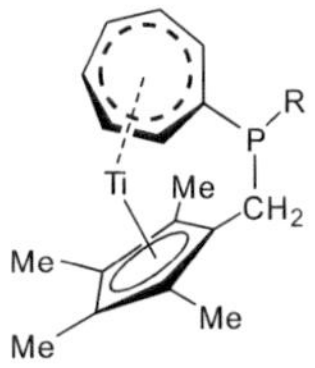

(13) (14) R = Ph or Cy (15) R = Ph or Mes

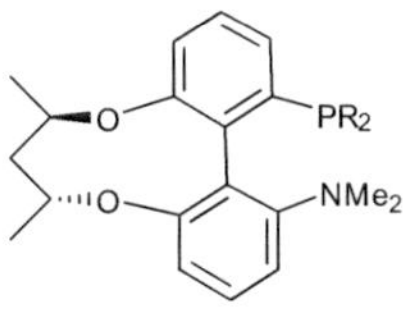

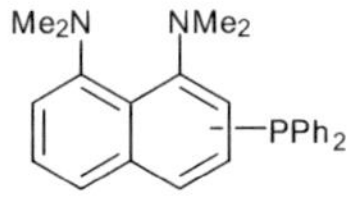

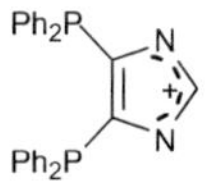

(16) R = Cy, Ph or Bu^t (17) (18)

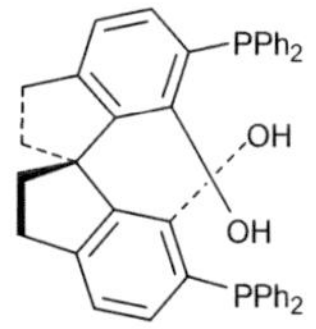

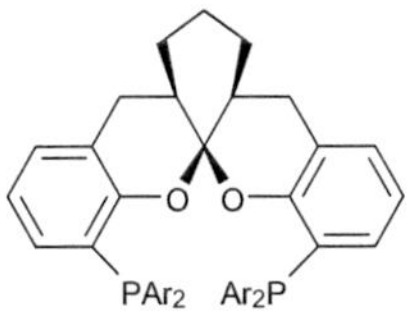

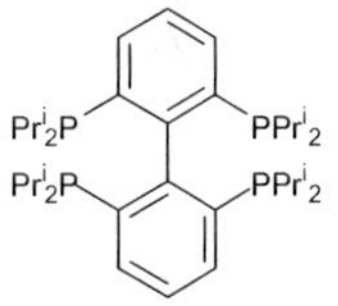

(19) (20) (21)

2.1.2 From metallated phosphines. This route has continued to find considerable use in the synthesis of new phosphines. Lithiophosphide reagents remain the most commonly used, sometimes as borane-protected systems, the borane group also providing protection against oxidation of the new phosphine during purification steps. New phosphines reported using these reagents in traditional procedures involving nucleophilic displacement reactions of sulfate esters or alkyl halides include the biarylmonophospholanes **(22)**,[35] various triangulane-based diphosphines, *e.g.*, **(23)**,[36] the 1,5-diphosphinodiethylether O{C$_2$H$_4$PH(SiPr$^i_3$)}$_2$,[37] novel chiral monophosphine ligands derived from isomannide and isosorbide, *e.g.*, (**24**),[38] and the chiral phosphine-phosphoramidite (**25**).[39] Ring-opening of a spirocyclic cyclopropane with lithium diphenylphosphide has given the phosphino-functional 4,5-diazafluorene (**26**), easily deprotonated to give the related phosphino-alkyldiazafluorenide anion.[40] Less-familiar routes involving lithiophosphide reagents include a stereoselective synthesis of *o*-bromo- or -iodo(aryl) P-chirogenic phosphines (**27**) based on the reaction of a lithiophosphide derived from a P-chirogenic secondary phosphine with the 1,2-dibromo- or 1,2-diiodo-arene, *via* addition to an intermediate aryne formed *in situ* in the presence of butyllithium.[41] The reactions of 2,3-dichloroquinoxaline with borane-protected lithiophosphides derived from both bulky and chiral secondary phosphines have provided routes to rigid diphosphinoquinoxalines, *e.g.*, (**28**). This work also disclosed new routes to enantiopure (*S*)- and (*R*)-*t*-butyl(methyl)phosphine-boranes.[42] Sodium and potassium-organophosphide reagents have also continued to find new applications in synthesis. The photostimulated reaction of sodium diphenylphosphide in liquid ammonia with the 2-chlorobenzoate ion provides a route to 2-(diphenylphosphino)benzoic acid (isolated as an ester), *via* an $S_{RN}1$ mechanism.[43] A new range of chiral platinum-phosphine catalysts has been developed for the asymmetric alkylation of the bis-secondary phosphine PhPH(CH$_2$)$_3$PHPh in the presence of NaSiOMe$_3$ in THF at room temperature.[44] Potassio-organophosphide reagents have been applied in the synthesis of the P,N,P-pincer ligand (**29**),[45] four new tripodal polyphosphine ligands of the type (Ar$_2$P)$_2$CHCH$_2$PAr$_2$ (Ar = Ph, *o*- and *p*-tolyl or *m*-xylyl),[46] and, by reaction with related *o*-fluoroaryloxazolines, the *o*-phosphinoaryloxazolines (**30**).[47] Less-familiar applications of potassio-organophosphide reagents include a series of nucleophilic substitution reactions at the nitrogen of sulfonamides that provide routes to phosphamides and amines,[48] and the characterisation of a heptaphosphaguanidine dianion derived from the reaction of [K(18-crown-6)]$_2$[HP$_7$] with a diarylcarbodiimide.[49] The synthesis of organophosphines (and other organophosphorus compounds) by activation of elemental phosphorus with superbase systems, *e.g.*, KOH-DMSO, has been reviewed.[50] Trofimov's group has also reported further work in this area in a series of papers describing routes to tris(2-pyridyl)phosphine, involving the reaction of red (or white) phosphorus with 2-bromopyridine in the presence of the superbase systems KOH/DMSO/(H$_2$O) or CsF/NaOH/DMSO, either on heating (70–100 °C)[51] or on microwave irradiation.[52] In related work, this group has also reported the synthesis of tris(2-pyridyl)phosphine in 50% yield by bubbling a mixture of PH$_3$ and H$_2$ through a mixture of 2-bromopyridine and KOH/DMSO at 70 °C.[53] Interest in the synthesis, structural characterisation and preparative uses of less common metalloorganophosphide systems has also continued. Alkali metal benzyl ether-substituted phosphido-borane complexes have been prepared and structurally-characterised.[54] Oxidation products derived from calcium- and strontium-bis(diphenylphosphides) on treatment with chalcogens have been investigated, phosphinite- and phosphonate-complexes, and some sulfur and selenium analogues having been characterised.[55] Work on organophosphido-complexes of the heavier main group 14 elements is included in a review of their coordination chemistry with anionic P-donor ligands of various types.[56] Further work has been reported on the chemistry of terminal organophosphido complexes of the heavier group 14 elements, including the formation of cage-phosphido systems derived from tin and lead,[57] a correlation of the structure of Ge, Sn and Pb-organophosphides with $|J_{MP}|$ NMR data,[58] and new work on diphospha-germylene - and -stannylene systems, $(R_2P)_2M$.[59,60] Among other metallo-organophosphide systems also investigated in the past year are those of zinc,[61,62] copper,[63,64] manganese,[65] molybdenum[66] and zirconium.[67] The use in synthesis of phosphine reagents metallated at atoms other than phosphorus has again continued to attract interest and further applications have been described. The usual starting point is a phosphine metallated at a carbon atom that is the site of subsequent transformations. Recent applications of C-lithiated phosphines in synthesis include routes to new ferrocenylphosphines, e.g., (31)[68] and (32),[69] and the 2-(4-diphenylphosphinophenyl)-1,3-benzodiazaborole (33).[70] The established asymmetric C-lithiation of a methyl group of borane-protected aryl- or t-butyl-dimethylphosphines in the presence of (−)-sparteine has found further application in routes to P-chirogenic α-alkoxyphosphine ligands, e.g., (34)[71] and the C_3-symmetric P-stereogenic triphosphine (35).[72] A similar asymmetric C-lithiation of borane-protected diphosphinoethanes of the type $R(Me)PCH_2CH_2P(Me)R$, followed by intramolecular oxidative coupling, has given a stereospecific synthesis of *trans*-1,4-diphosphacyclohexanes (36).[73] Interest has also continued in studies of the coordination chemistry of deprotonated phosphino-phenols[74] and -(secondary) amines.[75]

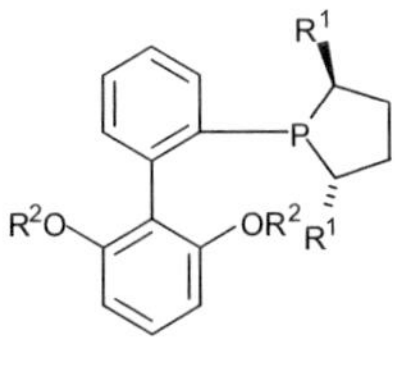

(22) R^1 = Et or Pr^i
R^2 = Me, Cy or Pr^i

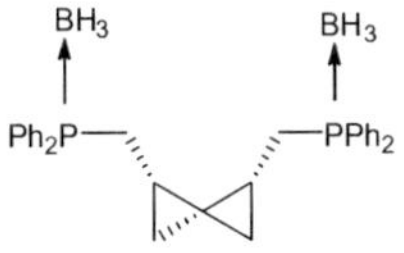

(23)

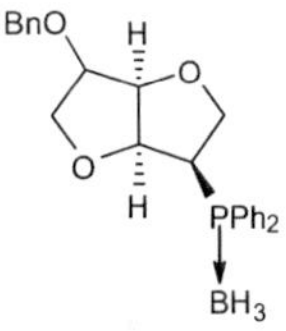

(24)

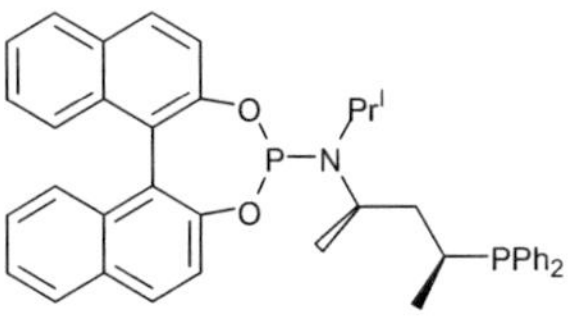

(25)

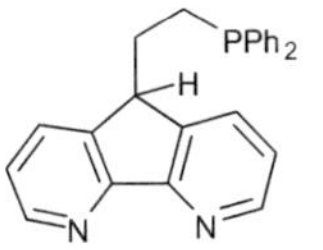

(26)

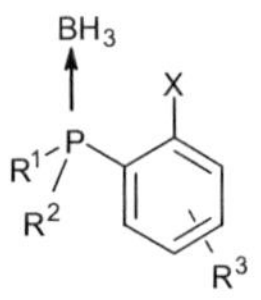

(27) R^1, R^2 = aryl or alkyl
R^3 = H or Me; X = Br or I

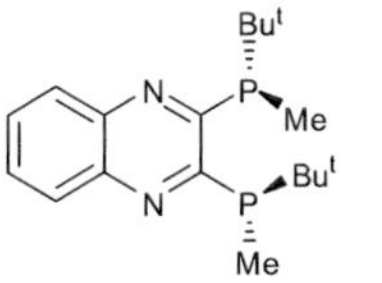

(28)

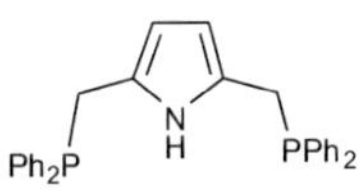

(29)

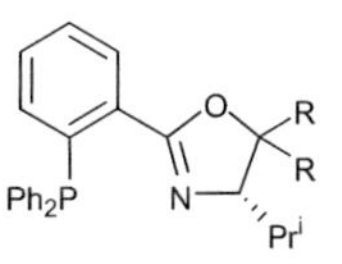

(30) R = Me, Et, Ph or m-tolyl

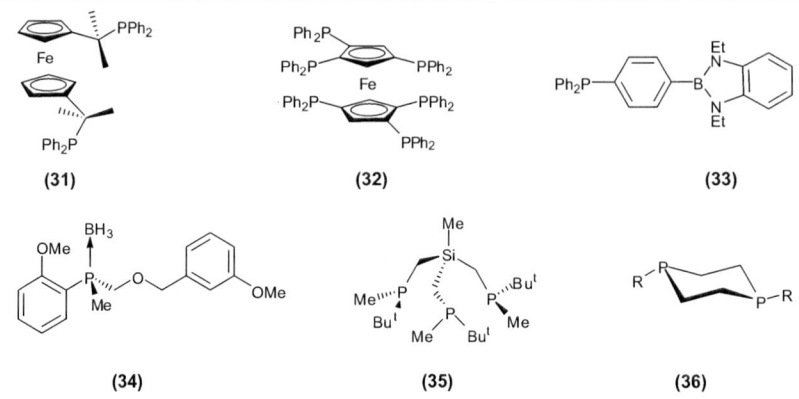

2.1.3 By the addition of P–H to unsaturated compounds.

This route has continued to find application, the number of papers published over the past year having increased a little compared to that in 2011. Addition of P–H bonds to unsaturated compounds has been used under a variety of conditions involving thermal-, radical (UV or AIBN)-, base- or metal complex-catalysed initiation in the synthesis of a range of new phosphines (and related chalcogenides). Anti-Markownikov free-radical addition of secondary phosphines (or their chalcogenides) to terminal alkenes has provided routes to a series of ω-phosphino-undecylphosphonium ionic liquids,[76] tripodal tetradentate phosphino-phosphine oxide ligands (37),[77] and α,ω-diphosphino alkane ether podands, e.g., (38).[78] In the past year, there has been considerable interest in metal-complex catalysed addition of P–H bonds to unsaturated compounds and this area has been covered in a recent review.[79] Examples reported commonly involve catalysis by palladium complexes, often involving chiral ligands, and include additions to enones, giving C* and P*-chiral tertiary phosphines, e.g., (39),[80] to methylidenemalonate esters, giving the functionalised phosphines (40),[81] and to nitroalkenes[82] and N-tosylimines.[83] Sequential palladium complex-catalysed additions of diphenylphosphine and primary or secondary amines to dimethyl acetylenedicarboxylate have provided a route to the chiral diester-substituted P,N-ligands (41).[84] An organopalladium-complex-catalysed hydroarsination of alcohol-functionalised vinyldiphenylphosphines to give the chiral arsinophosphines (42) has also been reported.[85] An iron complex-catalysed regioselective double hydrophosphination of terminal arylalkynes has given the phosphines $Ph_2P(Ar)CHCH_2PPh_2$.[86] Stereoselective hydrophosphination of dichloroacetylene with primary phosphines has been reported, both in the presence and absence of the primary phosphine-tungsten pentacarbonyl complex, to give mixtures of both mono- and di-addition products. Phosphination in the absence of the complex resulted in trans-addition whereas the complex-catalysed reaction gave the cis-products.[87] Chiral palladacycle-catalysed additions of diphenylphosphine to α,β-unsaturated imines in the presence of triethylamine provides a chemo- and enantio-selective route to enaminophosphines (43).[88] The same group has also reported a similar approach to the synthesis of chiral phosphorinanones by the addition of phenylphosphine to bis(enones).[89] Solvent- and catalyst-free regioselective hydrophosphinations of alkenes and alkynes have also been developed.[90]

As in recent years, P-H additions to unsaturated systems in the presence of a strong base have continued to find use in synthesis of new phosphines. Recent examples include routes to (cycloheptadienyl)-diphenylphosphine (44),[91] superhindered polydentate polyphosphine ligands, e.g., $P(CH_2CH_2PBut_2)_3$,[92] and a wide range of chiral aminoalkyl- and amidoalkyl-phosphines, e.g., (45), subsequently used as catalysts for enantioselective double-Michael indoline syntheses.[93] Trofimov's group has reported further examples of the superbase-red phosphorus generation of phosphine (PH_3) and its addition to alkenes, in this case 1-methoxy-4-vinylbenzene, to give, selectively, either the secondary phosphine $(4\text{-MeOC}_6H_4CH_2CH_2)_2PH$ or the tertiary phosphine $(4\text{-MeOC}_6H_4CH_2CH_2)_3P$.[94]

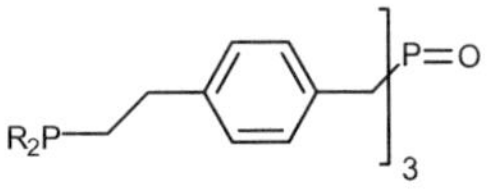

(37) R = e.g., $EtCH_2CH_2$ or $PhCH_2CH_2$

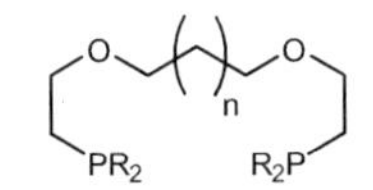

(38) n = 1–4; R = alkyl or aryl

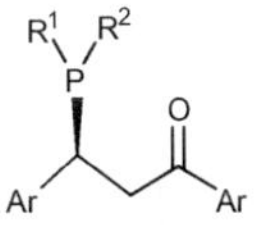

(39) R^1, R^2 = Ph or R^1 = Ph; R^2 = Me, Bu^t or Mes

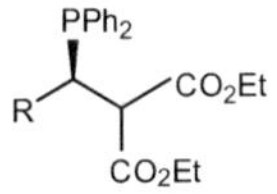

(40) R = Ph, OEt, Me, 2-furyl, p-anisyl, (E)-PhCH=CH or NH_2

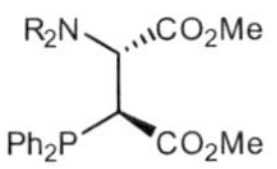

(41) R = Me, Et or Pr^i

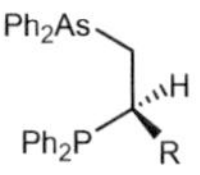

(42) R = CH_2CH_2OH or CH_2OH

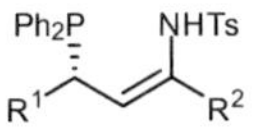

(43) R^1, R^2 = Ar

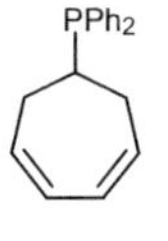

(44)

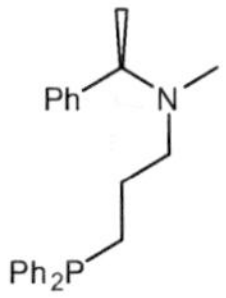

(45)

2.1.4 By the reduction of phosphine oxides and related compounds.

As in recent years, a wide range of reagents has been employed for the reduction of phosphine oxides, usually at the end of a multistage synthesis. Although silane-based reagents continue to be widely employed, before reviewing applications of these reagents it is important to note a recent report of a simple and unprecedented conversion of tertiary and secondary phosphine oxides and sulfides to phosphine-borane complexes using oxalyl chloride, followed by sodium borohydride, the reaction proceeding via reduction of an initially-formed chlorophosphonium chloride salt. Significantly, attempted reduction of P-stereogenic tertiary phosphine oxides under these conditions led to racemisation, probably due to the presence of the chloride anion. This problem was overcome by replacement of oxalyl chloride with triethyloxonium tetrafluoroborate, leading to initial alkylation at oxygen of the P=O bond, followed by borohydride reduction to give the chiral phosphine-borane with inversion of configuration at phosphorus. Further work is anticipated towards optimisation of this procedure.[95] In related work, the same group has shown that treatment of phosphine oxides and sulfides with oxalyl chloride, followed by reduction of the chlorophosphonium salt with lithium aluminium hydride, also regenerates the parent phosphines, thus providing a useful procedure for the removal of the phosphine oxide biproducts of Wittig and Appel reactions and recovery of the phosphine.[96]

Chlorosilanes, in particular trichlorosilane and hexachlorodisilane, continue to be widely used for the reduction of phosphine oxides and sulfides. In two significant papers, Krenske has addressed the underlying mechanistic background in order to understand the great variance in stereoselectivity observed in reductions of acyclic phosphine oxides and sulfides using different chlorosilane reagent systems. Thus, *e.g.*, hexachlorodisilane reduces phosphine oxides and sulfides to the corresponding phosphines, but with opposite stereoselectivities. Two theoretical studies have now shed some light on this issue. Phosphine oxide reduction is shown to proceed *via* the formation of conventional phosphorane intermediates (**46**), leading to inversion whereas the reduction of phosphine sulfides follows a totally different mechanism involving attack of the sulfur atom at the silicon atom of a donor-stabilised dichlorosilene (**47**), eventually leading to retention of configuration at phosphorus.[97] This general conclusion gained further support in a second study, which concluded that for reductions of phosphine oxides by trichlorosilane, the mechanism proposed originally by Horner applies, involving a four-centred frontside addition of the Si–H bond to O=P, resulting in transfer of hydride to phosphorus (and silicon to oxgen) to form an intermediate phosphorane, (**48**), leading eventually to retention of configuration at phosphorus in the resulting phosphine. Also rationalised in terms of phosphorane-based mechanisms are reductions of phosphine oxides by $HSiCl_3/Et_3N$ and Si_2Cl_6, all leading to inversions at phosphorus. Further support was also provided for the involvement of dichlorosilene intermediates in reductions of phosphine sulfides with Si_2Cl_6 that proceed with retention at phosphorus.[98]

(52) R = Me or Ph   **(53)**   **(54)** R = H, Me, PhCH$_2$ or CH$_2$(1-naphthyl)

(55) Ar = 3,5-(CF$_3$)$_2$C$_6$H$_3$ **(56)** R = ^{n}Bu, ^{t}Bu, or Ph   **(57)**

Trichlorosilane has been used in the absence of an amine base in the synthesis of (4-trimethylammoniophenyl)diphenylphosphine salts[99] and a combination of trichlorosilane and triphenylphosphine was used in the reduction of chiral, mesoporous cross-linked polymer-bound 2,2′-bis(diphenylphosphinoyl)-1,1′-binaphthyl (BINAPO) systems.[100] Among phosphines routinely accessed using trichlorosilane in the presence of a tertiary amine base are C-functionalised triarylphosphines bearing a perfluoroalkylester substituent in the 4-position of the benzene ring[101] and the chiral, atropisomeric phosphines (**49**),[102] (**50**)[103] and (**51**).[104] Among other silane-based reagent systems reported in the past year are Me$_3$SiCl-LiAlH$_4$, used for the reduction of an arylphosphonate ester to give the air-stable primary phosphine (**52**),[105] (EtO)$_2$MeSiH (or occasionally polymethylhydrosiloxane (PMHS), tetramethyldisiloxane (TMDS) or phenylsilanes) in the presence of catalytic amounts of specific Brønsted acids, commonly dialkyl- and diaryl-phosphates, enabling the reduction of phosphine oxides bearing sensitive carbonyl-based functional groups,[106] various combinations of TMDS with Lewis acids such as copper(II) triflate,[107] indium(III) bromide[108] or titanium(IV) isopropoxide (in an improved synthesis of enantiomers of 4,4′-diaminomethyl-BINAP via the reduction of 4,4′-dicyano-BINAPO,[109] and also in a route to P-chirogenic monodentate binaphthylphosphines (**53**)[110]), and combinations of PMHS with titanium(IV) isopropoxide in routes to the chiral phosphines (**54**)[111] and the phosphino-biferrocene (**55**).[112]

Finally, reduction of 1,1′-bis(chlorophosphino)ferrocenes with magnesium has provided a simple route to the diphospha[2]ferrocenophanes (**56**)[113] and phosphine oxides have been converted at room temperature into synthetically useful N-pyrazolylphosphonium salts using the highly charged, oxophilic phosphorus-centred trication (**57**). The pyrazolylphosphonium salts may then be reduced to the phosphines with cleavage of the pyrazole unit, using either LiAlH$_4$ or electrochemical methods.[114]

2.1.5 By miscellaneous methods.

Interest in methods for the activation of white phosphorus (P_4) has continued. Activation using compounds of the Group 14 elements in low oxidation states, providing routes for the synthesis of phosphines, polyphosphines and p_π-bonded phosphorus compounds, has been the subject of a recent overview.[115] Previously unknown polyphosphorus compounds have been obtained by the activation of P_4, coordinated to ruthenium complexes, using iodine and subsequent hydrolysis.[116] Weigand's group has reported several interesting approaches to the synthesis of new polyphosphines. The reactions of tris(3,5-dimethyl-1-pyrazolyl)phosphine with secondary phosphines and bis(secondary phosphines) at room temperature provide access to catenated and branched polyphosphine ligands, e.g., (**58**)[117] and (**59**),[118] via P-N/P-P bond metathesis steps. Also reported by this group is a one-pot synthesis of polyphosphorus frameworks involving two-, three- and four-coordinate phosphorus atoms by the reactions of the cationic system (**57**) with secondary phosphines.[119] Further examples of 2-phosphinomethyl(1-methyl)imidazoles, e.g., (**60**), have been obtained by the reactions of halogenophosphines with 1-methyl[2-(trimethylsilyl)methyl]-1H-imidazole.[120] The reaction of diphenylphosphine with 2-(dimethylaminomethyl)pyrroles provides a simple and direct route to the bis(diphosphinomethyl)pyrrole (**61**) and the PNNP system (**62**).[121] An alternative approach to (**61**) involves the initial reaction of pyrrole-2,5-dicarboxaldehyde with diphenylphosphine, followed by reduction of the initially-formed bis(phosphine oxide) using $LiAlH_4/CeCl_3/NaBH_4$, giving the related bis(phosphine-borane) adduct as a stable and easily purifiable compound, from which the free diphosphine can be obtained by heating under reflux with methanol.[122] The key step in the synthesis of a range of bis(perfluoroalkyl)phosphino-oxazolines (**63**) (and a related series of ferrocene ligands) is the reaction of a diphenyl 2-(oxazolinyl)arylphosphonite with a perfluoroalkyltrimethylsilane in the presence of CsF.[123] New phosphines prepared by the reactions of secondary phosphines with benzyl- or reactive alkyl-halides, followed by deprotonation of the initially-formed phosphonium salts, include the bisphosphinomethylsulfide $Bu^t_2PCH_2SCH_2PBu^t_2$[124] and a series of polystyrene-supported trialkylphosphine ligands.[125] Self-assembly of phosphine molecules has also continued to attract interest. A ligand system has been developed for a tandem rhodium-catalyzed hydroformylation-hydrogenation of alkenes which involves both 6-diarylphosphinopyridin-2-ones and 2-acylguanidino-5-phosphinopyrroles (**64**). The former undergo hydrogen-bonded dimerisations to give new self-assembled diphosphine ligands and the latter associate with the aldehyde component of the reaction mixture in a catalysed hydrogenation step.[126] Also of interest is the self-assembled reaction of isophthaloyl chloride and methyl-bis(trimethylsilyl)phosphine to give the first example of a diphosphametacyclophane (**65**)[127] and a strategy for the design of chiral phosphine-ligands, e.g., (**66**), by the ion-exchanged association of an achiral cationic ammonioalkyl-substituted arylphosphine with a chiral binaphthoate anion.[128] A new borataphosphine, (**67**), (essentially a complex of the di-t-butylphosphide anion with boratabenzene), has been prepared and its coordination chemistry with electronically unsaturated group 10 transition metals studied.[129] A series of acylphosphines of the type $R_2PC(O)(CH_2)_nCH_3$ (R = Ph, Cy, Pr^i or Bu^t; n = 4 or 13) has been prepared by the reaction of secondary phosphines and acyl halides in the presence of triethylamine and shown to undergo site-selective alkyl dehydrogenation when coordinated by phosphorus in an iridium complex.[130] Triarylphosphines involving a 9-anthracenyl group at the 4- or 5-position of each aryl ring, *e.g.*, (**68**), have been obtained lithiation of the related 9-bromophenylanthracenes followed by addition of triphenylphosphite. The *meta*-substituted triarylphosphines display extended fluorochromism, as also do the analogous triphenylamines.[131]

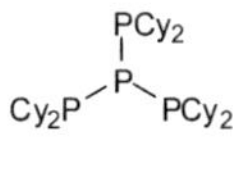

(**58**)

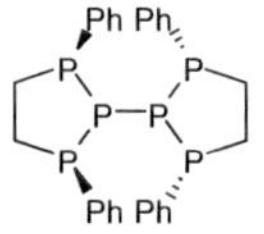

(**59**)

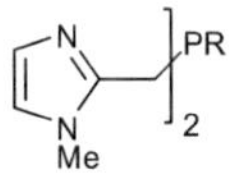

(**60**) R = Et or Ph

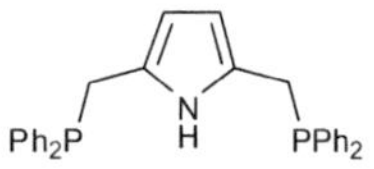

(**61**)

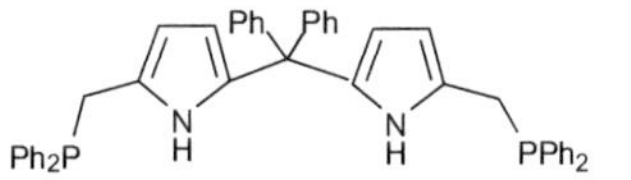

(**62**)

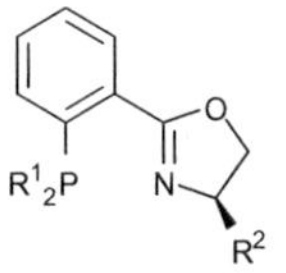

(**63**) R^1 = CF_3 or C_2F_5; R^2 = Ph, Bn, ^iPr or ^tBu

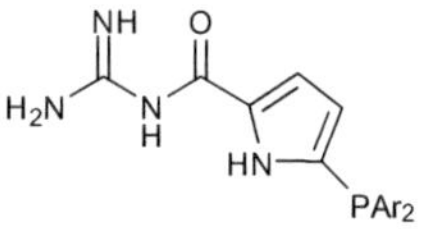

(**64**) Ar = Ph or p-$MeOC_6H_4$

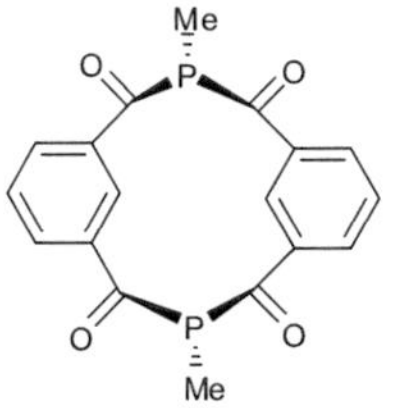

(**65**)

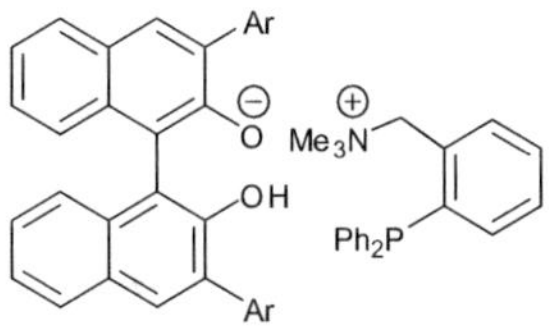

(**66**) Ar = Ph or 2-naphthyl

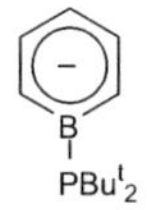

(**67**)

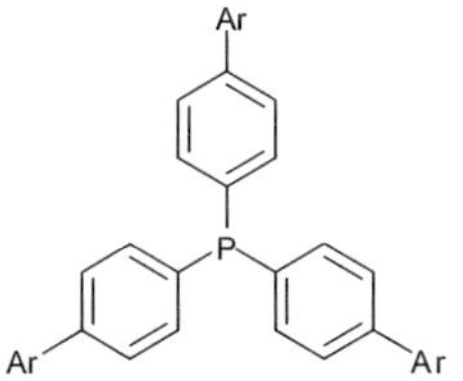

(**68**) Ar = 9-anthracenyl

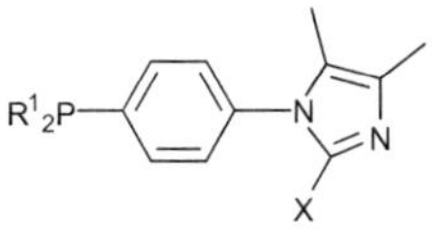

(**69**) R^1 = Ph or Cy; X = H or PR^2_2; R^2 = Ph, Cy or 2-furyl

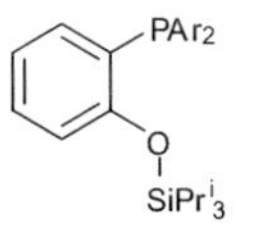

(**70**) Ar = 1-adamantyl

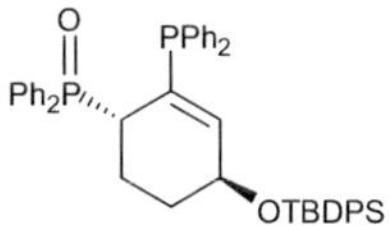

(**71**)

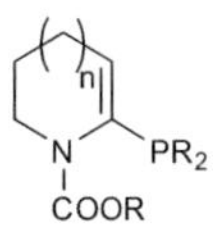

(**72**) n = 1 or 2

Applications of metal-catalysed routes for C–P bond formation in phosphine synthesis have also continued to appear. The usual approach is the reaction of a bromo- or iodo-arene with a secondary phosphine, catalysed by a palladium- or nickel-complex. Catalysis using Pd(OAc)$_2$ has been used recently to prepare phosphinoarylimidazoles, *e.g.*, (**69**),[132] the crowded silanoxyphenyl ligand (**70**),[133] 5-diphenylphosphinouracil and a related phosphinopyrimidine,[134] electron-rich 9-(B-connected)-phosphinocarboranes,[135] and the chiral phosphinocyclohexenyl-phosphine oxide (**71**).[136] Various palladium catalysts were used in a procedure involving a P–C cross-coupling between enol phosphates and secondary phosphine-borane complexes to give chiral and achiral αβ-alkenylphosphines bearing an amido group in the α-position to phosphorus, *e.g.*, (**72**).[137] Nickel(II)-bis(diphenylphosphinoalkane) complexes have been applied in routes to a wide range of aryl- and heteroaryl-phosphines,[138] and polymer-bound BINAP ligands.[139]

As in previous years, the elaboration of functional groups present in substituents at phosphorus has led to a wide range of new phosphines. A key step in the synthesis of a new water-soluble cyclodextrin-phosphine is the reaction of the borane-protected alkynyl-functional phosphine HC≡CCH$_2$PPh$_2$ with an azido-functional cyclodextrin to create a diphenylphosphinomethyltriazolyl linkage to the cyclodextrin.[140] Modifications at nitrogen of various pyrrolidinophosphines have given amino-acid- and imidazolium-tagged chiral pyrrolidinodiphosphines, *e.g.*, (**73**),[141] new amido- and N-heteroaryl-functional phosphines, *e.g.*, (**74**),[142] and a series of dendrimers, surface-capped with the chiral 3,4-bis(diphenylphosphino)pyrrolidino moiety.[143] Quaternisation at nitrogen of 1,3,5-triaza-7-phosphaadamantane (PTA) with various alkyl- and benzyl-halides has given further examples of cationic PTA systems, including a series of long chain N-alkyl derivatives[144] and a dense dendrimer, capped with PTA ligands.[145] A series of new phosphine-phosphite ligands has been prepared by phosphitylation of ω-hydroxyalkyldiphenylphosphines with chiral cyclic chlorophosphites derived from BINOL and the related partially reduced H$_8$-BINOL systems.[146]

Imine-formation from phosphinoarylaldehydes, and amide- and ester-formation from phosphinoarylcarboxylic acids have continued to be used in synthesis. New phosphines prepared by imination of o-diphenylphosphinobenzaldehyde include the chiral ligand (75),[147] the amidinium-bridged chiral diphosphine (76),[148] a series of chiral phosphinoarylhydrazones,[149] a phosphinopyrrolidinofullerene[150] and an imine derived from 2-aminobenzyl alcohol that exists in equilibrium with a phosphinoarylbenzoxazine tautomer.[151] The reaction of 4-(methoxycarbonyl)phenyldiphenylphosphine with sodium-3-aminopropane-sulfonate gave the anionic amidoalkanesulfonate-functionalised phosphine (77), subsequently grafted by electrostatic attraction to a cationic imidazolium-functionalised mesoporous silica support, this new material being of interest as a new solid-supported ligand for catalysis.[152] Modification of amino-functional arylphosphines has also been a useful approach to new ligands. N-phosphitylation of 2-methylaminophenyldiphenylphosphine, involving H_8-BINOL derivatives, has given new chiral phosphine-phosphoramidite ligands, e.g., (78).[153,154] Further examples of chiral thiourea-phosphine ligands have been prepared by treatment of 2-amino-2′-diphenylphosphino-1,1′-binaphthyls with arylisothiocyanates.[155] An aminomethyl-substituted BINAP has been used, via amide-formation, as a capping ligand in dendrimer synthesis.[156] Metallation of o-diphenylphosphinobenzenethiol, followed by treatment with chlorodiphenylarsine, provides a route to the multidonor ligand o-(Ph$_2$AsS)-C$_6$H$_4$PPh$_2$.[157] o-Lithiation of 2-bromophenyldiphenylphosphine has enabled the introduction of a difluoro(phenyl)silyl substituent into the phosphine.[158]

Mannich-type reactions involving ammonia, primary or secondary amines with hydroxymethylphosphonium salts, hydroxymethylphosphines or secondary phosphines (in the presence of formaldehyde) have continued to be used to generate new aminomethylphosphines. Among these are tris(diisopropylphosphinomethyl)amine, $N(CH_2PPr^i_2)_3$,[159] a series of carbosilane dendrons based on p-[bis(diphenylphosphinomethyl)amino]phenyltrialkylsilanes,[160] various phosphinoglycolates, $R_2PCH(OH)COO^-$ $R'_2NH_2^+$,[161] and 4-mono- and bis-(diphenylphosphinomethyl)aminopyridines.[162] Use of the phospha-Mannich reaction for the generation of the bis(diphenylphosphinomethyl)amino group has also been used as a strategy for the end-capping of dendrimers.[163] The reactions of rac/meso-1,2-bis(phenylphosphino)ethane with paraformaldehyde and primary amines bearing chiral substituents have given chiral 1-aza-3,6-diphosphacycloheptanes (79) as air-stable crystalline solids, with the prevailing formation of meso-phosphorus stereoisomers as kinetically-controlled products.[164] Further work has been reported on the formation of the 3-aza-7-phosphabicyclo[3,3,1]nonan-9-ones (80) from the reactions of 4-phosphacyclohexanones with amines and aldehydes.[165]

Side-chain functional group transformations of phosphinoferrocenes have again found application in the synthesis of new phosphines. A variety of side-chain modification routes have been used in routes to a series of phosphino-functionalised ferrocenyloxazolines.[166] Well-established nucleophilic displacements using primary or secondary

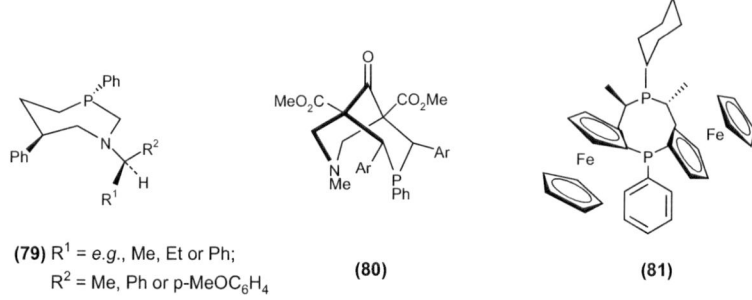

(79) R^1 = e.g., Me, Et or Ph;
R^2 = Me, Ph or p-MeOC$_6$H$_4$

(80)

(81)

phosphines and dimethylaminomethylferrocenes or ferrocenylmethyl acetate esters have received further application in the synthesis of chiral ferrocenylmethylphosphines,[167] new 2-phospha[3]ferrocenophanes,[168] conformationally-constrained phosphocin and 1,5-diphosphocins incorporating two ferrocenyl units, e.g., (**81**),[169] and for the introduction of a chiral P-phospholanylmethyl side-chain in a synthesis of Josiphos-type bis(phospholanyl)ferrocenes.[170] A 1-hydroxymethyl-2-phosphinoferrocene has been transformed into the corresponding 1-p-toluenesulfonamidomethyl system.[171] Štěpnička's group has described side-chain modifications of phosphinoferrocenyl carboxylic acids and esters to give a series of phosphinoferrocenyl carboxamides,[172,173] and has also reviewed the general chemistry of phosphino-carboxamide systems.[174] Other ferrocene side-chain modifications reported include the conversion of a 2-(aminomethyl)ferrocenylphosphine into a related chiral imidatomethyl-ferrocenylphosphine[175] and the reaction of a 1-lithio-2-(aminomethyl)ferrocene with phenyl- or adamantyl-bis(trifluoromethyl)-phosphine, resulting in displacement of a trifluoromethyl group from the phosphine and introduction of the organo(trifluoromethyl)phosphino moiety into the ferrocenyl system.[176]

2.2 Reactions

2.2.1 Nucleophilic attack at carbon.

The formation of zwitterionic phosphonium compounds by nucleophilic attack of phosphorus at unsaturated carbon and the subsequent engagement of such dipolar species in C–C and C–N bond-forming reactions continues to attract a great deal of attention. The long-running saga of the reactions of tertiary phosphines and acetylenedicarboxylic acid esters in the presence of a third reactant, a proton source that serves to protonate the initial dipolar species formed to give a vinylphosphonium salt, has continued. The vinylphosphonium salt then undergoes addition of the anion derived from the proton source to form a new phosphonium ylide. In many cases, these are stable, but some undergo intramolecular reactions to give new, non-phosphorus-containing products, often via a Wittig route. Further examples have also appeared of reactions of this type that lead to C–C bond formation with eventual reformation of the phosphine, the latter now assuming a catalytic role. The formation of stable ylides from the reactions of triarylphosphines, dialkyl acetylenedicarboxylates (DAAD) and various NH-, SH- and OH-acids has been investigated using proton sources such as a variety of NH-heterocyclic compounds,[177,178] a 2-mercaptopyrimidine,[179] 2,4-dinitrophenylacetic acid,[180] Meldrum's acid derivatives,[181,182] and thienylmethanols and other alcohols.[183] In some of the latter cases, the initially-formed alcohol-derived ylides undergo elimination of triphenylphosphine with formation of alkoxy-O-vinylated alkenes, making the reaction catalytic in the phosphine.[184] Other catalytic reactions involving α-halogenoketones as the proton source have resulted in the formation of halogenated αβ-unsaturated γ-butyrolactone derivatives.[185] Whereas the reaction of the Ph_3P-DAAD system with 2-nitro-trans-cinnamaldehyde results in a stable ylide, the related reaction of triphenylphosphine with an alkylpropiolate ester results in the formation of 2H-pyran derivatives with regeneration of the phosphine.[186] In other cases, the initially-formed ylides have been shown to undergo rearrangement or transformation to other ylides on heating in toluene[187,188] or collapse via elimination of triphenylphosphine oxide to form products derived from intramolecular Wittig reactions.[189,190] Stable ylides have also been obtained from the reactions of tris(dimethylamino)-phosphine, DAAD and NH-heterocyclic compounds.[191] Finally, it is interesting to note that formation of stable ylides derived from the reactions of a diversity of NH- and CH-acids with the Ph_3P-DAAD zwitterions can also be achieved under solvent-free conditions,[192] work in this area having previously involved the use of a halogenated or other solvent. Doubtless we will soon see further applications of these 'green' conditions in publications from these groups. Other papers on the reactions of t-phosphine-acetylenic ester systems report routes to dihydropyridine-fused benzopyrones from N-phenyl-chromonyl nitrones,[193] γ-lactones bearing an α-phosphorus ylide moiety from aldehydes,[194] fused nitrogen heterocycles from annulations of azomethine imines and ethyl 2-butynoate,[195] α-amino-γ-oxoacid derivatives of phthalimide,[196] and the formation of 5-(trifluoromethyl)-2,5-dihydro-1,2λ^5-oxaphospholes from aryl- or styryl-trifluoromethyl ketones.[197]

Tertiary phosphines have also been used as catalysts in promoting cycloaddition reactions of a much wider range of molecules bearing an alkyne functional group. Included among these are cycloaddition reactions of ynones with o-tosylamidobenzaldehydes (and related ketones) to give quinolines,[198] and with N-substituted isatins to give dihydropyrano[2,3-b]indoles (and related compounds),[199] and highly functionalised spiro-oxazolines.[200] Also reported are cycloadditions of dialkyl (E)-hex-2-en-4-ynedioates to [60]-fullerene to give a series of cyclopenteno-fullerenes bearing a phosphonio-ylide functionality,[201] cycloadditions of alkynyl esters to curcumin,[202] and phosphine-mediated cycloisomerisations of alkynyl hemiketals, providing access to spiroketals and dihydropyrazoles via tandem reactions.[203] Interest has again continued in the wider general synthetic applicability of tertiary phosphines in the nucleophilic catalysis of carbon-carbon bond formation as typified by the Morita-Bayliss-Hillman (MBH) and related aza-MBH reactions. Applications of new types of phosphine catalyst include the use of glucose-functionalised phosphinothioureas, *e.g.*, (82),[204] chiral phosphine-squaramides (83),[205] a (2-diphenylphosphinobenzoyl)-substituted calix[4]arene,[206] multifunctional binaphthylphosphines, *e.g.*, (84),[207] and a trialkylammonioalkyl-functionalised triarylphosphine, the insolubility of the latter in ether affording an easy separation of the desired products from the catalyst.[208] Further catalytic applications of 2-phosphino-2′-hydroxybiphenyls have also been reported.[209] Reports of the catalysis of MBH and related reactions by simple acyclic alkyl- and aryl-phosphines have also continued to appear.[210] Of some additional interest is a report of the use of tributylphosphine as both a nucleophilic catalyst and as a sacrificial reducing agent, involving water as the hydrogen source, in a domino-aza-MBH/reduction reaction of electron-poor dienes and N-tosylimines.[211]

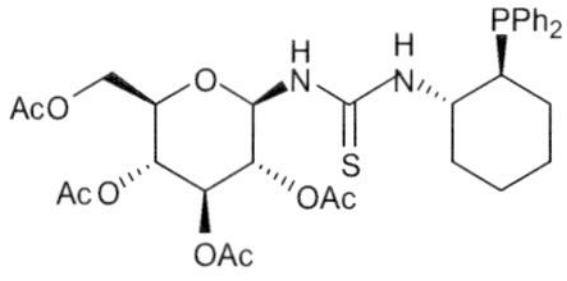

(82)

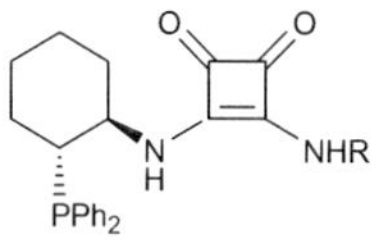

(83) R = alkyl or aryl

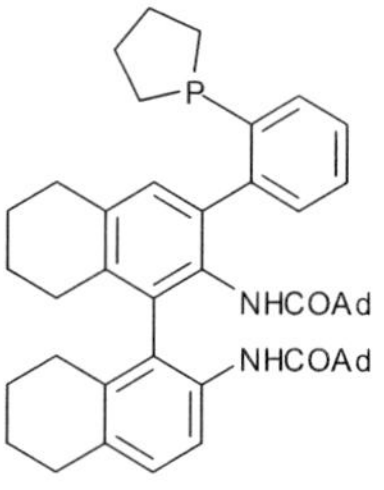

(84) Ad = 1-adamantyl

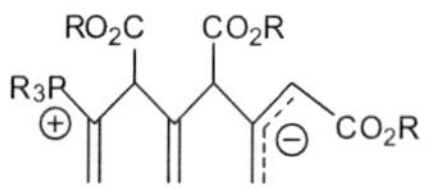

(85)

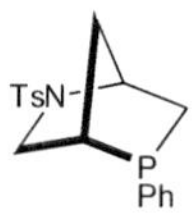

(86)

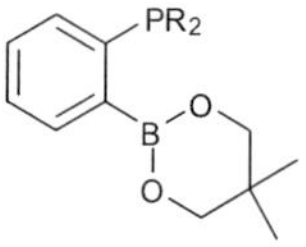

(87) R = Ph or Me

A host of other phosphine-catalysed reactions in which the initial step is the formation of a reactive phosphoniobetaine intermediate by addition to carbon-carbon multiple bonds, in particular α-substituted allenoates, have also been reported in the year under review. Chiral-phosphine-catalysed asymmetric annulations of allenes with electron-deficient alkenes and imines have been the subject of a feature article.[212] The mechanism of phosphine-catalysed [4+2] annulations of α-alkyl-allenoates has been investigated by computational methods.[213] Whereas the 1:1 zwitterionic adducts of phosphines with allenoates are well-studied, it has now been shown by a combined experimental and computational approach that higher order 1,n-zwitterionic intermediates ($n = 5$, 7 or 9), *e.g.*, (85), are real species that can be formed in phosphine-allenoate systems, and which can lead to the formation of medium- and large-ring compounds on coupling with appropriate electrophilic partners.[214] Although most of the cycloaddition reactions of allenoates noted have involved simple acyclic phosphines as catalysts, it is interesting to note the application of N-acylaminoalkylphosphines[215] and related dipeptide-functionalised phosphines,[216] and also the P-chiral bicyclic phosphine (86),[217] for the catalysis of enantioselective cycloaddition reactions. Among cycloaddition reactions of allenoates catalysed by tertiary phosphines are those leading to quaternary carbon-centered chromans,[218] dihydropyran- and cyclopenten-fused chromen-2-ones,[219] heterocyclic spiranes,[220] tetrahydro-pyridines[221] and -isoquinolines,[222] various functionalised 3-pyrrolines,[223] the indole alkaloid hirsutine,[224] conjugated 2,3-dihydrofurans,[225] nitrone-functionalised indolin-2-ones[226] and spirocyclic cyclopentenyl systems.[227] Phosphine-catalysed inverse conjugate additions of C- and N-nucleophiles to allenylphosphine oxides have also been described.[228] In addition to the reactions of allenes, many other systems involving phosphine-catalysed cycloadditions of electron-deficient alkenes, imines or carbonyl compounds have been reported, including a domino-benzannulation strategy leading to highly functionalised multiaryl skeletons,[229] and routes to tetrahydropyridines[230] and polysubstituted cyclohexanones.[231] Another major group of substrates shown to undergo phosphine-catalysed annulation and other reactions are suitably-modified Morita-Baylis Hillman adducts, which have been the subject of a recent review.[232] Recent papers describing such reactions include routes to N-fused pyrroles,[233] functionalised bicyclic imides,[234] saturated seven-membered 1,4-heterocyclic compounds,[235] spirocyclopentaneoxindoles,[236] 3-functionalised benzofuran-2(3H)-ones and oxindoles,[237] and various functionalised- and spirocyclic-cyclopentenes.[238,239] Among a miscellany of other papers describing catalysis by phosphines, it is interesting to note phosphine-promoted additions of o-iodobenzyl alcohols to methyl propiolate to give alkylidene phthalans, subsequently cyclised under Heck conditions to benzo-fused oxygen heterocycles,[240] the acylcyanation of α-substituted activated alkenes,[241] domino-routes to tetrasubstituted furans[242] and highly functionalised cyclopentenes[243] from simple terminally-activated alkenes, an efficient ββ-dimerisation of vinylphosphonates to afford linear dimers,[244] multicomponent routes to spirooxindoles[245] highly functionalised α-aminonitriles[246] and furo[3,2-c]coumarins,[247] and also for the Knoevenagel condensation of indole-carboxaldehydes with active methylene compounds.[248] Also of interest is the use of tris(2,4,6-trimethoxyphenyl)phosphine as a potent organocatalyst for the group transfer polymerisation of alkyl (meth)acrylates,[249] the use of phosphines in catalysing the β-boronation of αβ-unsaturated carbonyl compounds,[250] the application of o-diphenylphosphinoarylboronates, e.g., (87), as catalysts for Michael additions of malonate pronucleophiles to methyl vinyl ketone,[251] and catalysis by tributylphosphine of the ring-opening of aziridines with secondary amines.[252] The addition of the ambiphilic phosphinomethylalane $Me_2AlCH_2PMe_2$ to an allenyl vinyl ketone initiates Nazarov-cyclisations of the ketone but also allows the trapping and characterisation of a Nazarov intermediate, via phosphine-trapping of an oxoallyl cation.[253] Other reports describing initial nucleophilic attack by phosphorus at carbon to form ylidic or zwitterionic systems include further work on the interaction of triphenylphosphine with various benzofuroxanes and benzofurazanes bearing electron-withdrawing groups[254] and the formation of labile cyclic oxaphosphoranes in the reactions of triphenylphosphine and various phosphorus (III) esters with trichloropyruvate N-acylimines.[255] Less-familiar conventional reactions leading to phosphonium salt formation include alkylation of a phosphinoalkyne with a methylzirconocene cation, the alkynylphosphonium cation then becoming bound to zirconium *via* the alkynyl group,[256] the reactions of triphenylphosphine with benzylic ethers derived from the oxidative C–H activation of arylmethanes by DDQ,[257] a route to an [^{18}F]-labelled (2-(2-fluoroethoxy)ethyl)triphenylphosphonium cation,[258] and the alkylation of trioctylphosphine with the non-toxic reagent, dimethyl carbonate, *en-route* to a series of ionic liquids.[259]

2.2.2 Nucleophilic attack at halogen. Once again, relatively little new fundamental work has appeared, although phosphine-positive halogen systems have continued to attract interest as reagents in synthesis and some new procedures have been developed. Triphenylphosphine has been shown to react with thionyl chloride to give a mixture of [Ph$_3$PCl]Cl, Ph$_3$PS and Ph$_3$PS.[260] A series of Ar$_3$PI$_2$ adducts has been prepared from the 1:1 reactions of substituted arylphosphines with di-iodine. These compounds do not ionise to [Ar$_3$PI]I in CDCl$_3$, the familiar 'molecular spoke' structure being stable in solution, although the relative stabilities depend on the nature of the aryl group.[261] The reactions of *o-*, *m-* and *p*-tritolylphosphines (and related *o-* and *p*-tritolylarsines) with two equivalents of di-iodine or di-bromine give a series of complexes of the type R$_3$EX$_4$ (E = P or As; X = Br or I) which are largely ionic, consisting of halophosphonium or haloarsonium cations and trihalide anions, [R$_3$EX][X$_3$]. These compounds exhibit structural isomerism, existing either as simple 1:1 ion-pairs, [R$_3$EX][X$_3$] or as a 2:1 complex of the type [(R$_3$EX)$_2$X$_3$][X$_3$] in which the cation involves two [R$_3$EX]$^+$ ions interacting with one [X$_3$]$^-$ anion.[262] The reaction of the iodine(III) complex PhI(OTf)$_2$ with triphenylphosphine results in oxidation of the phosphorus, with formation of iodobenzene and [Ph$_3$POTf][OTf], the latter existing in equilibrium with Ph$_3$PO and Tf$_2$O.[263] Further work has been reported on the catalysis by triphenylphosphine oxide of Appel-type reactions using oxalyl chloride, which proceed *via* formation of the [Ph$_3$PCl]$^+$ cation as the effective reagent, this system now having been shown to promote the conversion of oximes to nitriles.[264] The phosphine oxide-oxalyl chloride system has also been applied to the development of a one-pot stereoselective synthesis of *P*-stereogenic phosphines and phosphine-boranes from racemic phosphine oxides that involves treatment of the phosphine oxide with oxalyl chloride in the presence of a chiral alcohol (menthol), followed by reduction of the alkoxy-phosphonium intermediate with either LiAlH$_4$ (to give the chiral phosphine) or NaBH$_4$ (to give the related phosphine-borane).[265]

New applications of phosphine-positive halogen systems include the use of the Ph$_3$P-C$_2$Cl$_6$ or CCl$_3$CN combinations for the *N*-protection of 2-amino-3-chloro-pyrazines as the iminotriphenylphosphorane derivatives in a synthesis of 3-(*N*-substituted)-(2-amino)pyrazines[266] and the amidation and esterification of phosphoric acids.[267] The CCl$_3$CN-Ph$_3$P and CBr$_4$-Ph$_3$P systems have been used for the conversion of N-heteroaromatic hydroxy compounds to the related N-heteroaromatic chlorides or bromides[268] and CBr_4-Ph_3P has been applied in a one-step synthesis of αω-dibromocompounds by the ring-opening of cyclic ethers.[269] The Ph_2PCl-I_2-NaN_3 system has been used in a one-pot direct synthesis of acyl azides from carboxylic acids.[270] The same group has also developed the use of a Ph_2PCl-I_2-imidazole system for the highly selective mono-N-alkylation and amidation of amines with alcohols or carboxylic acids.[271] A procedure for the mild and efficient esterification of alkylphosphonic acids with primary alcohols has been developed, which employs the use of a polymer-bound triarylphosphine with iodine and imidazole.[272]

2.2.3 Nucleophilic attack at other atoms.

The chemistry of phosphine-borane adducts has continued to generate interest, the main emphasis again being work on 'Frustrated Lewis Pair' (FLP) systems. Addition of sodium borohydride in water to a solution of 1,2-bis-(diphenylphosphino)ethane in THF provides a very simple route to the diphosphine bis(borane) complex, the crystal structure of which was also reported.[273] Treatment of 1,2-dibromo-1,2-dimesityldiborane, Mes(Br)BB(Br)Mes, with trialkylphosphines proceeds with rearrangement of the diborane, resulting in the formation of two constitutional isomers (**88**) of a monophosphine adduct, the relative ratios of the isomers depending on the size of the phosphine.[274] Whereas the reaction of the electron-deficient aminodiborane, μ-$H_2NB_2H_5$, with tricyclohexylphosphine at room temperature gives the known Cy_3PBH_3 adduct, at $-35\,°C$ the adduct (**89**) is formed.[275] The stabilty of the, as yet unknown, bis(phosphine)borylene complexes, $(R_3P)_2BH$, (BH being isoelectronic with carbon, for which carbodiphosphorane systems $R_3P=C=PR_3 \leftrightarrow R_3P \rightarrow C \leftarrow PR_3$ are well-known,) has been the subject of a theoretical study which suggests that such species, particularly those with bulky donors, may become isolable in condensed phases.[276] The association of frustrated phosphine-borane pairs in toluene has been investigated by solvent molecular dynamics simulations of the $^tBu_3P/B(C_6F_5)_3$ system.[277] A study of the reactivities of the FLP $^tBu_2PCH_2BPh_2$ towards H_2, CO_2 and isocyanates (together with a computational study and reactivity comparisons with other FLPs) supports the view that these systems exhibit increased reactivity when the donor and acceptor sites are perfectly aligned.[278] Structural investigations of intramolecular (mainly vinylene-linked) phosphine-borane adducts using solid state NMR techniques and DFT calculations have provided new insights into the behaviour of FLP systems. It was concluded that in such FLP systems, complete suppression of normal covalent interactions between Lewis acid and base centres is not apparent and that residual electron density between these centres may be a requirement for FLP behaviour.[279] The relative reactivities of the Lewis pairs $R_3P/B(C_6F_5)_3$ (R = Me, Ph, ^tBu or C_6F_5) towards addition of dihydrogen have been investigated by quantum chemical calculations, which indicate the importance of the Lewis basicity of the donor in achieving thermodynamic feasibility for the activation of H_2.[280] Steric and electronic effects of substituents in R_3P/BR'_3 systems have also been the subject of a second theoretical study that enables quantitative insights into the ease of their H_2-splitting ability. It was concluded that when the sum of the hydride affinity of the borane and the proton affinity of the phosphine is greater than 340 kcal/mol, the ion-pair [HPR_3][HBR'_3] can be observed. Below this figure, the ion-pair decomposes with release of H_2.[281] A recent study has shown that just because a given FLP system is apparently inactive towards H_2 at room temperature does not mean that, under different conditions, activity will never appear. Thus, e.g., the FLP (C_6F_5)PPh_2/B(C_6F_5)$_3$ is apparently inactive towards H_2 at room temperature, whereas at $-80°$ C, the ion-pair [(C_6F_5)PPh_2H][HB(C_6F_5)$_3$] is observed by NMR. Nevertheless, this FLP/H_2 system does, however, lead to the hydrogenation of 1,1-diphenylethene at room temperature, implying the rapid reversible formation of the ion-pair even at room temperature.[282] Developments in the application of FLP systems as metal-free hydrogenation catalysts have been reviewed.[283] New work on additions of FLP systems to H_2, CO_2, and simple alkenes includes the use of [2,2]paracyclophane-derived bisphosphines in the domino hydrosilylation/hydrogenation of enones,[284] reactions of tms$_3$P and B(C_6F_4H)$_3$ with H_2 and CO_2 which also lead to Me_3Si group transfers from either phosphonium to phosphine or from phosphonium to oxygen,[285] use of the chiral camphor-based system (90) for the asymmetric hydrogenation of prochiral-imines, the intermediate chiral phosphoniohydridoborate zwitterion also being characterised,[286] the reactions of modified intermolecular FLP systems involving a series of sterically crowded phosphines and B(C_6F_5)$_3$ with H_2, CO_2 and ethene,[287] and the reactions of FLP systems involving the alkenylboranes (91) and ^{t}Bu$_3$P with H_2 and for the hydrogenation of enones and imines.[288] Considerable attention has been paid to the reactions of FLP systems with alkenes and alkynes. Erker's group has reported studies of the addition of various FLP systems to 6-dimethylpentafulvene[289] and conjugated diynes.[290] In related work, this group has also shown that the reaction of the bis(phosphinoalkynes) (92) with B(C_6F_5)$_3$ give the new FLP systems (93) via the intermediacy of phosphireniumborate zwitterions. Increasing the steric bulk of the substituents at phosphorus to mesityl in the starting diphosphine results instead in the formation of a benzopentafulvene-based FLP (94).[291] FLP addition reactions to alkenes[292] and alkynes[293] have also been reported by Stephan's group. This group has also reported that the reaction of sterically-demanding phosphines with B(C_6F_5)$_3$/alkyl fluorides gives phosphonium fluoroborate salts.[294] Other new work on the reactions of FLP systems includes additions to enolisable conjugated ynones,[295] thionyl chloride and halogens,[296] azides,[297] hydrosilylation reactions,[298,299] CO_2,[300] oxides of nitrogen,[301,302] and complex formation with transition metal acceptors.[303] Further work has appeared on reactions of phosphines with aluminium(III) acceptors. A theoretical study of the reaction of primary phosphines with AlH_3 has concluded that the adducts RPH_2:AlH_3 are much stronger acids than the isolated phosphines and the analogous phosphine:borane complexes.[304] Both classical and frustrated Lewis pairs of the strong Lewis acid Al(C_6F_5)$_3$ with a variety of phosphines and other donors have been found to exhibit exceptional activity in the Lewis pair polymerisation of conjugated polar alkenes.[305] Both phosphine-boron and phosphine-aluminium FLP systems have been shown to promote C–H activation of isobutylene to form boron and aluminium σ-allyl complexes.[306] Hydroalumination of aryldialkynylphosphines RP(C≡CBu^t)$_2$ (R = Ph or Mes) with equimolar quantities of diethylaluminium hydride has given mixed alkenyl-alkynyl cyclic dimers in which the dative aluminium-phosphorus bonds are geminal to the exocyclic alkenyl groups. Subsequent treatment of these with triethylaluminium affords new cyclic phosphine-alane complexes which also behave as FLP systems that activate CO_2 and phenylisocyanate.[307] Adducts of a P/Al FLP system with alkali metal hydrides have also been characterised.[308] Also reported are hydrogen activation and catalytic amine-borane dehydrogenation reactions promoted by cationic titanium(IV) and neutral titanocene-phosphinoaryl oxide FLP systems.[309]

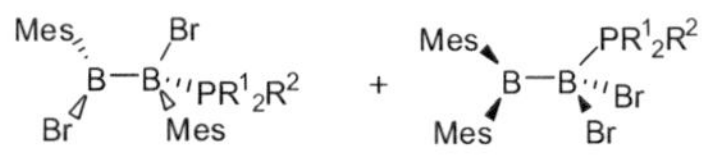

(88) R^1 = Et or Cy; R^2 = Et or Me

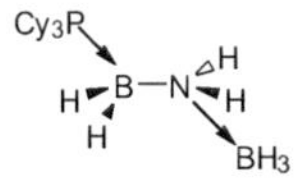

(89)

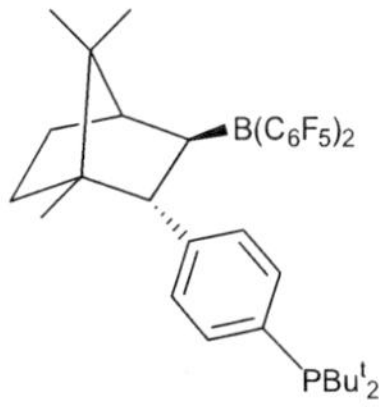

(90)

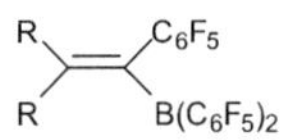

(91) R = H, Pr^n, Bu^n, Bu^t, Ph or p-Tol

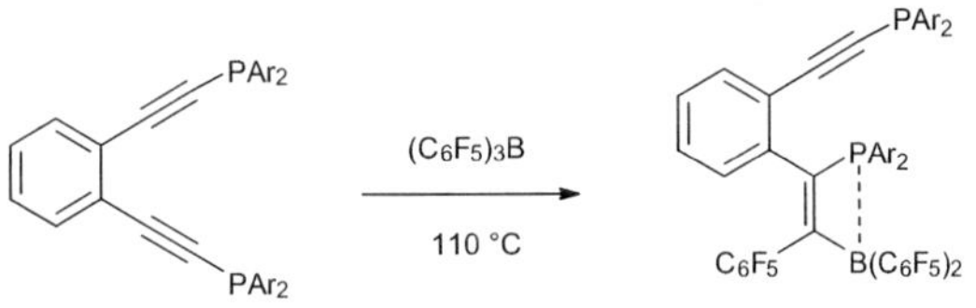

(92) Ar = Ph or o-Tolyl

(93)

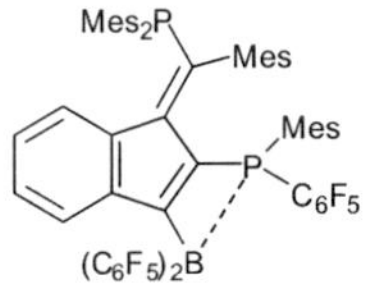

(94)

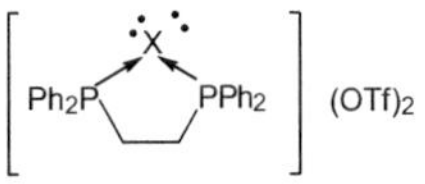

(95) X = Se or Te

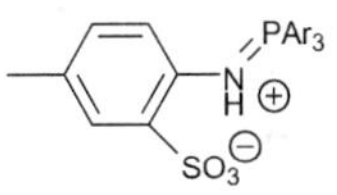

(96) Ar = Ph, m-Tolyl or p-Tolyl

The reactivity of phosphines towards oxygen, sulfur and selenium, and their compounds has also continued to generate interest, although at a very low level in the past year. Nitrene-formation in the reaction of tributylphosphine with o-nitrotoluene is the basis of a new procedure for the temperature-dependent formation of 3- and 7-methyl-3H-azepines in nucleophilic media.[310] Further applications of tris(2-carboxyethylphosphine) as a reagent for sulfur-sulfur bond cleavage in peptide chemistry have been described.[311,312] The cleavage of diphenyl-disulfide and -diselenide by triphenylphosphine in an ionic liquid solvent is the basis of a one-pot synthesis of unsymmetrical alkylaryl sulfides and selenides, via the reaction of the in situ generated phenyl-thiolate and -selenoate ions with alkyl halides.[313] Certain metal ions have been shown to promote sulfur-transfer reactions from phosphine sulfides to phosphines.[314]

Triphenylphosphine has been shown to add rapidly to a sulfur atom of molybdenum tris(dithiolene) complexes, with formation of the zwitterionic ligand $Ph_3P^+SC_6H_4S^-$.[315] The Trofimov-Gusarova group has reported further reactions between secondary phosphines and sulfur or selenium in the presence of a third reactant to give new thiophosphinate and diselenophosphinate esters.[316,317] Diphosphine- and related diarsine-stabilised complexes of Se(II) and Te(II) (**95**) have been obtained by a ligand exchange process by treatment of previously described diazabutadiene-chalcogen(II) complexes with the diphosphine or diarsine.[318] Also reported is the formation of stable phosphine and arsine complexes of the Te(II) cation, MesTe$^+$, of the type [MesTe(Ph$_3$E)]OTf (E = P or As).[319] In this context, it is interesting to note that similar attempts to stabilise related isoelectronic compounds of iodine(III) of the type PhI(OTf)$_2$ resulted only in oxidation of the phosphine ligand.[263]

Interest has also continued in the Mitsunobu and Staudinger reactions, in which nucleophilic attack by phosphorus at nitrogen is the initial step. Although there has again been no new fundamental mechanistic work on these reactions, their applications in synthetic chemistry have continued to develop, although at a much reduced rate compared to last year. New Mitsunobu procedures described, involving tertiary phosphine-diazoester combinations, include the formation of ethers from propargyl alcohols and α- or β-naphthols *en-route* to naphthofuran derivatives,[320] a route to pyrazolines from the reactions of ethenetricarboxylates, diethyl azodicarboxylate and triphenylphosphine,[321] a solvent-free approach for the synthesis of α-aminonitriles,[322] the synthesis of glycosylazetidines,[323] and procedures for the regioalkylation of hydrazines.[324] Heteroaromatic azo compounds such as 4,4′-azopyridine have been shown to be useful recyclable reagents for the direct conversion of aliphatic alcohols into symmetric disulfides under Mitsunobu conditions in combination with triphenylphosphine and ammonium thiocyanate, the resulting *N*-cyanohydrazine biproducts being easily oxidised to the corresponding azopyridine.[325] Conventional Staudinger reactions between phosphines and 2-azidobenzenesulfonates, followed by sulfonate ester deprotection, have afforded the zwitterions (**96**).[326] Staudinger ligation and aza-Wittig procedures have continued to be used in synthesis, recent applications including the development of a proline-based aminophosphine template for rapid Staudinger ligation reactions with azides,[327] a route to functionalised pyridines by an amidoalkylation/Staudinger/aza-Wittig sequence,[328] a synthesis of medium-sized cyclic peptides by a Staudinger-mediated ring-closure,[329] dendrimer-synthesis,[330] and routes to thiazolo-fused pyrimidinone-piperazines[331] and 1,4-oxazine- and 1,5-oxazocine-based sugar hybrids.[332] Two groups have reported the development of Staudinger procedures that are catalytic in the phosphine component, enabled by conducting the reaction in the presence of phenyltrisilane or diphenylsilane as reducing agent, providing obvious advantages in work-up and separation of phosphine oxide biproducts. These procedures have been applied in a simple reduction of azides to primary amines[333] and in the development of traceless Staudinger procedures that are also catalytic in the phosphine component.[334] The latter group has also reported a new traceless Staudinger ligation procedure involving a chlorophosphite reagent for the direct acyl substitution of carboxylic acids.[335]

Finally, a few papers have appeared describing aspects of the nucleophilic properties of phosphines towards the heavier group 14 elements. Equilibrium constants have been determined for the association of silylenes, R_2Si (R = Me, Ph or Mes), and the related germylenes, R_2Ge, with a range of oxygen, sulfur and phosphorus donors including triethylphosphine and tricyclohexylphosphine. The phosphines bind more strongly than the other donors, Me_2Si and Me_2Ge being borderline soft Lewis acids which are stronger and softer than trimethylborane.[336] Trimethylphosphine complexes of the Lewis acids $SnCl_4$, $SnCl_3^+$ and $SnCl_2^{2+}$ have also been characterised.[337] Also of interest is a theoretical study of carbodiphosphorane analogues $E(PPh_3)_2$ with E = silicon to lead. Although these are as yet unknown, the study reveals that these compounds should be sufficiently stable to be isolable in a condensed phase. The bonding model is viewed as being of the type $R_3P \rightarrow E \leftarrow PR_3$, involving donor-acceptor bonds to a bare central atom which retains its four valence electrons as two electron pairs,[338] a situation now agreed for the carbodiphosphoranes where E=P and for related Lewis base-stabilised dicarbon systems of the type $D \rightarrow C=C \leftarrow D$,[339] and also for related molecules of the type $D \rightarrow EE \leftarrow D$ where E is a heavier main group 14 or 15 element and ligand D is a phosphine or a carbene.[340]

2.2.4 Miscellaneous reactions. Interest in the electronic and other physicochemical properties of phosphines has continued with a study of the basicity of phosphines and diphosphines in acetonitrile, with pK_as recorded in the range from 4–16. The data have been compared with those for amines of similar structure, the phosphines displaying much smaller changes in basicity compared to amines when alkyl groups are replaced by aryl groups as a result of the much weaker resonance interaction between the aryl ring and phosphorus.[341] Triphenylphosphine acts as a base catalyst in the condensation reactions of the active methylene compound 'Meldrum's acid' with indole-3-carboxaldehydes, the initial step being proton abstraction from the acidic dicarbonyl system by the phosphine.[342] The use of phosphine-functionalised cyclodextrin systems as mass-transfer agents and ligands for aqueous organometallic catalysis has been reviewed.[343] A combined experimental and computational approach has shown that the mechanism of the reduction of activated carbonyl groups by trialkylphosphines in the presence of water proceeds either through proton transfer from the phosphine and cleavage by water during work-up, or *via* involvement of water at the initial stage and a double proton transfer to give the product, both pathways possibly operating simultaneously during the course of the reaction.[344] The effects of substituents, solvents and acids on the equilibrium between (*E*)-2-phosphinoazobenzenes and the inner phosphonium salts (**97**) have been investigated. The acceptor character of the solvent is mainly responsible for the solvent effects and acidic additives also influence the position of equilibrium depending on their ability to protonate at the nitrogen bearing the negative charge. Electron-withdrawing substituents at the 4- and 4′-positions also shift the equilibrium towards the salt and electron-donating substituents favour the phosphine. Inner phosphonium salts are not formed from the (Z)-isomer of the phosphine in the presence of acids but, on photoirradiation, the (E)-isomer is formed and enters into the equilibrium.[345]

Among a miscellany of papers reporting other aspects of phosphine reactivity is a review of 1,3-dipolar cycloadditions of nitrones to heterosubstituted alkenes that includes work on vinylphosphines and their derivatives,[346] the addition reactions of the chiral N-heterocyclic diphosphines (98) with activated alkynes or alkenes to give chiral diphosphines, e.g., (99),[347] the unexpected dimerisation of the monohydrazide of 2-diphenylphosphinobenzaldehyde to form an 8-membered ring diphosphonium system,[348] the effects of electrolysis conditions on the course of anodic oxidation of tertiary phosphines in the presence of camphene, involving the intermediacy of trialkylphosphine radical cations,[349] and the formation of 3-aza-7-phosphabicyclo[3,3,1]nonan-9-ones, e.g., (100), in the Mannich reactions of phosphorinanones, primary amines and formaldehyde.[350] The chemistry of hydroxymethylphosphines has continued to develop, the 1:1 reaction of tris(hydroxymethyl)phosphine with lignin-type (phenol functional) α,β-unsaturated aldehydes in various solvents giving tris(hydroxyalkyl)phosphonium phenolate zwitterions.[351] A study of the reaction of tris(hydroxymethyl)phosphine with $Na_2S_2O_4$ has led to the characterisation of the related secondary and primary (hydroxymethyl)phosphines and their oxides.[352] Boryl(phosphino)carbenes, B–Ë–P species (Ë=C), and their heavier group 14 analogues, (Ë=Si, Ge, Sn and Pb) have continued to attract attention. The recently prepared boryl(phosphino)carbene (101) has been shown to react at phosphorus with CO_2 to form the phosphacumulene ylide (102). With SO_2, the relatively unstable sulfine (103) is formed.[353] The reactivity of boryl(phosphino)carbenyl carbon analogues, B–Ë–P, (Ë=C, Si, Ge, Sn and Pb) has been considered using a theoretical approach, showing not only that carbenic reactivity decreases down the group but that the, as yet unknown, heavier carbene-like analogues should be isolable at room temperature.[354] P–P bond cleavage and formation reactions have also been of interest. In the presence of a chromium cyclopentadienyl β-diketiminate catalyst and an activated manganese powder reducing agent, tetraphenyldiphosphine undergoes cleavage, reacting with added chlorocyclohexane to give cyclohexyldiphenylphosphine via an intermediate cyclohexylchromium species, and providing a means of controlling the reactivity of secondary alkyl radicals.[355] Carbaborane-fused 1,2-diphosphetanes have been shown to undergo P–P bond cleavage reactions with lithium, followed by hydrogen chloride, to give secondary mono- and bis-(phosphino)carbaboranes. Treatment of the latter with formaldehyde and one equivalent of aniline gives a carbaborane-fused 1-aza-3,6-diphosphepane.[356] The newly prepared sterically-crowded diphosphine (104) has been shown to undergo homolytic P–P bond cleavage to form two phosphinyl radicals. Treatment of the diphosphine with chalcogens or P_4 gives products that involve insertion of elements into the P–P bond.[357] A two-coordinate palladium complex with two crowded dialkylphosphinyl radical ligands has also been reported.[358] The Menschutkin reactions of trimethylphosphine with alkyl- or aryl-chlorophosphines have been shown to yield phosphino-phosphonium salts, [Me$_3$P–PR$_2$]$^+$ Cl$^-$, providing a simple method for P–P bond formation which may have broader application for the formation of element-element bonds.[359] The favoured course of cleavage of the exo-cyclic R–P bond of oxaphosphirane complexes has been considered using a theoretical approach, showing that homolytic cleavage is disfavoured relative to heterolytic cleavage to give an oxaphosphiranide anion and R$^+$.[360]

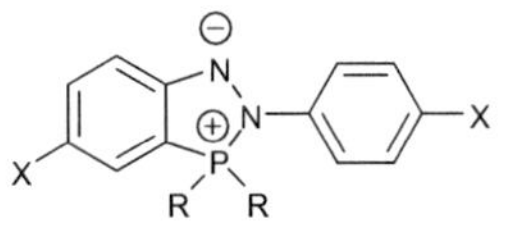

(97) X = electron-withdrawing or electron-donating group

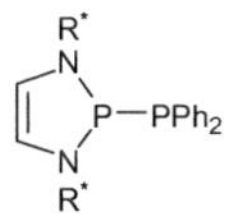

(98)

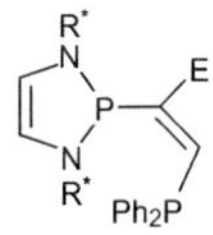

(99)

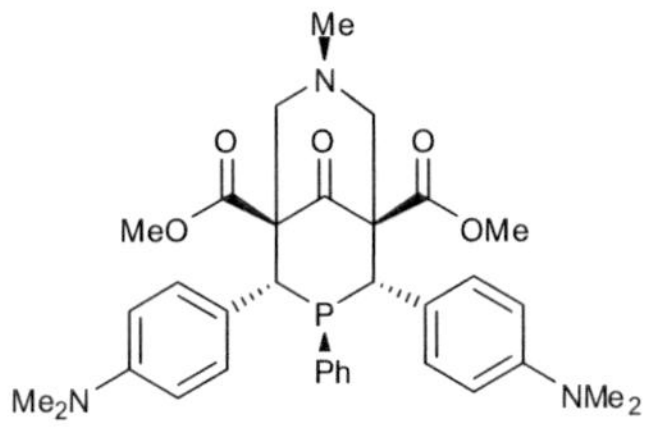

(100)

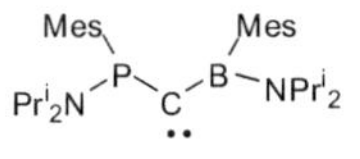

(101)

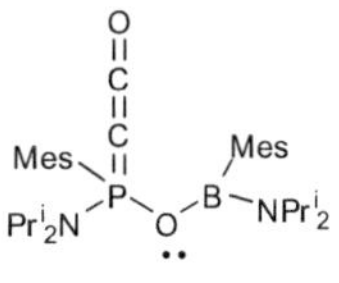

(102)

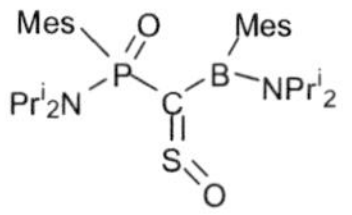

(103)

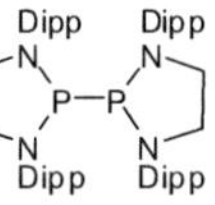

(104)

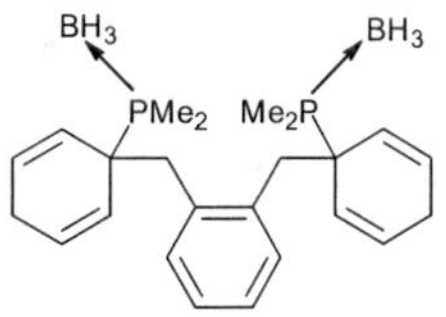

(105)

Phosphorus-element bond formation *via* dehydrocoupling reactions has continued to attract considerable interest, with particular reference to catalytic processes. The redox stability of the catalytic metal centre is the primary factor influencing catalytic *versus* stoichiometric behaviour of molecular main group compounds in homogeneous dehydrocoupling reactions.[361] Dehydrogenative processes occurring within alkylphosphines have been the subject of a recent feature article.[362] Active intermediates, product distributions and a suggested catalytic cycle have been reported for a rhodium-catalysed dehydrocoupling of the phosphine-borane H$_3$B·PBu$^t_3$ to give [HP^{t}Bu$_2$BH$_2$PBu$^t_2$BH$_3$].[363] Birch reductive de-aromatisation of aryldialkylphosphine-boranes (and related phosphine oxides), followed by treatment with reactive alkyl halides, provides a route to the corresponding α-functionalised (cyclohexa-1,4-dien-3-yl)phosphine derivatives. Thus, *e.g.*, Birch reduction of dimethyl(phenyl)phosphine-borane, followed by addition of *o*-di(bromomethyl)benzene, gave rise to the borane-protected diphosphine (**105**).[364] Symmetric sideways-on alkene-like ligand behavior has been observed for the coordination of phosphinoboranes $R_2PB(C_6F_5)_2$ to platinum.[365]

As ever, unusual phosphine ligands continue to be of interest in co-ordination chemistry. Included among many papers on phosphine complexes and the reactivity of coordinated phosphine ligands are studies of the coordination chemistry of derivatives of 2,3-bis(diphenylphosphino)maleic anhydride,[366] the solution structures and dynamics of P,N-containing heterocycles and their metal complexes,[367] intramolecular rearrangements of coordinated alkynylphosphines,[368] the photoelectronic properties of phosphametallacycle-linked coplanar porphyrin dimers,[369] and complexes of stable *N*-phosphino-1,2,4-triazol-5-ylidenes.[370] Further work has appeared on nucleophilic additions to chelated $(Ph_2P)_2C=CH_2$.[371] The application of 1,1′-bis(diphenylphosphino)ferrocene for the synthesis of functional molecular materials has been the subject of a review.[372] Finally, it is interesting to note interest in the ligand properties of stable fluorophosphines based on a phosphaadamantane cage or a phosphabicyclic structure.[373]

3 p$_\pi$-Bonded phosphorus compounds

Although activity in this area has increased compared to the previous year, work in specific areas *i.e.* the chemistry of phosphaalkenes and the less-developed groups of low coordination number phosphorus compounds, in particular phosphenium ions, phosphinidenes, and their metal complexes, has again dominated the area. Long-established topics such as the chemistry of diphosphenes and phosphaalkynes have continued to attract attention but to a much smaller extent. Interest in assessing the ability of the ferrocene system to stabilise a diphosphene has led to the synthesis of the stable 1,2-bis(ferrocenyl)diphosphene (**106**), in which the sterically-crowded ferrocene units provide the necessary kinetic stability, the unsubstituted parent Fc–P=P–Fc being, as yet, not fully isolated or characterised.[374] Also reported is the diphosphene-containing organometallic complex (**107**) derived from the heptaphosphide trianion,[375] a study of the photophysics of the diphosphene Dmp–P=P–Dmp (Dmp=2,6-dimesitylphenyl),[376] and a theoretical study of substituent effects on the relative energies of diphosphenes R–P=P–R and the, as yet unknown, isomeric diphosphinylidenes $R_2P=P$ that may lead to new preparative studies.[377] New routes to diphosphenes and phosphaalkenes have been developed from the reactions of lithium (2,4,6-tri-*t*-butylphenyl)silylphosphides with haloforms.[378] A new route to P=C double bond systems is provided by deoxygenation of oxaphosphirane complexes by *in situ*-generated Ti(III) complexes.[379] Alternative routes have been developed for the synthesis of *C*-monoacetylenic phosphaalkenes, *e.g.*, (**108**).[380] Among other new phosphaalkenes reported are enantiomerically pure phosphaalkenyl-oxazolines (**109**), accessed by the

(106) Dmp = 3,5-Mes$_2$C$_6$H$_3$ **(107)**   **(108)** R^1 = H, Me or C(OH)Ph$_2$
                                                                          R^2 = SiMe$_3$

reactions of a lithium silylphosphide with a chiral oxazolinyl-functionalised ketone (a phospha-Peterson reaction),[381] the bisphosphaalkene (**110**),[382] and the 4,5,6-triphospha[3]radialene (**111**).[383] Electrochemical reduction of the latter gave a related isolable, delocalised dianion which did not prove to be an aromatic system.[384] Phospharadialenes have also been the subject of an overview which comments on possible extensions of this area of work to include polyphospharadialenes based on larger central ring systems.[385] A range of W-shaped anions of the general formula [R$_2$C=P]$_2$E$^-$ (E=N, P, As or Sb) has been investigated by theoretical methods and a new member of the series (E=As) prepared. Structural studies on the known phosphorus and the recently prepared asenic-bridged anions support the theoretical prediction that a significant reduction in the P–E–P bond angle arises when the R$_2$C moiety is a strong π-acceptor, making the central P–E–P unit positively charged, giving a structure that is π-isoelectronic with the aromatic cyclopropylium cation and indicating an enhanced homoaromatic character.[386] UV-irradiation of the [4+2]cycloadduct (**112**) of [1-methyl-2-keto-1,2-dihydrophosphinine-W(CO)$_5$] with dimethyl acetylenedicarboxylate leads to the generation of the methylphosphaketene complex (**113**) which easily loses carbon monoxide to give the related methylphosphinidene complex. Both photolysis products can be trapped with reactive alcohols.[387] Gates *et al.* have shown that phosphaalkenes can be activated on treatment with electrophiles. With triflic acid, the phosphaalkenes RP=CHBu^t (R=Bu^t or Ad) give diphosphiranium salts (**114**) as two stereoisomers, whereas with methyl triflate, the diphosphetanium salts (**115**) are formed.[388] This group has also shown that treatment of phosphaalkenes with *N*-heterocyclic carbenes results in the formation of 4-phosphinocarbenes (**116**).[389] The phosphaalkene MesP=CPh$_2$ has been shown to add to arylacetylenes to give luminescent benzo-1,2-dihydrophosphinines (**117**).[390] The electrochemical oxidation of C-(dialkylamino)-substituted phosphaalkenes has also been investigated.[391] Other work relating to the reactivity of phosphaalkenes is a quantum chemical investigation of the products of their sulfurization,[392] a density functional study of 4π-electron-four-membered rings involving Si=P bonds,[393] and a theoretical study of the As=C bond in arsaallenes and arsaphosphaallenes.[394] New bis(phosphaalkenyl)germanium(II) and tin(II) systems (**118**) have been prepared and stabilised as their *N*-heterocyclic carbene complexes and their coordination chemistry studied.[395,396]

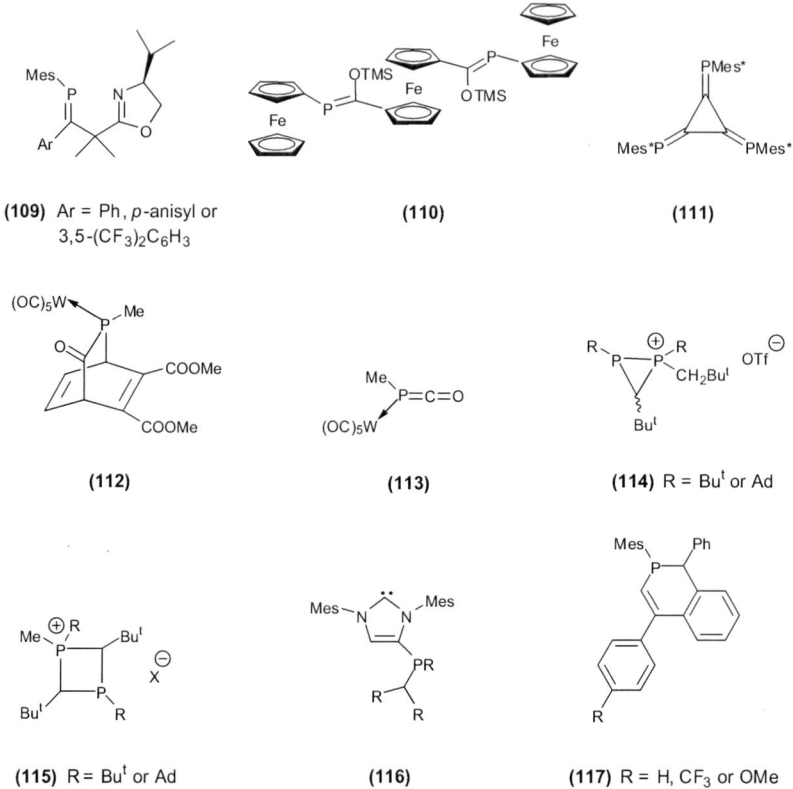

(109) Ar = Ph, p-anisyl or 3,5-(CF$_3$)$_2$C$_6$H$_3$

(110)

(111)

(112)

(113)

(114) R = Bu^t or Ad

(115) R = Bu^t or Ad

(116)

(117) R = H, CF$_3$ or OMe

As in the previous year, relatively little new work has appeared relating to p$_\pi$-bonded phosphorus compounds involving triple bonds from phosphorus to a group 14 or group 15 element. Interest in the chemistry of the phosphaethynolate anion (OC≡P)$^-$ has continued. A new route to it is provided by the reaction of an anionic borane-capped niobium phosphide complex with carbon monoxide in diethyl ether at room temperature.[397] The ligand properties of the phosphaethynolate ion have been compared with those of the isocyanate ion in studies of their rhenium complexes. Structural and theoretical work shows that while the ReNCO skeleton is linear, the RePCO skeleton is bent at phosphorus and is best described as a Re(I) metallaphosphaketene derivative.[398] The co-ordination chemistry of the phosphaalkyne Me$_3$SiC≡P has also been studied, showing that this molecule readily binds to a variety of transition metals in various coordination modes.[399] Complexes between the interhalogen FCl and P≡CX species have been investigated by theoretical methods.[400] Further work has been published on the formation (from Mes*C≡P and ^{t}BuLi) and reactivity of 1,3-diphosphacyclobutane-2,4-diyls.[401] The insertion reactions of phosphaalkynes into silicon-silicon bonds are included in a review of bis-silylene chemistry.[402] Finally, it is interesting to note a theoretical study of the nature of the carbon–arsenic bond in arsaalkynes that reveals that in most systems of this type, the bonds are essentially triple bonds with similar characteristics to those of the phosphaalkyne analogues.[403]

(118) E = Ge or Sn   **(119)** Ter = 2,6-bis(trimethylphenyl)phenyl   **(120)**

The chemistry of phosphenium ions (R_2P:$^+$ and RP:$^{2+}$, and related monophosphorus cationic species) and phosphinidenes (RP:) has also continued to attract attention. Fluoride ion affinity data have been used in a theoretical treatment of the Lewis acidity of phosphenium ions, showing that phosphenium ions are often stronger Lewis acids than neutral species but are generally less Lewis acidic than cations such as Me_3C^+ or Me_3Si^+.[404] The stabilisation of phosphenium ions by interaction with adjacent heteroatoms or intermolecular donors, usually nitrogen or phosphorus, has again been a major theme in the past year. The phosphenium salts (**119**) have been prepared, involving a range of different anions, structural studies of which have revealed significant steric interactions between the cation and anion which affect the ability of the anion to approach the cationic centre.[405] Variation of the nature of the potential anionic groups X and Y in the PNP pincer system (**120**) (accessed initially from the reaction of a bis(phosphino)diarylamine with PCl_3, followed by chloride exchange with other anions) has provided a range of neutral, mono- and di-cationic phosphorus (III) compounds.[406] The low-temperature reaction of $(Me_3Si)_2NPCl_2$ and $GaCl_3$ has given the disilylaminophosphenium salt $[R_2N=PCl]^+GaCl_4^-$, (R=Me_3Si), which decomposes at room temperature to give the stable species $[R–N\equiv P]^+$ $GaCl_4^-$. Treatment of the initial phosphenium ion with trimethylsilyl azide provides the highly reactive azidophosphenium salt $[R_2N=PN_3]^+$ $GaCl_4^-$, stable only below $-40\,°C$.[407] Phosphine complexes of an enantiomerically pure, atropisomeric arsenium cation have also been described.[408] Also reported are Lewis base-stabilised phosphenium oxides and sulfides $[D:\rightarrow P(X)R_2]^+$ OTf^- (X=O or S).[409,410] Considerable interest has been shown in studies of the coordination chemistry of stabilised phosphenium cations. Reviews have appeared of metal complexes of planar R_2P ligands with particular reference to their analogy with carbene ligands,[411] and also of metal complexes of carbene-stabilised phosphenium cations.[412] New work has been reported on the coordination chemistry of carbene-stabilised phosphenium cations towards Rh(I)[413] and Cu(I) acceptors.[414] Analogies have been drawn between the donor properties of nitrogen-stabilised heterocyclic phosphenium cations and nitrosyl ligands.[415] Platinum(II) complexes of cyclic diphosphine-stabilised triphosphenium cations have been characterised in solution by ^{31}P NMR spectroscopy.[416] Also reported is a DFT study of the coordination preferences of phosphenium ions with molybdenum in the presence of other phosphorus ligands.[417] Donor-stabilised complexes of P(I) cations have also continued to attract attention. Ragogna's group has reported new imino-based chelating ligand systems for the stabilisation of

P⁺ and other heavier Group 15 related cationic species.[418,419] Weigand and Feldman have published an overview of recent progress in the synthesis of multiply-charged P(III)-centred cations.[420] Weigand's group has also reported routes to cationic polyphosphorus-cage systems of the type [R₅PCl]⁺ via insertion of [RPCl]⁺ cations into P₄.[421,422] The Weigand and Burford groups have also reported routes to ligand-stabilised [P₄]²⁺ cations.[423] The relative stabilities of cyclic dicationic derivatives of diphosphines, involving three or four linked phosphorus atoms, have been investigated.[424] Carbene-stabilised trications of the type [L₃P]³⁺ have been used as ligands in some platinum(II)-catalysed reactions.[425]

Interest in the chemistry of phosphinidenes, RP:, and terminal phosphinidene complexes, RP:→[M], has again been maintained. The chemistry of phosphanylidene-σ^4-phosphoranes, RP=PR₃, (regarded as phosphine-complexed phosphinidenes, i.e., ArP←PR₃), has been reviewed.[426] The *peri*-acenaphthene phosphanylidene-phosphorane (**121**) has been isolated as a very air-sensitive red solid, stable at room temperature for several weeks, but which on treatment with Pd(Ph₃P)₄ gives the diphosphene complex (**122**).[427] Among new metal-complexed phosphinidenes reported is a tantalum-phosphinidene-complex, prepared by the reaction of a three-membered ring tantalaziridine complex with LiPHPh,[428] and side-on bonded phosphanylphosphinidene complexes of tungsten.[429] New aspects of the reactivity of terminal phosphinidene complexes include studies of their reactions with dihydrogen (initially giving secondary diphosphine complexes and then complexed primary phosphines),[430] the deoxygenation of carbon dioxide by electrophilic terminal phosphinidene complexes of chromium and tungsten,[431] insertion into the B–H bonds of amine- and phosphine-boranes,[432] and computational studies of copper-catalysed phosphinidene transfer to ethylene, acetylene and CO.[433] Ruiz et al. have reported a series of studies of the reactivity of phosphinidene-bridged metal complexes.[434] Also reported is a study of the effects of steric strain in the phosphiranes formed from the reactions of transient phosphinidene complexes with norbornene,[435] the involvement of phosphinidene (and arsinidene) complexes in the activation of 'non-innocent' Cp* substituents leading to novel cage compounds via cascade-like reactions[436] and a structure and bonding analysis of some cationic electrophilic phosphinidene complexes of iron, ruthenium and osmium.[437] The chemistry of lithium phosphinidenoid complexes of the type RPXLi[W(CO)₅] (X=halogen) has received further attention from Streubel's group. The reactivity of terminal phosphinidene versus Li-Cl phosphinidenoid complexes in cycloaddition chemistry has been compared.[438] The first room temperature-stable Li-Cl phosphinidenoid complex has been prepared, involving a P-trityl substituent.[439] Reactions of lithium phosphinidenoid complexes of the type RPXLi[W(CO)₅] (X=halogen) with TEMPO,[440] cyclic ketones having an oxygen atom in the ring system[441] and alkenyl-substituted aldehydes[442] have also been investigated. Although not a $\lambda^3\pi$-system, it is interesting to note here a report of the synthesis of a crystalline singlet phosphinonitrene (**123**), involving a very short phosphorus-nitrene bond, suggesting that it is a rare example of phosphorus $\sigma^3\lambda^5$-coordination.[443]

(121)   (122)   (123)

4 Phosphirenes, phospholes and phosphinines

Interest in potentially aromatic heterocyclic systems has continued at a similar level to the previous year, with most work again relating to the chemistry of fused ring phospholes and phospholyl anions. Activity in the phosphirene area, however, has again been minimal in the past year. Streubel's group has studied the thermal reactions of 2H-phosphirene complexes (**124**) with primary and secondary amine derivatives[444] and alcohols,[445] leading to the formation of aminophosphine and alkoxyphosphine complexes, so as to explore the boundaries between complexed phosphanide, phosphinidenoid and transient phosphinidene intermediates.

New synthetic work in phosphole chemistry continues to be driven by interest in the optical properties and molecular electronics potential of phospholes and their oxides and sulfides, particularly where the phosphole unit is part of a fused ring system or where phosphole rings are linked to other polarisable aromatic heterocyclic systems. Phosphole-based functional materials have been the subject of a recent review.[446] New phosphole systems reported which relate to this area of work include various phosphole-modified pentathienoacenes, e.g., (**125**),[447] carbazole-bridged dithienophospholes,[448] dithienophospholes bearing p-diphenyl-aminophenyl[449] or polyaromatic hydrocarbon[450] substituents at phosphorus, 2,2'-bis(benzo[b]phosphole) and 2,2'-benzo[b]phosphole-benzo[b]heterole hybrid systems,[451] bisphosphole-bridged ladder oligophenylenes,[452] the $\sigma^3\lambda^3$-dibenzophosphapentaphene (**126**),[453] the acenaphtho-fused phosphole (**127**)[454] and new P-chirogenic dibenzophospholes, e.g., (**128**).[455] Routes have also been developed for the synthesis of various phosphole-based phosphahelicenes, e.g., (**129**)[456] and (**130**),[457] the properties of which have so far been little studied. Also described is the synthesis of the bis(azido)-functional phosphole sulfide (**131**), subsequently used to prepare related phospholes bearing triazole-functionalised 2,5-diaryl substituents,[458] the 1,1-carboboration of bis(alkynyl)phosphines giving the phospholes (**132**),[459] the preparation of a series of 2,2'-biphospholes for tuning the HOMO-LUMO gap of π-systems,[460] and the synthesis of phospholes bearing a phosphinomethyl substituent at the phosphole phosphorus.[461] Other preparative work of interest is the synthesis of rhenium complexes of phosphole-pyridine chelate systems having large chiroptical properties,[462] and a series of highly substituted benzylphospholium salts derived from dithienophospholes, also having unusual photophysical properties.[463,464] Theoretical studies of phospholes have been few in number in the past year. The influence of electron-donating substituents on the conductivity of phosphole oligomers has been investigated by DFT methods[465] and the influence of complexation at phosphorus by the small polarising cations Li^+, Be^{2+} and Al^{3+} on the aromatic character of the parent phosphole system has been investigated by a combined MP2 and DFT approach, revealing that the aromatic character increases on complex formation.[466] A new aspect of the Diels-Alder reactivity of 2H-phospholes has emerged in the development of a facile stereoselective synthesis of P-chiral 1-phosphanorbornenes, utilising diastereotopic face differentiation.[467]

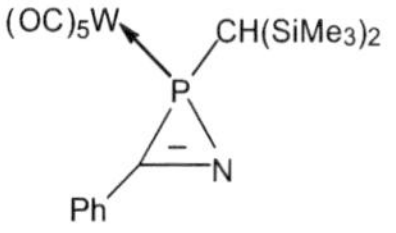

(124)

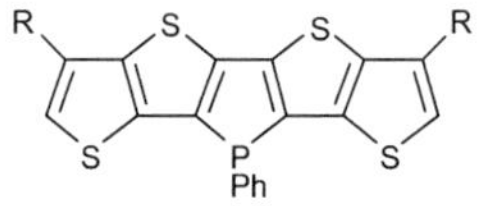

(125) R = Me or C_6H_{13}

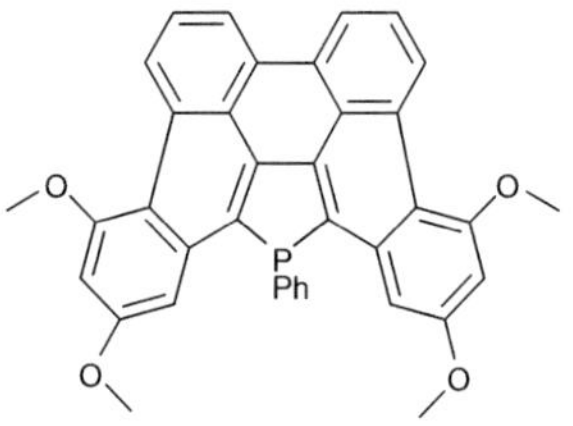

(126)

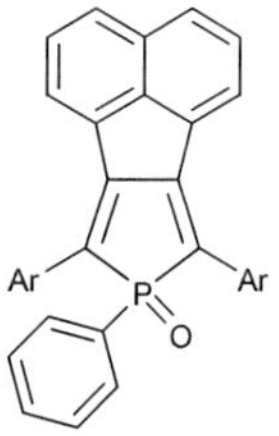

(127)

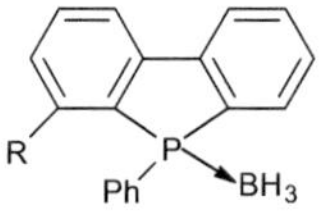

(128) R = CH(Ph)OMOM

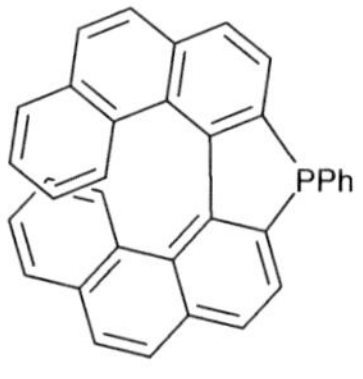

(129)

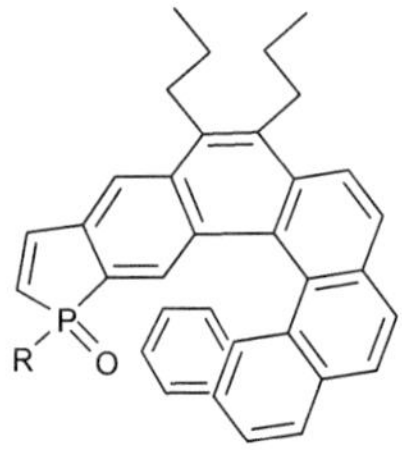

(130) R = Ph or menthyl

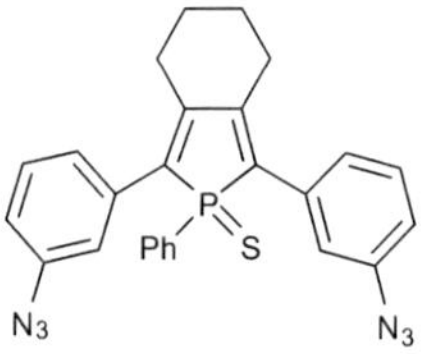

(131)

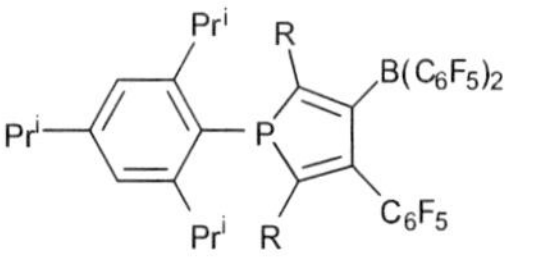

(132) R = Ph, Pr^n or $SiMe_3$

Interest in the chemistry of heterophospholes, possessing one or more additional heteroatoms as ring-members, has also continued. The reactions of 1-alkyl-1,2-diphospholes with diphenyldiazomethane proceed initially *via* addition to the P=C bond to give the bicyclic phosphiranes, *e.g.*, (133). In contrast, the related reactions with nitrones result initally in oxidation at phosphorus, followed by other cycloaddition steps.[468] A series of naphtho[2,3-d]-1,3-oxaphospholes has been prepared as new examples of fluorescent phospha-acenes[469] and a new route to highly-substituted 1,2-oxaphospholes (and some conventional phospholes) is provided by a series of cascade reactions starting from the phospha-Wittig-Horner reagent $(OC)_5WP(H)(Ph)P(O)(OEt)_2$ and dialkynylketones in the presence of LDA.[470] The first annelated 1,3-azaphospholoferrocenes have been prepared and structurally characterised[471] and a new route to N-methyl-1,3-benzazaphospholes has been developed.[472] Tautomeric equilibria and aromaticity of various types of phosphodiazole have been studied by *ab initio* theoretical methods, the results showing a clear preference for the tautomeric forms in which the mobile proton is connected to a nitrogen atom, the most stable being those possessing a pyrrole-like nitrogen and therefore having the greater aromatic stabilisation.[473] Also reported are studies of the reversible dissociation and subsequent reactivity of a sterically crowded tetrahydro-2,2'-bis(1,3,2-diazaphosphole).[474]

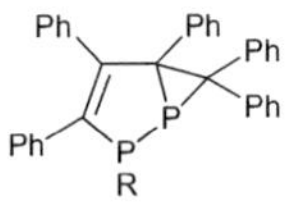

(133) R = alkyl

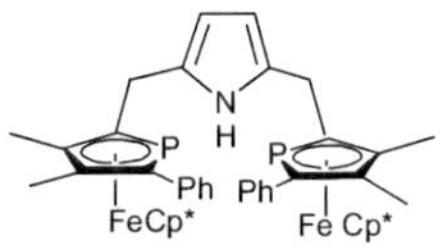

(134)

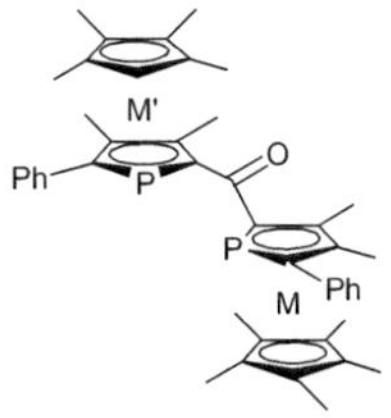

(135) M = M' = Fe or Ru; M = Fe, M' = Ru

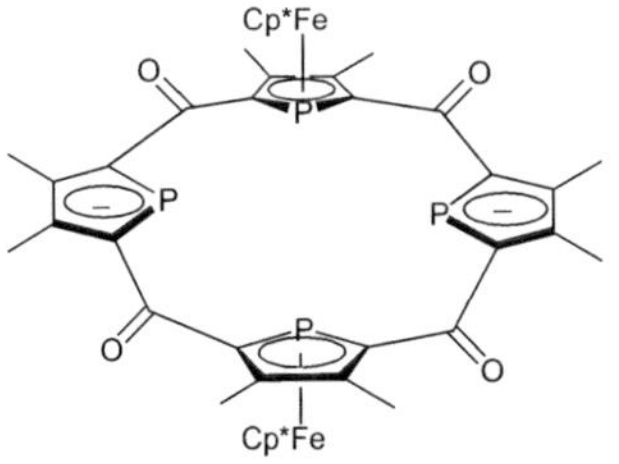

(136)

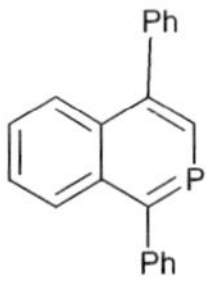

(137)

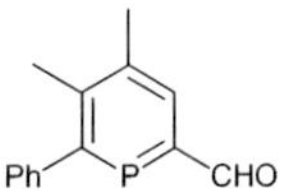

(138)

The chemistry of pholide and related anions and their metallocene complexes also continues to generate interest. Of note is a commentary on the impact of Marianne Baudler's work in 1987 leading to the discovery of the pentaphosphacyclopentadienide ion, $[P_5]^-$.[475] Recent work on the synthesis of pholide and related anions includes a route to 1,2,3-tripnictolide anions $[E_3C_2H_2]^-$ (E = P or As) by the reaction of DMF solutions of K_3E_7 with acetylene, and from which rhodium-metallocene complexes have also been obtained.[476] Also reported is the synthesis and structural characterisation of a 1,3-diphospholyl-magnesium(II) complex involving the anion $[1,3-P_2C_3Bu^t_3]^-$, generated by the reaction of the phosphaalkyne P≡CBu^t with sodium metal, followed by treatment with Mg(II)Cl$_2$.[477] New work on phosphaferrocenes includes the synthesis of the new phosphaferrocene-pyrrole-phosphaferrocene P-N-P pincer ligand (134),[478] a route to vinyl- and alkynyl-phosphaferrocenes (and their derivatives) starting from 2-formyl-3,4-dimethylphosphaferrocene,[479] a diastereoselective route to enantiopure keto-*bis*-(2-phosphametallocene)s, *e.g.*, (135),[480] a stepwise double ring-expansion methodology leading to the synthesis of *tri*- and *tetra*-phosphaporphyrinogens, *e.g.*, (136),[481] the immobilisation of a monophosphaferrocene on a silica support,[482] and the resolution and determination of the absolute configuration of 2,2′-diacetyl-3,3′,4,4′-tetramethyl-1,1′-diphosphaferrocenes.[483] Treatment of the pentaphosphaferrocene [Cp*Fe(η-P_5)] with Cu(I) halides in the presence of different templates has given novel fullerene-like spherical molecules that serve as hosts for the templates.[484] Other work on phosphametallocenes includes a theoretical study of various coordination modes in binuclear cobalt carbonyl complexes bridged by phospholyl ligands,[485] the insertion of a phospholide unit into a rhodium-rhodium bond,[486] a study of the influence of added ligands on the redox chemistry of cyclopentadienyl- and phospholyl-samarium complexes,[487] and the immobilisation onto periodic mesoporous silica SBA-15 of lanthanum- and neodymium-monophospholide complexes.[488]

Interest in the synthesis and reactivity of the six-membered, potentially aromatic, phosphinine ring system has also continued, but at a much lower level than in recent years. New synthetic work includes the application of the pyrylium salt route to phosphinine synthesis, this time starting from pyrylium salts bearing chiral substituents to give the related chiral phosphinines,[489] and the development of new routes to the 2-phosphanaphthalene (137)[490] and the phosphinine-2-aldehyde (138).[491] Also reported is an approach to the synthesis of 1,2-dihydro-1,4,2-benzodiazaphosphinines,[492] cationic gold(I) complexes of 2,4,6-tri-t-butyl-1,3,5-triphosphabenzene,[493] and the synthesis of some λ^5-phosphinines from phosphonium-iodonium ylides.[494]

References

1. O. I. Kolodiazhnyi, *Tetrahedron: Asymmetry*, 2012, **23**, 1.
2. G. J. Rowlands, *Isr. J. Chem.*, 2012, **52**, 60.
3. Y. Morisaki and Y. Chujo in *'Polymeric Chiral Catalyst Design and Chiral Polymer Synthesis'*, Ed. S. Itsuno, John Wiley & Sons, Inc.: Hoboken, NJ, 2011, 457–488.
4. E. Bernoud, R. Veillard, C. Alayrac and A. C. Gaumont, *Molecules*, 2012, **17**, 14573.
5. I. R. Butler, *Eur. J. Inorg. Chem.*, 2012, 4387.
6. A. Velian and C. C. Cummins, *J. Am. Chem. Soc.*, 2012, **134**, 13978.
7. Y. Dai, X. Feng, B. Wang, R. He and M. Bao, *J. Organomet. Chem.*, 2012, **696**, 4309.
8. P. A. Donets, T. Saget and N. Cramer, *Organometallics*, 2012, **31**, 8040.
9. S. C. Reynolds, R. P. Hughes, D. S. Glueck and A. L. Rheingold, *Org. Lett.*, 2012, **14**, 4238.
10. (a) S. Ueda, S. Ali, B. P. Fors and S. L. Buchwald, *J. Org. Chem.*, 2012, **77**, 2543; and (b) N. Hoshiya and S. L. Buchwald, *Adv. Synth. Catal.*, 2012, **354**, 2031.
11. (a) M. Su and S. L. Buchwald, *Angew. Chem. Int. Ed.*, 2012, **51**, 4710; and (b) L. Salvi, N. R. Davis, S. Z. Ali and S. L. Buchwald, *Org. Lett.*, 2012, **14**, 170.
12. B. Milde, D. Schaarschmidt, T. Rüffer and H. Lang, *Dalton Trans.*, 2012, **41**, 5377.

13 Y. Wang, M. J. P. Vaismaa, K. Rissanen and R. Franzén, *Eur. J. Org. Chem.*, 2012, 1569.
14 B. Vabre, Y. Canac, C. Duhayon, R. Chauvin and D. Zargarian, *Chem. Commun.*, 2012, **48**, 10446.
15 (a) T. Lovasz, E. Gal, L. Găină, I. Sas, C. Cristea and L. Silaghi-Dumitrescu, *Studia UBB. Chemia*, 2010, **55**, 248; *Chem. Abstr.*, 2012, **156**, 390224; and (b) T. Lovasz, E. Gala, C. Cristea and L. Silaghi-Dumitrescu, *Studia UBB. Chemia*, 2011, **56**, 191; *Chem. Abstr.*, 2012, **157**, 492775.
16 (a) S. Sauerbrey, P. K. Majhi, G. Schnakenburg, A. J. Arduengo (III) and R. Streubel, *Dalton Trans.*, 2012, **41**, 5368; and (b) S. Sauerbrey, P. K. Majhi, S. Schwieger, A. J. Arduengo (III) and R. Streubel, *Heteroatom Chem.*, 2012, **23**, 513.
17 (a) K. H. Chung, C. M. So, S. M. Wong, C. H. Luk, Z. Zhou, C. P. Lau and F. Y. Kwong, *Chem. Commun.*, 2012, **48**, 1967; and (b) S. M. Wong, C. M. So, K. H. Chung, C. H. Luk, C. P. Lau and F. Y. Kwong, *Tetrahedron Lett.*, 2012, **53**, 3754.
18 D. L. Dodds, M. D. K. Boele, G. P. F. van Strijdonk, J. G. de Vries, P. W. N. M. van Leeuwen and P. C. J. Kamer, *Eur. J. Inorg. Chem.*, 2012, 1660.
19 X. Hao, J. Huan, G.-A. Yu, M.-Q. Qiu, N.-F. She, Y. Sun, C. Zhao, S.-L. Mao, J. Jin and S.-H. Liu, *J. Organomet. Chem.*, 2012, **706–707**, 99.
20 W. Geng, W.-X. Zhang, W. Hao and Z. Xi, *J. Am. Chem. Soc.*, 2012, **134**, 20230.
21 B. Milde, D. Schaarschmidt, P. Ecorchard and H. Lang, *J. Organomet. Chem.*, 2012, **706–707**, 52.
22 A. C. T. Kuate, S. K. Mohapatra, C. G. Daniliuc, P. G. Jones and M. Tamm, *Organometallics*, 2012, **31**, 8544.
23 I. Philipova, G. Stavrakov and V. Dimitrov, *Tetrahedron: Asymmetry*, 2012, **23**, 927.
24 J. Xing, P. Cao and J. Liao, *Tetrahedron:Asymmetry*, 2012, **23**, 527.
25 L. Piche, J.-C. Daigle, G. Rehse and J. P. Claverie, *Chem. Eur. J.*, 2012, **18**, 3277.
26 W. Wu, S. Wang, Y. Zhou, Y. He, Y. Zhuang, L. Li, P. Wan, L. Wang, Z. Zhou and L. Qiu, *Adv. Synth. Catal.*, 2012, **354**, 2395.
27 N. J. Farrer, K. L. Vikse, R. McDonald and J. S. McIndoe, *Eur. J. Inorg. Chem.*, 2012, 733.
28 M. Lohan, B. Milde, S. Heider, J. M. Speck, S. Krausse, D. Schaarschmidt, T. Rüffer and H. Lang, *Organometallics*, 20102, **31**, 2310.
29 V. I. de Paula, C. A. Sato and R. G. Buffon, *J. Braz. Chem. Soc.*, 2012, **23**, 258.
30 J. Ruiz and A. F. Mesa, *Chem. Eur. J.*, 2012, **18**, 4485.
31 K. Endo, D. Hamada, S. Yakeishi, M. Ogawa and T. Shibata, *Org. Lett.*, 2012, **14**, 2342.
32 X. Wang, F. Meng, Y. Wang, Z. Han, Y.-J. Chen, L. Liu, Z. Wang and K. Ding, *Angew. Chem. Int. Ed.*, 2012, **51**, 9276.
33 C. E. Averre, M. P. Coles, I. R. Crossley and I. J. Day, *Dalton Trans.*, 2012, **41**, 278.
34 H. Petzold and A. I. S. Alrawashdeh, *Chem. Commun.*, 2012, **48**, 160.
35 T. Saget, S. J. Lemouzy and N. Cramer, *Angew. Chem. Int. Ed.*, 2012, **51**, 2238.
36 A. F. Khlebnikov, S. I. Kozhushkov, D. S. Yufit, H. Schill, M. Reggelin, V. Spohr and A. de Meijere, *Eur. J. Org. chem.*, 2012, 1530.
37 C. von Hänisch and A. Kracke, *Z. Anorg. Allg. Chem.*, 2012, **638**, 383.
38 H. Ibrahim, C. Bournaud, R. Guillot, M. Toffano and G. Vo-Thanh, *Tetrahedron Lett.*, 2012, **53**, 4900.
39 S. Balogh, G. Farkas, J. Madarász, A. Szöllősy, J. Kovács, F. Davas, L. Ürge and J. Bakos, *Green Chemistry*, 2012, **14**, 1146.

40   R. Tan, F. S. N. Chiu, A. Hadzovic and D. Song, *Organometallics*, 2012, **31**, 2184.
41   J. Bayardon, H. Laureano, V. Diemer, M. Dutartre, U. Das, Y. Rousselin, J.-C. Henry, F. Colobert, F. R. Leroux and S. Jugé, *J. Org. Chem.*, 2012, **77**, 5759.
42   (a) T. Imamoto, K. Tamura, Z. Zhang, Y. Horiuchi, M. Sugiya, K. Yoshida, A. Yanagisawa and I. D. Gridnev, *J. Am. Chem. Soc.*, 2012, **134**, 1754; and (b) Z. Zhang, K. Tamura, D. Mayama, M. Sugiya and T. Imamoto, *J. Org. Chem.*, 2012, **77**, 4184.
43   S. M. Barolo, S. E. Martín and R. A. Rossi, *ARKIVOC*, 2012, **viii**, 98.
44   M. A. Guino-o, A. H. Zureick, N. F. Blank, B. J. Anderson, T. W. Chapp, Y. Kim, D. S. Glueck and A. L. Rheingold, *Organometallics*, 2012, **31**, 6900.
45   N. Grüger, H. Wadepohl and L. H. Gade, *Dalton Trans.*, 2012, **41**, 14028.
46   P. E. Sues, A. J. Lough and R. H. Morris, *Organometallics*, 2012, **31**, 6589.
47   E. Bélanger, M.-F. Pouliot, M.-A. Courtemanche and J.-F. Paqin, *J. Org. Chem.*, 2012, **77**, 317.
48   S. Yoshida, K. Igawa and K. Tomooka, *J. Am. Chem. Soc.*, 2012, **134**, 19358.
49   (a) R. S. P. Turbervill and J. M. Goicoechea, *Chem. Commun.*, 2012, **48**, 1470; and (b) R. S. P. Turbervill and J. M. Goicoechea, *Organometallics*, 2012, **31**, 2452.
50   N. K. Gusarova, S. N. Arbuzova and B. A. Trofimov, *Pure Appl. Chem.*, 2012, **84**, 439.
51   (a) S. F. Malysheva, A. V. Artem'ev, N. A. Belogorlova, A. O. Korocheva, N. K. Gusarova and B. A. Trofimov, *Russ. J. Gen. Chem.*, 2012, **82**, 1307; (b) B. A. Trofimov, A. V. Artem'ev, S. F. Malysheva, N. K. Gusarova, N. A. Belogorlova, A. O. Korocheva, Y. V. Gatilov and V. I. Mamatyuk, *Tetrahedron Lett.*, 2012, **53**, 2424; and (c) S. F. Malysheva, A. O. Korocheva, N. A. Belogorlova, A. V. Artem'ev, N. K. Gusarova and B. A. Trofimov, *Doklady Chem.*, 2012, **445**, 164.
52   B. A. Trofimov, N. K. Gusarova, A. V. Artem'ev, S. F. Malysheva, N. A. Belogorlova, A. O. Korocheva, O. N. Kazheva, G. G. Alexandrov and O. A. Dyachenko, *Mendeleev Commun.*, 2012, **22**, 187.
53   B. A. Trofimov, N. K. Gusarova, A. V. Artem'ev, S. F. Malysheva, N. A. Belogorlova and A. O. Korocheva, *Heteroatom Chem.*, 2012, **23**, 411.
54   K. Izod, J. M. Watson, W. Clegg and R. W. Harrington, *Eur. J. Inorg. Chem.*, 2012, 1696.
55   T. M. A. Al-Shboul, G. Volland, H. Görls, S. Krieck and M. Westerhausen, *Inorg. Chem.*, 2012, **51**, 7903.
56   K. Izod, *Coord. Chem. Rev.*, 2012, **256**, 2972.
57   S. Almstätter, M. Eberl, G. Balázs, M. Bodensteiner and M. Scheer, *Z. Anorg. Allg. Chem.*, 2012, **638**, 1739.
58   E. C. Y. Tam, N. A. Maynard, D. C. Apperley, J. D. Smith, M. P. Coles and J. R. Fulton, *Inorg. Chem.*, 2012, **51**, 9403.
59   K. Izod, E. R. Clark, W. Clegg and R. W. Harrington, *Organometallics*, 2012, **31**, 246.
60   T. Řezníček, L. Dostál, A. Růžička and R. Jambor, *Eur. J. Inorg. Chem.*, 2012, 2983.
61   B. A. Vaughan, E. M. Arsenault, S. M. Chan and R. Waterman, *J. Organomet. Chem.*, 2012, **696**, 4327.
62   I. D'Auria, M. Lamberti, M. Mazzeo, S. Milione, G. Roviello and C. Pellecchia, *Chem. Eur. J.*, 2012, **18**, 2349.
63   P. J. Harford, J. Haywood, M. R. Smith, B. N. Bhawal, P. R. Raithby, M. Uchiyama and A. E. H. Wheatley, *Dalton Trans.*, 2012, **41**, 6148.
64   I. Abdellah, E. Bernoud, J.-F. Lohier, C. Alayrac, L. Toupet, C. Lepitit and A.-C. Gaumont, *Chem. Commun.*, 2012, **48**, 4088.

65 F. A. Stokes, R. J. Less, J. Haywood, R. L. Melen, R. I. Thompson, A. E. H. Wheatley and D. S. Wright, *Organometallics*, 2012, **31**, 23.
66 T. Kruczyński, R. Grubba, K. Baranowska and J. Pikies, *Polyhedron*, 2012, **39**, 25.
67 A. Łapczuk-Krygier, K. Baranowska, Ł. Ponikiewski, E. Matern and J. Pikies, *Inorg. Chim. Acta*, 2012, **387**, 361.
68 D. Roy, S. Mom, S. Royer, D. Lucas, J.-C. Hierso and H. Doucet, *ACS Catal.*, 2012, **2**, 1033.
69 R. V. Smaliy, M. Beupérin, A. Mielle, P. Richard, H. Cattey, A. N. Kostyuk and J.-C. Hierso, *Eur. J. Inorg. Chem.*, 2012, 1347.
70 L. Weber, H. Kuhtz, L. Böhling, A. Brockhinke, A. Chrostowska, A. Dargelos, A. Mazière, H.-G. Stammler and B. Neumann, *Dalton Trans.*, 2012, **41**, 10440.
71 M. J. Johansson, S. Berglund, Y. Hu, K. H. O. Andersson and N. Kann, *ACS Comb. Sci.*, 2012, **14**, 304.
72 M. F. Cain, D. S. Glueck, J. A. Golen and A. L. Rheingold, *Organometallics*, 2012, **31**, 775.
73 Y. Morisaki, H. Imoto, R. Kato, Y. Ouchi and Y. Chujo, *Heterocycles*, 2012, **85**, 2543.
74 L.-C. Liang, H.-Y. Shih, H.-S. Chen and S.-T. Lin, *Eur. J. Inorg. Chem.*, 2012, 298.
75 S. Schneider, J. Meiners and B. Askevold, *Eur. J. Inorg. Chem.*, 2012, 412.
76 K. L. Luska, K. Z. Demmans, S. A. Stratton and A. Moores, *Dalton Trans.*, 2012, **41**, 13533.
77 N. K. Gusarova, V. A. Kuimov, S. F. Malysheva, N. A. Belogorlova, A. I. Albanov and B. A. Trofimov, *Tetrahedron*, 2012, **68**, 9218.
78 L. A. Oparina, N. K. Gusarova, O. V. Vysotskaya, N. A. Kolyvanov, A. A. Artem'ev and B. A. Trofimov, *Synthesis*, 2012, **44**, 2938.
79 D. Zhao and R. Wang, *Chem. Soc. Rev.*, 2012, **41**, 2095.
80 Y. Huang, S. A. Pullarkat, Y. Li and P.-H. Leung, *Inorg. Chem.*, 2012, **51**, 2533.
81 C. Xu, G. J. H. Kennard, F. Hennersdorf, Y. Li, S. A. Pullarkat and P.-H. Leung, *Organometallics*, 2012, **31**, 3022.
82 J. J. Feng, M. Huang, Z.-Q. Lin and W.-L. Duan, *Adv. Synth. Catal.*, 2012, **354**, 3122.
83 M. Huang, C. Li, J. Huang, W.-L. Duan and S. Xu, *Chem. Commun.*, 2012, **48**, 11148.
84 K. Chen, S. A. Pullakat, M. Ma, Y. Li and P.-H. Leung, *Dalton Trans.*, 2012, **41**, 5391.
85 Y. L. Cheow, S. A. Pullarkat, Y. Li and P.-H. Leung, *J. Organomet. Chem.*, 2012, **696**, 4215.
86 M. Kamitani, M. Itazaki, C. Tamiya and H. Nakazawa, *J. Am. Chem. Soc.*, 2012, **134**, 11932.
87 (a) A. Benié, C. G. Kodjo, K. S. Traoré, J.-M. Denis and Y.-A. Békro, *Physical & Chemical News*, 2011, **59**, 73; and (b) *Chem. Abstr.*, 2012, **156**, 311129.
88 Y. Huang, R. J. Chew, S. A. Pullarkat, Y. Li and P.-H. Leung, *J. Org. Chem.*, 2012, **77**, 6849.
89 R. Huang, S. A. Pullarkat, S. Teong, R. J. Chew, X. Li and P.-H. Leung, *Organometallics*, 2012, **31**, 4871.
90 F. Alonso, Y. Moglie, G. Radivoy and M. Yus, *Green Chem.*, 2012, **14**, 2699.
91 A. Massard, V. Rampazzi, A. Perrier, E. Bodio, M. Picquet, P. Richard, J.-C. Hierso and P. Le Gendre, *Organometallics*, 2012, **31**, 947.
92 R. Gilbert-Wilson, L. D. Field and M. M. Bhadbhade, *Inorg. Chem.*, 2012, **51**, 3239.

93 S. N. Khong and O. Kwon, *Molecules*, 2012, **17**, 5626.
94 S. F. Malysheva, N. K. Gusarova, A. V. Artem'ev, N. A. Belogorlova, V. I. Smirnov, V. A. Shagun, V. A. Kuimov and B. A. Trofimov, *Synth. Commun.*, 2012, **42**, 1685.
95 K. V. Rajendran and D. G. Gilheany, *Chem. Commun.*, 2012, **48**, 817.
96 P. A. Byrne, K. V. Rajendran, J. Muldoon and D. G. Gilheany, *Org. Biomol. Chem.*, 2012, **10**, 3531.
97 E. H. Krenske, *J. Org. Chem.*, 2012, **77**, 1.
98 E. H. Krenske, *J. Org. Chem.*, 2012, **77**, 3969.
99 M. Dötterl, P. Thoma and H. G. Alt, *Adv. Synth. Catal.*, 2012, **354**, 389.
100 Q. Sun, X. Meng, X. Liu, X. Zhang, Y. Yang, Q. Yang and F.-S. Xiao, *Chem. Commun.*, 2012, **48**, 10505.
101 C. M. Friesen, C. D. Montgomery and S. A. J. U. Temple, *J. Fluorine Chem.*, 2012, **144**, 24.
102 X.-N. Zhang and M. Shi, *Eur. J. Org. Chem.*, 2012, 6271.
103 S. Wang, J. Li, T. Miao, W. Wu, Q. Li, Y. Zhuang, Z. Zhou and L. Qiu, *Org. Lett.*, 2012, **14**, 1966.
104 D.-Y. Zhang, C.-B. Yu, M.-C. Wang, K. Gao and Y.-G. Zhou, *Tetrahedron Lett.*, 2012, **53**, 2556.
105 L. H. Davies, B. Stewart, R. W. Harrington, W. Clegg and L. J. Higham, *Angew. Chem. Int. Ed.*, 2012, **51**, 4921.
106 Y. Li, L.-Q. Lu, S. Das, S. Pisiewicz, K. Junge and M. Beller, *J. Am. Chem. Soc.*, 2012, **134**, 18325.
107 Y. Li, S. Das, S. Zhou, K. Junge and M. Beller, *J. Am. Chem. Soc.*, 2012, **134**, 9727.
108 L. Pehlivan, E. Métay, D. Delbrayelle, G. Mignani and M. Lemaire, *Tetrahedron*, 2012, **68**, 3151.
109 W. Dayoub, A. Favre-Réguillon, M. Berthod, E. Jeanneau, G. Mignani and M. Lemaire, *Eur. J. Org. Chem.*, 2012, 3074.
110 M.-C. Duclos, Y. Singjunla, C. Petit, A. Favre-Réguillon, E. Jeanneau, F. Popowycz, E. Métay and M. Lemaire, *Tetrahedron Lett.*, 2012, **53**, 5984.
111 W. Tang, N. D. Patel, G. Xu, X. Xu, J. Savoie, S. Ma, M.-H. Hao, S. Keshipeddy, A. G. Capacci, X. Wei, Y. Zhang, J. J. Gao, W. Li, S. Rodriguez, B. Z. Lu, N. K. Kee and C. H. Senanayake, *Org. Lett.*, 2012, **14**, 2258.
112 M. A. Gross, K. Mereiter, Y. Wang and W. Weissensteiner, *J. Organomet. Chem.*, 2012, **716**, 32.
113 Y. Tanimoto, Y. Ishizu, K. Kubo, K. Miyoshi and T. Mizuta, *J. Organomet. Chem.*, 2012, **713**, 80.
114 K.-O. Feldmann, S. Schulz, F. Klotter and J. J. Weigand, *Chem. Sus. Chem.*, 2011, **4**, 1805; *Chem. Abstr.*, 2012, **156**, 122582.
115 S. Khan, S. S. Sen and H. W. Roesky, *Chem. Commun.*, 2012, **48**, 2169.
116 P. Barbaro, C. Bazzicalupi, M. Peruzzini, S. S. Constantini and P. Stoppioni, *Angew. Chem. Int. Ed.*, 2012, **51**, 8628.
117 K.-O. Feldmann, R. Fröhlich and J. J. Weigand, *Chem. Commun.*, 2012, **48**, 4296.
118 K.-O. Feldmann and J. J. Weigand, *J. Am. Chem. Soc.*, 2012, **134**, 15443.
119 K.-O. Feldmann and J. J. Weigand, *Angew. Chem. Int. Ed.*, 2012, **51**, 7545.
120 K. Junge, B. Wendt, F. A. Westerhaus, A. Spannenberg, H. Jiao and M. Beller, *Chem. Eur. J.*, 2012, **18**, 9011.
121 S. Kumar, G. Mani, S. Mondal and P. K. Chattaraj, *Inorg. Chem.*, 2012, **51**, 12527.
122 G. T. Venkanna, T. V. M. Ramos, H. D. Arman and Z. J. Tonzetich, *Inorg. Chem.*, 2012, **51**, 12789.

123 Z. Hu, Y. Li, K. Liu and Q. Shen, *J. Org. Chem.*, 2012, **77**, 7957.
124 M. Stickel, C. Maichle-Moessmer, L. Wesemann and H. A. Mayer, *Polyhedron*, 2012, **46**, 95.
125 E. Ullah, J. McNulty, M. Sliwinski and A. Robertson, *Tetrahedron Lett.*, 2012, **53**, 3990.
126 D. Fuchs, G. Rousseau, L. Diab, U. Gellrich and B. Breit, *Angew. Chem. Int. Ed.*, 2012, **51**, 2178.
127 A. J. Saunders, I. R. Crossley, M. P. Coles and S. M. Roe, *Chem. Commun.*, 2012, **48**, 5766.
128 K. Ohmatsu, M. Ito, T. Kunieda and T. Ooi, *Nature Chemistry*, 2012, **4**, 473.
129 B. B. Macha, J. Boudreau, L. Maron, T. Maris and F.-G. Fontaine, *Organometallics*, 2012, **31**, 6428.
130 S. M. Whittemore, R. J. Yoder and J. P. Stambuli, *Organometallics*, 2012, **31**, 6124.
131 Z. Li, H. Ishizuka, Y. Seri, M. Akita and M. Yoshizawa, *Chem. Asian J.*, 2012, **7**, 1789.
132 B. Milde, R. Packheiser, S. Hildebrandt, D. Schaarschmidt, T. Rüffer and H. Lang, *Organometallics*, 2012, **31**, 3661.
133 C. B. Lavery, R. McDonald and M. Stradiotto, *Chem. Commun.*, 2012, **48**, 7277.
134 T. D. Nixon, A. J. Gamble, R. J. Thatcher, A. C. Whitwood and J. M. Lynam, *Inorg. Chim. Acta*, 2012, **380**, 252.
135 A. M. Spokoyny, C. D. Lewis, G. Teverovskiy and S. L. Buchwald, *Organometallics*, 2012, **31**, 8478.
136 D. R. Boyd, M. Bell, K. S. Dunne, B. Kelly, P. J. Stevenson, J. F. Malone and C. C. R. Allen, *Org. Biomol. Chem.*, 2012, **10**, 1388.
137 M. Cieslikiewicz, A. Bouet, S. Jugé, M. Toffano, J. Bayardon, C. West, K. Lewinski and I. Gillaizeau, *Eur. J. Org. Chem.*, 2012, 1101.
138 Y.-L. Zhao, G.-J. Wu, Y. Li, L.-X. Gao and F.-S. Han, *Chem. Eur. J*, 2012, **18**, 9622.
139 B. H. Lipshutz, N. A. Isley, R. Moser, S. Ghorai, H. Leuser and B. R. Taft, *Adv. Synth. Catal.*, 2012, **354**, 3175.
140 D. N. Tran, F.-X. Legrand, S. Menuel, H. Bricout, S. Tilloy and E. Monflier, *Chem. Commun.*, 2012, **48**, 753.
141 X. Jin, X.-F. Xu and K. Zhao, *Tetrahedron: Asymmetry*, 2012, **23**, 1058.
142 D. Rageot and A. Pfaltz, *Helv. Chim. Acta*, 2012, **95**, 2176.
143 L. Zhao, J. Liu, Y. Feng, Y. He and Q. Fan, *Chin. J. Chem*, 2012, **30**, 2009.
144 P. Bergamini, L. Marvelli, A. Marchi, F. Vassanelli, M. Fogagnolo, P. Formaglio, T. Bernardi, R. Gavioli and F. Sforza, *Inorg. Chim. Acta*, 2012, **391**, 162.
145 P. Servin, R. Laurent, H. Dib, L. Gonsalvi, M. Peruzzini, J.-P. Majoral and A.-M. Caminade, *Tetrahedron Lett.*, 2012, **53**, 3876.
146 G. Farkas, S. Balogh, J. Madarász, Á. Szöllősy, F. Darvas, L. Ürge, M. Gouygou and J. Bakos, *Dalton Trans.*, 2012, **41**, 9493.
147 Y. Li, F. Liang, R. Wu, Q. Li, Q.-R. Wang, Y.-C. Xu and L. Jiang, *Synlett*, 2012, **23**, 1805.
148 P. D. Newman, K. J. Cavell and B. M. Kariuki, *Dalton Trans.*, 2012, **41**, 12395.
149 A. Ros, B. Estepa, A. Bermejo, E. Álvarez, R. Fernández and J. M. Lassaletta, *J. Org. Chem.*, 2012, **77**, 4740.
150 C.-H. Chen, C.-S. Chen, H.-F. Dai and W.-Y. Yeh, *Dalton Trans.*, 2012, **41**, 3030.
151 G. A. Ardizzoia, S. Brenna, S. Durini and B. Therrien, *Organometallics*, 2012, **31**, 5427.

152 L. Wang, D. Dehe, T. Philippi, A. Seifert, S. Ernst, Z. Zhou, M. Hartmann, R. N. Klupp Taylor, A. P. Singh, M. Jia and W. R. Thiel, *Catalysis Sci. Technol.*, 2012, **2**, 1188.
153 R. Bellini and J. N. H. Reek, *Chem. Eur. J.*, 2012, **18**, 13510.
154 C.-J. Hou, Y.-H. Wang, Z. Zheng, J. Xu and X.-P. Hu, *Org. Lett.*, 2012, **14**, 3554.
155 H.-P. Deng and M. Shi, *Eur. J. Org. Chem.*, 2012, 183.
156 J. Liu, Y. Feng, B. Ma, Y.-M. He and Q.-H. Fan, *Eur. J. Org. Chem.*, 2012, 6737.
157 I. Sárosi, A. Hildebrand, P. Lönnecke, L. Silaghi-Dumitrescu and E. Hey-Hawkins, *Dalton Trans.*, 2012, **41**, 5326.
158 T. Sanji, K. Naito, T. Kashiwabara and M. Tanaka, *Heteroatom Chem.*, 2012, **23**, 520.
159 P. Scherl, A. Kruckenberg, S. Mader, H. Wadepohl and L. H. Gade, *Organometallics*, 2012, **31**, 7024.
160 S. Vigo, R. Andrés, P. Gómez-Sal, J. de la Mata and E. de Jesús, *J. Organomet. Chem.*, 2012, **717**, 88.
161 N. Peulecke, M. K. Kindermann, M. Köckerling and J. Heinicke, *Polyhedron*, 2012, **41**, 61.
162 I. Angurell, E. Puig, O. Rossell, M. Seco and P. Gómez-Sal, *J. Organomet. Chem.*, 2012, **716**, 120.
163 J. Liu, Y. Feng, Y. He, N. Yang and Q.-H. Fan, *New J. Chem.*, 2012, **36**, 380.
164 E. I. Musina, A. A. Karasik, A. S. Balueva, I. D. Strelnik, T. I. Fesenko, A. B. Dobrynin, T. P. Gerasimova, S. A. Katsyuba, O. N. Kataeva, P. Lönnecke, E. Hey-Hawkins and O. G. Sinyashin, *Eur. J. Inorg. Chem.*, 2012, 1857.
165 A. I. Zayya and J. L. Spencer, *Tetrahedron*, 2012, **68**, 5109.
166 A. Zirakzadeh, R. Schuecker, N. Gorgas, K. Mereiter, F. Spindler and W. Weissensteiner, *Organometallics*, 2012, **31**, 4241.
167 O. N. Gorunova, M. V. Livantsov, Y. K. Grishin, N. A. Kataeva, K. A. Kochetkov, A. V. Churakov, L. G. Kuz'mina and V. V. Dunina, *Polyhedron*, 2012, **31**, 37.
168 M. Neel, A. Panossian, A. Voituriez and A. Marinetti, *J. Organomet. Chem.*, 2012, **716**, 187.
169 E. M. Barreiro, D. F. D. Broggini, L. A. Adrio, A. J. P. White, R. Schwenk, A. Togni and K. K. (M). Hii, *Organometallics*, 2012, **31**, 3745.
170 T. Hammerer, A. Dämbkes, W. Braun, A. Salzer, G. Franciò and W. Leitner, *Synthesis*, 2012, **44**, 2793.
171 M.-M. Wei, M. García-Melchor, J.-C. Daran, C. Audin, A. Lledós, R. Poli, E. Deydier and E. Manoury, *Organometallics*, 2012, **31**, 6669.
172 J. Schultz, I. Císařová and P. Štěpnička, *Organometallics*, 2012, **31**, 729.
173 J. Schultz, I. Císařová and P. Štěpnička, *Eur. J. Inorg. Chem.*, 2012, 5000.
174 P. Štěpnička, *Chem. Soc. Rev.*, 2012, **41**, 4273.
175 K. Bert, T. Noël, W. Kimpe, J. L. Goeman and J. Van der Eycken, *Org. Biomol. Chem.*, 2012, **10**, 8539.
176 J. F. Buergler, K. Niedermann and A. Togni, *Chem. Eur. J.*, 2012, **18**, 632.
177 M. Ziyaadini, M. T. Maghsoodlou, N. Hazeri and S. M. Habibi-Khorassani, *Monatsh. Chem.*, 2012, **143**, 1681.
178 A. Shaabani, S. Keshipour, M. Aghaei, M. H. Khodabandeh and M. Zahedi, *Chin. J. Chem.*, 2012, **30**, 1893.
179 (a) H. Ghasempour and G. Foroghi-Nematallahi, *Physical Chemistry (An Indian Journal)*, 2011, **6**(3); and (b) *Chem. Abstr.*, 2012, **156**, 588385.
180 A. Hassanabadi, M. H. Mosslemin, E. Abyar and M. Taleb-Malamiri, *J. Chem. Res.*, 2012, **36**, 497.

181 A. Habibi, H. Hosseinzadeh and S. M. Aghvami, *Phosphorus, Sulfur, and Silicon*, 2012, **187**, 409.
182 A. Habibi, K. Eskandari and A. Alizadeh, *Phosphorus, Sulfur, and Silicon*, 2012, **187**, 1109.
183 S. Salmanpour, A. Ramazani and Y. Ahmadi, *Bull. Chem. Soc. Ethiop.*, 2012, **26**, 153.
184 N. Shajari, A. Ramazani and Y. Ahmadi, *Bull. Chem. Soc. Ethiop.*, 2011, **25**, 91.
185 S. Asghari and A. K. Habibi, *Helv. Chim. Acta*, 2012, **95**, 810.
186 S. Asghari and M. Osia, *Phosphorus, Sulfur, and Silicon*, 2012, **187**, 1195.
187 M. Anary-Abbasinejad, A. Hassanabadi and Z. Khaksari, *Synthetic Commun.*, 2012, **42**, 204.
188 S. Asghari, A. Salimi and V. Taghipour, *Phosphorus, Sulfur, and Silicon*, 2012, **187**, 669.
189 M. Anary-Abbasinejad, H. D. Farashah, A. Hassanabadi, H. Anaraki-Ardakani and N. Shams, *Synthetic Commun.*, 2012, **42**, 1877.
190 H. Anaraki-Ardakani, M. Noei and A. Tabarzad, *Chin. Chem. Lett.*, 2012, **23**, 45.
191 M. Zyaadini, M. T. Maghsoodlou, N. Hazeri and S. M. Habibi-Khorassani, *Heteroatom Chem.*, 2012, **23**, 131.
192 G. Marandi, N. A. Torbati, R. Heydari, N. Hazeri, S. M. Habibi-Khorassani, M. T. Maghsoodlou, B. Adrom, B. W. Skelton and M. Makha, *Phosphorus, Sulfur, and Silicon*, 2012, **187**, 1450.
193 K. Wittstein, A. B. García, M. Schürmann and K. Kumar, *Synlett*, 2012, 227.
194 J.-C. Deng, F.-W. Chan, C.-W. Kuo, C.-A. Cheng, C.-Y. Huang and S.-C. Chuang, *Eur. J. Org. Chem.*, 2012, 5738.
195 J. Liu, H. Liu, R. Na, G. Wang, Z. Li, H. Yu, M. Wang, J. Zhong and H. Guo, *Chem. Lett.*, 2012, **41**, 218.
196 Y. Oe, T. Inoue, H. Kishimoto, M. Sasaki, T. Ohta and I. Furukawa, *Int. J. Org. Chem.*, 2012, **2**, 111.
197 H.-F. Fan, X.-W. Wang, J.-W. Zhao, X.-J. Li, J.-M. Gao and S.-Z. Zhu, *Synthesis*, 2012, **44**, 3315.
198 S. Khong and O. Kwon, *J. Org. Chem.*, 2012, **77**, 8257.
199 Z. Lian, Y. Wei and M. Shi, *Tetrahedron*, 2012, **68**, 2401.
200 (a) Z. Lian and M. Shi, *Eur. J. Org. Chem.*, 2012, 581; (b) Z. Lian and M. Shi, *Org. Biomol. Chem.*, 2012, **10**, 8048; and (c) L. Yang, P. Xie, E. Li, X. Li, Y. Huang and R. Chen, *Org. Biomol. Chem.*, 2012, **10**, 7628.
201 S.-C. Chuang, J.-C. Deng, F.-W. Chan, S.-Y. Chen, W.-J. Huang, L.-H. Lai and V. Rajeshkumar, *Eur. J. Org. Chem.*, 2012, 2606.
202 R. Amiri, P. Padyab, F. Chalabian and N. A. Rajabi, *J. Heterocycl. Chem.*, 2012, **49**, 683.
203 J. Saha, C. Lorenc, B. Surana and M. W. Peczuh, *J. Org. Chem.*, 2012, **77**, 3846.
204 W. Yang, F. Sha, X. Zhang, K. Yuan and X. Wu, *Chin. J. Chem.*, 2012, **30**, 2652.
205 J.-Y. Qian, C.-C. Wang, F. Sha and X.-Y. Wu, *RSC Adv.*, 2012, **2**, 6042.
206 Y. Shen, Q. Tiang, C. Zhang and W. Zhong, *Synlett.*, 2012, 741.
207 Y.-L. Yang, Y. Wei and M. Shi, *Org. Biomol. Chem.*, 2012, **10**, 7429.
208 Y. Imura, N. Shimojuh, K. Moriyama and H. Togo, *Tetrahedron*, 2012, **68**, 2319.
209 X. Meng, P. Xie, Y. Huang and R. Chen, *RSC Adv.*, 2012, **2**, 8104.
210 (a) C. Masusai, D. Soorukram, C. Kuhakarn, P. Tuchinda, C. Pakawatchai, V. Reutrakul and M. Pohmakotr, *Helv. Chim. Acta*, 2012, **95**, 1912; (b) Y.-J. Jang, S.-E. Syu, Y.-W. Jhang, Y.-T. Lee, C.-J. Lee, K.-W. Chen, U. Das and W. Lin, *Molecules*, 2012, **17**, 2529; and (c) C. Lindner, R. Tandon, Y. Liu, B. Maryasin and H. Zipse, *Org. Biomol. Chem.*, 2012, **10**, 3210.

211   D. Duvvuru, P. Retailleau, J.-F. Betzer and A. Marinetti, *Eur. J. Org. Chem.*, 2012, 897.
212   Q.-Y. Zhao, Z. Lian, Y. Wei and M. Shi, *Chem. Commun.*, 2012, **48**, 1724.
213   (a) L. Zhao, M. Wen and Z.-X. Wang, *Eur. J. Org. Chem.*, 2012, 3587; and (b) Y. Qiao and K.-L. Han, *Org. Biomol. Chem.*, 2012, **10**, 7689.
214   W. Meng, H.-T. Zhao, J. Nie, Y. Zheng, A. Fu and J.-A. Ma, *Chem. Sci.*, 2012, **3**, 3053.
215   (a) H. Xiao, Z. Chai, D. Cao, H. Wang, J. Chen and G. Zhao, *Org. Biomol. Chem.*, 2012, **10**, 3195; and (b) F. Zhong, X. Han, Y. Wang and Y. Lu, *Chem. Sci.*, 2012, **3**, 1231.
216   Q. Zhao, X. Han, Y. Wei, M. Shi and Y. Lu, *Chem. Commun.*, 2012, **48**, 970.
217   I. P. Andrews and O. Kwon, *Chem. Sci.*, 2012, **3**, 2510.
218   F. Hu, X. Guan and M. Shi, *Tetrahedron*, 2012, **68**, 4782.
219   Y. Wang, Z.-H. Yu, H.-F. Zheng and D.-Q. Shi, *Org. Biomol. Chem.*, 2012, **10**, 7739.
220   (a) D. Duvvuru, N. Pinto, C. Gomez, J.-F. Betzer, P. Retailleau, A. Voituriez and A. Marinetti, *Adv. Synth. Catal.*, 2012, **354**, 408; and (b) D. Wang, Y. Wei and M. Shi, *Chem. Commun.*, 2012, **48**, 2764.
221   X.-Y. Chen and S. Ye, *Eur. J. Org. Chem.*, 2012, 5723.
222   C. Jing, R. Na, B. Wang, H. Liu, L. Zhang, J. Liu, M. Wang, J. Zhong, O. Kwon and H. Guo, *Adv. Synth. Catal.*, 2012, **354**, 1023.
223   (a) I. P. Andrews, B. R. Blank and O. Kwon, *Chem. Commun.*, 2012, **48**, 5373; and (b) X. Han, F. Zhong, Y. Wang and Y. Lu, *Angew. Chem. Int. Ed.*, 2012, **51**, 767.
224   R. A. Villa, Q. Xu and O. Kwon, *Org. Lett.*, 2012, **14**, 4634.
225   P. Xie, W. Lai, Z. Geng, Y. Huang and R. Chen, *Chem. Asian J.*, 2012, **7**, 1533.
226   C.-K. Pei, Y. Jiang and M. Shi, *Eur. J. Org. Chem.*, 2012, 4206.
227   M. Steurer, K. L. Jensen, D. Worgull and K. A. Jørgensen, *Chem. Eur. J.*, 2012, **18**, 76.
228   J. M. de los Santos, Z. Ochoa and F. Palacios, *ARKIVOC*, 2012, **v**, 54.
229   P. Xie, Y. Huang and R. Chen, *Chem. Eur. J.*, 2012, **18**, 7362.
230   (a) Z. Jin, R. Yang, Y. Du, B. Tiwari, R. Ganguly and Y. R. Chi, *Org. Lett.*, 2012, **14**, 3226; (b) Z. Shi, P. Yu, T.-P. Lo and G. Zhong, *Angew. Chem. Int. Ed.*, 2012, **51**, 7825; and (c) Z. Shi, Q. Tong, W. W. Y. Leong and G. Zhong, *Chem. Eur. J.*, 2012, **18**, 9802.
231   R. Zhou, J. Wang, J. Tian and Z. He, *Org. Biomol. Chem.*, 2012, **10**, 773.
232   T.-Y. Liu, M. Xie and Y.-C. Chen, *Chem. Soc. Rev.*, 2012, **41**, 4101.
233   C. R. Reddy, M. D. Reddy and B. Srikanth, *Org. Biomol. Chem.*, 2012, **10**, 4280.
234   F. Zhong, G.-Y. Chen, X. Han, W. Yao and Y. Lu, *Org. Lett.*, 2012, **14**, 3764.
235   R. Zhou, J. Wang, C. Duan and Z. He, *Org. Lett.*, 2012, **14**, 6134.
236   Y. Wang, L. Liu, T. Zhang, N.-J. Zhong, D. Wang and Y.-J. Chen, *J. Org. Chem.*, 2012, **77**, 4143.
237   D. Wang, Y.-L. Yang, J.-J. Jiang and M. Shi, *Org. Biomol. Chem.*, 2012, **10**, 7158.
238   (a) X.-N. Zhang, H.-P. Deng, L. Huang, Y. Wei and M. Shi, *Chem. Commun.*, 2012, **48**, 8664; (b) H.-P. Deng, Y. Wei and M. Shi, *Adv. Synth. Catal.*, 2012, **354**, 783; and (c) H.-P. Deng, D. Wang, Y. Wei and M. Shi, *Beilstein J. Org. Chem.*, 2012, **8**, 1098.
239   F. Hu, Y. Wei and M. Shi, *Tetrahedron*, 2012, **68**, 7911.
240   Y. C. Fan and O. Kwon, *Org. Lett.*, 2012, **14**, 3264.
241   Z. Zhuang, J.-M. Chen, F. Pan and W.-W. Liao, *Org. Lett.*, 2012, **14**, 2354.
242   J. Wang, R. Zhou, Z.-R. He and Z. He, *Eur. J. Org. Chem.*, 2012, 6033.

243 C. Hu, Z. Geng, J. Ma, Y. Huang and R. Chen, *Chem. Asian J.*, 2012, **7**, 2032.
244 X.-B. Wang, Y. Saga, R. Shen, H. Fujino, M. Goto and L.-B. Han, *RSC. Adv.*, 2012, **2**, 5935.
245 S. Riaz, A. Naiduand and P. K. Dubey, *Letters in Org. Chem.*, 2012, **9**, 101.
246 J.-M. Chen, Y.-Z. Fang, Z-l. Wie and W.-W. Liao, *Synthesis*, 2012, **44**, 1849.
247 C.-J. Lee, Y.-J. Jang, Z.-Z. Wu and W. Lin, *Org. Lett.*, 2012, **14**, 1906.
248 M. Venkatanarayana and P. K. Dubey, *J. Chem. Sci.*, 2011, **123**, 609.
249 M. Fevre, J. Vignolle, V. Heroguez and D. Taton, *Macromolecules*, 2012, **45**, 7711.
250 C. Pubill-Ulldemolins, A. Bonet, H. Gulyás, C. Bo and E. Fernández, *Org. Biomol. Chem.*, 2012, **10**, 9677.
251 O. Baslé, S. Porcel, S. Ladeira, G. Bouhadir and D. Bourissou, *Chem. Commun.*, 2012, **48**, 4495.
252 W. X. Zhang, L. Su, W. G. Hu and J. Zhou, *Chinese Chem. Lett.*, 2012, **23**, 657.
253 J. Boudreau, M.-A. Courtemanche, V. M. Marx, D. J. Burnell and F.-G. Fontaine, *Chem. Commun.*, 2012, **48**, 11250.
254 (a) I. V. Galkina, E. V. Tudrii, E. A. Berdnikov, L. M. Yusupova, F. S. Levinson, D. B. Krivolapov, I. A. Litvinov, R. A. Cherkasov and V. I. Galkin, *Russ. J. Org. Chem.*, 2012, **48**, 721; (b) I. V. Galkina, M. P. Spiridonova, E. V. Tudrii, E. A. Berdnikov, L. M. Yusupova, R. A. Cherkasov and V. I. Galkin, *Russ. J. Org. Chem.*, 2012, **48**, 1133; and (c) I. V. Galkina, M. P. Spiridonova, F. S. Levinson, E. A. Berdnikov, R. A. Cherkasov and V. I. Galkin, *Russ. J. Gen. Chem.*, 2012, **82**, 1601.
255 Y. Y. Khomutnik, P. P. Onys'ko, Y. V. Rassukanaya, A. G. Vlasenko, A. N. Chernega, V. S. Brovarets, S. G. Pil'o and A. D. Sinitsa, *Russ. J. Gen. Chem.*, 2012, **82**, 1058.
256 X. Xu, R. Fröhlich, C. G. Daniliuc, G. Kehr and G. Erker, *Chem. Commun.*, 2012, **48**, 6109.
257 V. S. Batista, R. H. Crabtree, S. J. Konezny, O. R. Luca and J. M. Praetorius, *New J, Chem.*, 2012, **36**, 1141.
258 D.-Y. Kim, H.-J. Kim, K.-H. Yu and J.-J. Min, *Bioorg. Med. Chem. Lett.*, 2012, **22**, 319.
259 M. Selva, M. Noè, A. Perosa and M. Gottardo, *Org. Biomol. Chem.*, 2012, **10**, 6569.
260 H.-F. Klein, A. Kuhn, N. Kuhn, S. Laufer and M. Ströbele, *Z. Anorg. Allg. Chem.*, 2012, **638**, 1784.
261 N. A. Barnes, S. M. Godfrey, R. Z. Khan, A. Pierce and R. G. Pritchard, *Polyhedron*, 2012, **35**, 31.
262 F. B. Alhanash, N. A. Barnes, S. M. Godfrey, P. A. Hurst, A. Hutchinson, R. Z. Khan and R. G. Pritchard, *Dalton Trans.*, 2012, **41**, 7708.
263 T. P. Pell, S. A. Couchman, S. Ibrahim, D. J. D. Wilson, B. J. Smith, P. J. Barnard and J. L. Dutton, *Inorg. Chem.*, 2012, **51**, 13034.
264 R. M. Denton, J. An, P. Lindovska and W. Lewis, *Tetrahedron*, 2012, **68**, 2899.
265 (a) K. V. Rajendran and D. G. Gilheany, *Chem. Commun.*, 2012, **48**, 10040; and (b) K. V. Rajendran, J. S. Kudavalli, K. S. Dunne and D. G. Gilheany, *Eur. J. Org. Chem.*, 2012, 2720.
266 B. Delouvrié, H. Germain, C. S. Harris, M. Lamorlette, H. Lebraud, H. T. H. Nguyen, A. Noisier and G. Ouvry, *Tetrahedron Lett.*, 2012, **53**, 5380.
267 A. Kasemsuknimit, A. Satyender, W. Chavasiri and D. O. Jang, *Bull. Korean Chem. Soc.*, 2011, **32**, 3486.
268 W. Kijrungphaiboon, O. Chantarasriwong and W. Chavasiri, *Tetrahedron Lett.*, 2012, **53**, 674.
269 P. Billing and U. H. Brinker, *J. Org. Chem.*, 2012, **77**, 11227.

270   N. Nowrouzi and M. Z. Jonaghani, *Chinese Chem. Lett.*, 2012, **23**, 442.
271   N. Nowrouzi and M. Z. Jonaghani, *Canad. J. Chem.*, 2012, **90**, 498.
272   A. K. Purohit, D. Pardasani, V. Tak, A. Kumar, R. Jain and D. K. Dubey, *Tetrahedron Lett.*, 2012, **53**, 3795.
273   L. T. Yildirim, M. Masjedi and S. Özkar, *J. Crystallisation Process and Technology*, 2011, **1**, 1.
274   H. Braunschweig, A. Damme, J. O. C. Jiminez-Halla, T. Kupfer and K. Radacki, *Angew. Chem. Int. Ed.*, 2012, **51**, 6267.
275   A. C. Malcolm, K. J. Sabourin, R. McDonald, M. J. Ferguson and E. Rivard, *Inorg. Chem.*, 2012, **51**, 12905.
276   M. A. Celik, R. Sure, S. Klein, R. Kinjo, G. Bertrand and G. Frenking, *Chem. Eur. J.*, 2012, **18**, 5676.
277   I. Bakó, A. Stirling, S. Bálint and I. Pápai, *Dalton Trans.*, 2012, **41**, 9023.
278   F. Bertini, V. Lyaskovskyy, B. J. J. Timmer, F. J. J. de Kanter, M. Lutz, A. W. Ehlers, J. C. Slootweg and K. Lammertsma, *J. Amer. Chem. Soc.*, 2012, **134**, 201.
279   T. Wiegand, H. Eckert, O. Ekkert, R. Fröhlich, G. Kehr and S. Grimme, *J. Amer. Chem. Soc.*, 2012, **134**, 4236.
280   D. Wu, D. Jia, A. Liu, L. Liu and J. Guo, *Chem. Phys. Lett.*, 2012, **541**, 1.
281   S. Gao, W. Wu and Y. Mo, *Int. J. Quantum Chem.*, 2011, **111**, 3761.
282   L. Greb, P. Oña-Burgos, B. Schirmer, S. Grimme, D. W. Stephan and J. Paradies, *Angew. Chem. Int. Ed.*, 2012, **51**, 10164.
283   D. W. Stephan, *Org. Biomol. Chem.*, 2012, **10**, 5740.
284   L. Greb, P. Oña-Burgos, A. Kubas, F. C. Falk, F. Breher, K. Fink and J. Paradies, *Dalton Trans.*, 2012, **41**, 9056.
285   K. Takeuchi and D. W. Stephan, *Chem. Commun.*, 2012, **48**, 11304.
286   G. Ghattas, D. Chen, F. Pan and J. Klankermayer, *Dalton Trans.*, 2012, **41**, 9026.
287   M. Harhausen, R. Fröhlich, G. Kehr and G. Erker, *Organometallics*, 2012, **31**, 2801.
288   J. S. Reddy, B.-H. Xu, T. Mahdi, R. Fröhlich, G. Kehr, D. W. Stephan and G. Erker, *Organometallics*, 2012, **31**, 5638.
289   B.-H. Xu, C. M. Mömming, R. Fröhlich, G. Kehr and G. Erker, *Chem. Eur. J.*, 2012, **18**, 1826.
290   P. Feldhaus, B. Schirmer, B. Wibbeling, C. G. Daniliuc, R. Fröhlich, S. Grimme, G. Kehr and G. Erker, *Dalton Trans.*, 2012, **41**, 9135.
291   R. Liedtke, G. Kehr, R. Fröhlich, C. G. Daniliuc, B. Wibbeling, J. L. Petersen and G. Erker, *Helv. Chim. Acta*, 2012, **95**, 2515.
292   X. Zhao and D. W. Stephan, *Chem. Sci.*, 2012, **3**, 2123.
293   C. B. Caputo, S. J. Geier, E. Y. Ouyang, C. Kreitner and D. W. Stephan, *Dalton Trans.*, 2012, **41**, 237.
294   C. B. Caputo and D. W. Stephan, *Organometallics*, 2012, **31**, 27.
295   B.-H. Xu, R. A. A. Yanez, H. Nakatsuka, M. Kitamura, R. Fröhlich, G. Kehr and G. Erker, *Chem. Asian. J.*, 2012, **7**, 1347.
296   S. Frömel, R. Fröhlich, C. G. Daniliuc, G. Kehr and G. Erker, *Eur. J. Inorg. Chem.*, 2012, 3774.
297   A. Stute, L. Heletta, R. Fröhlich, C. G. Daniliuc, G. Kehr and G. Erker, *Chem. Commun.*, 2012, **48**, 11739.
298   D. Chen, V. Leich, F. Pan and J. Klankermayer, *Chem. Eur. J.*, 2012, **18**, 5184.
299   W. Lie, H. F. T. Klare, M. Oestreich, R. Fröhlich, G. Kehr and G. Erker, *Z. Naturforsch.*, 2012, **67b**, 987.
300   M. J. Sgro, J. Dömer and D. W. Stephan, *Chem. Commun.*, 2012, **48**, 7253.

301 M. Sajid, A. Stute, A. J. P. Cardenas, B. J. Culotta, J. A. M. Heperle, T. H. Warren, B. Schirmer, S. Grimme, A. Studer, C. G. Daniliuc, R. Fröhlich, G. Kehr and G. Erker, *J. Am. Chem. Soc.*, 2012, **134**, 10156.
302 T. M. Gilbert, *Dalton Trans.*, 2012, **41**, 9046.
303 S. L. Granville, G. C. Welch and D. W. Stephan, *Inorg. Chem.*, 2012, **51**, 4711.
304 A. Martín-Sómer, A. M. Lambsabhi, O. Mó and M. Yáñez, *J. Phys. Chem. A*, 2012, **116**, 6950.
305 Y. Zhang, G. M. Miyake, M. G. John, L. Falivene, L. Caporaso, L. Cavallo and E. Y.-X. Chen, *Dalton Trans.*, 2012, **41**, 9119.
306 G. Ménard and D. W. Stephan, *Angew. Chem. Int. Ed.*, 2012, **51**, 4409.
307 S. Roters, C. Appelt, H. Westenberg, A. Hepp, J. C. Slootweg, K. Lammertsma and W. Uhl, *Dalton Trans.*, 2012, **41**, 9033.
308 C. Appelt, J. C. Slootweg, K. Lammertsma and W. Uhl, *Angew. Chem. Int. Ed.*, 2012, **51**, 5911.
309 A. M. Chapman and D. F. Wass, *Dalton Trans.*, 2012, **41**, 9067.
310 S. M. Ulfa, H. Okamoto and K. Satake, *Chem. Lett.*, 2012, **41**, 400.
311 M. Monsó, W. Kowalczyk, D. Andreu and B. G. de la Torre, *Org. Biomol. Chem.*, 2012, **10**, 3116.
312 C. Zhang, Y. Li, M. Zhang and X. Li, *Tetrahedron*, 2012, **68**, 5152.
313 S. Banerjee, L. Adak and B. C. Ranu, *Tetrahedron Lett.*, 2012, **53**, 2149.
314 T. Tsukuda, R. Miyoshi, A. Esumi, A. Yamagiwa, A. Dairiki, K. Matsumoto and T. Tsubomura, *Inorg. Chim. Acta*, 2012, **384**, 149.
315 N. Nguyen, A. J. Lough and U. Fekl, *Inorg. Chem.*, 2012, **51**, 6446.
316 N. K. Gusarova, A. V. Artem'ev, L. A. Oparina, N. A. Kolyvanov, S. F. Malysheva, O. V. Vysotskaya and B. A. Trofimov, *Synthesis*, 2012, 431.
317 A. V. Artem'ev, N. K. Gusarova, S. F. Malysheva, Y. V. Gatilov and V. I. Mamatyuk, *Org. Prep. Proc. Int.*, 2012, **44**, 262.
318 J. W. Dube, M. M. Hänninen, J. L. Dutton, H. M. Tuononen and P. J. Ragogna, *Inorg. Chem.*, 2012, **51**, 8897.
319 J. Beckmann, J. Bolsinger, A. Duthie, P. Finke, E. Lork, C. Lüdtke, O. Mallow and S. Mebs, *Inorg. Chem.*, 2012, **51**, 12395.
320 V. S. P. R. Lingam, D. H. Dahale, K. Mukkanti, B. Gopalan and A. Thomas, *Tetrahedron Lett.*, 2012, **53**, 5695.
321 S. Yamazaki, Y. Maenaka, K. Fujinami and Y. Mikata, *RSC Adv.*, 2012, **2**, 8095.
322 D. Chaturvedi, A. K. Chaturvedi, N. Mishra and V. Mishra, *Tetrahedron Lett.*, 2012, **53**, 5398.
323 A. Singh, B. B. Mishra, R. R. Kale, D. Kushwaha and V. K. Tiwara, *Synth. Commun*, 2012, **42**, 3598.
324 D. G. Dunford, F. Chaudhry, B. Kariuki, D. W. Knight and R. C. Wheeler, *Tetrahedron Lett.*, 2012, **53**, 7006.
325 N. Iranpoor, H. Firouzabadi and D. Khalili, *Tetrahedron Lett.*, 2012, **53**, 6913.
326 C. T. Burns, S. Shang, R. Thapa and M. S. Mashuta, *Tetrahedron Lett.*, 2012, **53**, 4832.
327 C.-M. Park, W. Niu, C. Liu, T. D. Biggs, J. Guo and M. Xian, *Org. Lett.*, 2012, **14**, 4694.
328 A. A. Fesenko and A. D. Shutalev, *Tetrahedron Lett.*, 2012, **53**, 6261.
329 K. Ha, J.-C. M. Monbaliu, B. C. Williams, G. G. Pillai, C. E. Ocampo, M. Zeller, C. V. Stevens and A. R. Katritzky, *Org. Biomol. Chem.*, 2012, **10**, 8055.
330 S. C. Han, J. W. Lee and S.-H. Jin, *Macromol. Res.*, 2012, **20**, 1083.
331 Z. D. Fang and X. H. Wei, *J. Chem. Res.*, 2012, **36**, 612.
332 S. E. Kurhade, V. T. Salunkhe, V. Siddalah, D. Bhuniya and D. S. Reddy, *Synthesis*, 2011, 3523.

333 H. A. van Kalkeren, J. J. Bruins, F. P. J. T. Rutjes and F. L. van Delft, *Adv. Synth. Catal.*, 2012, **354**, 1417.
334 A. D. Kosal, E. E. Wilson and B. L. Ashfeld, *Angew. Chem. Int. Ed.*, 2012, **51**, 12036.
335 A. D. Kosal, E. E. Wilson and B. L. Ashfeld, *Chem. Eur. J.*, 2012, **18**, 14444.
336 S. S. Kostina, T. Singh and W. J. Leigh, *Organometallics*, 2012, **31**, 3755.
337 E. MacDonald, L. Doyle, S. S. Chitnis, U. Werner-Zwanziger, N. Burford and A. Decken, *Chem. Commun.*, 2012, **48**, 7922.
338 N. Takagi, R. Tonner and G. Frenking, *Chem. Eur. J.*, 2012, **18**, 1772.
339 J. L. Dutton and D. J. D. Wilson, *Angew. Chem. Int. Ed.*, 2012, **51**, 1477.
340 D. J. D. Wilson, S. A. Couchman and J. L. Dutton, *Inorg. Chem.*, 2012, **51**, 7657.
341 K. Haav, J. Saame, A. Kütt and I. Leito, *Eur. J. Org. Chem.*, 2012, 2167.
342 M. Venkatanarayana and P. K. Dubey, *Heteroatom Chem.*, 2012, **23**, 41.
343 (a) S. Tilloy, C. Binkowski-Machut, S. Menuel, H. Bricout and E. Monflier, *Molecules*, 2012, **17**, 13062; and (b) F. Hapiot, H. Bricout, S. Tilloy and E. Monflier, *Eur. J. Inorg. Chem.*, 2012, 1571.
344 Y. Wei, X.-G. Liu and M. Shi, *Eur. J. Org. Chem.*, 2012, 2386.
345 M. Yamamura, N. Kano and T. Kawashima, *Bull. Chem. Soc. Jpn.*, 2012, **85**, 110.
346 T. B. Nguyen, A. Martel, C. Gaulon-Nourry, R. Dhal and G. Dujardin, *Org. Prep. Proc. Int.*, 2012, **44**, 1.
347 D. Förster, I. Hartenbach, M. Nieger and D. Gudat, *Z. Naturforsch.*, 2012, **67b**, 765.
348 M. Milanković, B. Warżajtis, U. Rychlewska, D. Radanović, K Anđelković, T. Božić, M. Vujčić and D. Sladić, *Molecules*, 2012, **17**, 2567.
349 V. A. Zagumennov and N. A. Sizova, *Russ. J. Gen. Chem.*, 2012, **82**, 1368.
350 A. I. Zayya, R. Vagana, M. R. M. Nelson and J. L. Spencer, *Tetrahedron Lett.*, 2012, **53**, 923.
351 D. V. Moiseev, B. R. James and A. V. Gushchin, *Russ. J. Gen. Chem.*, 2012, **82**, 840.
352 D. V. Moiseev, B. R. James and T. Q. Hu, *Phosphorus, Sulfur, and Silicon*, 2012, **187**, 433.
353 F. Lavigne, E. Maerten, G. Alcaraz, V. Branchadell, N. Saffon-Merceron and A. Baceiredo, *Angew. Chem. Int. Ed.*, 2012, **51**, 2489.
354 C.-S. Wu and M.-D. Su, *Dalton Trans.*, 2012, **41**, 3253.
355 W. Zhou, K. C. MacLeod, B. O. Patrick and K. M. Smith, *Organometallics*, 2012, **31**, 7324.
356 A. Kreienbrink, P. Lönnecke, M. Findeisen and E. Hey-Hawkins, *Chem. Commun.*, 2012, **48**, 9385.
357 N. A. Giffin, A. D. Hendsbee, T. L. Roemmele, M. D. Lumsden, C. C. Pye and J. D. Masuda, *Inorg. Chem.*, 2012, **51**, 11837.
358 T. Iwamoto, F. Hirakawa and S. Ishida, *Angew. Chem. Int. Ed.*, 2012, **51**, 12111.
359 S. S. Chitnis, E. MacDonald, N. Burford, U. Werner-Zwanziger and R. McDonald, *Chem. Commun.*, 2012, **48**, 7359.
360 A. Espinosa and R. Streubel, *Chem. Eur. J.*, 2012, **18**, 13405.
361 R. J. Less, R. L. Melen and D. S. Wright, *RSC Adv.*, 2012, **2**, 2191.
362 M. Grellier and S. Sabo-Etienne, *Chem. Commun.*, 2012, **48**, 34.
363 M. A. Huertos and A. S. Weller, *Chem. Commun.*, 2012, **48**, 7185.
364 M. Stankevič, K. Wójcik, M. Jaklińska and K. M. Pietrusiewicz, *Eur. J. Org. Chem.*, 2102, 2521.
365 A. Amgoune, S. Ladeira, K. Miqueu and D. Bourissou, *J. Am. Chem. Soc.*, 2012, **134**, 6560.

366 W. Yu, O. Fuhr and D. Fenske, *J. Clust. Sci.*, 2012, **23**, 753.
367 S. K. Latypov, A. G. Strelnik, S. N. Ignatieva, E. Hey-Hawkins, A. S. Balueva, A. A. Karasik and O. G. Sinyashin, *J. Phys. Chem. A*, 2012, **116**, 3182.
368 K.-W. Tan, X.-Y. Yang, Y. Li, Y. Huang, S. A. Pullarkat and P.-H. Leung, *Organometallics*, 2012, **31**, 8407.
369 Y. Matano, K. Matsumoto, H. Hayashi, Y. Nakao, T. Kumpulainen, V. Chukharev, N. V. Tkachenko, H. Lemmetyinen, S. Shimizu, N. Kobayashi, D. Sakamaki, A. Ito, K. Tanaka and H. Imahori, *J. Am. Chem. Soc.*, 2012, **134**, 1825.
370 A. P. Marchenko, H. N. Koidan, E. V. Zarudnitskii, A. N. Hurieva, A. A. Kirilchuk, A. A. Yurchenko, A. Biffis and A. N. Kostyuk, *Organometallics*, 2012, **31**, 8257.
371 R. Mosteiro, A. Fernández, M. López-Torres, D. Vazquez-García, L. Naya, J. M. Ortigueira, J. M. Vila and J. J. Fernández, *J. Organomet. Chem.*, 2012, **720**, 30.
372 D. J. Young, S. W. Chien and T. S. A. Hor, *Dalton Trans.*, 2012, **41**, 12655.
373 N. Fey, M. Garland, J. P. Hopewell, C. L. McMullin, S. Mastroianni, A. G. Orpen and P. G. Pringle, *Angew. Chem. Int. Ed.*, 2012, **51**, 118.
374 T. Sasamori, M. Sakagami, M. Niwa, H. Sakai, Y. Furukawa and N. Tokitoh, *Chem. Commun.*, 2012, **48**, 8562.
375 C. M. Knapp, B. H. Westcott, M. A. C. Raybould, J. E. McGrady and J. M. Goicoechea, *Angew. Chem. Int. Ed.*, 2012, **51**, 9097.
376 H.-L. Peng, J. L. Payton, J. D. Protasiewicz and M. C. Simpson, *Dalton Trans.*, 2012, **41**, 13204.
377 S. Vogt-Geisse and H. F. Schaefer III, *J. Chem. Theory Comput.*, 2012, **8**, 1663.
378 S. Sasaki, M. Yoshifuji and N. Inamoto, *ARKIVOC*, 2012, **ii**, 15.
379 C. Albrecht, L. Shi, J. M. Pérez, M. van Gastel, S. Schwieger, F. Neese and R. Streubel, *Chem. Eur. J.*, 2012, **18**, 9780.
380 A. Orthaber, E. Öberg, R. T. Jane and S. Ott, *Z. Anorg. Allg. Chem.*, 2012, **638**, 2219.
381 J. Dugal-Tessier, S. C. Serin, E. B. Castillo-Contreras, E. D. Conrad, G. R. Dake and D. P. Gates, *Chem. Eur. J.*, 2012, **18**, 6349.
382 A. Orthaber, R. H. Herber and R. Pietschnig, *J. Organomet. Chem.*, 2012, **719**, 36.
383 H. Miyake, T. Sasamori and N. Tokitoh, *Angew. Chem. Int. Ed.*, 2012, **51**, 3458.
384 H. Miyake, T. Sasamori, J. I.-C. Wu, P. von, R. Schleyer and N. Tokitoh, *Chem. Commun.*, 2012, **48**, 11440.
385 H. Hopf, *Angew. Chem. Int. Ed.*, 2012, **51**, 11945.
386 A. B. Rhozhenko, A. Ruban, V. Thelen, M. Nieger, K. Airola, W. W. Schoeller and E. Niecke, *Eur. J. Inorg. Chem.*, 2012, 2502.
387 Y. Mao, Z. Wang, R. Ganguly and F. Mathey, *Organometallics*, 2012, **31**, 4786.
388 J. I. Bates and D. P. Gates, *Chem. Eur. J.*, 2012, **18**, 1674.
389 J. I. Bates and D. P. Gates, *Organometallics*, 2012, **31**, 4529.
390 L. C. Pavelka and K. M. Baines, *Dalton Trans.*, 2012, **41**, 3294.
391 V. A. Zagumennov, *Russ. J. Electrochemistry*, 2011, **47**, 1317.
392 K. V. Turcheniuk and A. B. Rozhenko, *Computational Chemistry*, 2012, **33**, 1023.
393 Y.-F. Yang, G.-J. Cheng, J. Zhu, X. Zhang, S. Inoue and Y.-D. Wu, *Chem. Eur. J*, 2012, **18**, 7516.
394 P. M. Petrar, G. Nemes and L. Silaghi-Dumitrescu, *Studia UBB Chemia*, 2010, **55**(part 1), 25; *Chem. Abstr.*, 2012, **156**, 421556.
395 D. Matioszek, T.-G. Kocsor, A. Castel, G. Nemes, J. Escudié and N. Saffon, *Chem. Commun.*, 2012, **48**, 3629.

396 T.-G. Kocsor, D. Matioszek, G. Nemes, A. Castel, J. Escudié, P. M. Petrar, N. Saffon and I. Haiduc, *Inorg. Chem.*, 2012, **51**, 7782.
397 I. Krummenacher and C. C. Cummins, *Polyhedron*, 2012, **32**, 10.
398 S. Alidori, D. Heift, G. Santiso-Quinones, Z. Benkő, H. Grützmacher, M. Caporalvi, L. Gonsalvi, A. Rossin and M. Peruzzini, *Chem. Eur. J.*, 2012, **18**, 14805.
399 S. M. Mansell, M. Green and C. A. Russell, *Dalton Trans.*, 2012, **41**, 14360.
400 I. Alkorta, G. Sanchez-Sanz and J. Elguero, *J. Phys. Chem. A*, 2012, **116**, 2300.
401 M. Yoshifuji, Y. Hirano, G. Schnakenburg, R. Streubel, E. Niecke and S. Ito, *Helv. Chim. Acta*, 2012, **95**, 1723.
402 S. S. Sen, S. Khan, S. Nagendran and H. W. Roesky, *Acc. Chem. Res.*, 2012, **45**, 578.
403 T. Marino, M. C. Michelini, N. Russo, E. Sicilia and M. Toscano, *Theor. Chem. Acc.*, 2012, **131**, 1141.
404 J. M. Slattery and S. Hussein, *Dalton Trans.*, 2012, **41**, 1808.
405 F. Reiss, A. Schulz and A. Villinger, *Eur. J. Inorg. Chem.*, 2012, 261.
406 D. E. Herbert, A. D. Miller and O. V. Ozerov, *Chem. Eur. J.*, 2012, **18**, 7696.
407 C. Hering, A. Schulz and A. Villinger, *Angew. Chem. Int. Ed.*, 2012, **51**, 6241.
408 N. L. Kilah and S. B. Wild, *Organometallics*, 2012, **31**, 2658.
409 A. D. Hendsbee, N. A. Giffin, Y. Zhang, C. C. Pye and J. D. Masuda, *Angew. Chem. Int. Ed.*, 2012, **51**, 10836.
410 C. Maaliki, C. Lepetit, C. Duhayon, Y. Canac and R. Chauvin, *Chem. Eur. J.*, 2012, **18**, 16153.
411 L. Rosenberg, *Coord. Chem. Rev.*, 2012, **256**, 606.
412 Y. Canac, C. Maaliki, I. Abdellah and R. Chauvin, *New J. Chem.*, 2012, **36**, 17.
413 C. Maaliki, C. Lepetit, Y. Canac, C. Bijani, C. Duhayon and R. Chauvin, *Chem. Eur. J.*, 2012, **18**, 7705.
414 E. Digard, J. Andrieu and H. Cattey, *Inorg. Chem. Commun.*, 2012, **25**, 39.
415 B. Pan, Z. Xu, M. W. Bezpalko, B. M. Foxman and C. M. Thomas, *Inorg. Chem.*, 2012, **51**, 4170.
416 P. K. Coffer (née Monks), R. M. K. Deng, K. B. Dillon, M. A. Fox and R. J. Olivey, *Inorg. Chem.*, 2012, **51**, 9799.
417 N. Tsuchida, M. Isoi, H. Nakazawa and K. Takano, *J. Organomet. Chem.*, 2012, **697**, 41.
418 A. L. Brazeau, N. D. Jones and P. J. Ragogna, *Dalton Trans.*, 2012, **41**, 7890.
419 E. Magdzinski, P. Gobbo, C. D. Martin, M. S. Workentin and P. J. Ragogna, *Inorg. Chem.*, 2012, **51**, 8425.
420 K.-O. Feldman and J. J. Weigand, *Angew. Chem. Int. Ed.*, 2012, **51**, 6566.
421 M. H. Holthausen, K.-O. Feldman, S. Schulz, A. Hepp and J. J. Weigand, *Inorg. Chem.*, 2012, **51**, 3374.
422 M. H. Holthausen and J. J. Weigand, *Z. Anorg. Allg. Chem.*, 2012, **638**, 1103.
423 M. Donath, E. Conrad, P. Jerabek, G. Frenking, R. Frölich, N. Burford and J. J. Weigand, *Angew. Chem. Int. Ed.*, 2012, **51**, 2964.
424 R. Bashforth, A. J. Boyall, P. K. Coffer (née Monks), K. B. Dillon, A. E. Goeta, J. A. K. Howard, A. M. Kenwright, M. R. Probert, H. J. Shepherd and A. L. Thompson, *Dalton Trans.*, 2012, **41**, 1165.
425 J. Carreras, M. Patil, W. Thiel and M. Alcarazo, *J. Am. Chem. Soc.*, 2012, **134**, 16753.
426 J. D. Protasiewicz, *Eur. J. Inorg. Chem.*, 2012, 4539.
427 B. A. Surgenor, M. Bühl, A. M. Z. Slawin, J. D. Woollins and P. Kilian, *Angew. Chem. Int. Ed.*, 2012, **51**, 10150.
428 M. A. Rankin and C. C. Cummins, *Dalton Trans.*, 2012, **41**, 9615.

429 R. Grubba, K. Baranowska, J. Chojnacki and J. Pikies, *Eur. J. Inorg. Chem.*, 2012, 3263.
430 M. P. Duffy, L. Yu Ting, L. Nicholls, Y. Li, R. Ganguly and F. Mathey, *Organometallics*, 2012, **31**, 2936.
431 G. Schulten, G. von Frantzius, G. Schnakenburg, A. Espinosa and R. Streubel, *Chem. Sci.*, 2012, **3**, 3526.
432 R. Tian and F. Mathey, *Chem. Eur. J.*, 2012, **18**, 11210.
433 M. J. Amme, A. B. Kazi and T. R. Cundari, *Int. J. Quantum Chem.*, 2010, **110**, 1702.
434 (a) M. A. Alvarez, M. E. García, R. González and M. A. Ruiz, *Dalton Trans.*, 2012, **42**, 14498; (b) M. A. Alvarez, M. E. García, D. García-Vivó, A. Ramos, M. A. Ruiz and J. Suárez, *Inorg. Chem.*, 2012, **51**, 34; (c) M. A. Alvarez, M. E. García, D. García-Vivó, A. Ramos and M. A. Ruiz, *Inorg. Chem.*, 2012, **51**, 3698; (d) B. Alvarez, M. A. Alvarez, I. Amor, M. E. García, D. García-Vivó, J. Suárez and M. A. Ruiz, *Inorg. Chem.*, 2012, **51**, 7810; and (e) M. A. Alvarez, I. Amor, M. E. García, D. García-Vivó, M. A. Ruiz and J. Suárez, *Organometallics*, 2012, **31**, 2749.
435 F. Ho, Y. Li and F. Mathey, *Organometallics*, 2012, **31**, 8456.
436 M. Stubenhofer, G. Lassandro, G. Balázs, A. Y. Timoshkin and M. Scheer, *Chem. Commun.*, 2012, **48**, 7262.
437 K. K. Pandey, P. Tiwari and P. Patidar, *J. Phys. Chem. A*, 2012, **116**, 11753.
438 R. Streubel, J. M. V. Franco, G. Schnakenburg and A. E. Ferao, *Chem. Commun.*, 2012, **48**, 5986.
439 V. Nesterov, G. Schnakenburg, A. Espinosa and R. Streubel, *Inorg. Chem.*, 2012, **51**, 12343.
440 V. Nesterov, S. Schwieger, G. Schnakenburg, S. Grimme and R. Streubel, *Organometallics*, 2012, **31**, 3457.
441 R. Streubel, E. Schneider and G. Schnakenburg, *Organometallics*, 2012, **31**, 4707.
442 R. Streubel, M. Klein and G. Schnakenburg, *Organometallics*, 2012, **31**, 4711.
443 F. Dielmann, O. Back, M. Henry-Ellinger, P. Jerabek, G. Frenking and G. Bertrand, *Science*, 2012, **337**, 1526.
444 L. Duan, G. Schnakenburg, J. Daniels and R. Streubel, *Eur. J. Inorg. Chem.*, 2012, 2314.
445 L. Duan, G. Schnakenburg, J. Daniels and R. Streubel, *Eur. J. Inorg. Chem.*, 2012, 3490.
446 Y. Ren and T. Baumgartner, *Dalton Trans.*, 2012, **41**, 7792.
447 J.-H. Wan, W.-F. Fang, Y.-B. Li, X.-Q. Xiao, L.-H. Zhang, Z. Xu, J.-J. Peng and G.-Q. Lai, *Org. Biomol. Chem.*, 2012, **10**, 1459.
448 M. Stolar and T. Baumgartner, *New J. Chem.*, 2012, **36**, 1153.
449 C. J. Chua, Y. Ren and T. Baumgartner, *Org. Lett.*, 2012, **14**, 1588.
450 C. J. Chua, Y. Ren and T. Baumgartner, *Organometallics*, 2012, **31**, 2425.
451 Y. Hayashi, Y. Matano, K. Suda, Y. Kimura, Y. Nakao and H. Imahori, *Chem. Eur. J.*, 2012, **18**, 15972.
452 D. Hanifi, A. Pun and Y. Liu, *Chem. Asian J.*, 2012, **7**, 2615.
453 P. A. Bouit, A. Escande, R. Szűcs, D. Szieberth, C. Lescop, L. Nyulászi, M. Hissler and R. Réau, *J. Amer. Chem. Soc.*, 2012, **134**, 6524.
454 Y. Matano, A. Saito, Y. Suzuki, T. Miyajima, S. Akiyama, S. Otsubo, E. Nakamoto, S. Aramaki and H. Imahori, *Chem. Asian J.*, 2012, **7**, 2305.
455 V. Diemer, A. Berthelot, J. Bayardon, S. Jugé, F. R. Leroux and F. Colobert, *J. Org. Chem.*, 2012, **77**, 6117.
456 K. Nakano, H. Oyama, Y. Nishimura, S. Nakasako and K. Nozaki, *Angew. Chem. Int. Ed.*, 2012, **51**, 695.

457 K. Yavari, S. Moussa, B. B. Hassine, P. Retailleau, A. Voituriez and A. Marinetti, *Angew. Chem. Int. Ed.*, 2012, **51**, 6748.
458 W. Weymiens, F. Hartl, M. Lutz, J. C. Slootweg, A. W. Ehlers, J. R. Mulder and K. Lammertsma, *Eur. J. Org. Chem.*, 2012, 6711.
459 J. Möbus, Q. Bonnin, K. Ueda, R. Fröhlich, K. Itami, G. Kehr and G. Erker, *Angew. Chem. Int. Ed.*, 2012, **51**, 1954.
460 H. Chen, W. Delaunay, L. Yu, D. Joly, Zuo. Wang, J. Li, C. Zisu Wang, D. Lescop, B. Tondelier, Z. Geffroy, M. Duan, F. Hissler, Mathey and R. Réau, *Angew. Chem. Int. Ed.*, 2012, **51**, 214.
461 D. H. Nguyen, J. Bayardon, C. Salomon-Bertrand, S. Jugé, P. Kalck, J.-C. Daran, M. Urrutigoity and M. Gouygou, *Organometallics*, 2012, **14**, 857.
462 E. Takacs, A. Escande, N. Vanthuyne, C. Roussel, C. Lescop, E. Guinard, C. Latouche, A. Boucekkine, J. Crassous, R. Réau and M. Hissler, *Chem. Commun.*, 2012, **48**, 6705.
463 Y. Ren and T. Baumgartner, *Inorg. Chem.*, 2012, **51**, 2669.
464 Y. Ren, W. H. Kan, V. Thangadurai and T. Baumgartner, *Angew. Chem. Int. Ed.*, 2012, **51**, 3964.
465 A. M. El-Nahas, A. H. Mangood and T. S. El-Shazly, *Computational and Theoretical Chem.*, 2012, **980**, 68.
466 A. Peña-Gallego, J. Rodríguez-Otero and E. M. Cabaleiro-Lago, *J. Mol. Model.*, 2012, **18**, 765.
467 T. Möller, M. B. Sárosi and E. Hey-Hawkins, *Chem. Eur. J.*, 2012, **18**, 16604.
468 A. Zagidullin, Y. Ganushevich, V. Miluykov, D. Krivolapov, O. Kataeva, O. Sinyashin and E. Hey-Hawkins, *Org. Biomol. Chem.*, 2012, **10**, 5298.
469 F. L. Laughin, A. L. Rheingold, N. Deligonul, B. J. Laughlin, R. C. Smith, L. J. Higham and J. D. Protasiewicz, *Dalton Trans.*, 2012, **41**, 12016.
470 A. I. Arkhypchuk, M.-P. Santoni and S. Ott, *Angew. Chem. Int. Ed.*, 2012, **51**, 7776.
471 A. Marchenko, A. Hurieva, H. Koidan, V. Rampazzi, H. Cattey, N. Pirio, A. N. Kostyuk and J.-C. Hierso, *Organometallics*, 2012, **31**, 5986.
472 M. Ghalib, B. Niaz, P. G. Jones and J. W. Heinicke, *Tetrahedron Lett.*, 2012, **53**, 5012.
473 W. P. Oziminski, *Computational and Theoretical Chem.*, 2012, **980**, 92.
474 D. Förster, H. Dilger, F. Ehret, M. Nieger and D. Gudat, *Eur. J. Inorg. Chem.*, 2012, 3989.
475 H. Grützmacher, *Z. Anorg. Allg. Chem.*, 2012, 1877.
476 R. S. P. Turbervill and J. M. Goicoechea, *Chem. Commun.*, 2012, **48**, 6100.
477 C. Jones and C. Schulten, *J. Chem. Crystallography*, 2012, **42**, 856.
478 R. Tian, X. Ng, R. Ganguly and F. Mathey, *Organometallics*, 2012, **31**, 2486.
479 J. Fassbender, W. Frank and C. Ganter, *Eur. J. Inorg. Chem.*, 2012, 4356.
480 D. Carmichael, X. F. Le Goff, E. Muller, L. Ricard and M. Stankevič, *Dalton Trans.*, 2012, **41**, 5155.
481 D. Carmichael, A. Escalle-Lewis, G. Frison, X. Le Goff and E. Muller, *Chem. Commun.*, 2012, **48**, 302.
482 S. K. Mallissery and D. Gudat, *Z. Anorg. Allg. Chem.*, 2012, **638**, 1141.
483 A. Kłys, *J. Organomet. Chem.*, 2012, **700**, 1.
484 A. Schindler, C. Heindl, G. Balázs, C. Gröger, A. V. Virovets, E. V. Peresypkina and M. Scheer, *Chem. Eur. J*, 2012, **18**, 829.
485 X. Chen, R. Jin, Q Du, H. Feng, Y. Xie and R. B. King, *J. Organomet. Chem.*, 2012, **701**, 1.
486 A. J. Arce, Y. De Sanctis, M. C. Goite, R. Machado, Y. Otero and T. Gonzalez, *Inorg. Chim. Acta*, 2012, **392**, 241.

487 S. Labouille, F. Nief, X.-F. Le Goff, L. Maron, D. R. Kindra, H. L. Houghton, J. W. Ziller and W. J. Evans, *Organometallics*, 2012, **31**, 5196.
488 E. Le Roux, Y. Liang, K. W. Törnroos, F. Nief and R. Anwander, *Organometallics*, 2012, **31**, 6526.
489 N. A. van der Velde, H. T. Korbitz and C. M. Garner, *Tetrahedron Lett.*, 2012, **53**, 5742.
490 Y. Mao, K. M. H. Lim, R. Ganguly and F. Mathey, *Org. Lett.*, 2012, **14**, 4974.
491 Y. Mao and F. Mathey, *Org. Lett.*, 2012, **14**, 1162.
492 A. P. Marchenko, G. N. Koidan, A. N. Huryeva, A. S. Merkulov and A. N. Kostyuk, *Heteroatom Chem.*, 2012, **23**, 309.
493 N. S. Townsend, M. Green and C. A. Russell, *Organometallics*, 2012, **31**, 2543.
494 E. D. Matveeva, T. A. Podrugina, M. A. Taranova, A. M. Ivanova, R. Gleiter and N. S. Zefirov, *J. Org. Chem.*, 2012, **77**, 5770.

Phosphine chalcogenides

G. Keglevich

DOI: 10.1039/9781782623977-00052

The acidity and hydrogen bond acceptor ability of different phosphine oxides was evaluated by MP2 computational methods. In the case of ring phosphine oxides, the ring strain influenced the properties.[1]

A bis(acyl)phosphine oxide, used as photoinitiator, was synthesized by the reaction of $NaPH_2(NaO^tBu)_x$ (x = 2.4–2.8; generated from elemental phosphorus, sodium and *tert*-butanol) and 2,4,6-trimethylbenzoyl chloride (mesitoyl chloride) to give sodium bis(mesitoyl)phosphide as an intermediate that was converted to the target compound by alkylation with a functionalized alkyl halide followed by oxidation (Scheme 1).[2]

Trofimov prepared a mixture of tris(2-pyridyl)phosphine and its phosphine oxide by the reaction of red or white phosphorus in a superbasic $KOH/DMSO/H_2O$ suspension at 100/75 °C (Scheme 2).[3]

The adduct of methyl propargyl ether with pinacolone was reacted with phosphorus trichloride. The ester so formed then underwent spontaneous rearrangement to the corresponding allene derivative. Chlorination of the allene led to an unstable chlorophosphonium salt that gave a dichlorophosphono-1,3-butadiene (**A**) undergoing cyclization to cyclobutene **B** on heating. Dialkyl- or diaryl-cyclobutenylphosphine oxides (**D**) were obtained by thermal electrocyclization of intermediate **C** formed from dichlorophosphonobutadiene **A** in a Grignard reaction. It was also possible to convert species **B** to target product **D** (Scheme 3).[4]

An analogous, but somewhat simpler reaction sequence was also described by the same authors (Scheme 4).[5]

A four-step synthesis of 4,6-bis(diphenylphosphinoylmethyl)-dibenzofuran including formylation, reduction, substitution and

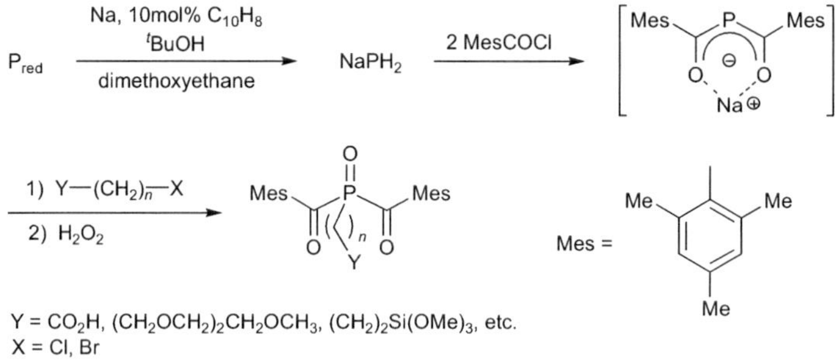

Scheme 1

Budapest University of Technology and Economics, Department of Organic Chemistry and Technology, 1521, Budapest, Hungary.
E-mail: gkeglevich@mail.bme.hu

Arbuzov reaction, together with the two-step preparation of 4,6-bis(diphenylphosphinoyl)dibenzofuran were reported (Scheme 5).[6]

In the multi-step total synthesis of Herboxidiene methyl ester, one step involved an Arbuzov reaction of an allyl bromide with the ethyl ester of diphenylphosphinous acid (Scheme 6).[7]

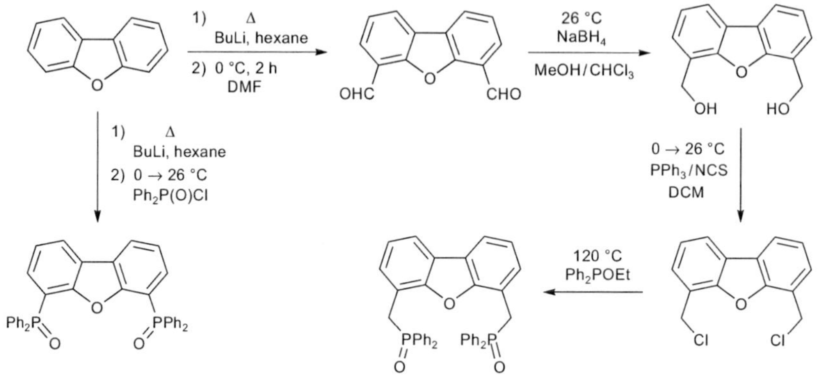

Scheme 5

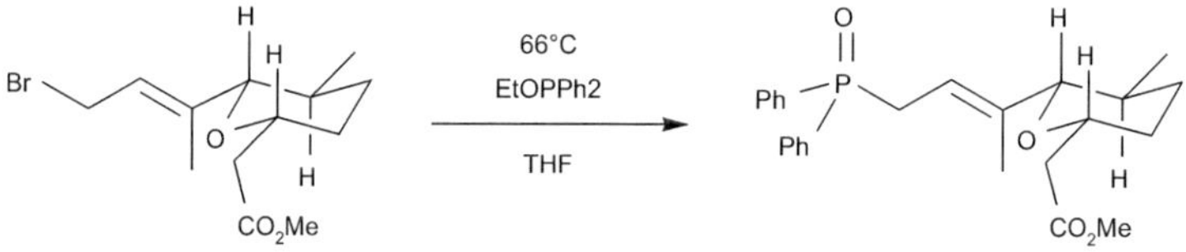

Scheme 6

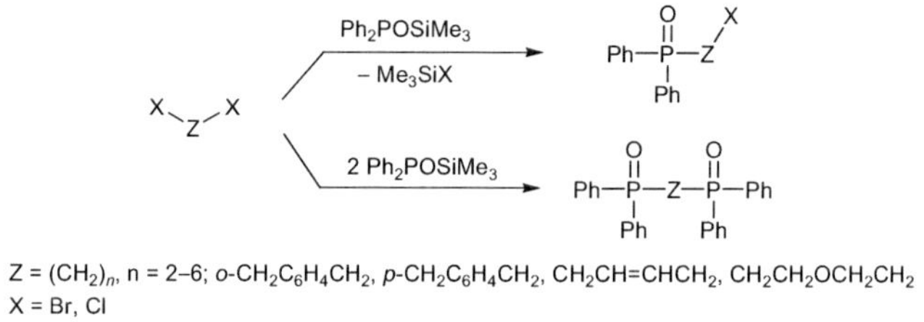

Scheme 7

Z = (CH$_2$)$_n$, n = 2–6; o-CH$_2$C$_6$H$_4$CH$_2$, p-CH$_2$C$_6$H$_4$CH$_2$, CH$_2$CH=CHCH$_2$, CH$_2$CH$_2$OCH$_2$CH$_2$
X = Br, Cl

α,ω-dihaloalkanes were also utilized in the Arbuzov reaction to furnish the corresponding ω-haloalkylphosphinoyl compounds in good or satisfactory yields along with the bis(phosphinoyl) derivatives as the by-products (Scheme 7).[8]

Phosphine oxide-functionalized imidazolium ionic liquids (ILs) were synthesized as tuneable ligands for lanthanide complexation. The task specific ILs were prepared in three steps, by reacting alkyl bromide-functionalized 1,2-dimethyl imidazolium ILs with (Tf$_2$N)$^-$ or (PF$_6$)$^-$ anions and potassium diphenylphosphide. The final step was oxidation of the phosphine to the P-oxide (Scheme 8).[9]

N-confused porphyrins have been involved in regioselective phosphinoylations. Thus, *e.g.*, the reaction of the Ag(III) complex with potassium diphenylphosphide provided a porphyrin derivative with a Ph$_2$P-moiety at

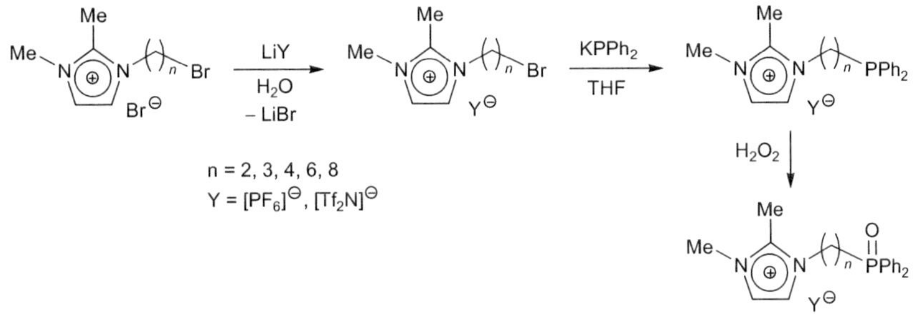

Scheme 8

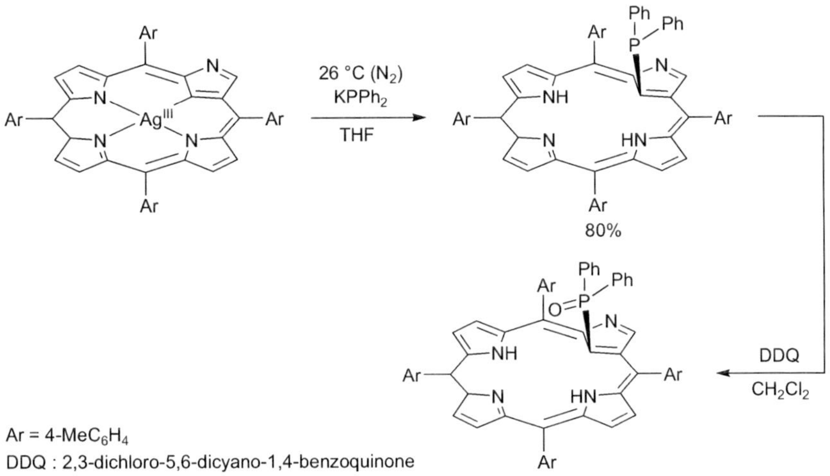

n = 2, 3, 4, 6, 8
Y = $[PF_6]^\ominus$, $[Tf_2N]^\ominus$

Scheme 9

Ar = 4-MeC$_6$H$_4$
DDQ : 2,3-dichloro-5,6-dicyano-1,4-benzoquinone

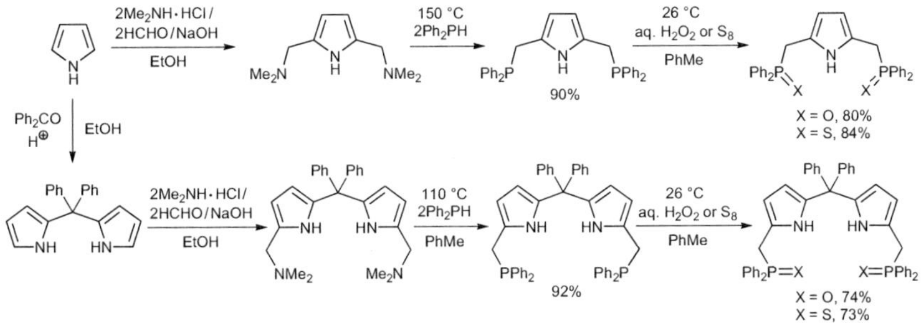

Scheme 10

the C(21) position. The phosphine was stabilized by oxidation (Scheme 9).[10]

Pyrrole-based PNP and PNNP pincer diphosphines were synthesized starting from pyrrole and dipyrrolyl-diphenylmethane utilizing a double Mannich reaction and then the reaction of the Mannich bases with two equivalents of Ph$_2$PH at 110–150 °C. It is noteworthy that there was no need to use Ph$_2$PLi in the substitution reaction. The bisphosphines were converted to the oxides and sulfides (Scheme 10).[11]

A mixed phosphine-phosphine oxide was prepared that was used as an organocatalyst in the asymmetric allylation of aldehydes and hydrogenation of alkenes. The protected bromo-dihydroxycyclohexene was coupled with Ph$_2$PH and the resulting phosphine was stabilized as the borane. Then, the double protecting group was removed and the homoallylic hydroxy group was silylated by the bulky TBDPSCl reactant to provide a selective protection. Treatment of the allylic alcohol with chlorodiphenylphosphine led, surprisingly, not to the expected allylic diphenylphosphinite, but to a rearranged species that was a phosphine oxide. The interesting conversion can be specified as an Arbuzov [2,3]-sigmatropic rearrangement. In the final step, either the protecting group was removed, or the phosphine borane function was deprotected (Scheme 11).[12]

A novel observation is that the reaction of aldehydes with Ph$_2$PI affords the corresponding phosphine oxides as shown in Scheme 12. In this instance, the Ph$_2$PI reagent, applied in a 4.5 equivalent quantity, acts as a reducing agent and, at the same time, as a phosphinoylating agent. The Ph$_2$PI was generated *in situ* by the reaction of Ph$_2$PCl and NaI.[13]

Phosphinoyl- and thiophosphinoyl-pyrazoles were prepared from β-enaminoesters (EAE) by a five-step procedure. The EAE was treated with diphenylchlorophosphine to afford the phosphino-iminoester. After blocking the tervalent P atom of the intermediate by oxygen or

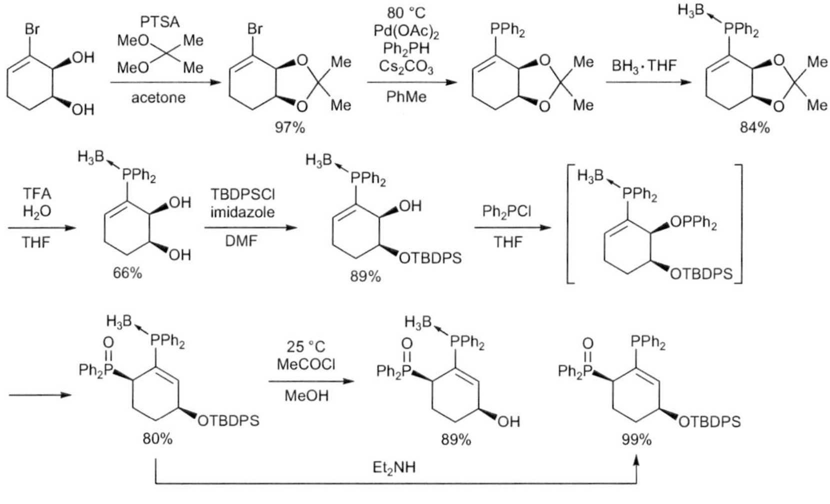

Scheme 11

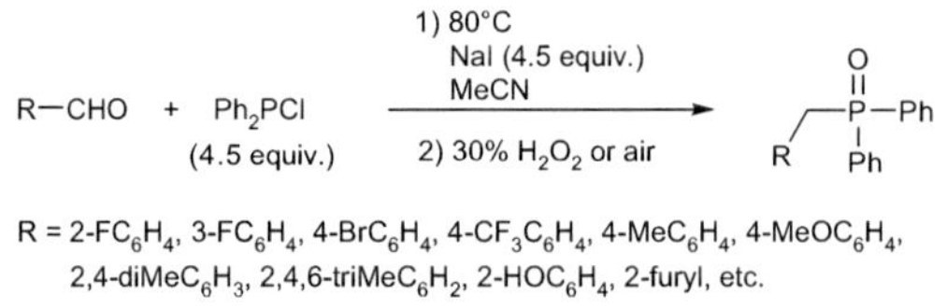

Scheme 12

sulfur, the imino function was converted to a carbonyl group. The resulting β-oxo ester was then cyclized by hydrazine in two steps to furnish the target compounds (Scheme 13).[14]

The reaction of aryl- or hetarylvinyl-methyl ketones with diphenylchlorophosphine in the presence of acetic acid led to diphenylphosphinoyl ketones (Scheme 14).[15]

A reaction sequence starting with the interaction of dialkylchloromethyleneiminium chloride with diphenylchlorophosphine, followed by the reaction of the adduct so formed with dialkylformamide to give (dialkylaminochloromethyl)phosphine oxides, was investigated in detail (Scheme 15).[16]

The precursor of a new atropisomeric bis(phosphine oxide) derivative was synthesized from 1-bromo-3-(trifluoromethoxy)benzene *via* P-functionalization, coupling and optical resolution (Scheme 16).[17]

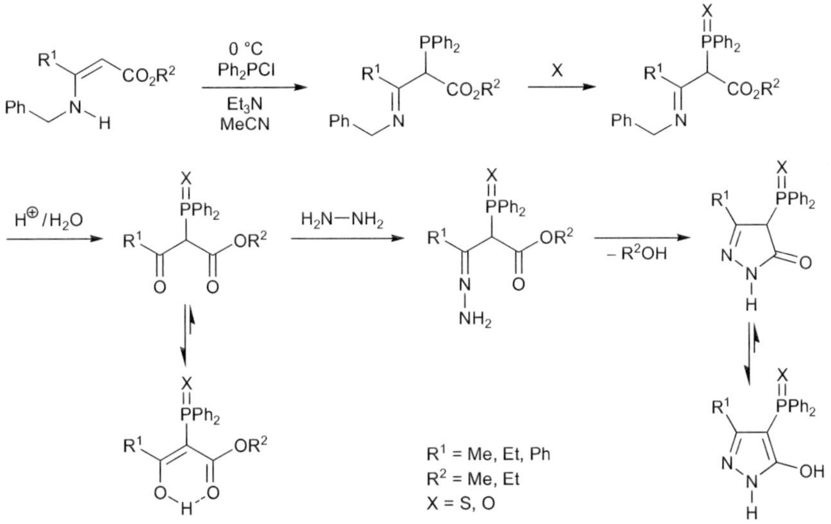

Scheme 13

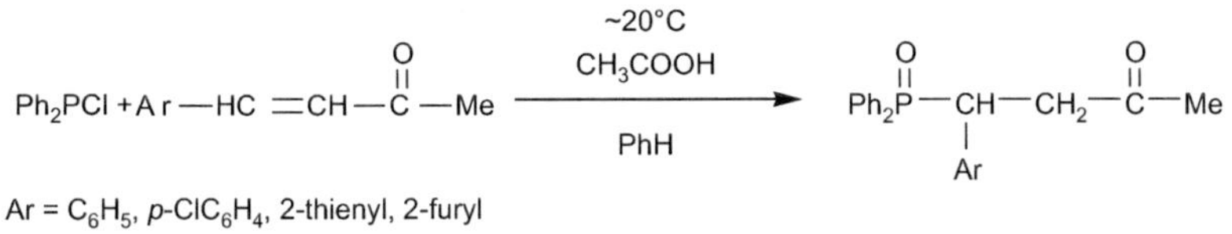

Scheme 14

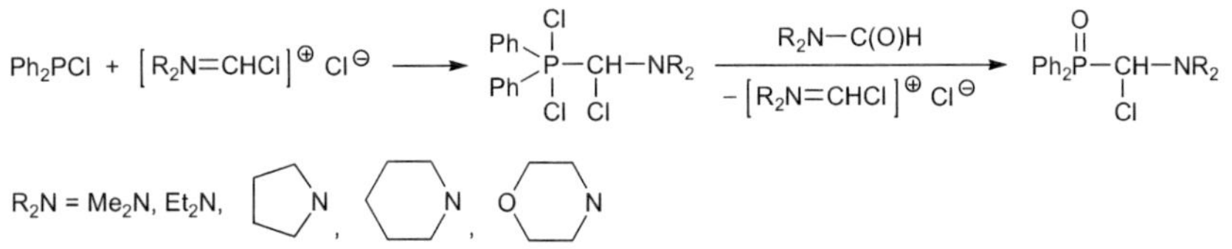

Scheme 15

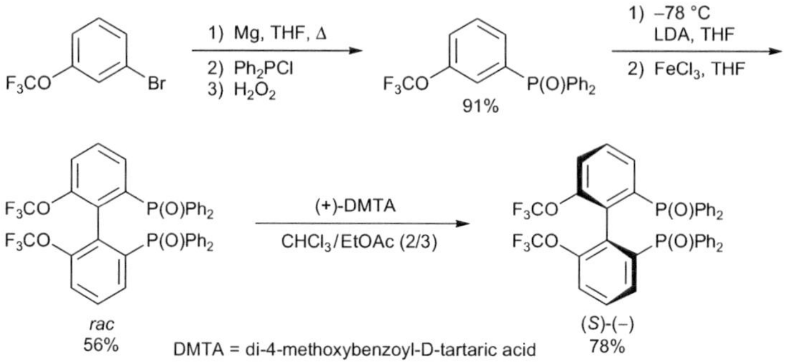

Scheme 16

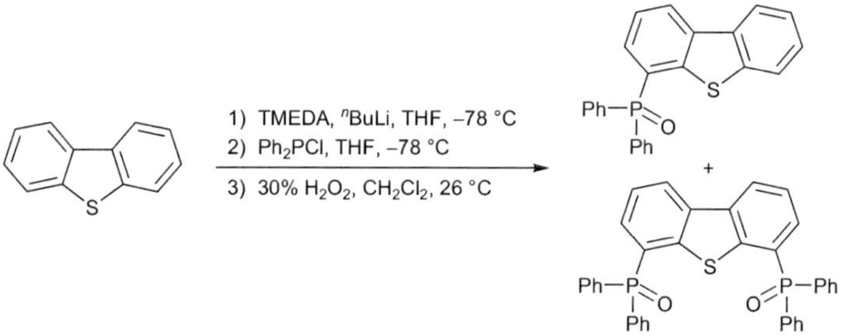

Scheme 17

Dibenzothiophene was P-functionalized *via* metallation, reaction with Ph$_2$PCl and oxidation. Both the mono- and the diphosphinoyl products were formed (Scheme 17).[18]

Bichromophoric dithieno[3,2-*b*:2′,3′-*d*]phospholes with polyaromatic hydrocarbon substituents were synthesized by the lithiation of dibromobithiophenes followed by the reaction of the intermediates so formed with arylphosphonous dichlorides. The phosphines obtained were oxidized to the corresponding P-oxides (Scheme 18).[19]

α,α′-Diarylacenaphtho[*c*]phosphole chalcogenides were prepared from 1,8-bis(trimethylsilylethynyl)naphthalene in 4/5 steps. In the first step, the bisacetylene derivative was reacted with the Sato-Urabe low-valent titanium reagent to provide a heterocyclic-titanium intermediate. This moiety was then treated with PhPCl$_2$ and a phosphine oxide was formed by oxidation. Then, the TMS groups were exchanged to bromine atoms and the resulting dibromophospholes were converted to the bisaryl derivatives by Stille coupling. Further functionalizations led to the corresponding phosphole oxides and sulfides (Scheme 19).[20]

Polycyclic compounds comprising carbazole- and/or fluorene units and one or two diphenylphosphinoyl moieties are of interest as components of phosphorescent light-emitting diodes. The functionalization of the skeleton may be achieved by reacting the 'Core' bromide with Ph$_2$PCl or Ph$_2$P(O)Cl after metallation by butyllithium (Scheme 20).[21] In the former case, the phosphine is oxidized to the phosphine oxide. A few possibilities for the stucture of the target compounds are shown in Fig. 1. 'Core' means a skeleton incorporating carbazole- and/or fluorene units.

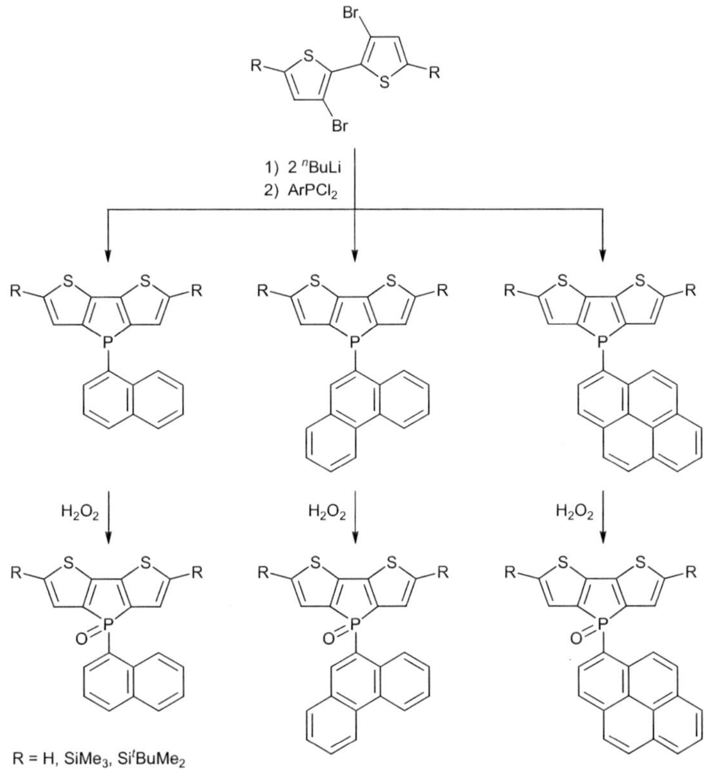

Scheme 18

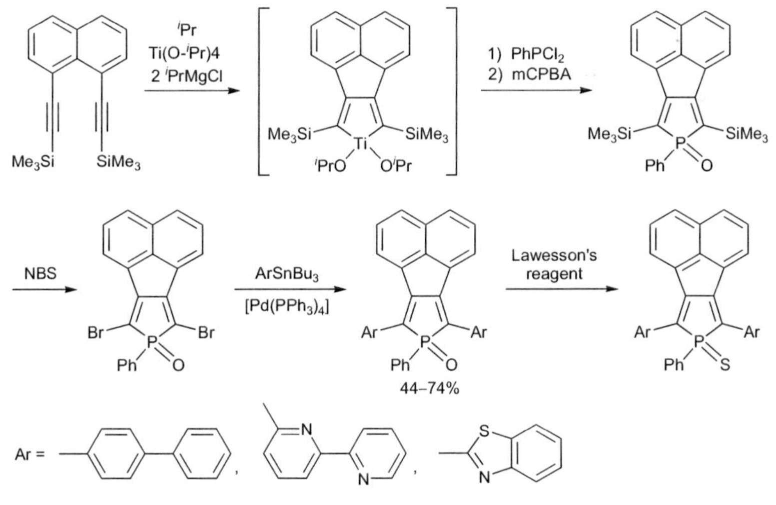

R = H, SiMe₃, Si'BuMe₂

Scheme 19

A novel rigid-rod type dianhydride monomer with a phosphine oxide function was prepared *via* the Suzuki coupling reaction of 4-(diphenylphosphinoyl)phenyl boronic acid, synthesized in two steps. The four methyl groups in the aromatic ring of the Suzuki product were then

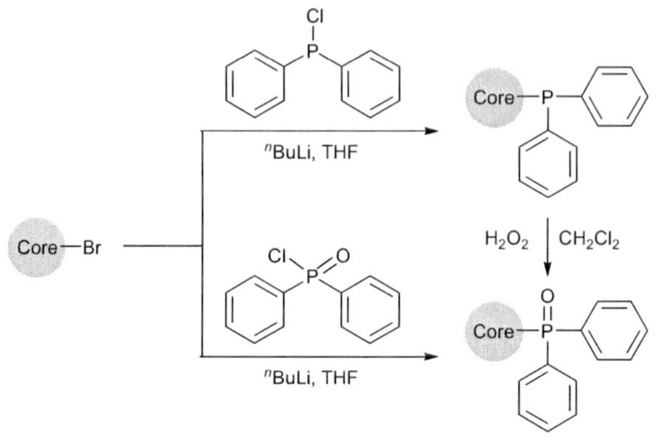

Scheme 20

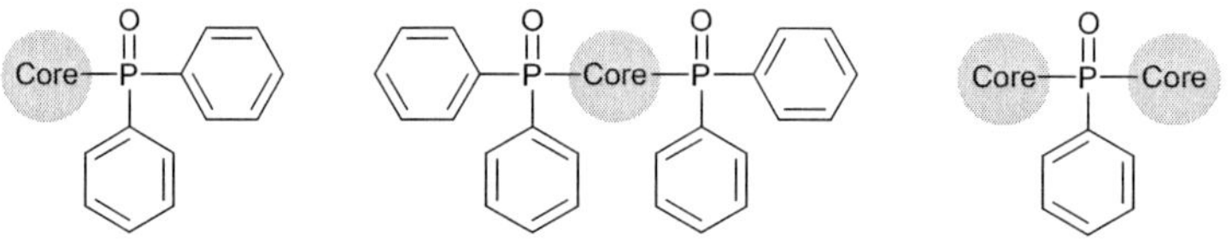

Fig. 1

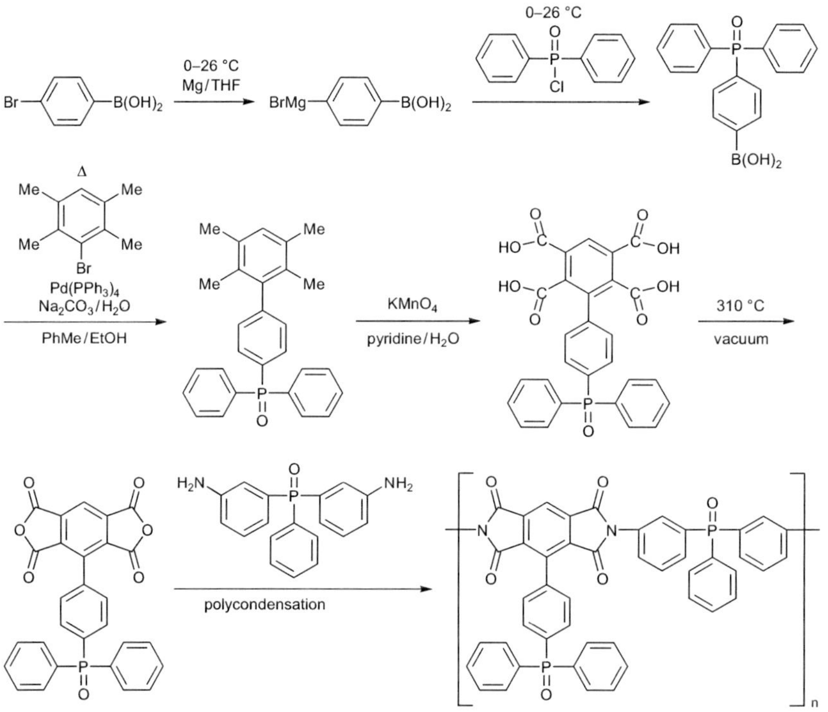

Scheme 21

oxidized to carboxyl functions and dehydrated to the corresponding bis(anhydride). This monomer was then used in polycondensation with bis(3-aminophenyl)phenylphosphine oxide (Scheme 21).[22]

Showing resemblance to the first two steps of the above synthesis, 2-oxazolyl-bromobenzenes were converted to the corresponding Grignard reagents that were then treated with a half equivalent of phenylphosphonic dichloride. The products may be regarded as phosphine oxide-linked bis(oxazolines) (Scheme 22).[23]

In a similar fashion, bis(4-fluoro-3-trifluoromethylphenyl-)phenylphosphine oxide was prepared and reacted with bisphenol derivatives to furnish poly(arylene ether phosphine oxides) (Scheme 23).[24]

The interaction of phenylphosphonic dichloride with perfluoroalkyl Grignard reagents afforded the corresponding phenyldi(perfluoroalkyl)phosphine oxides. It was observed that one of the electron-poor P–C bonds could be hydrolyzed to give the corresponding phenylperfluoroalkylphosphinic acid (Scheme 24).[25]

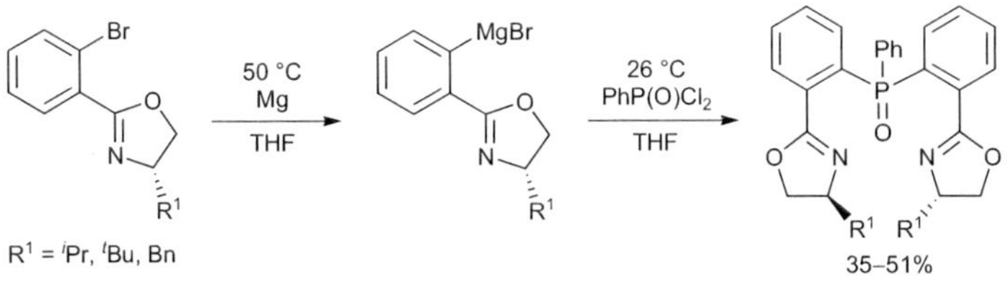

Scheme 22

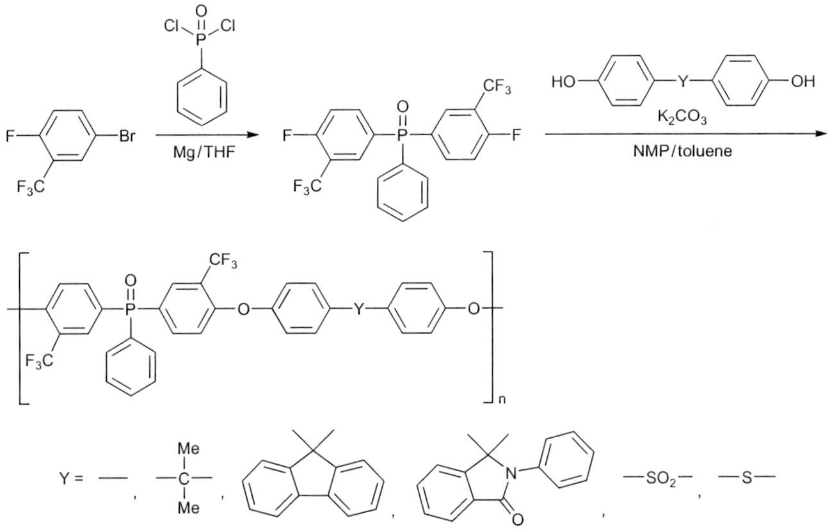

Scheme 23

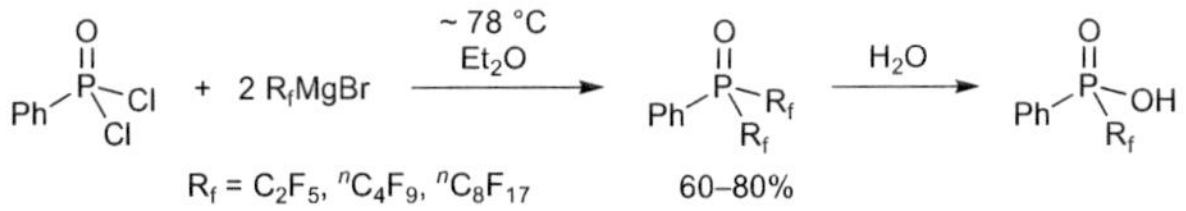

Scheme 24

Three homologues of dibutylaminoalkyl-diarylphosphine oxides were synthesized *via* the corresponding hydroxy and bromo intermediates as shown in Scheme 25. The starting materials/intermediates were diaryl-hydroxymethylphosphine oxides, as well as hydroxypropyl- and hydroxypentyl-diarylphosphine oxides. The latter two intermediates were prepared by the alkylation of the diarylphosphine oxides (Scheme 25).[26]

N-aryl-tetrahydroisoquinolines were the subject of a gold-catalyzed oxidative coupling with diarylphosphine oxides to provide the corresponding phosphinoyl derivatives (Scheme 26).[27]

Aryl group-modified DIOP dioxides (Ar-DIOPOs) were synthesed by the reaction of the corresponding 1,4-diiodo compound with two equivalents of secondary phosphine oxides in the presence of LHMDS as the base (Scheme 27-1).[28] A dibenzooxaphosphorine oxide was also used as the >P(O)H species to prepare a sterically crowded, more rigid DIOPO (Scheme 27-2).[28]

Morita-Baylis-Hillmann carbonates were involved in an asymmetric allylic substitution reaction with diphenylphosphine oxide in the presence of chiral thiourea-phosphine organocatalysts. The enantioselectivity was $\geq 96\%$ (Scheme 28).[29]

1,1,2-Tris(diarylphosphinoyl)ethanes were synthesized by the reaction of bromoacetaldehyde diethyl acetal with three equivalents of

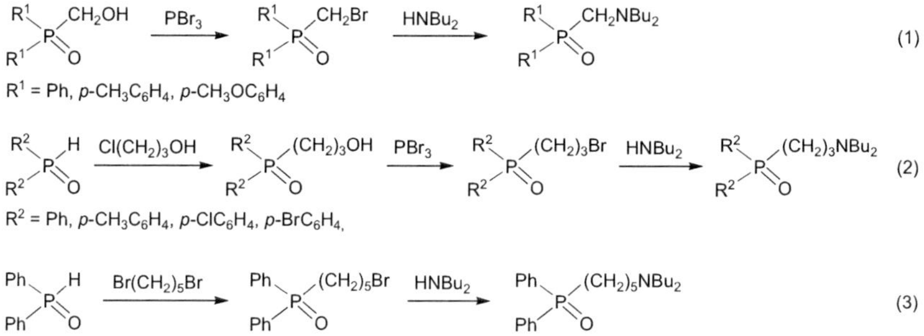

Scheme 25

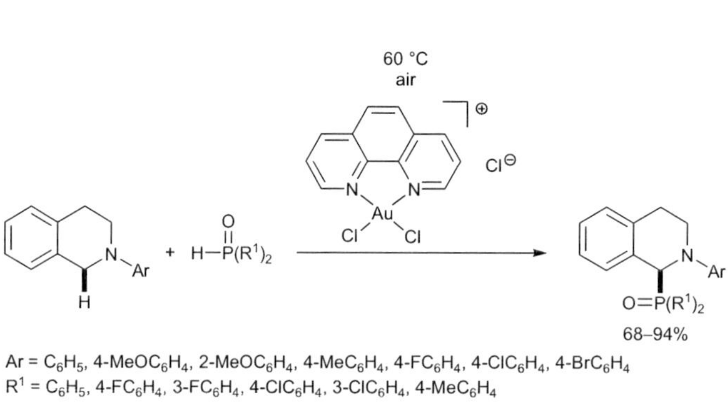

Scheme 26

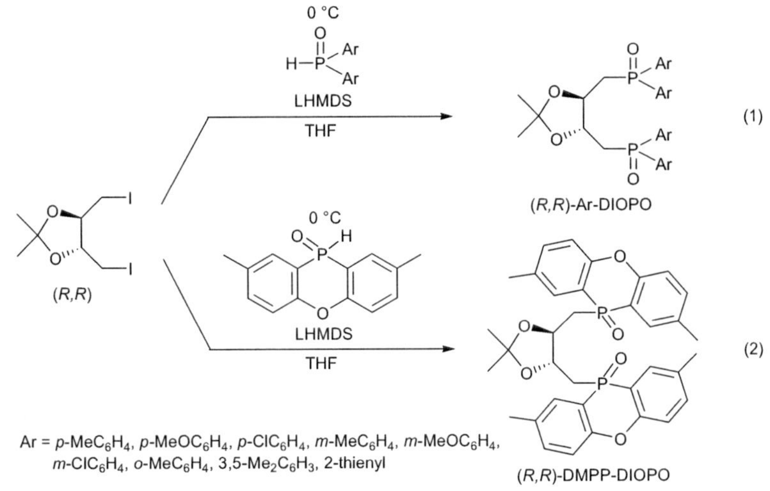

Scheme 27

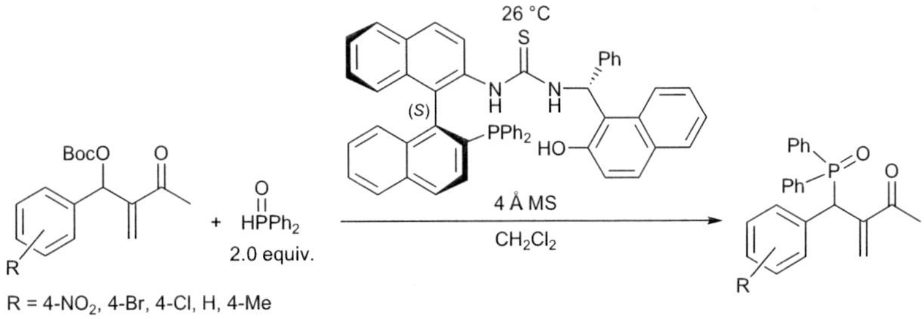

Ar = p-MeC$_6$H$_4$, p-MeOC$_6$H$_4$, p-ClC$_6$H$_4$, m-MeC$_6$H$_4$, m-MeOC$_6$H$_4$, m-ClC$_6$H$_4$, o-MeC$_6$H$_4$, 3,5-Me$_2$C$_6$H$_3$, 2-thienyl

Scheme 28

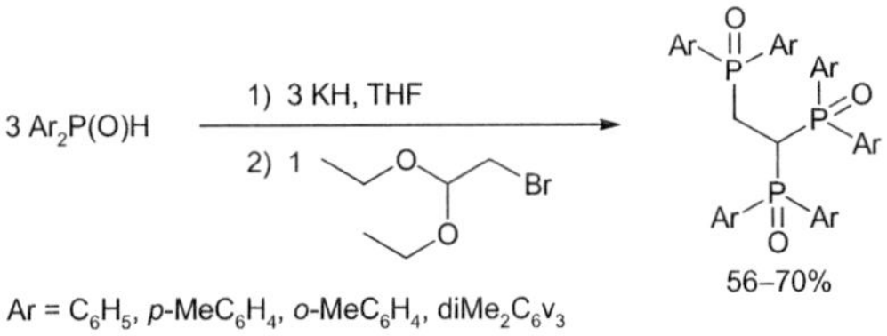

R = 4-NO$_2$, 4-Br, 4-Cl, H, 4-Me

Scheme 29

Ar = C$_6$H$_5$, p-MeC$_6$H$_4$, o-MeC$_6$H$_4$, diMe$_2$C$_6$v$_3$

diarylphosphine oxides in the presence of KH. The products are tripodal phosphine oxide ligands (Scheme 29).[30]

A uracil-substituted phosphine oxide was prepared by the Pd-catalyzed coupling of 5-iodouracil with diphenylphosphine in the presence of triethylamine, followed by oxidation of the initially-formed phosphine (Scheme 30).[31]

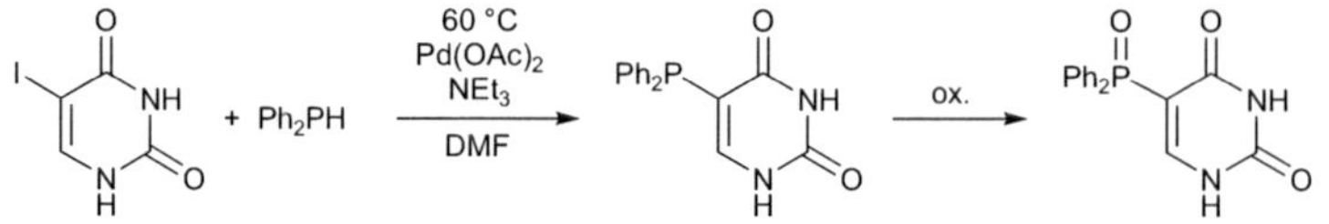

Scheme 30

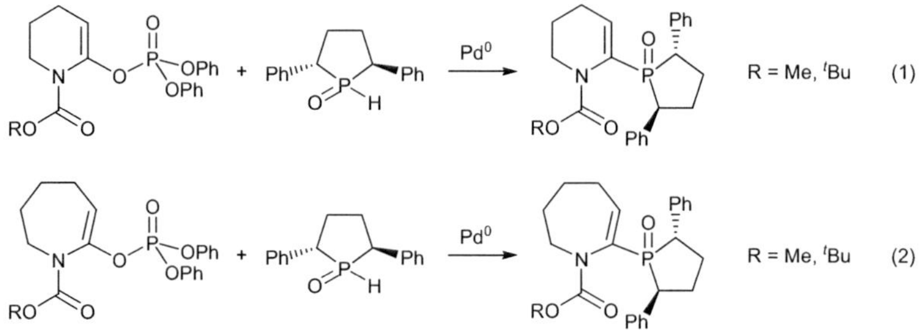

Scheme 31

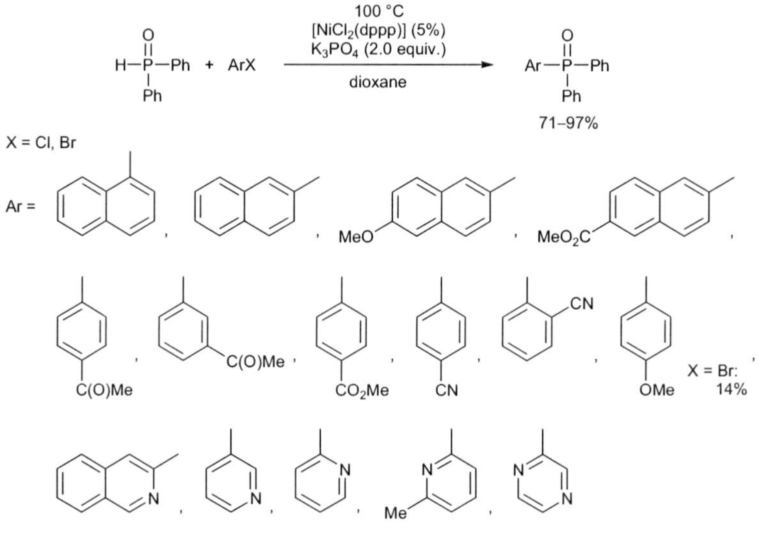

Scheme 32

The next scheme exemplifies the P–C cross-coupling of cyclic enamides and 2,5-diphenylphospholane 1-oxide in the presence of Pd^0 to furnish the corresponding α-enamido-phosphine oxides (Scheme 31).[32]

A series of aryl-diphenylphosphine oxides was prepared by the [NiCl$_2$(dppp)]-catalyzed cross-coupling of aryl halides with diphenylphosphine oxide (Scheme 32).[33]

In a similar way, but using NiCl$_2$(dppf), aryl mesylates and tosylates were coupled with diarylphosphine oxides (Scheme 33).[34]

In another series of P–C coupling reactions, involving a wide range of aryl iodides, a variety of secondary phosphine oxides was used in the presence of tris(dibenzylideneacetone)dipalladium(Pd$_2$dba$_3$) and Xantphos. (Scheme 34).[35]

It is noteworthy that the coupling of substituted halogenobenzoic acids with diphenylphosphine oxide could be accomplished under microwave conditions in water as the reaction medium (Scheme 35).[36]

A monomer for a fluorinated poly(arylene ether-phosphine oxide) polymer was prepared by the P–C coupling of a dioxoalkoxyaryl iodide with bis(3,4,5-trifluorophenyl)phosphine oxide under the classical conditions of the Hirao reaction (Scheme 36).[37]

The bistriflate of BINOL underwent selective monophosphinoylation in reaction with phenyl-alkylphosphine oxides (Scheme 37).[38]

An analogous precursor of a chiral-bridged atropisomeric monophosphine ligand was also described (Scheme 38).[39]

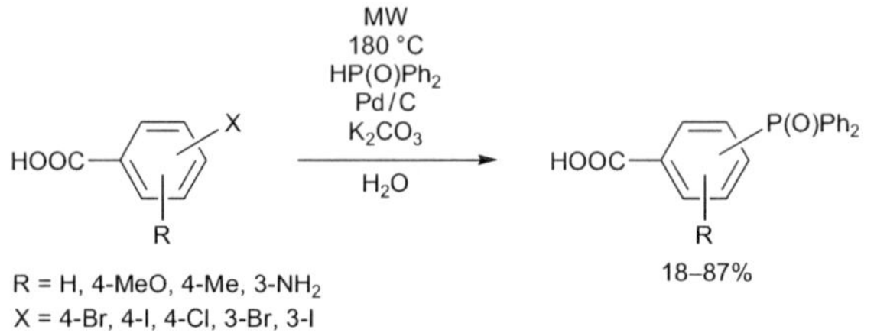

Scheme 35

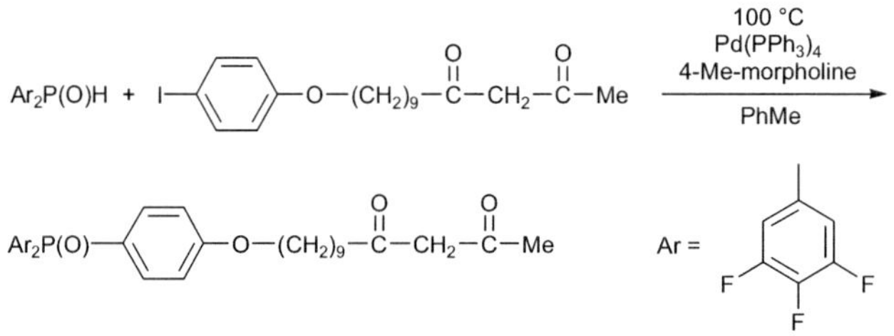

R = H, 4-MeO, 4-Me, 3-NH$_2$
X = 4-Br, 4-I, 4-Cl, 3-Br, 3-I

18–87%

Scheme 36

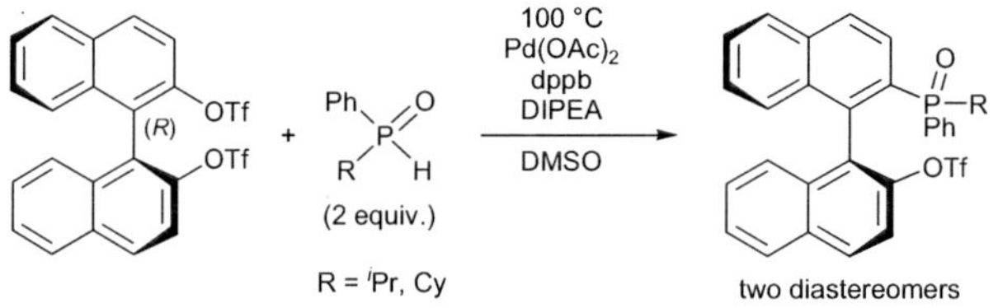

Scheme 37

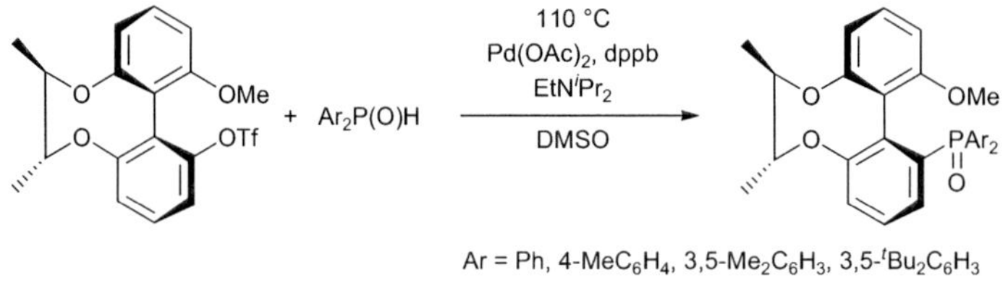

R = ^{i}Pr, Cy

two diastereomers

Scheme 38

Ar = Ph, 4-MeC$_6$H$_4$, 3,5-Me$_2$C$_6$H$_3$, 3,5-^{t}Bu$_2$C$_6$H$_3$

It is noteworthy that phenol derivatives could also be coupled with diphenylphosphine oxide. The catalyst was NiCl$_2$(dppp) and the activating reagent a special phosphonium salt (Scheme 39).[40]

Substituted chloropyrazines including a dichloro derivative were also the subject of P–C couplings using Pd(dppf)Cl$_2$ as the catalyst and DBU as the base to afford monophosphinoyl and diphosphinoyl products (Schemes 40-1 and 40-2).[41] A dithiophosphinoylpyrazine was also prepared via the diphosphine intermediate (Scheme 40-3).[41]

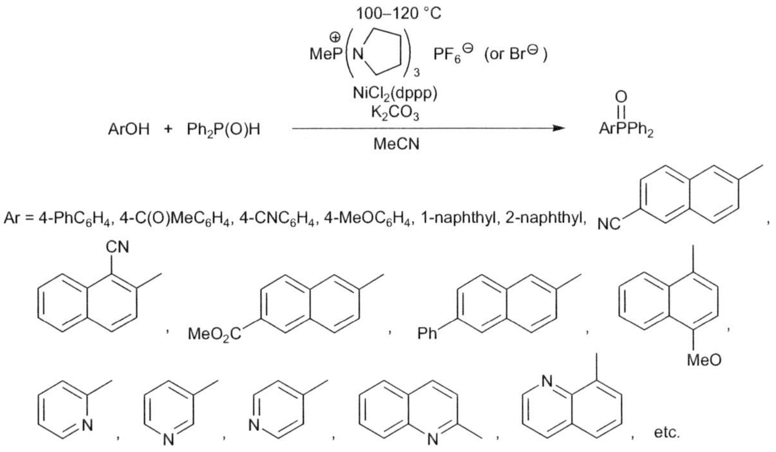

Scheme 39

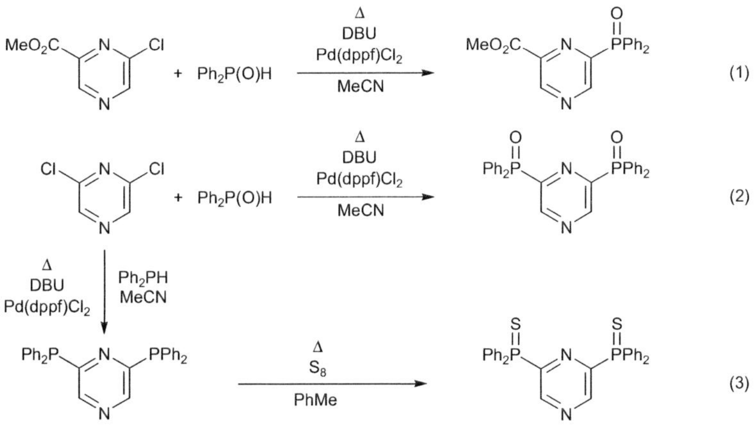

Scheme 40

A series of hetaryl chlorides and aryl chlorides with a heterocyclic substituent underwent coupling with diphenylphosphine oxide in the presence of a Ni-catalyst and K_2CO_3 (Scheme 41).[42]

α,β-Unsaturated ketones and diallylphosphine oxide reacted according to the phospha-Michael protocol using diethylzinc as the base and a special bispyrrole derivative as the chiral component. The adducts were formed in enantioselectivities ≥ 82 (Scheme 42-1).[43] Unsaturated acyl-pyrroles were also tested in the enantioselective Michael additions (Scheme 42-2).[43]

In another example, an enone derivative of methyl oleate was reacted with diphenylphosphine oxide (Scheme 43).[44] This reaction was then extended to analogous triglyceride derivatives.

The author of this chapter, together with co-workers, investigated the microwave(MS)-assisted addition of >P(O)H species, including diphenylphosphine oxide, to maleic acid derivatives. There was no need to use any catalyst or base (Scheme 44).[45]

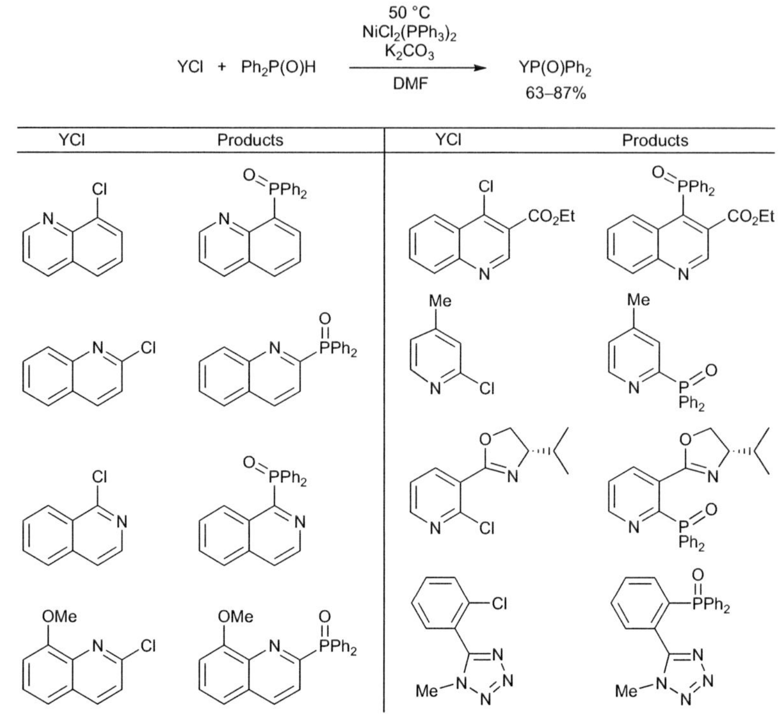

Scheme 41

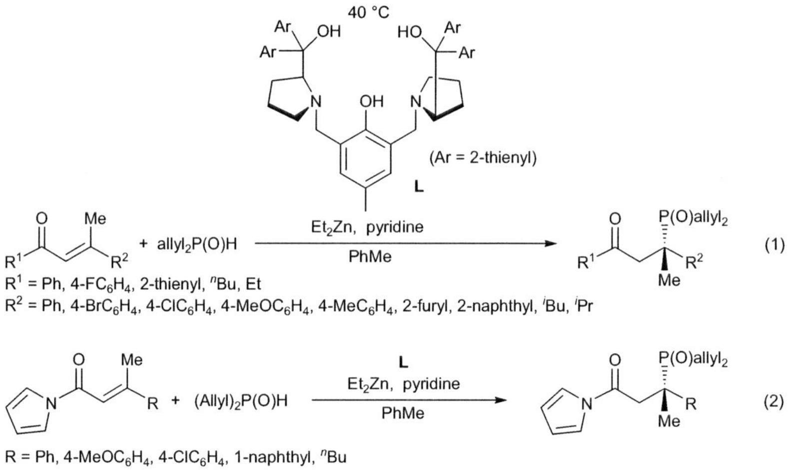

Scheme 42

The Michael addition of >P(O)H reagents including dialkyl phosphites H-phosphinates and diphenylphosphine oxide to the double-bond of 1-phenyl-2-phospholene oxide was also studied. In these cases, a traditional procedure involving the use of trimethylaluminum as the base was more appropriate than the MW-assisted variation due to better selectivities (Scheme 45).[46]

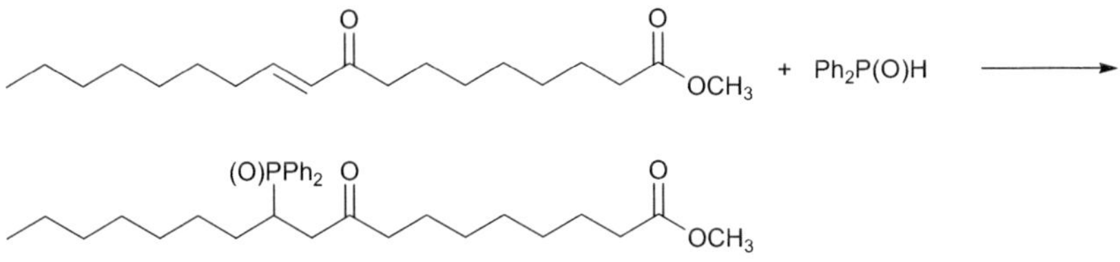

Scheme 43

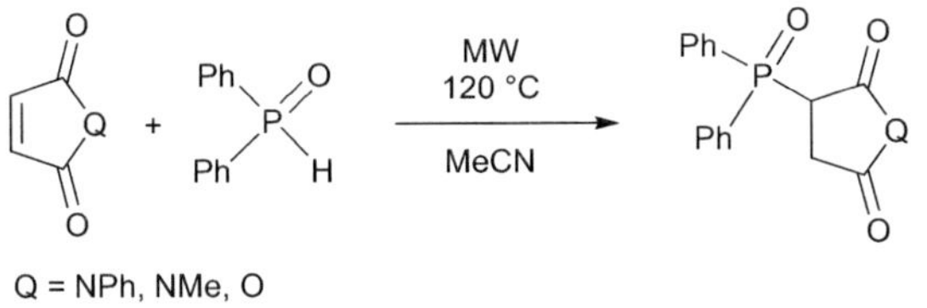

Scheme 44

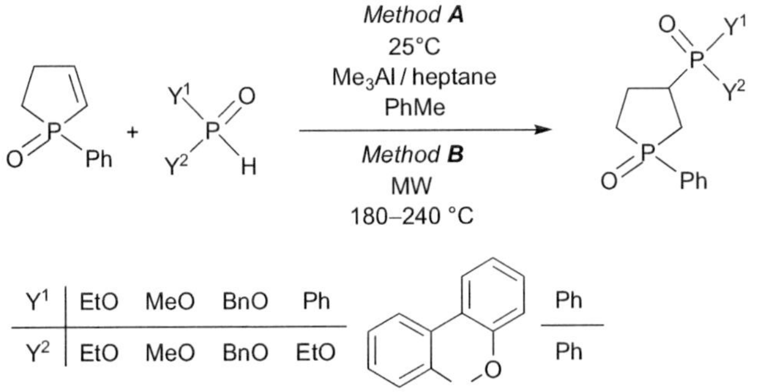

Q = NPh, NMe, O

Scheme 45

Trofimov et al. carried out the addition of dialkylphosphines and dialkylphosphine chalcogenides to tris(4-vinylbenzyl)phosphine oxide under radical conditions. The trisphosphines were converted to the stable oxides (Scheme 46).[47]

The addition of diphenylphosphine oxide to imines prepared from imidazole-2-carboxaldehyde and primary amines led to the corresponding α-aminophosphine oxides (Scheme 47).[48]

A new method involving the copper-catalyzed addition of diarylphosphine oxides to N-tosylhydrazones was elaborated for the synthesis of phosphine oxides (Scheme 48).[49]

A highly regio- and stereo-selective hydrophosphinylation of acetylenes in the presence of an immobilized rhodium-phosphine catalyst has been described (Scheme 49).[50]

In a ruthenium-catalyzed reaction with diphenylphosphine oxide, 1,6-diynes underwent a stereoselective and hydrophosphinylative cyclization to afford 1,3-dienylphosphine oxides (Scheme 50).[51]

In a novel reaction, terminal alkynes were phosphinylthiolated by reaction with butyl diphenylmonothiophosphinate using a Pd-PEt$_3$ catalyst system (Scheme 51).[52]

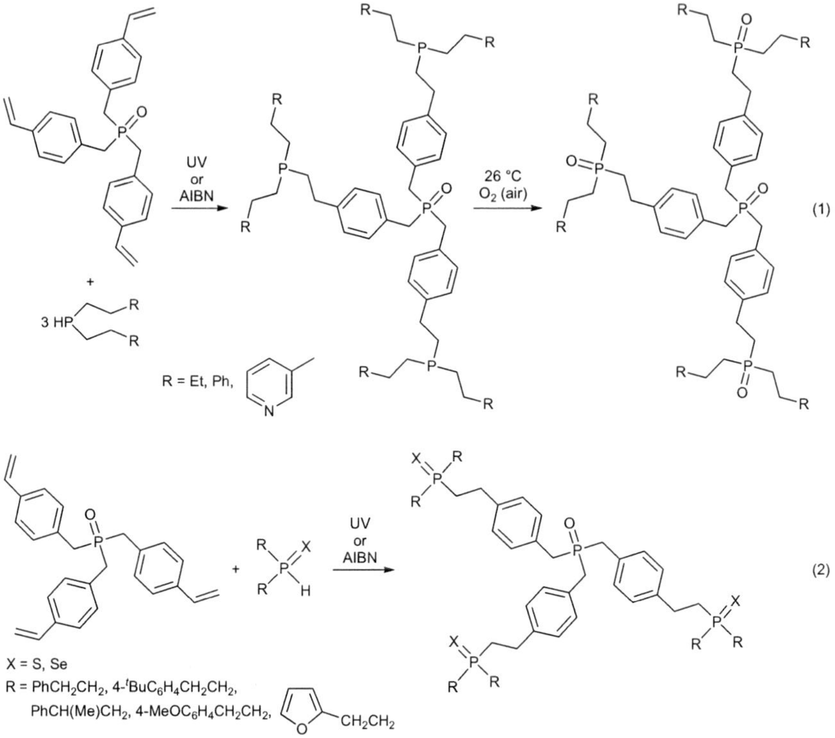

Scheme 46

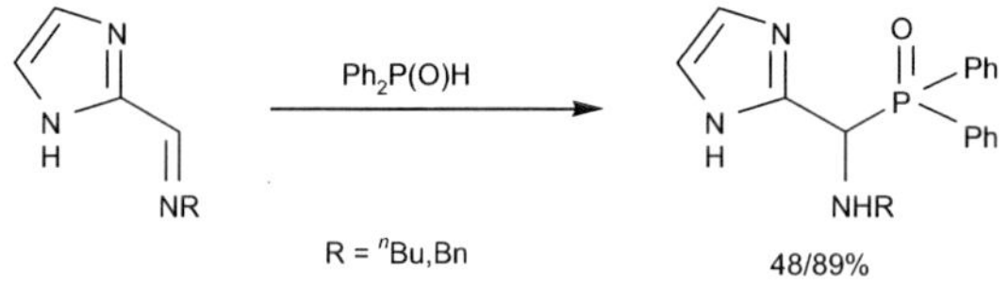

Scheme 47

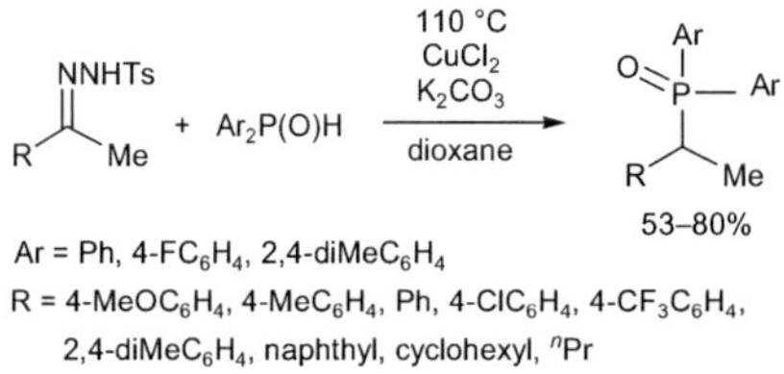

Scheme 48

The double Kabachnik-Fields (phospha-Mannich) reaction of primary amines, two equivalents of paraformaldehyde and the same amount of diphenylphosphine oxide gave bis(diphenylphosphinoylmethyl)amines that were precursors of bidentate P-ligands after double deoxygenation

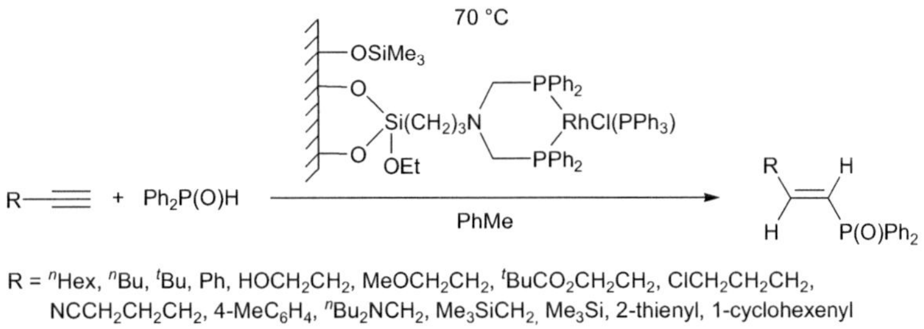

Scheme 49

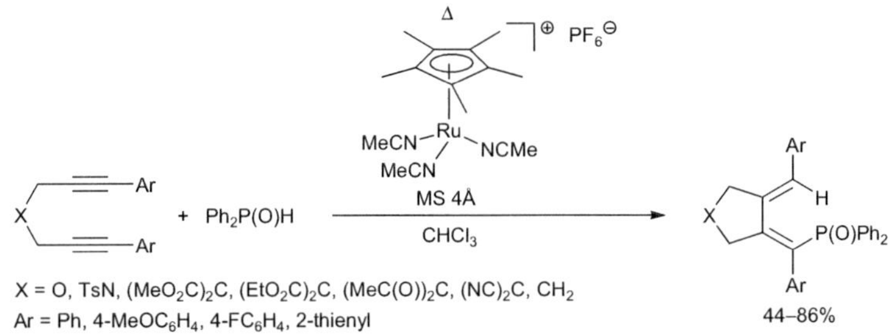

R≡ + Ph$_2$P(O)H → (R)(H)C=C(H)(P(O)Ph$_2$)

Conditions: supported Si(CH$_2$)$_3$N(PPh$_2$)$_2$RhCl(PPh$_3$), PhMe, 70 °C

R = ^{n}Hex, ^{n}Bu, ^{t}Bu, Ph, HOCH$_2$CH$_2$, MeOCH$_2$CH$_2$, ^{t}BuCO$_2$CH$_2$CH$_2$, ClCH$_2$CH$_2$CH$_2$, NCCH$_2$CH$_2$CH$_2$, 4-MeC$_6$H$_4$, ^{n}Bu$_2$NCH$_2$, Me$_3$SiCH$_2$, Me$_3$Si, 2-thienyl, 1-cyclohexenyl

Scheme 50

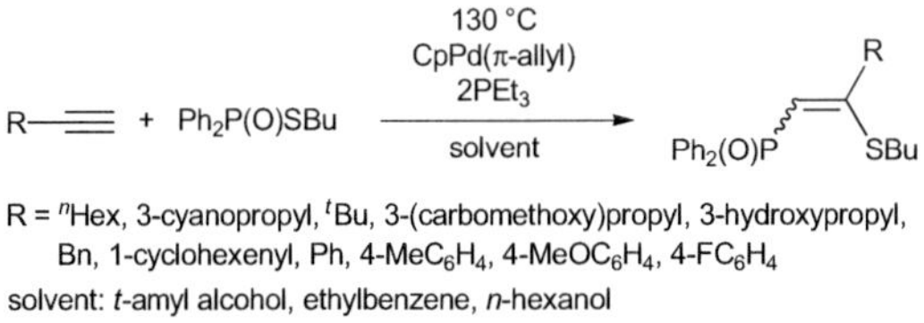

Ar–≡–X–≡–Ar + Ph$_2$P(O)H → cyclized diene product, 44–86%

Conditions: [Cp*Ru(MeCN)$_3$]$^+$PF$_6^-$, MS 4Å, CHCl$_3$, Δ

X = O, TsN, (MeO$_2$C)$_2$C, (EtO$_2$C)$_2$C, (MeC(O))$_2$C, (NC)$_2$C, CH$_2$
Ar = Ph, 4-MeOC$_6$H$_4$, 4-FC$_6$H$_4$, 2-thienyl

Scheme 51

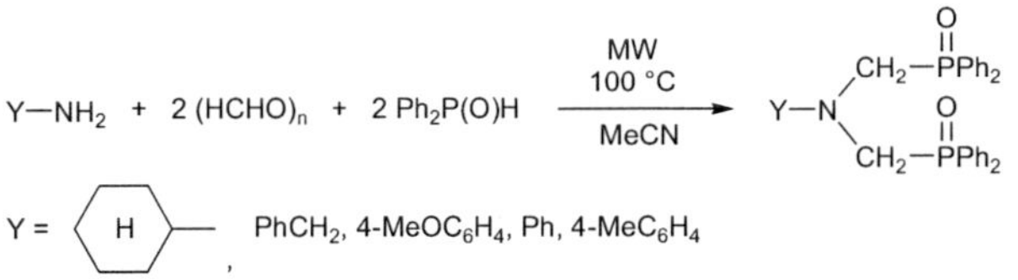

R≡ + Ph$_2$P(O)SBu → Ph$_2$(O)P–C(R)=CH–SBu

Conditions: CpPd(π-allyl), 2PEt$_3$, solvent, 130 °C

R = ^{n}Hex, 3-cyanopropyl, ^{t}Bu, 3-(carbomethoxy)propyl, 3-hydroxypropyl, Bn, 1-cyclohexenyl, Ph, 4-MeC$_6$H$_4$, 4-MeOC$_6$H$_4$, 4-FC$_6$H$_4$
solvent: t-amyl alcohol, ethylbenzene, n-hexanol

Scheme 52

Y–NH$_2$ + 2 (HCHO)$_n$ + 2 Ph$_2$P(O)H → Y–N(CH$_2$–P(O)Ph$_2$)$_2$

Conditions: MW, 100 °C, MeCN

Y = cyclohexyl, PhCH$_2$, 4-MeOC$_6$H$_4$, Ph, 4-MeC$_6$H$_4$

(Scheme 52).[53,54] Cherkasov *et al.* also studied the bis(Kabachnik-Fields) reaction[55] and recent results of the Kabachnik-Fields reaction including the preparation of α-aminophosphine oxides have been reviewed.[56]

β-Aminophosphine oxides were obtained by the addition of primary and secondary amines to diphenyl(vinyl)phosphine oxide in water as the reaction medium (Scheme 53).[57]

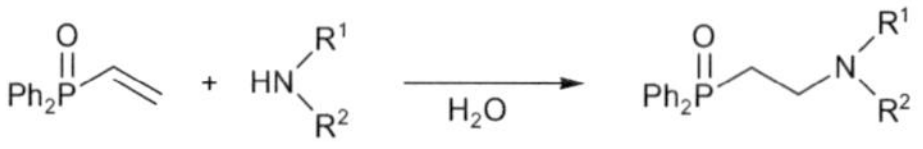

Scheme 53

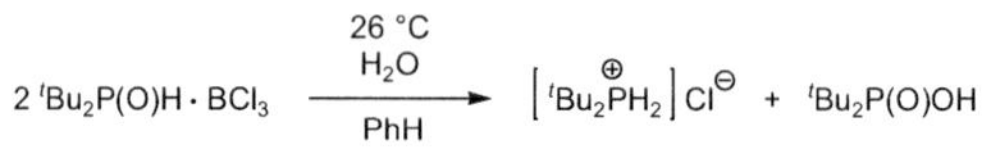

Scheme 54

Scheme 55

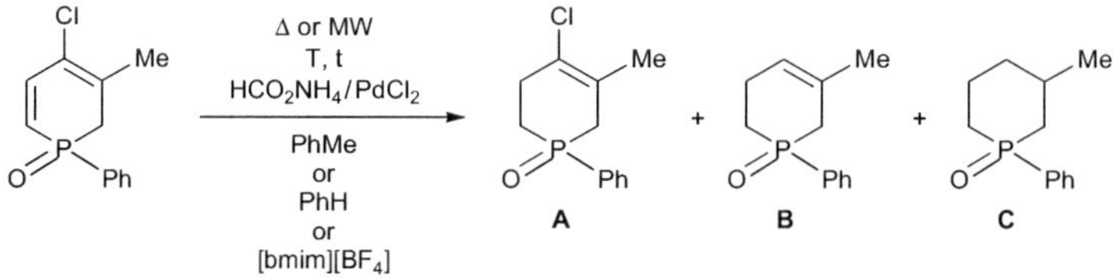

Mode of heating	Solvent	T (°C)	t (h)	Composition (%)			
				A	B	C	Other
Δ	PhMe	110	4	22	31	47	
MW	PhMe	110	2	11	79	5	5
Δ	[bmim][BF$_4$]	110	3	10	41	49	
MW	[bmim][BF$_4$]	110	2	2	59	39	
Δ	PhH	78	4	9	44	47	
Δ	[bmim][BF$_4$]	130	1.5	3	45	32	

^{t}BuP(O)H · BCl$_3$ underwent a kind of disproportionation reaction on exposure to moist air in benzene solution (Scheme 54).[58]

Recent developments in the field of metal-catalyzed asymmetric addition of >P(O)H species to C=O, C=N and C=C double-bonds have been reviewed by Zhao and Wang.[59] Under the conditions of transfer hydrogenation by ammonium formate, 4-chloro-1,2-dihydrophosphinine oxides were converted to a mixture of the corresponding 1,2,3,6-tetrahydrophosphinine oxide, its dechlorinated derivative and the respective 1,2,3,4,5,6-hexahydrophosphinine oxide (Scheme 55). The effects of variance in the reaction conditions are shown in Table below the Scheme.[60]

Pietrusiewicz and co-workers found that the dearomatization of aryldialkylphosphine oxides under Birch reduction conditions, followed by treatment with alkyl halides, provided the α-functionalized (cyclohexa-1,4-dien-3-yl)phosphine oxide derivatives (Scheme 56).[61]

(1,4-Cyclohexadien-3-yl)dimethylphosphine oxide, prepared by Birch reduction, was involved in the Michael addition of secondary phosphine oxides in the presence of a suitable base. Using diphenylphosphine oxide and n-butyllithium, the expected *trans* diphosphinoyl cyclohexene was formed (Scheme 57-1).[62] Applying other bases, the situation was more complex (Scheme 57-2).[62]

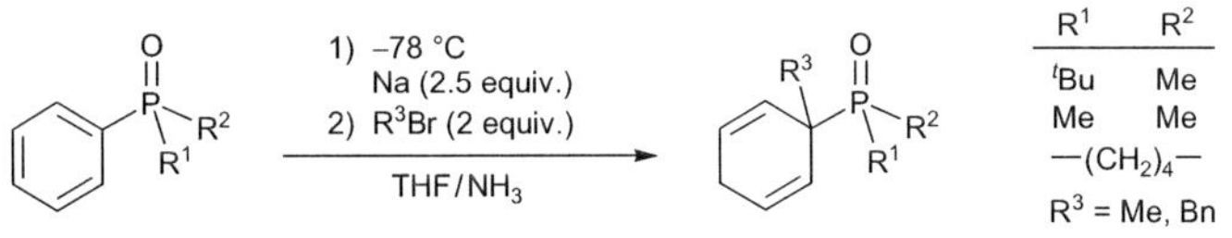

Scheme 56

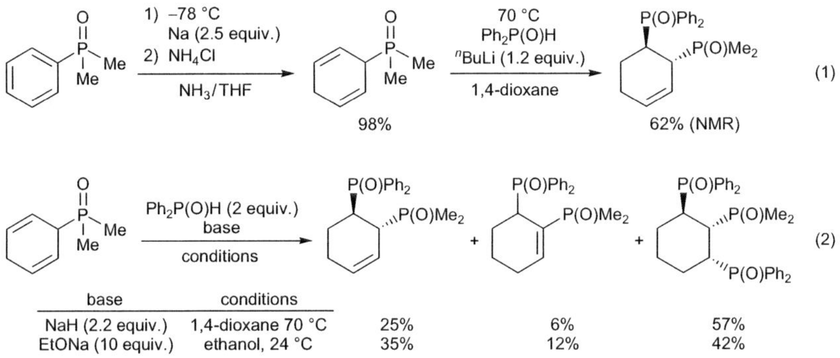

Scheme 57

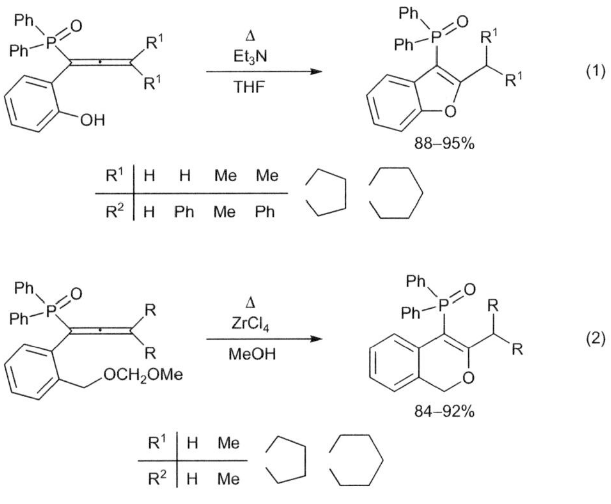

Scheme 58

Phenolic and benzylic allenylphosphine oxides were cyclized to phosphinoyl benzofurans and isochromenes, respectively. The conditions applied were, however, different (Schemes 58-1 and 58-2).[63]

The triphenylphosphine-catalyzed reaction of allenyl-diphenylphosphine oxide with N- and C-nucleophiles gave γ-substituted allylphosphine oxides. Scheme 59 shows the outcome of the reaction with phthalimide.[64]

The bisoxide of BINAP was modified by double bromination using NBS in an ionic liquid (hexylmethylimidazolium PF_6 or Ntf_2) (Scheme 60).[65]

The resolution of 2,2'-bis(diphenylphosphinoyl)-1,1'-binaphthyl was accomplished *via* an inclusion complex with chiral 2,2'-dihydroxy-1,1'-binaphthyl

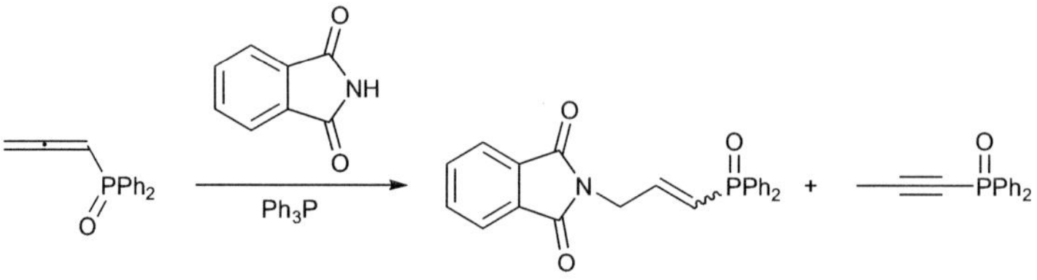

Scheme 59

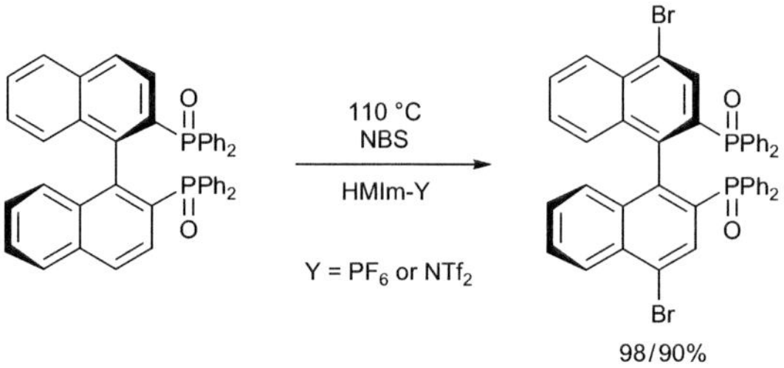

Scheme 60

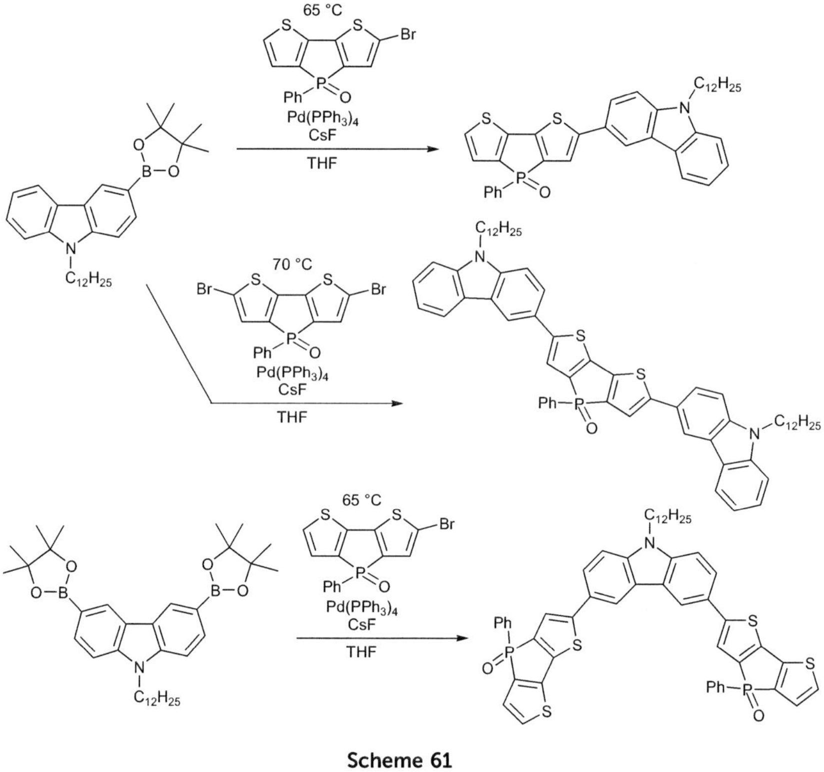

Scheme 61

(BINOL).[66] Carbazole-containing dithienophosphole oxides were prepared in 1:1, 2:1 and 1:2 'ratios' by different variations of the Suzuki-Miyaura protocol (Scheme 61).[67]

Diphenyl(4-methylphenyl)phosphine was oxidized by KMnO$_4$ to diphenyl(4-carboxyphenyl)phosphine oxide. Then, the carboxyl-function was converted to a perfluoroalkoxy ester group in two steps. After deoxygenation, the P-ligand was used in Rh-catalyst (Scheme 62).[68]

Nonisocyanate hybrid coatings were studied using a phosphine oxide-based bis(cyclic carbonate) prepared in three steps starting from bis(4-hydroxyphenyl)phenylphosphine oxide and epichlorohydrin as shown in Scheme 63.[69]

To make available electrophosphorescent polymers, fluorinated poly(arylene ether phosphine oxide)-type macromolecules were synthesized (Scheme 64).[70]

Tetraphenylsilane was modified with triphenylphosphine oxide and carbazole in order to make available blue-phosphorescent components.[71] Phosphine oxide-based poly(arylene ethers) with sulfonate functions were prepared for components of proton exchange membranes (Scheme 65).[72]

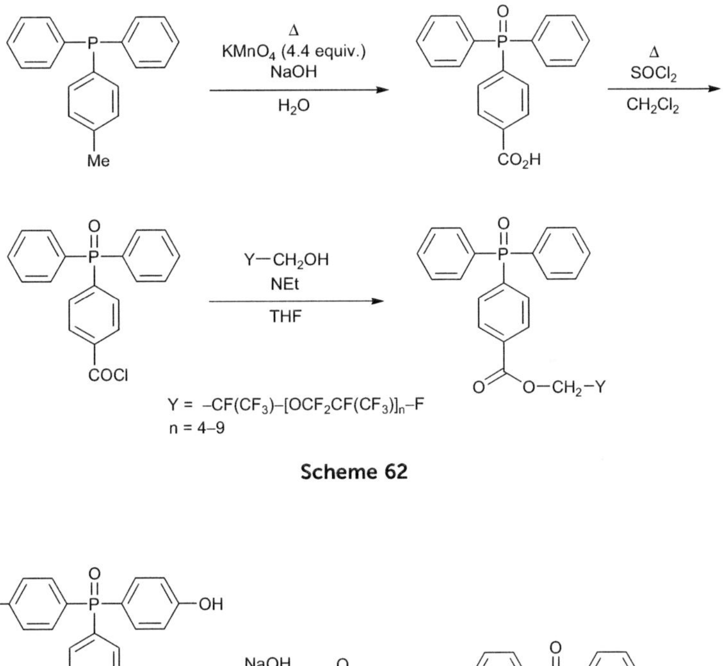

Scheme 62

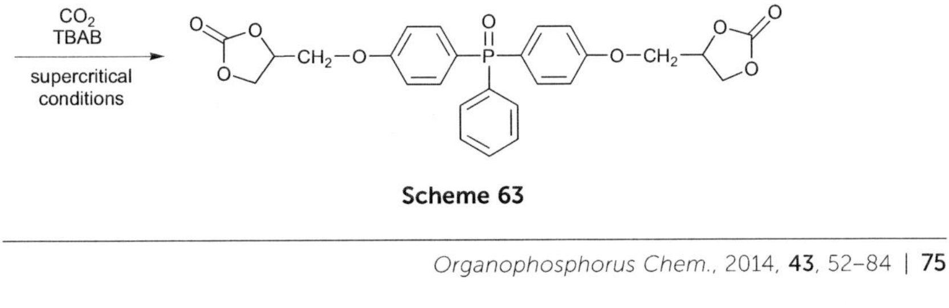

Scheme 63

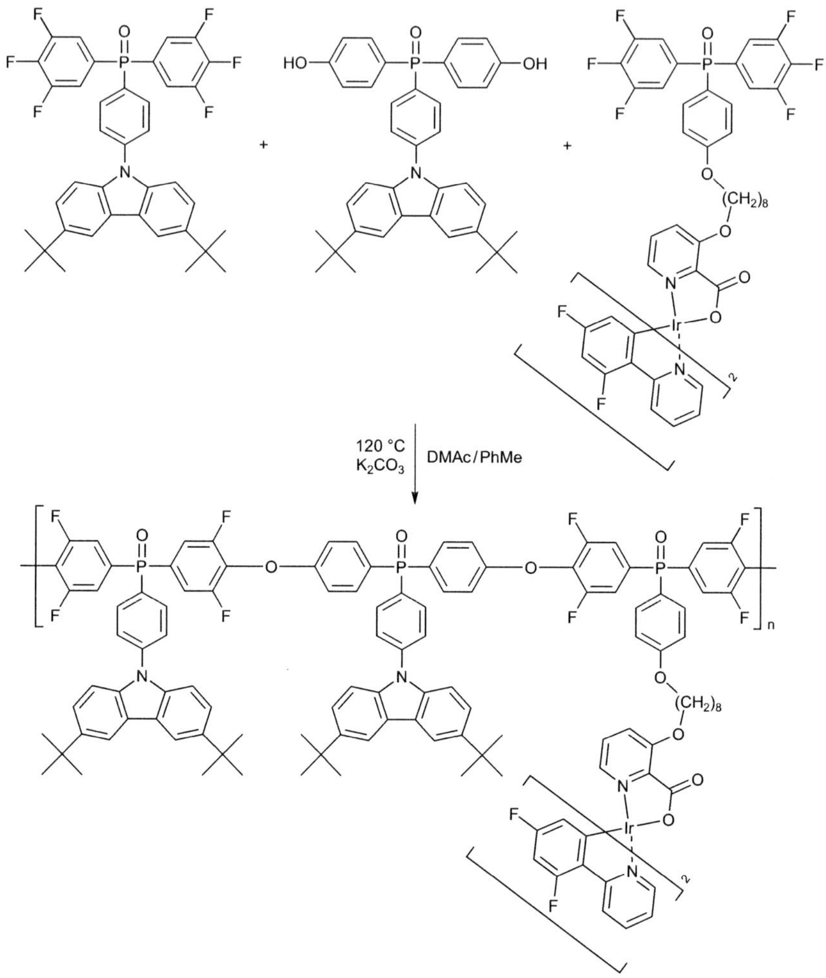

Scheme 64

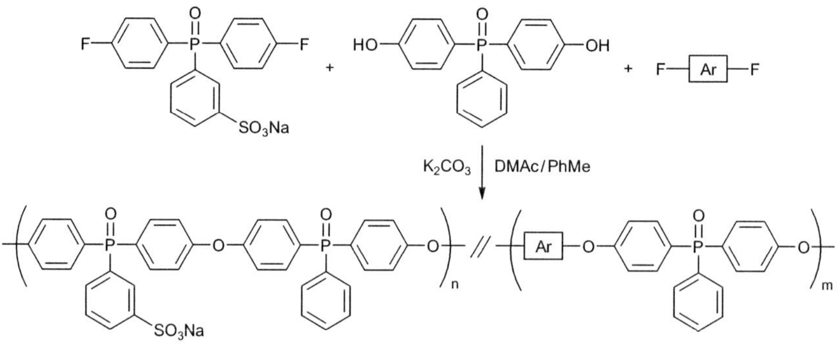

Scheme 65

A phosphine oxide-poly(trimethylene terephthalate) copolymer was also introduced (Scheme 66).[73]

Tris(glycidyloxy)phosphine oxide was used as a component in epoxy systems.[74] A benzylidene dichloride and diphenylphosphinous dichloride were converted via quaternization and treatment of the phosphonium salt so formed with sulfur dioxide to (4-hydroxy-3,5-di-*tert*-butylphenyl)-chloromethanediphenylphosphine oxide. This intermediate was utilized in substitution, elimination and Arbuzov reactions as shown in Scheme 67.[75]

Gilheany and co-workers transformed racemic phosphine oxides to phosphonium salts that on reaction with chiral non-racemic alcohols gave diastereomeric alkoxyphosphonium salts that were utilized in the synthesis of enantio-enriched P-stereogenic phosphine oxides, phosphines and phosphine boranes (Scheme 68).[76]

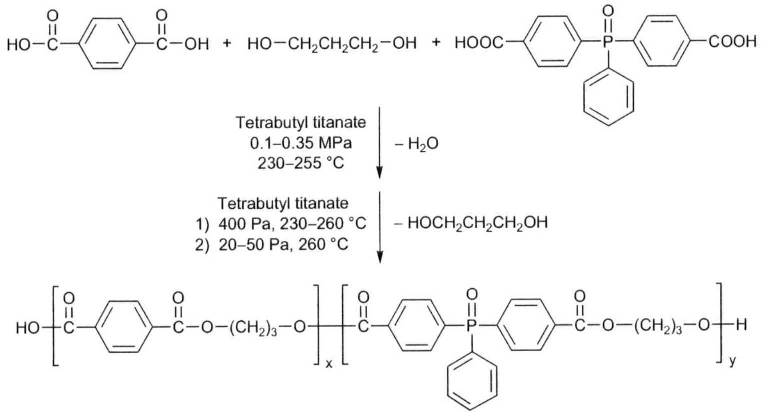

Scheme 66

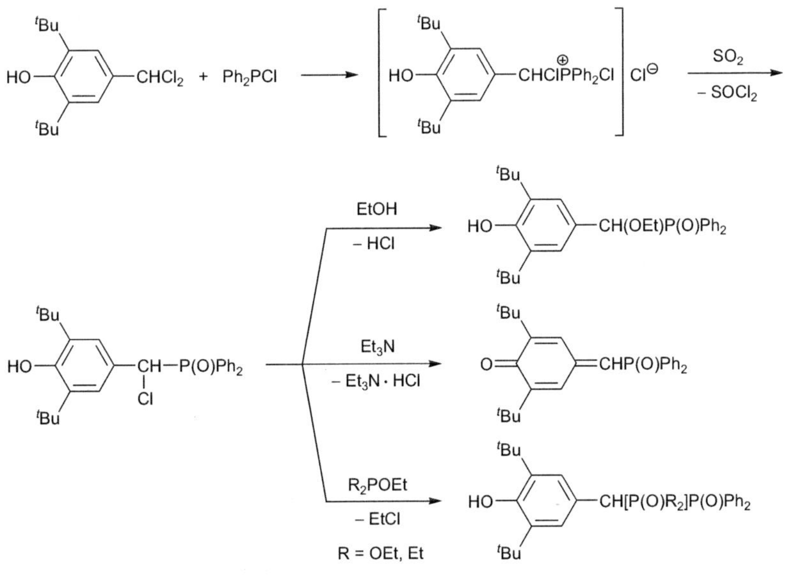

Scheme 67

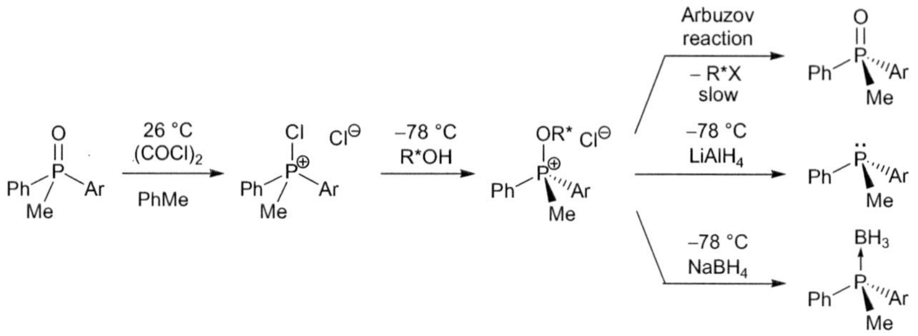

Scheme 68

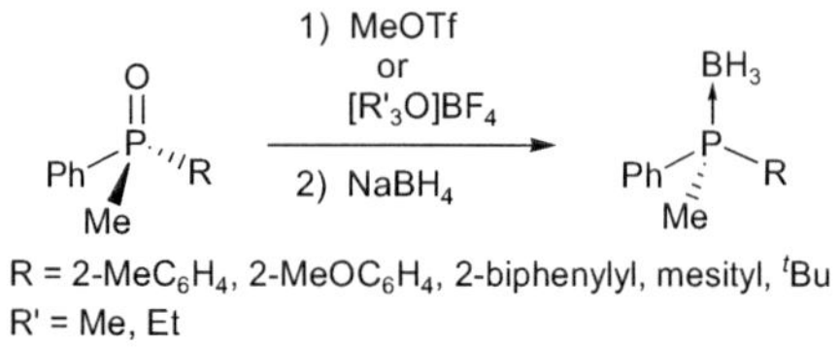

Scheme 69

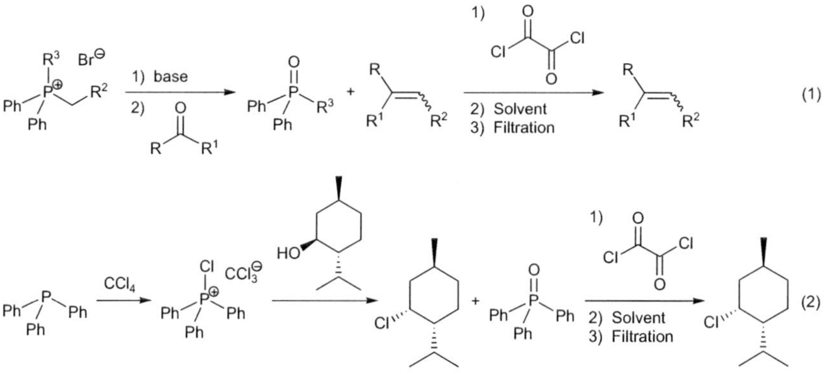

R = 2-MeC$_6$H$_4$, 2-MeOC$_6$H$_4$, 2-biphenylyl, mesityl, ^{t}Bu
R' = Me, Et

Scheme 70

In another variation, optically active phosphine oxides were converted to phosphonium salts with a non-nucleophilic counterion by alkylation of the P=O bond with methyl triflate or trialkoxonium tetrafluoroborate (Meerwein's salt). Hence, the Arbuzov fission could be avoided and the salt was reduced to the corresponding phosphine borane with inversion of configuration (Scheme 69).[77]

The conversion of the triphenylphosphine oxide by-product of the Wittig- and Appel reactions to chlorophosphonium salt (Scheme 70) by reaction with oxalyl chloride made possible an easy work-up of the reaction mixture, as the insoluble salt could be removed by filtration.[78]

The reduction of phosphine oxides to phosphines by (EtO)$_2$MeSiH or PMHS could be catalyzed by Brönsted acids. Scheme 71 shows examples

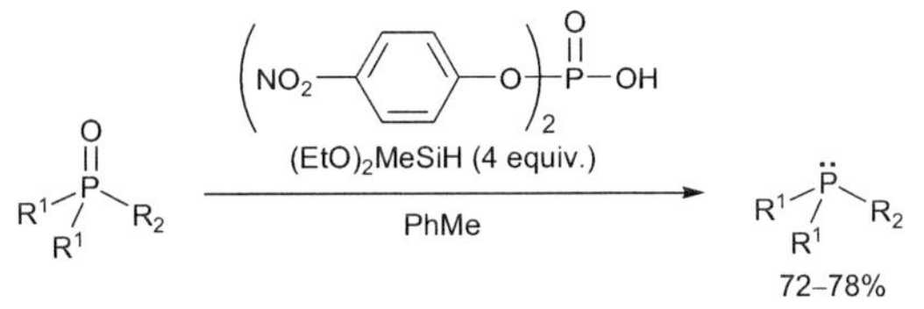

Scheme 71

R¹	4-MeC₆H₄	Ph	4-FC₆H₄	Ph	Ph	ᶜHexyl	
R²	4-MeC₆H₄	4-NH₂C₆H₄	Ph	2-(6-Me-pyridyl)	ᶜHexyl	ᶜHexyl	etc.

Scheme 72

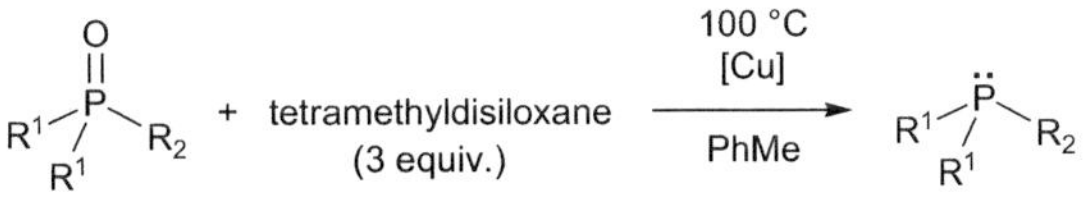

R¹	Ph	4-FC₆H₄	Ph	Octyl	Ph	Ph	Ph	Ph
R²	Ph	Ph	ᶜHexyl	Octyl	Cyclopropyl	Allyl	4-MeC(O)-C₆H₄	4-MeOC(O)-C₆H₄

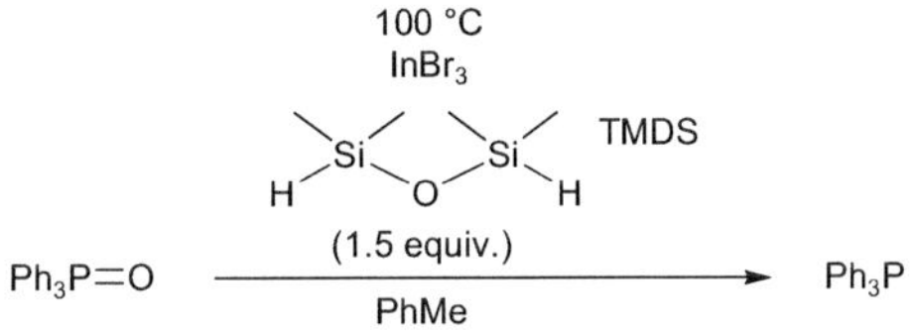

Scheme 73

for the reduction with (EtO)$_2$MeSiH in the presence of a diarylphosphate.[79]

Another observation by Beller et al. was that tetramethyldisiloxane is an efficient deoxygenating agent in the presence of copper catalysts (Scheme 72).[80]

It was also possible to deoxygenate phosphine oxides with the InBr$_3$/TMDS system (Scheme 73).[81] Krenske revisited the deoxygenation and desulfuration of phosphine chalcogenides by perchlorosilanes and chlorosilanes and evaluated the mechanisms and stereoselectivities by theoretical calculations.[82,83]

A mixed P=S/P=O-stabilized geminal dianion was generated and converted to the corresponding Ru-complexes.[84] In another field host-guest interactions between β-cyclodextrin and trioctylphosphine oxide were studied and utilized.[85] The reaction of yttrium anilido hydride [LY(NH(DIPP))(μ-H)]$_2$, where L is a ligand and DIPP is an aryl substituent, with triphenylphosphine oxide led to C–P bond cleavage, while its reaction with R$_2$MeP=O (R = Me, Ph) led to C–H bond cleavage.[86] It was found that the reaction of tertiary phosphines R$_3$P (R = Me, Bu, Oct, c-hexyl and Ph) with 35% aqueous H$_2$O$_2$ afforded the corresponding oxides as the H$_2$O$_2$ adducts (R$_3$P=O · (H$_2$O$_2$)x, where x = 0.5–1).[87] The reaction of N-aroyltrichloroethaneimines with triphenylphosphine oxide led first to labile monocyclic phosphoranes that were stabilized to afford trichloroazadienes (CCl_2=C(R)N=CClAr) and triphenylphosphine oxide.[88] Phosphine oxide intermediates were described in one of the catalytic versions of the Staudinger reaction.[89] Phosphine oxide additives were utilized in the synthesis of 2,2-disubstituted terminal epoxides by the catalytic asymmetric Corey-Chaykovsky epoxidation of ketones.[90] A trichlorosilyltriflate-promoted cross-aldol reaction between ketones in the presence of a chiral phosphine oxide as an organocatalyst was described.[91] Single crystal X-ray analysis proved that the molecules of (2,4,6-trimethylphenyl)boronic acid and triphenylphosphine oxide are held together by an O–H···O hydrogen bond.[92] It was found that methyl(diphenyl)phosphine oxide co-crystallized with 1,4-diodo-2,3,5,6-tetrafluorobenzene and that the molecules were held together by intermolecular P=O···I–C bonds.[93] Secondary phosphine oxides were reviewed as versatile ligands in transition metal complexes together with their possible utilization as catalysts in cross-coupling reactions.[94]

Transition metal complexes with electron-deficient phosphinous acids (($Rf)_2$POH) were prepared and used as catalysts in Heck cross-coupling reactions. The reaction of $(CF_3)_2$POH, $(C_2F_5)_2$POH, and [2,4-$(CF_3)_2C_6H_3]_2$POH with $PdCl_2$ yielded initially the mononuclear complexes [$PdCl_2${$(Rf)_2$POH}$_2$] that were condensed easily to neutral dinuclear complexes [$Pd_2(\mu$-Cl$)_2${[$(Rf)_2$PO]$_2$H}$_2$] with the liberation of HCl.[95] The coordination strength of phosphine oxides toward Pd-catalysts was investigated.[96] New complexes of lanthanide nitrates with tri(tertiary-butyl)-phosphine oxide[97] and those of uranyl and neodymium nitrates with (2-methyl-4-oxopent-2-yl)diphenylphosphine oxide[98] were described. A novel tris(thiocyanato)-tris(triphenylphosphinoxido)europium(III)-nitrato-bis(thiocyanato)-tris(triphenylphosphinoxido)europium(III) co-crystalline complex[99] and a new tris(thiocyanato)-tris(triphenylphosphinoxido)-terbium complex[100] were characterised. Phosphine oxide moieties were incorporated into polymers to establish 'coordination polymers' suitable for interaction with metal ions, such as Mn(II), Co(II), Ni(II), Pd(II), Au(I) and Rh(I).[101–103] Bis(2-diisopropylphosphinophenyl)phenylphosphine oxide was found to be an excellent tridentate ligand to coordinate with Pd(0).[104] Functionalized tertiary phosphine oxides, such as aminoalkylphosphine oxides[105,106] and acetyl-containing phosphine oxides[107] were used as extractants for rare earth elements, lanthanides, alkaline earth metals and actinides. Addition of 1–5 vol% amounts of IL-s increased the efficiency of Am(III) extraction by diphenyl(dibutylcarbamoylmethyl)phosphine oxide.[108]

An extractive spectrophotometric method was elaborated for the estimation of uranium in wastes using, among other extractants, tri-n-octylphosphine oxide.[109]

References

1 C. Trujillo, G. Sanchez-Sanz, I. Alkorta and J. Elguero, *Comp. Theor. Chem.*, 2012, **994**, 81.
2 L. Gonsalvi and M. Peruzzini, *Angew. Chem. Int. Ed.*, 2012, **51**, 7895.

3. B. A. Trofimov, A. V. Artemev, S. F. Malysheva, N. K. Gusarova, N. A. Belogorlova, A. O. Korocheva, Y. V. Gatilov and V. I. Mamatyuk, *Tetrahedron Lett.*, 2012, **53**, 2424.
4. A. S. Bogachenkov, M. M. Efremova and B. I. Ionin, *Tetrahedron Lett.*, 2012, **53**, 2100.
5. A. S. Bogachenkov and B. I. Ionin, *Russ. J. Gen. Chem.*, 2012, **82**, 2009.
6. D. Rosario-Amorin, E. N. Duesler, R. T. Paine, B. P. Hay, L. H. Delmau, S. D. Reilly, A. J. Gaunt and B. L. Scott, *Inorg. Chem.*, 2012, **51**, 6667.
7. R. Premraj, M. D. McLeod, G. W. Simpson and M. G. Banwell, *Heterocycles*, 2012, **85**, 2949.
8. V. V. Ragulin, *Russ. J. Gen. Chem.*, 2012, **82**, 1928.
9. J. A. Vicente, A. Mlonka, H. Q. N. Gunaratne, M. Swadzba-Kwasny and P. Nockemann, *Chem. Commun.*, 2012, **48**, 6115.
10. N. Grzegorzek, L. Latos-Grazynski and L. Szterenberg, *Org. Biomol. Chem.*, 2012, **10**, 8064.
11. S. Kumar, G. Mani, S. Mondal and P. K. Chattaraj, *Inorg. Chem.*, 2012, **51**, 12527.
12. D. R. Boyd, M. Bell, K. S. Dunne, B. Kelly, P. J. Stevenson, J. F. Malone and C. C. R. Allen, *Org. Biomol. Chem.*, 2012, **10**, 1388.
13. F. Wang, M. Qu, F. Chen, Q. Xu and M. Shi, *Chem. Commun.*, 2012, **48**, 8580.
14. H. Slimani and S. Touil, *Heterocycles*, 2012, **85**, 2987.
15. G. V. Bodrin, E. I. Goryunov, I. B. Goryunova, Yu. V. Nelyubina, P. V. Petrovskii, M. S. Grigor'ev, A. M. Safiulina, I. G. Tananaev and E. E. Nifant'ev, *Doklady Chem.*, 2012, **447**, 269.
16. V. P. Morgalyuk, T. V. Strelkova and E. E. Nifant'ev, *Russ. J. Gen. Chem.*, 2012, **82**, 1171.
17. D.-Y. Zhang, C.-B. Yu, M.-C. Wang, K. Gao and Y.-G. Zhou, *Tetrahedron Lett.*, 2012, **53**, 2556.
18. C. Han, Z. Zhang, H. Xu, S. Yue, J. Li, P. Yan, Z. Deng, Y. Zhao, P. Yan and S. Liu, *J. Am. Chem. Soc.*, 2012, **134**, 19179.
19. C. J. Chua, Y. Ren and T. Baumgartner, *Organometallics*, 2012, **31**, 2425.
20. Y. Matano, A. Saito, Y. Suzuki, T. Miyajima, S. Akiyama, S. Otsubo, E. Nakamoto, S. Aramaki and H. Imahori, *Chem. Asian J.*, 2012, **7**, 2305.
21. S. O. Jeon and J. Y. Lee, *J. Mater. Chem.*, 2012, **22**, 4233.
22. Y.-U. Bae and T.-H. Yoon, *J. Appl. Polym. Sci.*, 2012, **123**, 3298.
23. Y. Jin and D.-M. Du, *Tetrahedron*, 2012, **68**, 3633.
24. A. Ghosh, D. Bera, D.-Y. Wang, H. Komber, A. K. Mohanty, S. Banerjee and B. Voit, *Macromol. Mater. Eng.*, 2012, **297**, 145.
25. A. I. Hosein and A. J. M. Caffyn, *Dalton Trans.*, 2012, **41**, 13504.
26. A. N. Yarkevich, L. N. Petrova and S. O. Bachurin, *Russ. J. Gen. Chem.*, 2012, **82**, 1659.
27. J. Xie, H. Li, Q. Xue, Y. Cheng and C. Zhu, *Adv. Synth. Catal.*, 2012, **354**, 1646.
28. Y. Ohmaru, N. Sato, M. Mizutani, S. Kotani, M. Sugiura and M. Nakajima, *Org. Biomol. Chem.*, 2012, **10**, 4562.
29. H.-P. Deng and M. Shi, *Eur. J. Org. Chem.*, 2012, 183.
30. P. E. Sues, A. J. Lough and R. H. Morris, *Inorg. Chem.*, 2012, **51**, 9322.
31. T. D. Nixon, A. J. Gamble, R. J. Thatcher, A. C. Whitwood and J. M. Lynam, *Inorg. Chim. Acta*, 2012, **380**, 252.
32. M. Cieslikiewicz, A. Bouet, S. Juge, M. Toffano, J. Bayardon, C. West, K. Lewinski and I. Gillaizeau, *Eur. J. Org. Chem.*, 2012, 1101.
33. Y.-L. Zhao, G.-J. Wu, Y. Li, L.-X. Gao and F.-S. Han, *Chem. Eur. J.*, 2012, **18**, 9622.
34. C. Shen, G. Yang and W. Zhang, *Org. Biomol. Chem.*, 2012, **10**, 3500.

35 A. J. Bloomfield and S. B. Herzon, *Org. Lett.*, 2012, **14**, 4370.
36 S. M. Rummelt, M. Ranocchiari and J. A. van Bokhoven, *Org. Lett.*, 2012, **14**, 2188.
37 S. Shao, J. Ding, L. Wang, X. Jing and F. Wang, *J. Mater. Chem.*, 2012, **22**, 24848.
38 M.-C. Duclos, Y. Singjunla, C. Petit, A. Favre-Reguillon, E. Jeanneau, F. Popowycz, E. Metay and M. Lemaire, *Tetrahedron Lett.*, 2012, **53**, 5984.
39 S. Wang, J. Li, T. Miao, W. Wu, Q. Li, Y. Zhuang, Z. Zhou and L. Qiu, *Org. Lett.*, 2012, **14**, 1966.
40 Y.-L. Zhao, G.-J. Wu and F.-S. Han, *Chem. Commun.*, 2012, **48**, 5868.
41 N. I. Nikishkin, J. Huskens, J. Assenmacher, A. Wilden, G. Modolo and W. Verboom, *Org. Biomol. Chem.*, 2012, **10**, 5443.
42 H.-Y. Zhang, M. Sun, Y.-N. Ma, Q.-P. Tian and S.-D. Yang, *Org. Biomol. Chem.*, 2012, **10**, 9627.
43 D. Zhao, L. Mao, L. Wang, D. Yang and R. Wang, *Chem. Commun.*, 2012, **48**, 889.
44 M. Moreno, G. Lligadas, J. C. Ronda, M. Galia and V. Caldiz, *J. Polym. Sci. Pol. Chem.*, 2012, **50**, 3206.
45 E. Bálint, J. Takács, L. Drahos and G. Keglevich, *Heteroatom Chem.*, 2012, **23**, 235.
46 E. Jablonkai, L. Drahos, Z. Drzazga, K. M. Pietrusiewicz and G. Keglevich, *Heteroatom Chem.*, 2012, **23**, 539.
47 N. K. Gusarova, V. A. Kuimov, S. F. Malysheva, N. A. Belogorlova, A. I. Albanov and B. A. Trofimov, *Tetrahedron*, 2012, **68**, 9218.
48 B. Boduszek, T. K. Olszewski, W. Goldeman, K. Grzegolec and P. Blazejewska, *Tetrahedron*, 2012, **68**, 1223.
49 L. Wu, X. Zhang, Q.-Q. Chen and A.-K. Zhou, *Org. Biomol. Chem.*, 2012, **10**, 7859.
50 Y. Huang, W. Hao, G. Ding and M.-Z. Cai, *J. Organomet. Chem.*, 2012, **715**, 141.
51 Y. Yamamoto, K. Fukatsu and H. Nishiyama, *Chem. Commun.*, 2012, **48**, 7985.
52 N. Hoshi, T. Kashiwabara and M. Tanaka, *Tetrahedron Lett.*, 2012, **53**, 2078.
53 E. Bálint, E. Fazekas, G. Pintér, Á. Szöllősy, T. Holczbauer, M. Czugler, L. Drahos, T. Körtvélyesi and G. Keglevich, *Curr. Org. Chem.*, 2012, **16**, 547.
54 E. Bálint, E. Fazekas, P. Pongrácz, L. Kollár, L. Drahos, T. Holczbauer, M. Czugler and G. Keglevich, *J. Organomet. Chem.*, 2012, **717**, 75.
55 R. A. Cherkasov, A. R. Garifzyanov, S. A. Koshkin and N. V. Davletshina, *Russ. J. Gen. Chem.*, 2012, **82**, 1453.
56 G. Keglevich and E. Bálint, *Molecules*, 2012, **17**, 12821.
57 R. A. Cherkasov, A. R. Garifzyanov, N. V. Kurnosova, E. V. Matveeva and I. L. Odinets, *Russ. Chem. Bull. Int. Ed.*, 2012, **61**, 174.
58 I. Sänger, F. Schödel, M. Bolte and H.-W. Lerner, *J. Chem. Crystallogr*, 2012, **42**, 472.
59 D. Zhao and R. Wang, *Chem. Soc. Rev.*, 2012, **41**, 2095.
60 R. Kovács, G. T. Balogh, K. Ludányi, L. Drahos and G. Keglevich, *Phosphorus, Sulfur, Silicon*, 2012, **187**, 121.
61 M. Stankevic, K. Wojcik, M. Jaklinska and K. M. Pietrusiewicz, *Eur. J. Org. Chem.*, 2012, 2521.
62 M. Stankevic, M. Jaklinska and K. M. Pietrusiewicz, *J. Org. Chem.*, 2012, **77**, 1991.
63 K. V. Sajna and K. C. Kumara Swamy, *J. Org. Chem.*, 2012, **77**, 5345.
64 J. M. de los Santos, Z. Ochoa and F. Palacios, *ARKIVOC*, 2012, **iv**, 54.

65 W. Dayoub, A. Favre-Reguillon, M. Berthod, E. Jeanneau, G. Mignani and M. Lemaire, *Eur. J. Org. Chem.*, 2012, 3074.
66 B. Hatano, K. Hashimoto, H. Katagiri, T. Kijima, S. Murakami, S. Matsuba and M. Kusakari, *J. Org. Chem.*, 2012, **77**, 3595.
67 M. Stolar and T. Baumgartner, *New J. Chem.*, 2012, **36**, 1153.
68 C. M. Friesen, C. D. Montgomery and S. A. J. U. Temple, *J. Fluorine Chem.*, 2012, **144**, 24.
69 Z. Hosgor, N. Kayaman-Apohan, S. Karatas, A. Gungor and Y. Menceloglu, *Adv. Polym. Tech.*, 2012, **31**, 390.
70 S. Shao, J. Ding, L. Wang, X. Jing and F. Wang, *J. Am. Chem. Soc.*, 2012, **134**, 15189.
71 H. Liu, G. Cheng, D. Hu, F. Shen, Y. Lv, G. Sun, B. Yang, P. Lu and Y. Ma, *Adv. Funct. Mater.*, 2012, **22**, 2830.
72 L. Fu, H. Liao, G. Xiao and D. Yan, *J. Membrane Sci.*, 2012, **389**, 407.
73 H.-B. Chen, Y. Zhang, L. Chen, W. Wang, B. Zhao and Y.-Z. Wang, *Polym. Advan. Technol.*, 2012, **23**, 1276.
74 G. Durga and A. K. Narula, *J. Appl. Polym. Sci.*, 2012, **124**, 3685.
75 M. B. Gazizov, R. K. Ismagilov, L. P. Shamsutdinova, R. F. Karimova, R. Z. Musin, K. A. Nikitina, A. A. Bashkirtsev and O. G. Sinyashin, *Russ. J. Gen. Chem.*, 2012, **82**, 1587.
76 K. V. Rajendran and D. G. Gilheany, *Chem. Commun.*, 2012, **48**, 10040.
77 K. V. Rajendran and D. G. Gilheany, *Chem. Commun.*, 2012, **48**, 817.
78 P. A. Byrne, K. V. Rajendran, J. Muldoon and D. G. Gilheany, *Org. Biomol. Chem.*, 2012, **10**, 3531.
79 Y. Li, L.-Q. Lu, S. Das, S. Pisiewicz, K. Junge and M. Beller, *J. Am. Chem. Soc.*, 2012, **134**, 18325.
80 Y. Li, S. Das, S. Zhou, K. Junge and M. Beller, *J. Am. Chem. Soc.*, 2012, **134**, 9727.
81 L. Pehlivan, E. Metay, D. Delbrayelle, G. Mignani and M. Lemaire, *Tetrahedron*, 2012, **68**, 3151.
82 E. H. Krenske, *J. Org. Chem.*, 2012, **77**, 1.
83 E. H. Krenske, *J. Org. Chem.*, 2012, **77**, 3969.
84 H. Heuclin, X. F. Le Goff and N. Mezailles, *Chem. Eur. J.*, 2012, **18**, 16136.
85 H. Bavireddi and R. Kikkeri, *Analyst*, 2012, **137**, 5123.
86 E. Lu, Y. Chen, J. Zhou and X. Leng, *Organometallics*, 2012, **31**, 4574.
87 C. R. Hilliard, N. Bhuvanesh, J. A. Gladysz and J. Bluemel, *Dalton Trans.*, 2012, **41**, 1742.
88 Ya. Ya. Khomutnik, P. P. Onys'ko, Yu. V. Rassukanaya, A. G. Vlasenko, A. N. Chernega, V. S. Brovarets, S. G. Pil'o and A. D. Sinitsa, *Russ. J. Gen. Chem.*, 2012, **82**, 1058.
89 H. A. van Kalkeren, J. J. Bruins, F. P. J. T. Rutjes and F. L. van Delft, *Adv. Synth. Catal.*, 2012, **354**, 1417.
90 T. Sone, A. Yamaguchi, S. Matsunaga and M. Shibasaki, *Molecules*, 2012, **17**, 1617.
91 S. Aoki, S. Kotani, M. Sugiura and M. Nakajima, *Chem. Commun.*, 2012, **48**, 5524.
92 S. Rosca, M. Olaru and C. I. Rat, *Acta Crystallogr. E*, 2012, **68**, o31.
93 S. Y. Oh, C. W. Nickels, F. Garcia, W. Jones and T. Friscic, *CrystEngComm*, 2012, **14**, 6110.
94 T. M. Shaikh, C.-M. Weng and F.-E. Hong, *Coordin. Chem. Rev.*, 2012, **256**, 771.
95 B. Kurscheid, L. Belkoura and B. Hoge, *Organometallics*, 2012, **31**(4), 1329.
96 B. Neuwald, F. Oelscher, I. Goettker-Schnetmann and S. Mecking, *Organometallics*, 2012, **31**, 3128.

97 A. Bowden, S. J. Coles, M. B. Pitak and A. W. G. Platt, *Inorg. Chem.*, 2012, **51**, 4379.

98 A. G. Matveeva, M. S. Grigoriev, T. K. Dvoryanchikova, S. V. Matveev, A. M. Safiulina, O. A. Sinegribova, M. P. Passechnik, I. A. Godovikov, D. A. Tatarinov, V. F. Mironov and I. G. Tananaev, *Russ. Chem. Bull. Int. Ed.*, 2012, **61**, 399.

99 A. T. Thames, F. D. White, L. N. Pham, K. R. Xiang and R. E. Sykora, *Acta Crystallogr. E*, 2012, **68**, m1530.

100 L. N. Pham, A. T. Thames, F. D. White, K. R. Xiang and R. E. Sykora, *Acta Crystallogr. E*, 2012, **68**, m1531.

101 P. Schmiedel and H. Z. Krautscheid, *Anorg. Allg. Chem.*, 2012, **638**, 1839.

102 Q. Chen, F. Lian, F. Jiang, L. Chen and M. Hong, *Inorg. Chim. Acta*, 2012, **392**, 396.

103 F. Zhang, X. Yang, F. Zhu, J. Huang, W. He, W. Wang and H. Li, *Chem. Sci.*, 2012, **3**, 476.

104 E. J. Derrah, C. Martin, S. Ladeira, K. Miqueu, G. Bouhadir and D. Bourissou, *Dalton Trans.*, 2012, **41**, 14274.

105 R. A. Cherkasov, A. R. Garifzyanov, E. B. Bazanova, R. R. Davletshin and S. V. Leont'eva, *Russ. J. Gen. Chem.*, 2012, **82**, 33.

106 E. V. Matveeva, E. V. Sharova, A. N. Turanov, V. K. Karandashev and I. L. Odinets, *Cent. Eur. J. Chem.*, 2012, **10**, 1933.

107 A. M. Safiulina, A. G. Matveeva, T. K. Dvoryanchikova, O. A. Sinegribova, A. M. Tu, D. A. Tatarinov, A. A. Kostin, V. F. Mironov and I. G. Tananaev, *Russ. Chem. Bull.*, 2012, **61**, 392.

108 G. A. Pribylova, I. V. Smirnov and A. P. Novikov, *Radiochemistry*, 2012, **54**, 473.

109 S. Biswas, P. N. Pathak and S. B. Roy, *Spectrochim. Acta A*, 2012, **91**, 222.

Phosphonium salts and P-ylides

Maurizio Selva,* Alvise Perosa and Marco Noè

DOI: 10.1039/9781782623977-00085

1 Introduction

The present review was aimed at describing the state-of-the-art, for the period January 2011–December 2012, of two pillar classes of phosphorus-containing compounds, the phosphonium salts and ylides. For the Reader's convenience, topics are organized to offer an introductory survey on the methods of preparation and characterisation of both types of compounds, followed by an analysis of the most remarkable but also curiosity driven researches for their applications. A special section is devoted to phosphonium-based ionic liquids (PIls) due to the exceptional importance of this topic.

2 Phosphonium salts

2.1 Synthesis and characterisation

Quaternisation of the corresponding phosphine by reaction with an electrophile or a Brønsted acid is the most typical and simple procedure for the preparation of phosphonium salts. Besides this classic approach, new interesting synthetic pathways have been described in the period surveyed by this review.

A remarkable intramolecular cascade cyclisation was reported by Fukazawa and co-workers to yield a phosphonium-borate zwitterion[1] (Scheme 1). The so-formed polycyclic ladder π-conjugated structure served as an attractive fundamental skeleton for novel optoelectronic materials.

A phosphoniodefluorination reaction of 2,5,6,7,8-pentafluoro-1,4-naphthoquinone was described for the synthesis of 5,6,7,8-tetrafluoro-1,4-naphthoquinone phosphonium betaines which were recognized as potential inhibitors of tumoral cell growth and also as cell-protecting antioxidants (Scheme 2).[2]

Unconventional transformations were also discovered. The reaction of a phosphonium ylide such as tributyl[(trimethylsilyl)methylene]phosphorane with enolisable aldehydes did not afford the expected Wittig reaction product. Conversely, alkenylphosphonium salts were obtained in good yields (65–89%, Scheme 3).[3]

A method for single C–P bond formation at a tertiary ring-bridging position was claimed by Brown and co-workers[4] that, under mild conditions and without organolithium or Grignard reactants, was able to accomplish the reaction of adamantyl and diamantyl triflates with secondary phosphines in highly non-polar media (Scheme 4).

Dipartimento di Scienze Molecolari e Nanosistemi dell'Universita Ca' Foscari Venezia, Centre for Sustainable Technologies, calle Larga S. Marta, 2137-30123 Venezia, Italy. E-mail: selva@unive.it

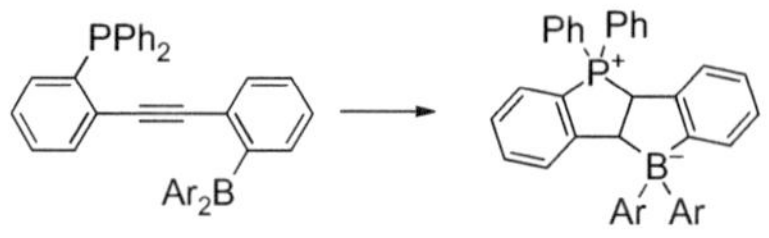

Scheme 1

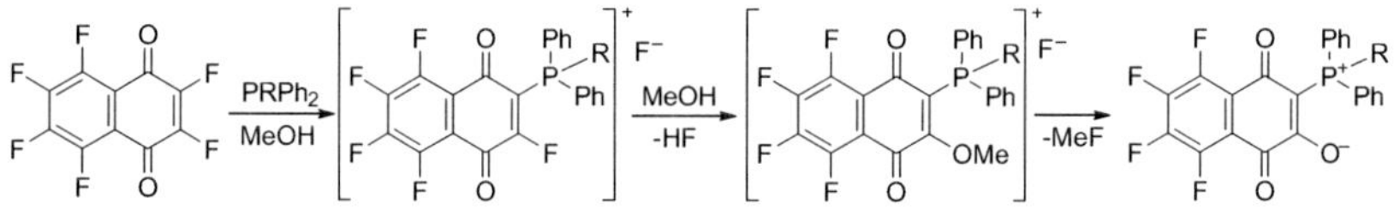

Scheme 2

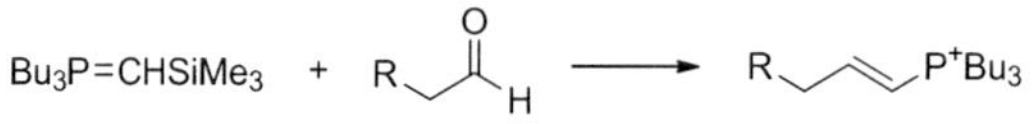

Scheme 3

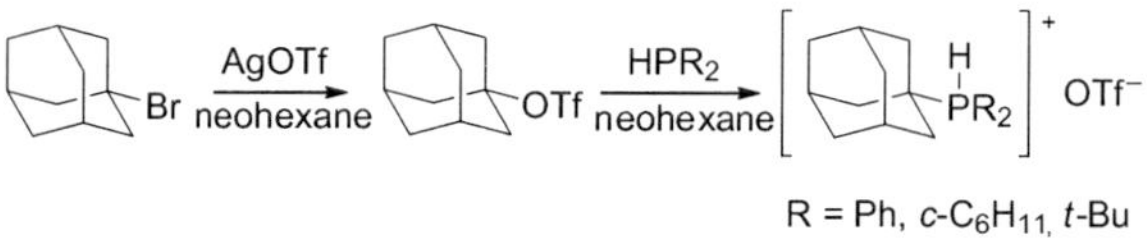

Scheme 4

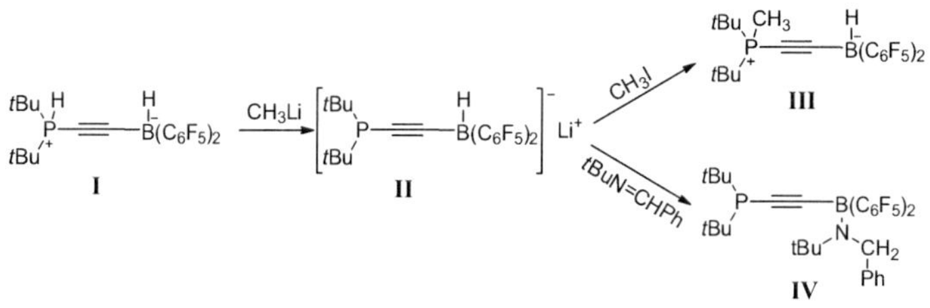

Scheme 5

In the field of Frustrated Lewis Pairs (FLP), new synthetic routes for the preparation of alkynyl-like phosphonium borates were reported[5] by Stephan and co-workers. An acetylene substituted phosphine was reacted with a chlorobis(pentafluorophenyl)borane yielding a phosphonium-borate zwitterion that, in turn, was converted to borohydride by the action of an excess of ClSiHMe$_2$, (Scheme 5, **I**). The resulting product was deprotonated by CH$_3$Li to obtain a phosphine-borate lithium salt (Scheme 5, **II**), the methylation of which afforded tBu$_2$-P(Me)C≡CB(H)(C$_6$F$_5$)$_2$ (Scheme 5, **III**). The phosphine-borate lithium salt also reacted with imines (*e.g.* tBuN=CHPh) to give adducts bearing N–B bonds (Scheme 5, **IV**).

The reaction of neutral phosphine-borane derivatives with 1-hexene, tetrahydrofuran and phenylacetylene produced an interesting set of innovative macrocyclic compounds (Scheme 6).

Preparations of a variety of new phosphonium salts are briefly reported below. The structures of the products are summarized in Fig. 1.

In order to synthesise new ligands for the preparation of Pt(II), Pd(II), and Ni(II) complexes, Sladic and co-workers attempted the reaction of 2-(diphenylphosphino)benzaldehyde and malonic acid dihydrazide in the presence of perchloric acid.[6] The desired transformation was not successful, but surprisingly, a bisimino bridged diphosphonium compound was obtained displaying substantial biological activity in the brine shrimp test and which cleaves plasmid DNA. The formation of the phosphonium moiety was also used for the preparation of new materials. A remarkable case was reported by Zhang et al. that used a nickel(0)-catalyzed Yamamoto-type cross-coupling reaction to prepare a microporous polymer.[7] The polymer networks proved stable toward water, base, and acid, and their apparent BET specific surface areas could be tuned (from 650 to 980 m^2 g^{-1}) by changing the counteranions (Br$^-$ to F$^-$). They also displayed high intrinsic catalytic activity for the reaction between epoxide and CO_2 and were also good supports for Pd nanoparticles for cross-coupling reactions. In other examples, phosphonium groups were used to modify the properties of already known molecules.

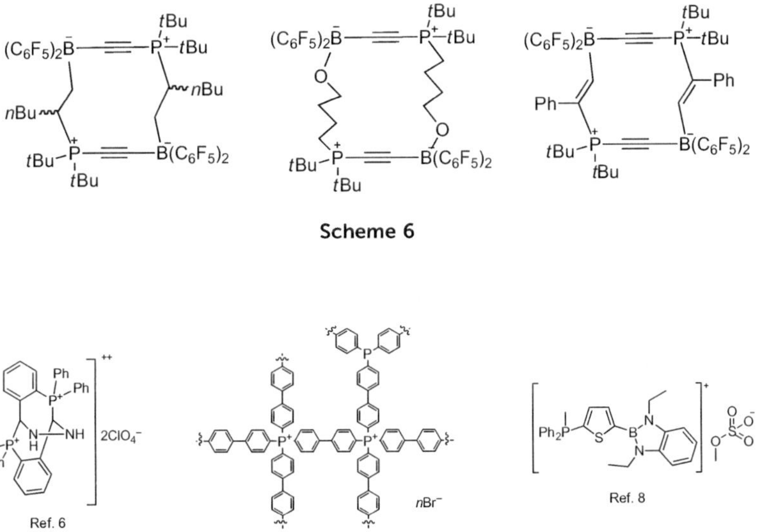

Scheme 6

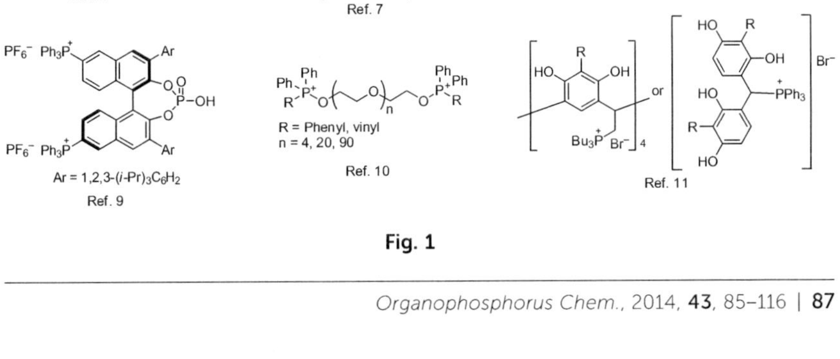

Fig. 1

A series of 2-(2′)thienyl-1,3,2-benzodiazaboroles and a phosphorus based end group [X = PPh$_2$, P(O)Ph$_2$, P(S)Ph$_2$, P(Se)Ph$_2$, P(AuCl)Ph$_2$ and P(Me)Ph$_2$] in the 2′ position were synthesised to study the influence of the phosphonium substituents on the fluorescence.[8] The attachment of phosphonium ion phase tags to 6,6′ position of chiral binapthyl-based phosphoric acids allowed to prepare excellent organocatalysts for the asymmetric Friedel–Crafts reactions.[9] Phosphonium-based bolaamphiphiles for the stabilisation of silicic acid aqueous solution were synthesised by the reaction of the appropriate glycol and a tertiary phosphine in the presence of hydrochloric acid.[10] New calix[4]resorcinols containing phosphonium groups at the lower rim were synthesised by the reaction of 2-tributylphosphonioacetaldehyde acetal with resorcinol and 2-methylresorcinol.[11] The use of a phosphonium counterion was also found to be useful in the synthesis of benzyl(triphenyl)phosphonium dichloroiodate (BnPh$_3$P$^+$(ICl$_2$)$^-$),[12] a versatile reagent for the efficient and selective iodination of organic substrates. The use of the phosphonium cation yielded crystals easily. Phosphonium derivatives have also found applications in supramolecular chemistry. For example, organoamino phosphonium cations [P(NHR)$_4$]$^+$ with four hydrogen bonding sites arranged in tetrahedral fashion around the central phosphorus atom, allowed the formation of supramolecular aggregates in the presence of chloride and several carboxylate anions.[13] X-ray structure analyses of these compounds provided evidence for hierarchical structures ranging from 1D chains to 2D sheets and 3D networks. Phosphonium salts prepared by the reaction of [P(NHR)$_4$]Cl [(R = cyclohexyl, iso-butyl, isopropyl)] with Na$_2$MoO$_4 \cdot$ 2H$_2$O/Na$_2$WO$_4 \cdot$ 2H$_2$O were used to obtain supramolecular aggregates of several iso- and hetero-polymetallate anions with topology of 2D-layers, grids, 3D-networks and -helical assemblies.[14] A chiral P-spiro-iminophosphorane and 3,5-dichlorophenol were also reported to assemble into three types of discrete molecular associations that depended on the stoichiometry of the substituted phenol (Scheme 7).[15] The resulting aggregates were successful catalysts for asymmetric conjugate addition of azlactones to cinnamoyl acylbenzotriazole. The enantiomeric excess increased with the number of phenol molecules bonded to the iminophosphorane centre.

Phosphonium compounds also act as guests in inclusion complexes. The interactions of Co(III) sarcophagine-type cage molecules, [Co(diCLsar)]$^{3+}$ or [Co(HONOsar)]$^{3+}$, with mono-phosphonium cations and sodium p-sulfonatocalix[4] arene were studied by Ling and co-workers.[16] They observed the formation of either 1 : 1 or 1 : 2 host-guest inclusion

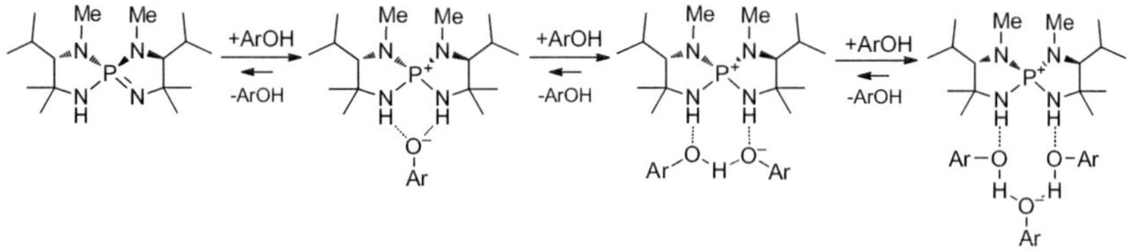

Scheme 7

complexes in the solid state yielding complex I [p-sulfonatocalix[4]-arene · Co(diCLsar) · 2[benzyltriphenyl phosphonium]], complex II [2[p-sulfonatocalix[4]arene] · Co(diCLsar) · 3[tetraphenyl phosphonium]] and complex III [p-sulfonatocalix[4]arene · Co(HONOsar) · tetraphenylphosphonium]. The behaviour of phosphonium surfactants in water was systematically investigated by Gainanova et al.[17] Critical micelle concentrations of alkyltriphenylphosphonium bromides (Ph_3PR^+ Br^-; $R = C_nH_{2n+1}$ where n = 8, 10, 12, 14, 16, 18) were found to decrease with the number of carbon atoms. Also, supramolecular polymeric assemblies were obtained by the complexation of various phosphonium monocations and dications with poly(acrylic acid) (PAA) as a multianion.[18] Remarkable studies were performed on the structural characterisation of phosphonium salts both in their pure form and in solution. Of note was the analysis by Yamamoto et al. of the nature of several phosphonium salts in the solid state and in solution and their equilibrium with the corresponding phosphorane form (Scheme 8).[19] Single crystal X-ray crystallography indicated that the geometry of fluoride compounds was almost an ideal trigonal bipyramid (TBP), while bromide and tetrafluoroborate compounds were essentially phosphonium salts with phosphorane character. This was confirmed by NMR investigations in solution, where poorly coordinating anions such as BF_4^- and OTf^- were merely counteranions of the phosphonium cation, while chloride and bromide anions allowed an equilibrium between the phosphorane and the phosphonium structures.

Solid-state NMR (SSNMR) techniques, including challenging [79/81]Br experiments, were employed for the characterisation of some triphenylphosphonium bromides.[20] Results were compared to single-crystal XRD information. NMR parameters allowed the detection of small structural changes accounting for the presence or absence of site symmetry elements in each compound. A noteworthy study on the structure of a series of wedge-shaped phosphonium salts having 3,4,5-tris(alkyloxy)benzyl moieties, was reported by Ichikawa et al.[21] The analysis showed the potential of nanostructured liquid crystals composed of ionic molecules, for the development of transportation nanochannel materials. Applications as media for lithium ion batteries were proposed. The reactivity of phosphonium salts in specific environments was also examined to prove various theories and assumptions. For example, photoinduced bond cleavage of a diphenylmethyltriphenylphosphonium salt in acetonitrile reverse micellar nanopools was studied by Pugliesi et al.[22] in order to evaluate the effects of encapsulation on the succeeding dynamics. Similarly, in work by Mayr,[23] the formation of benzhydryl cations Ar_2CH^+ and/or benzhydryl radicals Ar_2CH^- by UV irradiation of benzhydryl

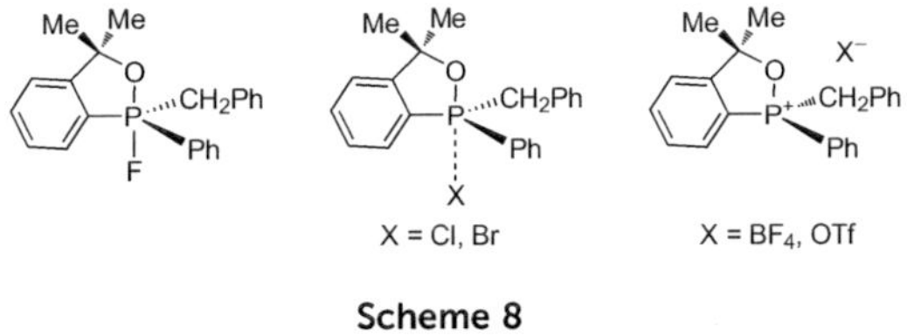

Scheme 8

triarylphosphonium salts $[Ar_2CH\text{-}PAr_3]^+X^-$ proved that the photochemistry of phosphonium salts is controlled by the degree of ion pairing, which in turns depends on the solvent and the concentration of the phosphonium salts.

2.2 Applications in synthesis

Phosphonium salts are successfully employed for a number of synthetical applications in organic chemistry. Of a particular note is their extensive and consolidated use as phase transfer catalysts (PTC). However, the development of new compounds and applications is always under investigation. For example, an innovative quaternary phosphonium salt (QPS)-type triphase catalyst (TPC) prepared by functionalisation of cross-linked polystyrene microspheres (CPS), was recently proposed as an efficient catalyst for the model reaction of phthalimide with 1-bromobutane in a biphase aqueous/organic phase.[24]

Moreover, an emerging area in the PTC sector deals with chiral phase-transfer catalysis mediated by phosphonium derivatives. This topic, which was mostly limited to quaternary ammonium salts, has been recently reviewed by Enders and Nguyen that described several examples of phosphonium salts as chiral phase transfer catalysts.[25]

In the past two years, many other applications of phosphonium salts as catalysts or intermediates of chemical reactions have also been described. Some of the most relevant ones are summarized below.

Tetrabutylphosphonium and triphenylchlorophosphonium chlorides were reported to act as catalysts for the addition of diethylzinc to aldehydes (aromatic, heteroaromatic or aliphatic) and the dehydration of aromatic and aliphatic aldoximes to nitriles, respectively.[26,27] Catalysis was so efficient that both types of reactions occurred at room temperature, even with phosphonium salts in polymer-bonded form.

Rodriguez-Zubiri and coworkers reported an extensive analysis of the effects of phosphonium salts on the catalytic activity of rhodium-triphenylphosphine complexes for the hydroamination of olefins.[28] A new ring expansion reaction based on the interaction between 1,2,3-triarylcyclopropenylphosphonium bromide[29] and sodium polyphosphides was developed by Bezkisko and co-workers. A number of sodium 3,4,5-triaryl-1,2-diphosphacyclopentadienides were obtained in high yields. A commercially available cyclopropyl phosphonium tetrafluoroborate (CPTB) was proposed to develop a new synthesis of 2-methylene-4-substituted ethyl butyrates (Scheme 9).[30] A variety of nucleophiles were able to react with CPTB to produce the corresponding ylides that in turn, coupled with several aldehydes to provide *E*-olefinic products.

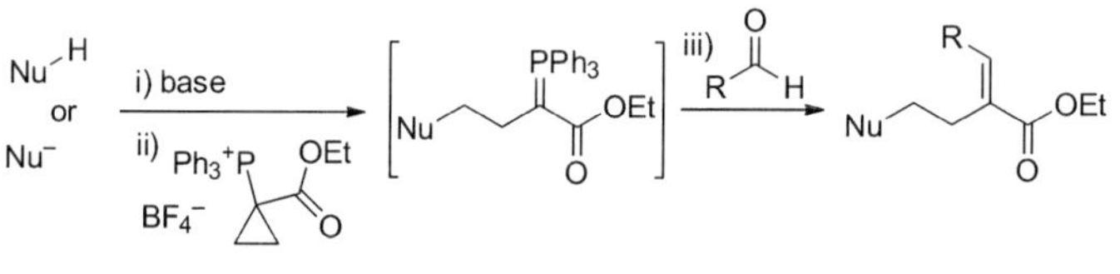

Scheme 9

The commercial BOP reagent ((Benzotriazolyloxy)tris(dimethylamino)-phosphonium hexafluorophosphate) was used as a coupling reagent for the synthesis of unnatural aminoacids (Scheme 10). Heras et al. described the incorporation of a range of nucleophiles including amino ester residues at position 4 of an electron-rich 2-amino-4(3H)-pyrimidinone.[31]

Another innovative application of the BOP reagent was reported by Wan and co-workers, in the cyclization of thioureas for a convenient synthesis of 2-aminobenzimidazoles.[32] Similarly, bromotripyrrolidino-phosphonium hexafluorophosphate (PyBroP) was used in an unprecedented Pd/Cu-catalysed direct heteroarylation of tautomerisable heterocycles.[33] This new coupling reaction afforded unsymmetrical as well as symmetrical biheterocycles. An intriguing example of reverse reactivity was reported by the use of a phosphonium intermediate in the reaction of a β ketoaldehyde (Scheme 11).[34] The overall strategy was based on the activation of an aldehyde functionality by a Lewis acid and its further reaction with triphenylphosphine to yield the corresponding phosphonium salt. The residual ketone group was therefore able to undergo subsequent nucleophilic additions.

Innovative methodologies for the synthesis and applications of phosphonium salts as organocatalysts have also been developed by our group. A green protocol for the methylation of trioctylphosphine with the non-toxic dimethylcarbonate was first implemented to obtain methyl-trioctylphosphonium methylcarbonate $[P_{8881}][MeOCO_2]$. This salt was then used not only as an excellent catalyst for the Henry condensation of nitroalkanes with aldehydes and ketones,[35] but also as a convenient source to synthesize a library of phosphonium derivatives in which the methyl trioctylphosphonium cation was coupled to weakly basic anions such as bicarbonate, acetate, and phenolate. These compounds proved to be efficient organocatalysts for the transesterification of dimethyl- and diethyl-carbonate with primary and secondary alcohols.[36] Moreover, the

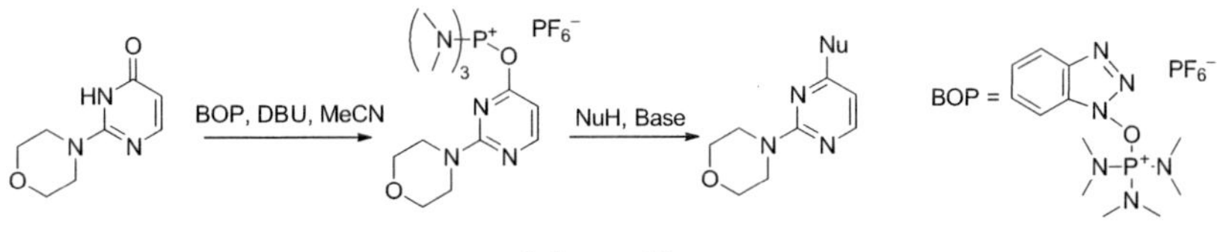

Scheme 10

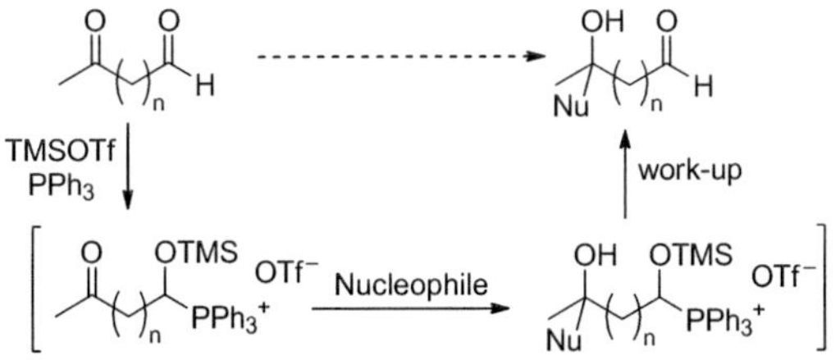

Scheme 11

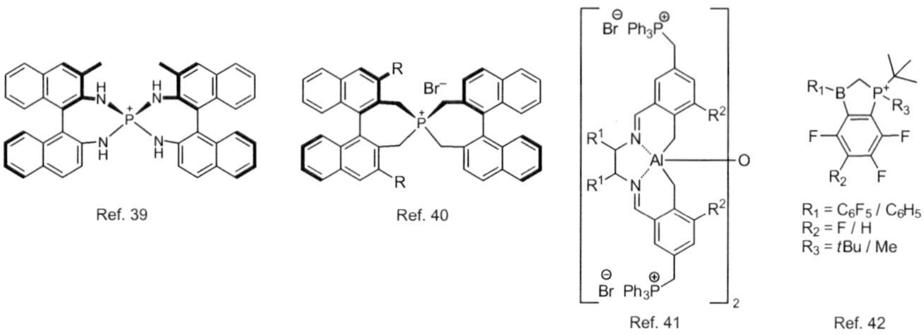

Ref. 39 Ref. 40 Ref. 41 Ref. 42

$R_1 = C_6F_5 / C_6H_5$
$R_2 = F / H$
$R_3 = tBu / Me$

Fig. 2

same onium salts also catalyzed the Baylis-Hillman condensation of cyclohexenone. A detailed kinetic analysis of such a reaction allowed us to demonstrate the occurrence of a cooperative effect between the anionic and the cationic partners of the catalysts which acted simultaneously as nucleophilic and electrophilic activators, respectively.[37] Meier et al.[38] also reported that in the presence of microwave irradiation, an easily accessible allylic phosphonium salt in aqueous $NaHCO_3$, promoted the synthesis of stereochemically defined α-methylalkenoic acids and esters, without the need of hydrides or other reducing agents. Ooi and coworkers[39] described the use of a P-spiro-heterochiral arylaminophosphonium barfate (tetrakis(3,5-bis(trifluoromethyl)phenyl)borate) (Fig. 2) as catalyst for enantioselective aza-Michael additions to conjugated nitroenynes. Similar P-spiro-quaternary phosphonium salts (Fig. 2) were prepared, and applied for the first highly enantioselective amination of benzofuranones.[40] In a study by North,[41] phosphonium groups were attached to the salen ligands of a bimetallic aluminium(salen) complex (Fig. 2) to form one-component catalysts for the cycloaddition of CO_2 to epoxides to produce cyclic carbonates. The catalytic activity proved to be heavily influenced by the solubility of the catalyst in the epoxide reagent. Phosphonium groups were also used to increase the Lewis acidity of organo boranes.[42] Examples are shown in Fig. 2; the reported rigid cyclic Lewis acids possess Gutmann acceptor numbers of AN = 87.3, 85.7, and 85.7, respectively, which are among the highest for organoboranes. This compounds were successfully employed as catalysts for the [4 + 2] cycloaddition reaction between 2,5-dimethyl-1,4-benzoquinone and cyclopentadiene.

2.3 Medical and biological applications

In medicinal chemistry, phosphonium salts were used mainly to obtain new agents for PET imaging based on labelled ^{18}F[43–52] or ^{64}Cu[53] compounds.

Quaternary phosphonium compounds were also found to be reversible inhibitors of cholinesterases of different animals and showed species-specificity of action depending on their inhibitor structure.[54] The use of phosphonium salt-containing materials as antimicrobial agents was also reported. For example, new 2,6-di-tert-butylphenol derivatives possessing simultaneously phosphonium and phosphonate fragments were reported to have a broad spectrum of antibacterial and antimycotic activity.[55,56] Also, a tetradecyltriphenylphosphonium bromide-functionalized few-layered graphite showed excellent thermal stability and long-acting antibacterial activity against E. coli and S. aureus.[57] A water-soluble brilliant blue/reduced graphene oxide/tetradecyltriphenylphosphonium bromide composite, with remarkable antibacterial properties, was also prepared by using non-covalent brilliant blue-functionalized reduced graphene oxide as the tetradecyltriphenylphosphonium bromide carrier.[58] In the field of protein-based hydrogels, tetrakis(hydroxymethyl)phosphonium chloride (THPC) was used as a covalent cross-linking agent. The identification of new, cytocompatible cross-linkers allowed for greater flexibility of hydrogel design. THPC was proposed as an inexpensive, aqueous cross-linker for 3D cell encapsulation in protein-based hydrogels. showing cytocompatibility with retention of cell growth and phenotype in an ELP hydrogel system.[59] Phosphonium cations are known to accumulate in the mitochondria of cells due to their liphophilic cationic nature.[60–64] This behaviour was employed in several studies as a tool for intracellular delivery of DNA or RNA[65–71] and in DNA differentiation applications.[72] Water-soluble phosphonium chitosan derivatives were also prepared to obtain a material for gene delivery or for antibacterial applications.[73]

2.4 Application in clay-modification technologies

Montmorillonites are used extensively as fillers in polymer nanocomposites. To attain compatibility of the hydrophilic montmorillonite with hydrophobic polymers, the cations on the clay surface were exchanged with long chain alkylammonium ions, thus providing intermixing between the organic and inorganic phases. In the processing of high-melting-point polymers, to circumvent the deleterious early decomposition effects of thermally unstable alkyl ammonium-treated montmorillonites, highly thermally-stable phosphonium salts are often used as intercalant agents. In this way, especially for sodium montmorillonite, ion-exchange reactions were reported to produce organo-montmorillonites with enhanced thermal stability.[74] For example, montmorillonites modified with phosphonium cations like tetra-octylphosphonium and octadecyl(tributyl)phosphonium, hexadecyl (triphenyl)phosphonium and tetraphenylphosphonium were, by far, more thermally stable than the alkylammonium montmorillonites.[75] The decomposition mechanism was studied in detail for montmorillonite intercalated with alkyl (methyl, ethyl, propyl, and dodecyl)triphenyl phosphonium cations.[76] Thermogravimetric analysis was applied to determine the thermal properties of polypropylene composites, and scanning electron microscopy was used to study the microstructures of chars.[77] The contemporary use of two different phosphonium salts as intercalant was also investigated.[78] Phosphonium salts were also used to prepare transparent clay films using tetraphenylphosphonium (TPP) modified smectites (TPP-SA, TPP-HE, and TPP-SA-HE).[79] Optical studies confirmed that the transparent clay films maintained their transparency in the visible range even after annealing at 350 °C for one hour.

Novel organoclays were synthesized which involve several kinds of phosphonium cations to improve the dispersibility in matrix resins of composites and accelerate the curing of matrix resin. The application of such epoxy/clay nanocomposites and their thermal, mechanical, and adhesive properties were investigated.[80] Interesting studies on the morphology and properties of polycarbonate (PC)/clay nanocomposites[81] and poly(ethylene terephthalate) (PET)/phosphonium vermiculite (P-VMT) composites[82] were also reported to prepare new thermally resistant materials.

2.5 Miscellaneous applications

2.5.1 Membranes.
Phosphonium-functionalised membranes have found a plethora of different applications in separation technologies, water treatment,[83] synthetic chemistry and more. Water swelling limits the development of a polymeric material for a membrane. To overcome this problem, Yan and co-workers[84] developed a self-crosslinking polymer functionalised with quaternary phosphonium-based hydroxide-exchange membranes. In a search for a thermally stable and high performance material, Stokes *et al.*[85] synthesised a trimethylphosphonium chloride functionalized polystyrene *via* reversible addition–fragmentation chain transfer (RAFT) polymerization. The polymer was characterised by a very high ionic content (up to 98 mol%) and excellent thermal stability. Also, the chemical inertness of a tetrakis(dialkylamino)phosphonium cation was evaluated as a functional group for alkaline anion exchange membranes.[86]

2.5.2 Leather tanning.
Another worthy application of phosphonium salts is in leather tanning. This also helps to reduce the environmental impact of this process. The use of tetrakis(hydroxymethyl)phosphonium sulfate (THPS) and chromium as a chrome-saver approach was reported by Fathima *et al.*[87] Environmental impact assessment showed that there was a significant reduction in the effluent (TS and sulfates) load when compared to conventional chrome tanning. Another study by Luo and co-workers,[88] investigated the use of phosphonium compounds in combination tannages with vegetable tannins and aluminium, in order to avoid the use of chromium.

2.5.3 New materials.
Phosphonium-type zwitterions synthesised through the quaternisation reaction of tri-n-hexylphosphine or tri-n-octylphosphine with 1,4-butanesultone were reacted with $HNTf_2$ to yield viscous homogeneous mixtures. These materials were studied by Ohno *et al.*,[89] emphasizing not only their gel properties, but more remarkably their lyotropic liquid-crystalline properties. The chemical and physical properties of diethyl(methyl)(isobutyl)phosphonium hexafluorophosphate ($[P_{i4\ 2\ 2\ 1}][PF_6]$) were studied by Jin and co-workers[90] in order to understand its ion transport behavior as organic ionic plastic crystals (OIPCs). These materials are very interesting since they could find applications as solid electrolytes in various electrochemical devices such as lithium batteries. Photoinduced bond cleavage can be used for preparative purposes. This was the case of carbene generation from 3-aryl-3-(trifluoromethyl)diazirine-functionalized highly fluorinated phosphonium

salts (HFPS), developed by Workentin and co-workers.[91] The reaction was employed to produce highly reactive carbene precursors for covalent attachment of the HFPS onto cotton/paper to impart hydrophobicity to these surfaces.

3 Phosphonium-based ionic liquids (PILs)

In the drive towards expanding the range of useful ionic liquids (ILs), with suitably low viscosities, sufficient stability, and conductivity for electrochemical uses, phosphonium-based ionic liquids (PILs) are exceptionally promising compounds. Due to the high thermal and electrochemical stabilities of PILs compared to ammonium-based ILs (alkylammonium, imidazolium, pyridinium, pyrrolidinium, and piperidinium compounds), phosphonium salts have been especially considered for applications as electrolytes. Their properties (both thermal and electrochemical stabilities) are in fact crucial to improve safety, durability, power and energy densities of electrochemical devices such as electric double layer capacitors.

Some phosphonium ILs also exhibit lower viscosity and higher conductivity than the corresponding ammonium ILs.

3.1 Preparation and characterisation

Studies of the properties of already known or even commercially available ILs have been reported extensively. Four independent studies (below summarized) compared the properties of protic phosphonium and ammonium ionic liquids of formula $[PHR_3]^+X^-$ and $[NHR_3]^+X^-$. Ali Rana and coworkers[92] did a comprehensive investigation using not only different cations but also different anions such as Ntf_2^-, OTf^-, $SO_3CH_3^-$ or NO_3^-. These authors demonstrated that phosphonium-based ILs were characterised not only by a higher thermal stability, but also a higher ionic conductivity and more facile proton reduction compared to the corresponding ammonium-based ILs. In a paper by Vankelecom,[93] only triflates were considered. This study focused on the ion conduction behavior of the ILs. The higher ionic conductivity observed for phosphonium-based PILs was attributed to their weaker hydrogen bond and Coulombic interactions. Also, a higher carrier ion concentration was indicated by infrared analysis and lattice potential energy estimation. Tsunashima *et al.*[94] studied the properties of the bistriflamide phosphonium salts. These authors ended up with the same observation about the higher conductivity, but they also highlighted that ionic liquids based on trialkylphosphonium cations were lower melting and of lower viscosity than the corresponding trialkylammonium ionic liquids. Analogous conclusions were drawn by Timperman and Anouti that focused[95] on the protic tributylphosphonium tetrafluoroborate ionic liquid in its pure form and in the mixture with acetonitrile. In a paper by Armel and co-workers,[96] the use of phosphonium cations with small alkyl chain substituents (methyl, ethyl, iso-butyl), in combination with a range of different anions (dicyanamide $N(CN)_2$; bistrifluoromethansulfonylamide, NTf_2; bis(fluorosulfonylamide), FSA), was reported to produce a variety of new halide free ionic liquids that were fluid, conductive and with sufficient thermal stability for a range of electrochemical applications. New density and viscosity data for ILs based on the tetradecyltrihexylphosphonium cation in combination with the bis(trifluoromethylsulfonyl)imide, bromide, chloride, decanoate, methanesulfonate, dicyanamide and bis(2,4,4-trimethylpentyl) phosphinate anions were determined by Coutinho et al.[97] For pure ILs, these authors predicted values of densities in good agreement with those experimentally determined. These decreased according to the anion trend: $[NTf_2]^- > Br^- > [CH_3SO_3]^- > [N(CN)_2]^- > Cl^- > [Phosphate]^- > [Decanoate]^-$. Instead, it was noticed that viscosity data correlated with the Vogel–Tammann–Fulcher method; accordingly, viscosity increased in the following order: $Br^- > Cl^- > [CH_3SO_3]^- > [Phosphate]^- \gg [Decanoate]^- > [N(CN)_2]^- > [NTf_2]^-$. These studies reinforced the significance of models directed towards thermophysical data prediction for which no experimental data are yet available. This was confirmed also by an investigation on high-pressure densities of phosphonium-based ionic liquids.[98] A detailed study on the chemical and physical properties of a series of alkyltributylphosphonium chloride ionic liquids was performed by the QUILL centre.[99] The focus was on the variability of the alkyl chain length and its impact on physical properties, such as melting points/glass transitions, thermal stability, density and viscosity. The same research group at QUILL investigated the effect of alkyltributylphosphonium chlorides $[P_{4\ 4\ 4\ n}]Cl$ (n = 4–8, 10, 12 or 14) on *A. nidulans conidia*. A relationship between the toxicity and the length of the alkyl substituent was observed.[100] These authors also provided a useful and robust toxicity screening method in order to stimulate the research for a conscious design on innovative ionic liquids. A very interesting nano-biphasic ionic liquid system composed of hydrophobic phosphonium salts and a hydrophilic ammonium salt was observed by Ohno and coworkers.[101] It was noticed that when a phosphonium zwitterion-lithium bis(trifluoromethanesulfonyl)imide complex was mixed with a hydrophilic ammonium salt, a nano-segregated liquid-crystalline matrix consisting of hydrophilic ionic liquid (IL) domains and hydrophobic IL domains was obtained. A TGA-MS study of the decomposition of commercially available trihexyl(tetradecyl)phosphonium decanoate and trihexyltetradecylphosphonium bis[(trifluoromethyl)sulfonyl] amide was performed by Keating and coworkers.[102] Among the other results, It was noted that the ionic liquid containing the NTf_2 anion decomposed by releasing HF. Hence, the use of the decanoate salt was preferable and safer for the environment. Several other studies on the physical and chemical characterisation of phosphonium ionic liquids were also conducted to gain information on properties of these compounds.[103–107] Also, new ILs were synthesised and their properties investigated with a view to proposing suitable applications for them. Fluorohydrogenate salts of quaternary phosphonium cations with alkyl and methoxy groups (tetraethylphosphonium $[P_{2\ 2\ 2\ 2}]^+$, triethyl-n-pentylphosphonium $[P_{5\ 2\ 2\ 2}]^+$, triethyl-n-octylphosphonium $[P_{2\ 2\ 2\ 8}]^+$, and triethylmethoxymethylphosphonium $[P_{2\ 2\ 2\ (101)}]^+$ were synthesized by Matsumoto et al.[108] These compounds were obtained by the action of anhydrous hydrogen fluoride on the corresponding phosphonium bromide or chloride precursors. Novel ionic liquids based on various chelated orthoborate anions with different phosphonium cations ($[P_{8\ 4\ 4\ 4}]$; $[P_{14\ 4\ 4\ 4}]$; $[P_{14\ 6\ 6\ 6}]$) were synthesised by Antzutkin and co-workers.[109] These compounds were characterised by the evaluation of several properties, including glass transition temperatures, density, viscosity and ionic conductivity. Due to their hydrophobicity and hydrolytical stability at room temperature, these hf-BILs were recommended for lubricant applications. Chiappe and co-workers also developed a series of ionic liquids based on trialkylglycerylphosphonium cations.[110] These salts were used as polar additives in a typical base-promoted Baylis-Hillman reaction. Long et al. reported thermal, rheological, and ion-transport properties of both commercially available triisobutylphosphonium tosylate and tri-n-butylethylphosphonium diethylphosphate, and other salts obtained by exchange reactions with BF_4, NTf_2 and DFOB (difluorooxalylborate) counteranions.[111] The latter anion (DFOB) was synthesized by the reaction of bis(trimethylsilyl)oxalate with BF_4^- ionic liquids.

3.2 Applications of phosphonium ionic liquids in synthesis

In synthetic chemistry, ionic liquids are very frequently used as solvents. Scammels and coworkers recently prepared tetraalkylphosphonium ionic liquids bearing an ether functionality and showed that these salts were excellent media for Grignard reactions.[112] Interestingly, while the reduction of starting aldehydes to the corresponding primary alcohol was the favored pathway in pyridinium ILs solvents, the same reactions carried out in the above mentioned phosphonium ILs afforded the expected Grignard products in good yield. The use of Grignard reagents in ILs was also reported by Kude and co-workers.[113] In this case, an iron-catalyzed homocoupling of aryl Grignard reagents was studied, and it was demonstrated that the reaction was extremely faster in phosphonium salt ionic liquids with respect to a common solvent like THF. Ionic liquid phases provide unique reaction environments and phosphonium ILs can act not only as solvents, but also as catalysts, co-catalysts and/or reactants. Trihexyl(tetradecyl)phosphonium tetrafluoroborate was reported to be an efficient phase-transfer catalyst for solid-liquid halex reactions, in particular for the introduction of fluoride ions by nucleophilic aromatic substitution. The ionic liquid was easy to handle and reuseable.[114] Another interesting example was observed during the dehydration of benzylic alcohols.[115] Under microwave irradiation, but in the absence of any metal catalyst, benzyl ethers and alkenes were obtained from primary and secondary benzylic alcohols in good to excellent yields using commercially available hydrophobic phosphonium ionic liquids as reaction media. A class of phosphonium amino acid ionic liquids was able to promote the copper-free and amine-free Sonogashira coupling reaction (Scheme 12).[116] A hydrophobic/lipophobic ionic liquid allowed an easy separation of the product from the ionic liquid and catalyst, and the recovered ionic liquid containing Pd catalyst could be reused.

Brønsted acidic (4-sulfobutyl)tris(4-sulfophenyl)phosphonium hydrogen sulfate was used by Shaterian and co-workers[117] to catalyse the

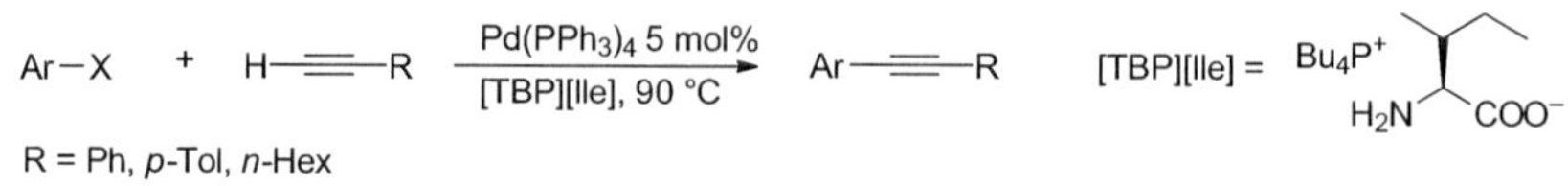

Scheme 12

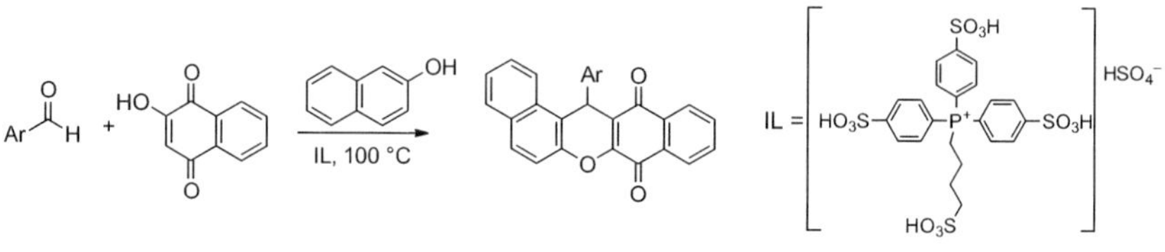

Scheme 13

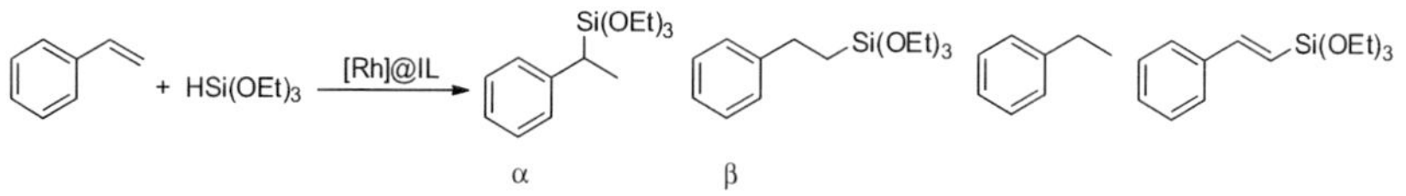

Scheme 14

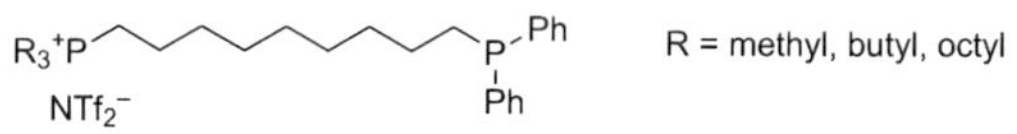

Fig. 3

synthesis of 14-aryl-14H-dibenzo[a,i]xanthene-8,13-dione, 3,4-dihydro-1H-benzo[b]xanthene-1,6,11(2H,12H)-trione, and aryl-5H-dibenzo[b,i]xanthene-5,7,12,14(13H)-tetraone derivatives (Scheme 13).

Metal-catalysed reactions were also reported in ILs in which the interaction between the metal complex and the IL led to very interesting effects. Wang et al.[118] prepared eleven tetrakis(dialkylamino)phosphonium cations and used them as catalytic supports for the hydrosilylation reaction of styrene with triethoxysilane catalyzed by Rh(PPh$_3$)$_3$Cl. Among the salts tested, the best catalytic activity and selectivity (94 %) in favor of the β-adduct was obtained in the presence of [[(C$_4$H$_9$)$_2$N]$_3$[(C$_8$H$_{17}$)$_2$N]P]PF$_6$ (Scheme 14). The metal complex-IL catalyst system could be reused more than 10 times without noticeable loss of catalytic activity and selectivity.

Another example of the use of metal complexes in ionic liquids was reported by Moores and co-workers.[119] In this case, besides the use of [P$_{4\ 4\ 4\ 14}$]NTf$_2$ as the solvent, different diphenylphosphino functionalised phosphonium salts (Fig. 3) acted as monodentate ligand for Rh complexes. These PFILs-Rh(I) complexes displayed very good activities and excellent selectivities in hydroformylation reactions.

3.3 Applications as nanoparticle stabilisers

Ionic liquids also find applications as nanoparticle (NP) stabilisers. The ability to stabilise NPs was explained by the combination of electrostatic interactions between the coordinatively unsaturated metal nanoparticle surface and the ionic-liquid anions and the steric protection offered by cations. In the case of strongly coordinating anions such as halides and bulky phosphonium cations, these effects are so strong that colloidal suspensions remain stable for months, as was observed by Scott et al.[120] In particular, the coordination of tetraalkylphosphonium halide ionic liquids by gold and palladium nanoparticles gave exceptionally stable colloidal solution.

De Vos and co-workers proved that a simple nickel(II) salt was sufficient as a catalyst for the aromatic amination of aryl chlorides when tetraoctylphosphonium bromide was used as the solvent.[121] The IL acted as a stabiliser for nickel nanoparticles that were formed *in situ* during reaction. In addition the phosphonium-based IL as a solvent made the metal catalytic system recyclable. Also, trihexyltetradecylphosphonium bis[(trifluoromethyl)sulfonyl]-imide, $[P_{6\,6\,6\,14}][NTf_2]$ was used as a solvent for the preparation of zinc oxide nanoparticles (ZnONPs) in the presence of microwave irradiation.[122] $[P_{6\,6\,6\,14}][NTf_2]$ had a low interface tension that made it able to enhance the nucleation rate and favor the formation of smaller ZnONPs. Moores et al.[123] investigated the dependence of the stability and catalytic activity of NPs on the nature of the IL used to generate them. $[P_{4\,4\,4\,1}]NTf_2$, $[P_{4\,4\,4\,8}]NTf_2$, $[P_{4\,4\,4\,14}]X$ (X= $-NTf_2$, $-OTf$, $-PF_6$), $[C_4mim]NTf_2$ and $[C_4dmim]NTf$ were compared and it was found that the ion associations, in terms of ion-pairs or supramolecular aggregates, played a key role in the stabilization of metal NPs by ILs. The use of trihexyl(tetradecyl)phosphonium-based ionic liquids was also reported for the formation of Pd(0) nanoparticles without the addition of reducing agents such as $NaBH_4$.[124] The particle formation process was highly dependent on the anion of the IL. Thus, *e.g.*, no nanoparticles were observed when the chloride IL was used. The authors proposed two mechanisms involving *i*) the formation of a trialkyl phosphine or *ii*) the dehydrogenation of alkyl chains of the phosphonium cation in the presence of palladium to yield molecular hydrogen. Particle shape could be controlled by varying the anion of the IL and the Pd(II) precursor used. Trihexyl(tetradecyl)phosphonium dicyanamide was used for the preparation of magnetic ionogels[125] by the reaction of organosilane-coated iron oxide nanoparticles and N-isopropylacrylamide in the ionic liquid. The silane-modified particles (Si-NPs) were immobilized in the ionogel and proved to be resistant to nanoparticle leaching. The Si-NPs also improved the mechanical stability of the ionogels with respect to ionogels synthesized with unmodified nanoparticles. The ionogels responded to external permanent magnets and were therefore prototypes of a new soft magnetic actuator.

3.4 Electrochemical applications

Ionic liquids have attracted a lot of interest as potential alternatives to conventional carbonate electrolytes in lithium-ion batteries. Major advantages of ILs are due to their intrinsic physical properties, such as low vapor pressure, non-flammability, high ionic conductivity and unlimited combinations of cations and anions. Zhang and co-workers[126]

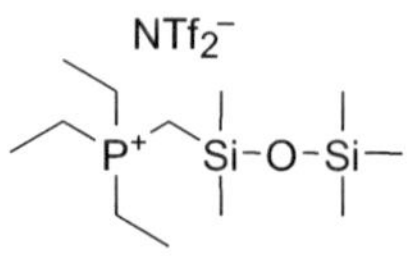

Fig. 4

explored a functionalised phosphonium salt bearing a disiloxane group (Fig. 4) This ionic liquid electrolyte showed good stability with a lithium transition metal oxide cathode and a graphite anode in lithium ion cells.

In the field of organic electrolytes for lithium secondary batteries, other phosphonium-based ionic liquids were proposed by Tsunashima et al.[127] These authors observed that cations bearing a methoxymethylene substituent bound to the phosphorus atom had very low viscosity when coupled with the NTf_2^- anion. These compounds exhibited very high thermal and electrochemical stability and were also proposed as electrolytes for dye-sensitized solar cells.[128] The physical properties of $[P_{6\ 6\ 6\ 14}][Cl]$ were studied with the addition of both water and metal salts[129] ($MgCl_2$ and LiCl) to verify their applicability as electrolyte for a Mg-air battery. The addition of water to the IL $[P_{6\ 6\ 6\ 14}][Cl]$ both reduced the viscosity and increased the conductivity. Hayyan et al.[130] demonstrated that phosphonium based ionic liquids were able to stabilise electrochemically-generated superoxide ion. Accordingly, the destruction of chlorobenzenes via the generation of O_2^- in $[P_{6\ 6\ 6\ 14}]NTf_2$ was proposed. Timperman et al.[131] described the use of a solution of the phosphonium protic ionic liquid $[Bu_3HP][BF_4]$ (3.4 mol L^{-1}) in acetonitrile as an electrolyte for carbon-based supercapacitors operating at a voltage (1.5 V) and capacities comparable to those of conventional aqueous electrolytes. The combination of good cycling abilities and an operating temperature ranging from $-40\ °C$ to $80\ °C$ allowed the realization of supercapacitors with an extended specific energy. Another application of phosphonium based ionic liquids in the electrochemistry field was as active components of solid-state iodide selective electrode[132] and as solvents for the electrodeposition of dysprosium.[133] A series of paper on the electrochemical reactivity and characterisation of different ILs also provided useful information for future applications.[134–137]

3.5 PIL based extraction and separation technologies

Several studies on the use of phosphonium ionic liquids for selective extraction of metal cations from aqueous solution were reported. Binnemans and co-workers described a green extraction process in which cobalt was separated as a tetrachlorocobaltate(ii) complex in a tri(hexyl)tetradecylphosphonium chloride ionic liquid phase.[138] This left behind nickel, magnesium and calcium in the aqueous phase. The overall process showed a very low environmental impact since no additional organic solvents were required. Other processes to extract zinc(II)[139,140] nickel(II) and cobalt(II),[141] cadmium (II),[142] palladium (II)[143,144] and Zn(II) through solid liquid extraction[145] were also described. Investigations on the partitioning of butanol and other fermentation broth components were carried out in phosphonium and ammonium-based ionic liquids.[146] This was a challenging issue in the case of organic materials and served also as an assessment on the toxicity of ILs for the microorganism involved in the process. Phosphonium-based ionic liquids (ILs) were examined for the extraction of ethanol from fermentation broths.[147] It was observed that such 'onium salts were highly efficient and a liquid-liquid extraction step coupled to an extractive fermentation process was proposed as an alternative to distillation. A study by Marrucho and co-workers[148] showed that phosphonium-based ILs were better extractants of short chain organic acids from aqueous dilute solutions than organic solvents traditionally used. The extraction efficiency of three phosphonium-based ILs showed different behaviors. Among them, tetradecyltrihexyl phosphonium bis(2,4,4-trimethylpentyl) phosphinate ([$P_{6\ 6\ 6\ 14}$][Phos]) was the best solvent for the investigated acids. A remarkable application of [$P_{6\ 6\ 6\ 14}$]Cl was reported as a surfactant for recovery of the oil from reservoirs that have lost their drive.[149] It was claimed that the studied IL was suitable to the purpose because of its ability to decrease the interfacial tension at the water–oil interface. The use of ionic liquid-based solvents was also described for separation processes in deep eutectic solvents.[150,151]

3.6 Miscellaneous applications

The phosphonium ILs (tetradecyl(trihexyl)phosphonium chloride, tetradecyl(trihexyl)phosphonium decanoate, tetradecyl(trihexyl)phosphonium dicyanimide, tetradecyl(trihexyl)phosphonium bis(triflamide)) have been considered as solvents for CO_2 capture using triethylendiamine (TEDA) by Clyburne et al.[152] Binary systems composed of two ILs were reported to generate crystalline ion pairs. Ohno and co-workers[153] used zwitterions to investigate the ion pair effect. The addition of a phosphonium type zwitterion (obtained by quaternisation of tri-n-octylphosphine with 1,4-butanesultone) to a typical amino acid ionic liquid such as 1-Ethyl-3-methylimidazolium leucine, gave a thermotropic and ion conductive gel.

Phosphonium ionic liquids were use also in the polymer chemistry field as monomers,[154–156] to yield new polyelectrolytes as well as modifiers for common polymeric materials.[157]

4 P-Ylides (Phosphoranes)

4.1 Preparation

Stable phosphonate ylides and phosphonate esters could be prepared *via* a one-pot reaction between activated acetylenes and triphenylphosphite in the presence of sulfonamides and heterocyclic NH-acids.[158,159] Ketenylidenetriphenylphosphorane ($Ph_3P=C=C=O$) was shown to react with protic reagents to yield novel carbonyl-stabilized ylides.[160] Lee reported the synthesis of triphenylphosphorane ylide precursors from alkyl halides utilizing a new sulfinyl reagent.[161] Phosphorus ylides were formed in the reaction of [(RNH)P-N(t-Bu)]$_2$ (R=t-Bu, i-Pr], with the allenylphosphine oxide $Ph_2P(O)C(Ph)=C=CH_2$.[162] A one-pot three-component reaction between hexamethyl phosphorous triamide and dimethyl

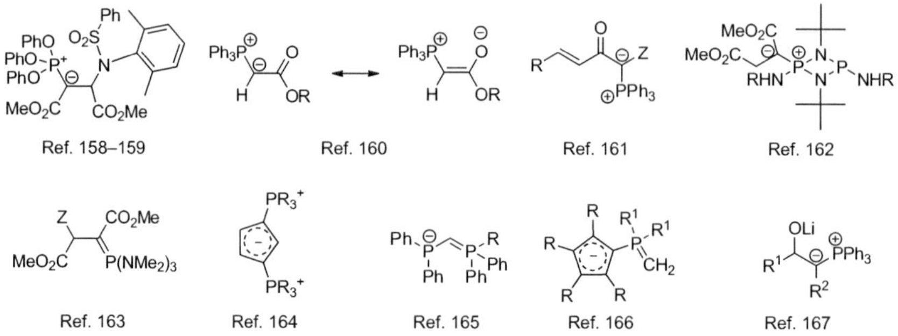

Fig. 5

acetylenedicarboxylate in the presence of NH-heterocyclic compounds yielded stable phosphorus ylides.[163] 1,3-Diphosphorus ylide cyclopentadienylium salts $(C_5H_3)(PPh_3)_2I$ and $(C_5H_3)[P(4-CH_3-Ph)_3]_2I$ were prepared from the reaction of 1,1′-dichloromercurioferrocene with $Pd(PPh_3)_4$ and with $Pd[P(4-CH_3-Ph)_3]_4$, respectively.[164] Phosphonium ylides as ligands could be obtained by the SN_2 reaction of dppm and haloalkanes, followed by deprotonation.[165] Phosphonium salts $[H_3C-PR_2-CpXH)]^+$ were starting materials for the synthesis of a set of lithium phosphonium diylides $Li[CH_2-PR_2-CpX]$,[166] and β-lithiooxyphosphonium ylides, used for Wittig–Schlosser and SCOOPY-type stereoselective olefination reactions, were synthesised directly from readily available terminal epoxides.[167] Figure 5 summarizes the structures of various ylides prepared through the above described methods.

Numerous reports detailed NMR and XRD studies of ylide structures and conformations.[168–173] In addition, the molecular interactions between phosphorus ylides and hypohalous acids HOX (X = F, Cl and Br) were investigated from a theoretical standpoint.[174]

The preferred conformations of ketodiester phosphonium ylides were investigated.[175] Acyclic diphosphonium bis-ylides from the corresponding diphosphonium precursors were systematically analysed with respect to the steric demand at the phosphorus centers and the substitution pattern of the phenylene bridge. In the o-phenylene series, bis-ylides were not observed due to the instability of the mono-ylide intermediates. In contrast, in the m-phenylene series, in which the two phosphonium centers were in remote positions, the bis-ylide was fully characterized.[176]

4.2 Applications in synthesis

A new synthetic approach toward the synthesis of flavones and pyranoflavone (Fig. 6, Ref. 177) was developed by light-induced intramolecular photochemical Wittig reactions in water with aryloxycarbonyl groups and suitably substituted phosphonium bromides without any phase-transfer catalyst or promoter.[177] Valla et al. reported the synthesis of retinoid-like structures using Wittig reactions[178] (Fig. 6, Ref. 178). In the synthesis of the antihistaminic drug olopatadine (Fig. 6, Ref. 179), the key step was a trans stereoselective Wittig olefination using a non-stabilized phosphorus ylide and a stereoselective Heck cyclization.[179]

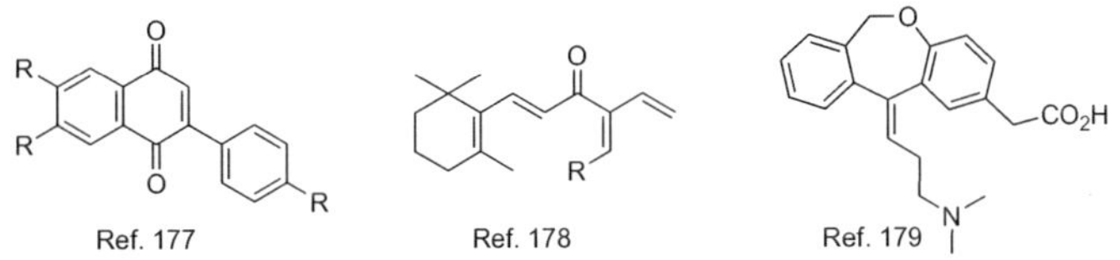

Ref. 177

Ref. 178

Ref. 179

Fig. 6

Scheme 15

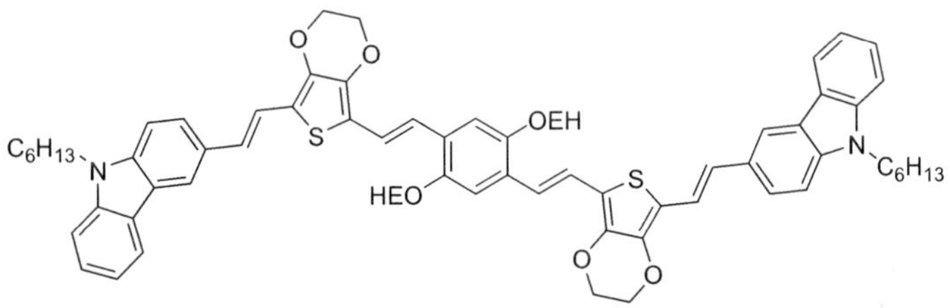

Fig. 7

New studies on the mechanism of the Wittig olefination reaction have uncovered an effect that is common to reactions of all ylide types, and that strongly argues for the operation of a common mechanism in all Li salt-free Wittig reactions.[180,181] The results reported by Byrne et al. showed that these effect were most easily explained by the [2+2] cycloaddition mechanism. The emerging consensus was that the mechanism centers on [2+2] cycloaddition of the ylide and carbonyl group to give the oxaphosphetane OPA directly (Scheme 15). The results, in tandem with the computational results of Aggarwal, Harvey, and co-workers, corroborated the cycloaddition mechanism proposed by Vedejs and strongly supported a common mechanism of the Wittig reaction for all ylide types which, ordinarily, operates under kinetic control.

Studies on the opening of a spiro-OPA led to the proposal of a stereomutation mechanism of the MB4 type for the OPA decomposition.[182]

Several oxidation-Wittig and oxidation Wittig-Horner one-pot procedures have also been developed for the efficient synthesis of a variety of substituted stilbenes.[183]

Phospha-Wittig-Horner type reactions are a common occurrence in the literature, and what follows is a listing of some of the most salient papers in the period considered by this review (2011–12).

π-Conjugated arylenevinylene oligomers (Fig. 7) could be prepared using Wittig–Horner click chemistry from bifunctional building blocks that could be interconnected to one another at any desired sequence.[184]

The reactions of P,P-dichlorophosphines and P,P-dibromophosphines with triethylphosphite under Michaelis-Arbuzov conditions were developed to give phosphinodiphosphonates in quantitative yields. After

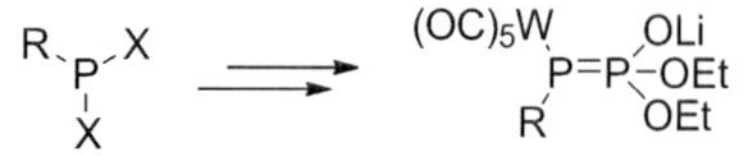

Fig. 8

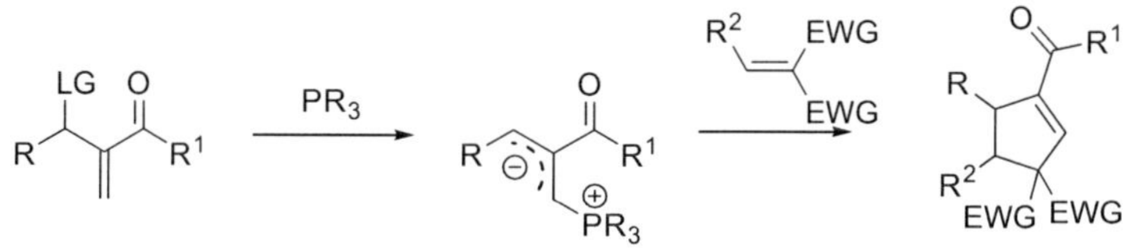

Scheme 16

complexation to W(CO)$_5$ and treatment with CH$_3$ONa, phospha–Wittig–Horner reagents (Fig. 8) were obtained on a multigram scale in good overall yield and were shown to react smoothly with acetone within minutes.[185]

Heteroatom Wittig chemistry also includes reactions of N-sulfonyl imines. It was demonstrated that these compounds underwent olefination reactions with nonstabilized phosphonium ylides under mild conditions to afford an array of both Z- and E-isomers of 1,2-disubstituted alkenes, allylic alcohols, and allylic amines.[186,187] Additionally, studies of the reactions of 5-bromo-4,6-dimethyl-2-thioxo-1,2-dihydropyridine-3-carbonitrile and thiazolidinone with phosphorus ylides have proved the formation of new phosphonium ylides.[188,189] Annulations via P-ylides are a common occurrence in the literature. For example, on photochemical irradiation, phosphonium-iodonium ylides were shown to undergo 1,3-dipolar cycloaddition reactions with triple bonds, via a carbene intermediate, to yield furans.[190–192] Even more common are the reactions of Morita-Baylis-Hillman (MBH) acetates and carbonates. Zhou et al. demonstrated that these substrates were able to generate very reactive 1,3-dipoles in the presence of tertiary phosphines; the dipoles then underwent cycloaddition reactions to yield annulation products (Scheme 16).[193]

This strategy was used to synthesize spirocyclopenteneoxindoles,[194] as well as stereodefined cyclopentenes.[195–197] Likewise, Hu reported a [3 + 2] annulation reaction of 2-arylideneindane-1,3-diones with MBH carbonates that proceeded smoothly in the presence of a multifunctional thiourea-phosphine catalysts to produce the corresponding quaternary carbon centered spirocyclic cyclopentenes.[198] A phosphine-triggered tandem [3 + 4] annulation reaction between MBH carbonates and 1,4-diheteroatom dinucleophiles was described, which provided access to saturated seven-membered 1,4-heterocycles.[199] Analogously, MBH carbonates were used as C3 synthons in asymmetric [3 + 2] annulation reactions for the asymmetric synthesis of 3-spirocyclopentene-2-oxindoles.[200] Another kind of annulation reaction for the construction of highly functionalized stereodefined cyclopentene skeletons involved a Rauhut–Currier Domino Reaction.[201] Finally, an efficient asymmetric [3 + 2] cycloaddition reaction between MBH carbonates of isatins and

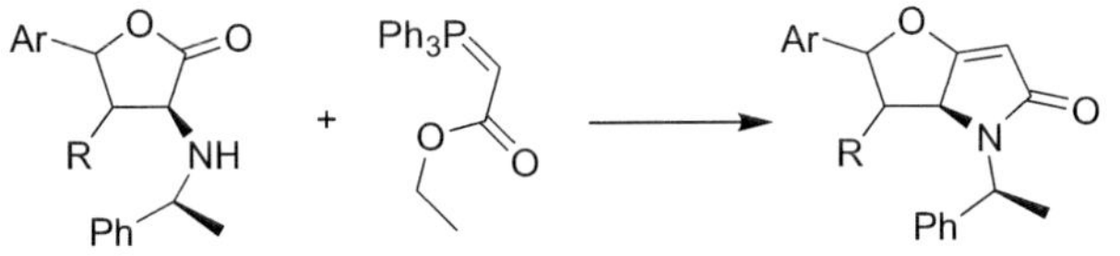

Scheme 17

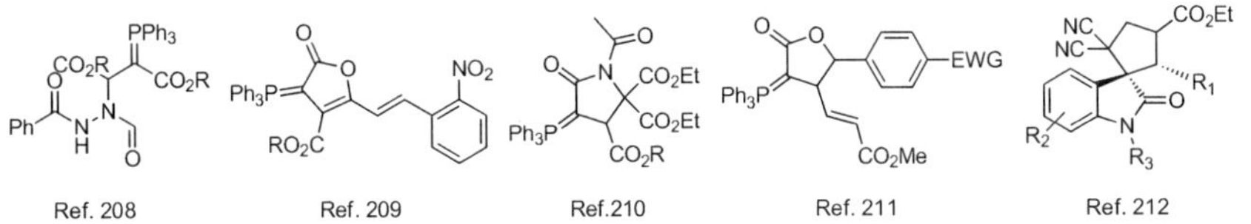

Fig. 9

Fig. 10

N-phenylmaleimide was developed for the construction of the spiro-cyclopentaneoxindole scaffold.[202] Duris described a new method for the synthesis of tetramic acids based on an intramolecular Wittig annulation procedure (Scheme 17).[203]

A phosphine-mediated benzannulation reaction between a β,γ-unsaturated α-ketoester and an allyic carbonate was used for the construction of multiaryl compounds.[204] A variety of carbocyclic and heterocyclic organic compounds are accessible *via* selective ring-closing intramolecular Wittig olefination, starting form phosphacumulene ylides.[205] An interesting phosphine-organocatalytic enantioselective reaction for the synthesis of spirocyclopenteneoxindoles was described, and the mechanistic hypothesis involves the intermediacy of a P-ylide.[206]

Multicomponent reactions involving phosphorus ylides were used to construct complex organic frameworks. For example, the reaction of triphenylphosphine, dialkyl acetylenedicarboxylates, and acids such as phenols, imides, amides, enols, oximes and alcohols was described to produce stabilized phosphorus ylides.[207] Other examples of products for multicomponent reactions involving P-ylides are shown in Fig. 9.

Miscellaneous P-ylide synthetic applications also included the hydrolysis of the zwitterionic adducts formed from Bu_3P, CS_2 and norbornene or norbornadiene to give the corresponding norbornane-fused 1,3-dithiolanes;[213] as well as hydrolysis using $HgCl_2$ and $CdCO_3$ for the formation of crystalline 1:1 $HgCl_2$ adducts. Based on this observation, a range of robust and readily accessible polymers containing norbornane-fused dithiolane units were evaluated for absorption of $HgCl_2$ from aqueous solutions (Fig. 10).

A large number of different syntheses involving phosphorus ylides were also reported. However, it is beyond the scope of the present chapter to illustrate them all. The interested reader can survey some of the following references.[214,215] A Wittig olefination reaction was used to derivatise ferrocenes with tetrazole rings.[216] π-Extended tetrathiafulvalene building blocks for new electroactive molecules have been prepared by Wittig olefination starting from a tetrathiafulvalene bearing a methyltriphenylphosphonium bromide moiety.[217] A trialkylphosphine-mediated Wittig reaction of trifluoromethylketones and allylic carbonates was used to prepare trifluoromethyl E-dienes in good yields.[218] Following their studies on allylic phosphorus ylides,[219] Xu et al. desribed a stereoselective PBu_3-mediated vinylogous Wittig olefination between α-methyl allenoates and a variety of aldehydes that represented the first example of a practical and synthetically useful vinylogous Wittig reaction (Scheme 18).[220]

The p-nitrobenzyltriphenylphosphonium ylide was used as an initiator for the terpolymerization of styrene and vinyl acetate with an electron-accepting monomer such as acrylonitrile. The mechanism is proposed to be based on the generation of a phenyl radical.[221,222] Also with a synthetic application perspective is an organic-inorganic hybrid 2D molecular space having a regular array of triphenylphosphine groups that has been investigated for its homogeneous catalytic potential for carbon-phosphorus ylide reactions.[223] Silylenes, i.e. P-ylides with a silicon atom in place of a carbon atom, are quite ubiquitous. For example a number of stable phosphonium sila-ylides, and their reactivity has been described,[224,225] including the first stable disilyne bis(phosphine) adduct.[226] The first N-donor-stabilized phosphasilene was synthesized in 87% yield. Remarkably, this compound was able to react with dichlorotriphenylphosphorane Ph_3PCl_2 to give the unprecedented 4π-electron Si_2P_2-cycloheterobutadiene $[(LSi)_2P_2]$ with two coordinate phosphorus atoms (Scheme 19).[227]

A different type of silylene having more electropositive carbon-based p-donating substituents, with the the carbanionic substituents additionally stabilized as part of a phosphorus ylide was also described (Scheme 20).[228,229]

Scheme 18

Scheme 19

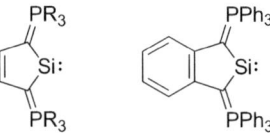

Scheme 20

4.3 Coordination chemistry

P-ylides can also act as ligands for metals. The following is a list ordered by the nature of the metallic element. Sabounchei et al. reported the reaction between MCl_4 (M = Ti, Zr and Hf) with α-ketophosphorus ylides to produce edge-shared $[M_2X_{10}]^{2-}$ complexes.[230] Bis(phosphaethenyl)pyridine iron complexes bearing novel ligand systems containing a phosphonium ylide structure were prepared by redox chemistry.[231] Several new nickel complexes were obtained by the treatment of the stabilized ylide benzoylmethylenetri(2-alkoxylphenyl)phosphorane with $Ni(cod)_2$ in the presence of PPh_3.[232] An yttrium phosphonium methylide complex – and its reaction with benzophenone – to form an excedingly rare hydrocarbon-soluble metal-coordinated betaine was also described. An X-ray crystal structure of the coordinated betaine was also reported.[233] Based on the phosphonium-1-indenylide (PHIN) ligands, triphenylphosphonium-1-indenylide, methyldiphenylphosphonium-1-indenylide, and dimethylphenylphosphonium-1-indenylide, a number of ruthenium(II) complexes were reported.[234,235] Grubbs described several new phosphonium alkylidene ruthenium metathesis catalysts incorporating different NHCs, that were used to generate ruthenacycles with the goal of rationalizing degenerate metathesis selectivity.[236] Another example was a photogenerated ruthenium-phosphonium-alkylidene complex used as a catalyst for ring-closing metathesis (RCM) and ring-opening metathesis polymerization (ROMP) reactions.

Möbius aromatic [28]hexaphyrin phosphonium adducts with palladium were also prepared and fully characterised,[237] as also were simpler Pd-ylide complexes.[238] The phosphorus ylides $Ph_3PCHC(O)C_6H_4R$ (R = 4-Me, 4-Br) react with $PdCl_2$ in equimolar ratios to give the C,C-orthopalladated $[Pd\{CHP(C_6H_4)Ph_2CO-C_6H_4-R)\} (\mu-Cl)]_2$.[239,240] Platinum(II) square-planar bis(ylide) complexes with nonstabilized ylide ligands, were generated by deprotonation of the tetra-n-butylphosphonium cation.[241] Phosphine attack at one carbene ligand on a Pt centre, to give an alkyl complex, better described as a stabilized ylide.[242] The reaction of a dinuclear gold(I) formamidinate with a phosphorus ylide yielded mixed-ligand gold(I) compounds featuring N and C connectivity from two dissimilar ligands, a formamidinate and a phosphorus ylide.[243] A mercury binuclear complex was also reported.[244] The synthesis of a uranium–ylide adduct was described[245] and its reactivity used for the synthesis of a uranium (V) terminal oxo complex and for the study of its chemistry.[246] α-Keto-stabilized phosphorus ylides of the type $Ph_3P=CHC(O)R$ (R = 2,4-dichlorophenyl (L1) and 4-isopropylphenyl (L2)) were used for the formation of complexes of structure $[Ag(L_2)_2(NO_3)]$ with $AgNO_3$ and $Hg(NO_3)_2 \cdot H_2O$.[247]

References

1. A. Fukazawa, E. Yamaguchi, E. Ito, H. Yamada, J. Wang, S. Irle and S. Yamaguchi, *Organometallics*, 2011, **30**, 3870.
2. L. I. Goryunov, S. I. Zhivetyeva, G. A. Nevinsky and V. D. Shteingarts, *Arkivoc*, 2011, 185.
3. P. Majewski and E. Lukaszewicz, *Collect. Czech. Chem. Commun.*, 2011, **76**, 1.
4. J. Prabagar, A. R. Cowley and J. M. Brown, *Synlett*, 2011, 2351.
5. X. Zhao, A. J. Lough and D. W. Stephan, *Chem. Eur. J.*, 2011, **17**, 6731.
6. M. Milenkovic, B. Warzajtis, U. Rychlewska, D. Radanovic, K. Andelkovic, T. Bozic, M. Vujcic and D. Sladic, *Molecules*, 2012, **17**, 2567.
7. Q. Zhang, S. Zhang and S. Li, *Macromolecules*, 2012, **45**, 2981.
8. L. Weber, H. Kuhtz, L. Bohling, A. Brockhinke, A. Chrostowska, A. Dargelos, A. Maziere, H.-G. Stammler and B. Neumann, *Dalton Trans.*, 2012, **41**, 10440.
9. J. Hermeke and P. H. Toy, *Tetrahedron*, 2011, **67**, 4103.
10. K. D. Demadis, A. Tsistraki, A. Popa, G. Ilia and A. Visa, *RSC Advances*, 2012, **2**, 631.
11. I. R. Knyazeva, A. R. Burilov and M. A. Pudovik, *Russ. Chem. Bull.*, 2011, **60**, 1956.
12. H. Imanieh, S. Ghammamy, M. M. A. Nikje, F. Hosseini, Z. S. Aghbolagh, H.-K. Fun, H. R. Khavasi and R. Kia, *Helv. Chim. Acta*, 2011, **94**, 2248.
13. A. K. Gupta, J. Nicholls, S. Debnath, I. Rosbottom, A. Steiner and R. Boomishankar, *Cryst. Growth Des.*, 2011, **11**, 555.
14. A. K. Gupta, A. Kalita and R. Boomishankar, *Inorg. Chim. Acta*, 2011, **372**, 152.
15. D. Uraguchi, Y. Ueki and T. Ooi, *Angew. Chem. Int. Ed.*, 2011, **50**, 3681.
16. I. Ling, Y. Alias, A. N. Sobolev, B. W. Skelton and C. L. Raston, *Dalton Trans.*, 2011, **40**, 10337.
17. G. A. Gainanova, G. I. Vagapova, V. V. Syakaev, A. R. Ibragimova, F. G. Valeeva, E. V. Tudriy, I. V. Galkina, O. N. Kataeva, L. Y. Zakharova, S. K. Latypov and A. I. Konovalov, *J. Colloid Interface Sci.*, 2012, **367**, 327.
18. X. Lin, L. Navailles, F. Nallet and M. W. Grinstaff, *Macromolecules*, 2012, **45**, 9500.
19. Y. Yamamoto, K. Nakao, T. Hashimoto, S. Matsukawa, N. Suzukawa, S. Kojima and K.-y. Akiba, *Heteroat. Chem.*, 2011, **22**, 523.
20. K. M. N. Burgess, I. Korobkov and D. L. Bryce, *Chem. Eur. J.*, 2012, **18**, 5748.
21. T. Ichikawa, M. Yoshio, A. Hamasaki, S. Taguchi, F. Liu, X.-b. Zeng, G. Ungar, H. Ohno and T. Kato, *J. Am. Chem. Soc.*, 2012, **134**, 2634.
22. C. F. Sailer, R. B. Singh, J. Ammer, E. Riedle and I. Pugliesi, *Chem. Phys. Lett.*, 2011, **512**, 60.
23. J. Ammer, C. F. Sailer, E. Riedle and H. Mayr, *J. Am. Chem. Soc.*, 2012, **134**, 11481.
24. B. Gao, L. Wang, R. Du and Y. Li, *Int. J. Chem. Kinet.*, 2011, **43**, 677.
25. D. Enders and T. V. Nguyen, *Org. Biomol. Chem.*, 2012, **10**, 5327.
26. T. Werner, A. M. Riahi and H. Schramm, *Synthesis*, 2011, **2011**, 3482.
27. R. M. Denton, J. An, P. Lindovska and W. Lewis, *Tetrahedron*, 2012, **68**, 2899.
28. M. Rodriguez-Zubiri, C. Baudequin, A. Béthegnies and J.-J. Brunet, *ChemPlusChem*, 2012, **77**, 445.
29. I. Bezkishko, V. Miluykov, O. Sinyashin and E. Hey-Hawkins, *Phosphorus, Sulfur, Silicon Relat. Elem.*, 2011, **186**, 657–659.
30. S. W. Chung, M. S. Plummer, L. A. McAllister, R. M. Oliver, J. A. Abramite, Y. Shen, J. Sun, D. P. Uccello, J. T. Arcari, L. M. Price and J. I. Montgomery, *Org. Lett.*, 2011, **13**, 5338.

31 A. ElMarrouni, J. M. Fabrellas and M. Heras, *Org. Biomol. Chem.*, 2011, **9**, 5967.
32 Z.-K. Wan, E. F. Ousman, N. Papaioannou and E. Saiah, *Tetrahedron Lett.*, 2011, **52**, 4149.
33 A. Sharma, D. Vachhani and E. Van der Eycken, *Org. Lett.*, 2012, **14**, 1854.
34 H. Fujioka, K. Yahata, O. Kubo, Y. Sawama, T. Hamada and T. Maegawa, *Angew. Chem. Int. Ed.*, 2011, **50**, 12232.
35 M. Fabris, M. Noè, A. Perosa, M. Selva and R. Ballini, *J. Org. Chem.*, 2012, **77**, 1805.
36 M. Selva, M. Noe, A. Perosa and M. Gottardo, *Org. Biomol. Chem.*, 2012, **10**, 6569.
37 V. Lucchini, M. Noe, M. Selva, M. Fabris and A. Perosa, *Chem. Commun.*, 2012, **48**, 5178.
38 L. Meier, M. Ferreira and M. M. Sá, *Heteroat. Chem*, 2012, **23**, 179.
39 D. Uraguchi, N. Kinoshita, T. Kizu and T. Ooi, *Synlett*, 2011, **2011**, 1265.
40 C.-L. Zhu, F.-G. Zhang, W. Meng, J. Nie, D. Cahard and J.-A. Ma, *Angew. Chem. Int. Ed.*, 2011, **50**, 5869.
41 M. North, P. Villuendas and C. Young, *Tetrahedron Lett.*, 2012, **53**, 2736.
42 A. Schnurr, M. Bolte, H.-W. Lerner and M. Wagner, *Eur. J. Inorg. Chem.*, 2012, 112.
43 D.-Y. Kim, H.-J. Kim, K.-H. Yu and J.-J. Min, *Bioconjugate Chem.*, 2012, **23**, 431.
44 D. Y. Kim, H. J. Kim, K. H. Yu and J. J. Min, *Bioorg. Med. Chem. Lett.*, 2012, **22**, 319.
45 Y. Sun, L. Zhu and W. He, *Eur. J. Nucl. Med. Mol. Imaging*, 2012, **39**, S398.
46 G. S. Gurm, S. B. Danik, T. M. Shoup, S. Weise, K. Takahashi, S. Laferrier, D. R. Elmaleh and H. Gewirtz, *Jacc-Cardiovascular Imaging*, 2012, **5**, 285.
47 A. Haslop, C. Marzano, N. Long, A. Gee and C. Plisson, *J. Labelled Comp. Radiopharm.*, 2011, **54**, S454.
48 T. Higuchi, K. Fukushima, C. Rischpler, T. Isoda, M. S. Javadi, H. Ravert, D. P. Holt, R. F. Dannals, I. Madar and F. M. Bengel, *J. Nucl. Med.*, 2011, **52**, 965.
49 D. Y. Kim, H. S. Kim, U. N. Le, S. N. Jiang, H. J. Kim, K. C. Lee, S. K. Woo, J. Chung, H. S. Kim, H. S. Bom, K. H. Yu and J. J. Min, *J. Nucl. Med.*, 2012, **53**, 1779.
50 I. Madar, T. Isoda, P. Finley, J. Angle and R. Wahl, *J. Nucl. Med.*, 2011, **52**, 808.
51 T. M. Shoup, D. R. Elmaleh, A. L. Brownell, A. Zhu, J. L. Guerrero and A. J. Fischman, *Mol. Imaging Biol.*, 2011, **13**, 511.
52 D. Y. Kim, H. J. Kim, K. H. Yu and J. J. Min, *Nucl. Med. Biol.*, 2012, **39**, 1093.
53 Y. Zhou and S. Liu, *Bioconjugate Chem.*, 2011, **22**, 1459.
54 N. E. Basova, E. V. Rozengart and A. A. Suvorov, *J. Evol. Biochem. Physiol.*, 2011, **47**, 420.
55 V. V. Andriyashin, Y. V. Bakhtiyarova, R. A. Cherkasov, V. I. Galkin and I. V. Galkina, *Russ. J. Org. Chem.*, 2012, **48**, 1574.
56 V. V. Andriyashin, Y. V. Bakhtiyarova, R. A. Cherkasov, V. I. Galkin and I. V. Galkina, *Russ. J. Org. Chem.*, 2012, **48**, 1576.
57 A. G. Xie, X. Cai, M. S. Lin, T. Wu, X. J. Zhang, Z. D. Lin and S. Z. Tan, *Mat. Sci. Eng. B-Solid*, 2011, **176**, 1222.
58 X. Cai, S. Z. Tan, M. S. Lin, A. Xie, W. J. Mai, X. J. Zhang, Z. D. Lin, T. Wu and Y. L. Liu, *Langmuir*, 2011, **27**, 7828.
59 C. Chung, K. J. Lampe and S. C. Heilshorn, *Biomacromolecules*, 2012, **13**, 3912.
60 N. Kolevzon, U. Kuflik, M. Shmuel, S. Benhamron, I. Ringel and E. Yavin, *Pharm. Res.*, 2011, **28**, 2780.

61 S. Biswas, N. S. Dodwadkar, A. Piroyan and V. P. Torchilin, *Biomaterials*, 2012, **33**, 4773.
62 Y. Ju-Nam, Y. S. Chen, J. J. Ojeda, D. W. Allen, N. A. Cross, P. H. E. Gardiner and N. Bricklebank, *RSC Advances*, 2012, **2**, 10345.
63 T. E. Long, X. Lu, M. Galizzi, R. Docampo, J. Gut and P. J. Rosenthal, *Bioorg. Med. Chem. Lett.*, 2012, **22**, 2976.
64 A. Taladriz, A. Healy, E. J. F. Perez, V. H. Garcia, C. R. Martinez, A. A. M. Alkhaldi, A. A. Eze, M. Kaiser, H. P. de Koning, A. Chana and C. Dardonville, *J. Med. Chem.*, 2012, **55**, 2606.
65 S. T. Hemp, M. H. Allen, M. D. Green and T. E. Long, *Biomacromolecules*, 2012, **13**, 231.
66 S. T. Hemp, A. E. Smith, M. H. Allen, J. M. Bryson and T. E. Long, *Mol. Ther.*, 2012, **20**, S248.
67 S. T. Hemp, A. E. Smith, J. M. Bryson, M. H. Allen and T. E. Long, *Biomacromolecules*, 2012, **13**, 2439.
68 A. E. Smith, S. T. Hemp, M. H. Allen, J. M. Bryson and T. E. Long, *Mol. Ther.*, 2012, **20**, S246.
69 R. Bansal, S. K. Tripathi, K. C. Gupta and P. Kumar, *J. Mater. Chem.*, 2012, **22**, 25427.
70 M. Li, G. M. Ganea, C. Lu, S. L. De Rooy, B. El-Zahab, V. E. Fernand, R. Jin, S. Aggarwal and I. M. Warner, *J. Inorg. Biochem.*, 2012, **107**, 40.
71 C. Ornelas-Megiatto, P. R. Wich and J. M. J. Fréchet, *J. Am. Chem. Soc.*, 2012, **134**, 1902.
72 L.-M. Tumir, I. Crnolatac, T. Deligeorgiev, A. Vasilev, S. Kaloyanova, M. Grabar Branilović, S. Tomić and I. Piantanida, *Chem. Eur. J.*, 2012, **18**, 3859.
73 L. Wang, X. Xu, S. Guo, Z. Peng and T. Tang, *Int. J. Biol. Macromol.*, 2011, **48**, 375.
74 W. Abdallah and U. Yilmazer, *Thermochim. Acta*, 2011, **525**, 129.
75 V. Mittal, *Applied Clay Science*, 2012, **56**, 103.
76 S. Ganguly, K. Dana, T. K. Mukhopadhyay and S. Ghatak, *J. Therm. Anal. Calorim.*, 2011, **105**, 199.
77 B. Yu, X. S. Zeng, G. J. Dai, X. A. Cai, Z. D. Lin, X. J. Zhang, S. Z. Tan and Y. L. Liu, *Asian J. Chem.*, 2011, **23**, 2069.
78 S. Ganguly, K. Dana, T. K. Mukhopadhyay and S. Ghatak, *Clays Clay Miner.*, 2011, **59**, 13.
79 K. Kawasaki, T. Ebina, T. Hanaoka, H. Tsuda and K. Motegi, *Jpn. J. Appl. Phys.*, 2011, **50**, 121601.
80 K. Saitoh, K. Ohashi, T. Oyama, A. Takahashi, J. Kadota, H. Hirano and K. Hasegawa, *J. Appl. Polym. Sci.*, 2011, **122**, 666.
81 S. Suin and B. B. Khatua, *Ind. Eng. Chem. Res.*, 2012, **51**, 15096.
82 X. S. Zeng, D. M. Cai, Z. D. Lin, X. Cai, X. J. Zhang, S. Z. Tan and Y. B. Xu, *J. Appl. Polym. Sci.*, 2012, **126**, 601.
83 K. Wang, Y. Zeng, L. He, J. Yao, A. K. Suresh, J. Bellare, T. Sridhar and H. Wang, *Desalination*, 2012, **292**, 119.
84 S. Gu, R. Cai and Y. Yan, *Chem. Commun.*, 2011, **47**, 2856.
85 K. K. Stokes, J. A. Orlicki and F. L. Beyer, *Polym. Chem*, 2011, **2**, 80.
86 K. J. T. Noonan, K. M. Hugar, H. A. Kostalik, E. B. Lobkovsky, H. D. Abruna and G. W. Coates, *J. Am. Chem. Soc.*, 2012, **134**, 18161.
87 N. N. Fathima, J. R. Rao and B. U. Nair, *J. Am. Leather Chem. Assoc.*, 2011, **106**, 249.
88 J. X. Luo, Y. J. Feng and Z. H. Shan, *J. Soc. Leather Technol. Chem.*, 2011, **95**, 215.
89 S. Ueda, J. Kagimoto, T. Ichikawa, T. Kato and H. Ohno, *Adv. Mater.*, 2011, **23**, 3071.

90 L. Jin, K. M. Nairn, C. M. Forsyth, A. J. Seeber, D. R. MacFarlane, P. C. Howlett, M. Forsyth and J. M. Pringle, *J. Am. Chem. Soc.*, 2012, **134**, 9688.
91 S. Ghiassian, H. Ismaili, B. D. W. Lubbock, J. W. Dube, P. J. Ragogna and M. S. Workentin, *Langmuir*, 2012, **28**, 12326.
92 U. A. Rana, R. Vijayaraghavan, M. Walther, J. Sun, A. A. J. Torriero, M. Forsyth and D. R. MacFarlane, *Chem. Commun.*, 2011, **47**, 11612.
93 J. Luo, O. Conrad and I. F. J. Vankelecom, *J. Mater. Chem.*, 2012, **22**, 20574.
94 K. Tsunashima, M. Fukushima and M. Matsumiya, *Electrochemistry*, 2012, **80**, 904.
95 L. Timperman and M. Anouti, *Ind. Eng. Chem. Res.*, 2012, **51**, 3170.
96 V. Armel, D. Velayutham, J. Sun, P. C. Howlett, M. Forsyth, D. R. MacFarlane and J. M. Pringle, *J. Mater. Chem.*, 2011, **21**, 7640.
97 C. M. S. S. Neves, P. J. Carvalho, M. G. Freire and J. A. P. Coutinho, *J. Chem. Thermodyn.*, 2011, **43**, 948.
98 L. I. N. Tome, R. L. Gardas, P. J. Carvalho, M. J. Pastoriza-Gallego, M. M. Piñeiro and J. o. A. P Coutinho, *J. Chem. Eng. Data*, 2011, **56**, 2205.
99 G. Adamova, R. L. Gardas, M. Nieuwenhuyzen, A. V. Puga, L. P. N. Rebelo, A. J. Robertson and K. R. Seddon, *Dalton Trans.*, 2012, **41**, 8316.
100 M. Petkovic, D. O. Hartmann, G. Adamova, K. R. Seddon, L. P. N. Rebelo and C. S. Pereira, *New J. Chem.*, 2012, **36**, 563.
101 S. Taguchi, T. Ichikawa, T. Kato and H. Ohno, *Chem. Commun.*, 2012, **48**, 5271.
102 M. Keating, F. Gao and J. Ramsey, *J. Therm. Anal. Calorim.*, 2011, **106**, 207.
103 D. Cholico-Gonzalez, M. Avila-Rodriguez, J. A. Reyes-Aguilera, G. Cote and A. Chagnes, *J. Mol. Liq.*, 2012, **169**, 27.
104 A. F. Ferreira, P. N. Simões and A. G. M. Ferreira, *J. Chem. Thermodyn.*, 2012, **45**, 16.
105 C. E. Ferreira, N. M. C. Talavera-Prieto, I. M. A. Fonseca, A. T. G. Portugal and A. G. M. Ferreira, *J. Chem. Thermodyn.*, 2012, **47**, 183.
106 X. Liu, Y. Zhao, X. Zhang, G. Zhou and S. Zhang, *J. Phys. Chem. B*, 2012 **116**, 4934.
107 M. S. Manic, E. A. Macedo and V. Najdanovic-Visak, *Fluid Phase Equilib.*, 2012, **324**, 82.
108 T. Enomoto, S. Kanematsu, K. Tsunashima, K. Matsumoto and R. Hagiwara, *PCC*, 2011, **13**, 12536.
109 F. U. Shah, S. Glavatskih, D. R. MacFarlane, A. Somers, M. Forsyth and O. N. Antzutkin, *PCCP*, 2011, **13**, 12865.
110 F. Bellina, C. Chiappe and M. Lessi, *Green Chem.*, 2012, **14**, 148.
111 M. D. Green, C. Schreiner and T. E. Long, *J. Phys. Chem. A*, 2011, **115**, 13829.
112 L. Ford, F. Atefi, R. D. Singer and P. J. Scammells, *Eur. J. Org. Chem.*, 2011, **2011**, 942.
113 K. Kude, S. Hayase, M. Kawatsura and T. Itoh, *Heteroat. Chem*, 2011, **22**, 397.
114 A. Fan, G.-K. Chuah and S. Jaenicke, *Catal. Today*, 2012, **198**, 300.
115 H. A. Kalviri and F. M. Kerton, *Adv. Synth. Catal.*, 2011, **353**, 3178.
116 T. Fukuyama, M. T. Rahman, S. Maetani and I. Ryu, *Chem. Lett.*, 2011, **40**, 1027.
117 H. R. Shaterian, M. Ranjbar and K. Azizi, *J. Mol. Liq.*, 2011, **162**, 95.
118 D. Wang, J. Li, J. Peng, Y. Bai and G. Lai, *Phosphorus, Sulfur Silicon Relat. Elem.*, 2011, **186**, 2258.
119 K. L. Luska, K. Z. Demmans, S. A. Stratton and A. Moores, *Dalton Trans.*, 2012, **41**, 13533.
120 A. Banerjee, R. Theron and R. W. J. Scott, *ChemSusChem*, 2012, **5**, 109.
121 I. Geukens, J. Fransaer and D. E. De Vos, *ChemCatChem*, 2011, **3**, 1431.

122   E. K. Goharshadi, M. Abareshi, R. Mehrkhah, S. Samiee, M. Moosavi, A. Youssefi and P. Nancarrow, *Mater. Sci. Semicond. Process.*, 2011, **14**, 692.
123   K. L. Luska and A. Moores, *Green Chem.*, 2012, **14**, 1736.
124   H. A. Kalviri and F. M. Kerton, *Green Chem.*, 2011, **13**, 681.
125   B. Ziółkowski, K. Bleek, B. Twamley, K. J. Fraser, R. Byrne, D. Diamond and A. Taubert, *Eur. J. Inorg. Chem.*, 2012, **2012**, 5245.
126   W. Weng, Z. Zhang, J. Lu and K. Amine, *Chem. Commun.*, 2011, **47**, 11969.
127   K. Tsunashima, F. Yonekawa, M. Kikuchi and M. Sugiya, *Electrochemistry*, 2011, **79**, 453.
128   Y. Kunugi, N. Hamada, S. Kodama, M. Sugiya and K. Tsunashima, *Electrochemistry*, 2011, **79**, 810.
129   P. M. Bayley, J. Novak, T. Khoo, M. M. Britton, P. C. Howlett, D. R. Macfarlane and M. Forsyth, *Aust. J. Chem.*, 2012, **65**, 1542.
130   M. Hayyan, F. S. Mjalli, M. A. Hashim, I. M. AlNashef, S. M. Al-Zahrani and K. L. Chooi, *J. Electroanal. Chem.*, 2012, **664**, 26.
131   L. Timperman, H. Galiano, D. Lemordant and M. Anouti, *Electrochem. Commun.*, 2011, **13**, 1112.
132   N. V. Shvedene, A. V. Rzhevskaia and I. V. Pletnev, *Talanta*, 2012 **102**, 123.
133   A. Kurachi, M. Matsumiya, K. Tsunashima and S. Kodama, *J. Appl. Electrochem.*, 2012, **42**, 961.
134   J.-A. Latham, P. C. Howlett, D. R. MacFarlane and M. Forsyth, *Electrochim. Acta*, 2011, **56**, 5328.
135   M. Matsumiya, K. Tsunashima and S. Kodama, *Zeitschrift Fur Naturforschung Section a-a Journal of Physical Sciences*, 2011, **66**, 668.
136   K. Tsunashima, Y. Ono and M. Sugiya, *Electrochim. Acta*, 2011, **56**, 4351.
137   J. E. F. Weaver, D. Breadner, F. Deng, B. Ramjee, P. J. Ragogna and R. W. Murray, *The Journal of Physical Chemistry C*, 2011, **115**, 19379.
138   S. Wellens, B. Thijs and K. Binnemans, *Green Chem.*, 2012, **14**, 1657.
139   M. Regel-Rosocka and M. Wisniewski, *Hydrometallurgy*, 2011, **110**, 85.
140   M. Regel-Rosocka, Ł. Nowak and M. Wiśniewski, *Sep. Purif. Technol.*, 2012, **97**, 158.
141   P. Rybka and M. Regel-Rosocka, *Sep. Sci. Technol.*, 2012, **47**, 1296.
142   A. Arias, I. Saucedo, R. Navarro, V. Gallardo, M. Martinez and E. Guibal, *React. Funct. Polym.*, 2011, **71**, 1059.
143   A. Cieszynska and M. Wisniewski, *Sep. Purif. Technol.*, 2011, **80**, 385.
144   R. Navarro, I. Saucedo, C. Gonzalez and E. Guibal, *Chem. Eng. J*, 2012, **185–186**, 226.
145   D. Kogelnig, A. Regelsberger, A. Stojanovic, F. Jirsa, R. Krachler and B. Keppler, *Monatsh. Chem.*, 2011, **142**, 769.
146   H. R. Cascon, S. K. Choudhari, G. M. Nisola, E. L. Vivas, D.-J. Lee and W.-J. Chung, *Sep. Purif. Technol.*, 2011, **78**, 164.
147   C. M. S. S. Neves, J. F. O. Granjo, M. G. Freire, A. Robertson, N. M. C. Oliveira and J. A. P. Coutinho, *Green Chem.*, 2011, **13**, 1517.
148   F. S. Oliveira, J. M. M. Araújo, R. Ferreira, L. P. N. Rebolo and I. M. Marrucho, *Sep. Purif. Technol.*, 2012, **85**, 137.
149   S. Lago, H. Rodriguez, M. K. Khoshkbarchi, A. Soto and A. Arce, *RSC Advances*, 2012, **2**, 9392.
150   M. A. Kareem, F. S. Mjalli, M. A. Hashim and I. M. AlNashef, *Fluid Phase Equilib.*, 2012, **314**, 52.
151   M. A. Kareem, F. S. Mjalli, M. A. Hashim, M. K. O. Hadj-Kali, F. S. G. Bagh and I. M. Alnashef, *Fluid Phase Equilib*, 2012, **333**, 47.

152 N. D. Harper, K. D. Nizio, A. D. Hendsbee, J. D. Masuda, K. N. Robertson, L. J. Murphy, M. B. Johnson, C. C. Pye and J. A. C. Clyburne, *Ind. Eng. Chem. Res.*, 2011, **50**, 2822.

153 S. Taguchi, T. Matsumoto, T. Ichikawa, T. Kato and H. Ohno, *Chem. Commun.*, 2011, **47**, 11342.

154 S. Cheng, F. L. Beyer, B. D. Mather, R. B. Moore and T. E. Long, *Macromolecules*, 2011, **44**, 6509.

155 Y. P. Borguet and N. V. Tsarevsky, *Polym. Chem*, 2012, **3**, 2487.

156 S. Cheng, M. Zhang, T. Wu, S. T. Hemp, B. D. Mather, R. B. Moore and T. E. Long, *J. Polym. Sci., Part A: Polym. Chem*, 2012, **50**, 166.

157 A. M. A. Dias, S. Marceneiro, M. E. M. Braga, J. F. J. Coelho, A. G. M. Ferreira, P. N. Simões, H. I. M. Veiga, L. C. Tomé, I. M. Marrucho, J. M. S. S. Esperança, A. A. Matias, C. M. M. Duarte, L. P. N. Rebelo and H. C. de Sousa, *Acta Biomaterialia*, 2012, **8**, 1366.

158 F. R. Charati, M. T. Maghsoodlou, S. M. Habibi-Khorassani, N. Hazeri, A. Ebrahimi, Z. Hossaini, P. Ebrahimi, N. Maleki, S. R. Adhamdoust, F. Vasheghani-Farahani, G. Marandi, A. Sobolev and M. Makha, *Comb. Chem. High Throughput Screening*, 2011, **14**, 2159.

159 F. R. Charati, M. Moghimi, M. T. Maghsoodlou, S. M. Habibi-Khorassani, Z. Hossaini, N. Maleki, B. W. Skelton and M. Makha, *Phosphorus, Sulfur Silicon Relat. Elem.*, 2011, **186**, 1428.

160 M. G. Makramalla, M. T. Diab, C. N. Sherren, A. D. Hendsbee, J. D. Masuda, K. N. Robertson and J. A. C. Clyburne, *Phosphorus, Sulfur Silicon Relat. Elem.*, 2012, **187**, 1537.

161 K. Lee, *Bull. Korean Chem. Soc.*, 2011, **32**, 3477.

162 K. C. Kumara Swamy, G. Gangadhararao, V. Srinivas, N. N. Bhuvan Kumar, E. Balaraman and M. Chakravarty, *Inorg. Chim. Acta*, 2011, **372**, 374.

163 M. Ziyaadini, M. T. Maghsoodlou, N. Hazeri and S. M. Habibi-Khorassani, *Heteroat. Chem*, 2012, **23**, 131.

164 C. Xu, Z.-Q. Wang, Z. Li, W.-Z. Wang, X.-Q. Hao, W.-J. Fu, J.-F. Gong, B.-M. Ji and M.-P. Song, *Organometallics*, 2012, **31**, 798.

165 J. Langer, S. Meyer, F. Dundar, B. Schowtka, H. Gorls and M. Westerhausen, *Arkivoc*, 2012, 210.

166 C. Lichtenberg, N. S. Hillesheim, M. Elfferding, B. Oelkers and J. Sundermeyer, *Organometallics*, 2012, **31**, 4259.

167 D. M. Hodgson and R. S. D. Persaud, *Beilstein J. Org. Chem.*, 2012, **8**, 1896.

168 A. Habibi, K. Eskandari and A. Alizadeh, *Phosphorus, Sulfur Silicon Relat. Elem.*, 2012, **187**, 1109.

169 S. M. Habibi-Khorassani, A. Ebrahimi, M. T. Maghsoodlou, S. Same-Salari, S. Nasiri and H. Ghasempour, *Magn. Reson. Chem.*, 2011, **49**, 213.

170 S. M. Habibi-Khorassani, M. T. Maghsoodlou, A. Ebrahimi, M. Mohammadi, M. Shahraki and E. Aghdaei, *J. Phys. Org. Chem.*, 2012, **25**, 1328.

171 S. M. Habibi-Khorassani, M. T. Maghsoodlou, E. Aghdaei and M. Shahraki, *Progress in Reaction Kinetics and Mechanism*, 2012, **37**, 301.

172 M. Rostamizadeh, M. T. Maghsoodlou, S. M. Habibi-Khorassani, N. Hazeri, F. R. Charati, M. A. Kazemian, B. W. Skelton and M. Makha, *Heteroat. Chem*, 2011, **22**, 715.

173 M. Rostamizadeh, M. T. Maghsoodlou, S. M. Habibi-Khorassani, N. Hazeri, S. S. Sajadikhah, F. R. Charati, M. A. Kazemian, B. W. Skelton and M. Makha, *Heteroat. Chem.*, 2011, **22**, 36.

174 A. Zabaradsti, A. Kakanejadifard and M. Ghasemian, *Comp. Theor. Chem.*, 2012, **989**, 1.

175 F. Castañeda, P. Silva, M. Teresa Garland, A. Shirazi and C. A. Bunton, *J. Mol. Struct.*, 2011, **1004**, 284.
176 C. Maaliki, M. Abdalilah, C. Barthes, C. Duhayon, Y. Canac and R. Chauvin, *Eur. J. Inorg. Chem.*, 2012, **2012**, 4057.
177 J. Das and S. Ghosh, *Tetrahedron Lett.*, 2011, **52**, 7189.
178 A. Valla, N. Meheux, D. Cartier, B. Valla, L. Dufosse and R. Labia, *Synth. Commun.*, 2011, **41**, 184.
179 J. Bosch, J. Bachs, A. M. Gómez, R. Griera, M. Écija and M. Amat, *J. Org. Chem.*, 2012, **77**, 6340.
180 P. A. Byrne and D. G. Gilheany, *J. Am. Chem. Soc.*, 2012, **134**, 9225.
181 P. A. Byrne, L. J. Higham, P. McGovern and D. G. Gilheany, *Tetrahedron Lett.*, 2012, **53**, 6701.
182 J. García López, A. Morán Ramallal, J. González, L. Roces, S. García-Granda, M. J. Iglesias, P. Oña-Burgos and F. López Ortiz, *J. Am. Chem. Soc.*, 2012, **134**, 19504.
183 A. S. Saiyed, K. N. Patel, B. V. Kamath and A. V. Bedekar, *Tetrahedron Lett.*, 2012, **53**, 4692.
184 K. N. Shivananda, I. Cohen, E. Borzin, Y. Gerchikov, M. Firstenberg, O. Solomeshch, N. Tessler and Y. Eichen, *Adv. Funct. Mater.*, 2012, **22**, 1489.
185 A. I. Arkhypchuk, M.-P. Santoni and S. Ott, *Organometallics*, 2012, **31**, 1118.
186 D. J. Dong, Y. A. Li, J. Q. Wang and S. K. Tian, *Chem. Commun.*, 2011, **47**, 2158.
187 F. Fang, Y. A. Li and S. K. Tian, *Eur. J. Org. Chem.*, 2011, 1084.
188 H. A. Abdel-Malek, M. S. Salem and L. S. Boulos, *Life Science Journal-Acta Zhengzhou University Overseas Edition*, 2012, **9**, 695.
189 H. A. Abdel-Malek, *Phosphorus, Sulfur Silicon Relat. Elem.*, 2012, **187**, 506.
190 E. D. Matveeva, T. A. Podrugina, M. A. Taranova, A. A. Borisenko, A. V. Mironov, R. Gleiter and N. S. Zefirov, *J. Org. Chem.*, 2011, **76**, 566.
191 E. D. Matveeva, T. A. Podrugina, M. A. Taranova, A. M. Ivanova, R. Gleiter and N. S. Zefirov, *J. Org. Chem.*, 2012, **77**, 5770.
192 T. D. Nekipelova, M. A. Taranova, E. D. Matveeva, T. A. Podrugina, V. A. Kuzmin and N. S. Zefirov, *Doklady Chemistry*, 2012, **447**, 262.
193 R. Zhou, J. Wang, H. Song and Z. He, *Org. Lett.*, 2011, **13**, 580.
194 H. P. Deng, Y. Wei and M. Shi, *Org. Lett.*, 2011, **13**, 3348.
195 H.-P. Deng, D. Wang, Y. Wei and M. Shi, *Beilstein J. Org. Chem.*, 2012, **8**, 1098.
196 H.-P. Deng, Y. Wei and M. Shi, *Adv. Synth. Catal.*, 2012, **354**, 783.
197 X.-n. Zhang, H.-P. Deng, L. Huang, Y. Wei and M. Shi, *Chem. Commun.*, 2012, **48**, 8664.
198 F. Hu, Y. Wei and M. Shi, *Tetrahedron*, 2012, **68**, 7911.
199 R. Zhou, J. Wang, C. Duan and Z. He, *Org. Lett.*, 2012, **14**, 6134.
200 F. Zhong, X. Han, Y. Wang and Y. Lu, *Angew. Chem. Int. Ed.*, 2011, **50**, 7837.
201 C. Hu, Z. Geng, J. Ma, Y. Huang and R. Chen, *Chemistry – An Asian Journal*, 2012, **7**, 2032.
202 Y. Wang, L. Liu, T. Zhang, N.-J. Zhong, D. Wang and Y.-J. Chen, *J. Org. Chem.*, 2012, **77**, 4143.
203 A. Duriš, A. Daïch and D. Berkeš, *Synlett*, 2011, **2011**, 1631.
204 P. Xie, Y. Huang and R. Chen, *Chem. Eur. J.*, 2012, **18**, 7362.
205 S. S. Maigali, M. H. Arief, M. El-Hussieny and F. M. Soliman, *Phosphorus, Sulfur Silicon Relat. Elem*, 2012, 187.
206 B. Tan, N. R. Candeias and C. F. Barbas, *J. Am. Chem. Soc.*, 2011, **133**, 4672.
207 A. Ramazani and A. R. Kazemizadeh, *Curr. Org. Chem.*, 2011, **15**, 3986.

208 A. Hassanabadi, M. H. Mosslemin, S. Salari and M. M. Landi, *Synth. Commun.*, 2011, **42**, 2309.
209 S. Asghari and M. Osia, *Phosphorus, Sulfur Silicon Relat. Elem.*, 2012, **187**, 1195.
210 S. Asghari, A. Salimi and V. Taghipour, *Phosphorus, Sulfur Silicon Relat. Elem.*, 2012, **187**, 669.
211 J. C. Deng and S. C. Chuang, *Org. Lett.*, 2011, **13**, 2248.
212 J.-C. Deng, F.-W. Chan, C.-W. Kuo, C.-A. Cheng, C.-Y. Huang and S.-C. Chuang, *Eur. J. Org. Chem.*, 2012, **2012**, 5738.
213 R. A. Aitken, K. M. Aitken, S. Lambert, R. Playfair and N. J. Wilson, *Heterocycles*, 2012, **84**, 1113.
214 L. S. Boulos, H. A. Abdel-Malek, N. F. El-Sayed and M. E. Moharam, *Phosphorus, Sulfur Silicon Relat. Elem.*, 2012, **187**, 225.
215 L. S. Boulos, H. A. Abdel-Malek, N. F. El-Sayed and M. E. Moharam, *Phosphorus, Sulfur Silicon Relat. Elem.*, 2012, **187**, 697.
216 J. Grodner and T. Sałaciński, *Synthesis*, 2012, **44**, 3071.
217 A. Molina-Ontoria, R. García, A. Gouloumis, F. Giacalone, M. R. Torres and N. Martín, *Eur. J. Org. Chem.*, 2012, **2012**, 3581.
218 T. Wang, L.-T. Shen and S. Ye, *Synthesis*, 2011, **2011**, 3359.
219 S. Xu and Z. He, *Chin. J. Org. Chem.*, 2012, **32**, 1159.
220 S. Xu, R. Chen and Z. He, *J. Org. Chem.*, 2011, **76**, 7528.
221 K. Prajapati and A. Varshney, *Asian J. Chem.*, 2011, **23**, 2361.
222 K. Prajapati and A. Varshney, *International Journal of Organic Chemistry*, 2011, **1**, 57.
223 C. Chen, X. Shao, K. Yao, J. Yuan, W. Shangguan, T. Kawaguchi and K. Shimazu, *Langmuir*, 2011, **27**, 11958.
224 D. Gau, A. Baceiredo and T. Kato, *Actualite Chimique*, 2012, **369**, 24.
225 R. Azhakar, S. P. Sarish, H. W. Roesky, J. Hey and D. Stalke, *Organometallics*, 2011, **30**, 2897.
226 D. Gau, R. Rodriguez, T. Kato, N. Saffon-Merceron, A. de Cozar, F. P. Cossio and A. Baceiredo, *Angew. Chem. Int. Ed.*, 2011, **50**, 1092.
227 S. Inoue, W. Wang, C. Präsang, M. Asay, E. Irran and M. Driess, *J. Am. Chem. Soc.*, 2011, **133**, 2868.
228 M. Asay, S. Inoue and M. Driess, *Angew. Chem. Int. Ed.*, 2011, **50**, 9589.
229 M. Karni and Y. Apeloig, *Organometallics*, 2012, **31**, 2403.
230 S. J. Sabounchei, M. Ahmadi and P. Shahriary, *Asian J. Chem.*, 2012, **24**, 1333.
231 Y. Nakajima and F. Ozawa, *Organometallics*, 2012, **31**, 2009.
232 D.-W. Wan, Y.-S. Gao, J.-F. Li, Q. Shen, X.-L. Sun and Y. Tang, *Dalton Trans.*, 2012, **41**, 4552.
233 M. R. Crimmin and A. J. P. White, *Chem. Commun.*, 2012, **48**, 1745.
234 K. G. Fowler, S. L. Littlefield, M. C. Baird and P. H. M. Budzelaar, *Organometallics*, 2011, **30**, 6098.
235 R. Hussain, K. G. Fowler, F. Sauriol, M. C. Baird and P. H. M. Budzelaar, *Organometallics*, 2012, **31**, 6926.
236 B. K. Keitz and R. H. Grubbs, *J. Am. Chem. Soc.*, 2011, **133**, 16277.
237 M. Inoue, T. Yoneda, K. Youfu, N. Aratani and A. Osuka, *Chem. Eur. J.*, 2011, **17**, 9028.
238 K. Karami and M. Salah, *Transition Met. Chem.*, 2011, **36**, 363.
239 S. J. Sabounchei, F. A. Bagherjeri, A. dolatkhah, J. Lipkowski and M. Khalaj, *J. Organomet. Chem.*, 2011, **696**, 3521.
240 A. Y. Khalimon, E. M. Leitao and W. E. Piers, *Organometallics*, 2012, **31**, 5634.

241 P. A. Dub, A. Béthegnies, J.-C. Daran and R. Poli, *Organometallics*, 2012, **31**, 3081.
242 R. C. Klet, J. A. Labinger and J. E. Bercaw, *Organometallics*, 2012, **31**, 6652.
243 D. Y. Melgarejo, G. M. Chiarella and J. P. Fackler, *Organometallics*, 2011, **30**, 5374.
244 K. Karami and O. Buyukgungor, *J. Chem. Crystallogr.*, 2012, **42**, 38.
245 S. Fortier, J. R. Walensky, G. Wu and T. W. Hayton, *J. Am. Chem. Soc.*, 2011, **133**, 6894.
246 S. Fortier, N. Kaltsoyannis, G. Wu and T. W. Hayton, *J. Am. Chem. Soc.*, 2011, **133**, 14224.
247 S. J. Sabounchei, M. Sarlakifar, S. Salehzadeh, M. Bayat, M. Pourshahbaz and H. R. Khavasi, *Polyhedron*, 2012, **38**, 131.

Nucleotides and oligonucleotides: mononucleotides

Yao-Zhong Xu[*a] and Raman Narukulla[b]

DOI: 10.1039/9781782623977-00117

Since the publication of the double-helical DNA structure by Watson and Crick in 1953, research on DNA and related subjects has never paused. In fact, surprising findings emerge every so often, and even today more and more exciting discoveries and applications are being made. 2013 marks the 60th anniversary of that milestone. This chapter will cover some research papers published in 2012 regarding the chemical synthesis and biological, particularly medicinal, applications of nucleotides and oligonucleotides.

Nucleotides are the building blocks of nucleic acids and play vital roles in many biological processes. A nucleotide comprises a nitrogen-containing unit (base) linked to a sugar (ribose or deoxyribose) and a phosphate group. Each of the four bases in a nucleotide has its distinctive role in metabolism and a molecule similar to these compounds could be used as a potential therapeutic agent. Indeed, chemically modified nucleosides and nucleotides can act as therapeutic agents in the process of nucleic acid metabolism. For instance, the first anti-cancer drug methotrexate acted on both thymidylate synthase and on *de novo* purine synthesis. The chemistry of these analogues has now been well established and led to effective drugs such as anti-cancer and anti-viral agents. The importance of these moieties has created great interest and resulted in the formation of a new field of chemistry of nucleic acid components, *i.e.*, modified nucleotides and oligonucleotides. Relevant research articles on this subject will be summarised and discussed in the sections below.

1 Nucleoside monophosphates

Nucleoside monophosphates, also called mononucleotides, are moieties consisting of one (occasionally two) nucleosides and a single phosphate. They are generally prepared from mono-phosphorylation of nucleosides.

Modified nucleoside monophosphates. The number of naturally occurring nucleotides is relatively few. However, synthetic chemistry is now capable of offering an unlimited number of modified nucleoside monophosphates for biological studies or medicinal exploitations. Based upon their chemical moieties (base, sugar and phosphate), it is natural for this paper to discuss all of the three corresponding modifications. In this section the focus will be on base-modified and sugar-modified mononucleotides. Phosphate-modified mononucleotides will be dealt with in Sections 2 and 3.

---

[a]*Department of Life, Health and Chemical Sciences, the Open University, Walton Hall, Milton Keynes, MK7 6AA, UK. E-mail: yao.xu@open.ac.uk*
[b]*Argenta, a Galapagos Company, 8-9 Spire Green Centre, Flex Meadow, Harlow, Essex, CM19 5TR, UK*

1.1 Base-modified nucleoside monophosphates

Since the bases encode genetic information, any base modification would have significant biological consequences and offer great synthetic opportunities for novel mononucleotides with useful chemical or physical properties. For instance, base-modified nucleotides are reported as potent medicinal or diagnostic agents.

1.1.1 Modifications to the atoms not directly involved with hydrogen-bonding.
As the bases are involved in base-pairings through H-bonding, certain modifications are designed to alter the base-pairing while others have only a minimal effect. For the latter, modification at the 5-position of pyrimidines (particularly thymine) is often the choice. Kore et al.[1] reviewed recent developments in the synthesis and applications of C5-substituted pyrimidine nucleosides and nucleotides. These compounds have potential therapeutic applications. The attachment of a suitable functional group onto the 5-position of a pyrimidine is commonly accomplished *via* a Suzuki reaction. The Suzuki reaction (also called Suzuki-Miyaura reaction or Suzuki coupling) is a widely used reaction in which an aryl- or vinyl-boronic acid (R-BY$_2$) reacts with an aryl- or vinyl-halide or iodo-nucleoside (R$_1$-X) in the presence of a palladium complex to synthesize poly-olefins, styrenes, substituted biphenyls as well as modified nucleosides and nucleotides (R-R$_1$) as illustrated in Scheme 1.

Fresneau et al.[2] described the synthesis of unprotected C5-aryl/(hetero)aryl-2′-deoxyuridine (**1**) *via* a Suzuki reaction in aqueous media in the presence of palladium acetate and triphenylphosphine. Although the reaction took place at the nucleoside level, it is worth noting that the reaction in aqueous solution is particularly advantageous as it can be exploited at the nucleotide and oligomer level. Similarly Lalut et al.[3] reported the synthesis of C3-arylated-3-deazauridine derivatives (**2**) using Suzuki coupling conditions. These compounds showed anti-herpetic activity against HSV-1, with activity comparable to acyclovir.

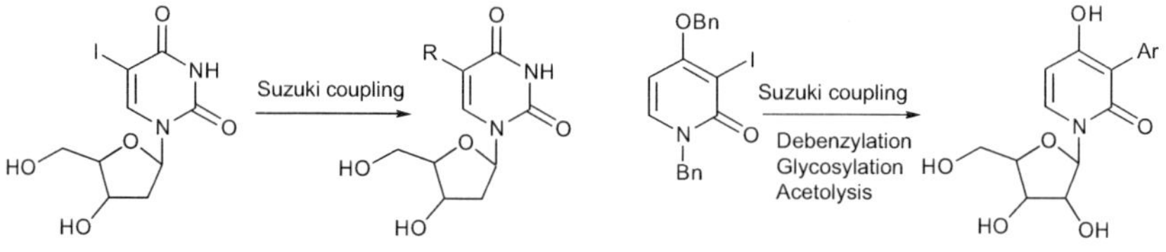

1a: R = aryl
1b: R = heteroaryl

2: Ar = Substituted Aryl

Kalachova et al.[4] reported the synthesis of nucleoside mono- and triphosphates bearing oligo-pyridine ligands. The synthesis involves a

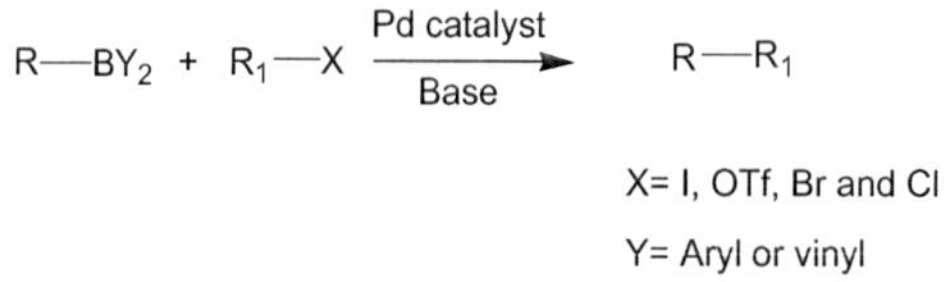

X= I, OTf, Br and Cl
Y= Aryl or vinyl

Scheme 1 Suzuki coupling reaction.

single-step Suzuki coupling of 5-iodo-dCMP (or -dCTP) (**3**) and 7-iodo-7-deaza-dAMP (or -dATP) (**4**) with the corresponding bipyridine- or terpyridine-linked acetylenes. As with the 5-position of pyrimidines, the 7-position (*i.e.* 7-deaza) of purines is also frequently used to attach a functional group onto purine ring. These modified nucleoside triphosphates were successfully incorporated into oligonucleotides by primer extension experiments using different DNA polymerases and the resultant products were used for post-synthetic complexation with Fe^{2+}.

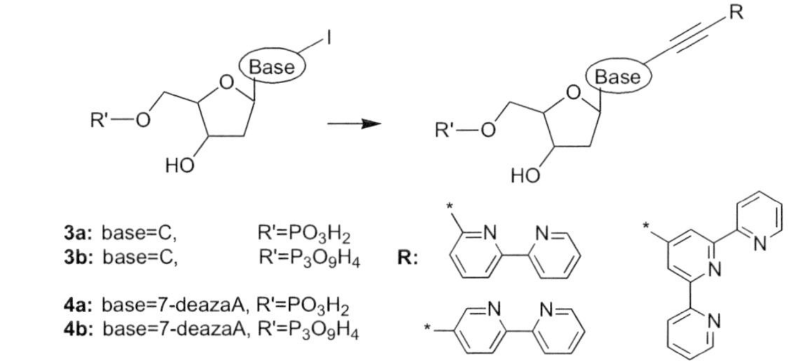

3a: base=C,   R'=PO_3H_2
3b: base=C,   R'=$P_3O_9H_4$
4a: base=7-deazaA, R'=PO_3H_2
4b: base=7-deazaA, R'=$P_3O_9H_4$

Raindlova *et al.*[5] reported the synthesis of aldehyde-linked nucleotides (**5a**), triphosphate (**5b**), oligomers and their bio-conjugations with lysine and peptides through reductive amination as shown in Scheme 2.

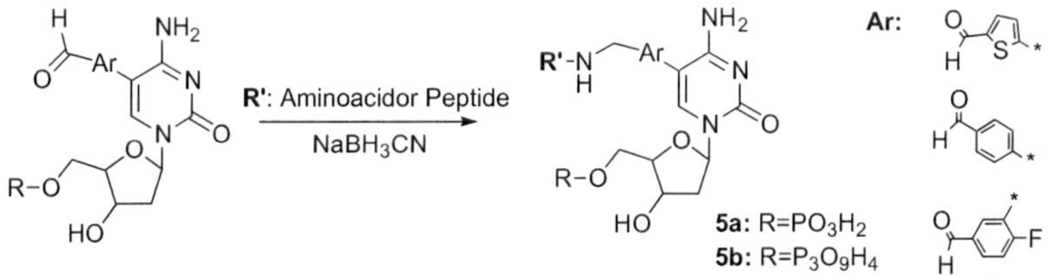

R': Aminoacid or Peptide
$NaBH_3CN$

5a: R=PO_3H_2
5b: R=$P_3O_9H_4$

Scheme 2 Reductive amination with aminoacids or peptides.

Nucleoside and nucleotide analogues have been widely used in the treatment of various diseases. Koegler *et al.*[6] reported the synthesis of 6-aza-2'-deoxyuridine monophosphate analogues (**6**). These compounds can inhibit thymidylate synthases, and also act as substrates or inhibitors of thymidine monophosphate kinase in *mycobacterium tuberculosis*.

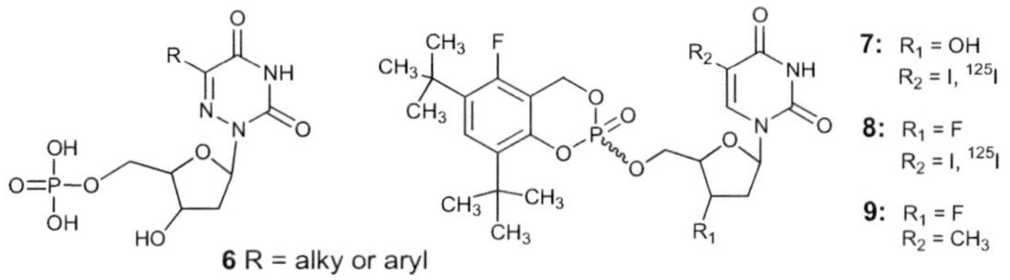

6 R = alky or aryl

7: R_1 = OH
     R_2 = I, ^{125}I
8: R_1 = F
     R_2 = I, ^{125}I
9: R_1 = F
     R_2 = CH_3

Kortylewicz *et al.*[7] reported the synthesis of *cyclo*-saligenyl monophosphates of radio-labelled 5-iodo-2'-deoxyuridine (**7**), 5-iodo-3'-fluoro-2',3'-dideoxyuridine (**8**), and 3-fluorothymidine (**9**). The synthesised diastereomers have been well separated on HPLC. All these ^{125}I-radio-labelled compounds are suitable for imaging and molecular radiotherapy. Shen et al.[8] reported the total synthesis of triciribine (trycyclic nucleoside) and its monophosphate (**10**), both of which were showed antineoplastic activity. Their bioavailability and antineoplastic activity were significantly improved by linking with amino acids.

1.1.2 Modifications to the atoms directly involved with hydrogen-bonding.
Kopina et al.[9] reported a novel and efficient preparation of 2,4-diaminopyrimidine nucleosides from cytidine in six steps with good yields. The reagents LiCl and DBU (1,8-diazabicyclo[5,4,0]undec-7-ene) used in the amine addition step allowed the generation of a single tautomeric product. By utilising this route, the modified tRNA nucleosides lysidine (**11**) and agmatidine (**12**) have been prepared.

Lakshman et al.[10] reported the synthesis and biological properties of C2 triazolylinosine derivatives including **13** and **14**. The chemistry involves O^6-(benzotriazol-1H-yl)guanosine and its 2′-deoxy analogue being converted to the O^6-allyl derivatives followed by diazotization, cyclization and removal of protecting groups which then afforded C2 triazolylinosine derivatives. These derivatives showed modest inhibitory activity against cytomegalovirus.

Martín-Ortíz et al.[11] reported the preparation of purine-derived metallo-nucleosides (**15a**), metallo-nucleotides (**15b**), and metallo-dinucleotides containing Ir and Rh. This method would be applicable for the preparation of functionalized oligonucleotides.

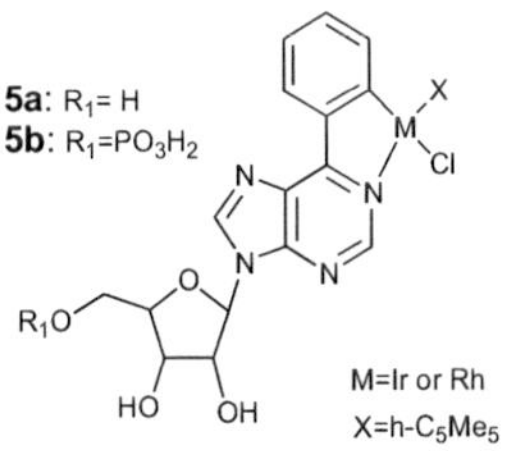

15a: R_1= H
15b: R_1=PO_3H_2

M=Ir or Rh
X=h-C_5Me_5

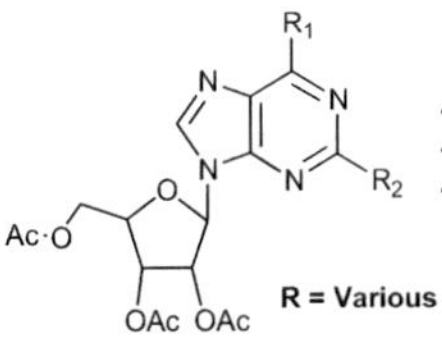

16: R_1 = Cl, R_2 = R
17: R_1 = R, R_2 = Cl
18: R_1 = R, R_2 = R

R = Various amides, carbamates

Bosch *et al.*[12] reported the Pd-catalysed amidation of 2,6-dihalopurine nucleosides (**16–18**). The authors compared the reactivity of 2-Cl, 2-Br and 2-I derivatives of a 6-chloropurine ribose-protected nucleoside with various amides and carbamates, under Buchwald conditions, between 0 and 110 °C. The reactivity order was 2-I > 2-Br > 6-Cl ≫ 2-Cl. The 2-iodo substituent could be replaced even at 0 °C while the replacement of 2-Cl substituent required 110 °C. Adamska *et al.*[13] reported the synthesis of N^6-derivatives of adenine and N^4-substituted derivatives of cytosine and their 2-deoxyribosides using unprotected nucleobases (adenine and cytosine) or unprotected 2-deoxyribosides with high yields.

1.1.3 Other base-modifications. Hassan and co-workers[14] reported a novel synthesis of pyrazolone nucleosides and their antimicrobial activities. Mata and his group[15] reported the stereoselective *N*-glycosylation of 2-deoxythioribosides. This facile synthesis (shown in Scheme 3) involves *N*-2-deoxyribosylation of a modified nucleobase with 2-deoxythioriboside in the presence of bis-(trimethylsilyl)acetamide, *N*-iodosuccinimide and trimethylsilyl trifluoromethanesulfonate to afford stereoselective nucleosides (**19**) in excellent yields.

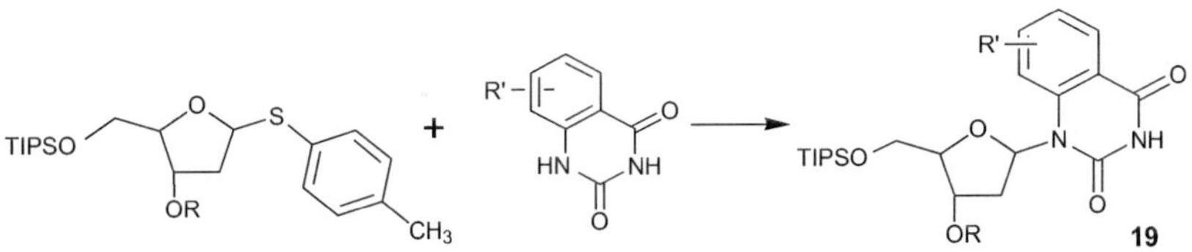

19

TIPS = triisopropylsilyl
R, R' = substituents

Scheme 3 Stereoselective N-glycosylation of 2′-dexoythioribosides.

Parmenopoulou *et al.*[16] reported the synthesis of triazole pyrimidine nucleosides (**20**). The triazolo group between the sugar and the base has an important role in efficient inhibition of Ribonuclease A.

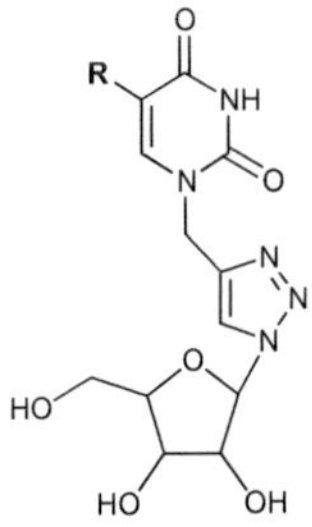

20a: R = H
20b: R = CH_3
20c: R = F
20d: R = Cl
20e: R = Br

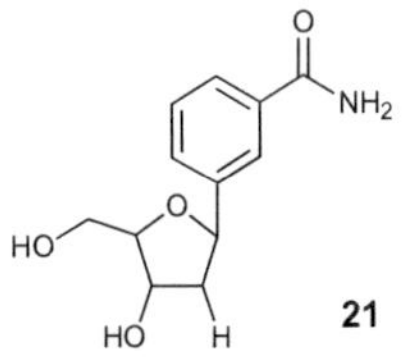

21

Midtkandal et al.[17] reported a facile and new synthetic route to a C-nucleoside, 2-deoxy benzamide riboside (**21**). The synthesised target was found to be active against a number of cancer cell lines. Similar ribose-based C-nucleosides are known as antiviral and antiproliferative drugs, such as formycin, tiazofurin and benzamide riboside.

1.2 Sugar-modified nucleoside monophosphates

Common modifications to the sugar are those in which a hydrogen atom or hydroxyl group is replaced by other atoms or groups. A fluorine atom is frequently used in this case, since the size of F is similar to that of H and the electronegativity of F is close to that of the O atom of an OH group.

1.2.1 Modification by fluorine atom.

Viral diseases have caused many problems for humans, but have also provoked a search for new antiviral agents. As a result of this, a number of modified nucleosides have been designed and synthesized. Fluorinated nucleosides, such as sugar-fluorinated nucleosides (**22**) and base-fluorinated nucleosides (**23**), are an important type of modified nucleosides and used as antiviral drugs.[18] In the cases of **22b–d**, the fluorine atom is used to replace and mimic an OH group in the sugar due to their similar electronegativities and bond lengths (C–F bond length 1.35 Å and C–O bond length 1.43 Å). The fluorine atom also mimics hydrogen (**22a** and **23**) without causing much structural deviation. These modified nucleosides can be converted to their corresponding mononucleotides for other medicinal purposes.

Fluorinated derivatives of Tubercidin

Perlikova et al.[19] synthesised a number of 2'-deoxy-2'-fluoro-ribonucleosides (**24a**) and 2'-deoxy-2',2'-difluororibonucleosides (**24b**) derived from 6-(het)aryl-7-deazapurines using palladium catalysed cross-coupling reactions.

R = aryl group
24a: R_1 = H; **24b**: R_1=F

1.2.2 Modification by other atoms or groups.
Besides fluorine atoms, other atoms (such as C, N and O) and groups are also used to produce novel sugar-modified nucleoside monophosphates. 2-C-Methyl nucleosides are potential candidates with activity against Hepatitis C virus. Leisvuori et al.[20] prepared 3,5-cyclic phosphates (**25a**) and 3,5-cyclic thiophosphates (**25b**) of 2-C-methyl ribonucleosides as potential antiviral drugs.

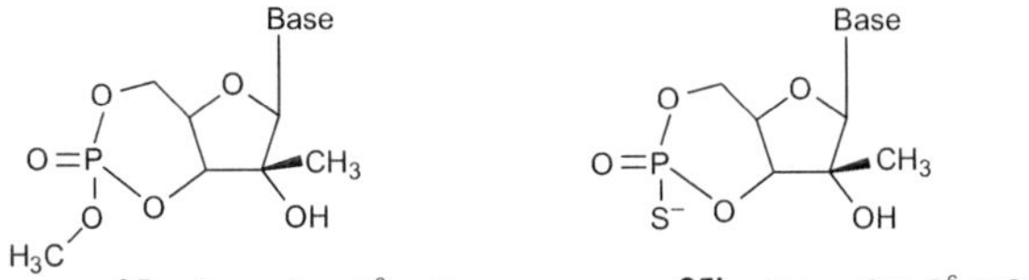

25a Base = G or O^6-meG

25b Base = G or O^6-meG

Kumar et al.[21] reported the synthesis of pyrene-modified nucleotides and their effects in secondary nucleic acid structures. Pyrene was attached to the 5′-position of thymidine (**26a**) or to the 2′-position of 2′-deoxyuridine via a triazolomethylene linker (**26b**), or via a triazole linker (**26c**). These nucleosides were converted into their phosphoramidites and then incorporated into oligonucleotides, and analysed by thermal stability and fluorescence studies. These experiments indicated that these oligonucleotides can act as specific recognition probes.

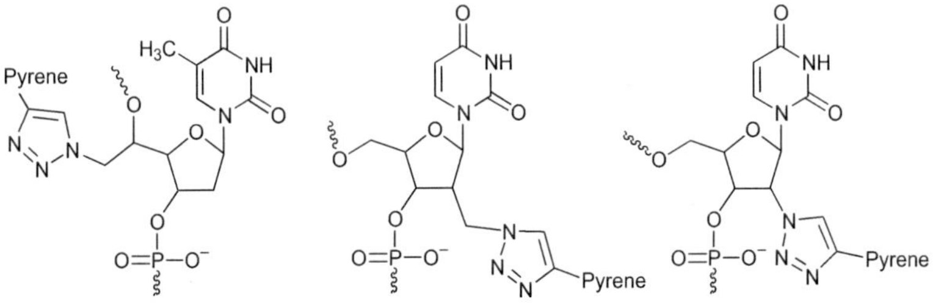

Casu et al.[22] reported the synthesis of 2′-substituted inosine analogues (**27**) by masking the 6-hydroxyl group. The corresponding analogues were prepared from 6-chloropurine riboside via 6-dimethylaminopurine or 6-benzyloxypurine intermediates.

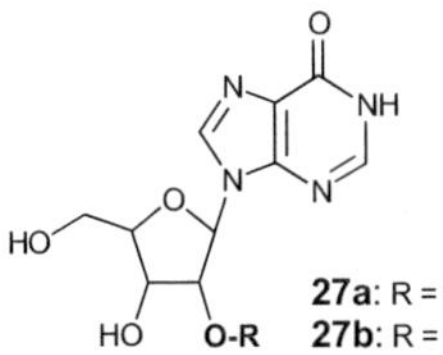

27a: R = CH$_2$COOH
27b: R = C(O)CH$_2$CH$_2$COOH

Cui et al.[23] reported the synthesis of α-thymidine and β-thymidine analogues. The authors prepared and optimised a series of 5′-urea-α- and β-thymidine analogues. 5′-urea-α-thymidines (**28**) have particularly potent antimalarial activity. These findings may be useful for future structure-based drug discovery programmes.

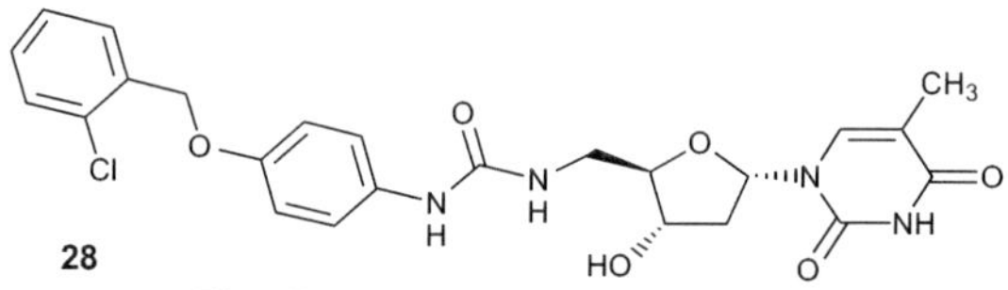

28

EC$_{50}$: 28 nm

Gotkowska et al.[24] reported the synthesis of novel isoxazolidine analogues of homonucleosides (**29** and **30**). The facile synthesis and the potential biological activity of isoxazolidine nucleoside analogues may open a new window to optimise lead molecules in this series.

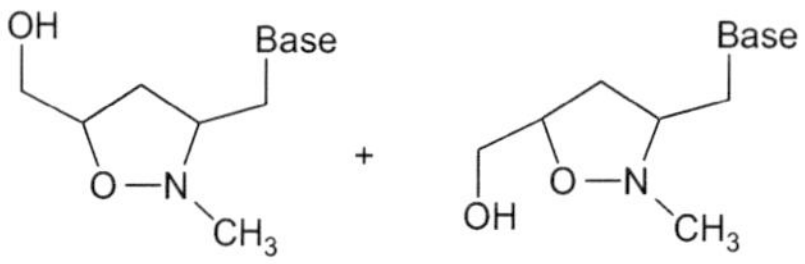

29   **30**

Kiritsis et al.[25] synthesised a new series of pyrimidine pyranonucleosides (**31**). These compounds have a cyano group on the pyranose moiety which results in a significant change in biological activity and physical properties. The synthesised compounds with cyano substitution showed antiviral and cytostatic activities in cell culture.

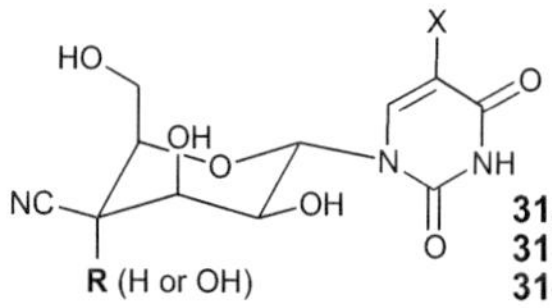

31a: X=H
31b: X=F
31c: X=CH$_3$

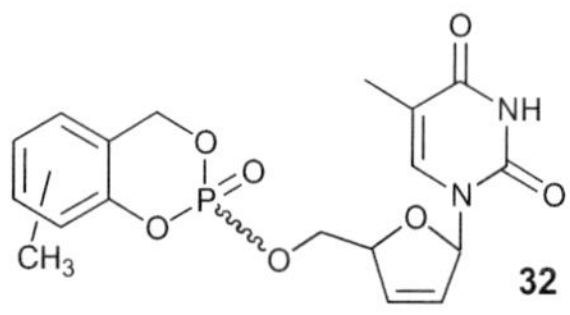

32

Morales et al.[26] described the synthesis of methyl-substituted *cyclo*-Salpronucleotides of stavudine (d4T) monophosphates (**32**) with a high stereo-selectivity and in good yields. Each diastereoisomer was tested against HIV and showed significant antiviral activity. Interestingly all *S*-configurated diastereomers showed a higher antiviral activity against HIV-1 and HIV-2 in wild-type cells and in mutant thymidine kinase-deficient cells than the *R*-configurated counterparts.

Locked nucleic acids (LNA) are a modified RNA in which the ribose moiety is modified with an extra bridge connecting the 2′-O and 4′-C atoms. Thus LNA nucleosides can be regarded as sugar-modified nucleosides. Madsen et al.[27] synthesized the 2′-amino-LNA phosphoramidites of thymidine (**33a**) and 5-methylcytidine (**33b**) on a large scale with high yields.

33a: Base = thymine
33b: Base = 5-methylcytosine

Schmucker et al.[28] reviewed the importance of chromophores in DNA functionalization. These chromophores can be DNA bases or their substitutes. The authors summarized the synthetic chemistry published over the last ten years through which they have modified DNA with various chromophores, fluorescent probes, and metal-ion ligands by using three different approaches, i.e., replacement of DNA bases, modifications of DNA bases, and sugar modifications at the 2′-position via either the phosphoramidite chemistry or a post-synthetic modification. These modifications could play a useful role in the development of functional nanomaterials.

1.3 Doubly-modified nucleoside monophosphates

Modifications are not limited to the single moiety of either the base or the sugar. Sometimes both moieties are modified to produce a novel nucleoside monophosphate with new properties. A couple of examples are described as follows. Naus et al.[29] reported a series of base- and sugar-modified derivatives of 7-(hetero)aryl-7-deazaadenosines including 2′-C-methylribonucleosides (**34a**), 2′-deoxy-2′-fluoroarabinonucleosides (**34b**), arabinonucleosides (**34c**) and 2′-deoxyribonucleosides (**34d**). The syntheses involve Suzuki couplings of sugar-modified 7-iodo-7-deazaadenine nucleosides with (hetero)arylboronic acids or Stille couplings with (hetero)arylstannanes. These compounds showed anti-HCV activities, sub-micromolar anti-proliferative activitiy and cytostatic effects.

R = Aryl groups
34a: R_1 = CH_3, R_2 = OH
34b: R_1 = F, R_2 = H
34c: R_1 = OH, R_2 = H
34d: R_1 = H, R_2 = H

Di Francesco et al.[30,31] synthesised a series of 2′-C-methyl-ribose nucleosides and triphosphates containing the nucleobases 7-aryl-7-deazaadenine (**35**) and 7-(hetero)aryl-7-deazaadenine (**36**) and examined their inhibitory properties towards Hepatitis C NS5B polymerase. These analogues showed IC_{50} values of 0.1 µM in cellular assays and this potency is similar to the current lead molecules.

35a: R=H
35b: R=P$_3$O$_9$H$_4$

36a: R=H
36b: R=P$_3$O$_9$H$_4$

2 Nucleoside phosphoramidites and phosphoramidates

2.1 Nucleoside phosphoramidites

Nucleoside phosphoramidites are the monomers of choice when carrying out the chemical synthesis of oligonucleotides by an automated synthesizer. Various synthetic methods discussed above for modified, particularly base-modified, nucleoside monophosphates can be adapted to produce their corresponding phosphoramidites.

2.1.1 Base-modified nucleoside phosphoramidites. Szombati et al.[32] described the synthesis of C8-arylamine-modified 2′-deoxyadenosine phosphoramidites (**37**), their incorporation into oligonucleotides and their use as a biological probe. The key step in the synthesis of these compounds involves a palladium-catalyzed C–N cross coupling.

Yaren et al.[33] described the synthesis of carbocyclic *exo*-amino nucleoside phosphoramidites (**38**) and the incorporation of them into oligomers. Their base-pairing properties were investigated and the results showed such analogues engage in specific interactions with natural bases in DNA duplexes. In addition they can form self-pairs that match or exceed the stability of Watson–Crick base-pairs. Therefore, these nucleosides may have applications in biotechnology.

Dziuba et al.[34] synthesised a universal nucleoside with strong two-band switchable fluorescence. This nucleoside analogue (**39a**) contains 3-hydroxychromone as a fluorophore and as a nucleobase. The nucleoside has been converted into its phosphoramidite (**39b**) and incorporated into oligonucleotides. This molecule contains good environmental sensitivity switching between two fluorescence bands without altering the conformation and exhibits high fluorescence properties.

39a: R₁ = H, R₂ = H
39b: R₁ = DMT, R₂ = −P(N(iPr)₂)(O−CH₂CH₂CN)

Barrois et al.[35] reported the synthesis of 5-diarylethene-2′-deoxyuridines (**40**) and their phosphoramidites and the incorporation of these monomers into oligomers. Photochemical studies showed that nucleosides (**40** and **41**) are photochemically stable and the switching of optical properties in DNA can be achieved selectively at 310 nm (forward) and 450 nm (backward).

40a: R = Br
40b: R = alkynyl

41a: R = Br
41b: R = alkynyl

Ibarra-Soza et al.[36] reported the synthesis of new 7-substituted 8-aza-7-deaza-adenosine ribonucleoside phosphoramidites (**42**). The authors described various 7-position modifications of deazaadenosine and used these in generating major groove-modified duplex RNAs. The analogues with 7-position modification maintain base pairing specificity.

43: n=0; R=CH₃
44: n=2; R=H

Harusawa et al.[37] synthesised C5-linked C₀- and C₂-ribonucleoside phosphoramidites via tetrazole C-nucleosides for probing RNA catalysis. The nitrogen in the tetrazole was protected by methyl-substituted pivaloyloxymethyl (**43**) and pivaloyloxymethyl (**44**) groups which can easily be removed under basic conditions.

McManus et al.[38] synthesised O^4-alkyl-2′-deoxythymidine phosphoramidites (**45** and **46**) to produce a self-complementary duplex containing a site-specific interstrand cross-link which was used for probe recognition and repair by O^6-alkylguanine DNA alkyltransferases.

Tarashima et al.[39] described a post synthetic modification of oligodeoxynucleotides. 4,7-Diaminoimidazo-[5′,4′:4,5]-pyrido[2,3-d]pyrimidine (**47**) was incorporated into oligomers using phosphoramidite chemistry. The p-bromobenzoyl group was used as a protecting group for the 4-amino group. After synthesis, the controlled pore glass (CPG) support bearing the oligomer was treated with ammonium hydroxide at 55 °C to remove the protecting groups and convert the 7-chloro group into the required amino group, giving an oligomer containing the modified base (**48**).

2.1.2 **Sugar-modified nucleoside phosphoramidites.** Similar to base-modified nucleoside monophosphates, modifications can also be made to the sugar moiety of their phosphoramidites. In the following two examples, one is to add a functional group on 2′-O atom of ribose sugar, while the other is to use a different sugar instead of ribose. Brzezinska et al.[40] synthesised phosphoramidites of 2′-O-guanidinopropyl-modified nucleoside (**49**) and incorporated them into siRNAs. During chemical synthesis, the guanidinopropyl group (in Boc-protected form) was used as the protecting group for the 2′-OH group. Biological studies indicated that modification of siRNAs with guanidinopropyl groups may be useful for the creation of siRNA activators targeting the hepatitis B virus.

Seth and co-workers[41] reported the synthesis of 2′-fluorocyclohexenyl pyrimidine phosphoramidite (**50**) in multigram quantities using methyl-D-mannose pyranoside and carried out structural and biological evaluation of modified oligonucleotides. Biophysical evaluation showed that the above modified nucleotide behaves similarly to a 2′-modified nucleotide and that the mannose-containing oligonucleotides are more stable towards digestion by snake venom phosphodiesterases when compared to unmodified DNA and 2′-fluoro RNA, and thus would be useful for antisense applications.

2.2 Nucleoside phosphoramidates

Nucleoside phosphoramidates are the oxidised form of nucleoside phosphoramidites, but they are designed for different purposes. Nucleoside phosphoramidites are very reactive and used as the monomers for chemical synthesis of oligomers while nucleoside phosphoramidates are relatively stable and used primarily as prodrugs since the amidate moiety is often constructed by linking with a natural amino acid to enhance the cellular uptake of a phosphoramidate which is then hydrolysed to generate the desired nucleoside monophosphate. Cho et al.[42] developed an efficient method for the synthesis of exo-N-carbamoyl nucleosides and their phosphoramidate prodrugs (51) in excellent yields.

Meneghesso et al.[43] reported the synthesis of pyrimidine nucleoside phosphoramidate prodrugs (52) which showed improved antiviral activity against influenza virus. Jain et al.[44] developed a one-step synthesis of cyclic pronucleotide derivatives of 5-fluoro-2′-deoxyuridine using a novel phosphoramidating reagent (bis(4-nitrophenyl)-methoxyglycinyl phosphoramidate). These compounds (53) were found to be active against cancer cells. This chemistry would be applicable to other nucleoside analogues for modification of the 5′- and 3′-OH groups. It is worth pointing out that the phosphorus atom in 53 is a chiral centre; once hydrolysed the cyclic phosphoramidate becomes a phosphate and the chirality on the phosphorus no longer exists.

3 Nucleoside phosphonates

Nucleoside phosphonates are a type of nucleotide in which the sugar is linked with a phosphorus atom *via* a stable P–C bond instead of a hydrolysable P–O bond. The moiety containing such a carbon-bonded phosphorus atom is called a phosphonate which is very different from the phosphate within natural nucleic acids and is highly resistant to enzymatic degradation. Nucleoside phosphonates, in particular their acyclic analogues, have been successfully developed as antiviral agents with potent and selective activity *in vitro* and *in vivo*. For instance, the drugs tenofovir, adefovir and cidofovir are some members of the nucleoside phosphonate family. Efforts are continuously being made to design and develop newer and better nucleoside phosphonate drugs against viruses or other infectious pathogens. However, the phosphonates are poorly cell-permeable; therefore suitable functional groups are often used to increase their bioavailability. Pertusati *et al.*[45] reviewed the medicinal chemistry of nucleoside phosphonate prodrugs for antiviral therapy and presented an overview of acyclic and cyclic nucleoside phosphonates acting as antiviral agents. Vertuani *et al.*[46] reported the synthesis and *in vitro* stability of nucleoside 5′-phosphonate derivatives (adenosine, inosine, guanosine and thioadenosine analogues). These derivatives were tested to investigate their inhibitory effects on platelet aggregation. The adenosine-5′-phosphonate derivatives were found to have relatively high *in vivo* stability.

For nucleoside phosphonates, modifications to the sugar are the common approach to altering chemical properties of the phosphonates. In terms of the absence or presence of the sugar ring, two such types of modifications are presented as follows.

3.1 Ring-opened: acyclic phosphonates

Brel[47] reported an efficient synthesis of a series of new purine and pyrimidine nucleotides containing acyclic phosphonate analogues with a triple or double bond (**54**) in the carbon skeleton.

Pradere *et al.*[48] described the synthesis of bis(pivaloyloxymethyl) prodrugs of (*E*)-[4′-phosphonobut-2′-en-1′-yl]purine nucleosides (**55**) using a one-step Mitsunobu reaction. The authors evaluated these compounds for their antiviral activities against a number of DNA and RNA viruses and some of the compounds showed submicromolar activity. Modification of the side chain or base in these analogues may lead to the discovery of new antiviral agents.

Hexadecyloxypropyl prodrugs of acyclic nucleoside phosphonates (**56** and **57**) were also synthesized and evaluated for their antiviral activities.[49] Among them, **57b** was found most potent against DNA viruses, but none of these compounds to be inhibitory against RNA viruses. The highly lipophilic hexadecycloxypropyl moiety is believed to enhance antiviral activities due to the better cellular permeability of the prodrug.

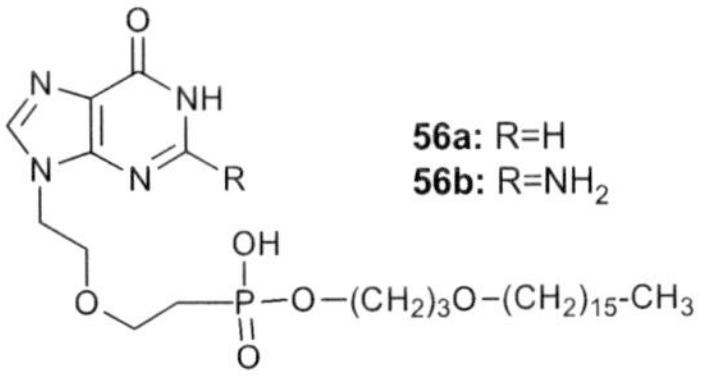

56a: R=H
56b: R=NH$_2$

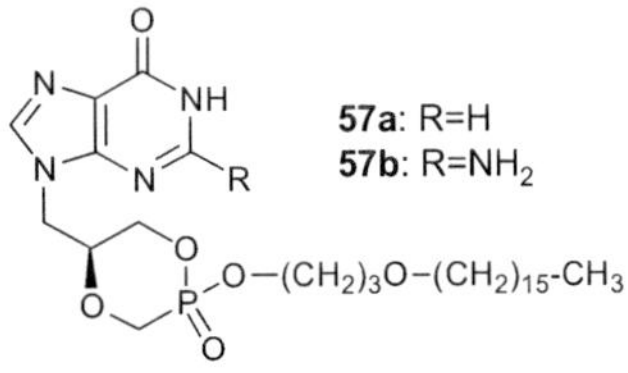

57a: R=H
57b: R=NH$_2$

Diab et al.[50] reported a series of fluorophosphonylated pyrimidine nucleosides (**58**). These analogues act as thymidine phosphorylase multisubstrate inhibitors.

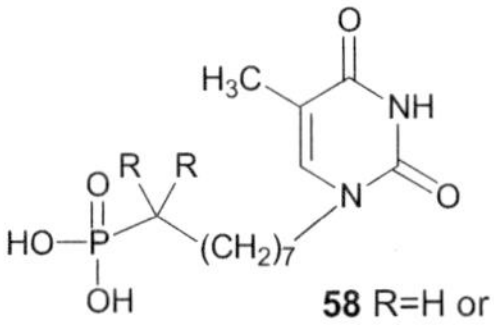

58 R=H or F

Cesnek et al.[51] reported the synthesis of a series of linear alkyl and alkyloxyalkyl phosphonates (9-phosphonoalkyl and 9-phosphonoalkoxyalkyl) purines (**59**). Studies showed that these compounds act as inhibitors of *Plasmodium falciparum*, *Plasmodium vivax* and human hypoxanthine–guanine–(xanthine) phosphoribosyltransferases.

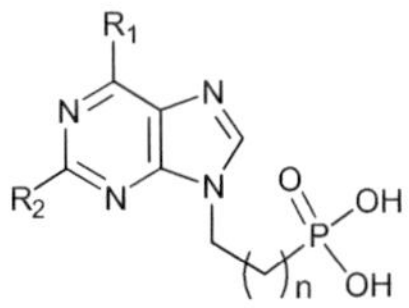

59a R$_1$ = OH, Cl, Br; R$_2$ = H, NH$_2$, OH; n = 0–6

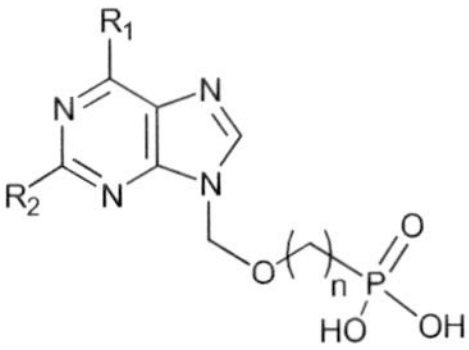

59b R$_1$ = OH, Cl; R$_2$ = H, NH$_2$, OH; n = 2–4

Hockova et al.[52] described the synthesis of novel N-branched acyclic nucleoside phosphonates (**60**). These analogues are potent inhibitors of human *plasmodium falciparum* and *plasmodium vivax* 6-oxopurine phosphoribosyltransferases.

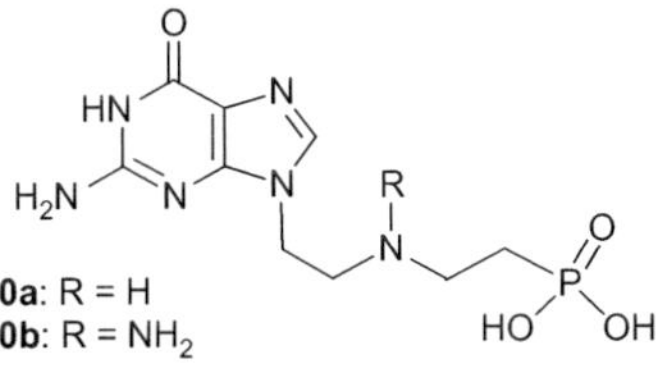

60a: R = H
60b: R = NH$_2$
R = Alkyl groups

3.2 Ring-retained nucleoside phosphonates

Pradere et al.[53] reported the synthesis of 5′-methylene-phosphonate furanonucleoside prodrugs. The synthesis involves a Horner–Wadsworth–Emmons reaction, starting with protected 5′-aldehydic nucleosides.

This chemistry was successfully applied to the synthesis of 2′-deoxy-2′-α-fluoro-2′-β-C-methyl nucleosides (**61**).

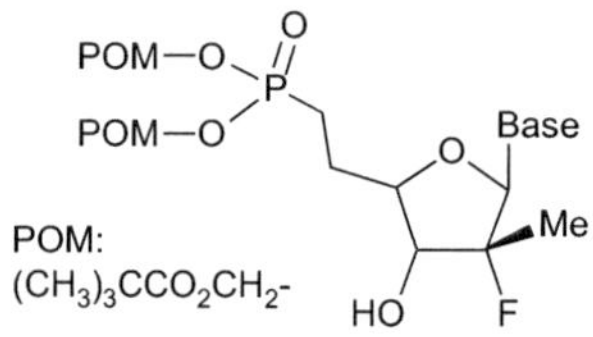

POM:
(CH$_3$)$_3$CCO$_2$CH$_2$-

61

Shen et al.[54] reported the synthesis and conformational study of novel 3′-branched threosyl-5′-deoxyphosphonic acid nucleoside analogues (**62**). The chemistry involves multiple steps starting from acetol. These analogues showed potent anti-HIV activities, demonstrating that 5′-deoxy versions of threosyl nucleoside phosphonates are worth further exploitation. They[55] also synthesised 3′-hydroxymethyl 5′-deoxythreosyl nucleoside phosphonate (**63**) and related analogues which showed potent anti-human immunodeficiency virus (anti-HIV) activity.

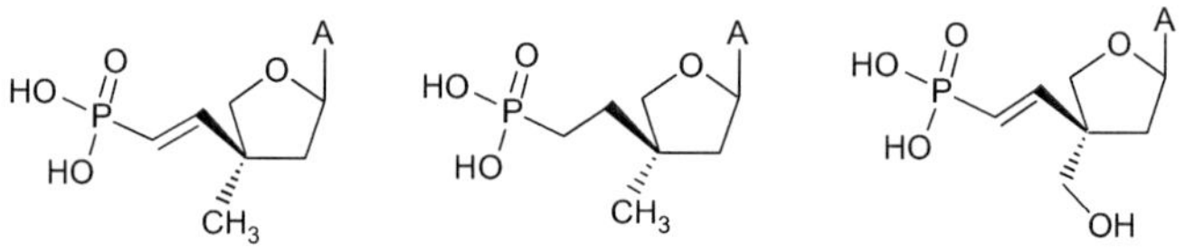

62a   **62b**   **63**

Kang et al.[56] reported an efficient synthetic route to 4′-fluorinated 5′-deoxythreosyl nucleoside phosphonates (**64**). The chemistry involves multiple steps starting from D-glyceraldehyde and mainly glycosylation of the nucleobases, phosphorylation and hydrolysis to yield the desired analogues. These analogues showed significant anti-HIV activity.

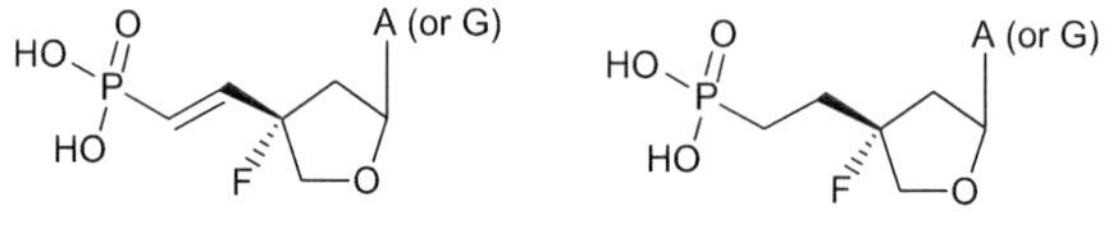

64a   **64b**

Hladezuk et al.[57] developed rhodium catalysed O–H insertion reactions (shown in Scheme 4) using α-diazophosphonates reacting with uridine, thymidine and protected adenosine derivatives for the preparation of

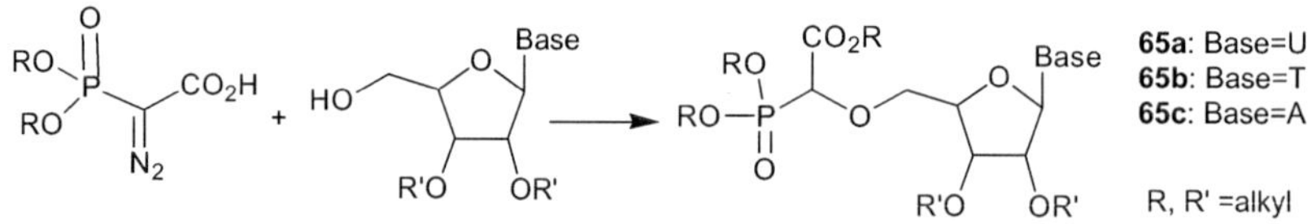

65a: Base=U
65b: Base=T
65c: Base=A

R, R′ =alkyl

Scheme 4

phosphononucleoside derivatives (**65**) containing a carboxylic acid moiety adjacent to the phosphonate group.

Van Poecke et al.[58] prepared a series of 5-(hetero)-aryl-modified nucleoside phosphonates (**66**) by an eight-step synthetic route involving a Wittig reaction and Suzuki couplings, protections and deprotections. These analogues showed promising antiviral activity.

Rejman et al.[59] reported the N-phosphonocarbonylpyrrolidine derivatives of guanine (**67**). These compounds can act as bisubstrate inhibitors that occupy both the phosphate and nucleoside binding sites of human purine nucleoside phosphorylase.

4 Dinucleoside phosphates and other nucleotides

4.1 Dinucleoside phosphates

Dinucleoside phosphates contain *two* nucleosides linked by one or more phosphate group(s) and play a vital role in physiological activities such as platelet aggregation, neurotransmission, glycogen breakdown and activation of nucleotide receptors. Joshi et al.[60] developed the synthesis of dinucleoside monophosphates without the use of protecting groups. This facile chemistry, with montmorillonite as the catalyst, was used to synthesise the dimers (**68b** and **68c**) from the activated nucleotides (**68a**) by reacting with nucleosides. It can also be applied to the synthesis of related dimers from activated nucleotides by reacting with nucleoside 5′-monophosphates or from activated nucleotides with themselves.

Some of the nucleotides acting as agonists consist of a nucleotide scaffold, but have a stability issue. To overcome this instability it is necessary to prepare dinucleotides which are metabolically more stable than the corresponding nucleotides. Yelovitch et al.[61] reported the synthesis of boranophosphate dinucleoside polyphosphates and monitored the metabolic stability of those analogues. These studies showed that uridine containing dinucleoside analogues are potent, stable, and highly selective receptor agonists.

4.2 Other nucleotides

In this sub-section, selected publications on nucleotide-like compounds are discussed. Tanaka et al.[62] described a synthesis of sugar-nucleoside diphosphates. This novel chemistry involves a one-pot synthesis by a chemical coupling in water (shown in Scheme 5). 2-Imidazolyl-1,3-dimethylimidazolinium chloride was used as an agent to activate the nucleoside 5′-monophosphate in water to form a phosphorimidazolide intermediate. Reacting this intermediate with sugar-1-phosphate in water produced sugar-nucleoside diphosphates (**69**) in good yield. These products can be used directly for the preparation of oligosaccharides by glycosyltransferases.

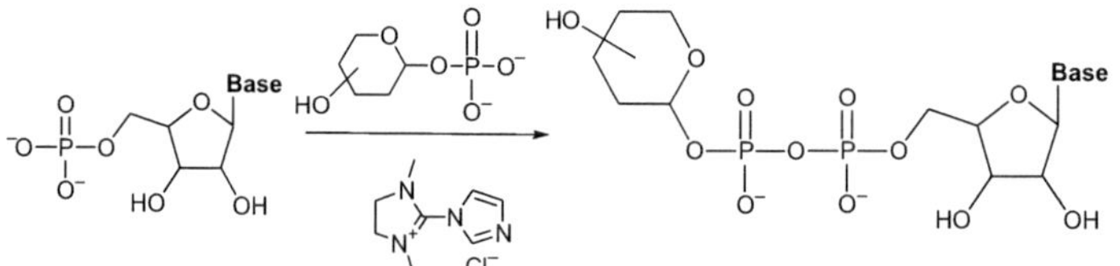

Scheme 5 Synthesis of sugar-nucleoside diphosphates.

Wolf et al.[63] reported the synthesis of cytidine-5′-monophosphate-sialic acid derivatives (**70**). The synthesis involves the reaction of *cyclo*-Sal-nucleotides with different phosphate nucleophiles. *Cyclo*-Sal-nucleotides were also used to prepare monophosphate-linked sugar nucleotides in high yields. These compounds were used for enzymatic studies (bacterial polysialyltransferase) and the enzyme showed marked differences in utilising these compounds.

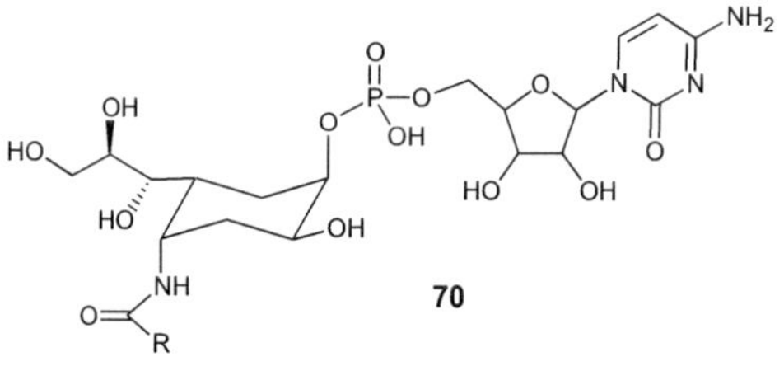

Kowalska et al.[64] reported an efficient and scalable procedure for the synthesis of nucleoside 5′-phosphosulfates (**71**) and the 2′,3′-cyclophosphate containing adenine (**72**). These compounds are important intermediates in the metabolism of sulphur in living organisms.

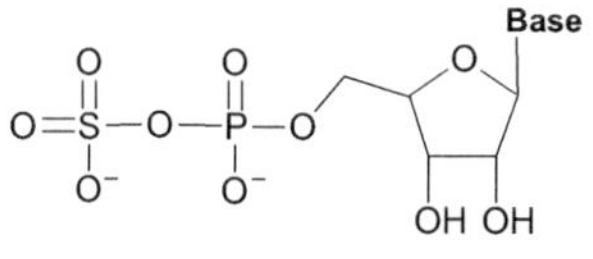

71 Base: A, G, N^7-MeG, C, U

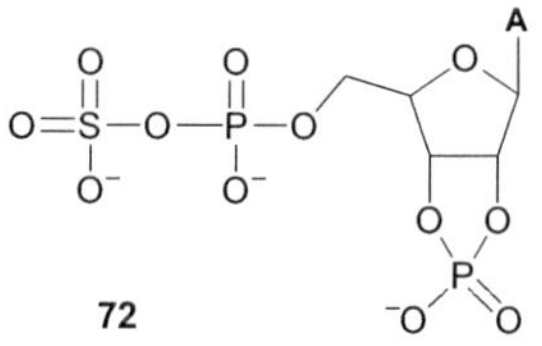

Kim et al.[65] reported the synthesis of pyrimidine nucleoside analogues containing a boronic acid in place of the 3′-hydroxyl group of deoxyribose (73). Unfortunately, these nucleoside analogues were found to be inactive against human hepatocarcinoma cells due to decomposition at neutral pH with liberation of free pyrimidine base.

Peyrat and Xie[66] reported the synthesis of thymidine dimers from 5′-O-aminothymidine. These dinucleosides (74 and 75) can be readily converted to phosphoramidite monomers, then incorporated into oligonucleotides to study their base-pairing properties with complementary DNA or RNA. The synthetic method should be applicable to the synthesis of various O-amino dinucleosides.

5 Tri- and poly-phosphates

5.1 Modified nucleoside triphosphates

The nucleoside phosphoramidites discussed in Section 2 are used for *chemical* synthesis of oligomers, nucleoside triphosphates (NTPs) and their deoxy-analogues (dNTP) are used primarily for *enzymatic* synthesis of nucleic acids. Therefore the availability of triphosphates of modified nucleosides would allow one to generate, using enzymes, oligomers with novel properties. Such properties can be influenced by a number of factors such as the nature of the substituent and its position on the nucleoside of a modified NTP or dNTP. Since oligomers and nucleic acids are involved in various biological processes such as replication, transcription and translation, nucleoside triphosphates bearing functional groups can also offer probes to explore nucleic acid structures and functions. Therefore further investigations are needed to generate novel and useful modified nucleoside triphosphates. A review,[67] that can be used as a starting point, summarises recent developments on the synthesis of modified nucleoside triphosphates and the understanding of the mechanisms underlying their acceptance by polymerases.

The commonly used syntheses of nucleoside triphosphates are the Yoshikawa method and the Ludwig-Eckstein method. The Yoshikawa method[67], the simplest one, is shown in the upper reaction in Scheme 6. The chemistry involves first the 5′-monophosphorylation of an unprotected nucleoside with phosphorous oxychloride ($POCl_3$), which affords a highly reactive intermediate (76). This intermediate is then reacted with pyrophosphate to yield the cyclic triphosphate (77), followed by hydrolysis to the target (80). This method allows the selective

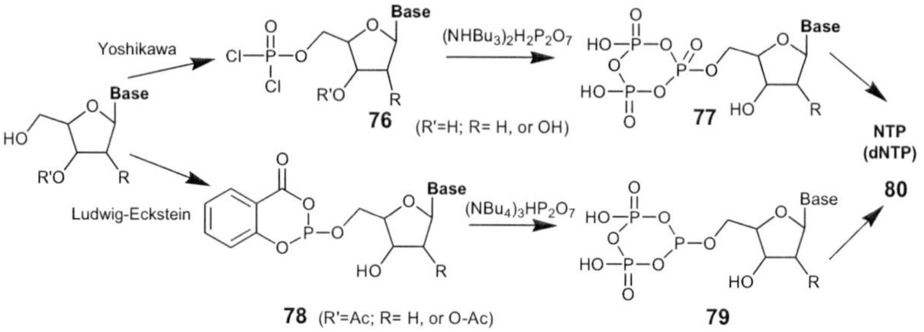

Scheme 6 Synthesis of nucleoside triphosphates.

phosphorylation at 5′-position to afford the 5′-regioisomer. However, it is not compatible with certain nucleosides and can result in a quantity of undesirable by-products.

Another popular reaction is the Ludwig-Eckstein's method[67] (the lower reaction in Scheme 6), which involves the protection of a 3′-O (and 2′-O in the case for NTPs) modified nucleoside and the reaction of it with a phosphitylating agent bearing a good leaving group (in this case, salicyl phosphorochlorite) to afford an activated phosphite intermediate (**78**). This intermediate then undergoes a nucleophilic substitution by tris(tetra-n-butylammonium) hydrogen pyrophosphate, resulting in the cyclic intermediate (**79**). The iodine-mediated oxidation (and hydrolysis) of **79** yields the corresponding nucleoside triphosphate (**80**). The advantage of this method is the formation of fewer by-products which simplifies the purification.

The search for better methods is still on-going. Kore et al.[68] reported a facile procedure for the gram-scale synthesis of 2′-deoxynucleoside triphosphates, including dGTP, dATP, dCTP, TTP and dUTP with good yields (around 70%) starting from the corresponding nucleoside. This synthesis involved mono-phosphorylation of nucleoside followed by reaction with tributylammonium pyrophosphate and the hydrolysis of resulting cyclic intermediate to provide the corresponding dNTP. The authors noted the use of almost substoichiometric amounts of pyrophosphate (0.85 equivalents) for the triphosphorylation step is one of the key contributing factors to the pure product. Strenkowska et al.[69] described a method for phosphate-modified nucleoside triphosphates (**81**) by reacting with cyanoethyl phosphorimidazolides using microwave irradiation. The chemistry involves phosphorylation and deprotection in one pot without the need for isolation and purification of the intermediate.

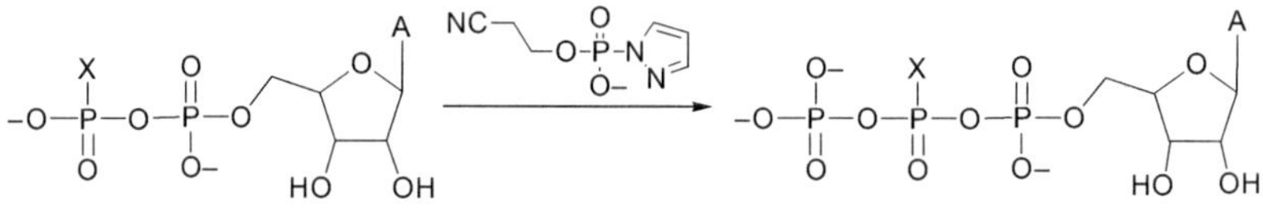

Hacker et al.[70] synthesised a number of phosphate-modified ATP analogues (**82**) and examined their stability in aqueous solutions. It was found that the nitrogen-linked ATP analogue is less stable, whereas the oxygen- and carbon-linked tri- and tetra-phosphate analogues are stable in the pH range of 3 to 12.

Nawale et al.[71] synthesised 4′-C-aminomethyl-2′-O-methylthymidine 5′-triphosphate (**83**) and incorporated it into DNA by means of thermophilic DNA polymerases. The dual sugar modification increases the synthesized DNA's resistance to exonuclease when compared with the single 2′-O-methyl modification.

Fluorescent nucleotide analogues are widely used as molecular probes for studying DNA-protein interaction. Riedl et al.[72] synthesised the triphosphates of modified deoxycytidine bearing fluorophores 3,5-difluoro-4-hydroxybenzylidene-imidazolinone (**84a**) and 3,5-bis(methoxy)-4-hydroxy-benzylideneimidazolinone (**84b**), and used them for the synthesis of oligonucleotides. These DNA probes showed fluorescence increment 2 to 3.2 times upon interaction with a protein.

Holzberger et al.[73] synthesised 8-vinyl- and 8-styryl-2′-deoxyguanosine modified DNA as novel fluorescent molecular probes. Guanosine 5′-triphosphate analogues (**85**) were processed by various DNA polymerases to create fluorescent DNA and these synthesised compounds have potential as probes for fluorescence spectroscopy.

Bergen et al.[74] synthesised C5-modified pyrimidine and C7-modified 7-deazapurine nucleoside triphosphates (**86a–d**). The authors reported crystal structures of modified dNTPs carrying modifications at the C5 positions of pyrimidines or at the C7 positions of 7-deazapurines in complex with a DNA polymerase, which provides useful insights into the mechanism of incorporation of modified dNTPs into DNA. Santner et al.[75] described the synthesis of 2′-methylseleno adenosine and guanosine 5′- triphosphates (**87**). These compounds are useful for enzymatic RNA synthesis by RNA polymerases.

Seela et al.[76] reported the synthesis of 7-deazapurine (pyrrolo[2,3-d]pyrimidine) 2′-deoxyribonucleosides (**88**), including β-D- and β-L-enantiomers, fluoro-derivatives and 2′,3′-dideoxyribonucleosides with well-defined stereochemistry. These compounds have potential applications as antiviral or anticancer agents (7-deazapurine nucleosides), and the 7-deazapurine nucleoside triphosphates are used in Sanger dideoxy DNA-sequencing.

Caton-Williams et al.[77] reported a one-pot synthesis of nucleoside 5′-(α-P-thio) triphosphates (**89**) and phosphorothioate nucleic acids without protecting any nucleoside functionalities. The chemistry of this facile synthesis involves the treatment of unprotected nucleosides with a mild phosphitylating reagent (salicyl phosphorochloridite and pyrophosphate) followed by sulfurization and hydrolysis to afford the corresponding analogues. These compounds can be easily purified by ion-exchange chromatography.

Mohamady et al.[78] described an efficient preparation of nucleoside polyphosphates and their conjugates such as nucleoside triphosphates, symmetrical and unsymmetrical dinucleoside polyphosphates and sugar nucleotides using sulfonyl imidazolium salts as coupling reagents in excellent yields.

Kore et al.[79] developed a one-pot chemical synthesis of 2′-deoxynucleoside-5′-tetraphosphates (**90**). The synthetic strategy involves monophosphorylation of the nucleoside, subsequent reaction of the intermediate with tris-(tri-n-butylammonium) triphosphate and hydrolysis of the cyclic tetrametaphosphate intermediate to afford the corresponding 2′-deoxynucleoside-5′-tetraphosphate in good yields. This method has been utilized for both purine and pyrimidine deoxynucleotides. In a separate paper, they[80] also reported a convenient and efficient synthesis of pyrimidine 2′-deoxynucleoside-5′-tetraphosphates. The facile chemistry involves monophosphorylation of 2′-deoxynucleoside, conversion of 5′-monophosphate into an imidazolide salt, and finally conversion of this intermediate to the 2′-deoxynucleoside-5′-tetraphosphate in excellent yields.

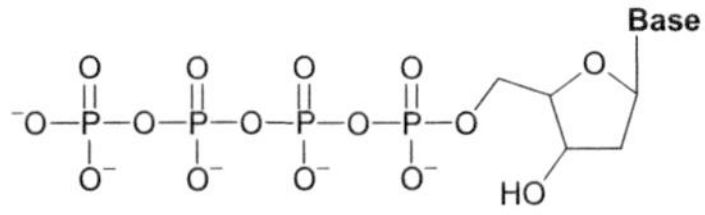

Base = A, G, C, T   90

5.2 Modified oligonucleotides

We have discussed modified nucleoside phosphoramidites (Section 2.1) and modified nucleoside triphosphates (Section 5.1) and their uses as the monomers for chemical or enzymatic synthesis of modified oligomers (oligonucleotides and oligodeoxynucleotides). In this section, our emphasis is primarily on interesting properties of modified oligomers.

Insertion of modified nucleotides into DNA strands often induces conformational changes or increases steric hindrance, consequently altering base pairing properties. Synthetic oligonucleotides have been used to affect gene expression for research and therapeutic purposes over the last few decades. Free radical DNA damage is another interesting topic in the area of modified oligonucleotide chemistry, since reactive species can cause extensive damage to cellular DNA. In the majority of nucleotides, a hydroxyl radical attacks the sugar unit of DNA at H5′, resulting in the formation of a C5′ radical, which may cause intramolecular attack at various positions of the nucleobase, and could generate a cyclic base-sugar adduct and lead to DNA damage. This type of lesion has also been identified in human cellular DNA. To improve the performance of oligonucleotides with the aim of avoiding the above lesions, various modified nucleotides have been developed. Nakano et al.[81] synthesised ureido-linked phenyl-2′-deoxycytidine derivatives (**91**).

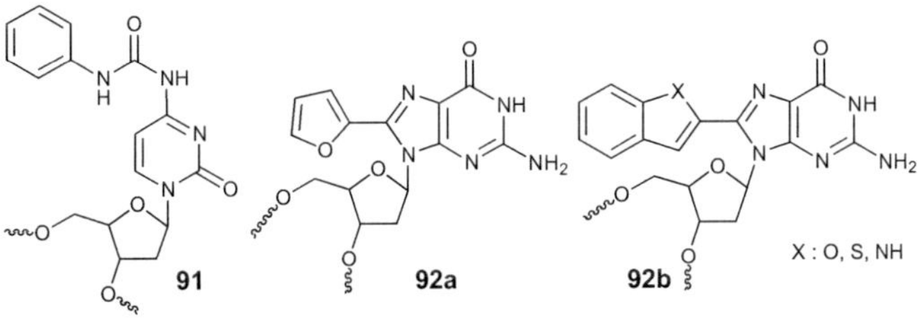

X : O, S, NH

91   92a   92b

Base-pairing studies of these derivatives in DNA showed the selectivity of the carbamoylated cytosine lesions produced in cells and possible applications of the 2′-deoxycytidine derivatives in medical technologies and molecular biology. Modified oligonucleotide analogues, especially at the 2′-position of the sugar moiety, have been useful as antisense oligonucleotides.[82] Fluorescent nucleoside analogues have many potential applications in understanding nucleic acid structures and functions, DNA damage and repair, protein-DNA interactions among others. Rankin et al.[83] prepared C8-(hetero)aryl-2′-deoxy-guanosine adducts (92) as conformational fluorescent probes. Various physical experiments showed that these probes demonstrated sensitivity to the opposite base in the duplex and illustrated the utility of these compounds in fluorescence spectroscopy to monitor probe conformation.

Kaura et al.[84] described the synthesis and hybridization properties of oligonucleotides containing 5-(1-aryl-1,2,3-triazol-4-yl)-2′-deoxyuridines (93). These modifications showed strong thermal affinity and binding specificity towards RNA targets, due to formation of chromophore arrays in the major groove.

Torigoe et al.[85] described the chemical modification of oligonucleotides. The authors used 2′-O,4′-C-ethylene (instead of 2′-O,4′-C-methylene used in LNA) bridged oligonucleotides (94) for gene targeting. These compounds have increased ability to form duplex and triplex structures and have higher nuclease resistance. Such a modification increases the thermal stability of the duplexes. Further studies on these bridged oligonucleotides showed that the sugar moiety with a polar atom-containing alkyl chain increases the rigidity and the degree of hydration which is vital to form duplexes and triplexes in vivo.

Khomyakova et al.[86] reported the synthesis of 2′-modified uridine oligoribonucleotides containing 1, 2-diol group (96) by phosphoramidite approach. The 1,2-diol was converted by a post-synthetic periodate oxidation to an aldehyde group which can react with other functional groups. These oligomers were studied for their applications in the affinity modification of RNA recognizing proteins.

97a: Base = Adenine
97b: Base = Thymine

Zhang et al.[87] reported the synthesis of novel L-isonucleoside (**97**) modified oligonucleotides containing 5′-CH$_2$-extended chains at the sugar moiety and isoadenine or isothymine as the nucleobase. These were incorporated into DNA and siRNA. These modified oligonucleotides can form stable double helical structures with their complementary DNA and RNA and are stable towards nucleases when compared with the unmodified DNA.

6 Summaries

This chapter reviewed selected papers published in 2012 on the chemical synthesis and biological, particularly medicinal, applications of nucleotides and oligonucleotides. The term "nucleotide" is the central point of the chapter. As a nucleotide has a nucleoside in its make-up and plays its functional roles as part of oligonucleotides, some relevant aspects of nucleosides and oligonucleotides are also included in this chapter. For its intended functions, a nucleotide can be presented in several forms, such as phosphoramidite, phosphoramidate, phosphonate and triphosphate, their relationships can be summarized in the Scheme below.

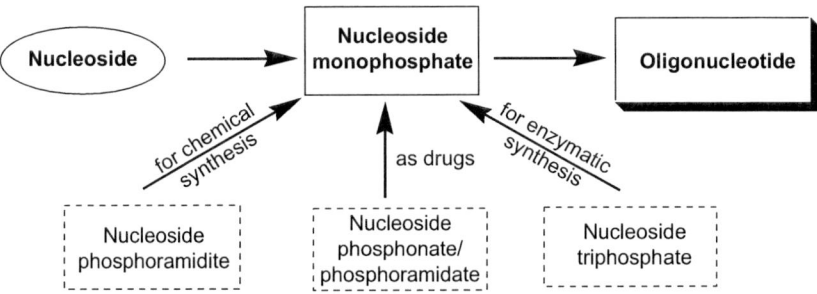

As the nucleoside is the key moiety within a nucleotide, any modification made to the nucleoside would also be worth investigating at the nucleotide level. Indeed, many synthetic methods for modified nucleosides have been successfully extended to their corresponding nucleotides. Nucleoside phosphoramidites and triphosphates are primarily employed as nucleotide monomers during chemical and enzymatic synthesis of oligonucleotides respectively. Nucleoside phosphonates and phosphoramidates have been developed as potential drugs against viral and other pathogens. Phosphonates can be regarded as a stable form of the nucleotides due to their resistance to nucleases while phosphoramidates can be viewed as a masked form of nucleotides to increase their bioavailability. Recently chemical methods for the synthesis of modified nucleotides and oligonucleotides have undergone dramatic improvements. Nucleotides and oligonucleotides are of interest both due to their biological significance and for their synthetic challenges. These analogues have the necessary attributes for pharmaceutical use. They provide an opportunity for the design of therapeutic molecules with highly potent properties and meet the pharmacological requirements for possible drug development. As discussed above, nucleotides and oligonucleotides have potential biological applications and are well-suited for understanding and prediction of the effects of molecular structure on biological activities. The above-illustrated synthetic routes and applications are useful to generate molecules with robust biological stability and may lead to a successful DNA-based therapeutics. These molecules could be useful for understanding of the events that occur in living cells. Subsequently more exciting discoveries could be made of DNA's yet unknown roles and functions in biological systems. Along with methodological advances in chemistry, particular in synthetic chemistry, many more novel nucleosides and nucleotides would be prepared as effective drugs and/or incorporated into oligonucleotides for biological and medicinal exploitation. There is no doubt that research papers on nucleotides and oligonucleotides will be published for many years to come.

References

1 A. R. Kore and I. Charles, *Curr. Org. Chem.*, 2012, **16**, 1996–2013.
2 N. Fresneau, M. A. Hiebel, L. A. Agrofoglio and S. Berteina-Raboin, *Molecules*, 2012, **17**, 14409–14417.
3 J. Lalut, L. Tripoteau, C. Marty, H. Bares, N. Bourgougnon and F. X. Felpin, *Bioorg. Med. Chem. Lett.*, 2012, **22**, 7461–7464.
4 L. Kalachova, R. Pohl and M. Hocek, *Org. Biomol. Chem.*, 2012, **10**, 49–55.
5 V. Raindlova, R. Pohl and M. Hocek, *Chem. Eur. J*, 2012, **18**, 4080–4087.
6 M. Kogler, R. Busson, S. De Jonghe, J. Rozenski, K. Van Belle, T. Louat, H. Munier-Lehmann and P. Herdewijn, *Chem. and Biodiversity*, 2012, **9**, 536–556.
7 Z. P. Kortylewicz, Y. Kimura, K. Inoue, E. Mack and J. Baranowska-Kortylewicz, *J. Med. Chem.*, 2012, **55**, 2649–2671.
8 W. Shen, J. S. Kim and J. Hilfinger, *Synth. Commun.*, 2012, **42**, 358–374.
9 B. J. Kopina and C. T. Lauhon, *Org. Lett.*, 2012, **14**, 4118–4121.
10 M. K. Lakshman, A. Kumar, R. Balachandran, B. W. Day, G. Andrei, R. Snoeck and J. Balzarini, *J. Org. Chem.*, 2012, **77**, 5870–5883.
11 M. Martin-Ortiz, M. Gomez-Gallego, C. R. de Arellano and M. A. Sierra, *Chem. Eur. J*, 2012, **18**, 12603–12608.
12 L. Bosch, I. Cialicu, J. Caner, X. Ariza, A. M. Costa, M. Terrazas and J. Vilarrasa, *Tet. Lett.*, 2012, **53**, 1358–1362.
13 E. Adamska, J. Barciszewski and W. T. Markiewicz, *Nucleosides Nucleotides and Nucl. Acids*, 2012, **31**, 861–871.
14 A. E. A. Hassan, A. H. Moustafa, M. M. Tolbah, H. F. Zohdy and A. Z. Haikal, *Nucleosides Nucleotides and Nucl. Acids*, 2012, **31**, 783–800.
15 G. Mata and N. W. Luedtke, *J. Org. Chem.*, 2012, **77**, 9006–9017.
16 V. Parmenopoulou, D. S. M. Chatzileontiadou, S. Manta, S. Bougiatioti, P. Maragozidis, D. N. Gkaragkouni, E. Kaffesaki, A. L. Kantsadi, V. T. Skamnaki, S. E. Zographos, P. Zounpoulakis, N. A. A. Balatsos, D. Komiotis and D. D. Leonidas, *Bioorg. Med. Chem.*, 2012, **20**, 7184–7193.
17 R. R. Midtkandal, P. Redpath, S. A. J. Trammell, S. J. F. Macdonald, C. Brenner and M. E. Migaud, *Bioorg. Med. Chem. Lett.*, 2012, **22**, 5204–5207.
18 H. Wojtowicz-Rajchel, *J. Fluorine Chem.*, 2012, **143**, 11–48.
19 P. Perlikova, N. J. Martinez, L. Slavetinska and M. Hocek, *Tetrahedron*, 2012, **68**, 8300–8310.
20 A. Leisvuori, Z. Ahmed, M. Ora, L. Beigelman, L. Blatt and H. Lonnberg, *Helv. Chim. Acta*, 2012, **95**, 1512–1520.

21 P. Kumar, K. I. Shaikh, A. S. Jorgensen, S. Kumar and P. Nielsen, *J. Org. Chem.*, 2012, **77**, 9562–9573.
22 F. Casu, R. K. Harston, M. A. Chiacchio and G. Gumina, *Nucleosides Nucleotides and Nucl. Acids*, 2012, **31**, 224–235.
23 H. Q. Cui, J. Carrero-Lerida, A. P. G. Silva, J. L. Whittingham, J. A. Brannigan, L. M. Ruiz-Perez, K. D. Read, K. S. Wilson, D. Gonzalez-Pacanowska and I. H. Gilbert, *J. Med. Chem.*, 2012, **55**, 10948–10957.
24 J. Gotkowska, J. Balzarini and D. G. Piotrowska, *Tet. Lett.*, 2012, **53**, 7097–7100.
25 C. Kiritsis, S. Manta, V. Parmenopoulou, A. Dimopoulou, N. Kollatos, I. Papasotiriou, J. Balzarini and D. Komiotis, *Carbohydrate Res.*, 2012, **364**, 8–14.
26 E. H. R. Morales, J. Balzarini and C. Meier, *J. Med. Chem.*, 2012, **55**, 7245–7252.
27 A. S. Madsen, A. S. Jorgensen, T. B. Jensen and J. Wengel, *J. Org. Chem.*, 2012, **77**, 10718–10728.
28 W. Schmucker and H. A. Wagenknecht, *Synlett*, 2012, 2435–2448.
29 P. Naus, P. Perlikova, A. Bourderioux, R. Pohl, L. Slavetinska, I. Votruba, G. Bahador, G. Birkus, T. Cihlar and M. Hocek, *Bioorg. Med. Chem.*, 2012, **20**, 5202–5214.
30 M. E. Di Francesco, S. Avolio, G. Dessole, U. Koch, M. Pompei, V. Pucci, M. Rowley and V. Summa, *Nucleosides Nucleotides and Nucl. Acids*, 2012, **31**, 592–607.
31 M. E. Di Francesco, S. Avolio, M. Pompei, S. Pesci, E. Monteagudo, V. Pucci, C. Giuliano, F. Fiore, M. Rowley and V. Summa, *Bioorg. Med. Chem.*, 2012, **20**, 4801–4811.
32 Z. Szombati, S. Baerns, A. Marx and C. Meier, *Chembiochem*, 2012, **13**, 700–712.
33 O. Yaren, P. Rothlisberger and C. J. Leumann, *Synthesis-Stuttgart*, 2012, **44**, 1011–1025.
34 D. Dziuba, V. Y. Postupalenko, M. Spadafora, A. S. Klymchenko, V. Guerineau, Y. Mely, R. Benhida and A. Burger, *J. Am. Chem. Soc.*, 2012, **134**, 10209–10213.
35 S. Barrois and H. A. Wagenknecht, *Beilstein J. Org. Chem.*, 2012, **8**, 905–914.
36 J. M. Ibarra-Soza, A. A. Morris, P. Jayalath, H. Peacock, W. E. Conrad, M. B. Donald, M. J. Kurth and P. A. Beal, *Org. Biomol. Chem.*, 2012, **10**, 6491–6497.
37 S. Harusawa, H. Yoneyama, D. Fujisue, M. Nishiura, M. Fujitake, Y. Usami, Z. Y. Zhao, S. A. McPhee, T. J. Wilson and D. M. J. Lilley, *Tet. Lett*, 2012, **53**, 5891–5894.
38 F. P. McManus, D. K. O'Flaherty, A. M. Noronha and C. J. Wilds, *Org. Biomol. Chem.*, 2012, **10**, 7078–7090.
39 N. Tarashima, Y. Higuchi, Y. Komatsu and N. Minakawa, *Bioorg. Med. Chem.*, 2012, **20**, 7095–7100.
40 J. Brzezinska, J. D'Onofrio, M. C. R. Buff, J. Hean, A. Ely, M. Marimani, P. Arbuthnot and J. W. Engels, *Bioorg. Med. Chem.*, 2012, **20**, 1594–1606.
41 P. P. Seth, J. H. Yu, A. Jazayeri, P. S. Pallan, C. R. Allerson, M. E. Ostergaard, F. W. Liu, P. Herdewijn, M. Egli and E. E. Swayze, *J. Org. Chem.*, 2012, **77**, 5074–5085.
42 J. H. Cho, S. J. Coats and R. F. Schinazi, *Org. Lett.*, 2012, **14**, 2488–2491.
43 S. Meneghesso, E. Vanderlinden, A. Stevaert, C. McGuigan, J. Balzarini and L. Naesens, *Antiviral Res.*, 2012, **94**, 35–43.
44 H. V. Jain and T. I. Kalman, *Bioorg. Med. Chem. Lett.*, 2012, **22**, 4497–4501.

45 F. Pertusati, M. Serpi and C. McGuigan, *Antiviral Chem. Chemotherapy*, 2012, **22**, 24.
46 S. Vertuani, A. Baldisserotto, K. Varani, P. A. Borea, B. D. M. Cruz, L. Ferraro, S. Manfredini and A. Dalpiaz, *Eur. J. Med. Chem.*, 2012, **54**, 202–209.
47 V. K. Brel, *Synthesis-Stuttgart*, 2012, **44**, 2359–2364.
48 U. Pradere, V. Roy, A. Montagu, O. Sari, M. Hamada, J. Balzarini, R. Snoeck, G. Andrei and L. A. Agrofoglio, *Eur. J. Med. Chem.*, 2012, **57**, 126–133.
49 T. Tichy, G. Andrei, R. Snoeck, J. Balzarini, M. Dracinsky and M. Krecmerova, *Eur. J. Med. Chem.*, 2012, **55**, 307–314.
50 S. A. Diab, C. De Schutter, M. Muzard, R. Plantier-Royon, E. Pfund and T. Lequeux, *J. Med. Chem.*, 2012, **55**, 2758–2768.
51 M. Cesnek, D. Hockova, A. Holy, M. Dracinsky, O. Baszczynski, J. de Jersey, D. T. Keough and L. W. Guddat, *Bioorg. Med. Chem.*, 2012, **20**, 1076–1089.
52 D. Hockova, D. T. Keough, Z. Janeba, T. H. Wang, J. de Jersey and L. W. Guddat, *J. Med. Chem.*, 2012, **55**, 6209–6223.
53 U. Pradere, F. Amblard, S. J. Coats and R. F. Schinazi, *Org. Lett.*, 2012, **14**, 4426–4429.
54 G. H. Shen, L. Kang, E. Kim, W. Lee and J. H. Hong, *Bull. Korean Chem. Soc.*, 2012, **33**, 2574–2580.
55 G. H. Shen, L. Kang, E. Kim and J. H. Hong, *Nucleosides Nucleotides and Nucl. Acids*, 2012, **31**, 720–735.
56 L. Kang, E. Kim, E. J. Choi, J. C. Yoo, W. Lee and J. H. Hong, *Bull. Korean Chem. Soc.*, 2012, **33**, 4007–4014.
57 I. Hladezuk, V. Chastagner, S. G. Collins, S. J. Plunkett, A. Ford, S. Debarge and A. R. Maguire, *Tetrahedron*, 2012, **68**, 1894–1909.
58 S. Van Poecke, D. Sinnaeve, J. C. Martins, J. Balzarini and S. Van Calenbergh, *Nucleosides Nucleotides and Nucl. Acids*, 2012, **31**, 256–272.
59 D. Rejman, N. Panova, P. Klener, B. Maswabi, R. Pohl and I. Rosenberg, *J. Med. Chem.*, 2012, **55**, 1612–1621.
60 P. C. Joshi, M. F. Aldersley, D. V. Zagorevskii and J. P. Ferris, *Nucleosides Nucleotides and Nucl. Acids*, 2012, **31**, 536–566.
61 S. Yelovitch, J. Camden and G. Weisman, *J. Med. Chem.*, 2012, **55**, 437–448.
62 H. Tanaka, Y. Yoshimura, M. R. Jorgensen, J. A. Cuesta-Seijo and O. Hindsgaul, *Angew. Chem. Int. Ed.*, 2012, **51**, 11531–11534.
63 S. Wolf, S. Warnecke, J. Ehrit, F. Freiberger, R. Gerardy-Schahn and C. Meier, *Chembiochem*, 2012, **13**, 2605–2615.
64 J. Kowalska, A. Osowniak, J. Zuberek and J. Jemielity, *Bioorg. Med. Chem. Lett.*, 2012, **22**, 3661–3664.
65 B. J. Kim, J. H. Zhang, S. L. Tan, D. S. Matteson, W. H. Prusoff and Y. C. Cheng, *Org. Biomol. Chem.*, 2012, **10**, 9349–9358.
66 S. Peyrat and J. Xie, *Synthesis-Stuttgart*, 2012, **44**, 1718–1724.
67 M. Hollenstein, *Molecules*, 2012, **17**, 13569–13591.
68 A. R. Kore, M. Shanmugasundaram, A. Senthilvelan and B. Srinivasan, *Nucleosides Nucleotides and Nucl. Acids*, 2012, **31**, 423–431.
69 M. Strenkowska, P. Wanat, M. Ziemniak, J. Jemielity and J. Kowalska, *Org. Lett.*, 2012, **14**, 4782–4785.
70 S. M. Hacker, M. Mex and A. Marx, *J. Org. Chem.*, 2012, **77**, 10450–10454.
71 G. N. Nawale, K. R. Gore, C. Hobartner and P. I. Pradeepkumar, *Chem. Comm*, 2012, **48**, 9619–9621.
72 J. Riedl, P. Menova, R. Pohl, P. Orsag, M. Fojta and M. Hocek, *J. Org. Chem.*, 2012, **77**, 8287–8293.
73 B. Holzberger, J. Strohmeier, V. Siegmund, U. Diederichsen and A. Marx, *Bioorg. Med. Chem. Lett.*, 2012, **22**, 3136–3139.

74 K. Bergen, A. L. Steck, S. Strutt, A. Baccaro, W. Welte, K. Diederichs and A. Marx, *J. Am. Chem. Soc.*, 2012, **134**, 11840–11843.
75 T. Santner, V. Siegmund, A. Marx and R. Micura, *Bioorg. Med. Chem.*, 2012, **20**, 2416–2418.
76 F. Seela, S. Budow and X. H. Peng, *Curr. Org. Chem.*, 2012, **16**, 161–223.
77 J. Caton-Williams, B. Fiaz, R. Hoxhaj, M. Smith and Z. Huang, *Science China-Chem*, 2012, **55**, 80–89.
78 S. Mohamady, A. Desoky and S. D. Taylor, *Org. Lett.*, 2012, **14**, 402–405.
79 A. R. Kore, A. Senthilvelan and M. Shanmugasundaram, *Tet. Lett*, 2012, **53**, 5868–5870.
80 A. R. Kore, Z. J. Xiao, A. Senthilvelan, I. Charles, M. Shanmugasundaram, S. Mukundarajan and B. Srinivasan, *Nucleosides Nucleotides and Nucl. Acids*, 2012, **31**, 567–573.
81 S. Nakano, H. Oka, D. Yamaguchi, M. Fujii and N. Sugimoto, *Org. Biomol. Chem.*, 2012, **10**, 9664–9670.
82 T. P. Prakash, *Chem. and Biodiversity*, 2011, **8**, 1616–1641.
83 K. M. Rankin, M. Sproviero, K. Rankin, P. Sharma, S. D. Wetmore and R. A. Manderville, *J. Org. Chem.*, 2012, **77**, 10498–10508.
84 M. Kaura, P. Kumar and P. J. Hrdlicka, *Org. Biomol. Chem.*, 2012, **10**, 8575–8578.
85 H. Torigoe, N. Sato and N. Nagasawa, *J. Biochem.*, 2012, **152**, 17–26.
86 E. A. Khomyakova, E. M. Zubin, L. V. Pavlova, E. V. Kazanova, I. P. Smirnov, G. E. Pozmogova, S. Muller, N. G. Dolinnaya, E. A. Kubareva, R. K. Hartmann and T. S. Oretskaya, *Russian J. Bioorg. Chem.*, 2012, **38**, 488–499.
87 J. Zhang, Y. Chen, Y. Huang, H. W. Jin, R. P. Qiao, L. Xing, L. R. Zhang, Z. J. Yang and L. H. Zhang, *Org. Biomol. Chem.*, 2012, **10**, 7566–7577.

Nucleotides and nucleic acids; oligo- and poly-nucleotides

David Loakes

DOI: 10.1039/9781782623977-00146

1 Introduction

The synthesis of modified oligonucleotides is, as in previous years, divided into sections describing updates on synthesis, followed by modifications to the backbone, sugar and nucleobase. Modifications to the nucleobase continues to be the largest section with a broad range of nucleobase analogues described, whilst updates to oligonucleotide synthesis continue to decline due primarily to the fact that oligonucleotide synthesis is fully automated. The section on aptamers and aptazymes is a two year review, and, as in previous years, the number of publications in this field continues to grow. The range of aptamer targets ranges from small molecules, such as adenine, to aptamers that bind to cell surface proteins, the latter being particularly aimed at identifying cancer cells. The majority of publications in aptamers and aptazymes relate to uses as sensors, with a broad range of methods of detection of the aptamer target. The section described as oligonucleotide conjugates is again the largest section, with topics such as fluorophores that deal with a broad range of subjects including not only the different dyes that have been added to oligonucleotides but to the continually increasing area of single molecule studies. Another area that has seen continual growth is oligonucleotide nanostructures and nanodevices, where there have been many new publications concerning the number of different self-assembly structures and many more nanodevices, such as logic gates and mechanical systems described. Oligonucleotide structure determination has also continued to generate many publications, and in this volume the sections on X-ray crystallography and NMR are a review of the past two years. There have been a large number of complex structures reported describing oligonucleotide-protein interactions in particular, and this has not been confined to crystallography as advances in NMR studies allow solution structures of more complex systems. Finally there are a number of other methods used for the study of oligonucleotide systems, primarily using electron microscopy, but which also includes a number of other methods, such as atomic force spectroscopy, surface plasmon spectroscopy and Raman spectroscopy. All of these methodologies add to our understanding of oligonucleotides in living systems and in artificial constructs.

Medical Research Council, Laboratory of Molecular Biology, Francis Crick Avenue, Cambridge Biomedical Campus, Cambridge, CB2 0QH, UK.
E-mail: dml@mrc-lmb.cam.ac.uk

1.1 Oligonucleotide synthesis

1.1.1 Recent developments in oligonucleotide synthesis.
There have been few reports on advances in the synthesis of oligonucleotides during this review period. A review of methods for the synthesis of mixed oligonucleotides for random mutagenesis has been reported,[1] and a novel protocol for automated solid-phase synthesis of RNA-nucleopeptides using silyl-protected amino acids has been described.[2] A protocol for parallel purification of oligonucleotides using a 5′-hexa-His tag has been reported that is suitable up to mM scale purification.[3] Solution phase synthesis of siRNA has been reported using a 2-(azidomethyl)benzoyl protecting group for the 3′-hydroxyl that is cleaved on treatment with triphenylphosphine.[4] A new reagent (1) has been synthesised for the stereocontrolled (Rp) solid-phase synthesis of phosphorothioate oligonucleotides with diastereoselectivity $\geq 98:2$.[5] Also, a building block suitable for the synthesis of non-bridging phosphorodithioate RNA oligomers has been described.[6]

A reagent-free protocol for solid-phase synthesis of 5′-phosphate oligonucleotides has been devised. The synthesis requires synthesis of oligonucleotides with an additional nucleotide which is then subjected to oxidation to the 5′-aldehyde, and a number of different oxidising reagents were examined. Finally, under basic deprotection the 5′-nucleoside undergoes β-elimination leaving the oligonucleotide with a 5′-phosphate group.[7] Post-synthetic on column labelling of RNA containing either 2-iodoadenosine or 5-iodouridine with a range of organostannanes *via* Stille coupling reactions has been reported.[8] Richert and co-workers have described different polyaromatic cores for the synthesis of four- or six-arm DNA hybrid structures.[9,10] A novel polymer-bound support has been synthesised for the preparation of DNA bearing a 3′-carboxyalkyl group suitable for reaction with epoxy-coated surfaces for the construction of biochips.[11]

1

1.1.2 Oligonucleotide microarrays.
Oligonucleotide microarrays are now widely-used in a variety of applications, and this section deals with some novel synthetic and analytical publications, as well as some novel applications. A method for the synthesis of microarrays on silicon surfaces is described using photolytic thiol-ene chemistry.[12,13] An analysis of artifactual probe signals using multiple samples, using qPCR as validation, has been reported,[14] and hybridisation parameters on microarrays have been examined using a nearest-neighbour model.[15] A review of statistical considerations for microarray meta-analysis has been

reported.[16] A scanometric analysis of DNA microarrays using an optical scanner or by naked eye, in which an intercalated gold nanoparticle is used for detection, has been reported. By this method a hemaglutinin subtyping array was developed.[17]

A few new applications have been described in this review period. Arrays are reported that contain a T7 promoter sequence at the 5′-end of the DNA such that transcription yields many RNA copies of the probe sequences that self-assemble yielding full length transcripts which may be used for gene assembly using RT-PCR.[18] Quantitative analysis of microRNAs from tissue has been reported on arrays using *in situ* hybridisation.[19] Arrays have been reported for expression analysis of neural stem cells,[20] for the detection of single nucleotide polymorphisms (SNPs) to allow for identification of highly pathogenic influenza virus genes,[21] for the synthesis of protein microarrays using coupled *in vitro* transcription-translation systems,[22] and a carbohydrate microarray has been described in which the DNA probes are conjugated with fucosylated glycoclusters to identify novel binders toward *Pseudomonas aeruginosa* lectins.[23]

1.2 The synthesis of modified oligonucleotides

The following section deals with modifications to oligonucleotides dealing with modifications to the internucleotide linkage, the sugar and the nucleobase as separate sections. As in previous years the modification to the nucleobase remains the major part of this review as the number of alternate nucleobase structures continues to increase. In addition there are many un-natural nucleobases described aimed at introducing some particular functionality into oligonucleotides or for the introduction of some spectral property.

1.2.1 Oligonucleotides containing modified phosphodiester linkages.

A variety of backbone modifications that have been described recently are discussed in this section, and a review of modified oligonucleotides suitable for use in gene silencing has also been described.[24] Certain backbone modifications that have been in use for many years, such as the phosphorothioate linkage, are not described unless there is some new development. One of the simplest modifications is the use of triesters rather than diesters, which includes cyclic, lariat and catenane-modified oligonucleotides. 2′,5′-Adenylic acids (2−5A) occur naturally: a prostate-specific membrane antigen (RBI1033)[25] has been modified by inclusion of 2−5A linkages where it exhibited up to ten times greater affinity than the parent compound, and activity was decreased when the 2−5A linkage was removed.[26] Lonnberg has examined whether phosphate-branched RNA is stable under physiological conditions and demonstrated that when embedded in an oligonucleotide it is stable with a half-life of several hours.[27] New photocaged analogues, such as (2), have been described to provide control of transcription.[28] The phosphate group of DNA has been modified to a triester by addition of dodecyl units to provide amphiphilic sites in the oligonucleotide and it was shown that this stabilised duplex DNA by formation of a hydrophobic zipper.[29] The synthesis of catenanes has been described using ssDNA[30] and quadruplex DNA,[31] and two methods for the formation of DNA rotaxane structures have been reported.[32]

The phosphorothioate modification is well known and most reports describing its use are excluded from this review. A method has been reported for the synthesis of non-bridging phosphorodithioates during the solid-phase synthesis of RNA.[6] A phosphorothioate modification has been used for the introduction of a spin label, which was then used to study the global structure of a three-way junction in the Phi29 packaging RNA dimer.[33] Phosphorothioate oligonucleotides have been used to conjugate various amine substituents to introduce cationic groups into DNA. It was found that the cationic DNA showed improved thermal stability in duplex and triplex structures compared to the unmodified oligonucleotides.[34] Phosphorothioate oligonucleotides bearing a 5'-maleimide group were found to be compatible with either reaction of a thiol compound or with dienes in a Diels-Alder reaction, leaving the phosphorothioate unmodified.[35]

Alkyl phosphonates are a neutral modification of oligonucleotides, which have been used for a number of years, primarily in antisense strategies, but suffer from the problem that oligonucleotides tend to be less water-soluble. A series of 2'- and 5'-O-methylphosphonate tetramers in various sugar conformations (*i.e.* ribo-, arabino- and xylo-) were investigated as substrates for RNase L, and each was found to be as good a substrate as the normal phosphodiester analogues.[36] The alkynyl phosphonate backbone (3) was used as a versatile DNA backbone suitable for modification using Click chemistry.[37] The synthesis of RNA containing a phosphonoacetate or thiophosphonoacetate (4) has been reported and it was found that the presence of the modification was generally stabilising in duplex stability, and ssRNA containing the modification was readily taken up into HeLa cells without transfection agents.[38] Another known modification that has often been used in antisense strategies is the boranophosphate. Wada and co-workers have devised a method to synthesise stereocontrolled boranophosphates using the intermediate (5) in either *R*p or *S*p configuration, and using acid labile protecting groups for exocyclic amines.[39] Caruthers and co-workers have used boranophosphonates as reducing agents of a number of metal ions yielding metal nanoparticles.[40]

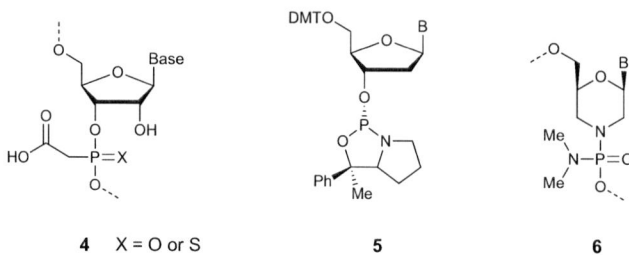

4 X = O or S   **5**   **6**

Another common backbone modification is the phosphoramidate, which includes the morpholino-modification (**6**). The non-enzymatic template-dependent synthesis of DNA has been described using an amino group and an activated phosphate at either 3′- or 5′-positions of the building blocks, allowing for DNA synthesis in either direction.[41,42] A phosphoramidate internucleotide linkage has been described in which propargylamine has been attached to phosphorus in a non-bridging mode that allows for elaboration of the backbone with azides using Click chemistry.[43] The other major class of phosphoramidate linkage is the morpholino-derivative (**6**) as noted above, and this class of oligomer is primarily used as antisense agents. An improved method for the synthesis of morpholino-oligonucleotides has been described in which the yield of oligomers could be improved by the addition of inorganic salts, such as LiBr.[44] Antisense oligonucleotides derived from (**6**) have been used to target the poly(A) tail junction of maternal mRNA in zebrafish.[45,46] A photo-activatable morpholino-derivative has been reported to allow for embryonic gene control by UV irradiation.[47] Non-enzymatic ligation of DNA has been described using periodate cleavage of a terminal ribose nucleotide followed by ligation with an amino-modified DNA strand.[48]

A method for the synthesis of RNA oligonucleotides having the guanidinium linkage (**7**) has been reported.[49] The chimeric building block between a PNA-like (see below) and RNA dimer (**8**) has been used to introduce an amide bond linkage into siRNA. It was found that the presence of the amide linkage in the siRNA was well tolerated in the RNAi pathway when incorporated at either internal positions or at the 3′-end.[50] There are several reports that use nucleosides modified at 3′- and 5′-positions by an alkyne and an azide to form oligomers bearing a triazole linkage by Click chemistry. An improvement on the synthesis of these oligomers has been reported, with yields of up to 90%.[51] The incorporation of triazole linkages in an otherwise mixed DNA sequence was reported to be destabilising in contrast to previous reports.[52] The incorporation of up to two triazole linkages into DNA was assessed in a transcription-translation system in *E. coli*, where it was reported that the backbone modification had little effect on overall efficacy.[53] Triazole linkages in siRNA have also been examined, and again the modification was well tolerated in positions approaching the 3′-end of the siRNA.[54] A triazole linkage has also been used to form an interstrand crosslink in RNA using a propargyl unit attached at O2′ and crosslinking with a bis(azide) using Click chemistry.[55]

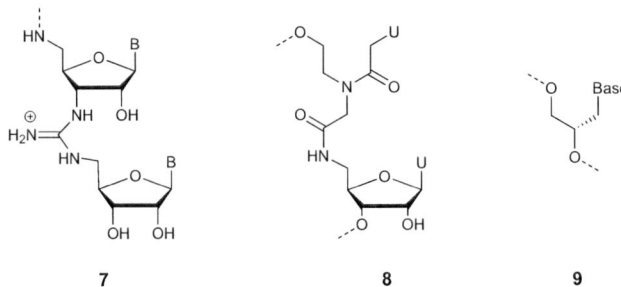

Other backbone modifications that have been reported are oligo(dT) in which the entire phosphate group is replaced by ferrocene for electrochemical studies,[56] and a crystal structure of the known glycol nucleic acid (GNA) (**9**) which was found to form a N-type conformational structure.[57]

Peptide nucleic acids (PNA) (**10**) were first described by Nielsen and co-workers,[58] and consist of a nucleobase attached to an aminoethylglycine backbone. PNA exhibits very good binding to natural DNA and RNA, and is able to strand-invade oligonucleotide duplexes. PNA has a neutral backbone and therefore is more readily taken up into cells where it can bind to complementary oligonucleotides, but does suffer from the drawback that it is frequently insoluble in water and is there often tagged with cationic amino acid residues to aid cellular uptake.

A PNA-RNA dimer unit has been synthesised and incorporated into an RNA oligonucleotide. It was found that the presence of the PNA unit was destabilising.[59] PNA (**10**), which is non-chiral, shows no directional preference in binding with oligonucleotides other than by complementary base pairing, and therefore PNA can be used to probe cooperativity in formation of duplexes with DNA or RNA, and this has been reviewed by Totsingan et al.[60] PNA has been used in microarrays to detect DNA from olive leaves, though PNA modified by an arginine side chain showed greater selectivity.[61] DNA-encoded synthesis (see section 3.2) has received a lot of interest of the past few years, and this has now been extended to PNA using Mtt-protection groups orthogonal to the usual Fmoc chemistry used in PNA synthesis.[62] A PNA-encoded library has also been used to identify small molecule binders to streptavidin.[63] A G-rich PNA oligomer has been shown to be able to bind to a DNA quadruplex, such as the *c-myc* oncogenic promoter sequence, to form a PNA/DNA hybrid quadruplex structure, and a single PNA strand will bind to both the DNA strands of the G-rich DNA sequence.[64]

One of the principal applications of PNA once it had been introduced was in antisense and antigene therapy, as PNA is not readily degraded. The effect of arginine-modified PNA on transfection volume and cell density has been studied, where it was observed that antisense activity increased with increasing volume.[65] PNA-DNA hybrids have been used to target NFκB transcriptions factors in cystic fibrosis IB3-1 cells without use of a transfection agent.[66] PNA and PNA hybrids have been used to target microRNA[67] and pre-miRNA,[68] the latter being shown to interfere with microRNA maturation. Fluorescent peptide-PNA chimeras have been used to image monoamine oxidase A mRNA in neuronal cells,[69] and PNA displayed on polymer nanoparticles have been used to inhibit mRNA sequences for inducible nitric oxide synthase (iNOS) to treat inflammation and acute lung injury.[70] The antisense activity of PNA was found to be dramatically enhanced when conjugated to cholesterol or cholic acid.[71]

PNA probes having two fluorophores have been used for detection of DNA and RNA,[72] exemplified by the dual colour imaging of two mRNA targets in influenza H1N1 infected cells.[73] PNA has been modified with a quinone methide for the site-specific alkylation of complementary DNA.[74] A PNA probe has been used in a label-free electrochemical sensor with picomolar sensitivity for profiling the methylation states of the p53 tumour suppressor gene.[75] A lysine-tagged PNA oligomer has been developed for use as an amperometric sensor.[76]

A few PNA conjugates have been prepared; the effect of conjugation of a ruthenocene and a ferrocene derivative has been examined to assist cellular uptake of the PNA. It was found that not only was the ruthenocene more stable than the ferrocene, but also cellular uptake was considerably better, with a final concentration of up to 4mM found on incubation with 50μM of the conjugate.[77] Another ruthenium-PNA conjugate is described for use in a template-directed synthesis such that when a second PNA strand conjugated to an azido-dye binds adjacent to the ruthenium-PNA conjugate on DNA, the azide group is reduced in the presence of ascorbate to yield the amino-dye derivative, the latter now exhibiting fluorescence.[78] Flavin conjugates of PNA were shown to have enhanced cellular uptake by an endocytotic delivery process.[79] A PNA-aminosugar conjugate, targeting the transactivation response element of HIV-1, has been described that exhibited high bioavailability in human cell lines and strongly inhibited Tat-mediated transactivation of HIV-1 transcription.[80]

In addition to the normal aminoethylglycine PNA derivatives (**10**) a number of base and backbone modifications have been described. Amongst the base modifications described are 5-azidomethyluracil, which was used to introduce the azide group as a masked amine for the introduction of amide-linked functional groups for the generation of a chemical library.[81] 2-Aminopyridine has been introduced into PNA to aid triplex formation with DNA under physiological conditions.[82] 8-Vinyl-guanine has been introduced into PNA for use as a fluorescent analogue.[83]

11, **12**, **13**, **14**, **15**, **16** n = 0, 1, 2

PNA (**10**) is achiral and exhibits little preference (other than base complementarity) for binding to nucleic acids, binding in both parallel and anti-parallel modes. The introduction of substituents into the backbone creates a chiral centre and induces some preference for directional binding, and also frequently shows a preference for either DNA or RNA. The introduction of a methyl group onto the γ-position of (**10**) results in a more rigid structure when bound to DNA. A comparison of charge transfer through PNA (**10**) and the γ-methyl derivative revealed that charge transfer rate constant was more than double through (**10**) than through the methylated analogue due to the greater flexibility as demonstrated in silico.[84] α,α-Dimethylglycine has also been used as a backbone which showed enhanced binding compared to (**10**), and also showed a preference for binding to DNA.[85]

Various ligands, e.g. (**11**) have been synthesised and conjugated to one of the termini of PNA: the ligands were designed to bind cerium ions such that they would selectively cleave the target DNA.[86] Guanidine-modified PNA (**12**) has been used to enhance cellular uptake of PNA. Unlike normal PNA (**10**), that forms a 1:1 triplex with RNA, it was found that the guanidine-modified PNA (**12**) preferentially formed a 2:1 strand invasion complex with dsRNA.[87] Aminomethylene-modified PNA, at either the α or γ positions, have been synthesised and shown that they form more stable duplexes with DNA, in particular the γ(S) modification (**13**).[88] γ(R)-Modified PNA, bearing a mini-PEGylated side chain (**14**), have been shown to exhibit enhanced binding to DNA, in particular in that they are able to strand-invade any of the DNA duplexes examined, and have superior water-solubility compared with normal PNA (**10**).[89] N^{γ}-ω-Carboxyalkyl PNA (**15**), where the alkyl chain is either one or five in length, have been described and examined for its ability to hybridise with DNA. It was found that the longer alkyl chain caused significant improvement in duplex stability, whilst with a single methylene group there was a slight destabilisation.[90] Pyrrolidinyl-PNA modifications have been previously described, *e.g.* (**16**)[91] and exhibit enhanced stability in duplexes with DNA and RNA. Modification of the cycloalkyl ring to the cyclobutyl or cyclohexyl derivatives had a significant effect on duplex stability, with the highest stability observed with the cyclobutyl ring and destabilisation with the six-membered analogue.[92] The analogue (**16**) with a cyclopentane ring bearing an azide group on the pyrrolidine ring has also been described for attachment of a pyrene moiety using Click chemistry to prepare fluorescent PNA.[93]

1.2.2 Oligonucleotides containing modified sugars.

A large number of modifications to the sugar portion of oligonucleotides have been reported, and these have been incorporated either to introduce some additional functionality or to aid oligonucleotide stability. A review of oligonucleotide modifications suitable in gene silencing has been reported.[24]

The most common point of modification of the (deoxy)ribose ring is at the 2′-position, and the most widely used modification is 2′-*O*-methyl derivatives, and as such is excluded from this review as properties of 2′-*O*-methyl nucleosides have been well established. Selective 2′-hydroxyl acylation analysed by primer extension (SHAPE) is a technique that has been used to measure oligonucleotide structure and dynamics, and this has been further investigated by Weeks and co-workers.[94,95] A phosphoramidite building block bearing a 2′-*O*-phthalimidooxymethyl substituent has been described suitable for modifying oligonucleotides with various aldehydes.[96] 2′-*O*-Mesyloxyethyl ribothymidine has been incorporated into DNA which can be converted to the corresponding azidooxyethyl derivative post-synthetically. The resultant azide was then used to label DNA with various alkyne-modified fluorescent dyes using Click chemistry.[97] Azide and alkyne modifications have introduced to the 2′-hydroxyl group for crosslinking in a copper-free Click reaction.[98] 2′-*O*-Methoxyethyl (MOE) derivatives have been incorporated into RNA where they were found to potently and allele-selectively inhibit mutant Huntingtin expression.[99] Uridine and ribothymidine modified by MOE have also been used to activate RNAi in mice.[100] The 2′-aminoethoxy-modification has also been used to aid triplex stabilisation when the modification is used in the third strand.[101] Multiple 2′-*O*-pyrenylmethyl derivatives in an RNA duplex result in higher ordered pyrene π-stacking and enhanced excimer fluorescence.[102] Cationic residues have also been incorporated into siRNA; using the guanidine derivative (**17**) siRNA targeting hepatitis B virus was synthesised and found that the cationic siRNA exhibited improved silencing activity.[103] Cationic 2′-*O*-[*N*-(4-aminobutyl)carbamoyl]uridine formed more stable complementary and mismatched RNA duplexes, as well as increasing nuclease resistance.[104]

A number of 2′-fluoro-nucleotide analogues have been reported: using NMR and thermodynamic data it has been shown the 2′-fluoro RNA exhibits increased Watson-Crick hydrogen bonding and stacking interactions.[105] Incorporation of 2′-fluoro nucleotides into RNA has been shown to greatly improve RNAi[106] and splice correction[107] activities. To identify the specificity of polymerases for either dNTP's or rNTP's, Holliger and co-workers have evolved mutant polymerases capable of recognising 2′-fluoro- or 2′-azido-modified RNA and identified a point mutation (Y409G) controlling the specificity.[108] Nucleoside lesions may be stabilised to repair enzymes by use of 2′-fluoro analogues, which act as inhibitors, allowing for studies of the mechanism of repair. The 2′-fluoro analogue of thymine glycol has thus been used to examine its interactions with the DNA glycosylase NEIL1,[109] as has the 2′-ara-fluoro derivative of 5-carboxylcytosine with the human DNA glycosylase hTDG.[110] 2′-Fluoro- and 2′-ara-fluoro-guanosine have been used to control the conformation of G-quadruplexes as the fluoro-derivatives show a preference for *anti*-conformation of the guanosine residues.[111]

There have been many attempts over previous years to design a universal base, *i.e.* a nucleoside that binds to all natural nucleobases without discrimination, but most such nucleosides have been non-hydrogen bonding which usually compromises thermal stability. Starting from 2′-azido-uridine, a series of triazole-linked pyrene analogues (**18**) formed by Click chemistry have been synthesised and incorporated into oligonucleotides. It was shown that the resulting analogues behaved as a universal base with both complementary DNA or RNA.[112] An analogue related to (**18**, R=Py), derived from $C^{2'}$-azidomethyl-uridine or –thymidine has been used to stabilise secondary structures, in particular bulges and three-way junctions, and, due to stacking between two pyrene moieties in the secondary structure, results in excimer fluorescence, and may therefore be used as a recognition probe.[113] A 2′-*O*-allyl derivative has been used in a Click reaction to incorporate a styryl dye suitable for use in energy transfer interactions through duplex DNA.[114] By screening a library of mutant T7 RNA polymerases, a polymerase was identified that would efficiently incorporate 2′-methylseleno-dUTP.[115] The effect of incorporation of a thymine base attached to the sugar through a carbon linkage at either C2′ or C5′ (**19**) to provide additional base pairing sites has been examined. NMR and molecular modelling studies suggest an unwinding of duplex DNA at the site of incorporation, with both analogues slightly depressing the duplex stability.[116] 2′-Trifluoromethylthiouridine has also been incorporated into RNA where it had utility for probing RNA structures using ^{19}F-NMR spectroscopy.[117]

A solid support has been reported suitable for the synthesis of oligonucleotides bearing a 3′-carboxyalkyl group.[11] These oligonucleotides were reacted with epoxy-modified surfaces for construction of diagnostic biochips as exemplified by detection of the bacterial disease meningitis. 3′-Amino-modified oligonucleotides have been used for the synthesis of oligonucleotide-peptide conjugates *via* native chemical ligation.[118]

Substitution of various nucleotides by 4′-thio-ribonucleotides into an siRNA sequence led to increased thermal and nuclease resistance, resulting in a longer term RNAi effect.[119] A series of thymidine derivatives has been reported bearing C4′-carboxyl-, methoxycarbonyl-, carbamoyl- and methylcarbamoyl-substituents and evaluated for hybridisation properties with complementary DNA and RNA. Of note was a C4′-carboxy-derivative that adopted an *S*-conformation resulting in an increase in thermal stability with RNA.[120] 4′-*C*-Aminomethyl-2′-*O*-methylthymidine triphosphate has been shown to be incorporated into DNA using the thermophilic DNA polymerases Pfu and Terminator III. The incorporation of the C4′-modification resulted in an increased resistance towards exonuclease activity compared with the known 2′-*O*-methyl modification.[121,122]

5′-*O*-Aminooxymethyl nucleosides have been used during chemical synthesis of DNA for modification by various aldehydes. The monomer was found to be suitable to the conditions of chemical synthesis, in particular to the treatment with TBAF.[96] The 2′-*O*-MOE-5′-vinylphosphonate (**20**) (and its reduced form) have been incorporated into siRNA resulting in a potent RNAi effect in animal models.[100]

In addition to the (deoxy)ribose sugar portion of the natural oligonucleotides, a large number of alternate sugars have been applied to nucleotides. These include not only pentofuranose sugars but also a number of hexose sugars. The final part of this section describes a number of locked nucleic acid sugar structures based largely on the now common LNA but also includes a number of other conformationally locked sugar systems.

A series of isonucleoside derivatives has been prepared and incorporated into ssDNA or ssRNA, where it was found that duplexes with either DNA or RNA were more stable than their native counterparts. In addition, the isonucleoside (**21**) could be used in siRNA and retain gene silencing, and was also a substrate for RNase H.[123]

A number of other sugar modifications have been examined. A self-complementary duplex of fully-modified deoxy-xylo nucleic acid has been prepared, and its structure solved by NMR. It was found that the duplex formed a ladder structure, and is the first example of a furanose nucleic acid complex to adopt this structure.[124] Threose nucleic acid (TNA), (**22**), has been used to evolve aptamers binding to thrombin,[125] demonstrating that TNA is capable of adopting tertiary structures similar to DNA and RNA. Two epimeric 1,4-dioxane-based monomers (**23**, 5*R*-isomer shown) have been incorporated into the centre of a DNA duplex where they were found to be destabilising, in particular the 5S-isomer.[126]

A few hexose-based sugar modifications have been applied to oligonucleotides. 2′-Fluoro cyclohexenyl nucleic acids (CeNA) (**24**) have been incorporated into a DNA duplex where it was slightly destabilising but exhibited enhanced nuclease stability.[127] A crystal structure of RNA with (**24**) showed that it formed an A-type duplex not unlike that found with RNA:RNA duplexes.[128] A crystal structure of a duplex between RNA and altritol nucleic acids (**25**) has been reported and the hybrid structure formed an A-type helix analogous to a fully RNA duplex.[129] Aptamers have been evolved using hexitol nucleic acids (HNA) and CeNA that were able to fold into defined structures just like DNA and RNA,[130] and crystal structures of the S- and R-6′-methyl-3′-fluorohexitol systems have been compared.[131]

Acyclic nucleosides have also been examined. The acyclic analogue (**26**) was incorporated into DNA where it functioned as a fluorescent probe for detection of SNPs.[132] The pyrene modified acyclic derivative (**27**) was incorporated into ssDNA and it was observed that three such substitutions resulted in excimer fluorescence in the ssDNA, which disappeared when the DNA was bound to its target DNA.[133]

Locked nucleic acids (**28**) were first introduced by Imanishi[134] and Wengel,[135] and contain a methylene bridge between C4′ and O2′ that locks the sugar into an *endo* conformation. Oligonucleotides containing (**28**) exhibit enhanced binding towards both DNA and RNA as well as enhanced nuclease resistance. Due to the enhanced binding and nuclease resistance LNA has frequently applied to antisense and antigene strategies as well as used for probes particularly in cells.

During this review period LNA has been used to aid stabilisation of hairpin oligonucleotides,[136] as probes to detect the *E. coli* O157:H7 pathogen on gold nanoparticles,[137] and to probe the mode of action of the *B. subtilis* σ^A RNA polymerase.[138] LNA has been used to knockdown gene expression of three miRNAs at low concentration and toxicity,[139] to stabilise siRNA in rainbow trout against VHSV,[140] and to protect DNA cleavage by a number of restriction endonucleases.[141] LNA has also been used to stabilise triplex strands in a thermodynamic study.[142] LNA has also been used in studies of charge transport through DNA, dealt with in more detail in section 3.4.[143–145] A few analogues of (28) have been described. The α-L-LNA analogue has been previously reported, but in this review period modification of α-L-LNA by substitution at the 5′-position has been described. A comparison of the two isomers showed that the *R*-5′-Me analogue had little effect on duplex stability, but that the *S*-5′-Me derivative was destabilising.[146] α-L-LNA has also been shown to be incorporated into DNA as its 5′-triphosphate by various DNA polymerases and HIV-RT.[147] Replacement of the oxygen in the C4′-O2′ bridge by nitrogen allows for attachment to the new amino group, and various pyrene derivatives have been attached to the bridging amino group where they were used as fluorescent probes for the detection of SNPs.[148] 2′-Amino-LNA has also been incorporated into DNA as its 5′-triphosphate by a number of DNA polymerases.[149] The amide-bridging analogue (29) has been reported and found to have similar stabilising effects as (28) but with enhanced nuclease resistance.[150] Both isomers of 4′-methoxyethyl- and ethyl-modified (28) have been described and compared with LNA in a thermodynamic study.[151]

Other locked nucleic acids have been described: the bicyclic analogue (30) has been prepared in both the ribose and arabinose conformations where it was found that the ribose analogue was destabilising with complementary DNA and RNA, whilst the arabinose derivative was Tm neutral.[152] The tricyclic analogue (31, R=H), which has been previously described, has been used in an antisense study compared with 2′-O-methoxyethyl-DNA, where it was observed that oligomers of (31, R=H) exhibited better antisense activity.[153] The effect of substitution at C6′ (31, R=esters) has been reported where it was found to favourably stabilise duplex DNA.[154]

The so-called W-shaped nucleotide analogue (32) has been used in a triplex-forming oligonucleotide for antigene activity targeting the Bcl-2 or survivin gene, where it was shown to inhibit cell proliferation and induce caspase-dependent apoptosis.[155] The analogue (32) has also been synthesised with a methyl group at each position of the phenyl ring, and it was found that when used in a triplex-forming oligonucleotide that a methyl group in the *meta* or *para* positions stabilised the triplexes at TA or GC interruption sites of the duplex.[156] The *North* and *South* methanocarba-dT analogues (**33**, *N*-isomer shown) have been incorporated into a DNA hairpin where it was found that (**33**) adopted an unexpected C4'-*exo* conformation.[157] A crystal structure has been reported of the DNA polymerase Dpo4 with the dATP derivative of the *North* (**33**) analogue in the polymerase active site.[158,159] A number of highly-strained sugar analogues have been reported having *spiro*-modifications.[160]

32   **33**

1.2.3 Oligonucleotides containing modified bases.

This section describes the range of modified nucleobases incorporated into oligonucleotides and their applications and deals with modified pyrimidines, purines and finally a number of modified nucleobases ranging from abasic sites to large polyaromatic hydrocarbon *C*-nucleosides, in addition to a number of both pyrimidine and purine adducts of environmental mutagens. tRNAs contain a number of modified nucleotides, and a review of them and their function has been reported.[161]

A series of novel geranylated uridine analogues has been identified in a range of different bacterial species, including *E. coli*, *Enterobacter*, and *Salmonella*, all of which have geranyl groups attached through a 2-thio group, such as (**34**).[162] 2-Selenouridine has been identified as being able to selectively form a wobble base pair with adenosine rather than a wobble pair with guanosine.[163] The synthesis of the *E. coli* tRNA^{Glu2} anticodon arm, containing the modified nucleotide 5-methylaminomethyl-2-thiouridine has been described.[164]

In order to study the effects of interstrand crosslinks in DNA a protocol for the synthesis of *N*3-thymidine-butylene-*N*3-thymidine has been described using allyloxycarbonyl for the protection of the 5'-hydroxyl group and a silyl protection for the 3'-hydroxyl group.[165] To control the gene silencing effects in miRNA, a caged uridine with *N*3 protected with 6-nitropiperonyloxymethyl has been reported, allowing for photochemical removal on demand.[166]

To study the mechanism of flap endonucleases an interstrand crosslink disulphide linkage has been introduced using 4-thiouridine and 6-thioguanosine, where it was observed that specificity was abolished using hFEN1.[167] To study repair by DNA alkyltransferases a duplex DNA containing an interstrand crosslink between two thymidine residues having variable length alkyl chains between the two O4 groups of thymidine has been synthesised. It was found that the repair enzyme hAGT bound to the thymidine lesion with similar affinity to O^6-methylguanosine.[168] The synthesis of the *E.coli* tRNA^{Glu2} anticodon arm containing the modified nucleotide 5-methylaminomethyl-2-thiouridine has been reported.[164] The structure of the human tRNA^{Lys3} containing the modified 5-methoxycarbonyl-2-thiouridine has been described.[169] Incorporation of 5-aminomethyl-dU into ssDNA has been shown to form stable antiparallel triplexes with either DNA or RNA targets.[170] 5-Propargylamino-dU has been used in triplex forming oligonucleotides to aid stabilisation of the third strand.[101]

The C5 position of pyrimidines is the most common point for attachment of a broad range of modifications. To identify the localisation of DNA replication in *Sulfolobus* archaea, 5-bromodUTP and 5-ethynyldUTP were incorporated, followed by immunolocalisation analysis to reveal peripheral localisation of replisomes within the cell.[171] 5-Bromouracil will undergo an intrastrand photoaddition reaction to an adjacent uracil forming a C5-C5 linkage, and the mechanism of this reaction has been described.[172]

There are a few reports concerning the 5-hydroxymethyluracil modification during this review period. There are crystal structures of thymine DNA glycosylase[173,174] and methyl-binding protein domain 4 (MBD4)[175] bound to DNA containing 5-hydroxymethyl-dU, and a report that germ line ablation of SMUG1 DNA glycosylase results in loss of 5-hydroxymethyluracil excision.[176] 5-Hydroxymethyl-dU and dC triphosphates have been described bearing photolabile protecting groups suitable for use in Next-Generation Sequencing technologies.[177] A crystal structure of RNA containing 5-methoxyuridine showed the presence of hydrogen bonding between the methoxyl group and the phosphate backbone.[178]

Various C5-modifed dU analogues have been reported bearing either an alkyne or an azide for use in Click reactions. 5-Ethynyluridine has been incorporated into RNA for modification with azidofluorophores.[179] The diethynylbenzene derivative (35) has been used to probe non-polar interactions within a polymerase active site.[180] C5-Ethynyl-dU has been modified with varying size aryl azides and shown to aid stabilisation of duplexes,[181] and a long-chain alkyne has been incorporated onto DNA to act as an end-cap after modification by Click chemistry.[182] DNA containing an alkynyl-modified dU in each strand has been crosslinked with 1,4-bis-azidomethylbenzene.[183] C5-Azidopropyluridine has been introduced into RNA for modification by Click chemistry or the Staudinger reaction.[184] Alkyne- and azide-modified dU analogues suitable for DNA crosslinking in a copper-free Click reaction have also been prepared.[98] Various arylsulfonamide groups have been attached to ethynyl-dU and were found to aid stabilisation of duplex structures.[185] C5-Phenylethynyl-dU has been incorporated into DNA where it acts as a modulator of charge transport.[186]

Different reporter groups have been attached to C5 of uracil, *e.g.*, a number of fluorophores including naphthalene,[187] pyrene,[188] fluorene,[189] benzofuran[190] and the biaryl derivative (36) that also has an NMR label (^{19}F).[191] The structure of an RNA-DNA hybrid has been examined by EPR using a spin label attached to C5 of dU.[192] A terpyridine moiety has been attached to C5 to chelate metal ions.[193] Different carbohydrates have been attached *via* a linker to C5 of dU and the resultant DNA was hybridised to form a three-way junction with the carbohydrate displayed on each arm to probe binding to lectin.[194] Two nitroimidazole derivatives attached to C5 *via* a methylene bridge have been described where they form interstrand crosslinks under hypoxic conditions.[195] The acyclic pyrimidine analogue (37) has been incorporated into the thrombin-binding aptamer where it was found to increase the efficacy of the aptamer.[196]

The dihydrouracil analogue (38) has been introduced into RNA as its phosphoramidite derivative in order to study strand scission in dsRNA *via* generation of a C5-pyrimidine radical.[197] The dihydrouracil analogue (39) has been introduced into RNA as a photolabile moiety to study RNA folding. The presence of the arylsulfide group prevents proper RNA folding but is removed in less than a microsecond upon irradiation at 350 nm, regenerating ribothymidine.[198] (6-^{13}C)-Pyrimidine nucleotides have been synthesised for use as a spin label in NMR studies.[199]

37   38   39

There are a few lesions for the pyrimidine nucleotides, the most commonly-studied being the family of photoproducts. Perhaps the most common of the photoproducts is the cyclobutane pyrimidine dimer (40), and there are repair enzymes for this lesion. The mechanism of action of DNA photolyase in the repair of (40) has been investigated and in particular the role of the adenine in the FAD co-factor.[200] Another photoproduct is the (6→4) derivative (**41**), and the mechanism for alkaline degradation of it has been reported.[201] The translesion bypass of both of these blocking lesions has been monitored in *E. coli*.[202] Another photoproduct is the spore photoproduct (**42**), and is repaired by a radical SAM enzyme spore photoproduct lyase. A crystal structure of this enzyme with DNA containing (**42**) has been described.[203] The (6→4) derivative (**41**) will also undergo UV-induced rearrangement to the Dewar lesion (**43**), the mechanism of which has been examined.[204]

Another thymine lesion is thymine glycol, and the yield of this lesion in pUC18 films following ionising radiation have been measured.[205] The lesion is repaired by DNA glycosylases, and the binding of two such enzymes, NEIL1 and Endo III, to DNA containing 2′-fluoro-thymine glycol with the 2′-fluoro group in either the ribose or arabinose conformation has been examined. Whilst NEIL1 appears to show no preference for binding to either isomer, Endo III shows a preference for binding to the analogue in the ribose conformation.[109] Two crosslinked lesions between thymine and guanine are described, both formed by one electron oxidation of guanine. The dynamics and energetics of DNA containing a guanine(C8)-thymine(N3) lesion have been examined,[206] whilst formation and repair of an intrastrand crosslink between guanine(C8) and the methyl group of thymine have been described.[207]

There are fewer cytosine analogues than uracil analogues described, and the main site of modification is at C5. The role of 2-thiocytidine in the anticodon loop of *E. coli* tRNA^{Arg1,2} has been reported.[208] Incorporation of a phenyl group *via* a ureido-linkage to N4 of cytosine was found to decrease the selectivity of the analogue base pairing with guanosine under PEG-mediated osmotic stress conditions.[209] There are a number of reports concerning the analogues 5-formyl- and 5-carboxycytosine, which are oxidation products from 5-methylcytosine, and these analogues are repaired by thymine DNA glycosylases (TDG). Crystal structures of hTDG in complex with 5-carboxycytosine showing how it is recognised by the enzyme have been reported.[173,210,211] Transcription of 5-methyl, 5-formyl- and 5-carboxycytosine by mammalian and yeast RNA polymerase II have been compared.[212] Oligonucleotides containing 2′-ara-fluoro-5-carboxyl-dC have been synthesised and found to be resistant to repair by hTDG.[110] The mechanism by which 5-carboxy-dC undergoes decarboxylation in stem cells has been reported.[213]

5-Hydroxycytosine is another oxidative lesion that is mutagenic by replicative polymerases. Thermodynamic studies of DNA containing the lesion show that it is destabilising and suggest that there is a low energy barrier for unstacking for repair by a DNA glycosylase.[214] 5-Dimethyl-amino-dC may be used for a hole trap in charge transport through DNA, and the synthesis of oligonucleotides containing the analogue has been reported.[215] Hocek and co-workers have described the synthesis of a range of C5-modified cytosine analogues for further post-synthesis modification[216,217] and for the introduction of metal ions into the DNA duplex.[218]

5-Hydroxymethylcytosine (5-HMC) is the most widely-reported analogue; it is an important epigenetic marker and is abundant in mammalian cells, and has sometimes been referred to as the sixth base in the genetic code. It has been reported to be abundant in mammalian brains in synaptic genes, showing differences in abundance at the exon-intron boundary.[219] A method for sequencing DNA containing 5-HMC has been reported involving oxidative bisulfite treatment, where it was found that in mouse embryonic stem cells there is 3.3% hydroxymethylation in CpG islands.[220] The loss of 5-HMC has been reported as an epigenetic marker for melanoma.[221] RNAi-mediated knockdown of Tet1 enzyme resulted in depletion of hydroxymethylation in mouse embryonic stem cells.[222] Using genome mapping it was found that hydroxymethylation is associated with transcription factor binding to distal regulatory sites during neural differentiation.[223] It has been shown that the APOBEC deaminase system does not deaminate 5-HMC,[224] and the methyltransferase DNMT1 binds to 5-HMC over 60-fold more weakly than to 5-methylC.[225] Addition of glucose to the hydroxymethyl group leads to a nucleoside known as "J", that is found in kinetoplastids and T-even phage. The biochemical synthesis of J by β-glucosyltransferase and uridine diphosphoglucose has been described,[226] and binding by J-binding protein to DNA containing glucosylated 5-HMC shows rapid conformational change upon binding.[227]

44   45   46   47

5-Azacytidine is a chemotherapeutic agent, but is also mutagenic. It rapidly breaks down in water to give 2′-deoxyriboguanylurea (**44**) which is also mutagenic. The analogue (**44**) has been introduced into oligonucleotides as its phosphoramidite and shown to be a potent inhibitor of both human and bacterial methyltransferases and can induce 5-azaC-type fragile sites, and that the pharmacokinetic properties of 5-azaC are also demonstrated by its breakdown product (**44**).[228] Pyrrolocytosine is often used in place of cytosine as it is fluorescent, and has been used to probe the mechanism of repair of formamidopyrimidine-DNA glycosylase.[229] It has been used to probe for a mismatch with cytosine mediated by Ag(I) ions where the formation of the base pair quenches the fluorescence of pyrrolocytosine.[230] The modified pyrrolo-dC (**45**) has been used for the detection of mismatches, and it exhibits better fluorescent properties, especially when it is further modified with benzylazide using Click chemistry.[231] The pyrrolocytosine analogue (**46**) has been used as a G-clamp as it forms stronger base pairs with both guanosine and isoguanosine.[232] Another fluorescent cytosine analogue is (**47**), and this has been used in a crystal structure of RB69 (Y567A) DNA polymerase where it was observed that it could form a wobble pair with an incoming dATP.[233] The analogue (**47**) has been further modified by addition of a spin label onto the aromatic ring where it can be used to study conformational dynamics using EPR,[234,235] and a crystal structure revealed that the analogue did not perturb the native duplex structure in contrast to the crystal structure described above for the duplex containing (**47**).[236]

The main N2-modified adenosine used is 2-aminopurine, which is frequently used as an adenine surrogate because it maintains hydrogen bonding with thymine and is fluorescent. Applications of this analogue are described in section 3.5. A synthesis of the *E. coli* tRNA^{Glu2} anticodon arm containing 2-methyladenosine has been reported,[164] and the role of 2-methyladenosine in the anticodon loop of *E. coli* tRNA^{Arg1,2} has been reported.[208] A spin label has been attached to the C2-position of adenine to probe oligonucleotide structures by EPR.[192]

Using antibody-capture and parallel sequencing it has been reported that there are more than 12000 N^6-methyladenosine sites in the mouse genome.[237] The structure of the human tRNA^{Lys3} containing the modified 2-methylthio-N^6-threonylcarbamoyladenosine has been described.[169] Using the repair dioxygenase FTO, a library of small molecule inhibitors of the demethylase enzyme have been identified.[238] N^6-Propynyladenosine has been used for monitoring RNA synthesis by labelling with a fluorophore using Click chemistry.[239] Quantum dots have been attached to N6 of adenosine to probe for DNA-binding proteins.[240] The diazo compound (**48**) has been shown to site-specifically alkylate the N6-position of adenine when the adenine residue is in an overhanging position in a DNA duplex.[241] The C6-modified purine (**49**) has been incorporated into DNA to introduce a metal-binding site for functionalisation of nucleic acids.[242] The incorporation of either N^6-carbamoylcyclohexyl-A or the N^2-G analogue onto the 5′-end of an oligonucleotide duplex has been reported to aid stabilisation of an RNA duplex.[243]

48

49

Incorporation of C8-arylamino-dA analogues into the recognition sequence of the restriction endonuclease *Eco*R1 has been shown to dramatically reduce DNA cleavage.[244] 8-Aminoadenosine has been used as a triplex-forming oligonucleotide that forms a stable triplex at physiological pH.[245] Pairs of C8-ethynylpyrene adenosine nucleotides have been introduced into *i*-motif forming sequences where the *i*-motif could be identified due to excimer fluorescence of the pyrene moieties.[246] The introduction of a C8-alkoxyadenosine residue into siRNA was found to be well tolerated leading to efficient gene knock down, but multiple substitutions were less effective.[247] A C8-azidoadenosine triphosphate has been used for labelling of RNA following Click chemistry with an alkynyl-fluorophore.[248]

Of the natural nucleobases, guanine is the most susceptible to adduct formation by environmental mutagens, but adenine will also form adducts. The most common adduct is 1,N^6-ethenoadenine (εA) and derivatives, and various repair enzymes exist to remove this lesion. ALKBH2 is one such repair enzyme, and a crystal structure of it containing duplex DNA with the εA lesion has been reported.[249] Li *et al.* have examined the mechanism of repair by AlkB repair enzymes and have shown that the exocyclic carbons adjacent to N^6-nitrogen are the primary targets for oxidation leading to repair of the lesion.[250] Attack of DNA by glyoxal or methylglyoxal gives rise to the εA derivatives (**50**, R = H or Me, respectively), and the mechanism of crosslinking of this adduct has been examined.[251] Another common lesion of adenosine is the 8,5'-cyclonucleoside, and a synthesis of both diastereoisomers and incorporation into DNA has been described.[252] The presence of these two diastereoisomers in human urine has been reported as a biomarker for the detection of atherosclerosis.[253] The presence of either cyclonucleoside of dA or dG has been shown to strongly inhibit transcription *in vivo*.[254] The exocyclic amino group of adenine is susceptible to alkylation by environmental polyaromatic hydrocarbons, and the repair of a range of such lesions has been examined.[255] Exposure to aristocholic acid gives rise to N^6-dA and N^2-dG adducts, and the repair of these lesions has been examined, showing that a lack of recognition of the adducts leads to low levels of repair unless they are in a mismatch position.[256]

The modified nucleosides N2-methylG and 5-methylC have been shown to be important for bacterial fitness by modulating the early stages of initiation by stabilising the binding of fMet-tRNA^{fMet} to the 30S pre-initiation complex.[257]

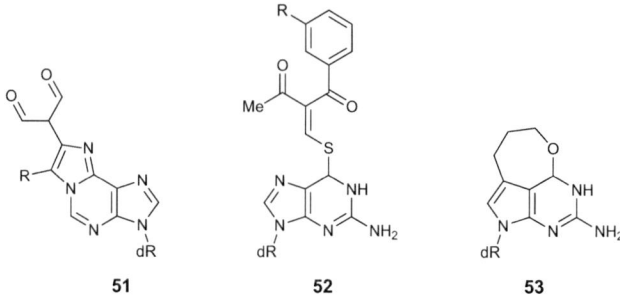

51 **52** **53**

Another common lesion found in nucleic acids are O^6-alkylguanine, and again there are specific enzymes evolved for the repair of such lesions. The alkyltransferase-like proteins (ATL) in *Schizosaccharomyces pombe* (Atl1) and *T. thermophilus* (TTHA1564) have been shown to be involved in flagging O^6-alkylG lesions by examining a number of O^6-alkylG analogues, with Atl1 in particular being able to distinguish between guanine and its alkylated derivatives.[258] Human O^6-alkylguanine DNA alkyltransferase has been shown to form a specific complex with duplex DNA containing an O^6-alkylguanine lesion in a stoichiometry of 2 : 1, with no binding to natural DNA or ssDNA containing a O^6-alkylguanine lesion.[259] It has also been shown to bind co-operatively to O^6-alkylguanine-containing DNA, and forms clusters of up to 11 proteins as determined by AFM.[260] To study the mechanism of flap endonucleases, an interstrand crosslink disulphide linkage has been introduced using 4-thiouridine and 6-thioguanosine, where it was observed that specificity was abolished using hFEN1.[167]

A method for introducing O^6-methylG into DNA by solid phase synthesis has been described using the analogue (**51**) which results in transfer of the thiol-protecting group onto the exocyclic amine at alkaline pH, and is further enhanced in the presence of $NiCl_2$. When R in (**51**) is an alkynyl group, this allows for further functionalisation of the guanine residue using Click chemistry.[261] The conformationally-locked *anti* O^6-alkylguanine analogue (**52**) has been incorporated into DNA where it was found to be a poor substrate for human O^6-methylguanine DNA methyltransferase and Atl1, suggesting that the *syn* conformation is required for recognition by these enzymes.[262] O^6-CarboxymethylG and O^6-(4-oxo-(3-pyridyl)butyl)G are lesions formed by bile acids and are associated with red meat. A method for the incorporation of these two analogues into DNA has been described, and the resultant DNA was shown to be a good substrate for *S. pombe* alkyltransferase-like protein Atl1.[263] The photocaged guanine derivative (**53**) has been incorporated into DNA where, after photolysis at 385 nm, it forms an interstrand crosslink between the guanine N1 and a cytosine N3 *via* an ethyl linkage.[264]

A few C8-modifications have been described, though the main C8-guanine modification is 8-oxoguanine. A C8-(2-pyridyl)guanine analogue has been reported that forms selective binding with Cu(II), Ni(II), Cd(II) and Zn(II) ions *via* coordination with the guanine N7.[265] An unstable mutagenic lesion, 8-nitroG, has been incorporated into DNA as its 2′-O-methyl derivative. It has been shown to adopt a *syn* conformation, and forms a stable base pair with dG. Templating with either AMV-RT or human DNA polymerase β resulted in preferential incorporation of dA and dC respectively.[266] C8-Arylethynylated guanine derivatives have been used as fluorescent labels,[267] as has 8-vinylguanine and 8-styrylguanine.[268]

53

54

syn 8-oxoG

As guanine has the lowest oxidation potential of all the nucleobases, it is the most susceptible to oxidation, and the 8-oxoguanine lesion is another common lesion found in nucleic acids. 8-Oxoguanine forms *syn* and *anti* conformations resulting in base pairing with either adenine or cytosine, and these conformations have been observed in a crystal structure of human DNA polymerase β in complex with DNA containing 8-oxoG.[269] The *syn* conformation is stabilised by minor groove hydrogen bonding between an Arg and the C8-oxygen. An adenine analogue has been described (**54**) that selectively binds to 8-oxoG through an extended hydrogen bonding network when 8-oxoG is in its *syn* conformation, and when it binds the resultant complex is fluorescent.[270] The replicative DNA polymerase PolB1 from *S. solfataricus* has been shown to stall upon encountering the lesion, whereas the lesion bypass polymerase Dpo4 faithfully replicates 8-oxoguanine by insertion of dCMP.[271] Correct translesion synthesis of 8-oxoG by DNA polymerase λ is coordinated with the glycosylase enzyme MutYH and involves phosphorylation and ubiquitination of pol λ.[272] Replication by DNA polymerase δ results in stalling of replication, and replication is continued in the presence of DNA polymerase λ.[273]

There are also a few reports on repair of 8-oxoG lesions. By using formaldehyde crosslinking it has been shown that the Werner syndrome helicase-exonuclease (a MutYH homologue) is recruited in the repair of 8-oxoG lesions in conjunction with DNA polymerase λ.[274] Repair of 8-oxoG by formamidopyrimidine-DNA glycosylase has been examined by stopped-flow kinetics using the fluorescent pyrrolocytosine.[229] Also, repair of 8-oxoG lesions in dinucleosomes has been shown to be repaired by the glycosylase OGG1 in the absence of histone H1.[275] A non-polar isostere of 8-oxoG, 2-chloro-4-fluoroindole (**55**), has been examined under single-turner kinetic conditions with the glycosylases Fpg and hOGG1 where it was found that (**55**) was an even better substrate for Fpg glycosylase than the native 8-oxoG, and also a substrate for hOGG1, though removed less efficiently.[276]

The generation of 8-oxoG lesions by singlet oxidation through photolysis of an anthraquinone-labelled oligonucleotide has been reported.[277] The mechanism by which cell death by the actions of bacterial antibiotics has been examined, revealing that an overproduction of the DNA polymerase DinB arises by the overproduction of 8-oxo-dGTP in the nucleotide pool.[278] The effect of local sequence for oxidation of guanine to 8-oxoguanine has been examined using a crosslinking assay with various restriction endonucleases.[279] A thermodynamic assay of guanine oxidation in trinucleotide repeat sequences showed that guanine oxidation preferentially occurs at sites of stem loop structures that place the 8-oxoguanine residue at or near the top of the loop.[280]

55   **56**   **57**

Oxidation of guanine is complex and can give rise to more than 10 products following oxidation to guanine. Lim *et al* have reported a mass spectral analysis of guanine oxidation using either nitrosoperoxycarbonate- or riboflavin-mediated photo-oxidation, and found that with both oxidants the main oxidation product was dehydroguanidinohydantoin (**56**).[281] Unzipping kinetics of duplex DNA containing a range of guanine oxidation products has been examined by translocation through an α-hemolysin nanopore.[282] The lifetime of guanine radical cations has been examined at varying pH by pulsed laser in the presence of two oxidants and shown that it persists for up to 3ms at acidic pH, but for up to 70ms at alkaline pH.[283] In aqueous solutions, oligonucleotides containing 8-oxoguanine were susceptible to further oxidation leading primarily to guanidinohydantoin products at low pH.[284] An additional 8-oxoguanine oxidation product has been identified, namely the guanidinoformimine (**57**), formed by oxidation with riboflavin, being mutagenic and able to base pair with A, G or C.[285]

Guanine is also known to form a number of adducts with environmental mutagens. Reaction of guanine with vinyl halides gives rise to lesions including N^2,3-ethenoguanine (**58**), the 2′-*ara*-fluoro-derivative of which has been shown to give rise to GC→AT mutations.[286] The mutagens 2-aminofluorene and 2-acetylaminofluorene give rise to C8 adducts of guanine (**59**, R=H or acetyl respectively). Sequence effects for the translesion synthesis past (**59**, R=H, X=F) have been carried out and shown that when the 5′-nucleotide is C it gives rise to insertion of dAMP or dCMP in addition to stalling at the n − 1 position.[287] Repair of the same analogue by NarI occurs in a sequence specific manner, and is repaired more efficiently than the acetylated derivative (**59**, R=acetyl, X=F).[288] This is in contrast to the results observed by another group using the same NER system for the analogue (**59**, R=H or acetyl,

X=H).²⁸⁹ DNA polymerase κ is involved in translesion synthesis past bulky lesions, and a quantum mechanical analysis of translesion past 10S(+)-*trans-anti*-B[*a*]P-*N*²-dG has been reported.²⁹⁰ Quantum mechanical analysis for the mitomycin C adduct are also described.²⁹¹

58

59 R = H or Ac
       X = H or F

60

A few aza- and deaza-purine analogues have been described during this review period. Crystal structures of KlenTaq DNA polymerase about to incorporate 7-deazapurine analogues are described,²⁹² as is the structure of RB69 DNA polymerase with DNA containing 3-deaza-dA to probe minor groove interactions.²⁹³ A range of 7-deaza-7-oligopyridine derivatives have been incorporated into DNA for functionalisation using metal ions.²¹⁸ 7-Deazapurine and 7-deaza-8-azapurine analogues have been reported suitable for modification by Click chemistry for the addition of pyrene,²⁹⁴ as well as for forming branched DNA complexes,²⁹⁵ and for modification in the major groove of siRNA²⁹⁶ and for interstrand crosslinking DNA duplexes,¹⁸³ also using Click chemistry. 7-Deaza-7-ethynyl-dA and -dG have been used for labelling genomic DNA by Click chemistry *in vivo*.²⁹⁷ 7-Deaza- and 7-deaza-8-aza-dA and -dG have been used to probe RNA editing by adenosine deaminases.²⁹⁸ A novel quadracyclic adenine analogue (**60**) has been synthesised that shows specific base pairing properties with thymine, and is fluorescent at >300 nm, allowing for selective excitation.²⁹⁹ Novel C7-hydroxymethylated 7-deaza-dA and dG triphosphates have been reported bearing photocleavable 2-nitrobenzyl protecting groups for use in Next-Generation Sequencing technologies.¹⁷⁷

The abasic site is one of the most common lesions found in DNA, and as such there are various enzymes evolved to specifically repair and/or replicate the abasic site. A model for the translocation step of replication with the bacteriophage phi29 DNA polymerase has been reported for individual DNA-polymerase complexes by capture by a α-hemolysin nanopore. In this case the copying of an abasic site is not investigated, rather a series of consecutive abasic sites in the template acts as a reporter for the polymerase translocation.³⁰⁰ The effect of base stacking for primer extension at and beyond an abasic site with RB69 DNA polymerase has been investigated by varying the penultimate base pair using all 16 possible base pair combinations, and by solving the crystal structures of many of these variants. It was found that incorporation opposite an abasic site varied over two orders of magnitude depending upon the neighbouring base pair.[301] By use of NMR spectroscopy it has been observed that the KlenTaq DNA polymerase adopts unique recognition states when encountering an abasic template site.[302]

A number of enzymes are involved in the repair of abasic sites. A crystal structure of the human methyl-binding domain IV glycosylase that binds at CpG sites, bound to an abasic site opposite a guanine showed no obvious recognition features suggesting the recognition must occur at a methylated cytosine residue.[303] Two exonuclease III family enzymes have been found to be involved in strand excision at abasic sites with the obligate pathogen *Neisseria meningitidis*,[304] and this has been further elaborated by a crystallographic structure.[305] A new endonuclease independent repair pathway has been described from *S. pombe* involving the DNA glycosylate Nth1 followed by tyrosyl phosphodiesterase (Tdp1).[306] One of the main repair enzyme types are endonucleases involved in base excision repair (BER). Various mechanistic studies have been reported, including unwinding of the DNA by the primosomal enzyme DnaD in *B. subtilis*,[307] the mechanism of histone-catalysed cleavage of abasic sites in nucleosome core particles,[308] and conformational dynamics of APE1 using stopped-flow analyses.[309]

An abasic site has been synthesised at the 3′-end of siRNA, where it was found that it enhanced the nuclease resistance of the siRNA.[310] Various techniques for the detection of abasic sites have been described. Treatment of DNA containing an abasic site with amines, followed by reduction with cyanoborohydride, results in the furanose ring being opened and with the amine attached to the C1′ position. These modified oligonucleotides may then be detected by translocation through an α-hemolysin nanopore, where it was reported that the best results were obtained when a crown-ether was attached to the abasic site.[311,312] A number of ligands have been described that bind selectively in an abasic site. Nucleobase-polyamine conjugates (attached *via* N1 of pyrimidines or N9 of purines) have been described such that the nucleobase forms a normal base pair with the opposing nucleobase and the polyamine anchors the ligand to the duplex.[313] The pyrazine diuretic drug Amiloride (**61**) has previously been shown to bind in DNA at an abasic site opposite a thymine, and has now been shown to also bind in RNA with an abasic site opposite uracil.[314] Another class of compound that binds at an abasic site by base pairing with the opposing nucleobase is 2-amino-1,8-naphthyridine (AMND) and its derivatives, which are fluorescent and hence readily detected when bound in an abasic site. AMND can form base pairs with either uracil or cytosine, but introduction of the electron-withdrawing trifluoromethyl group to the 5-position makes it highly selective for base pairing with cytosine.[315] The 5,6,7-trimethyl AMND derivative has been used to detect depurination events by ribosome-inactivating protein toxins.[316] A further analogue having a benzodiazole derivative attached to the 2-amino group has been found to be environment-sensitive and used to detect SNPs.[317] Perylenediimide has been used to intercalate into DNA having opposing abasic sites, and is proposed to be of used in ID arrays.[318] The spin label cytosine derivative (**62**) has also been used for the detection of abasic sites opposite a guanine using EPR. The effect of modification at N3 of (62) has been examined and it was found that a PEG derivative showed the greatest affinity.[319] The effect of the neighbouring base pair has also been reported, where it was found that best binding of the (62) derivative was at 5'd(G-T) sites.[320]

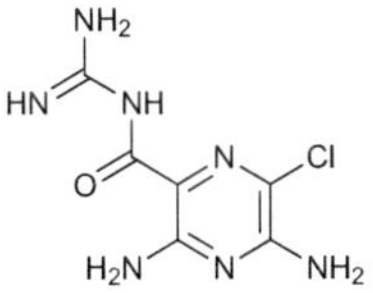

61 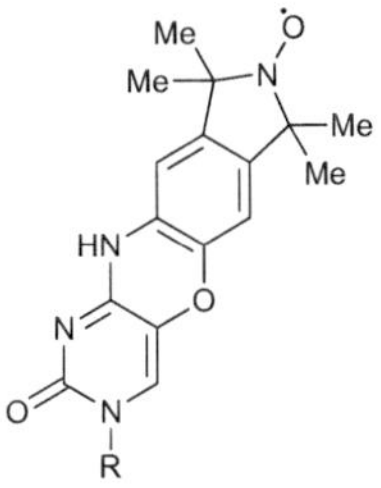 62 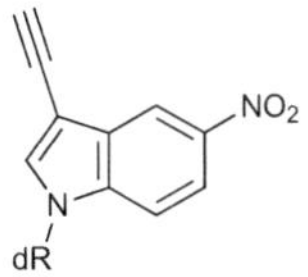 63

Another method used for the detection of abasic sites is modified nucleosides. A thymidine derivative having a benzofuran moiety attached to C5 has been used as a fluorescent probe for abasic sites in RNA.[190] A pyrene C-nucleoside and an imidazophenanthrene nucleoside have been incorporated into a triplex-forming oligonucleotide for the detection of an abasic site.[321] The non-hydrogen bonding nucleoside 3-ethynyl-5-nitroindole (63) has been shown to be highly selective for incorporation (as its triphosphate) opposite an abasic site, being selectively incorporated even in the presence of 50-fold excess of the natural dNTPs. The analogue (63) may then be further modified with an azido-fluorophore using Click chemistry for detection of the abasic site.[322] The sugar-modified nucleoside (18) has also been used to detect abasic sites with fluorescent detection from the pyrene moiety.[112]

Other abasic sites that have been reported are oxidised abasic sites. The mechanism of action of C1027, an enediyne antitumor antibiotic, has been examined in detail. It leads to the formation of an abasic site but also to oxidised abasic sites, such as the 2-deoxyribonolactone (64) and the C4'-oxidised abasic site (65), which then forms interstrand crosslinks, and ultimately results in cell death.[323] The mechanism by which histone-catalysed cleavage of nucleosomal DNA containing 2-deoxyribonolactone has been reported.[324] Another C4'-oxidised basic site analogue has been synthesised, the 2',2''-difluoro-derivative (66), rather than leading to amine derivatives on treatment with amines, undergoes a rearrangement to (67).[325]

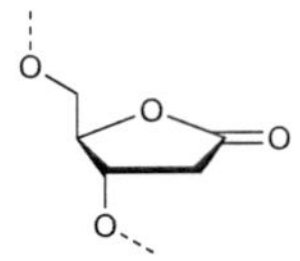

64 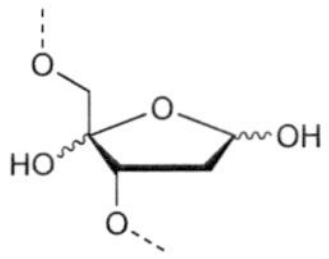 65

66 **67** **68**

A broad range of nucleobase analogues has been described, ranging from nucleobase isosteres to polyaromatic C-nucleosides. One of the widely-studied nucleoside isosteres in recent years is the difluorotoluene C-nucleoside (**68**), which is an isostere of thymidine, but possesses no conventional hydrogen bonding potential, a matter that has also come under considerable debate since (**68**) was first reported. Experimental and theoretical calculations have shown that the interaction between (**68**) and adenosine is 28% that of the normal T-A base pair, and that the electronegative fluorine does not form any hydrogen bonding, which arise rather from CH-N interactions.[326] In contrast to these findings, crystal structures of (**68**) incorporation into DNA by the RB69 DNA polymerase appear to suggest that the two fluorine atoms are forming weak hydrogen bonds with the opposing adenosine.[327] Transcription of DNA containing (**68**), as well as each of the other halogen analogues of it, by RNA polymerase II revealed that whilst incorporation of (**68**) was poor, transcription extension beyond it proceeded efficiently.[328] An isostere of the guanine oxidation lesion 8-oxoG, (**55**) was examined for its substrate activity with two DNA glycoylases and shown to be an even better substrate for Fpg glycosylase than 8-oxoG.[276]

A series of imidazole and tetrazole C-nucleosides with and without an ethyl linker, e.g., (**69**) and (**70**), have been synthesised and incorporated into RNA to examine their role in general acid-base catalysis in the VS ribozyme.[329] The modified base (**71**) has been designed for use in a triplex-forming oligonucleotide that will recognise a pseudouridine within a duplex.[330] Incorporation of the universal base analogue 5-nitroindole into siRNA has been examined. It was found that incorporation into the guide strand had little effect on potency when part of the RISC complex, but at position 15 of the passenger strand RNAi was greatly reduced, and that therefore the analogue could be used to reduce off-target effects.[331] The fluorescent, acyclic, ring-expanded analogue (**26**) and the corresponding diaminopurine derivative have been used to detect SNPs in RNA. When used in a matched duplex the analogue is extrahelical and fluorescent in the visible region, but at a mismatch site it is intercalated into the duplex and the fluorescence is quenched.[332] Building blocks for the incorporation of hydrazones and oximes into oligonucleotides have been described, for example 3-formylindole nucleoside, which when reacted with aromatic residues (in the case of the formylindole a hydrazine or oxyamine) are designed to quench a fluorophore opposing it in the complementary strand of a duplex.[333]

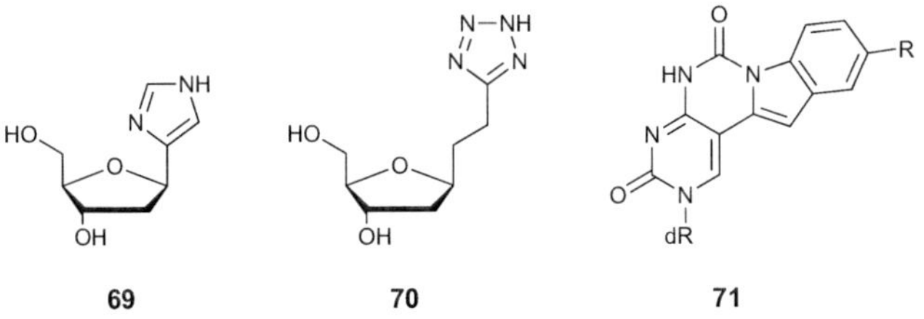

69   **70**   **71**

Various nucleosides have been introduced for crosslinking studies. Two vinylpurine analogues have been described, 2-amino-6-vinylpurine and 6-amino-2-vinylpurine, each having been used to form interstrand crosslinks in duplex DNA. 6-Amino-2-vinylpurine was shown to form crosslinks with an opposing cytosine,[334] whilst the other analogue when incorporated into RNA formed a sequence specific crosslink resulting in blocking of ribosomal activity.[335] 6-Vinylpurine has also been used to form specific interstrand crosslinks to an opposing cytosine residue.[336] 3-Cyanovinylcarbazole nucleoside (**72**) forms crosslinks with pyrimidines in the opposing strand of a duplex, and has been used for labelling plasmid DNA[337] and to arrest RNA transcription.[338] The acyclic furan derivative (**73**) undergoes methylene blue induced oxidation, (**73**) having been shown to then form a crosslink with N3 and N4 of cytidine.[339]

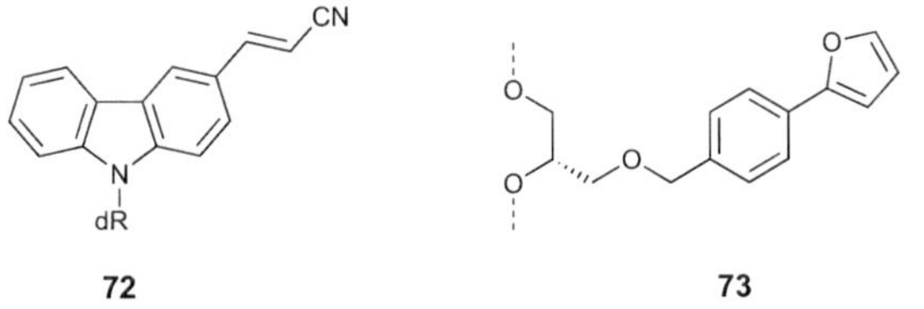

72   **73**

A number of aromatic residues have been incorporated into nucleic acids to induce charge transport through DNA and these are dealt with in section 3.4. Incorporation of an aromatic moiety, such as azobenzene, Methyl Red or thiazole orange, *via* a D-threoninol linkage onto the 5′-end of the sense of siRNA has been shown to greatly improve RNAi activity and strand selectivity.[340] Three or four aromatic rings connected by acetylene bridges form a stiff conjugated system but with sufficient conformational flexibility to be able to link together two strands of a DNA clamp in either duplex or triplex structures.[341] Single-stranded cyclic DNA containing the aromatic spacer (**74**) has previously been shown to form a triangular structure. When combined with ssDNA, generated from rolling circle amplification using Phi29 DNA polymerase containing complementary regions to the triangular DNA, they come together to form DNA nanotubes.[342] Anthrone, attached *via* a L-threoninol linker has been incorporated into a DNA probe for the detection of single nucleotide polymorphisms. When the anthracene residue is adjacent to a mismatch site it exhibits fluorescence but the fluorescence is quenched when adjacent to a matched base pair.[343] The same analogue when incorporated twice into ssDNA, spaced by varying length natural nucleotides, undergoes a reversible photodimerisation to induce a kink into the DNA strand.[344]

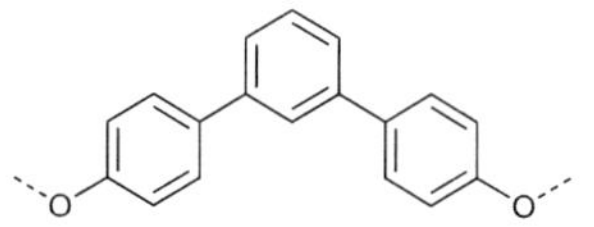

74

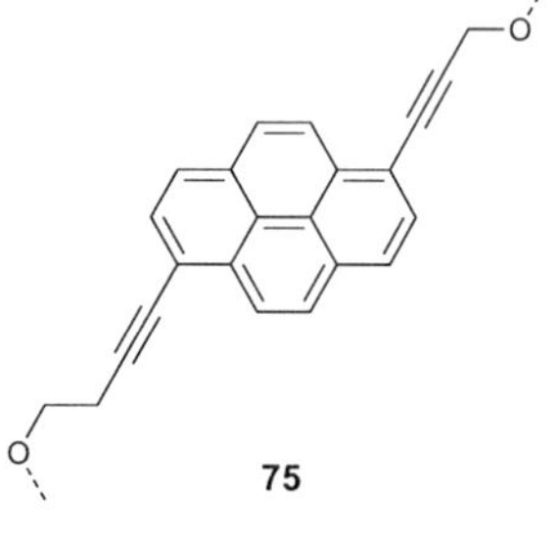

75

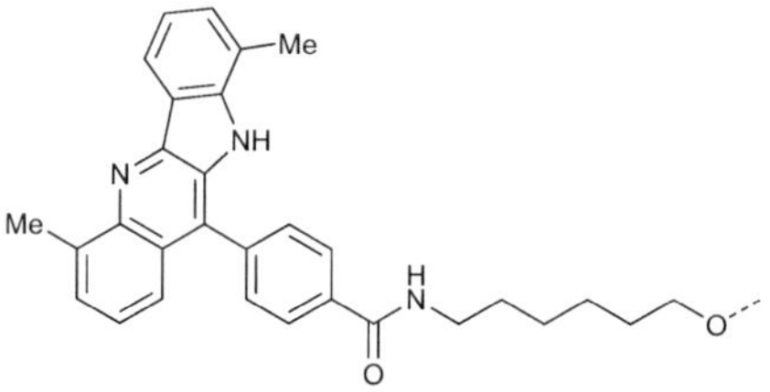

76

A number of reports describe the use of pyrene derivatives incorporated into oligonucleotides. A pyrene C-nucleotide has been incorporated into the centre of a DNA duplex, and when multiple consecutive substitutions of it are made, the duplex exhibits excimer fluorescence.[345] Three isomeric pyrene C-nucleoside triphosphates, in which the pyrene is attached to the sugar at different positions of the pyrene moiety, have been examined as substrates for DNA polymerases. It was found that they could be well tolerated by Klenow fragment DNA polymerase, though chain termination occurred after five subsequent dNTP insertions.[346] Pyrene, attached by a propane-1,2-diol spacer, was inserted into i-motif forming sequences where it was observed that the i-motif could form at higher pH.[347] The pyrene spacer (**75**), as well as the analogous perylene diimide spacer, have been inserted into DNA to assist formation of three-way junctions.[348] A DNA-based light-harvesting system has been reported that involves incorporation of phenanthrene (light connecting chromophore) and the excimer-forming pyrene (energy collection centre) through an extended π-stacked array.[349,350] A range of polyaromatic hydrocarbon residues has been described for intercalation into nucleic acid structures, and used to aid stabilisation of triplexes and of three-way junctions.[351] The quinoline derivative (**76**) has been tethered to the end of a pyrimidine strand to determine its effect at stabilising triplex structures.[352] It was found that the intercalator (**76**) had greatest stabilisation of the triplex strand when it followed the TAT triplet. The attachment of a poly(trihydroxymethylphenyl) residue to the end of ssDNA formed a dendron-like structure that was shown to be effective at forming micelles.[353]

A variety of base pairing systems have been examined during this review period. It has been shown that 5-methyl-2-thiocytidine forms a specific base pair with inosine in either DNA or 2′-O-methyl RNA.[354] Benner and co-workers have examined a number of novel base pairing systems having alternative donor/acceptor sites and have recently described the synthesis and stability of the new base pair (**77**):(**78**).[355] The synthesis of two base pairing systems, (**79**:**80** and **81**:**82**), each capable of forming four hydrogen bonds but with different donor/acceptor patterns, has been described.[356] Hirao and co-workers have previously described the base pairing system (**83**):(**84**, X=NO$_2$), and have now used these modified nucleotides in PCR. They demonstrated that the novel base pair was incorporated during PCR with extremely high PCR efficiency and very low rates of misincorporation.[357] They have also introduced a variant on this system (**84**, X=CHO, R=terminal alkyne) in place of (**84**, X=NO$_2$) in a transcription system that allows for modification of (**84**, X=CHO, R=terminal alkyne) using Click chemistry.[358] Romesberg and co-workers have also examined a number of potential base pairing systems, in particular (**85**):(**86**),[359] have also shown that this base pair functions well during PCR,[360] and have reported a crystal structure of the base pair within KlenTaq DNA polymerase.[361]

The final class of modified base pairing system is metal-mediated base pairs. It has previously been shown that a T-T or a C-C mismatch can be stabilised by Hg(II) or Ag(I) ions respectively. It has been shown that the T-Hg(II)-T base pair can be detected by Raman spectroscopy, having a specific Raman band at 1586 cm^{-1}.[362] Formation of Hg(II)-mediated base pairs between two thymine residues has been monitored by FRET.[363] Primer extension reactions with DNA polymerases and Ag(I) ions have been performed and a new base pair identified being the Ag(I)-mediated mispair between cytidine and adenosine.[364] It has also been shown that Ag(I) ions mediate a base pair between two imidazole nucleotide residues,[365] as well as between pyrroloytosine and cytosine.[230]

85 86 87

2 Aptamers and (deoxy)ribozymes

The field of aptamer and aptazyme research has seen an explosion of publications over the last few years, and there have been reviews published during this review period,[366–372] together with some method development advances.[373–378] As in previous years most aptamers and aptazymes are evolved using DNA rather than RNA, despite the fact that RNA can adopt more structures and many non-canonical base pairs, though of course this is probably a reflection of the fact that RNA is far more susceptible to degradation. During this review period there are also examples of aptamers constructed from non-native oligonucleotides.

The two major targets that DNA aptamers have been evolved for are thrombin and adenosine/ATP. Thrombin aptamers are usually G-quadruplex structures, and crystal structures of a thrombin aptamer binding to thrombin has been reported,[379,380] as well as a thermodynamic study revealing the importance of hydration of the aptamer for binding to its target.[381] A study of the effect of sugar conformation using the locked nucleic acid (87) in the G-quadruplex loops has also been reported.[382] Use of a single acyclic nucleotide has been reported as giving enhanced binding of the aptamer to its target,[383] and substitution of the thymidines by the bicyclic analogue (37) was found to improve anticoagulant activity.[196] Elongation of the thrombin-binding aptamer at its 5′-end gave a structure that varied depending upon the nature of the monovalent ion used to stabilise the G-quadruplex.[384] Replacement of the loops of the G-quadruplex structure by polyaromatic moieties, e.g., pyrene, was found to improve the binding of the aptamer to its target.[385] The binding of a thrombin-binding aptamer to thrombin has been examined at the single-molecule level.[386] A PCR-based assay has been used to measure thrombin using a split-aptamer,[387] and a dual aptamer has been described that binds to thrombin and docetaxel targeting tumour cell proliferation.[388] Attachment of a thrombin-binding aptamer to gold nanoparticles has been reported to yield enhanced binding to thrombin,[389] and a multiplex system for the detection of adenosine, thrombin and cocaine has been reported using the aptamers attached to gold nanoparticles.[390]

Thrombin-binding aptamers have been used to profile thrombin in human blood,[391] to detect thrombin by attachment of the aptamer inside a nanopore,[392] and in an assay that uses DNaseI as a nicking enzyme.[393] Assays have been developed to detect thrombin using its aptamers in a variety of methods, including fluorescence,[394–400] chemiluminescence,[401] electrochemiluminescence,[402] and plasmon resonance.[403,404] A photo-electrochemical detection system has been described in which the aptamer is attached to graphene with quantum dot nanolayers,[405] and graphene oxide has been used in an electrochemiluminescent assay,[406] and the aptamers have been attached to single-walled carbon nanotubes for electrochemical sensing.[407] Perylene/hemin nanocomposites have been used as redox probes for the detection of thrombin,[408] and stable enzyme precipitate coatings have also been described as a method for detecting thrombin.[409] The aptamer has also been used as a logic device.[410]

A second major group of aptamers that has been examined during this review period are those that recognise adenine, adenosine or its 5'-phosphate derivatives. Barbu and Stojanovic have used a range of adenosine analogues to examine adenosine-binding motifs.[411] Adenosine aptasensors have been evolved that can be used for the detection of abasic sites,[412] that respond to the activity of glucose oxidase,[413] and a dual aptamer has been reported that can bind to both a small molecule (adenosine) and a protein (lysozyme).[414] Many different methods of detection have been described for adenosine (and derivatives) aptamers, including fluorescence,[415,416] attached to gold surfaces for electrochemical[417–419] or electrochemiluminescent[420] detection, luminescence resonance transfer (LRET),[421] AFM,[422] and nanoplasmonic detection.[423,424] Adenine aptamers have been attached to carbon nanotubes,[425,426] titanium nanotubes,[427] silica nanoparticles,[428,429] and a gadolinium complex for MRI imaging.[430] Adenine aptamers have been used as nanodevices for drug delivery,[431] and logic devices.[432,433]

In addition to the above, aptamers have been evolved and used as sensors for Pb(II),[434] Hg(II),[435,436] As(III)[437] and uranyl[438] ions, as well as for various small molecules, including 8-oxoguanine,[439] melamine,[440] theophylline,[441] cocaine,[442–445] L-argininamide,[446] steroids,[447,448] polyaromatic foldamers,[449] and hemin.[450] Aptamers have been evolved for the peptide hormone systemin,[451] antibodies,[452–454] and various proteins including streptavidin,[455] MUC1,[456] ricin,[457] and a particularly interesting application for an aptamer evolved to bind to lysozyme that was used for visualisation of latent fingerprints.[458] Aptamers have also been evolved binding to histones,[459,460] and to cellular components[461] and cell surface proteins,[462–466] particularly to act as sensors of cancer cells.[467–470] Aptamers have been evolved for the detection of H1N1 influenza A virus,[471] and for the detection of HIV-1.[472,473] Various other proteins have been the target for aptamer evolution, including polyphosphatase kinase 2 from *M. tuberculosis*,[474] the oncogenic protein Shp2,[475] as inhibitors of 2-oxoglutarate-dependent oxygenases,[476] as well as for protein modification.[477,478]

As in previous years, there are far fewer RNA aptamers reported, despite the fact that RNA can adopt a far greater range of structures and base-pairing than DNA. A number of methodology reports have been described for the selection and use of RNA aptamers.[479–487] Locked nucleic acids (LNA) have also been used in the evolution of aptamers against HIV-1-RT.[488] A broad range of RNA aptamers has been described binding to small molecules, such as adenosine,[489] the cyclic dinucleotide c-di-GMP,[490,491] theophylline,[492] tetracycline[493] and helix-threading peptides.[494] Various reports have described the use of a GTP aptamer that has been modified by addition of fluorophores as a means of detection of the analyte.[495–497] There are also further reports on the use of the known green fluorescent protein aptamer[498,499] and thermodynamic studies of the malachite green aptamer.[500–502]

A number of proteins have been used to evolve RNA aptamers, including the tetracycline repressor TetR,[503] the cytosolic receptor RIG-1,[504] a family of GTPases,[505] a ^{99}Tc-labelled aptamer against the human matrix metalloprotease 9 for imaging tumour cells,[506] and against the heat shock transcription factor from D. melanogaster.[507] A prostate-specific cell-surface antigen (PSMA) aptamer has been radiolabelled with ^{64}Cu for imaging,[508] and an aptamer against HER2$^+$ breast cancer cells has been used for the targeted delivery of siRNA.[509] A report describes the development of an aptamer containing two binding sites against the HCV ($-$)-IRES domain I.[510]

In addition to DNA and RNA, aptamers have been evolved using nucleoside analogues. Thrombin binding aptamers have been evolved using threose nucleic acids (TNA, 22) as a potential progenitor to RNA,[125] and the known thrombin binding aptamer has been synthesised containing acyclic nucleic acids (termed unlocked nucleic acids, UNA) albeit with reduced affinity to thrombin.[511]

A number of different DNA aptamers with catalytic activity (DNAzymes) have been reported, though there are two classes of DNAzyme that have received the most attention, namely RNA-cleaving DNAzymes and DNA peroxidase catalysts. There are a number of metal-dependent RNA cleaving DNAzymes reported that rely of the presence of specific metal ions, including Pb(II),[512–516] Co(II),[517] Cu(II),[518,519] Zn(II) or Mn(II)[520–523] and uranyl[524] ions. Fluorogenic DNAzymes that cleave RNA in the presence of Mg(II) ions are also reported,[525,526] as well as DNAzymes that cleave RNA only in the presence of HIV-1-RT,[527] and one report that uses a DNA hairpin, the overall product of which is the formation of overlapping DNA strands forming DNA wires.[528] Other RNA-cleaving DNAzymes rely on the presence of histidine for cleavage,[529,530] or a C8-imidazolyl-modified adenosine at the catalytic centre.[531]

The other major class of DNAzyme reported mimics the action of enzymes like horseradish peroxidase, utilising in the process hydrogen peroxide as oxidant. Various reports describe mechanistic details of this class of DNAzyme,[532–535] and a number of modifications have also been described. These include the modification by addition of PEG to the DNAzyme to make the DNAzyme active in organic solvents,[536] the effect of backbone modifications,[537] and a variant of the DNAzyme that requires Pb(II) ions for activity.[538] A variant of the DNAzyme has been reported that mimics a NADH oxidase,[539] and variants are reported that act as logic-controlled biofuel cells,[540,541] and for the generation of DNA nanostructures.[542] Publications have been described that use the peroxidase DNAzyme as a colourimetric assay for the detection of PCR products,[543] for a fluorimetric assay of genetically-modified organisms,[544] and in electrochemical[545,546] and chemiluminescent assays.[547,548] Various reports describe amplification of DNA using the DNAzyme as sensor.[549-553]

There have been further reports on mechanistic details of the known 8-17 DNAzyme,[554,555] and further studies on the 10–23 DNAzyme in which modified nucleosides are used.[556-558] Other DNAzymes that have been reported include ligases,[559-563] peptide modifying DNAzymes,[564-566] an RNA-branching DNAzyme[567] and a DNAzyme that catalyses a reductive amination.[568] A DNAzyme is reported as a molecular device that converts input from a small molecule (FMN) into an RNA output,[569] and DNAzymes have been used as logic devices.[570,571]

There have been a number of reports for naturally occurring ribozymes, including the *glm*S ribozyme,[572-574] the hairpin ribozyme,[575,576] the hammerhead ribozyme,[577,578] the VS ribozyme,[329,579] the HDV ribozyme,[580-582] the *Azoarus* ribozyme,[583] as well as Group I[584-587] and Group II[588-591] introns. A number of other RNAzymes have been described, including a self-replicating L-RNA ribozyme,[592] and a self-replicating D-RNA ribozyme,[593] a ligase[594] including a crystal structure of a previously described RNA ligase ribozyme,[595] an RNA-cleaving[596] and a tetracyclin-dependent self-cleaving ribozyme,[597] a thiolesterase ribozyme,[598] a kinase,[599] an aminoacylation RNAzyme,[600] and mechanistic studies of the known Diels-Alderase ribozyme.[601,602]

Riboswitches are *cis*-acting elements that regulate gene expression by affecting transcriptional termination or translational initiation in response to binding of an analyte. A number of riboswitches are described which are listed for inclusion without comment. These include adenine,[603,604] ATP,[605] guanine,[606] purine,[607] preQ$_1$,[198,608-611] cyclic di-GMP,[612] theophylline,[613-615] SAM-I,[616-619] SAM-II,[620] SAM-III,[621] adenosylcobalamine,[622] thiamine pyrophosphate,[623] FMN,[624,625] lysine,[626-629] fluoride,[630] T-box,[631] M6,[632] and the btuB riboswitch from *E. coli*.[633]

3 Oligonucleotide conjugates

This section deals with a broad range of cargoes that have been attached to oligonucleotides, and includes a number of simple (novel) amino acids attached to proteins, saccharides, nanoparticles and solid surfaces. The larger part of this section deals with a range of fluorophores that enable visualisation of oligonucleotides, but in particular which have been been used in a vast range of single molecule studies. The section also includes advances in topics including template-directed synthesis, metal conjugates, charge transport, self-assembly constructs and finally a miscellaneous section describing a broad range of cargoes that have been attached to oligonucleotides.

3.1 Oligonucleotide-peptide conjugates

Recent developments in the synthesis of oligonucleotide-peptide conjugates include the synthesis of an RNA-peptide conjugate related to viral protein genome-linked (VPg) protein. The synthesis was carried out by first synthesising the peptide portion using Fmoc chemistry and using amino acid side chain protection with acid-labile or silyl protecting groups. The RNA was then synthesised by conjugation to the 5′-end of a nucleotide monomer, using reverse synthesis building blocks, *via* a tyrosine hydroxyl group.[2] An improved method for conjugation of peptides to oligonucleotides *via* a phosphoramidate linkage has been reported,[634] and 3′-peptidyl-tRNA mimics have been synthesised using native chemical ligation followed by a desulfurisation step involving treatment with TCEP, a radical initiator and glutathione.[118]

A number of publications describe various tRNA's with un-natural amino acids conjugated to them such that following translation the un-natural amino acid is incorporated into proteins. Amino acid-tRNA conjugates described include allyl-glycine, methyl-serine and biotinyl-lysine,[635] proline analogues,[636] and incorporation of crotonyl-lysine into histones.[637] *p*-Acetyl-lysine was also incorporated into antibodies such that the antibodies had dual specificities.[638,639] Various other amino acid-tRNA conjugates have been described for specific labelling of proteins *via* the modified amino acid. Various norbornene analogues of amino acids have been genetically encoded into proteins for site-specific labelling of the proteins *via* conjugation by Click chemistry with tetrazines,[640–642] and by Diels-Alder reactions.[643] Conversely, a tetrazine amino acid has been genetically encoded into proteins for conjugation by Click chemistry,[644] and bicyclononynes and *trans*-cyclooctenes have been genetically encoded for modification by Diels-Alder reactions.[645] Other genetically-encoded modified amino acids have been incorporated into proteins for modification by Click chemistry including a cyclopropene,[646] a bicyclononyne,[647] and a phenyl azide and allyl carbamate derivative.[648]

Oligonucleotide-peptide conjugates are frequently used now to aid cellular uptake, and this section will not describe all such publications during this review period. Some recent examples include delivery of siRNA to cancer cells using gastrin-releasing peptides,[649] and cellular uptake of DNA using the membrane-modifying peptide peptaibol.[650] Two oligonucleotides conjugated to collagen-like peptide have been examined at the single-molecule level for their translocation through a α-hemolysin nanopore.[651] Not only was translocation time dependent upon oligonucleotide length, but the collagen-like peptide unfolds corresponding to the rate of the entire conjugate translocation. Peptides have been used as scaffolds for pre-arrangement of oligonucleotide structures. A cyclic peptide has been used to assemble four DNA strands using either oxime formation or Click chemistry to prepare the conjugate. By this means a parallel-stranded G-quadruplex and an antiparallel *i*-motif were constructed.[652] A triplex structure has been designed using two dT$_{10}$ oligomers *via* a peptide containing melamine (**88**).[653] DNA oligonucleotides conjugated to diphenylalanine dipeptides have been shown to undergo a morphological transition from fibrillar to vesicular structures, and it is suggested this may be used as a delivery vehicle as it undergoes a pH-dependent change.[654] Oligonucleotide-peptide conjugates with a photocleavable group (**89**) attached to the 5′-end of the DNA has been used in a photolytic assay to study protein kinases using MALDI-TOF MS.[655] Oligonucleotides and peptides undergo a photolytic crosslink between tryptophan amino groups and guanine residues in a photo-induced electron transfer (PET) in the presence of Ru(II)-1,4,5,8-tetra-azaphenanthrine (Ru(II)-TAP) complexes.[656]

88

89

Oligonucleotides have also been conjugated to proteins, and a protocol has been described for the synthesis of such conjugates using azide-modified proteins and alkyne-modified DNA in a copper-free Click reaction.[657] Gene A* protein from the bacteriophage phi X174 is known to cleave ssDNA at its recognition site followed by formation of a stable phosphotyrosine linkage to the oligonucleotide. Using recombinant proteins containing the gene A* protein, any protein may be conjugated to DNA.[658] Triplex-forming ssDNA conjugated to a restriction endonuclease subunit has been examined, such that on a DNA duplex bearing two triplex-forming sites, it controls the dimerisation of the nuclease which then performs its catalytic cleavage.[659] Conjugation of up to four 48.5kbp of lambda phage DNA to a β-lactamase enzyme was found to control the enzyme activity. Compaction in the presence of spermine caused a decrease in enzyme activity, which could be increased beyond the activity of the enzyme alone by treatment with sodium chloride.[660] Conjugation of oligonucleotides to antibodies has been carried out and shown to enhance cellular targeting,[661] to be of use as small molecule carriers,[662] and to be of use by immuno-PCR[663] and in an immuno-fluoescence assay.[664]

3.2 Template-directed synthesis

Template-directed synthesis is an area of research that has been of interest for a number of years, although the number of publications is small. The non-enzymatic synthesis of DNA has been described using a 3′-amino DNA primer with an activated 5′-monophosphate,[665] as well as the converse allowing for synthesis of DNA in either direction.[41] The principle of template-directed synthesis relies on bringing together two DNA strands on a complementary sequence, with the two strands bearing functional groups such that once in proximity the two functional groups react, forming new bonds. There have been a few examples of this during this review period. One of the more common reactions is the Staudinger reaction in which a phosphine reduces an azide-modified fluorophore to the amine which is then fluorescent, indicating that a reaction has occurred.[78,321,666] There have been reports of alkylation using a 5′-iodide and a 3′-thiophosphate,[667] Wittig reactions,[668] peptide bond formation,[669]

Diels-Alder and Friedel-Crafts reactions,[670] and amidation and reductive amination.[671] The template-dependent synthesis of a palladium catalyst has been reported,[672] and multi-step syntheses have been described[673] including the synthesis of macromolecules.[674,675]

3.3 Oligonucleotide-metal conjugates

There have been many reports describing nucleic acid-gold conjugates, with the two main areas of research involving either gold electrodes or gold nanoparticles (gold and silver nanoparticles have been used in SERS, which is dealt with in section 4.3). The reversible melting thermodynamics of duplex DNA bound with netropsin has been examined,[676] and other hybridisation assays have been described using stem-loop probes[677] and DNA concatamers.[678] A number of reports describe the use of nucleic acids bound to a gold electrode as an assay for a specific target, such as exosomal microRNAs,[679] human telomeric DNA,[680] gene-specific DNA hypermethylation,[681] the breast cancer gene BRAC1,[682] and for quantification of transcription factor binding using cellular extracts.[683] Methylene blue labelled DNA bound to a gold electrode has been used as a molecular beacon for the detection of Y-shaped DNA or DNA junctions.[684] Other assays based on interactions of methylene blue and gold electrodes for the detection of Y-shaped DNA have been described based on restriction endonuclease-aided target recycling,[685] or for distinguishing chiral metallosupramolecular complexes.[686]

There are a few reports describing improvements for the synthesis of nucleic acid-functionalised gold nanoparticles. Gold nanoparticles as small as 83 nm have been functionalised with DNA in only one region using template electrodeposition, with detection limits in the pM region.[687] Gold nanoparticles with defined numbers of oligonucleotides attached can be separated using gel electrophoresis.[688] Synthesis of DNA-gold nanoparticles with defined orientation and conformation has been achieved using diblock oligonucleotides free from modification; a poly(A) tail was found to act as a good anchoring block for preferential binding with the gold surface.[689] Synthesis of gold nanoparticles with quantitative functionalisation by thiolated DNA may be performed in a matter of minutes using a pH 3 citrate buffer.[690] A protocol for the assembly of DNA-gold nanoparticle conjugates into defined shapes (*e.g.* triangles, squares) has been reported.[691] Conjugation efficiency of oligonucleotides has been improved (30–50%) using poly(dT) spacers.[692]

Nucleic acid-Au nanoparticles have been used in a number of applications, including cellular delivery. siRNA has been shown to have improved uptake when conjugated to Au nanoparticles,[693,694] a multi-colour fluorescent nucleic acid has been used as a multi-colour probe for the detection of tumour-related mRNA in living cells,[695] and DNA-capped nanoparticles have been used for the delivery of the anticancer agents doxorubicin and actinomycin D.[696] Various sensors have been reported using nucleic acids conjugated to gold nanoparticles, including a pH colourimetric sensor based on A-motif formation of poly(dA) at low pH,[697] a colourimetric sensor for the detection of SNPs,[698] an ultrasensitive detection of DNA using real-time ligation chain reaction,[699] as well as a method using isothermal strand displacement PCR,[700] and a method for assaying proteins using DNA conjugated to an affinity ligand.[701] A glucometer has been devised based on hybridisation of HIV DNA to probes bound to gold nanoparticles which in the presence of invertase converts sucrose to glucose to give positive readouts.[702] Other applications include a molecular beacon used for the detection of microRNA,[703] DNA-Au conjugates containing CpG islands for cytokine induction,[704] and DNA-Au conjugates for formation of DNA origami structures.[705]

Other nucleic acid-gold conjugates include a photochemical method for nanolithography using ssDNA bound to a gold surface *via* oligo(ethylene glycol) linkers,[706] an oxidative nanolithography method based on oxidation of Au(0) to Au(III) using *N*-bromosuccinimide,[707] the use of gold nanorods for formation of DNA assemblies detected using reversible plasmonic CD that claims detection limits down to 75nM,[708] and the effect of polycations (spermine) on mismatched DNA bound to single Au(II) crystals has been examined.[709] An aptamer bound to an Au(III)-iron oxide nanoparticle has been used as an electrochemical sensor based on DNase I-catalysed target recycling.[419]

Hairpin DNA, labelled with dye-functionalised silver nanoparticles, bound to a gold surface have been used in a SERS assay to recognise a third strand, in this example a sequence from HIV – 1 U5 LTR, capable of forming triplexes with the hairpin DNA.[710] Stable silver nanoparticles have been functionalised with thiolated DNA with high density at pH 3; adsorption was slower than onto gold nanoparticles, but considerably more efficient than at neutral pH.[711] A number of applications of nucleic acid-silver nanoclusters have been reported over the past few years. These conjugates are prepared by treating the nucleic acids with silver ions and reducing with borohydride. The properties and past applications of these conjugates have been reviewed by Latorre and Somoza.[712] Novel applications that have been reported during this review period include the detection of SNPs wherein there is a substantial shift in the fluorescence maxima depending on the environment of the silver nanocluster.[713] Using endonuclease-assisted target recycling, a sensitive method for the detection of Y-shaped DNA structures has been devised using silver nanoparticle-tagged carbon nanospheres.[714] Silver nanoclusters have also been used as signal transducers in DNA logic gates (see section 3.6).[715] Site-specific nanoclusters have been generated using triplex DNA, where the nanocluster formed specifically at a $CG.C^+$ site.[716] G-Rich DNA has been metallised with silver, which prevents formation of G-quadruplexes, even in the presence of quadruplex-stabilising ligands such as porphyrins. However, the same construct in the presence of a thiol undergoes a change in which the thiol sequesters the silver clusters, allowing for formation of the stabilised quadruplex, and this has been used as a method for detection of biological thiols (*e.g.* cysteine, glutathione).[717] Near infrared-emitting DNA conjugates have been formed by silver metallisation using HPLC-MS with in-line spectroscopy, where it was found that 10–20 silver atoms were required to shift the emission spectra into the infrared region.[718] Silver- and mercury-mediated base pairs have also been described, but are discussed in section 1.2.3.

Another metal that is widely-studied as conjugates with nucleic acids is platinum. An electrochemical method for the detection of DNA has been described in which thiolated DNA is bound to platinum nanoparticles, with the complementary strand bound to gold nanoparticles.[719,720] Such constructs, it has been suggested, allow for detection of single biomolecules. However, the majority of publications of platinum-nucleic acid conjugates concern the use of platinum species as anti-cancer agents, especially of *cis*-platin. A fluorescent array-based system has been reported for analysis of a range of platinum species related to *cis*-platin,[721] and the effect of mono- and bifunctional adducts of platinum species and their interaction with DNA has been examined.[722] The repair of platinated DNA by the Y family polymerase Pol η has been examined,[723] and a crystal structures of platinated DNA within human DNA Pol η have been reported.[724,725] Interactions of platinated DNA by the HMG proteins HMGB1[726] and HMGB4,[727] whilst repair of platinated DNA by the repair factors XPA, XPF and XPG[728] are also described.

The generation of reactive oxygen species by *cis*-platinated DNA has been examined using scanning electrochemical microscopy.[729] Platinated DNA has been shown to affect the conformation of zinc fingers,[730] and the effect of a platinum crosslink in the hammerhead ribozyme has been examined.[731] The repair of DNA by an azolated-bridged Pt(II) complex has been examined as novel antitumor agents,[732] and the mechanism by which *trans*-platinum species act as chemotherapeutic agents has also been reported.[733]

A few other nucleic acid-metal conjugates/complexes have been reported. A G-quadruplex stabilised with various divalent metal ions were assessed for their ability to catalyse Diels-Alder reactions. It was found that Cu(II) ions gave the highest yield of product, with *ee* up to *ca.* 75%, and that the absolute configuration of the Diels-Alder product could be reversed by switching between parallel and anti-parallel G-quadruplexes.[734] The same group has investigated Friedel-Crafts reactions with Cu(II)-G-quadruplexes, and has reported similar *ee* and absolute configuration as the Diels-Alder reaction above.[735] Enantioselective intermolecular oxa-Michael addition of achiral alcohols to enones, with *ee* up to 86% have been reported using duplex DNA with Cu(II) complexes of intercalated bidentate nitrogen ligands.[736] Incorporation of terpyridine-linked C5-dU (**90**) into duplex DNA resulted in DNA self-assembly of higher order structures with various divalent metal ions, but in particular with Ni(II) and Fe(II) ions.[193] A range of 7-deaza-7-oligopyridine derivatives has been incorporated into DNA for functionalisation using metal ions.[218] A novel stem-loop structure has been devised such that a nucleoside is replaced by either a terpyridine or dipicolylamine moiety such that the chelating groups are opposed at the neck of the duplex region of the hairpin. In the presence of transition metal ions, in particular Zn(II) ions the whole structure is stabilised by chelation of a metal ion.[737] Duplex DNA has also been stabilised by incorporation of Zn(II)-porphyrin residues at the ends of the duplex,[738,739] and a Zn(II)-porphyrin attached at C5 of dU has been used as a molecular switch.[740]

90

DNA conjugated to gadolinium phosphate nanoparticles have been reported as MRI contrast agents.[741] The modified purine analogue (**49**, M = Ir or Rh, L = ligand) has been synthesised for the introduction of iridium or rhodium species into nucleic acid structures.[242] The metal ion chelators DOTA and NOTA have been conjugated to the 3′-and of oligonucleotides for chelation of ^{68}Ga for *in vivo* imaging by PET.[742]

3.4 Charge transport

Charge transfer through DNA is an essential process in some biological systems associated with DNA damage and repair. The repair of damaged DNA by the XPD helicase containing an iron-sulphur cluster has been examined at the single molecule level using AFM.[743] Charge transport has been carried out to detect a C-A mismatch using haemoglobin as electron sink.[744] The transport of positive charge, or holes, through DNA has been examined in conjunction with histones where it was found to modulate DNA damage of guanine residues in the nucleosome core particle.[745] Charge transport has also been examined in a three-dimensional origami DNA structure.[746] Charge transport has been studied using PNA, where it was observed that the rate of transfer is twice as fast through PNA (**10**) than it was through PNA bearing a methyl group at the γ-position.[84] Substituting LNA into the DNA duplex causes a modest decrease in charge transport.[143–145]

Various modified nucleobases have been examined in charge transport. The incorporation of C5-phenylethynyl-dU into DNA drastically enhances charge transport,[186] as does incorporation of 5-methylzebularine.[144] A number of C5-modified dU analogues have been incorporated into DNA where they were found to enhance charge transport, possible by reducing the reverse rate of charge transport.[747] Terthiophene incorporated into a DNA hairpin structure has been used as an electron donor for charge transport,[748] as has pyrene.[749] The dynamics of charge transport through DNA hairpins having an anthraquinone end-cap have been examined.[750] Charge transport through an aptamer has been used as an electronic biomarker of early-stage lung cancer,[751] and gold electrodes have been used as a redox label for charge transport through DNA.[752]

3.5 Fluorescence

Probably the most common class of nucleic acid conjugate are fluorophores which are used for a range of applications and are the basis of a number of single molecule studies. A number of known and some novel fluorophores have been used during this review period. A review of methods for studying RNA interactions using fluorophores has been published.[753] The modified base 2-aminopurine (2AP) is often used as a fluorophore as it can be considered a surrogate for adenine, causing minimal perturbations when substituted into oligonucleotides. 2AP has been used to study breathing fluctuations in DNA,[754] to study G-quadruplex structures[755] and to study pH-induced conformational changes in a RNA helix from the 23S ribosomal RNA.[756] TAMRA has been used to monitor the dynamics of DNA-based reaction circuits,[757] and cleavage of G-quadruplexes,[758] and Cy dyes have been used to study RNA folding.[759] Thymidine has been doubly-modified by substitution at C5 with thiazole orange to study duplex formation,[760] and used in qPCR to minimise false positive signals.[761] Novel Cy dyes have been reported,[762] and a novel fluorescein dye has similarly been studied.[763] The fluorophore 6-methylisoxanthopterin (91) has been used as a G surrogate to study G-quadruplexes[764] and to study DNA-protein interactions.[765]

A number of fluorescent groups have been attached to the natural nucleobases, with pyrene being a common example, and pyrene-modified thymidine has been used to study G-quadruplex formation.[188] Fluorene has been attached to thymidine to study *i*-motif structures,[189] naphthalene has been attached to thymidine as an excimer fluorophore,[187] the analogue (92) has been described as a GFP-like fluorophore,[766] and the silole derivative (93) has been reported attached to C5-allylamino-dUTP.[767] Kool and co-workers have described a range of aromatic *C*-nucleosides and have used them for real-time monitoring of biological systems.[768] A 3-hydroxychromone universal nucleoside has been synthesised for investigating DNA interactions.[769]

A number of applications have been described relying on fluorophores, and a strategy for high-throughput analysis of DNA-protein interactions has been reported.[770] Fluorescence has been used to monitor DNA base excision repair,[771] to monitor the activity of the endonuclease XPF-ERCC1,[772] as hybridisation probes,[773,774] for the detection of protein interactions,[775] in PCR[776] and in fluorescence correlation spectroscopy.[777,778]

Molecular beacons (MBs) are frequently used as an optical method for the detection of target oligonucleotides, and a review of luminescent detection of DNA-binding proteins has been reported.[779] A Raman MB has been reported for efficient detection of DNA followed by silver deposition onto target DNA bound to magnetic nanoparticles.[780] MBs attached to a gold electrode have been used for electrochemical detection,[684] and an L-DNA MB has been reported that acts as a DNA thermometer in living cells.[781] Nucleotides modified by photocleavable nitrophenyl moieties have been described for light-activatable molecular beacons.[782] Coralyne promotes the formation of stable duplexes of poly(dA), and this has been used to design a molecular beacon based on conformational changes induced by coralyne.[783] Use of the dye 7-hydroxycoumarin allows for the use of a molecular beacon responsive to pH as the fluorescence of the dye is quenched upon protonation.[784] 2-Ethynylfluorenone moieties attached at C5 of dU have been used in a MB as a quencher-free system for the detection of SNPs.[785] 2′-Azidoribothymidine has been used for introduction of alkynyl dyes onto oligonucleotides for use as MBs.[97]

Applications of molecular beacons described in this review period include monitoring of the endonuclease activity of mammalian Ago2 protein,[786] quantitation of epithelial tumor marker mucin 1,[787] *in vitro* quantitation of specific miRNAs,[788] detection of G-quadruplexes,[789] and detection of K(I) ions using a G-quadruplex based MB.[790] Molecular beacons have been used as biosensors designed to detect allosteric inhibitors and activators,[791] and as a logic gate.[792]

Quantum dots may also be attached to oligonucleotides for fluorescence monitoring, and they have been used to monitor DNA-binding proteins,[240] for cancer cell imaging,[793] and for monitoring *in vivo* delivery of oligonucleotides.[794] A novel CdTe quantum dot has been described and applied as a biosensor for the detection of DNA.[795] A method for immobilisation of quantum dot-modified oligonucleotides onto solid supports has been described,[796] and a three-colour FRET system using quantum dots and two fluorophores has been reported.[797]

The majority of publications in which FRET is used are single molecule applications (see below), but there remain a number of 'bulk' applications of FRET, and a review on advances on FRET-based methods has been published.[798] A new three-colour FRET system has also been described.[799] FRET has been used to study conformational changes of the 30S ribosomal unit during mRNA translocation,[800] to study DNA packaging during viral packaging,[801] DNA bending by the *E. coli* integration host factor (IHF),[802] DNA loop formation by Cre-mediated recombination[803] and by the Lac repressor,[804] triplex DNA formation,[805] and during the formation of hairpin DNA formed by T:T mispairs in the presence of Hg(II) ions.[363]

Other uses of FRET include a method to study gene expression,[806] characterisation of the interaction between a tRNA and small binding protein B (SmpB),[807] examination of the mechanism of action of the FokI endonuclease,[808] and real-time monitoring of RAG-catalysed DNA cleavage during V(D)J recombination.[809] The binding of small molecules to various DNA structures has been assessed using FRET,[810] as has the binding of a neomycin dimer to HIV-1 TAR RNA,[811] and hairpin loop structures containing porphyrin-modified uracil moieties that act as a molecular switch.[740]

There have been many developments in the study of single molecules over the past year, and this has seen a vast expansion in publications in this field. A number of review articles and methodologies have been reported.[812–817] Systems that have been examined at the single molecule level include studies on the ribosome,[818–820] DNA[821–827] and RNA[828–830] polymerases, repair enzymes,[831–837] a telomerase,[838] helicases,[743,839] topoisomerase,[840] DNA methylases,[841,842] translocases[843,844] the hTERT promotor unfolding and refolding a G-quadruplex,[845] a Cre-mediated recombinase,[846] TATA binding protein binding to and bending DNA[847] and an ATPase motor winding ssDNA.[848] The binding of thrombin to its aptamer has been described.[386]

Other single molecule studies include single molecule sequencing,[849,850] nuclear export of mRNA,[851] translocation of DNA through a nanopore,[312,651,852] and dynamics of minor-groove binders binding to DNA.[853,854] The dynamics of Holliday junctions[855] and three-way junctions[856] has been examined as has the bending of DNA,[857] the dynamics of DNA supercoils,[858,859] the binding of Na(I) and Mg(II) ions to RNA,[860,861] binding of DNA intercalators,[862] the switching of DNA containing an azobenzene moiety,[863] and binding of actinomycin D to DNA.[864] Dynamics of unwinding and rewinding of the nucleosome has been examined,[865] as has the dynamics of overstretched DNA,[866,867] and the observation of growth of a DNA origami tile,[868] the docking of an RNA tetraloop to its receptor,[869] and conformational switching of DNA on a gold electrode.[870]

3.6 Nanostructures and nanodevices

There are two types of oligonucleotide barcodes that have been described. The first is a defined sequence that may be used to identify the presence of a given organism. For example, the nuclear ribosomal internal transcribed spacer region has been identified as a biobarcode for fungi,[871] and a biobarcode has been identified for the antibiotics found in soil Fasamycin A and B[872] and for uropathogen specific species.[873] The second type of barcode here is the addition of a known oligonucleotide sequence to a target to allow for specific recognition of that target. In this manner a chemical library has been constructed with 30,000 drug-like small molecules tagged with unique DNA sequences and from this library interleukin-2 inhibitors were identified.[874] An assay has been reported that uses pyrosequencing of DNA-tagged reporter groups that allowed for multiple sequences to be identified without any pre-separation.[875] A photocleavable barcode-antibody conjugate system has been reported allowing for multiplexed protein analysis in single cells.[876] Barcodes have been conjugated to gold nanoparticles to allow for electronic detection of disease-related proteins.[877] Fluorescently-labelled DNA has been self-assembled into DNA nanorods bearing 216 distinct barcodes which were used to tag yeast surface receptors.[878] dNTPs have been tagged with barcodes for identification after incorporation into DNA by a DNA polymerase.[879] A new method for RNA sequencing has been described that uses single-molecule barcodes and which was used for transcriptome profiling of *E. coli* more accurately than conventional RNA sequencing methods.[880]

The major part of this section describes the various ways that oligonucleotides can be constructed into what has become known as DNA origami. DNA can be controlled to adopt a large range of defined structures either through hydrogen bonding or by use of a small molecule conjugated to the oligonucleotides, and two reviews of DNA origami have been published.[881,882] Also, a 'genetic code' for preparing DNA-gold nanoparticles of defined morphology has been reported.[883] Cryo-EM has been used to visualise a DNA origami structure about twice the size of a prokaryotic ribosome in high detail due to the dense packing of oligonucleotides.[884] A DNA-functionalised quantum dot has been synthesised that is suitable for use in DNA origami.[885]

A number of well-defined structures have been reported, including tetrahedral,[886,887] spheres,[888] rotaxanes,[889] tubes[890,891] prisms[892] and rectangles.[705] A DNA tetrahedron has been used for *in vivo* siRNA delivery.[893] Folding and thermal denaturation of a DNA rectangle has been visualised using AFM,[894] as well as folding and unfolding of a DNA cuboid.[895] DNA hexagonal nanostructures have been formed linked together with a photoresponsive azo-unit to control the assembly.[896] Barrel shaped DNA bearing cholesterol moieties has been formed that penetrated and spanned a lipid membrane to act as an ion channel.[897] Three dimensional structures have been formed using DNA bricks,[898] and using DNA-conjugated gold nanoparticles and silver nanoparticles shapes resembling snowmen have been reported.[899]

Various DNA tiles and tube structures have been reported. Nanotubes and two dimensional arrays have been made using multi-helix bundles,[900,901] and various two dimensional structures have been formed using rigid organic linkers.[902] DNA arrays have been formed using triplexes, crosslinking the structures with psoralen attached to the triplex strand.[903] Long chains of duplex DNA with overhanging chains have been self-assembled and monitored on a crystal quartz microbalance.[904] DNA has been folded to form parallel left-handed helices,[905] DNA conjugated to gold nanoparticles have been assembled into helices of defined handedness.[906]

Conducting polymers have been formed containing 2,5-bis(2-thienyl)pyrrole attached to the exocyclic amino group of cytosine.[907] Two DNA duplexes containing bubble regions can be designed such that the bubbles interact through hydrogen bonding to form self-assembly structures.[908] Insertion of an *i*-motif-forming sequence into DNA allows for pH-responsive self-assembly structures.[909] DNA self-assembly on near-infrared-responsive gold nanoparticles has been utilised for the efficient delivery of cancer agents that are released by NIR irradiation,[910] and DNA origami has also been used as a carrier of doxorubicin.[911]

DNA origami has been formed on modified graphene,[912] and novel nanomaterials have been formed by assembly of DNA-conjugated gold nanoparticles bound to a DNA tile array with quantum dots, both being bound at defined positions on the DNA scaffold.[913] DNA has also been used in conjunction with single-walled carbon nanotubes for the detection of Sr(II) ions,[914] for ultrasensitive detection of DNA[915] and for the growth of palladium nanocrystals as catalytic agents.[916] DNA assemblies have been formed using pyrrole-imidazole polyamides[917] and helical structures have been formed using silica mineralisation.[918] A DNA nanostructure has been used as a scaffold to organise block copolymer chains to generate DNA-block copolymer cages.[919] DNA nanostructures have also been formed around various proteins.[920–923] DNA crosslinked through Click chemistry followed by metallisation generates conducting nanowires.[924] DNA nanoplates laid down onto SiN nanopores act as ion channels which it is suggested could be used for single molecule studies,[925] whilst a DNA surface has been used as a platform to transfer protein ensembles to a single molecule microscope[926] and origami structures containing zeptolitre volume sites have been created for single molecule studies.[927]

A set of 10^{12} random 20-mer oligonucleotides were found to produce a hierarchical sequence of structured self-assembly resulting in liquid crystal phases.[928] Long chains of DNA generated by a polymerase were found to form a meta-hydrogel. It has liquid-like properties when out of water but solid-like properties in water and reforms its original shape.[929] Addition of the bacterial motor protein FtsK50C to a DNA hydrogel causes the gel to stiffen and creates stochastic contractile events.[930]

Various DNA-based nanodevices have been described including photoresponsive,[931] pH[932] and Hg(I) ion[933] responsive switches. A large number of logic gates have been reported,[934–936] including those with electrochemical,[937] fluorescent[938–941] or colourimetric[942] output, or responding to light[943] or Ag(I) ions.[715] There are also DNA walking devices,[944] DNA transporters,[945,946] a DNA motor that can navigate a network of tracks,[947] a light-driven actuator,[948] and DNA-based machines.[949–951]

3.7 Miscellaneous conjugates

In addition to the examples described in earlier sections, there have been a number of other oligonucleotide conjugates that have been prepared using Click chemistry. A protocol using tris(triphenylphosphine)copper(I) bromide has been described for Click reactions to 5′-alkyne-modified RNA.[952] Bis- and tris-alkyne phosphoramidites have been synthesised for multiple 5′-labeling of oligonucleotides.[953] DNA has been doubly-labelled in a one pot reaction using orthogonal cycloaddition chemistry. Using a C5-alkynylated dU for Click chemistry and a 5′-*trans*-cyclooctene for inverse electron-demand Diels-Alder reactions the DNA was labelled with two fluorophores suitable for use in FRET.[954]

RNA has been 5′-labelled with a cyclooctyne group for use in strain-promoted (copper-free) Click reactions whilst still attached to the solid support.[955] DNA modified by a *trans* cyclooctyne derivative have also been used to attach to azide-modified cell-surface glycans to help control cell adhesion.[956] 5′-Alkyne-modified siRNA has been modified with azide derivatives of anandamide (an arachidonic acid derivative), folate and cholesterol to assist in receptor-mediated uptake for gene silencing.[957] Alkynyl terminally-modified RNA has also been used to introduce ethylene glycol linkages, such as HEG, used at splice correction sites, but was found to be inhibitory possibly due to the triazole linkage.[958] By use of H-phosphonate chemistry DNA has been site-specifically modified at internucleotide linkages *via* a phosphoramidate linkage bearing an alkyne.[43] The authors carried out Click reactions with various sized azidodiamondoids and observed that thermal stability of duplexes increased with the size of the diamondoid.

Different lipid structures have been conjugated to oligonucleotides, frequently to assist cellular uptake, which has been recently reviewed.[959] siRNA modified with both an amine-modifier and palmitic acid led to greatly enhanced inhibition of renilla luciferase,[960] and siRNA has been modified with a number of lipids where it was found that cholesterol gave the best cellular uptake and gene silencing.[961] DNA has been modified with three anabolic androgen steroids and bound to gold chips for detection using antibodies by SPR.[962]

Another common oligonucleotide conjugate involves various carbohydrates. A DNA-based microarray has been prepared in which the DNA is conjugated to a library of fucosylated glycoclusters to identify binders toward *Pseudomonas aeruginosa* lectin B.[23] DNA has been used to barcode glycosoaminoglycans (GAGs); using chondroitin sulfate and sialylated N-glycans as controls, following conjugation to DNA the GAG could be identified using qPCR.[963] Duplex DNA has been end-capped using apolar derivatives of glucose and cellobiose which stack onto the ends of the duplex greatly enhancing duplex stability.[964]

A protected maleimide building block (**94**) has been synthesised for site-specific conjugation with thiol-containing compounds. The protecting group is removed in a retro-Diels-Alder reaction by boiling in toluene for four hours.[965] DNA containing a terminal thiol group has been conjugated to a bis-maleimide, which is then reacted with a free amino group from a cell-surface glycan to control cell adhesion processes.[956] Other thiol-conjugated oligonucleotides have been described to covalently bind to cysteine in a single cysteine mutant of MutS,[966] and for the formation of DNA amphiphiles in which the DNA forms the hydrophilic portion and a hydrocarbon disulfide-linked chain the hydrophilic part. The amphiphiles are able to trap water-soluble drug-like compounds, and as the disulfide link is radiation-sensitive, X-irradiation results in release of the drug.[967] siRNA has been conjugated to polymeric materials connected through a disulphide linkage forming hydrogel nanoparticles that under reducing conditions slowly release siRNA intracellularly.[968] Photolabile adenosine-based transcription initiators have been reported and used to synthesise RNA using T7 phi2.5

promoter. The resultant RNA has a free thiol group and a photolabile-protected amino group that can be used for further modification by, for example, biotin.[969] DNA has also been synthesised with a terminal thiol linker that is protected by a 2-nitrobenzyl group.[970]

Destabilising modifications near the centre of an siRNA duplex in the RISC complex frequently enhances the efficacy of gene silencing. This has been further tested by introducing non-scissile alkyl linkages in the centre of the siRNA, where it was found that potent RNAi activity was retained. An i-motif oligonucleotide has been conjugated with poly(propylene oxide) and found that it underwent a pH-dependent morphology change shifting from spherical micelles to nanofibres.[971]

Oligonucleotides have been attached to a variety of different surfaces. DNA has been attached to a glass prism coated with a photochemical monolayer in a density gradient to study DNA occupancy using evanescent fluorescence.[972] The dynamics of DNA reorientation on a glass electrode has been measured using fluorescence quenching.[973] RNA has been bound to a quartz crystal microbalance to measure the traveling time of a translating ribosome along mRNA.[974] Graphene is proving to be an ideal candidate for development of novel nanodevices, and a protocol for efficient attachment of DNA to graphene surfaces has been described. Using a sulfo-NHS ester of carboxylated graphene, amino-modified DNA rapidly undergoes efficient carbodiimide coupling to the graphene surface, where it may have applications in the field of biosensing.[975]

94

95

A photolabile, protected chloroaldehyde (**95**) has been conjugated to the end of DNA for the generation of site-specific crosslinks with complementary oligonucleotides after irradiation at 365 nm for one minute.[976] Su et al. have reported a general strategy for photocaging oligonucleotides and some labelling applications.[977] A building block for the introduction of a squaraine derivative (**96**) has been described and incorporated into DNA. The squaraine moiety exhibited high molar absorptivity, environment-sensitive fluorescence properties and intense response in CD making it useful for DNA-based optical probes.[978] The translesion synthesis with DNA polymerases β, κ and ι of DNA containing a psoralen crosslink has been examined *in vitro*.[979]

96

4 Nucleic acid structures

The number of nucleic acid structures described in this review period continues the trend of increasing number in spite of the fact that this review period for X-ray crystallographic and NMR structures covers a two year period. In addition, the number of other structural methods has continued to increase, especially in areas involving all forms of electron microscopy. There are also some newer developing areas described, such as an increase in applications of EPR and SERRS. Small-angle X-ray scattering (SAXS) is also dealt with in section 4.3.

4.1 Crystal structures

There has continued to be an increase in the number of complex systems involving protein-nucleic acid interactions, and structures of the ribosome, and many of these structures are too complex to discuss in any detail, but are included as they form the major structural work described during this review period. Complex structures include the ribosome,[980–990] nucleosome,[991,992] DNA[993–1001] and RNA[1002–1009] polymerases, (including incorporation of RNTPs by human DNA pol λ,[1010] incorporation of an unnatural base pair[361] and 7-deazapurines or C5-modified pyrimidines[292] by KlenTaq DNA polymerase, structures of human DNA pol η forming a phosphodiester bond,[1011] the structure of RB69 DNA polymerase with DNA containing 3-deaza-dA,[293] and the mutant RB69 (Y567A) DNA polymerase forming a wobble pair between (47) and dATP,[233] and translesion synthesis past cis-platinated DNA,[724,725] past the thymine isostere difluorotoluene (68),[327] the polymerase Dpo4 with an incoming triphosphate of (33),[158,159] human DNA pol β bypassing 8-oxoguanine[269] and in ternary complex with (S)-α,β-fluoromethylene-dATP,[1012] and human DNA pol ι bypassing a locked DNA analogue,[158] reverse transcriptases[1013] and clamp loader complexes,[1014,1015] tRNAs,[1016–1018] transcription and related factors[1019–1031] including zinc fingers,[1032,1033] and terminators,[1034,1035] receptors,[1036,1037] various complexes associated with miRNA processing,[1038–1045] telomerases[1046–1048] a translocase,[1049] topoisomerases[1050–1052] and helicases.[1053–1059] Structures include complexes with various repair enzymes including endo-[305,1060–1069] and exo-nucleases,[1070–1074] photolyases,[203,1075,1076] for example, repair of the spore photoproduct (42) and tRNA modifying enzymes,[1077,1078] and the P. falciparum single-stranded binding protein wrapped around DNA.[1079] Complexes with modifying enzymes include phosphatases and kinases,[1080–1083] methyltransferases,[249,258,1077,1084–1087] ligases,[1088,1089] glycosylases,[173–175,210,211,303,1090,1091] pseudouridylation,[1092] a ubiquitin ligase,[1093] adenylation[1094] and deadenylation.[1095] Structures also include ATPases[1096,1097] and GTPases,[1098] visualisation of the mechanism of Group II intron splicing,[588,591] and other spliceosome complexes,[1099,1100] RNA packaging,[1101,1102] and viral RNA recognition[1103–1106] and packaging.[1107]

The crystal structure of the duplex d(CCAGGCCTGG) has been reported where there are multiple competing positions of the entire base pairing system.[1108] A high resolution structure of d(CGCGCG)$_2$ has been described that forms a Z-DNA duplex.[1109] A crystal structure of d(CCGGTACCGG) revealed that it forms a four-way junction.[1110] There has been a number of crystal structures of G-quadruplexes,[1111–1114] including structures of the thrombin-binding aptamer bound to thrombin.[379,380]

There has been a number of crystal structures reported that contain either modified DNA or a ligand bound to the DNA. For example, two reports describe the mode of intercalation of the ruthenium ligand Δ-[Ru(bpy)$_2$dppz]$^{2+}$ with matched[1115] and mismatched[1116] steps in the DNA duplex, as well as to a structure of the ligand intercalating into a DNA kink.[1117] The structure of the duplex d(CGTACG) bound to the isoquinoline alkaloids berberine and sanguinarine has also been described.[1118] A crystal structure of a duplex containing the spin-label derivative of (47) opposite to dG showed that the analogue did not perturb the duplex structure.[236] The cationic platinum species TriplatinNC-A has been showed to act as a phosphate clamp within a duplex, binding in the duplex minor groove.[1119] The structure of a DNA duplex containing dU modified at C5 by substitution with various hydrophobic groups, bound to its target protein (PDGF-BB) showed that it folded into a compact structure with a hydrophobic binding surface.[1120] The structure of a glycol nucleic acid duplex (9, (S)-isomer) showed that the duplex folded into an N-type conformation.[57]

Of the RNA crystal structures reported there is the structure of the known RNA ligase ribozyme,[595] an RNA aptamer binding to its bacterial protein target Hfq,[1121] myotonic dystrophy type 1 (DM1) RNA that showed internal UU loop structures,[1122] and the structure of the CUG helix from DM1,[1123] and a structure of the Fragile X syndrome repeating r(CGG) transcript.[1124] A number of crystal structures of riboswitches have been reported, including dG,[1125] cyclic-di-GMP,[1126,1127] glycine,[1128] lysine[1129] and M-box[1130] riboswitches.

The crystal structures of RNA containing a number of modified nucleotides have been reported. An RNA duplex containing 5-methoxyuridine revealed a hydrogen bond between the methoxyl group and the phosphate backbone.[178] The structure of the CGG repeats derived from the fragile X-associated tremor ataxia syndrome containing 8-bromoG has been reported.[1131] The structure of an RNA G-quadruplex with an acridine ligand intercalating into the structure has been described.[1132] A few sugar-modified RNA structures have been reported, including a structure containing the locked nucleic acid (28),[1133] a duplex between RNA and altritol nucleic acid (25),[129] a duplex between RNA and cyclohexene nucleic acid (24),[128] and a comparison of the duplexes with RNA of S- and R-6′methyl-3′-fluorohexitol nucleic acids have been reported.[131]

4.2 NMR structures

Most complex structures are solved by X-ray crystallography, but recent advances in NMR spectroscopy have allowed for several complex solution structures too. During this review period these include bent DNA being bound by the protein HMGB1,[1134] and binding of the high mobility group nucleosomal protein-2 to DNA,[1135] binding of DNA regulatory factors to B- and to Z-DNA,[1136] binding of human PARP-1 to DNA,[1137] binding of DNA to cyanobacterial sensory rhodopsin transducer,[1138] full-length p53 bound to DNA,[1139] chicken methylcytosine binding protein,[1140] cTAR DNA bound to the HIV-1 nucleocapsid protein,[1141] gyrase bound DNA,[1142] and ligase bound DNA,[1143] DNA bound in the active site of DNA Pol β,[1144,1145] ssDNA bound to the human ssDNA-binding protein Sμbp-2,[1146] and solid-state NMR of DNA from the Pf1 bacteriophage.[1147]

Methodologies in NMR have included methods to define ion co-ordination sites in DNA,[1148] and methods for examining DNA folding.[1149] There have been a large number of solution structures of G-quadruplexes,[1150–1159] including one containing an A-tetrad,[1160] a structure containing isoguanosine,[1161] and a G-quartet with an intercalating quindoline.[1162] In addition there have been a few reports of i-motif structures.[1163–1166] The solution structure of the Dickerson dodecamer containing a single riboguanosine derivative showed that whilst locally the nucleotide adopted an A-form conformation this was insufficient to perturb the overall B-DNA duplex structure.[1167] The structure of a B-Z junction from the Z-DNA binding domain of human ADAR1 has also been reported.[1168] NMR studies have also been used to observe the existence of transient Hoogsteen base pairs within a DNA duplex.[1169,1170]

There are a number of reports of DNA structures with small molecule binders or intercalators, including an indoloquinolone bound to dsDNA,[352] binding of the metabolite nemorubicin to short duplex DNA,[1171] threading of a bis-naphthalene macrocycle into a T-T mismatch,[1172] binding of a diaminonaphthyridine to DNA,[1173] intercalation of benzo[kl]lignans,[1174] binding of the topoisomerase inhibitor genistein to nicked DNA,[1175] binding of the antitumor agent adozelesin,[1176] and binding of Cidofovir to DNA.[1177] There are also structures of duplex DNA containing the *S. capreolus* nucleotide analogue oxanine (97),[1178] and a duplex structure containing an interstrand disulphide linkage between two thiophenol C-nucleotides,[1179] as well as a duplex with an interstrand crosslink formed by the antitumor antibiotic SJG-136.[1180] The solution structure of a DNA duplex containing a single triazole internucleotide linkage showed little perturbation from the usual B-form structure.[1181] The solution structure of a duplex derived from deoxyxylose nucleic acids revealed a ladderlike structure.[124]

97

A few advances in NMR methodology have been reported in the context of RNA oligonucleotides.[1182–1186] As with solution structures of DNA there have been a number of more complex structures for RNA. These include the binding of siRNA with argonaute 2 PAZ domain,[1187] binding of RNA to the SR protein involved in splicing of pre-mRNA,[1188] binding of RNA to the pluripotency factor Lin28,[1189] the structure of the mammalian RNA-binding protein Musashi1 with its RNA target,[1190] binding of purine-rich RNA to the RNA recognition motif Tra2-β1,[1191,1192] and tRNA structures.[1193,1194] Further complex structures involve binding of the fragile X mental retardation protein (FMRP) to an RNA duplex-quadruplex junction,[1195] conformational dynamics in the cystoviral RNA-directed RNA polymerase P2,[1196] the structure of the *tyrS*-T-box leader RNA,[1197] the splice protein SRSF2,[1198] the HIV-1 5′-leader RNA,[1199] the *Mesoplasma florum* dG riboswitch,[1200] and the structures of RNA with the chaperones from the DEAD-box[1201] and Hfq.[1202]

Other RNA structures include the RNA claw from the DNA packaging motor from bacteriophage φ29,[1203] the conformation of an RNA structural switch,[1204] structures of miRNA,[1205] the structure of the A730 loop from the *Neurospora* VS ribozyme,[1206] structures or rRNA,[1207] the structure of RNA containing an internal UU loop from myotonic dystrophy type 1,[1208] base pair opening found in the adenine tract of an RNA helix,[1209] and a conformational analysis of TAR RNA in response to binding of drug-like ligands.[1210] The solution structure of a duplex between arabino nucleic acid and 2′-fluoroarabinonucleic acid is also reported.[1211]

4.3 Other structural methods

A few other methods have been described to examine nucleic acid structures, the major method being the various forms of electron microscopy (EM). Various nucleic acid-protein complexes have been observed using EM, including a B DNA polymerase (PCNA)-DNA ternary complex,[1212] a DNA-dependent protein kinase[1213] and its complex with PARP1,[1214] the eukaryotic helicase Mcm2-7 activation,[1215] the global structures of eukaryotic RNase P and RNase MRP,[1216] and the recombinatorial repair enzyme Rad51 in complex with DNA.[1217] The structure of the UV-damaged DNA-binding protein involved in the nucleotide excision repair pathway has been solved by X-ray crystallography and the structure validated using EM.[1093] Other nucleic acid-protein structures examined using EM are the THO complex, involved in co-transcriptional formation of messenger ribonucleoprotein particles,[1218] the melanoma differentiation-association protein 5 (MDA5) in complex with dsRNA,[1219] the protection of telomeres 1 (POT1) protein complexed with ssDNA,[1220] and different binding modes of p53 complexes with DNA.[1221]

The main variation of EM that has been reported is cryo-electron microscopy (CEM). Of the reports on the use of CEM during this review period the most-studied structure is the ribosome (from various organisms) and its associated machinery.[1222–1236] There have also been a number of studies on various viruses.[1237–1242] Other studies that have been reported include ribonucleoproteins,[1243,1244] the RNA guide surveillance complex from *E.coli*,[1245] a p53 tetramer bound to a DNA-encoding transcription factor response element,[1246] human RXR/VDR nuclear receptor in complex with its DR3 target DNA,[1247] a complex (UPF-EJC) from eukaryotic nonsense-mediated mRNA decay,[1248] and of a DNA origami structure.[884]

Another form of EM is tunnelling-EM (TEM), and this has been used to observe cooperative folding of long DNA template strands into complex nanoscale structures,[1249] assembly of phosphorothioate-modified G-quadruplex-gold nanoparticle conjugates into flower-shaped structures,[1250] DNA-peptide conjugate structures,[654] and DNA/micelle self-assembly structures that mimic chromatin compaction.[1251] Gold nanoparticles with 10's of oligonucleotides attached have been shown to self-assemble into flower-shaped particles as observed by TEM.[688] Nucleic acid-gold nanoparticle assemblies of defined shapes, such as triangles and squares, have been characterised using TEM.[691] Poly-(propylene oxide) conjugates with an *i*-motif oligonucleotide undergo a pH-dependent change in morphology between micelles and nanofibres as observed using TEM.[971]

Small-angle X-ray scattering (SAXS) is a technique where the scattering of X-rays by an inhomogeneous sample is measured at very low angles, typically up to 10°. The information obtained from this technique concerns shape and size of macromolecules. It has advantages over X-ray crystallography in that a crystalline material is not required. The effect of Mg(II) ion concentration on the docking of two hairpin loops from an adenine riboswitch has been examined by SAXS to show two distinct free energy pathways of tertiary interactions.[1252] Typically, SAXS has been used to study nucleic acid-protein interactions. The chaperone effects of RNA-binding protein Hfq from *E.coli*[1202] and *Vibrio cholerae*[1253] with their cognate RNA have been reported, as have the mRNA-binding proteins TIAR and HuR.[1254] The specificity of the dsRNA-binding domain of the protein kinase PKR has been shown to straighten bulged dsRNA regions,[1255] and the structural basis for the recognition of RNA by the immune receptor RIG-I[1256] are both studied using SAXS. Various DNA-protein interactions have also been reported including the recognition of the J base by a DNA-binding domain in JBP1,[1257] evidence of DNA compaction in the mycobacterial chromosomal ParB,[1258] and the identification of a common architecture of nuclear hormone receptors binding to DNA repeat elements.[1259]

Many studies include analysis of nucleic acids, or their interaction with proteins using atomic force microscopy, and this is an area that has seen many publications, some of which are included below. 5′-Single-stranded loops on ssDNA have been used as polarity markers in single molecule AFM studies,[1260] and the generation of backbone self-crossing structures in intrinsically curved DNA has been observed.[1261] Using dsDNA with an overhanging ssDNA region, the binding of single-stranded DNA-binding protein has been studied and shown to adopt different conformations depending upon ionic strength.[1262] Intrinsic curvature of DNA has been observed,[1263] as well as induced curvature by DNA intercalating agents.[1264] DNA catenane structures[30] as well as G-quadruplex structures have also been observed by use of AFM.[1265,1266] Analysis of HCV viral RNA[1267] and a TAR RNA dimer in the presence or absence of gold colloids have been described.[1268] Various nucleic nanostructures have been observed using AFM, including rectangular tiles,[894] cubes,[895] circular[1269,1270] and toroidal DNA.[1271] The transition of B-DNA to Z-DNA in

AT-rich sequences induced by a ruthenium complex has also been observed.[1272] AFM has also been used for controlling the fabrication of thiolated DNA nanostructures onto gold surfaces.[1273] A method for photochemical nanolithography based on gold-bound ssDNA via oligo(ethylene glycol) linkers has been reported and observed using AFM.[706]

More complex structures that have been observed using AFM include the binding domain of ankyrin-R complexed with the folded ribosome in which the characteristic horse-shoe shape of nascent ankyrin-R may be observed emerging from the ribosome.[1274] There are also 'snap-shots' of RNA polymerases, yeast RNA pol II[1275] and T7 RNA polymerase,[1276] during transcription, observation of MutS during DNA mismatch repair,[1277] and changes in DNA packing density upon addition of Mg(II) or cationic spermine with bacteriophage λ.[1278] Single molecule high-speed AFM has been used to directly visualise the dynamics of the cytosine deaminase APOBEC3G in complex with ssDNA.[1279]

A number of nucleic acid structures have been examined using electroparamagnetic resonance spectroscopy (EPR) or variants of the technique, and during this review period a review of methods for spin labelling of nucleic acids has been reported.[1280] The method involves a spin-label that is frequently attached to a nucleotide or to a protein target, though in some cases extraneous spin-labels are involved that become included into the nucleic acid complex. Examples include conformational changes in DNA upon binding to T7 Endonuclease I binding to a DNA four-way junction,[1281] dynamics of binding of a series of myosins using a spin-labelled ADP,[1282] interactions of the helicase UvrB in complex with DNA,[1057] the global structure of a Phi29 packaging RNA dimer,[33] and distance measurements in cells using pulsed EPR.[1283,1284] The TEMPO derivative (98) has been used as an external spin label for study of DNA nanostructures.[1285] PELDOR has been used to measure conformational flexibility in duplex DNA,[234] and to examine DNA-RNA hybrid structures.[192] Radical formation induced by Fenton chemistry[1286] and in the presence of a Cu(II) nitroxide derivative are also reported.[1287]

98

Surface plasmon resonance (SPR) is a method for measuring adsorption of materials onto planar (frequently gold or silver) surfaces or to the surface of metal nanoparticles. Surface plasmon resonance is observed when the frequency of photons matches the frequency of oscillation of the bound metal electrons. SPR can be used in a number of colour-based biosensor applications as well as lab-on-a chip sensors. SPR has been used to follow the rate of release of DNA III polymerase holoenzyme following gap filling between Ozaki fragments where it was observed that the off-rate is too fast to support lagging strand synthesis.[1288] The specificity for binding to C-rich ssDNA and ssRNA by the K-homology single-stranded binding proteins has been reported using SPR.[1289] Protein microarrays have been developed and observed from gold-bound DNA microarrays using a coupled *in vitro* transcription-translation system.[22] Gold-bound DNA has been conjugated by three androgenic steroids with detection using steroid-specific antibodies visualised using SPR.[962] Gold nanoparticles have been used for nano-plasmonic detection of DNA hybridisation,[1290] whilst the effects SPR using various modified oligonucleotides (e.g. PNA, LNA, aptamers) bound to gold nanoparticles has also been studied.[369]

Raman spectroscopy is a technique used to observe vibrational and rotational modes in a system. It relies on inelastic or Raman scattering from monochromatic light in the near infrared through to near ultraviolet. The (usually) laser light interacts with phonons or other excitations in the system resulting in a shift in energy of the laser photons, yielding information regarding vibrational modes in the system of interest. A review of imaging living systems using coherent anti-Stokes Raman scattering microscopy (CARS) has been reported.[1291] Time-resolved events for the transcription of a RNA polymerase bound to a DNA template in a crystal have been measured using Raman spectroscopy,[1292] as well as detection of sub-picomolar DNA polymerisation on magnetic nanoparticles.[780] Raman spectroscopy has also been used to detect T-Hg(II)-T base pairs (see also sections 1.2.3 and 3.3 for further details of Hg(II) mediated base pairs).[362]

A variation of Raman spectroscopy is surface enhanced Raman spectroscopy (SERS) in which the target molecules are adsorbed onto a rough metal surface. The sensitivity enhancement by this technique can be as high as 10^{10}–10^{11},[1293] which means that even single molecules can be detected. Label-free detection of micro-RNA by hybridisation has been improved using a least-squares analysis approach.[1294] Using a silver colloid, label-free detection of RNA (polyA/G/C/U) has been described by observing specific marker bands in the SERS spectra.[1295] Five different strains of *E. coli* could be differentiated using SERS on both gold and silver colloids.[1296] Silver colloids have also been used in a label-free detection of single-base mismatches in short DNA duplexes.[1297] The same authors have shown that DNA attached to gold nanoparticles *via* a thiol linker are orientated in an extended position (*i.e.* standing up) which shows an enhancement in the adenine breathing SERS band.[1298] Novel methods of oligonucleotide detection using SERS include fluorescently-labelled DNA-silver nanoparticle conjugates using a series of RAMAN reporter probes,[710,1299] and dye-labelled DNA bound to positively-charged silver-spermine nanoparticles.[1300] Other charged SERS active nanoparticles that have been reported include gold or silver citrate and silver-EDTA conjugates with dye-labelled DNA.[1301] Ultra-sensitive SERS biosensors have also been reported using gold dendrimer nanoparticles.[1302] Other methods developed for DNA detection using SERS include Ramos cell aptamers for detection of cancer cells,[1303] and a protocol involving signal amplification using lambda exonuclease.[1304]

Other methods for studying nucleic acids and their complexes include infrared spectroscopy. By use of surface-enhanced infrared absorption spectroscopy, *in situ* hybridisation kinetics have been measured,[1305] and detection of formation of G-quadruplexes has been studied using a combination of IR and vibrational circular dichroism.[1306]

Summary

This chapter reviews the literature published in 2012 relating to modified oligonucleotides, in addition to literature from the previous year concerning aptamers and aptamzymes and nucleic acid X-ray and NMR structures. This being the 60th anniversary of the publication of the first crystal structure of DNA, it is clear how crucial that work was in the development of nucleic acid chemistry. The literature continues the trend of previous years in that modified oligonucleotides continue to play a key role in many biochemical and structural studies. Key areas of research include modified nucleobases, aptamers/aptazymes, nucleic acid conjugates and nucleic acid structures. Nucleic acid structures, including single-molecule studies, are one of the most studied areas, and the systems described becoming more complex as novel techniques emerge.

References

1. T. S. Arunachalam, C. Wichert, B. Appel and S. Muller, *Org. Biomol. Chem.*, 2012, **10**, 4641.
2. G. J. van der Heden van Noort, P. v. Delft, N. J. Meeuwenoord, H. S. Overkleeft, G. A. van der Marel and D. V. Filippov, *Chem. Commun.*, 2012, **48**, 8093.
3. K. T. York, R. C. Smith, R. Yang, P. C. Melnyk, M. M. Wiley, C. M. Turk, M. Ronaghi, K. L. Gunderson and F. J. Steemers, *Nucl. Acids Res.*, 2012, **40**, e4.
4. J. Huang and Z. Xi, *Tetrahedron Letters*, 2012, **53**, 3654.
5. Y. Nukaga, K. Yamada, T. Ogata, N. Oka and T. Wada, *J. Org. Chem.*, 2012, **77**, 7913.
6. N.-S. Li, J. K. Frederiksen and J. A. Piccirilli, *J. Org. Chem.*, 2012, **77**, 9889.
7. C. Sallamand, A. Miscioscia, R. Lartia and E. Defrancq, *Org. Lett.*, 2012, **14**, 2030.
8. L. Wicke and J. W. Engels, *Bioconjugate Chem.*, 2012, **23**, 627.
9. H. Griesser, M. Tolev, A. Singh, T. Sabirov, C. Gerlach and C. Richert, *J. Org. Chem.*, 2012, **77**, 2703.
10. A. Singh, M. Tolev, C. I. Schilling, S. Brase, H. Griesser and C. Richert, *J. Org. Chem.*, 2012, **77**, 2718.
11. S. Patnaik, S. K. Dash, D. Sethi, A. Kumar, K. C. Gupta and P. Kumar, *Bioconjugate Chem.*, 2012, **23**, 664.
12. J. Escorihuela, M. J. Banuls, R. Puchades and A. Maquieira, *Chem. Commun.*, 2012, **48**, 2116.
13. J. Escorihuela, M.-J. Banuls, R. Puchades and A. Maquieira, *Bioconjugate Chem.*, 2012, **23**, 2121.
14. S. Mao, A. L. Souza, R. J. Goodrich and S. A. Krawetz, *Biotechniques*, 2012, **53**, 91.
15. W. W. Hadiwikarta, J.-C. Walter, J. Hooyberghs and E. Carlon, *Nucl. Acids Res.*, 2012, **40**, e138.
16. G. C. Tseng, D. Ghosh and E. Feingold, *Nucl. Acids Res.*, 2012, **40**, 3785.

17 H. Cho, J. Jung and B. H. Chung, *Chem. Commun.*, 2012, **48**, 7601.
18 C.-H. Wu, M. R. Lockett and L. M. Smith, *Angew. Chem. Int. Ed.*, 2012, **51**, 4628.
19 J. A. Hanna, H. Wimberly, S. Kumar, F. Slack, S. Agarwal and D. L. Rimm, *Biotechniques*, 2012, **52**, 235.
20 M. Li, S.-T. Choi, K.-S. Tsang, P.-C. Shaw and K.-F. Lau, *ChemBioChem*, 2012, **13**, 1286.
21 V.-T. Nguyen, S. B. Nimse, K.-S. Song, J. Kim, V.-T. Ta, H. W. Sung and T. Kim, *Chem. Commun.*, 2012, **48**, 4582.
22 T. H. Seefeld, A. R. Halpern and R. M. Corn, *J. Am. Chem. Soc.*, 2012, **134**, 12358.
23 B. Gerland, A. Goudot, G. Pourceau, A. Meyer, V. Dugas, S. Cecioni, S. Vidal, E. Souteyrand, J.-J. Vasseur, Y. Chevolot and F. Morvan, *Bioconjugate Chem.*, 2012, **23**, 1534.
24 G. F. Deleavey and M. J. Damha, *Chem. Biol.*, 2012, **19**, 937.
25 H. Cramer, J. R. Okicki, T. Rho, X. Wang, R. H. Silverman and W. D. Weston, *Nucleosides, Nucleotides & Nucl. Acids*, 2007, **26**, 1471.
26 X. Wang, H. Tian, Z. Lee and W. D. Heston, *Nucleosides, Nucleotides & Nucl. Acids*, 2012, **31**, 432.
27 T. A. Lonnberg, *ChemBioChem*, 2012, **13**, 2690.
28 L. M. Ceo and J. T. Koh, *ChemBioChem*, 2012, **13**, 511.
29 C. Dohno, T. Shibata, M. Okazaki, S. Makishi and K. Nakatani, *Eur. J. Org. Chem.*, 2012, 5317.
30 J. Elbaz, Z.-G. Wang, F. Wang and I. Willner, *Angew. Chem. Int. Ed.*, 2012, **51**, 2349.
31 Y. Sannohe and H. Sugiyama, *Bioorg. Med. Chem.*, 2012, **20**, 2030.
32 F. Lohmann, D. Ackermann and M. Famulok, *J. Am. Chem. Soc.*, 2012, **134**, 11884.
33 X. Zhang, C.-S. Tung, G. Z. Sowa, M. m. M. Hatmal, I. S. Haworth and P. Z. Qin, *J. Am. Chem. Soc.*, 2012, **134**, 2644.
34 S. M. A. Rahman, T. Baba, T. Kodama, M. A. Islam and S. Obika, *Bioorg. Med. Chem.*, 2012, **20**, 4098.
35 A. Sanchez, E. Pedroso and A. Grandas, *Bioconjugate Chem.*, 2012, **23**, 300.
36 O. Pav, N. Panova, J. Snasel, E. Zborníkova and I. Rosenberg, *Bioorg. Med. Chem. Lett.*, 2012, **22**, 181.
37 H. Krishna and M. H. Caruthers, *J. Am. Chem. Soc.*, 2012, **134**, 11618.
38 R. N. Threlfall, A. G. Torres, A. Krivenko, M. J. Gait and M. H. Caruthers, *Org. Biomol. Chem.*, 2012, **10**, 746.
39 N. Iwamoto, N. Oka and T. Wada, *Tetrahedron Letters*, 2012, **53**, 4361.
40 S. Roy, M. Olesiak, P. Padar, H. McCuen and M. H. Caruthers, *Org. Biomol. Chem.*, 2012, **10**, 9130.
41 A. Kaiser, S. Spies, T. Lommel and C. Richert, *Angew. Chem. Int. Ed.*, 2012, **51**, 8299.
42 H. Vogel and C. Richert, *ChemBioChem*, 2012, **13**, 1474.
43 J. B. Crumpton and W. L. Santos, *Chem. Commun.*, 2012, **48**, 2018.
44 T. Harakawa, H. Tsunoda, A. Ohkubo, K. Seio and M. Sekine, *Bioorg. Med. Chem. Lett.*, 2012, **22**, 1445.
45 T. Wada, M. Hara, T. Taneda, C. Qingfu, R. Takata, K. Moro, K. Takeda, T. Kishimoto and H. Handa, *Nucl. Acids Res.*, 2012, **40**, e173.
46 Y. Wang, L. Wu, P. Wang, C. Lv, Z. Yang and X. Tang, *Nucl. Acids Res.*, 2012, **40**, 11155.
47 S. Yamazoe, I. A. Shestopalov, E. Provost, S. D. Leach and J. K. Chen, *Angew. Chem. Int. Ed.*, 2012, **51**, 6908.

48. A. E. Engelhart, B. J. Cafferty, C. D. Okafor, M. C. Chen, L. D. Williams, D. G. Lynn and N. V. Hud, *ChemBioChem*, 2012, **13**, 1121.
49. A. M. Awad, M. J. Collazo, K. Carpio, C. Flores and T. C. Bruice, *Tetrahedron Letters*, 2012, **53**, 3792.
50. W. Gong and J.-P. Desaulniers, *Bioorg. Med. Chem. Lett.*, 2012, **22**, 6934.
51. T. Fujino, N. Yamazaki, A. Hasome, K. Endo and H. Isobe, *Tetrahedron Letters*, 2012, **53**, 868.
52. A. Varizhuk, A. Chizhov, I. Smirnov, D. Kaluzhny and V. Florentiev, *Eur. J. Org. Chem.*, 2012, 2173.
53. A. P. Sanzone, A. H. El-Sagheer, T. Brown and A. Tavassoli, *Nucl. Acids Res.*, 2012, **40**, 10567.
54. T. C. Efthymiou, V. Huynh, J. Oentoro, B. Peel and J.-P. Desaulniers, *Bioorg. Med. Chem. Lett.*, 2012, **22**, 1722.
55. S. S. Pujari and F. Seela, *J. Org. Chem.*, 2012, **77**, 4460.
56. H. V. Nguyen, Z.-y. Zhao, A. Sallustrau, S. L. Horswell, L. Male, A. Mulas and J. H. R. Tucker, *Chem. Commun.*, 2012, **48**, 12165.
57. A. T. Johnson, M. K. Schlegel, E. Meggers, L.-O. Essen and O. Wiest, *J. Org. Chem.*, 2011, **76**, 7964.
58. P. E. Nielsen, M. Egholm, R. H. Berg and O. Buchardt, *Science*, 1991, **254**, 1487.
59. W. Gong and J.-P. Desaulniers, *Nucleosides, Nucleotides & Nucl. Acids*, 2012, **31**, 389.
60. F. Totsingan, V. Jain and M. M. Green, *Artificial DNA: PNA & XNA*, 2012, **3**, 31.
61. S. Rossi, A. Calabretta, T. Tedeschi, S. Sforza, S. Arcioni, L. Baldoni, R. Corradini and R. Marchelli, *Artificial DNA: PNA & XNA*, 2012, **3**, 63.
62. D. Chouikhi, M. Ciobanu, C. Zambaldo, V. Duplan, S. Barluenga and N. Winssinger, *Chem. Eur. J.*, 2012, **18**, 12698.
63. J.-P. Daguer, M. Ciobanu, S. Barluenga and N. Winssinger, *Org. Biomol. Chem.*, 2012, **10**, 1502.
64. T. Ishizuka, J. Yang, M. Komiyama and Y. Xu, *Angew. Chem. Int. Ed.*, 2012, **51**, 7198.
65. L. Llovera, P. Berthold, P. E. Nielsen and T. Shiraishi, *Artificial DNA: PNA & XNA*, 2012, **3**, 22.
66. A. Finotti, M. Borgatti, V. Bezzerri, E. Nicolis, I. Lampronti, M. Dechecchi, I. Mancini, G. Cabrini, M. Saviano, C. Avitabile, A. Romanelli and R. Gambari, *Artificial DNA: PNA & XNA*, 2012, **3**, 97.
67. A. G. Torres, M. M. Fabani, E. Vigorito, D. Williams, N. Al-Obaidi, F. Wojciechowski, R. H. E. Hudson, O. Seitz and M. J. Gait, *Nucl. Acids Res.*, 2012, **40**, 2152.
68. C. Avitabile, M. Saviano, L. D'Andrea, N. Bianchi, E. Fabbri, E. Brognara, R. Gambari and A. Romanelli, *Artificial DNA: PNA & XNA*, 2012, **3**, 88.
69. D. Sethi, C.-P. Chen, R.-Y. Jing, M. L. Thakur and E. Wickstrom, *Bioconjugate Chem.*, 2012, **23**, 158.
70. R. Shrestha, Y. Shen, K. A. Pollack, J.-S. A. Taylor and K. L. Wooley, *Bioconjugate Chem.*, 2012, **23**, 574.
71. T. Shiraishi and P. E. Nielsen, *Bioconjugate Chem.*, 2012, **23**, 196.
72. E. Socher, A. Knoll and O. Seitz, *Org. Biomol. Chem.*, 2012, **10**, 7363.
73. S. Kummer, A. Knoll, E. Socher, L. Bethge, A. Herrmann and O. Seitz, *Bioconjugate Chem.*, 2012, **23**, 2051.
74. Y. Liu and S. E. Rokita, *Biochemistry*, 2012, **51**, 1020.
75. P. Wang, H. Wu, Z. Dai and X. Zou, *Chem. Commun.*, 2012, **48**, 10754.

76　C. Zanardi, F. Terzi, R. Seeber, C. Baldoli, E. Licandro and S. Maiorana, *Artificial DNA: PNA & XNA*, 2012, **3**, 80.

77　A. Gross, N. Husken, J. Schur, L. Raszeja, I. Ott and N. Metzler-Nolte, *Bioconjugate Chem.*, 2012, **23**, 1764.

78　M. Rothlingshofer, K. Gorska and N. Winssinger, *Org. Lett.*, 2012, **14**, 482.

79　F. Marlin, P. Simon, S. Bonneau, P. Alberti, C. Cordier, C. Boix, L. Perrouault, A. Fossey, T. Saison-Behmoaras, M. Fontecave and C. Giovannangeli, *ChemBioChem*, 2012, **13**, 2593.

80　I. Das, J. Desire, D. Manvar, I. Baussanne, V. N. Pandey and J.-L. Décout, *J. Med. Chem.*, 2012, **55**, 6021.

81　A. Manicardi, A. Accetta, T. Tedeschi, S. Sforza, R. Marchelli and R. Corradini, *Artificial DNA: PNA & XNA*, 2012, **3**, 53.

82　T. Zengeya, P. Gupta and E. Rozners, *Angew. Chem. Int. Ed.*, 2012, **51**, 12593.

83　S. Mullar, J. Strohmeier and U. Diederichsen, *Org. Lett.*, 2012, **14**, 1382.

84　E. Wierzbinski, A. de Leon, X. Yin, A. Balaeff, K. L. Davis, S. Reppireddy, R. Venkatramani, S. Keinan, D. H. Ly, M. Madrid, D. N. Beratan, C. Achim and D. H. Waldeck, *J. Am. Chem. Soc.*, 2012, **134**, 9335.

85　A. Gourishankar and K. N. Ganesh, *Artificial DNA: PNA & XNA*, 2012, **3**, 5.

86　Y. Aiba, Y. Honda, Y. Han and M. Komiyama, *Artificial DNA: PNA & XNA*, 2012, **3**, 73.

87　P. Gupta, O. Muse and E. Rozners, *Biochemistry*, 2011, **51**, 63.

88　R. Mitra and K. N. Ganesh, *J. Org. Chem.*, 2012, **77**, 5696.

89　R. Bahal, B. Sahu, S. Rapireddy, C.-M. Lee and D. H. Ly, *ChemBioChem*, 2012, **13**, 56.

90　C. De Cola, A. Manicardi, R. Corradini, I. Izzo and F. De Riccardis, *Tetrahedron*, 2012, **68**, 499.

91　T. Vilaivan and C. Srisuwannaket, *Org. Lett.*, 2006, **8**, 1897.

92　W. Mansawat, C. Vilaivan, A. Balazs, D. J. Aitken and T. Vilaivan, *Org. Lett.*, 2012, **14**, 1440.

93　W. Mansawat, C. Boonlua, K. Siriwong and T. Vilaivan, *Tetrahedron*, 2012, **68**, 3988.

94　J. L. McGinnis, J. A. Dunkle, J. H. D. Cate and K. M. Weeks, *J. Am. Chem. Soc.*, 2012, **134**, 6617.

95　K.-A. Steen, G. M. Rice and K. M. Weeks, *J. Am. Chem. Soc.*, 2012, **134**, 13160.

96　J. Cieslak, A. Grajkowski, C. Ausín, A. Gapeev and S. L. Beaucage, *Nucl. Acids Res.*, 2012, **40**, 2312.

97　M. Gerowska, L. Hall, J. Richardson, M. Shelbourne and T. Brown, *Tetrahedron*, 2012, **68**, 857.

98　M. Shelbourne, T. Brown, A. H. El-Sagheer and T. Brown, *Chem. Commun.*, 2012, **48**, 11184.

99　D. Yu, H. Pendergraff, J. Liu, H. B. Kordasiewicz, D. W. Cleveland, E. E. Swayze, W. F. Lima, S. T. Crooke, T. P. Prakash and D. R. Corey, *Cell*, 2012, **150**, 895.

100　W. F. Lima, T. P. Prakash, H. M. Murray, G. A. Kinberger, W. Li, A. E. Chappell, C. S. Li, S. F. Murray, H. Gaus, P. P. Seth, Eric E. Swayze and S. T. Crooke, *Cell*, 2012, **150**, 883.

101　A. S. Cardew, T. Brown and K. R. Fox, *Nucl. Acids Res.*, 2012, **40**, 3753.

102　M. Nakamura, M. Fukuda, T. Takada and K. Yamana, *Org. Biomol. Chem.*, 2012, **10**, 9620.

103　J. Brzezinska, J. D'Onofrio, M. C. R. Buff, J. Hean, A. Ely, M. Marimani, P. Arbuthnot and J. W. Engels, *Bioorg. Med. Chem.*, 2012, **20**, 1594.

104 K. Seio, M. Tokugawa, T. Kanamori, H. Tsunoda, A. Ohkubo and M. Sekine, *Bioorg. Med. Chem. Lett.*, 2012, **22**, 2470.
105 A. Patra, M. Paolillo, K. Charisse, M. Manoharan, E. Rozners and M. Egli, *Angew. Chem. Int. Ed.*, 2012, **51**, 11863.
106 H. J. Haringsma, J. J. Li, F. Soriano, D. M. Kenski, W. M. Flanagan and A. T. Willingham, *Nucl. Acids Res.*, 2012, **40**, 4125.
107 F. Rigo, Y. Hua, S. J. Chun, T. P. Prakash, A. R. Krainer and C. F. Bennett, *Nature Chem. Biol.*, 2012, **8**, 555.
108 C. Cozens, V. B. Pinheiro, A. Vaisman, R. Woodgate and P. Holliger, *Proc. Natl. Acad. Sci. USA*, 2012, **109**, 8067.
109 K. Onizuka, J. Yeo, S. S. David and P. A. Beal, *ChemBioChem*, 2012, **13**, 1338.
110 Q. Dai, X. Lu, L. Zhang and C. He, *Tetrahedron*, 2012, **68**, 5145.
111 C. J. Lech, Z. Li, B. Heddi and A. T. Phan, *Chem. Commun.*, 2012, **48**, 11425.
112 S. P. Sau and P. J. Hrdlicka, *J. Org. Chem.*, 2012, **77**, 5.
113 P. Kumar, K. I. Shaikh, A. S. Jorgensen, S. Kumar and P. Nielsen, *J. Org. Chem.*, 2012, **77**, 9562.
114 M. M. Rubner, C. Holzhauser, P. R. Bohlander and H.-A. Wagenknecht, *Chem. Eur. J.*, 2012, **18**, 1299.
115 V. Siegmund, T. Santner, R. Micura and A. Marx, *Chem. Commun.*, 2012, **48**, 9870.
116 C. S. Madsen, S. Witzke, P. Kumar, K. Negi, P. K. Sharma, M. Petersen and P. Nielsen, *Chem. Eur. J.*, 2012, **18**, 7434.
117 K. Fauster, C. Kreutz and R. Micura, *Angew. Chem. Int. Ed.*, 2012, **51**, 13080.
118 A.-S. Geiermann and R. Micura, *ChemBioChem*, 2012, **13**, 1742.
119 M. Takahashi, C. Nagai, H. Hatakeyama, N. Minakawa, H. Harashima and A. Matsuda, *Nucl. Acids Res.*, 2012, **40**, 5787.
120 Y. Hari, T. Osawa and S. Obika, *Org. Biomol. Chem.*, 2012, **10**, 9639.
121 G. N. Nawale, K. R. Gore, C. Hobartner and P. I. Pradeepkumar, *Chem. Commun.*, 2012, **48**, 9619.
122 K. R. Gore, G. N. Nawale, S. Harikrishna, V. G. Chittoor, S. K. Pandey, C. Hobartner, S. Patankar and P. I. Pradeepkumar, *J. Org. Chem.*, 2012, **77**, 3233.
123 J. Zhang, Y. Chen, Y. Huang, H.-W. Jin, R.-P. Qiao, L. Xing, L.-R. Zhang, Z.-J. Yang and L.-H. Zhang, *Org. Biomol. Chem.*, 2012, **10**, 7566.
124 M. Maiti, V. Siegmund, M. Abramov, E. Lescrinier, H. Rosemeyer, M. Froeyen, A. Ramaswamy, A. Ceulemans, A. Marx and P. Herdewijn, *Chem. Eur. J.*, 2012, **18**, 869.
125 H. Yu, S. Zhang and J. C. Chaput, *Nature Chem.*, 2012, **4**, 183.
126 A. S. Madsen and J. Wengel, *J. Org. Chem.*, 2012, **77**, 3878.
127 P. P. Seth, J. Yu, A. Jazayeri, P. S. Pallan, C. R. Allerson, M. E. Ostergaard, F. Liu, P. Herdewijn, M. Egli and E. E. Swayze, *J. Org. Chem.*, 2012, **77**, 5074.
128 M. Ovaere, P. Herdewijn and L. Van Meervelt, *Chem. Eur. J.*, 2011, **17**, 7823.
129 M. Ovaere, J. Sponer, J. E. Sponer, P. Herdewijn and L. Van Meervelt, *Nucl. Acids Res.*, 2012, **40**, 7573.
130 V. B. Pinheiro, A. I. Taylor, C. Cozens, M. Abramov, M. Renders, S. Zhang, J. C. Chaput, J. Wengel, S.-Y. Peak-Chew, S. H. McLaughlin, P. Herdewijn and P. Holliger, *Science*, 2012, **336**, 341.
131 P. S. Pallan, J. Yu, C. R. Allerson, E. E. Swayze, P. Seth and M. Egli, *Biochemistry*, 2011, **51**, 7.
132 K. Furukawa, M. Hattori, T. Ohki, Y. Kitamura, Y. Kitade and Y. Ueno, *Bioorg. Med. Chem.*, 2012, **20**, 16.
133 K. K. Karlsen, A. Pasternak, T. B. Jensen and J. Wengel, *ChemBioChem*, 2012, **13**, 590.

134 S. Obika, K. Morio, D. Nanbu and T. Imanishi, *Chem. Commun.*, 1987, 1643.
135 S. K. Singh, P. E. Nielsen, A. A. Koshkin and J. Wengel, *Chem. Commun.*, 1998, 455.
136 C. Hull, C. Szewcyk and P. M. St. John, *Nucleosides, Nucleotides & Nucl. Acids*, 2012, **31**, 28.
137 S. K. Rastogi, C. M. Gibson, J. R. Branen, D. Eric Aston, A. Larry Branen and P. J. Hrdlicka, *Chem. Commun.*, 2012, **48**, 7714.
138 B. M. Beckmann, P. G. Hoch, M. Marz, D. K. Willkomm, M. Salas and R. K. Hartmann, *EMBO J.*, 2012, **31**, 1727.
139 H. Suryawanshi, M. K. Lalwani, S. Ramasamy, R. Rana, V. Scaria, S. Sivasubbu and S. Maiti, *ChemBioChem*, 2012, **13**, 584.
140 B. D. Schyth, J. B. Bramsen, M. M. Pakula, S. Larashati, J. Kjems, J. Wengel and N. Lorenzen, *Nucl. Acids Res.*, 2012, **40**, 4653.
141 L. Crouzier, C. Dubois, J. Wengel and R. N. Veedu, *Bioorg. Med. Chem. Lett.*, 2012, **22**, 4836.
142 S. P. Sau, P. Kumar, P. K. Sharma and P. J. Hrdlicka, *Nucl. Acids Res.*, 2012, **40**, e162.
143 A. K. Thazhathveetil, J. Vura-Weis, A. Trifonov, M. R. Wasielewski and F. D. Lewis, *J. Am. Chem. Soc.*, 2012, **134**, 16434.
144 K. Kawai, M. Hayashi and T. Majima, *J. Am. Chem. Soc.*, 2012, **134**, 9406.
145 U. Wenge, J. Wengel and H.-A. Wagenknecht, *Angew. Chem. Int. Ed.*, 2012, **51**, 10026.
146 P. P. Seth, C. R. Allerson, M. E. Ostergaard and E. E. Swayze, *Bioorg. Med. Chem. Lett.*, 2012, **22**, 296.
147 T. Hojland, R. N. Veedu, B. Vester and J. Wengel, *Artificial DNA: PNA & XNA*, 2012, **3**, 14.
148 I. K. Astakhova, E. Samokhina, B. R. Babu and J. Wengel, *ChemBioChem*, 2012, **13**, 1509.
149 M. W. Johannsen, R. N. Veedu, A. S. Madsen and J. Wengel, *Bioorg. Med. Chem. Lett.*, 2012, **22**, 3522.
150 A. Yahara, A. R. Shrestha, T. Yamamoto, Y. Hari, T. Osawa, M. Yamaguchi, M. Nishida, T. Kodama and S. Obika, *ChemBioChem*, 2012, **13**, 2513.
151 P. S. Pallan, C. R. Allerson, A. Berdeja, P. P. Seth, E. E. Swayze, T. P. Prakash and M. Egli, *Chem. Commun.*, 2012, **48**, 8195.
152 A. I. Haziri and C. J. Leumann, *J. Org. Chem.*, 2012, **77**, 5861.
153 S. Murray, D. Ittig, E. Koller, A. Berdeja, A. Chappell, T. P. Prakash, M. Norrbom, E. E. Swayze, C. J. Leumann and P. P. Seth, *Nucl. Acids Res.*, 2012, **40**, 6135.
154 J. Lietard and C. J. Leumann, *J. Org. Chem.*, 2012, **77**, 4566.
155 Y. Taniguchi and S. Sasaki, *Org. Biomol. Chem.*, 2012, **10**, 8336.
156 E. Aoki, Y. Taniguchi, Y. Wada and S. Sasaki, *ChemBioChem*, 2012, **13**, 1152.
157 P. S. Pallan, V. E. Marquez and M. Egli, *Biochemistry*, 2012, **51**, 2639.
158 A. Ketkar, M. K. Zafar, S. Banerjee, V. E. Marquez, M. Egli and R. L. Eoff, *J. Am. Chem. Soc.*, 2012, **134**, 10698.
159 A. Ketkar, M. K. Zafar, S. Banerjee, V. E. Marquez, M. Egli and R. L. Eoff, *Biochemistry*, 2012, **51**, 9234.
160 S. Hanessian, B. R. Schroeder, R. D. Giacometti, B. L. Merner, M. Ostergaard, E. E. Swayze and P. P. Seth, *Angew. Chem. Int. Ed.*, 2012, **51**, 11242.
161 T. Carell, C. Brandmayr, A. Hienzsch, M. Müller, D. Pearson, V. Reiter, I. Thoma, P. Thumbs and M. Wagner, *Angew. Chem. Int. Ed.*, 2012, **51**, 7110.

162 C. E. Dumelin, Y. Chen, A. M. Leconte, Y. G. Chen and D. R. Liu, *Nature Chem. Biol.*, 2012, **8**, 913.
163 H. Sun, J. Sheng, A. E. A. Hassan, S. Jiang, J. Gan and Z. Huang, *Nucl. Acids Res.*, 2012, **40**, 5171.
164 G. Leszczynska, J. Pieta, P. Leonczak, A. Tomaszewska and A. Malkiewicz, *Tetrahedron Letters*, 2012, **53**, 1214.
165 G. Sun, A. Noronha and C. Wilds, *Tetrahedron*, 2012, **68**, 7787.
166 C. M. Connelly, R. Uprety, J. Hemphill and A. Deiters, *Mol. BioSys.*, 2012, **8**, 2987.
167 A. Beddows, N. Patel, L. D. Finger, J. M. Atack, D. M. Williams and J. A. Grasby, *Chem. Commun.*, 2012, **48**, 8895.
168 F. P. McManus, D. K. O'Flaherty, A. M. Noronha and C. J. Wilds, *Org. Biomol. Chem.*, 2012, **10**, 7078.
169 F. A. P. Vendeix, F. V. Murphy, Iv, W. A. Cantara, G. Leszczynska, E. M. Gustilo, B. Sproat, A. Malkiewicz and P. F. Agris, *J. Mol. Biol.*, 2012, **416**, 467.
170 A. Shibata, Y. Ueno, M. Iwata, H. Wakita, A. Matsuda and Y. Kitade, *Bioorg. Med. Chem. Lett.*, 2012, **22**, 2681.
171 T. Gristwood, I. G. Duggin, M. Wagner, S. V. Albers and S. D. Bell, *Nucl. Acids Res.*, 2012, **40**, 5487.
172 J. Lepczynska, K. Komodzinski, J. Milecki, R. Kierzek, Z. Gdaniec, S. Franzen and B. Skalski, *J. Org. Chem.*, 2012, **77**, 11362.
173 H. Hashimoto, S. Hong, A. S. Bhagwat, X. Zhang and X. Cheng, *Nucl. Acids Res.*, 2012, **40**, 10203.
174 H. Hashimoto, X. Zhang and X. Cheng, *Nucl. Acids Res.*, 2012, **40**, 8276.
175 S. Morera, I. Grin, A. Vigouroux, S. Couve, V. Henriot, M. Saparbaev and A. A. Ishchenko, *Nucl. Acids Res.*, 2012, **40**, 9917.
176 K. Kemmerich, F. A. Dingler, C. Rada and M. S. Neuberger, *Nucl. Acids Res.*, 2012, **40**, 6016.
177 B. P. Stupi, H. Li, J. Wang, W. Wu, S. E. Morris, V. A. Litosh, J. Muniz, M. N. Hersh and M. L. Metzker, *Angew. Chem. Int. Ed.*, 2012, **51**, 1724.
178 J. Sheng, W. Zhang, A. E. A. Hassan, J. Gan, A. S. Soares, S. Geng, Y. Ren and Z. Huang, *Nucl. Acids Res.*, 2012, **40**, 8111.
179 C. Uttamapinant, A. Tangpeerachaikul, S. Grecian, S. Clarke, U. Singh, P. Slade, K. R. Gee and A. Y. Ting, *Angew. Chem. Int. Ed.*, 2012, **51**, 5852.
180 S. Obeid, H. Buskamp, W. Welte, K. Diederichs and A. Marx, *Chem. Commun.*, 2012, **48**, 8320.
181 M. Kaura, P. Kumar and P. J. Hrdlicka, *Org. Biomol. Chem.*, 2012, **10**, 8575.
182 B. M. Laing and D. E. Bergstrom, *Bioconjugate Chem.*, 2012, **23**, 683.
183 H. Xiong and F. Seela, *Bioconjugate Chem.*, 2012, **23**, 1230.
184 H. Rao, A. A. Sawant, A. A. Tanpure and S. G. Srivatsan, *Chem. Commun.*, 2012, **48**, 498.
185 P. Kumar, N. Chandak, P. Nielsen and P. K. Sharma, *Bioorg. Med. Chem.*, 2012, **20**, 3843.
186 M. Tanaka, K. Oguma, Y. Saito and I. Saito, *Chem. Commun.*, 2012, **48**, 9394.
187 M. Tanaka, K. Oguma, Y. Saito and I. Saito, *Bioorg. Med. Chem. Lett.*, 2012, **22**, 4103.
188 D. Musumeci, G. Oliviero, G. N. Roviello, E. M. Bucci and G. Piccialli, *Bioconjugate Chem.*, 2012, **23**, 382.
189 I. J. Lee, M. Park, T. Joo and B. H. Kim, *Mol. BioSys.*, 2012, **8**, 486.
190 A. A. Tanpure and S. G. Srivatsan, *ChemBioChem*, 2012, **13**, 2392.
191 J. Riedl, R. Pohl, L. Rulisek and M. Hocek, *J. Org. Chem.*, 2012, **77**, 1026.

192 O. Romainczyk, B. Endeward, T. F. Prisner and J. W. Engels, *Mol. BioSys.*, 2011, **7**, 1050.
193 T. Ehrenschwender, A. Barth, H. Puchta and H.-A. Wagenknecht, *Org. Biomol. Chem.*, 2012, **10**, 46.
194 M. Matsui and Y. Ebara, *Bioorg. Med. Chem. Lett.*, 2012, **22**, 6139.
195 Y. Kuang, H. Sun, J. C. Blain and X. Peng, *Chem. Eur. J.*, 2012, **18**, 12609.
196 N. Borbone, M. Bucci, G. Oliviero, E. Morelli, J. Amato, V. D'Atri, S. D'Errico, V. Vellecco, G. Cirino, G. Piccialli, C. Fattorusso, M. Varra, L. Mayol, M. Persico and M. Scuotto, *J. Med. Chem.*, 2012, **55**, 10716.
197 M. J. E. Resendiz, V. Pottiboyina, M. D. Sevilla and M. M. Greenberg, *J. Am. Chem. Soc.*, 2012, **134**, 3917.
198 M. J. E. Resendiz, A. Schon, E. Freire and M. M. Greenberg, *J. Am. Chem. Soc.*, 2012, **134**, 12478.
199 C. H. Wunderlich, R. Spitzer, T. Santner, K. Fauster, M. Tollinger and C. Kreutz, *J. Am. Chem. Soc.*, 2012, **134**, 7558.
200 Z. Liu, X. Guo, C. Tan, J. Li, Y.-T. Kao, L. Wang, A. Sancar and D. Zhong, *J. Am. Chem. Soc.*, 2012, **134**, 8104.
201 N. Arichi, A. Inase, S. Eto, T. Mizukoshi, J. Yamamoto and S. Iwai, *Org. Biomol. Chem.*, 2012, **10**, 2318.
202 V. Pages, G. Mazon, K. Naiman, G. Philippin and R. P. Fuchs, *Nucl. Acids Res.*, 2012, **40**, 9036.
203 A. Benjdia, K. Heil, T. R. M. Barends, T. Carell and I. Schlichting, *Nucl. Acids Res.*, 2012, **40**, 9308.
204 K. Haiser, B. P. Fingerhut, K. Heil, A. Glas, T. T. Herzog, B. M. Pilles, W. J. Schreier, W. Zinth, R. de Vivie-Riedle and T. Carell, *Angew. Chem. Int. Ed.*, 2012, **51**, 408.
205 A. R. Peoples, J. Lee, M. Weinfeld, J. R. Milligan and W. A. Bernhard, *Nucl. Acids Res.*, 2012, **40**, 6060.
206 S. Ding, K. Kropachev, Y. Cai, M. Kolbanovskiy, S. A. Durandina, Z. Liu, V. Shafirovich, S. Broyde and N. E. Geacintov, *Nucl. Acids Res.*, 2012, **40**, 2506.
207 J. Wang, H. Cao, C. You, B. Yuan, R. Bahde, S. Gupta, C. Nishigori, L. J. Niedernhofer, P. J. Brooks and Y. Wang, *Nucl. Acids Res.*, 2012, **40**, 7368.
208 W. A. Cantara, Y. Bilbille, J. Kim, R. Kaiser, G. Leszczynska, A. Malkiewicz and P. F. Agris, *J. Mol. Biol.*, 2012, **416**, 579.
209 S.-i. Nakano, H. Oka, D. Yamaguchi, M. Fujii and N. Sugimoto, *Org. Biomol. Chem.*, 2012, **10**, 9664.
210 L. Zhang, X. Lu, J. Lu, H. Liang, Q. Dai, G.-L. Xu, C. Luo, H. Jiang and C. He, *Nature Chem. Biol.*, 2012, **8**, 328.
211 A. Maiti, M. S. Noon, A. D. MacKerell, E. Pozharski and A. C. Drohat, *Proc. Natl. Acad. Sci. USA*, 2012, **109**, 8091.
212 M. W. Kellinger, C.-X. Song, J. Chong, X.-Y. Lu, C. He and D. Wang, *Nature Struct. Mol. Biol.*, 2012, **19**, 831.
213 S. Schiesser, B. Hackner, T. Pfaffeneder, M. Muller, C. Hagemeier, M. Truss and T. Carell, *Angew. Chem. Int. Ed.*, 2012, **51**, 6516.
214 M. Ganguly, M. W. Szulik, P. S. Donahue, K. Clancy, M. P. Stone and B. Gold, *Biochemistry*, 2012, **51**, 2018.
215 H. Yamada, M. Kurata, K. Tanabe, T. Ito and S.-i. Nishimoto, *Org. Biomol. Chem.*, 2012, **10**, 2035.
216 P. Menova and M. Hocek, *Chem. Commun.*, 2012, **48**, 6921.
217 V. Raindlova, R. Pohl and M. Hocek, *Chem. Eur. J.*, 2012, **18**, 4080.
218 L. Kalachova, R. Pohl and M. Hocek, *Org. Biomol. Chem.*, 2012, **10**, 49.
219 T. Khare, S. Pai, K. Koncevicius, M. Pal, E. Kriukiene, Z. Liutkeviciute, M. Irimia, P. Jia, C. Ptak, M. Xia, R. Tice, M. Tochigi, S. Morera, A. Nazarians,

D. Belsham, A. H. C. Wong, B. J. Blencowe, S. C. Wang, P. Kapranov, R. Kustra, V. Labrie, S. Klimasauskas and A. Petronis, *Nature Struct. Mol. Biol.*, 2012, **19**, 1037.

220 M. J. Booth, M. R. Branco, G. Ficz, D. Oxley, F. Krueger, W. Reik and S. Balasubramanian, *Science*, 2012, **336**, 934.

221 C. G. Lian, Y. Xu, C. Ceol, F. Wu, A. Larson, K. Dresser, W. Xu, L. Tan, Y. Hu, Q. Zhan, C.-w. Lee, D. Hu, Bill Q. Lian, S. Kleffel, Y. Yang, J. Neiswender, A. J. Khorasani, R. Fang, C. Lezcano, L. M. Duncan, R. A. Scolyer, J. F. Thompson, H. Kakavand, Y. Houvras, L. I. Zon, M. C. Mihm, U. B. Kaiser, T. Schatton, B. A. Woda, G. F. Murphy and Y. G. Shi, *Cell*, 2012, **150**, 1135.

222 J. M. Freudenberg, S. Ghosh, B. L. Lackford, S. Yellaboina, X. Zheng, R. Li, S. Cuddapah, P. A. Wade, G. Hu and R. Jothi, *Nucl. Acids Res.*, 2012, **40**, 3364.

223 A. A. Serandour, S. Avner, F. Oger, M. Bizot, F. Percevault, C. Lucchetti-Miganeh, G. Palierne, C. Gheeraert, F. Barloy-Hubler, C. L. Peron, T. Madigou, E. Durand, P. Froguel, B. Staels, P. Lefebvre, R. Metivier, J. Eeckhoute and G. Salbert, *Nucl. Acids Res.*, 2012, **40**, 8255.

224 C. S. Nabel, H. Jia, Y. Ye, L. Shen, H. L. Goldschmidt, J. T. Stivers, Y. Zhang and R. M. Kohli, *Nature Chem. Biol.*, 2012, **8**, 751.

225 H. Hashimoto, Y. Liu, A. K. Upadhyay, Y. Chang, S. B. Howerton, P. M. Vertino, X. Zhang and X. Cheng, *Nucl. Acids Res.*, 2012, **40**, 4841.

226 J. Terragni, J. Bitinaite, Y. Zheng and S. Pradhan, *Biochemistry*, 2012, **51**, 1009.

227 T. Heidebrecht, A. Fish, E. von Castelmur, K. A. Johnson, G. Zaccai, P. Borst and A. Perrakis, *J. Am. Chem. Soc.*, 2012, **134**, 13357.

228 K. Lamparska, J. Clark, G. Babilonia, V. Bedell, W. Yip and S. S. Smith, *Nucl. Acids Res.*, 2012, **40**, 9788.

229 N. A. Kuznetsov, Y. N. Vorobjev, L. N. Krasnoperov and O. S. Fedorova, *Nucl. Acids Res.*, 2012, **40**, 7384.

230 K. S. Park, J. Y. Lee and H. G. Park, *Chem. Commun.*, 2012, **48**, 4549.

231 X. Ming and F. Seela, *Chem. Eur. J.*, 2012, **18**, 9590.

232 X. Ming, P. Ding, P. Leonard, S. Budow and F. Seela, *Org. Biomol. Chem.*, 2012, **10**, 1861.

233 S. Xia, J. Beckman, J. Wang and W. H. Konigsberg, *Biochemistry*, 2012, **51**, 4609.

234 A. Marko, V. Denysenkov, D. Margraf, P. Cekan, O. Schiemann, S. T. Sigurdsson and T. F. Prisner, *J. Am. Chem. Soc.*, 2011, **133**, 13375.

235 C. Hobartner, G. Sicoli, F. Wachowius, D. B. Gophane and S. T. Sigurdsson, *J. Org. Chem.*, 2012, **77**, 7749.

236 T. E. Edwards, P. Cekan, G. W. Reginsson, S. A. Shelke, A. R. Ferre-D'Amare, O. Schiemann and S. T. Sigurdsson, *Nucl. Acids Res.*, 2011, **39**, 4419.

237 D. Dominissini, S. Moshitch-Moshkovitz, S. Schwartz, M. Salmon-Divon, L. Ungar, S. Osenberg, K. Cesarkas, J. Jacob-Hirsch, N. Amariglio, M. Kupiec, R. Sorek and G. Rechavi, *Nature*, 2012, **485**, 201.

238 B. Chen, F. Ye, L. Yu, G. Jia, X. Huang, X. Zhang, S. Peng, K. Chen, M. Wang, S. Gong, R. Zhang, J. Yin, H. Li, Y. Yang, H. Liu, J. Zhang, H. Zhang, A. Zhang, H. Jiang, C. Luo and C.-G. Yang, *J. Am. Chem. Soc.*, 2012, **134**, 17963.

239 M. Grammel, H. Hang and N. K. Conrad, *ChemBioChem*, 2012, **13**, 1112.

240 S. Kim, A. Gottfried, R. R. Lin, T. Dertinger, A. S. Kim, S. Chung, R. A. Colyer, E. Weinhold, S. Weiss and Y. Ebenstein, *Angew. Chem. Int. Ed.*, 2012, **51**, 3578.

241　K. Tishinov, K. Schmidt, D. Haussinger and D. G. Gillingham, *Angew. Chem. Int. Ed.*, 2012, **51**, 12000.
242　M. Martín-Ortiz, M. Gomez-Gallego, C. Ramírez de Arellano and M. A. Sierra, *Chem. Eur. J.*, 2012, **18**, 12603.
243　K. Seio, S. Kurohagi, E. Kodama, Y. Masaki, H. Tsunoda, A. Ohkubo and M. Sekine, *Org. Biomol. Chem.*, 2012, **10**, 994.
244　Z. Szombati, S. Baerns, A. Marx and C. Meier, *ChemBioChem*, 2012, **13**, 700.
245　L. G. Carrascosa, S. Gomez-Montes, A. Avino, A. Nadal, M. Pla, R. Eritja and L. M. Lechuga, *Nucl. Acids Res.*, 2012, **40**, e56.
246　I. J. Lee and B. H. Kim, *Chem. Commun.*, 2012, **48**, 2074.
247　U. Ghanty, E. Fostvedt, R. Valenzuela, P. A. Beal and C. J. Burrows, *J. Am. Chem. Soc.*, 2012, **134**, 17643.
248　M.-L. Winz, A. Samanta, D. Benzinger and A. Jaschke, *Nucl. Acids Res.*, 2012, **40**, e78.
249　C. Yi, B. Chen, B. Qi, W. Zhang, G. Jia, L. Zhang, C. J. Li, A. R. Dinner, C.-G. Yang and C. He, *Nature Struct. Mol. Biol.*, 2012, **19**, 671.
250　D. Li, J. C. Delaney, C. M. Page, X. Yang, A. S. Chen, C. Wong, C. L. Drennan and J. M. Essigmann, *J. Am. Chem. Soc.*, 2012, **134**, 8896.
251　D. Pluskota-Karwatka, D. Matysiak, M. Makarewicz and L. Kronberg, *Eur. J. Org. Chem.*, 2012, 4797.
252　H. Yueh, H. Yu, C. S. Theile, A. Pal, A. Horhota, N. Greco, C. V. Christianson and L. W. McLaughlin, *Nucleosides, Nucleotides & Nucl. Acids*, 2012, **31**, 661.
253　P. Jaruga, R. Rozalski, A. Jawien, A. Migdalski, R. Olinski and M. Dizdaroglu, *Biochemistry*, 2012, **51**, 1822.
254　C. You, X. Dai, B. Yuan, J. Wang, J. Wang, P. J. Brooks, L. J. Niedernhofer and Y. Wang, *Nature Chem. Biol.*, 2012, **8**, 817.
255　Y. Cai, N. E. Geacintov and S. Broyde, *Biochemistry*, 2012, **51**, 1486.
256　V. S. Sidorenko, J.-E. Yeo, R. R. Bonala, F. Johnson, O. D. Schärer and A. P. Grollman, *Nucl. Acids Res.*, 2012, **40**, 2494.
257　D. E. Burakovsky, I. V. Prokhorova, P. V. Sergiev, P. Milon, O. V. Sergeeva, A. A. Bogdanov, M. V. Rodnina and O. A. Dontsova, *Nucl. Acids Res.*, 2012, **40**, 7885.
258　O. J. Wilkinson, V. Latypov, J. L. Tubbs, C. L. Millington, R. Morita, H. Blackburn, A. Marriott, G. McGown, M. Thorncroft, A. J. Watson, B. A. Connolly, J. A. Grasby, R. Masui, C. A. Hunter, J. A. Tainer, G. P. Margison and D. M. Williams, *Proc. Natl. Acad. Sci. USA*, 2012, **109**, 18755.
259　M. Melikishvili and M. G. Fried, *Nucl. Acids Res.*, 2012, **40**, 9060.
260　I. Tessmer, M. Melikishvili and M. G. Fried, *Nucl. Acids Res.*, 2012, **40**, 8296.
261　K. Onizuka, T. Nishioka, Z. Li, D. Jitsuzaki, Y. Taniguchi and S. Sasaki, *Chem. Commun.*, 2012, **48**, 3969.
262　K. Abdu, M. K. Aiertza, O. J. Wilkinson, J. A. Grasby, P. Senthong, A. C. Povey, G. P. Margison and D. M. Williams, *Chem. Commun.*, 2012, **48**, 11214.
263　C. L. Millington, A. J. Watson, A. S. Marriott, G. P. Margison, A. C. Povey and D. M. Williams, *Nucleosides, Nucleotides & Nucl. Acids*, 2012, **31**, 328.
264　S. Hentschel, J. Alzeer, T. Angelov, O. D. Scharer and N. W. Luedtke, *Angew. Chem. Int. Ed.*, 2012, **51**, 3466.
265　A. Dumas and N. W. Luedtke, *Chem. Eur. J.*, 2012, **18**, 245.
266　I. Bhamra, P. Compagnone-Post, I. A. O'Neil, L. A. Iwanejko, A. D. Bates and R. Cosstick, *Nucl. Acids Res.*, 2012, **40**, 11126.
267　Y. Saito, K. Kugenuma, M. Tanaka, A. Suzuki and I. Saito, *Bioorg. Med. Chem. Lett.*, 2012, **22**, 3723.

268   B. Holzberger, J. Strohmeier, V. Siegmund, U. Diederichsen and A. Marx, *Bioorg. Med. Chem. Lett.*, 2012, **22**, 3136.
269   V. K. Batra, D. D. Shock, W. A. Beard, C. E. McKenna and S. H. Wilson, *Proc. Natl. Acad. Sci. USA*, 2012, **109**, 113.
270   Y. Taniguchi, Y. Koga, K. Fukabori, R. Kawaguchi and S. Sasaki, *Bioorg. Med. Chem. Lett.*, 2012, **22**, 543.
271   B. A. Maxwell and Z. Suo, *Biochemistry*, 2012, **51**, 3485.
272   E. Markkanen, B. van Loon, E. Ferrari, J. L. Parsons, G. L. Dianov and U. Hubscher, *Proc. Natl. Acad. Sci. USA*, 2012, **109**, 437.
273   E. Markkanen, B. Castrec, G. Villani and U. Hubscher, *Proc. Natl. Acad. Sci. USA*, 2012, **109**, 20401.
274   R. Kanagaraj, P. Parasuraman, B. Mihaljevic, B. van Loon, K. Burdova, C. Konig, A. Furrer, V. A. Bohr, U. Hubscher and P. Janscak, *Nucl. Acids Res.*, 2012, **40**, 8449.
275   H. Menoni, M. S. Shukla, V. Gerson, S. Dimitrov and D. Angelov, *Nucl. Acids Res.*, 2012, **40**, 692.
276   P. L. McKibbin, A. Kobori, Y. Taniguchi, E. T. Kool and S. S. David, *J. Am. Chem. Soc.*, 2011, **134**, 1653.
277   Y. Osakada, K. Kawai, T. Tachikawa, M. Fujitsuka, K. Tainaka, S. Tero-Kubota and T. Majima, *Chem. Eur. J.*, 2012, **18**, 1060.
278   J. J. Foti, B. Devadoss, J. A. Winkler, J. J. Collins and G. C. Walker, *Science*, 2012, **336**, 315.
279   A. L. Madison, Z. A. Perez, P. To, T. Maisonet, E. V. Rios, Y. Trejo, C. Ochoa-Paniagua, A. Reno and E. D. A. Stemp, *Biochemistry*, 2011, **51**, 362.
280   C. B. Volle, D. A. Jarem and S. Delaney, *Biochemistry*, 2011, **51**, 52.
281   K. S. Lim, L. Cui, K. Taghizadeh, J. S. Wishnok, W. Chan, M. S. DeMott, I. R. Babu, S. R. Tannenbaum and P. C. Dedon, *J. Am. Chem. Soc.*, 2012, **134**, 18053.
282   Q. Jin, A. M. Fleming, C. J. Burrows and H. S. White, *J. Am. Chem. Soc.*, 2012, **134**, 11006.
283   Y. Rokhlenko, N. E. Geacintov and V. Shafirovich, *J. Am. Chem. Soc.*, 2012, **134**, 4955.
284   A. M. Fleming, J. G. Muller, A. C. Dlouhy and C. J. Burrows, *J. Am. Chem. Soc.*, 2012, **134**, 15091.
285   D. Stathis, U. Lischke, S. C. Koch, C. A. Deiml and T. Carell, *J. Am. Chem. Soc.*, 2012, **134**, 4925.
286   L. Zhao, P. P. Christov, I. D. Kozekov, M. G. Pence, P. S. Pallan, C. J. Rizzo, M. Egli and F. P. Guengerich, *Angew. Chem. Int. Ed.*, 2012, **51**, 5466.
287   V. G. Vaidyanathan and B. P. Cho, *Biochemistry*, 2012, **51**, 1983.
288   V. Jain, B. Hilton, S. Patnaik, Y. Zou, M. P. Chiarelli and B. P. Cho, *Nucl. Acids Res.*, 2012, **40**, 3939.
289   H. Mu, K. Kropachev, L. Wang, L. Zhang, A. Kolbanovskiy, M. Kolbanovskiy, N. E. Geacintov and S. Broyde, *Nucl. Acids Res.*, 2012, **40**, 9675.
290   L. Lior-Hoffmann, L. Wang, S. Wang, N. E. Geacintov, S. Broyde and Y. Zhang, *Nucl. Acids Res.*, 2012, **40**, 9193.
291   J. A. Bueren-Calabuig, A. Negri, A. Morreale and F. Gago, *Org. Biomol. Chem.*, 2012, **10**, 1543.
292   K. Bergen, A.-L. Steck, S. Strütt, A. Baccaro, W. Welte, K. Diederichs and A. Marx, *J. Am. Chem. Soc.*, 2012, **134**, 11840.
293   S. Xia, T. D. Christian, J. Wang and W. H. Konigsberg, *Biochemistry*, 2012, **51**, 4343.
294   S. A. Ingale, S. S. Pujari, V. R. Sirivolu, P. Ding, H. Xiong, H. Mei and F. Seela, *J. Org. Chem.*, 2012, **77**, 188.

295 H. Xiong, P. Leonard and F. Seela, *Bioconjugate Chem.*, 2012, **23**, 856.
296 J. M. Ibarra-Soza, A. A. Morris, P. Jayalath, H. Peacock, W. E. Conrad, M. B. Donald, M. J. Kurth and P. A. Beal, *Org. Biomol. Chem.*, 2012, **10**, 6491.
297 A. B. Neef, F. Samain and N. W. Luedtke, *ChemBioChem*, 2012, **13**, 1750.
298 R. A. Mizrahi, K. J. Phelps, A. Y. Ching and P. A. Beal, *Nucl. Acids Res.*, 2012, **40**, 9825.
299 A. Dierckx, F.-A. Miannay, N. Ben Gaied, S. Preus, M. Bjorck, T. Brown and L. M. Wilhelmsson, *Chem. Eur. J.*, 2012, **18**, 5987.
300 K. R. Lieberman, J. M. Dahl, A. H. Mai, M. Akeson and H. Wang, *J. Am. Chem. Soc.*, 2012, **134**, 18816.
301 S. Xia, A. Vashishtha, D. Bulkley, S. H. Eom, J. Wang and W. H. Konigsberg, *Biochemistry*, 2012, **51**, 4922.
302 B. Holzberger, M. G. Pszolla, A. Marx and H. M. Moller, *ChemBioChem*, 2012, **13**, 635.
303 B. A. Manvilla, A. Maiti, M. C. Begley, E. A. Toth and A. C. Drohat, *J. Mol. Biol.*, 2012, **420**, 164.
304 J. Silhan, K. Nagorska, Q. Zhao, K. Jensen, P. S. Freemont, C. M. Tang and G. S. Baldwin, *Nucl. Acids Res.*, 2012, **40**, 2065.
305 D. Lu, J. Silhan, J. T. MacDonald, E. P. Carpenter, K. Jensen, C. M. Tang, G. S. Baldwin and P. S. Freemont, *Proc. Natl. Acad. Sci. USA*, 2012, **109**, 16852.
306 L. Nilsen, R. J. Forstrom, M. Bjoras and I. Alseth, *Nucl. Acids Res.*, 2012, **40**, 2000.
307 C. Collier, C. Machon, G. S. Briggs, W. K. Smits and P. Soultanas, *Nucl. Acids Res.*, 2012, **40**, 739.
308 C. Zhou, J. T. Sczepanski and M. M. Greenberg, *J. Am. Chem. Soc.*, 2012, **134**, 16734.
309 L. Y. Kanazhevskaya, V. V. Koval, Y. N. Vorobjev and O. S. Fedorova, *Biochemistry*, 2012, **51**, 1306.
310 K. Taniho, R. Nakashima, M. Kandeel, Y. Kitamura and Y. Kitade, *Bioorg. Med. Chem. Lett.*, 2012, **22**, 2518.
311 N. An, H. S. White and C. J. Burrows, *Chem. Commun.*, 2012, **48**, 11410.
312 N. An, A. M. Fleming, H. S. White and C. J. Burrows, *Proc. Natl. Acad. Sci. USA*, 2012, **109**, 11504.
313 Y. Abe, O. Nakagawa, R. Yamaguchi and S. Sasaki, *Bioorg. Med. Chem.*, 2012, **20**, 3470.
314 Y. Sato, T. Ichihashi, S. Nishizawa and N. Teramae, *Angew. Chem. Int. Ed.*, 2012, **51**, 6369.
315 Y. Sato, Y. Zhang, T. Seino, T. Sugimoto, S. Nishizawa and N. Teramae, *Org. Biomol. Chem.*, 2012, **10**, 4003.
316 A. A. Tanpure, P. Patheja and S. G. Srivatsan, *Chem. Commun.*, 2012, **48**, 501.
317 C.-x. Wang, Y. Sato, M. Kudo, S. Nishizawa and N. Teramae, *Chem. Eur. J.*, 2012, **18**, 9481.
318 T. Takada, Y. Otsuka, M. Nakamura and K. Yamana, *Chem. Eur. J.*, 2012, **18**, 9300.
319 S. A. Shelke and S. T. Sigurdsson, *ChemBioChem*, 2012, **13**, 684.
320 S. A. Shelke and S. T. Sigurdsson, *Nucl. Acids Res.*, 2012, **40**, 3732.
321 S. H. Lee, S. Wang and E. T. Kool, *Chem. Commun.*, 2012, **48**, 8069.
322 E. A. Motea, I. Lee and A. J. Berdis, *Nucl. Acids Res.*, 2012, **40**, 2357.
323 J. M. N. San Pedro, T. A. Beerman and M. M. Greenberg, *Bioorg. Med. Chem.*, 2012, **20**, 4744.
324 C. Zhou and M. M. Greenberg, *J. Am. Chem. Soc.*, 2012, **134**, 8090.
325 B. Yang, A. Jinnouchi, H. Suemune and M. Aso, *Org. Lett.*, 2012, **14**, 5852.

326 O. Khakshoor, S. E. Wheeler, K. N. Houk and E. T. Kool, *J. Am. Chem. Soc.*, 2012, **134**, 3154.
327 S. Xia, S. H. Eom, W. H. Konigsberg and J. Wang, *Biochemistry*, 2012, **51**, 1476.
328 M. W. Kellinger, S. Ulrich, J. Chong, E. T. Kool and D. Wang, *J. Am. Chem. Soc.*, 2012, **134**, 8231.
329 S. Harusawa, H. Yoneyama, D. Fujisue, M. Nishiura, M. Fujitake, Y. Usami, Z.-y. Zhao, S. A. McPhee, T. J. Wilson and D. M. J. Lilley, *Tetrahedron Letts.*, 2012, **53**, 5891.
330 T. Kanamori, Y. Masaki, M. Mizuta, H. Tsunoda, A. Ohkubo, M. Sekine and K. Seio, *Org. Biomol. Chem.*, 2012, **10**, 1007.
331 J. Zhang, J. Zheng, C. Lu, Q. Du, Z. Liang and Z. Xi, *ChemBioChem*, 2012, **13**, 1940.
332 M. Hattori, T. Ohki, E. Yanase and Y. Ueno, *Bioorg. Med. Chem. Lett.*, 2012, **22**, 253.
333 P. Crisalli, A. R. Hernandez and E. T. Kool, *Bioconjugate Chem.*, 2012, **23**, 1969.
334 S. Kusano, T. Sakuraba, S. Hagihara and F. Nagatsugi, *Bioorg. Med. Chem. Lett.*, 2012, **22**, 6957.
335 S. Hagihara, S. Kusano, W.-C. Lin, X.-g. Chao, T. Hori, S. Imoto and F. Nagatsugi, *Bioorg. Med. Chem. Lett.*, 2012, **22**, 3870.
336 S. Imoto, T. Chikuni, H. Kansui, T. Kunieda and F. Nagatsugi, *Nucleosides, Nucleotides & Nucl. Acids*, 2012, **31**, 752.
337 K. Fujimoto, K. Hiratsuka-Konishi, T. Sakamoto, T. Ohtake, K.-i. Shinohara and Y. Yoshimura, *Mol. BioSys.*, 2012, **8**, 491.
338 A. Shigeno, T. Sakamoto, Y. Yoshimura and K. Fujimoto, *Org. Biomol. Chem.*, 2012, **10**, 7820.
339 M. Op de Beeck and A. Madder, *J. Am. Chem. Soc.*, 2012, **134**, 10737.
340 H. Ito, M. Urushihara, X. Liang and H. Asanuma, *ChemBioChem*, 2012, **13**, 311.
341 M. I. Fatthalla and E. B. Pedersen, *Helv. Chim. Acta*, 2012, **95**, 1538.
342 G. D. Hamblin, K. M. M. Carneiro, J. F. Fakhoury, K. E. Bujold and H. F. Sleiman, *J. Am. Chem. Soc.*, 2012, **134**, 2888.
343 Z.-y. Zhao, M. San, J.-L. H. A. Duprey, J. R. Arrand, J. S. Vyle and J. H. R. Tucker, *Bioorg. Med. Chem. Lett.*, 2012, **22**, 129.
344 J. Manchester, D. M. Bassani, J.-L. H. A. Duprey, L. Giordano, J. S. Vyle, Z.-y. Zhao and J. H. R. Tucker, *J. Am. Chem. Soc.*, 2012, **134**, 10791.
345 F. Wojciechowski, J. Lietard and C. J. Leumann, *Org. Lett.*, 2012, **14**, 5176.
346 M. Hollenstein, F. Wojciechowski and C. J. Leumann, *Bioorg. Med. Chem. Lett.*, 2012, **22**, 4428.
347 A. A. El-Sayed, E. B. Pedersen and N. A. Khaireldin, *Nucleosides, Nucleotides & Nucl. Acids*, 2012, **31**, 872.
348 M. Probst, D. Wenger, S. M. Biner and R. Haner, *Org. Biomol. Chem.*, 2012, **10**, 755.
349 F. Garo and R. Haner, *Angew. Chem. Int. Ed.*, 2012, **51**, 916.
350 F. Garo and R. Haner, *Bioconjugate Chem.*, 2012, **23**, 2105.
351 M. I. Fatthalla, Y. M. Elkholy, N. S. Abbas, A. H. Mandour, P. T. Jorgensen, N. Bomholt and E. B. Pedersen, *Bioorg. Med. Chem.*, 2012, **20**, 207.
352 A. Eick, F. Riechert-Krause and K. Weisz, *Bioconjugate Chem.*, 2012, **23**, 1127.
353 L. Wang, Y. Feng, Z. Yang, Y.-M. He, Q.-H. Fan and D. Liu, *Chem. Commun.*, 2012, **48**, 3715.
354 A. Ohkubo, Y. Nishino, Y. Ito, H. Tsunoda, K. Seio and M. Sekine, *Org. Biomol. Chem.*, 2012, **10**, 2008.

355 H.-J. Kim, F. Chen and S. A. Benner, *J. Org. Chem.*, 2012, **77**, 3664.
356 N. Tarashima, Y. Higuchi, Y. Komatsu and N. Minakawa, *Bioorg. Med. Chem.*, 2012, **20**, 7095.
357 R. Yamashige, M. Kimoto, Y. Takezawa, A. Sato, T. Mitsui, S. Yokoyama and I. Hirao, *Nucl. Acids Res.*, 2012, **40**, 2793.
358 T. Ishizuka, M. Kimoto, A. Sato and I. Hirao, *Chem. Commun.*, 2012, **48**, 10835.
359 T. Lavergne, D. A. Malyshev and F. E. Romesberg, *Chem. Eur. J.*, 2012, **18**, 1231.
360 D. A. Malyshev, K. Dhami, H. T. Quach, T. Lavergne, P. Ordoukhanian, A. Torkamani and F. E. Romesberg, *Proc. Natl. Acad. Sci. USA*, 2012, **109**, 12005.
361 K. Betz, D. A. Malyshev, T. Lavergne, W. Welte, K. Diederichs, T. J. Dwyer, P. Ordoukhanian, F. E. Romesberg and A. Marx, *Nature Chem. Biol.*, 2012, **8**, 612.
362 T. Uchiyama, T. Miura, H. Takeuchi, T. Dairaku, T. Komuro, T. Kawamura, Y. Kondo, L. Benda, V. Sychrovsky, P. Bour, I. Okamoto, A. Ono and Y. Tanaka, *Nucl. Acids Res.*, 2012, **40**, 5766.
363 M. M. Kiy, Z. E. Jacobi and J. Liu, *Chem. Eur. J.*, 2012, **18**, 1202.
364 T. Funai, Y. Miyazaki, M. Aotani, E. Yamaguchi, O. Nakagawa, S.-i. Wada, H. Torigoe, A. Ono and H. Urata, *Angew. Chem. Int. Ed.*, 2012, **51**, 6464.
365 K. Petrovec, B. J. Ravoo and J. Muller, *Chem. Commun.*, 2012, **48**, 11844.
366 S. Hoon, B. Zhou, K. D. Janda, S. Brenner and J. Scolnick, *Biotechniques*, 2011, **51**, 413.
367 J. L. Vinkenborg, N. Karnowski and M. Famulok, *Nature Chem. Biol.*, 2011, **7**, 519.
368 L. H. Lauridsen, J. A. Rothnagel and R. N. Veedu, *ChemBioChem*, 2012, **13**, 19.
369 R. D'Agata and G. Spoto, *Artificial DNA: PNA & XNA*, 2012, **3**, 45.
370 F. Mancin, P. Scrimin and P. Tecilla, *Chem. Commun.*, 2012, **48**, 5545.
371 G. Zhu, M. Ye, M. J. Donovan, E. Song, Z. Zhao and W. Tan, *Chem. Commun.*, 2012, **48**, 10472.
372 M. Mascini, I. Palchetti and S. Tombelli, *Angew. Chem. Int. Ed.*, 2012, **51**, 1316.
373 L. Tang, X. Yang, L. W. Dobrucki, I. Chaudhury, Q. Yin, C. Yao, S. Lezmi, W. G. Helferich, T. M. Fan and J. Cheng, *Angew. Chem. Int. Ed.*, 2012, **51**, 12721.
374 J.-W. Park, R. Tatavarty, D. W. Kim, H.-T. Jung and M. B. Gu, *Chem. Commun.*, 2012, **48**, 2071.
375 W. He, M.-A. Elizondo-Riojas, X. Li, G. L. R. Lokesh, A. Somasunderam, V. Thiviyanathan, D. E. Volk, R. H. Durland, J. Englehardt, C. N. Cavasotto and D. G. Gorenstein, *Biochemistry*, 2012, **51**, 8321.
376 X. Zhang, S. Zhu, C. Deng and X. Zhang, *Chem. Commun.*, 2012, **48**, 2689.
377 Y. Huang, S. Zhao, Z.-F. Chen, M. Shi, J. Chen and H. Liang, *Chem. Commun.*, 2012, **48**, 11877.
378 J. Das, K. B. Cederquist, A. A. Zaragoza, P. E. Lee, E. H. Sargent and S. O. Kelley, *Nature Chem.*, 2012, **4**, 642.
379 I. Russo Krauss, A. Merlino, C. Giancola, A. Randazzo, L. Mazzarella and F. Sica, *Nucl. Acids Res.*, 2011, **39**, 7858.
380 I. Russo Krauss, A. Merlino, A. Randazzo, E. Novellino, L. Mazzarella and F. Sica, *Nucl. Acids Res.*, 2012, **40**, 8119.
381 S. Nagatoishi, N. Isono, K. Tsumoto and N. Sugimoto, *ChemBioChem*, 2011, **12**, 1822.

382   A. Avino, S. Mazzini, R. Ferreira, R. Gargallo, V. E. Marquez and R. Eritja, *Bioorg. Med. Chem.*, 2012, **20**, 4186.
383   A. Pasternak, F. J. Hernandez, L. M. Rasmussen, B. Vester and J. Wengel, *Nucl. Acids Res.*, 2011, **39**, 1155.
384   A. De Rache, I. Kejnovska, M. Vorlickova and C. Buess-Herman, *Chem. Eur. J.*, 2012, **18**, 4392.
385   F. Rohrbach, M. I. Fatthalla, T. Kupper, B. Potzsch, J. Muller, M. Petersen, E. B. Pedersen and G. Mayer, *ChemBioChem*, 2012, **13**, 631.
386   S. Liu, X. Zhang, W. Luo, Z. Wang, X. Guo, M. L. Steigerwald and X. Fang, *Angew. Chem. Int. Ed.*, 2011, **50**, 2496.
387   C. Ma, L. Cao, C. Shi and N. Ye, *Chem. Commun.*, 2011, **47**, 11303.
388   L. Gao, Y. Cui, Q. He, Y. Yang, J. Fei and J. Li, *Chem. Eur. J.*, 2011, **17**, 13170.
389   Y.-C. Shiang, C.-L. Hsu, C.-C. Huang and H.-T. Chang, *Angew. Chem. Int. Ed.*, 2011, **50**, 7660.
390   F. Luo, L. Zheng, S. Chen, Q. Cai, Z. Lin, B. Qiu and G. Chen, *Chem. Commun.*, 2012, **48**, 6387.
391   J. Muller, T. Becher, J. Braunstein, P. Berdel, S. Gravius, F. Rohrbach, J. Oldenburg, G. Mayer and B. Pötzsch, *Angew. Chem. Int. Ed.*, 2011, **50**, 6075.
392   D. Rotem, L. Jayasinghe, M. Salichou and H. Bayley, *J. Am. Chem. Soc.*, 2012, **134**, 2781.
393   A.-X. Zheng, J.-R. Wang, J. Li, X.-R. Song, G.-N. Chen and H.-H. Yang, *Chem. Commun.*, 2012, **48**, 374.
394   S. Yan, R. Huang, Y. Zhou, M. Zhang, M. Deng, X. Wang, X. Weng and X. Zhou, *Chem. Commun.*, 2011, **47**, 1273.
395   C.-L. Hsu, H.-T. Chang, C.-T. Chen, S.-C. Wei, Y.-C. Shiang and C.-C. Huang, *Chem. Eur. J.*, 2011, **17**, 10994.
396   Y. Zhang and X. Sun, *Chem. Commun.*, 2011, **47**, 3927.
397   J. Liu, J. Li, Y. Jiang, S. Yang, W. Tan and R. Yang, *Chem. Commun.*, 2011, **47**, 11321.
398   D. Zhang, M. Lu and H. Wang, *J. Am. Chem. Soc.*, 2011, **133**, 9188.
399   J.-W. Jian and C.-C. Huang, *Chem. Eur. J.*, 2011, **17**, 2374.
400   J. Sharma, H.-C. Yeh, H. Yoo, J. H. Werner and J. S. Martinez, *Chem. Commun.*, 2011, **47**, 2294.
401   Y. Qi and B. Li, *Chem. Eur. J.*, 2011, **17**, 1642.
402   Y. Chen, B. Jiang, Y. Xiang, Y. Chai and R. Yuan, *Chem. Commun.*, 2011, **47**, 7758.
403   L. Guo and D.-H. Kim, *Chem. Commun.*, 2011, **47**, 7125.
404   L. Wang, C. Zhu, L. Han, L. Jin, M. Zhou and S. Dong, *Chem. Commun.*, 2011, **47**, 7794.
405   X. Zhang, S. Li, X. Jin and S. Zhang, *Chem. Commun.*, 2011, **47**, 4929.
406   Y. Guo, X. Jia and S. Zhang, *Chem. Commun.*, 2011, **47**, 725.
407   Y. Fu, T. Wang, L. Bu, Q. Xie, P. Li, J. Chen and S. Yao, *Chem. Commun.*, 2011, **47**, 2637.
408   Y. Yuan, R. Yuan, Y. Chai, Y. Zhuo, X. Gan and L. Bai, *Chem. Eur. J.*, 2012, **18**, 14186.
409   H.-J. Lee, B. C. Kim, M.-K. Oh and J. Kim, *Chem. Commun.*, 2012, **48**, 5971.
410   D. Han, Z. Zhu, C. Wu, L. Peng, L. Zhou, B. Gulbakan, G. Zhu, K. R. Williams and W. Tan, *J. Am. Chem. Soc.*, 2012, **134**, 20797.
411   M. Barbu and M. N. Stojanovic, *ChemBioChem*, 2012, **13**, 658.
412   Y. Pang, Z. Xu, Y. Sato, S. Nishizawa and N. Teramae, *ChemBioChem*, 2012, **13**, 436.
413   S. Sitaula, S. D. Branch and M. F. Ali, *Chem. Commun.*, 2012, **48**, 9284.

414 Y. Xiang, X. Qian, B. Jiang, Y. Chai and R. Yuan, *Chem. Commun.*, 2011, **47**, 4733.
415 C. Holzhauser and H.-A. Wagenknecht, *ChemBioChem*, 2012, **13**, 1136.
416 X. Liu, R. Freeman and I. Willner, *Chem. Eur. J.*, 2012, **18**, 2207.
417 W. Wang, T. Yan, S. Cui and J. Wan, *Chem. Commun.*, 2012, **48**, 10228.
418 H. Liu, Y. Xiang, Y. Lu and R. M. Crooks, *Angew. Chem. Int. Ed.*, 2012, **51**, 6925.
419 B. Liu, Y. Cui, D. Tang, H. Yang and G. Chen, *Chem. Commun.*, 2012, **48**, 2624.
420 Z. Lin, F. Luo, Q. Liu, L. Chen, B. Qiu, Z. Cai and G. Chen, *Chem. Commun.*, 2011, **47**, 8064.
421 K. Song, X. Kong, X. Liu, Y. Zhang, Q. Zeng, L. Tu, Z. Shi and H. Zhang, *Chem. Commun.*, 2012, **48**, 1156.
422 T.-H. Nguyen, L. J. Steinbock, H.-J. Butt, M. Helm and R. Berger, *J. Am. Chem. Soc.*, 2011, **133**, 2025.
423 Q. Liu, C. Jing, X. Zheng, Z. Gu, D. Li, D.-W. Li, Q. Huang, Y.-T. Long and C. Fan, *Chem. Commun.*, 2012, **48**, 9574.
424 Z. F. Liu, J. Ge and S. Zhao, *Chem. Commun.*, 2011, **47**, 4956.
425 B. K. Das, C. Tlili, S. Badhulika, L. N. Cella, W. Chen and A. Mulchandani, *Chem. Commun.*, 2011, **47**, 3793.
426 J. Tang, D. Tang, J. Zhou, H. Yang and G. Chen, *Chem. Commun.*, 2012, **48**, 2627.
427 C.-Y. Tian, J.-J. Xu and H.-Y. Chen, *Chem. Commun.*, 2012, **48**, 8234.
428 N. Dave and J. Liu, *Chem. Commun.*, 2012, **48**, 3718.
429 Y. Huang, S. Zhao, Z.-F. Chen, M. Shi and H. Liang, *Chem. Commun.*, 2012, **48**, 7480.
430 W. Xu and Y. Lu, *Chem. Commun.*, 2011, **47**, 4998.
431 V. C. Ozalp and T. Schäfer, *Chem. Eur. J.*, 2011, **17**, 9893.
432 M. B. Serrano-Santos, E. Llobet, V. C. Ozalp and T. Schafer, *Chem. Commun.*, 2012, **48**, 10087.
433 Y. Jiang, N. Liu, W. Guo, F. Xia and L. Jiang, *J. Am. Chem. Soc.*, 2012, **134**, 15395.
434 Y. Liu, J. Ren, Y. Qin, J. Li, J. Liu and E. Wang, *Chem. Commun.*, 2012, **48**, 802.
435 Y. Wu, S. Zhan, L. Xu, W. Shi, T. Xi, X. Zhan and P. Zhou, *Chem. Commun.*, 2011, **47**, 6027.
436 G. Pelossof, R. Tel-Vered, X.-Q. Liu and I. Willner, *Chem. Eur. J.*, 2011, **17**, 8904.
437 Y. Wu, S. Zhan, F. Wang, L. He, W. Zhi and P. Zhou, *Chem. Commun.*, 2012, **48**, 4459.
438 J. Kim, M. Y. Kim, H. S. Kim and S. S. Hah, *Bioorg. Med. Chem. Lett.*, 2011, **21**, 4020.
439 J. Roy, P. Chirania, S. Ganguly and H. Huang, *Bioorg. Med. Chem. Lett.*, 2012, **22**, 863.
440 Z. Jiang, L. Zhou and A. Liang, *Chem. Commun.*, 2011, **47**, 3162.
441 Y. Sato, Y. Zhang, S. Nishizawa, T. Seino, K. Nakamura, M. Li and N. Teramae, *Chem. Eur. J.*, 2012, **18**, 12719.
442 Y. Zhao, X.-W. He and X.-B. Yin, *Chem. Commun.*, 2011, **47**, 6419.
443 R. Kawano, T. Osaki, H. Sasaki, M. Takinoue, S. Yoshizawa and S. Takeuchi, *J. Am. Chem. Soc.*, 2011, **133**, 8474.
444 A. K. Sharma and J. M. Heemstra, *J. Am. Chem. Soc.*, 2011, **133**, 12426.
445 A. Porchetta, A. Vallee-Belisle, K. W. Plaxco and F. Ricci, *J. Am. Chem. Soc.*, 2012, **134**, 20601.

446 Z. Zhu, C. Yang, X. Zhou and J. Qin, *Chem. Commun.*, 2011, **47**, 3192.
447 K.-A. Yang, R. Pei, D. Stefanovic and M. N. Stojanovic, *J. Am. Chem. Soc.*, 2011, **134**, 1642.
448 O. Reinstein, M. A. D. Neves, M. Saad, S. N. Boodram, S. Lombardo, S. A. Beckham, J. Brouwer, G. F. Audette, P. Groves, M. C. J. Wilce and P. E. Johnson, *Biochemistry*, 2011, **50**, 9368.
449 L. Delauriere, Z. Dong, K. Laxmi-Reddy, F. Godde, J.-J. Toulme and I. Huc, *Angew. Chem. Int. Ed.*, 2012, **51**, 473.
450 Y. Shi, W. T. Huang, H. Q. Luo and N. B. Li, *Chem. Commun.*, 2011, **47**, 4676.
451 T. Bing, T. Chang, X. Yang, H. Mei, X. Liu and D. Shangguan, *Bioorg. Med. Chem.*, 2011, **19**, 4211.
452 D. Muharemagic, M. Labib, S. M. Ghobadloo, A. S. Zamay, J. C. Bell and M. V. Berezovski, *J. Am. Chem. Soc.*, 2012, **134**, 17168.
453 X. Luo, I. Lee, J. Huang, M. Yun and X. T. Cui, *Chem. Commun.*, 2011, **47**, 6368.
454 M. R. Battig, B. Soontornworajit and Y. Wang, *J. Am. Chem. Soc.*, 2012, **134**, 12410.
455 V. J. B. Ruigrok, E. van Duijn, A. Barendregt, K. Dyer, J. A. Tainer, R. Stoltenburg, B. Strehlitz, M. Levisson, H. Smidt and J. van der Oost, *ChemBioChem*, 2012, **13**, 829.
456 C. Da Pieve, E. Blackshaw, S. Missailidis and A. C. Perkins, *Bioconjugate Chem.*, 2012, **23**, 1377.
457 B. Wang, C. Guo, G. Chen, B. Park and B. Xu, *Chem. Commun.*, 2012, **48**, 1644.
458 M. Wood, P. Maynard, X. Spindler, C. Lennard and C. Roux, *Angew. Chem. Int. Ed.*, 2012, **51**, 12272.
459 N. Shao, K. Zhang, Y. Chen, X. He and Y. Zhang, *Chem. Commun.*, 2012, **48**, 6684.
460 H. Yu, B. Jiang and J. C. Chaput, *ChemBioChem*, 2011, **12**, 2659.
461 Z. Sun, Y. Wang, Y. Wei, R. Liu, H. Zhu, Y. Cui, Y. Zhao and X. Gao, *Chem. Commun.*, 2011, **47**, 11960.
462 P. R. Mallikaratchy, A. Ruggiero, J. R. Gardner, V. Kuryavyi, W. F. Maguire, M. L. Heaney, M. R. McDevitt, D. J. Patel and D. A. Scheinberg, *Nucl. Acids Res.*, 2011, **39**, 2458.
463 J. Song, F. Lv, G. Yang, L. Liu, Q. Yang and S. Wang, *Chem. Commun.*, 2012, **48**, 7465.
464 T. Tokunaga, S. Namiki, K. Yamada, T. Imaishi, H. Nonaka, K. Hirose and S. Sando, *J. Am. Chem. Soc.*, 2012, **134**, 9561.
465 J. A. Phillips, H. Liu, M. B. O,ÄôDonoghue, X. Xiong, R. Wang, M. You, K. Sefah and W. Tan, *Bioconjugate Chem.*, 2011, **22**, 282.
466 L. Yang, L. Meng, X. Zhang, Y. Chen, G. Zhu, H. Liu, X. Xiong, K. Sefah and W. Tan, *J. Am. Chem. Soc.*, 2011, **133**, 13380.
467 K. Wang, M. You, Y. Chen, D. Han, Z. Zhu, J. Huang, K. Williams, C. J. Yang and W. Tan, *Angew. Chem. Int. Ed.*, 2011, **50**, 6098.
468 H. Shi, X. He, K. Wang, X. Wu, X. Ye, Q. Guo, W. Tan, Z. Qing, X. Yang and B. Zhou, *Proc. Natl. Acad. Sci. USA*, 2011, **108**, 3900.
469 H. Zhong, Q. Zhang and S. Zhang, *Chem. Eur. J.*, 2011, **17**, 8388.
470 C. Ding, S. Wei and H. Liu, *Chem. Eur. J.*, 2012, **18**, 7263.
471 Y. Song, L. Cui, J. Wu, W. Zhang, W. Y. Zhang, H. Kang and C. J. Yang, *Chem. Eur. J.*, 2011, **17**, 9042.
472 M. A. Ditzler, D. Bose, N. Shkriabai, B. Marchand, S. G. Sarafianos, M. Kvaratskhelia and D. H. Burke, *Nucl. Acids Res.*, 2011, **39**, 8237.

473. V. D'Atri, G. Oliviero, J. Amato, N. Borbone, S. D'Errico, L. Mayol, V. Piccialli, S. Haider, B. Hoorelbeke, J. Balzarini and G. Piccialli, *Chem. Commun.*, 2012, **48**, 9516.
474. K. T. Shum, E. L. H. Lui, S. C. K. Wong, P. Yeung, L. Sam, Y. Wang, R. M. Watt and J. A. Tanner, *Biochemistry*, 2011, **50**, 3261.
475. J. Hu, J. Wu, C. Li, L. Zhu, W. Y. Zhang, G. Kong, Z. Lu and C. J. Yang, *ChemBioChem*, 2011, **12**, 424.
476. S. M. Krylova, V. Koshkin, E. Bagg, C. J. Schofield and S. N. Krylov, *J. Med. Chem.*, 2012, **55**, 3546.
477. J. L. Vinkenborg, G. Mayer and M. Famulok, *Angew. Chem. Int. Ed.*, 2012, **51**, 9176.
478. J. Shimada, T. Maruyama, M. Kitaoka, H. Yoshinaga, K. Nakano, N. Kamiya and M. Goto, *Chem. Commun.*, 2012, **48**, 6226.
479. H. Iioka, D. Loiselle, T. A. Haystead and I. G. Macara, *Nucl. Acids Res.*, 2011, **39**, e53.
480. L. Qi, J. B. Lucks, C. C. Liu, V. K. Mutalik and A. P. Arkin, *Nucl. Acids Res.*, 2012, **40**, 5775.
481. C. Annoni, E. Nakata, T. Tamura, F. F. Liew, S. Nakano, M. L. Gelmi and T. Morii, *Org. Biomol. Chem.*, 2012, **10**, 8767.
482. B. Klauser, A. Saragliadis, S. Auslander, M. Wieland, M. R. Berthold and J. S. Hartig, *Mol. BioSys.*, 2012, **8**, 2242.
483. A. A. Bastian, A. Marcozzi and A. Herrmann, *Nature Chem.*, 2012, **4**, 789.
484. G.-H. Lee and S. S. Hah, *Bioorg. Med. Chem. Lett.*, 2012, **22**, 1520.
485. C. Carrasquilla, P. S. Lau, Y. Li and J. D. Brennan, *J. Am. Chem. Soc.*, 2012, **134**, 10998.
486. A. Murata, S.-i. Sato, Y. Kawazoe and M. Uesugi, *Chem. Commun.*, 2011, **47**, 4712.
487. J. T. Sczepanski and G. F. Joyce, *Chem. Biol.*, 2012, **19**, 1324.
488. S. Stewart, A. Syrett, A. Pothukuchy, S. Bhadra, A. Ellington and E. Anslyn, *ChemBioChem*, 2011, **12**, 2021.
489. M. M. K. Vu, N. E. Jameson, S. J. Masuda, D. Lin, R. Larralde-Ridaura and A. Luptak, *Chem. Biol.*, 2012, **19**, 1247.
490. K. D. Smith, S. V. Lipchock and S. A. Strobel, *Biochemistry*, 2011, **51**, 425.
491. S. Nakayama, Y. Luo, J. Zhou, T. K. Dayie and H. O. Sintim, *Chem. Commun.*, 2012, **48**, 9059.
492. T. Endoh and N. Sugimoto, *ChemBioChem*, 2011, **12**, 1174.
493. U. Forster, J. E. Weigand, P. Trojanowski, B. Suess and J. Wachtveitl, *Nucl. Acids Res.*, 2012, **40**, 1807.
494. H. Peacock, R. Bachu and P. A. Beal, *Bioorg. Med. Chem. Lett.*, 2011, **21**, 5002.
495. T. S. Singh, B. J. Rao and G. Krishnamoorthy, *Biochemistry*, 2012, **51**, 9260.
496. S. Nakano, E. Nakata and T. Morii, *Bioorg. Med. Chem. Lett.*, 2011, **21**, 4503.
497. F. F. Liew, H. Hayashi, S. Nakano, E. Nakata and T. Morii, *Bioorg. Med. Chem.*, 2011, **19**, 5771.
498. B. Shui, A. Ozer, W. Zipfel, N. Sahu, A. Singh, J. T. Lis, H. Shi and M. I. Kotlikoff, *Nucl. Acids Res.*, 2012, **40**, e39.
499. J. S. Paige, K. Y. Wu and S. R. Jaffrey, *Science*, 2011, **333**, 642.
500. J. E. Sokoloski, S. E. Dombrowski and P. C. Bevilacqua, *Biochemistry*, 2011, **51**, 565.
501. J. Lux, E. J. Pena, F. Bolze, M. Heinlein and J.-F. Nicoud, *ChemBioChem*, 2012, **13**, 1206.
502. J. Bernard Da Costa and T. Dieckmann, *Mol. BioSys.*, 2011, **7**, 2156.

503 M. Steber, A. Arora, J. Hofmann, B. Brutschy and B. Suess, *ChemBioChem*, 2011, **12**, 2608.
504 S.-Y. Hwang, H.-Y. Sun, K.-H. Lee, B.-H. Oh, Y. J. Cha, B. H. Kim and J.-Y. Yoo, *Nucl. Acids Res.*, 2012, **40**, 2724.
505 M. Hons, B. Niebel, N. Karnowski, B. Weiche and M. Famulok, *ChemBioChem.*, 2012, **13**, 1433.
506 S. Da Rocha Gomes, J. Miguel, L. Azema, S. Eimer, C. Ries, E. Dausse, H. Loiseau, M. Allard and J.-J. Toulme, *Bioconjugate Chem.*, 2012, **23**, 2192.
507 H. H. Salamanca, N. Fuda, H. Shi and J. T. Lis, *Nucl. Acids Res.*, 2011, **39**, 6729.
508 W. M. Rockey, L. Huang, K. C. Kloepping, N. J. Baumhover, P. H. Giangrande and M. K. Schultz, *Bioorg. Med. Chem.*, 2011, **19**, 4080.
509 K. W. Thiel, L. I. Hernandez, J. P. Dassie, W. H. Thiel, X. Liu, K. R. Stockdale, A. M. Rothman, F. J. Hernandez, J. O. McNamara and P. H. Giangrande, *Nucl. Acids Res.*, 2012, **40**, 6319.
510 K. Konno, M. Iizuka, S. Fujita, S. Nishikawa, T. Hasegawa and K. Fukuda, *Nucleosides, Nucleotides Nucl. Acids*, 2011, **30**, 185.
511 T. B. Jensen, J. R. Henriksen, B. E. Rasmussen, L. M. Rasmussen, T. L. Andresen, J. Wengel and A. Pasternak, *Bioorg. Med. Chem.*, 2011, **19**, 4739.
512 D. Nie, H. Wu, Q. Zheng, L. Guo, P. Ye, Y. Hao, Y. Li, F. Fu and Y. Guo, *Chem. Commun.*, 2012, **48**, 1150.
513 X. Miao, L. Ling and X. Shuai, *Chem. Commun.*, 2011, **47**, 4192.
514 Y. Wang and J. Irudayaraj, *Chem. Commun.*, 2011, **47**, 4394.
515 Y. Wen, C. Peng, D. Li, L. Zhuo, S. He, L. Wang, Q. Huang, Q.-H. Xu and C. Fan, *Chem. Commun.*, 2011, **47**, 6278.
516 L. Zhang, B. Han, T. Li and E. Wang, *Chem. Commun.*, 2011, **47**, 3099.
517 K. E. Nelson, H. E. Ihms, D. Mazumdar, P. J. Bruesehoff and Y. Lu, *ChemBioChem*, 2012, **13**, 381.
518 M. Liu, H. Zhao, S. Chen, H. Yu, Y. Zhang and X. Quan, *Chem. Commun.*, 2011, **47**, 7749.
519 H. Lin, Y. Zou, Y. Huang, J. Chen, W. Y. Zhang, Z. Zhuang, G. Jenkins and C. J. Yang, *Chem. Commun.*, 2011, **47**, 9312.
520 Y. Xiao, E. C. Allen and S. K. Silverman, *Chem. Commun.*, 2011, **47**, 1749.
521 Y. Xiao, R. J. Wehrmann, N. A. Ibrahim and S. K. Silverman, *Nucl. Acids Res.*, 2012, **40**, 1778.
522 C. H. Lam, C. J. Hipolito, M. Hollenstein and D. M. Perrin, *Org. Biomol. Chem.*, 2011, **9**, 6949.
523 A. Ariza-Mateos, S. Prieto-Vega, R. Díaz-Toledano, A. Birk, H. Szeto, I. Mena, A. Berzal-Herranz and J. Gomez, *Nucl. Acids Res.*, 2012, **40**, 1748.
524 Y. He and Y. Lu, *Chem. Eur. J.*, 2011, **17**, 13732.
525 M. M. Ali, S. D. Aguirre, H. Lazim and Y. Li, *Angew. Chem. Int. Ed.*, 2011, **50**, 3751.
526 M.-P. Chien, M. P. Thompson and N. C. Gianneschi, *Chem. Commun.*, 2011, **47**, 167.
527 R. Sugiyama, M. Hayafune, Y. Habu, N. Yamamoto and H. Takaku, *Nucl. Acids Res.*, 2011, **39**, 589.
528 F. Wang, J. Elbaz, R. Orbach, N. Magen and I. Willner, *J. Am. Chem. Soc.*, 2011, **133**, 17149.
529 J. Liang, Z. Chen, L. Guo and L. Li, *Chem. Commun.*, 2011, **47**, 5476.
530 R. Hu, T. Fu, X.-B. Zhang, R.-M. Kong, L.-P. Qiu, Y.-R. Liu, X.-T. Liang, W. Tan, G.-L. Shen and R.-Q. Yu, *Chem. Commun.*, 2012, **48**, 9507.
531 C. J. Hipolito, M. Hollenstein, C. H. Lam and D. M. Perrin, *Org. Biomol. Chem.*, 2011, **9**, 2266.

532   L. Stefan, H.-J. Xu, C. P. Gros, F. Denat and D. Monchaud, *Chem. Eur. J.*, 2011, **17**, 10857.
533   S. Nakayama, J. Wang and H. O. Sintim, *Chem. Eur. J.*, 2011, **17**, 5691.
534   L. Stefan, F. Denat and D. Monchaud, *Nucl. Acids Res.*, 2012, **40**, 8759.
535   L. Stefan, F. Denat and D. Monchaud, *J. Am. Chem. Soc.*, 2011, **133**, 20405.
536   H. Abe, N. Abe, A. Shibata, K. Ito, Y. Tanaka, M. Ito, H. Saneyoshi, S. Shuto and Y. Ito, *Angew. Chem. Int. Ed.*, 2012, **51**, 6475.
537   C. Li, L. Zhu, Z. Zhu, H. Fu, G. Jenkins, C. Wang, Y. Zou, X. Lu and C. J. Yang, *Chem. Commun.*, 2012, **48**, 8347.
538   X. Zhu, X. Gao, Q. Liu, Z. Lin, B. Qiu and G. Chen, *Chem. Commun.*, 2011, **47**, 7437.
539   E. Golub, R. Freeman and I. Willner, *Angew. Chem. Int. Ed.*, 2011, **50**, 11710.
540   M. Zhou, F. Kuralay, J. R. Windmiller and J. Wang, *Chem. Commun.*, 2012, **48**, 3815.
541   M. Zhang, S. Xu, S. D. Minteer and D. A. Baum, *J. Am. Chem. Soc.*, 2011, **133**, 15890.
542   S. Shimron, N. Magen, J. Elbaz and I. Willner, *Chem. Commun.*, 2011, **47**, 8787.
543   F. Du and Z. Tang, *ChemBioChem*, 2011, **12**, 43.
544   B. Qiu, Z.-Z. Zheng, Y.-J. Lu, Z.-Y. Lin, K.-Y. Wong and G.-N. Chen, *Chem. Commun.*, 2011, **47**, 1437.
545   J. Chen, J. Zhang, Y. Guo, J. Li, F. Fu, H.-H. Yang and G. Chen, *Chem. Commun.*, 2011, **47**, 8004.
546   J. Tang, L. Hou, D. Tang, B. Zhang, J. Zhou and G. Chen, *Chem. Commun.*, 2012, **48**, 8180.
547   R. Freeman, X. Liu and I. Willner, *J. Am. Chem. Soc.*, 2011, **133**, 11597.
548   M. Luo, X. Chen, G. Zhou, X. Xiang, L. Chen, X. Ji and Z. He, *Chem. Commun.*, 2012, **48**, 1126.
549   F. Wang, J. Elbaz, C. Teller and I. Willner, *Angew. Chem. Int. Ed.*, 2011, **50**, 295.
550   A.-X. Zheng, J. Li, J.-R. Wang, X.-R. Song, G.-N. Chen and H.-H. Yang, *Chem. Commun.*, 2012, **48**, 3112.
551   N. Wang, D.-M. Kong and H.-X. Shen, *Chem. Commun.*, 2011, **47**, 1728.
552   D. M. Koster, D. Haselbach, H. Lehrach and H. Seitz, *Mol. BioSys.*, 2011, **7**, 2882.
553   R. Fu, K. Jeon, C. Jung and H. G. Park, *Chem. Commun.*, 2011, **47**, 9876.
554   J. C. F. Lam, S. O. Kwan and Y. Li, *Mol. BioSys*, 2011, **7**, 2139.
555   W. Rong, L. Xu, Y. Liu, J. Yu, Y. Zhou, K. Liu and J. He, *Bioorg. Med. Chem. Lett.*, 2012, **22**, 4238.
556   J. He, D. Zhang, Q. Wang, X. Wei, M. Cheng and K. Liu, *Org. Biomol. Chem.*, 2011, **9**, 5728.
557   L. Robaldo, F. Izzo, M. Dellafiore, C. Proietti, P. V. Elizalde, J. M. Montserrat and A. M. Iribarren, *Bioorg. Med. Chem.*, 2012, **20**, 2581.
558   A. A. Fokina, M. I. Meschaninova, T. Durfort, A. G. Venyaminova and J.-C. Francois, *Biochemistry*, 2012, **51**, 2181.
559   L.-M. Lu, X.-B. Zhang, R.-M. Kong, B. Yang and W. Tan, *J. Am. Chem. Soc.*, 2011, **133**, 11686.
560   K. He, W. Li, Z. Nie, Y. Huang, Z. Liu, L. Nie and S. Yao, *Chem. Eur. J.*, 2012, **18**, 3992.
561   F. Wachowius and C. Hobartner, *J. Am. Chem. Soc.*, 2011, **133**, 14888.
562   F. Wang, J. Elbaz and I. Willner, *J. Am. Chem. Soc.*, 2012, **134**, 5504.
563   C.-H. Lu, F. Wang and I. Willner, *J. Am. Chem. Soc.*, 2012, **134**, 10651.
564   O. Y. Wong, P. I. Pradeepkumar and S. K. Silverman, *Biochemistry*, 2011, **50**, 4741.

565 A. Sachdeva and S. K. Silverman, *Org. Biomol. Chem.*, 2012, **10**, 122.
566 A. Sachdeva, M. Chandra, J. Chandrasekar and S. K. Silverman, *ChemBioChem.*, 2012, **13**, 654.
567 K. O. Alila and D. A. Baum, *Chem. Commun.*, 2011, **47**, 3227.
568 O. Y. Wong, A. E. Mulcrone and S. K. Silverman, *Angew. Chem. Int. Ed.*, 2011, **50**, 11679.
569 S. Ayukawa, Y. Sakai and D. Kiga, *Chem. Commun.*, 2012, **48**, 7556.
570 R. Orbach, L. Mostinski, F. Wang and I. Willner, *Chem. Eur. J.*, 2012, **18**, 14689.
571 R. Orbach, F. Remacle, R. D. Levine and I. Willner, *Proc. Natl. Acad. Sci. USA*, 2012, **109**, 21228.
572 K. M. Brooks and K. J. Hampel, *Biochemistry*, 2011, **50**, 2424.
573 J. Viladoms and M. J. Fedor, *J. Am. Chem. Soc.*, 2012, **134**, 19043.
574 J. Viladoms, L. G. Scott and M. J. Fedor, *J. Am. Chem. Soc.*, 2011, **133**, 18388.
575 S. Kath-Schorr, T. J. Wilson, N.-S. Li, J. Lu, J. A. Piccirilli and D. M. J. Lilley, *J. Am. Chem. Soc.*, 2012, **134**, 16717.
576 J. A. Liberman, M. Guo, J. L. Jenkins, J. Krucinska, Y. Chen, P. R. Carey and J. E. Wedekind, *J. Am. Chem. Soc.*, 2012, **134**, 16933.
577 A. Carbonell, R. Flores and S. Gago, *Nucl. Acids Res.*, 2011, **39**, 2432.
578 C. A. Strulson, R. C. Molden, C. D. Keating and P. C. Bevilacqua, *Nature Chem.*, 2012, **4**, 941.
579 J. Lacroix-Labonte, N. Girard, S. Lemieux and P. Legault, *Nucl. Acids Res.*, 2012, **40**, 2284.
580 B. L. Golden, *Biochemistry*, 2011, **50**, 9424.
581 N. Veeraraghavan, A. Ganguly, J.-H. Chen, P. C. Bevilacqua, S. Hammes-Schiffer and B. L. Golden, *Biochemistry*, 2011, **50**, 2672.
582 F. J. Sanchez-Luque, M. C. Lopez, F. Macias, C. Alonso and M. C. Thomas, *Nucl. Acids Res.*, 2011, **39**, 8065.
583 N. Vaidya, M. L. Manapat, I. A. Chen, R. Xulvi-Brunet, E. J. Hayden and N. Lehman, *Nature*, 2012, **491**, 72.
584 T. Lonnberg, *Chem. Eur. J.*, 2011, **17**, 7140.
585 M. Forconi, R. H. Porecha, J. A. Piccirilli and D. Herschlag, *J. Am. Chem. Soc.*, 2011, **133**, 7791.
586 X. Shi, S. V. Solomatin and D. Herschlag, *J. Am. Chem. Soc.*, 2012, **134**, 1910.
587 T. Lonnberg and K.-M. Kero, *Org. Biomol. Chem.*, 2012, **10**, 569.
588 M. Marcia and A. M. Pyle, *Cell*, 2012, **151**, 497.
589 C.-F. Li, M. Costa and F. Michel, *EMBO J.*, 2011, **30**, 3040.
590 T. Huang, T. R. Shaikh, K. Gupta, L. M. Contreras-Martin, R. A. Grassucci, G. D. Van Duyne, J. Frank and M. Belfort, *Nucl. Acids Res.*, 2011, **39**, 2845.
591 R. T. Chan, A. R. Robart, K. R. Rajashankar, A. M. Pyle and N. Toor, *Nature Struct. Mol. Biol.*, 2012, **19**, 555.
592 C. Olea, D. P. Horning and G. F. Joyce, *J. Am. Chem. Soc.*, 2012, **134**, 8050.
593 A. Wochner, J. Attwater, A. Coulson and P. Holliger, *Science*, 2011, **332**, 209.
594 G. M. Giambasu, T.-S. Lee, W. G. Scott and D. M. York, *J. Mol. Biol.*, 2012, **423**, 106.
595 D. M. Shechner and D. P. Bartel, *Nature Struct. Mol. Biol.*, 2011, **18**, 1036.
596 A. Ogawa, *Bioorg. Med. Chem. Lett.*, 2011, **21**, 155.
597 A. Wittmann and B. Suess, *Mol. BioSys*, 2011, **7**, 2419.
598 T.-P. Wang, Y.-C. Su, Y. Chen, Y.-M. Liou, K.-L. Lin, E.-C. Wang, L.-C. Hwang, Y.-M. Wang and Y.-H. Chen, *Biochemistry*, 2011, **51**, 496.
599 E. Biondi, A. W. R. Maxwell and D. H. Burke, *Nucl. Acids Res.*, 2012, **40**, 7528.

600 R. M. Turk, M. Illangasekare and M. Yarus, *J. Am. Chem. Soc.*, 2011, **133**, 6044.
601 S. Kraut, D. Bebenroth, A. Nierth, A. Y. Kobitski, G. U. Nienhaus and A. Jaschke, *Nucl. Acids Res.*, 2012, **40**, 1318.
602 B. Strauss, A. Nierth, M. Singer and A. Jäschke, *Nucl. Acids Res.*, 2012, **40**, 861.
603 V. Kumar, T. Endoh, K. Murakami and N. Sugimoto, *Chem. Commun.*, 2012, **48**, 9684.
604 K. L. Frieda and S. M. Block, *Science*, 2012, **338**, 397.
605 P. Y. Watson and M. J. Fedor, *Nature Chem. Biol.*, 2012, **8**, 963.
606 J. Buck, A. Wacker, E. Warkentin, J. Wohnert, J. Wirmer-Bartoschek and H. Schwalbe, *Nucl. Acids Res.*, 2011, **39**, 9768.
607 P. Daldrop, F. E. Reyes, D. A. Robinson, C. M. Hammond, D. M. Lilley, R. T. Batey and R. Brenk, *Chem. Biol.*, 2011, **18**, 324.
608 J. Feng, N. G. Walter and C. L. Brooks, *J. Am. Chem. Soc.*, 2011, **133**, 4196.
609 Q. Zhang, M. Kang, R. D. Peterson and J. Feigon, *J. Am. Chem. Soc.*, 2011, **133**, 5190.
610 T. Santner, U. Rieder, C. Kreutz and R. Micura, *J. Am. Chem. Soc.*, 2012, **134**, 11928.
611 C. D. Eichhorn, J. Feng, K. C. Suddala, N. G. Walter, C. L. Brooks and H. M. Al-Hashimi, *Nucl. Acids Res.*, 2012, **40**, 1345.
612 Y. Luo, J. Zhou, S. K. Watt, V. T. Lee, T. K. Dayie and H. O. Sintim, *Mol. BioSys*, 2012, **8**, 772.
613 A. Ogawa, *Bioorg. Med. Chem. Lett.*, 2012, **22**, 1639.
614 L. Martini and S. S. Mansy, *Chem. Commun.*, 2011, **47**, 10734.
615 J.-J. Jo, J.-H. Kim and J.-S. Shin, *ChemBioChem*, 2012, **13**, 2048.
616 S. P. Hennelly and K. Y. Sanbonmatsu, *Nucl. Acids Res.*, 2011, **39**, 2416.
617 W. Huang, J. Kim, S. Jha and F. Aboul-ela, *J. Mol. Biol.*, 2012, **418**, 331.
618 R. L. Hayes, J. K. Noel, U. Mohanty, P. C. Whitford, S. P. Hennelly, J. N. Onuchic and K. Y. Sanbonmatsu, *J. Am. Chem. Soc.*, 2012, **134**, 12043.
619 R. C. Wilson, A. M. Smith, R. T. Fuchs, I. R. Kleckner, T. M. Henkin and M. P. Foster, *J. Mol. Biol.*, 2011, **405**, 926.
620 B. Chen, X. Zuo, Y.-X. Wang and T. K. Dayie, *Nucl. Acids Res.*, 2012, **40**, 3117.
621 C. Lu, A. M. Smith, F. Ding, A. Chowdhury, T. M. Henkin and A. Ke, *J. Mol. Biol.*, 2011, **409**, 786.
622 A. Peselis and A. Serganov, *Nature Struct. Mol. Biol.*, 2012, **19**, 1182.
623 P. C. Anthony, C. F. Perez, C. Garcia-Garcia and S. M. Block, *Proc. Natl. Acad. Sci. USA*, 2012, **109**, 1485.
624 Q. Vicens, E. Mondragon and R. T. Batey, *Nucl. Acids Res.*, 2011, **39**, 8586.
625 D. B. Pedrolli, A. Matern, J. Wang, M. Ester, K. Siedler, R. Breaker and M. Mack, *Nucl. Acids Res.*, 2012, **40**, 8662.
626 P. Budhathoki, L. F. Bernal-Perez, O. Annunziata and Y. Ryu, *Org. Biomol. Chem.*, 2012, **10**, 7872.
627 S. N. Wilson-Mitchell, F. J. Grundy and T. M. Henkin, *Nucl. Acids Res.*, 2012, **40**, 5706.
628 L. R. Fiegland, A. D. Garst, R. T. Batey and D. J. Nesbitt, *Biochemistry*, 2012, **51**, 9223.
629 M.-P. Caron, L. Bastet, A. Lussier, M. Simoneau-Roy, E. Masse and D. A. Lafontaine, *Proc. Natl. Acad. Sci. USA*, 2012, **109**, E3444.
630 A. Ren, K. R. Rajashankar and D. J. Patel, *Nature*, 2012, **486**, 85.
631 S. Zhou, J. A. Means, G. Acquaah-Harrison, S. C. Bergmeier and J. V. Hines, *Bioorg. Med. Chem.*, 2012, **20**, 1298.

632 N. Dixon, C. J. Robinson, T. Geerlings, J. N. Duncan, S. P. Drummond and J. Micklefield, *Angew. Chem. Int. Ed.*, 2012, **51**, 3620.
633 G. A. Perdrizet, I. Artsimovitch, R. Furman, T. R. Sosnick and T. Pan, *Proc. Natl. Acad. Sci. USA*, 2012, **109**, 3323.
634 T.-P. Wang, N. C. Ko, Y.-C. Su, E.-C. Wang, S. Severance, C.-C. Hwang, Y. T. Shih, M. H. Wu and Y.-H. Chen, *Bioconjugate Chem.*, 2012, **23**, 2417.
635 K.-W. Ieong, M. Y. Pavlov, M. Kwiatkowski, A. C. Forster and M. Ehrenberg, *J. Am. Chem. Soc.*, 2012, **134**, 17955.
636 A. Chatterjee, H. Xiao and P. G. Schultz, *Proc. Natl. Acad. Sci. USA*, 2012, **109**, 14841.
637 C. H. Kim, M. Kang, H. J. Kim, A. Chatterjee and P. G. Schultz, *Angew. Chem. Int. Ed.*, 2012, **51**, 7246.
638 C. H. Kim, J. Y. Axup, A. Dubrovska, S. A. Kazane, B. A. Hutchins, E. D. Wold, V. V. Smider and P. G. Schultz, *J. Am. Chem. Soc.*, 2012, **134**, 9918.
639 J. Y. Axup, K. M. Bajjuri, M. Ritland, B. M. Hutchins, C. H. Kim, S. A. Kazane, R. Halder, J. S. Forsyth, A. F. Santidrian, K. Stafin, Y. Lu, H. Tran, A. J. Seller, S. L. Biroc, A. Szydlik, J. K. Pinkstaff, F. Tian, S. C. Sinha, B. Felding-Habermann, V. V. Smider and P. G. Schultz, *Proc. Natl. Acad. Sci. USA*, 2012, **109**, 16101.
640 E. Kaya, M. Vrabel, C. Deiml, S. Prill, V. S. Fluxa and T. Carell, *Angew. Chem. Int. Ed.*, 2012, **51**, 4466.
641 A. Bianco, F. M. Townsley, S. Greiss, K. Lang and J. W. Chin, *Nature Chem. Biol.*, 2012, **8**, 748.
642 K. Lang, L. Davis, J. Torres-Kolbus, C. Chou, A. Deiters and J. W. Chin, *Nature Chem.*, 2012, **4**, 298.
643 T. Plass, S. Milles, C. Koehler, J. Szymanski, R. Mueller, M. Wiesler, C. Schultz and E. A. Lemke, *Angew. Chem. Int. Ed.*, 2012, **51**, 4166.
644 J. L. Seitchik, J. C. Peeler, M. T. Taylor, M. L. Blackman, T. W. Rhoads, R. B. Cooley, C. Refakis, J. M. Fox and R. A. Mehl, *J. Am. Chem. Soc.*, 2012, **134**, 2898.
645 K. Lang, L. Davis, S. Wallace, M. Mahesh, D. J. Cox, M. L. Blackman, J. M. Fox and J. W. Chin, *J. Am. Chem. Soc.*, 2012, **134**, 10317.
646 Z. Yu, Y. Pan, Z. Wang, J. Wang and Q. Lin, *Angew. Chem. Int. Ed.*, 2012, **51**, 10600.
647 A. Borrmann, S. Milles, T. Plass, J. Dommerholt, J. M. M. Verkade, M. Wiesler, C. Schultz, J. C. M. van Hest, F. L. van Delft and E. A. Lemke, *ChemBioChem*, 2012, **13**, 2094.
648 B. Wu, Z. Wang, Y. Huang and W. R. Liu, *ChemBioChem*, 2012, **13**, 1405.
649 M. Sioud and A. Mobergslien, *Bioconjugate Chem.*, 2012, **23**, 1040.
650 S.-i. Wada, Y. Hitora, S. Yokoe, O. Nakagawa and H. Urata, *Bioorg. Med. Chem.*, 2012, **20**, 3219.
651 Y.-L. Ying, D.-W. Li, Y. Liu, S. K. Dey, H.-B. Kraatz and Y.-T. Long, *Chem. Commun.*, 2012, **48**, 8784.
652 R. Bonnet, P. Murat, N. Spinelli and E. Defrancq, *Chem. Commun.*, 2012, **48**, 5992.
653 Y. Zeng, Y. Pratumyot, X. Piao and D. Bong, *J. Am. Chem. Soc.*, 2011, **134**, 832.
654 N. Gour, D. Kedracki, I. Safir, K. X. Ngo and C. Vebert-Nardin, *Chem. Commun.*, 2012, **48**, 5440.
655 G. Zhou, F. Khan, Q. Dai, J. E. Sylvester and S. J. Kron, *Mol. BioSys*, 2012, **8**, 2395.
656 J. Ghesquiere, N. Gauthier, J. De Winter, P. Gerbaux, C. Moucheron, E. Defrancq and A. Kirsch-De Mesmaeker, *Chem. Eur. J.*, 2012, **18**, 355.

657 S. L. Khatwani, J. S. Kang, D. G. Mullen, M. A. Hast, L. S. Beese, M. D. Distefano and T. A. Taton, *Bioorg. Med. Chem.*, 2012, **20**, 4532.
658 Y. Mashimo, H. Maeda, M. Mie and E. Kobatake, *Bioconjugate Chem.*, 2012, **23**, 1349.
659 A. Silanskas, M. Zaremba, G. Sasnauskas and V. Siksnys, *Bioconjugate Chem.*, 2012, **23**, 203.
660 S. Rudiuk, A. Venancio-Marques and D. Baigl, *Angew. Chem. Int. Ed.*, 2012, **51**, 12694.
661 K. Zhang, L. Hao, S. J. Hurst and C. A. Mirkin, *J. Am. Chem. Soc.*, 2012, **134**, 16488.
662 A. Gangar, A. Fegan, S. C. Kumarapperuma and C. R. Wagner, *J. Am. Chem. Soc.*, 2012, **134**, 2895.
663 S. A. Kazane, D. Sok, E. H. Cho, M. L. Uson, P. Kuhn, P. G. Schultz and V. V. Smider, *Proc. Natl. Acad. Sci. USA*, 2012, **109**, 3731.
664 R. M. Schweller, J. Zimak, D. Y. Duose, A. A. Qutub, W. N. Hittelman and M. R. Diehl, *Angew. Chem. Int. Ed.*, 2012, **51**, 9292.
665 J. L. Cape, J. B. Edson, L. P. Spencer, M. S. DeClue, H.-J. Ziock, S. Maurer, S. Rasmussen, P.-A. Monnard and J. M. Boncella, *Bioconjugate Chem.*, 2012, **23**, 2014.
666 Y. Tamura, K. Furukawa, R. Yoshimoto, Y. Kawai, M. Yoshida, S. Tsuneda, Y. Ito and H. Abe, *Bioorg. Med. Chem. Lett.*, 2012, **22**, 7248.
667 E. M. Harcourt and E. T. Kool, *Nucl. Acids Res.*, 2012, **40**, e65.
668 X.-H. Chen, A. Roloff and O. Seitz, *Angew. Chem. Int. Ed.*, 2012, **51**, 4479.
669 C. Zhang, Y. Li, M. Zhang and X. Li, *Tetrahedron*, 2012, **68**, 5152.
670 A. J. Boersma, B. de Bruin, B. L. Feringa and G. Roelfes, *Chem. Commun.*, 2012, **48**, 2394.
671 Y. Li, M. Zhang, C. Zhang and X. Li, *Chem. Commun.*, 2012, **48**, 9513.
672 D. K. Prusty, M. Kwak, J. Wildeman and A. Herrmann, *Angew. Chem. Int. Ed.*, 2012, **51**, 11894.
673 M. L. McKee, P. J. Milnes, J. Bath, E. Stulz, R. K. O'Reilly and A. J. Turberfield, *J. Am. Chem. Soc.*, 2012, **134**, 1446.
674 P. J. Milnes, M. L. McKee, J. Bath, L. Song, E. Stulz, A. J. Turberfield and R. K. O'Reilly, *Chem. Commun.*, 2012, **48**, 5614.
675 G. Georghiou, R. E. Kleiner, M. Pulkoski-Gross, D. R. Liu and M. A. Seeliger, *Nature Chem. Biol.*, 2012, **8**, 366.
676 I. Belozerova and R. Levicky, *J. Am. Chem. Soc.*, 2012, **134**, 18667.
677 Z.-g. Yu and R. Y. Lai, *Chem. Commun.*, 2012, **48**, 10523.
678 B.-C. Yin, Y.-M. Guan and B.-C. Ye, *Chem. Commun.*, 2012, **48**, 4208.
679 T. Goda, K. Masuno, J. Nishida, N. Kosaka, T. Ochiya, A. Matsumoto and Y. Miyahara, *Chem. Commun.*, 2012, **48**, 11942.
680 L. Feng, B. Xu, J. Ren, C. Zhao and X. Qu, *Chem. Commun.*, 2012, **48**, 9068.
681 Z. Dai, X. Hu, H. Wu and X. Zou, *Chem. Commun.*, 2012, **48**, 1769.
682 H. Xu, L. Wang, H. Ye, L. Yu, X. Zhu, Z. Lin, G. Wu, X. Li, X. Liu and G. Chen, *Chem. Commun.*, 2012, **48**, 6390.
683 A. J. Bonham, K. Hsieh, B. S. Ferguson, A. Vallee-Belisle, F. Ricci, H. T. Soh and K. W. Plaxco, *J. Am. Chem. Soc.*, 2012, **134**, 3346.
684 Z. Shen, S. Nakayama, S. Semancik and H. O. Sintim, *Chem. Commun.*, 2012, **48**, 7580.
685 Q. Wang, L. Yang, X. Yang, K. Wang, L. He, J. Zhu and T. Su, *Chem. Commun.*, 2012, **48**, 2982.
686 L. Feng, C. Zhao, Y. Xiao, L. Wu, J. Ren and X. Qu, *Chem. Commun.*, 2012, **48**, 6900.

687   E. Spain, B. Miner, T. E. Keyes and R. J. Forster, *Chem. Commun.*, 2012, **48**, 838.
688   N. Borovok, E. Gillon and A. Kotlyar, *Bioconjugate Chem.*, 2012, **23**, 916.
689   H. Pei, F. Li, Y. Wan, M. Wei, H. Liu, Y. Su, N. Chen, Q. Huang and C. Fan, *J. Am. Chem. Soc.*, 2012, **134**, 11876.
690   X. Zhang, M. R. Servos and J. Liu, *J. Am. Chem. Soc.*, 2012, **134**, 7266.
691   Y. Wen, L. Chen, W. Wang, L. Xu, H. Du, Z. Zhang, X. Zhang and Y. Song, *Chem. Commun.*, 2012, **48**, 3963.
692   A. Barchanski, N. Hashimoto, S. Petersen, C. L. Sajti and S. Barcikowski, *Bioconjugate Chem.*, 2012, **23**, 908.
693   Z. Wang, H. Liu, S. H. Yang, T. Wang, C. Liu and Y. C. Cao, *Proc. Natl. Acad. Sci. USA*, 2012, **109**, 12387.
694   D. Zheng, D. A. Giljohann, D. L. Chen, M. D. Massich, X.-Q. Wang, H. Iordanov, C. A. Mirkin and A. S. Paller, *Proc. Natl. Acad. Sci. USA*, 2012, **109**, 11975.
695   N. Li, C. Chang, W. Pan and B. Tang, *Angew. Chem. Int. Ed.*, 2012, **51**, 7426.
696   C. M. Alexander, J. C. Dabrowiak and M. M. Maye, *Bioconjugate Chem.*, 2012, **23**, 2061.
697   S. Saha, K. Chakraborty and Y. Krishnan, *Chem. Commun.*, 2012, **48**, 2513.
698   W. Shen, H. Deng, A. K. L. Teo and Z. Gao, *Chem. Commun.*, 2012, **48**, 10225.
699   W. Shen, H. Deng and Z. Gao, *J. Am. Chem. Soc.*, 2012, **134**, 14678.
700   P. Lie, J. Liu, Z. Fang, B. Dun and L. Zeng, *Chem. Commun.*, 2012, **48**, 236.
701   F. Li, H. Zhang, C. Lai, X.-F. Li and X. C. Le, *Angew. Chem. Int. Ed.*, 2012, **51**, 9317.
702   J. Xu, B. Jiang, J. Xie, Y. Xiang, R. Yuan and Y. Chai, *Chem. Commun.*, 2012, **48**, 10733.
703   Y. Tu, P. Wu, H. Zhang and C. Cai, *Chem. Commun.*, 2012, **48**, 10718.
704   M. Wei, N. Chen, J. Li, M. Yin, L. Liang, Y. He, H. Song, C. Fan and Q. Huang, *Angew. Chem. Int. Ed.*, 2012, **51**, 1202.
705   X. Shen, C. Song, J. Wang, D. Shi, Z. Wang, N. Liu and B. Ding, *J. Am. Chem. Soc.*, 2011, **134**, 146.
706   M. N. Khan, V. Tjong, A. Chilkoti and M. Zharnikov, *Angew. Chem. Int. Ed.*, 2012, **51**, 10303.
707   P. Yang and X. Zhang, *Chem. Commun.*, 2012, **48**, 8787.
708   Z. Li, Z. Zhu, W. Liu, Y. Zhou, B. Han, Y. Gao and Z. Tang, *J. Am. Chem. Soc.*, 2012, **134**, 3322.
709   P. Salvatore, K. K. Karlsen, A. G. Hansen, J. Zhang, R. J. Nichols and J. Ulstrup, *J. Am. Chem. Soc.*, 2012, **134**, 19092.
710   J. Zheng, A. Jiao, R. Yang, H. Li, J. Li, M. Shi, C. Ma, Y. Jiang, L. Deng and W. Tan, *J. Am. Chem. Soc.*, 2012, **134**, 19957.
711   X. Zhang, M. R. Servos and J. Liu, *Chem. Commun.*, 2012, **48**, 10114.
712   A. Latorre and Á. Somoza, *ChemBioChem*, 2012, **13**, 951.
713   H.-C. Yeh, J. Sharma, I.-M. Shih, D. M. Vu, J. S. Martinez and J. H. Werner, *J. Am. Chem. Soc.*, 2012, **134**, 11550.
714   Z. Zhu, F. Gao, J. Lei, H. Dong and H. Ju, *Chem. Eur. J.*, 2012, **18**, 13871.
715   Z. Huang, Y. Tao, F. Pu, J. Ren and X. Qu, *Chem. Eur. J.*, 2012, **18**, 6663.
716   L. Feng, Z. Huang, J. Ren and X. Qu, *Nucl. Acids Res.*, 2012, **40**, e122.
717   Z. Chen, Y. Lin, C. Zhao, J. Ren and X. Qu, *Chem. Commun.*, 2012, **48**, 11428.
718   D. Schultz and E. G. Gwinn, *Chem. Commun.*, 2012, **48**, 5748.
719   S. J. Kwon and A. J. Bard, *J. Am. Chem. Soc.*, 2012, **134**, 10777.
720   Y. Li, Y. Zheng, M. Gong and Z. Deng, *Chem. Commun.*, 2012, **48**, 3727.

721 M. Gonzalez, R. Bartolome, S. Matarraz, E. Rodriguez-Fernandez, J. L. Manzano, M. Perez-Andres, A. Orfao, M. Fuentes and J. J. Criado, *J. Inorg. Biochem.*, 2012, **106**, 43.
722 D. Y. Lando, E. N. Galyuk, C.-L. Chang and C.-K. Hu, *J. Inorg. Biochem.*, 2012, **117**, 164.
723 V. Brabec, J. Malina, N. Margiotta, G. Natile and J. Kasparkova, *Chem. Eur. J.*, 2012, **18**, 15439.
724 Y. Zhao, C. Biertümpfel, M. T. Gregory, Y.-J. Hua, F. Hanaoka and W. Yang, *Proc. Natl. Acad. Sci. USA*, 2012, **109**, 7269.
725 A. Ummat, O. Rechkoblit, R. Jain, J. Roy Choudhury, R. E. Johnson, T. D. Silverstein, A. Buku, S. Lone, L. Prakash, S. Prakash and A. K. Aggarwal, *Nature Struct. Mol. Biol.*, 2012, **19**, 628.
726 S. Ramachandran, B. Temple, A. N. Alexandrova, S. G. Chaney and N. V. Dokholyan, *Biochemistry*, 2012, **51**, 7608.
727 S. Park and S. J. Lippard, *Biochemistry*, 2012, **51**, 6728.
728 M. Enoiu, J. Jiricny and O. D. Schärer, *Nucl. Acids Res.*, 2012, **40**, 8953.
729 M. M. N. Zhang, Y.-T. Long and Z. Ding, *J. Inorg. Biochem.*, 2012, **108**, 115.
730 S. Quintal, A. Viegas, S. Erhardt, E. J. Cabrita and N. P. Farrell, *Biochemistry*, 2012, **51**, 1752.
731 E. G. Chapman and V. J. DeRose, *J. Am. Chem. Soc.*, 2011, **134**, 256.
732 J. Mlcouskova, J. Kasparkova, T. Suchankova, S. Komeda and V. Brabec, *J. Inorg. Biochem.*, 2012, **114**, 15.
733 A. G. Quiroga, *J. Inorg. Biochem.*, 2012, **114**, 106.
734 C. Wang, G. Jia, J. Zhou, Y. Li, Y. Liu, S. Lu and C. Li, *Angew. Chem. Int. Ed.*, 2012, **51**, 9352.
735 C. Wang, Y. Li, G. Jia, Y. Liu, S. Lu and C. Li, *Chem. Commun.*, 2012, **48**, 6232.
736 R. P. Megens and G. Roelfes, *Chem. Commun.*, 2012, **48**, 6366.
737 J. R. Morgan, D. V. X. Nguyen, A. R. Frohman, S. R. Rybka and J. A. Zebala, *Bioconjugate Chem.*, 2012, **23**, 2020.
738 G. Sargsyan and M. Balaz, *Org. Biomol. Chem.*, 2012, **10**, 5533.
739 G. Sargsyan, B. L. MacLeod, U. Tohgha and M. Balaz, *Tetrahedron*, 2012, **68**, 2093.
740 J. R. Burns, S. Preus, D. G. Singleton and E. Stulz, *Chem. Commun.*, 2012, **48**, 11088.
741 M. F. Dumont, C. Baligand, Y. Li, E. S. Knowles, M. W. Meisel, G. A. Walter and D. R. Talham, *Bioconjugate Chem.*, 2012, **23**, 951.
742 A. Kiviniemi, J. Makela, J. Makila, T. Saanijoki, H. Liljenback, P. Poijarvi-Virta, H. Lonnberg, T. Laitala-Leinonen, A. Roivainen and P. Virta, *Bioconjugate Chem.*, 2012, **23**, 1981.
743 P. A. Sontz, T. P. Mui, J. O. Fuss, J. A. Tainer and J. K. Barton, *Proc. Natl. Acad. Sci. USA*, 2012, **109**, 1856.
744 C. G. Pheeney, L. F. Guerra and J. K. Barton, *Proc. Natl. Acad. Sci. USA*, 2012, **109**, 11528.
745 W. B. Davis, C. C. Bjorklund and M. Deline, *Biochemistry*, 2012, **51**, 3129.
746 N. Lu, H. Pei, Z. Ge, C. R. Simmons, H. Yan and C. Fan, *J. Am. Chem. Soc.*, 2012, **134**, 13148.
747 T. Ito, Y. Hamaguchi, K. Tanabe, H. Yamada and S.-i. Nishimoto, *Angew. Chem. Int. Ed.*, 2012, **51**, 7558.
748 M. J. Park, M. Fujitsuka, K. Kawai and T. Majima, *Chem. Eur. J.*, 2012, **18**, 2056.
749 M. J. Park, M. Fujitsuka, H. Nishitera, K. Kawai and T. Majima, *Chem. Commun.*, 2012, **48**, 11008.

750 R. Carmieli, A. L. Smeigh, S. M. Mickley Conron, A. K. Thazhathveetil, M. Fuki, Y. Kobori, F. D. Lewis and M. R. Wasielewski, *J. Am. Chem. Soc.*, 2012, **134**, 11251.

751 J. M. Thomas, B. Chakraborty, D. Sen and H.-Z. Yu, *J. Am. Chem. Soc.*, 2012, **134**, 13823.

752 A. Abi and E. E. Ferapontova, *J. Am. Chem. Soc.*, 2012, **134**, 14499.

753 B. D. Blakeley, S. M. DePorter, U. Mohan, R. Burai, B. S. Tolbert and B. R. McNaughton, *Tetrahedron*, 2012, **68**, 8837.

754 D. Jose, S. E. Weitzel and P. H. von Hippel, *Proc. Natl. Acad. Sci. USA*, 2012, **109**, 14428.

755 R. Buscaglia, D. M. Jameson and J. B. Chaires, *Nucl. Acids Res.*, 2012, **40**, 4203.

756 Y. Sakakibara, S. C. Abeysirigunawardena, A.-C. E. Duc, D. N. Dremann and C. S. Chow, *Angew. Chem. Int. Ed.*, 2012, **51**, 12095.

757 A. Padirac, T. Fujii and Y. Rondelez, *Nucl. Acids Res.*, 2012, **40**, e118.

758 M. Schoonover and S. M. Kerwin, *Bioorg. Med. Chem.*, 2012, **20**, 6904.

759 E. D. Holmstrom, J. L. Fiore and D. J. Nesbitt, *Biochemistry*, 2012, **51**, 3732.

760 Y. Kimura, T. Hanami, Y. Tanaka, M. J. L. de Hoon, T. Soma, M. Harbers, A. Lezhava, Y. Hayashizaki and K. Usui, *Biochemistry*, 2012, **51**, 6056.

761 F. Hovelmann, L. Bethge and O. Seitz, *ChemBioChem*, 2012, **13**, 2072.

762 L. M. Hall, M. Gerowska and T. Brown, *Nucl. Acids Res.*, 2012, **40**, e108.

763 D. A. Tsybulsky, M. V. Kvach, I. A. Stepanova, V. A. Korshun and V. V. Shmanai, *J. Org. Chem.*, 2012, **77**, 977.

764 K. Datta, N. P. Johnson, G. Villani, A. H. Marcus and P. H. von Hippel, *Nucl. Acids Res.*, 2012, **40**, 1191.

765 A. Moreno, J. Knee and I. Mukerji, *Biochemistry*, 2012, **51**, 6847.

766 J. Riedl, P. Menova, R. Pohl, P. Orsag, M. Fojta and M. Hocek, *J. Org. Chem.*, 2012, **77**, 8287.

767 Y. Yu, J. Liu, Z. Zhao, K. M. Ng, K. Q. Luo and B. Z. Tang, *Chem. Commun.*, 2012, **48**, 6360.

768 S. Wang, J. Guo, T. Ono and E. T. Kool, *Angew. Chem. Int. Ed.*, 2012, **51**, 7176.

769 D. Dziuba, V. Y. Postupalenko, M. Spadafora, A. S. Klymchenko, V. Guerineau, Y. Mely, R. Benhida and A. Burger, *J. Am. Chem. Soc.*, 2012, **134**, 10209.

770 A. R. Hieb, S. D'Arcy, M. A. Kramer, A. E. White and K. Luger, *Nucl. Acids Res.*, 2012, **40**, e33.

771 T. Ono, S. Wang, C.-K. Koo, L. Engstrom, S. S. David and E. T. Kool, *Angew. Chem. Int. Ed.*, 2012, **51**, 1689.

772 M. Bowles, J. Lally, A. J. Fadden, S. Mouilleron, T. Hammonds and N. Q. McDonald, *Nucl. Acids Res.*, 2012, **40**, e101.

773 J. F. Swennenhuis, B. Foulk, F. A. W. Coumans and L. W. M. M. Terstappen, *Nucl. Acids Res.*, 2012, **40**, e20.

774 X. Shu, Y. Liu and J. Zhu, *Angew. Chem. Int. Ed.*, 2012, **51**, 11006.

775 Q. Xue, Z. Wang, L. Wang and W. Jiang, *Bioconjugate Chem.*, 2012, **23**, 734.

776 J. E. Rice, A. H. Reis, L. M. Rice, R. K. Carver-Brown and L. J. Wangh, *Nucl. Acids Res.*, 2012, **40**, e164.

777 A. Borodavka, R. Tuma and P. G. Stockley, *Proc. Natl. Acad. Sci. USA*, 2012, **109**, 15769.

778 W. K. Ridgeway, D. P. Millar and J. R. Williamson, *Proc. Natl. Acad. Sci. USA*, 2012, **109**, 13614.

779 C.-H. Leung, D. S.-H. Chan, H.-Z. He, Z. Cheng, H. Yang and D.-L. Ma, *Nucl. Acids Res.*, 2012, **40**, 941.

780   F. Gao, Z. Zhu, J. Lei and H. Ju, *Chem. Commun.*, 2012, **48**, 10603.
781   G. Ke, C. Wang, Y. Ge, N. Zheng, Z. Zhu and C. J. Yang, *J. Am. Chem. Soc.*, 2012, **134**, 18908.
782   K. B. Joshi, A. Vlachos, V. Mikat, T. Deller and A. Heckel, *Chem. Commun.*, 2012, **48**, 2746.
783   Y.-H. Lin and W.-L. Tseng, *Chem. Commun.*, 2012, **48**, 6262.
784   H. Kashida, K. Yamaguchi, Y. Hara and H. Asanuma, *Bioorg. Med. Chem.*, 2012, **20**, 4310.
785   J. H. Ryu, J. Y. Heo, E.-K. Bang, G. T. Hwang and B. H. Kim, *Tetrahedron*, 2012, **68**, 72.
786   F. Li, P. Li, L. Yang and B. Tang, *Chem. Commun.*, 2012, **48**, 12192.
787   S. Shin, H. Y. Nam, E. J. Lee, W. Jung and S. S. Hah, *Bioorg. Med. Chem. Lett.*, 2012, **22**, 6081.
788   M. B. Baker, G. Bao and C. D. Searles, *Nucl. Acids Res.*, 2012, **40**, e13.
789   H. Zhou, S.-J. Xie, J.-S. Li, Z.-S. Wu and G.-L. Shen, *Chem. Commun.*, 2012, **48**, 10760.
790   B. Kim, I. H. Jung, M. Kang, H.-K. Shim and H. Y. Woo, *J. Am. Chem. Soc.*, 2012, **134**, 3133.
791   F. Ricci, A. Vallee-Belisle, A. Porchetta and K. W. Plaxco, *J. Am. Chem. Soc.*, 2012, **134**, 15177.
792   M. Zhang, H.-N. Le, P. Wang and B.-C. Ye, *Chem. Commun.*, 2012, **48**, 10004.
793   H. Zhong, R. Zhang, H. Zhang and S. Zhang, *Chem. Commun.*, 2012, **48**, 6277.
794   J. H. Kim, Y.-W. Noh, M. B. Heo, M. Y. Cho and Y. T. Lim, *Angew. Chem. Int. Ed.*, 2012, **51**, 9670.
795   C. Zhang, J. Xu, S. Zhang, X. Ji and Z. He, *Chem. Eur. J.*, 2012, **18**, 8296.
796   Q. Tian, W. Wong, Y. Xu, Y. Chan, H. K. Ho, G. Pastorin and W. H. Ang, *Chem. Commun.*, 2012, **48**, 5467.
797   H. Zhang and D. Zhou, *Chem. Commun.*, 2012, **48**, 5097.
798   S. Preus and L. M. Wilhelmsson, *ChemBioChem*, 2012, **13**, 1990.
799   A. Altevogt nee Kienzler, R. Flehr, S. Gehne, M. U. Kumke and W. Bannwarth, *Helv. Chim. Acta*, 2012, **95**, 543.
800   Z. Guo and H. F. Noller, *Proc. Natl. Acad. Sci. USA*, 2012, **109**, 20391.
801   A. B. Dixit, K. Ray and L. W. Black, *Proc. Natl. Acad. Sci. USA*, 2012, **109**, 20419.
802   P. Vivas, Y. Velmurugu, S. V. Kuznetsov, P. A. Rice and A. Ansari, *J. Mol. Biol.*, 2012, **418**, 300.
803   M. J. Shoura, A. A. Vetcher, S. M. Giovan, F. Bardai, A. Bharadwaj, M. R. Kesinger and S. D. Levene, *Nucl. Acids Res.*, 2012, **40**, 7452.
804   A. R. Haeusler, K. A. Goodson, T. D. Lillian, X. Wang, S. Goyal, N. C. Perkins and J. D. Kahn, *Nucl. Acids Res.*, 2012, **40**, 4432.
805   N. A. Kolganova, A. K. Shchyolkina, A. V. Chudinov, A. S. Zasedatelev, V. L. Florentiev and E. N. Timofeev, *Nucl. Acids Res.*, 2012, **40**, 8175.
806   M. Schifferer and O. Griesbeck, *J. Am. Chem. Soc.*, 2012, **134**, 15185.
807   M. Daher and D. Rueda, *Biochemistry*, 2012, **51**, 3531.
808   C. Pernstich and S. E. Halford, *Nucl. Acids Res.*, 2012, **40**, 1203.
809   G. Wang, K. Dhar, P. C. Swanson, M. Levitus and Y. Chang, *Nucl. Acids Res.*, 2012, **40**, 6082.
810   L. Stefan, B. Bertrand, P. Richard, P. Le Gendre, F. Denat, M. Picquet and D. Monchaud, *ChemBioChem*, 2012, **13**, 1905.
811   S. Kumar, P. Kellish, W. E. Robinson, D. Wang, D. H. Appella and D. P. Arya, *Biochemistry*, 2012, **51**, 2331.

812 A. Rajendran, M. Endo and H. Sugiyama, *Angew. Chem. Int. Ed.*, 2012, **51**, 874.
813 T. Plenat, C. Tardin, P. Rousseau and L. Salome, *Nucl. Acids Res.*, 2012, **40**, e89.
814 T. Fessl, F. Adamec, T. Polivka, S. Foldynova-Trantirkova, F. Vacha and L. Trantirek, *Nucl. Acids Res.*, 2012, **40**, e121.
815 I. V. Gopich and A. Szabo, *Proc. Natl. Acad. Sci. USA*, 2012, **109**, 7747.
816 H. Bai, J. E. Kath, F. M. Zorgiebel, M. Sun, P. Ghosh, G. F. Hatfull, N. D. F. Grindley and J. F. Marko, *Proc. Natl. Acad. Sci. USA*, 2012, **109**, 16546.
817 I. I. Cisse, H. Kim and T. Ha, *Nature Struct. Mol. Biol.*, 2012, **19**, 623.
818 X. Qu, L. Lancaster, H. F. Noller, C. Bustamante and I. Tinoco, *Proc. Natl. Acad. Sci. USA*, 2012, **109**, 14458.
819 T. Masuda, A. N. Petrov, R. Iizuka, T. Funatsu, J. D. Puglisi and S. Uemura, *Proc. Natl. Acad. Sci. USA*, 2012, **109**, 4881.
820 L. Wang, A. Pulk, M. R. Wasserman, M. B. Feldman, R. B. Altman, J. H. D. Cate and S. C. Blanchard, *Nature Struct. Mol. Biol.*, 2012, **19**, 957.
821 J. A. Morin, F. J. Cao, J. M. Lazaro, J. R. Arias-Gonzalez, J. M. Valpuesta, J. L. Carrascosa, M. Salas and B. Ibarra, *Proc. Natl. Acad. Sci. USA*, 2012, **109**, 8115.
822 S. Y. Berezhna, J. P. Gill, R. Lamichhane and D. P. Millar, *J. Am. Chem. Soc.*, 2012, **134**, 11261.
823 R. E. Georgescu, I. Kurth and M. E. O'Donnell, *Nature Struct. Mol. Biol.*, 2012, **19**, 113.
824 E. A. Manrao, I. M. Derrington, A. H. Laszlo, K. W. Langford, M. K. Hopper, N. Gillgren, M. Pavlenok, M. Niederweis and J. H. Gundlach, *Nature Biotechnol.*, 2012, **30**, 349.
825 G. Lia, B. Michel and J.-F. Allemand, *Science*, 2012, **335**, 328.
826 R. P. Markiewicz, K. B. Vrtis, D. Rueda and L. J. Romano, *Nucl. Acids Res.*, 2012, **40**, 7975.
827 S. Zorman, H. Seitz, B. Sclavi and T. R. Strick, *Nucl. Acids Res.*, 2012, **40**, 7375.
828 A. Chakraborty, D. Wang, Y. W. Ebright, Y. Korlann, E. Kortkhonjia, T. Kim, S. Chowdhury, S. Wigneshweraraj, H. Irschik, R. Jansen, B. T. Nixon, J. Knight, S. Weiss and R. H. Ebright, *Science*, 2012, **337**, 591.
829 B. Zamft, L. Bintu, T. Ishibashi and C. Bustamante, *Proc. Natl. Acad. Sci. USA*, 2012, **109**, 8948.
830 M. H. Larson, J. Zhou, C. D. Kaplan, M. Palangat, R. D. Kornberg, R. Landick and S. M. Block, *Proc. Natl. Acad. Sci. USA*, 2012, **109**, 6555.
831 J. Gorman, F. Wang, S. Redding, A. J. Plys, T. Fazio, S. Wind, E. E. Alani and E. C. Greene, *Proc. Natl. Acad. Sci. USA*, 2012, **109**, E3074.
832 A. L. Forget and S. C. Kowalczykowski, *Nature*, 2012, **482**, 423.
833 G. Lee, M. A. Bratkowski, F. Ding, A. Ke and T. Ha, *Science*, 2012, **336**, 1726.
834 M. Cristovao, E. Sisamakis, M. M. Hingorani, A. D. Marx, C. P. Jung, P. J. Rothwell, C. A. M. Seidel and P. Friedhoff, *Nucl. Acids Res.*, 2012, **40**, 5448.
835 T. A. Clark, I. A. Murray, R. D. Morgan, A. O. Kislyuk, K. E. Spittle, M. Boitano, A. Fomenkov, R. J. Roberts and J. Korlach, *Nucl. Acids Res.*, 2012, **40**, e29.
836 H. Yardimci, X. Wang, A. B. Loveland, I. Tappin, D. Z. Rudner, J. Hurwitz, A. M. van Oijen and J. C. Walter, *Nature*, 2012, **492**, 205.
837 R. Qiu, V. C. DeRocco, C. Harris, A. Sharma, M. M. Hingorani, D. A. Erie and K. R. Weninger, *EMBO J.*, 2012, **31**, 2528.
838 M. Hengesbach, N.-K. Kim, J. Feigon and M. D. Stone, *Angew. Chem. Int. Ed.*, 2012, **51**, 5876.

839　A. K. Byrd, D. L. Matlock, D. Bagchi, S. Aarattuthodiyil, D. Harrison, V. Croquette and K. D. Raney, *J. Mol. Biol.*, 2012, **420**, 141.
840　S. Lee, S.-R. Jung, K. Heo, J. A. W. Byl, J. E. Deweese, N. Osheroff and S. Hohng, *Proc. Natl. Acad. Sci. USA*, 2012, **109**, 2925.
841　J. Y. Lee and T.-H. Lee, *J. Am. Chem. Soc.*, 2011, **134**, 173.
842　B. R. Cipriany, P. J. Murphy, J. A. Hagarman, A. Cerf, D. Latulippe, S. L. Levy, J. J. Benítez, C. P. Tan, J. Topolancik, P. D. Soloway and H. G. Craighead, *Proc. Natl. Acad. Sci. USA*, 2012, **109**, 8477.
843　J. Y. Lee, I. J. Finkelstein, E. Crozat, D. J. Sherratt and E. C. Greene, *Proc. Natl. Acad. Sci. USA*, 2012, **109**, 6531.
844　V. I. Kottadiel, V. B. Rao and Y. R. Chemla, *Proc. Natl. Acad. Sci. USA*, 2012, **109**, 20000.
845　Z. Yu, V. Gaerig, Y. Cui, H. Kang, V. Gokhale, Y. Zhao, L. H. Hurley and H. Mao, *J. Am. Chem. Soc.*, 2012, **134**, 5157.
846　H.-F. Fan, *Nucl. Acids Res.*, 2012, **40**, 6208.
847　R. H. Blair, J. A. Goodrich and J. F. Kugel, *Biochemistry*, 2012, **51**, 7444.
848　H. You, R. Iino, R. Watanabe and H. Noji, *Nucl. Acids Res.*, 2012, **40**, e151.
849　S. Koren, M. C. Schatz, B. P. Walenz, J. Martin, J. T. Howard, G. Ganapathy, Z. Wang, D. A. Rasko, W. R. McCombie, E. D. Jarvis and A. M. Phillippy, *Nature Biotechnol.*, 2012, **30**, 693.
850　K. E. Ocwieja, S. Sherrill-Mix, R. Mukherjee, R. Custers-Allen, P. David, M. Brown, S. Wang, D. R. Link, J. Olson, K. Travers, E. Schadt and F. D. Bushman, *Nucl. Acids Res.*, 2012, **40**, 10345.
851　J. P. Siebrasse, T. Kaminski and U. Kubitscheck, *Proc. Natl. Acad. Sci. USA*, 2012, **109**, 9426.
852　G. M. Cherf, K. R. Lieberman, H. Rashid, C. E. Lam, K. Karplus and M. Akeson, *Nature Biotechnol.*, 2012, **30**, 344.
853　J. Bordello, M. I. Sanchez, M. E. Vazquez, J. L. Mascarenas, W. Al-Soufi and M. Novo, *Angew. Chem. Int. Ed.*, 2012, **51**, 7541.
854　T. Yoshidome, M. Endo, G. Kashiwazaki, K. Hidaka, T. Bando and H. Sugiyama, *J. Am. Chem. Soc.*, 2012, **134**, 4654.
855　C. Hyeon, J. Lee, J. Yoon, S. Hohng and D. Thirumalai, *Nature Chem.*, 2012, **4**, 907.
856　T. Sabir, A. Toulmin, L. Ma, A. C. Jones, P. McGlynn, G. F. Schröder and S. W. Magennis, *J. Am. Chem. Soc.*, 2012, **134**, 6280.
857　R. Vafabakhsh and T. Ha, *Science*, 2012, **337**, 1097.
858　M. T. J. van Loenhout, M. V. de Grunt and C. Dekker, *Science*, 2012, **338**, 94.
859　F. C. Oberstrass, L. E. Fernandes and Z. Bryant, *Proc. Natl. Acad. Sci. USA*, 2012, **109**, 6106.
860　C. V. Bizarro, A. Alemany and F. Ritort, *Nucl. Acids Res.*, 2012, **40**, 6922.
861　J. L. Fiore, E. D. Holmstrom and D. J. Nesbitt, *Proc. Natl. Acad. Sci. USA*, 2012, **109**, 2902.
862　D. H. Paik and T. T. Perkins, *Angew. Chem. Int. Ed.*, 2012, **51**, 1811.
863　M. Endo, Y. Yang, Y. Suzuki, K. Hidaka and H. Sugiyama, *Angew. Chem. Int. Ed.*, 2012, **51**, 10518.
864　T. Paramanathan, I. Vladescu, M. J. McCauley, I. Rouzina and M. C. Williams, *Nucl. Acids Res.*, 2012, **40**, 4925.
865　A. H. Mack, D. J. Schlingman, R. P. Ilagan, L. Regan and S. G. J. Mochrie, *J. Mol. Biol.*, 2012, **423**, 687.
866　X. Zhang, H. Chen, H. Fu, P. S. Doyle and J. Yan, *Proc. Natl. Acad. Sci. USA*, 2012, **109**, 8103.
867　N. Bosaeus, A. H. El-Sagheer, T. Brown, S. B. Smith, B. Åkerman, C. Bustamante and B. Norden, *Proc. Natl. Acad. Sci. USA*, 2012, **109**, 15179.

868   C. G. Evans, R. F. Hariadi and E. Winfree, *J. Am. Chem. Soc.*, 2012, **134**, 10485.
869   J. L. Fiore, E. D. Holmstrom, L. R. Fiegland, J. H. Hodak and D. J. Nesbitt, *J. Mol. Biol.*, 2012, **423**, 198.
870   E. A. Josephs and T. Ye, *J. Am. Chem. Soc.*, 2012, **134**, 10021.
871   C. L. Schoch, K. A. Seifert, S. Huhndorf, V. Robert, J. L. Spouge, C. A. Levesque, W. Chen and F. B. Consortium, *Proc. Natl. Acad. Sci. USA*, 2012, **109**, 6241.
872   Z. Feng, D. Chakraborty, S. B. Dewell, B. V. B. Reddy and S. F. Brady, *J. Am. Chem. Soc.*, 2012, **134**, 2981.
873   J. Xu, B. Jiang, J. Su, Y. Xiang, R. Yuan and Y. Chai, *Chem. Commun.*, 2012, **48**, 3309.
874   M. Leimbacher, Y. Zhang, L. Mannocci, M. Stravs, T. Geppert, J. Scheuermann, G. Schneider and D. Neri, *Chem. Eur. J.*, 2012, **18**, 7729.
875   Z. Chen, X. Fu, X. Zhang, X. Liu, B. Zou, H. Wu, Q. Song, J. Li, T. Kajiyama, H. Kambara and G. Zhou, *Chem. Commun.*, 2012, **48**, 2445.
876   S. S. Agasti, M. Liong, V. M. Peterson, H. Lee and R. Weissleder, *J. Am. Chem. Soc.*, 2012, **134**, 18499.
877   J. Zhou, M. Xu, D. Tang, Z. Gao, J. Tang and G. Chen, *Chem. Commun.*, 2012, **48**, 12207.
878   C. Lin, R. Jungmann, A. M. Leifer, C. Li, D. Levner, G. M. Church, W. M. Shih and P. Yin, *Nature Chem.*, 2012, **4**, 832.
879   A. Baccaro, A.-L. Steck and A. Marx, *Angew. Chem. Int. Ed.*, 2012, **51**, 254.
880   K. Shiroguchi, T. Z. Jia, P. A. Sims and X. S. Xie, *Proc. Natl. Acad. Sci. USA*, 2012, **109**, 1347.
881   E. Stulz, *Chem. Eur. J.*, 2012, **18**, 4456.
882   B. Sacca and C. M. Niemeyer, *Angew. Chem. Int. Ed.*, 2012, **51**, 58.
883   Z. Wang, L. Tang, L. H. Tan, J. Li and Y. Lu, *Angew. Chem. Int. Ed.*, 2012, **51**, 9078.
884   X.-c. Bai, T. G. Martin, S. H. W. Scheres and H. Dietz, *Proc. Natl. Acad. Sci. USA*, 2012, **109**, 20012.
885   Z. Deng, A. Samanta, J. Nangreave, H. Yan and Y. Liu, *J. Am. Chem. Soc.*, 2012, **134**, 17424.
886   C. Zhang, C. Tian, X. Li, H. Qian, C. Hao, W. Jiang and C. Mao, *J. Am. Chem. Soc.*, 2012, **134**, 11998.
887   H. Pei, L. Liang, G. Yao, J. Li, Q. Huang and C. Fan, *Angew. Chem. Int. Ed.*, 2012, **51**, 9020.
888   J. I. Cutler, E. Auyeung and C. A. Mirkin, *J. Am. Chem. Soc.*, 2012, **134**, 1376.
889   D. Ackermann, S.-S. Jester and M. Famulok, *Angew. Chem. Int. Ed.*, 2012, **51**, 6771.
890   X. Shen, Q. Jiang, J. Wang, L. Dai, G. Zou, Z.-G. Wang, W.-Q. Chen, W. Jiang and B. Ding, *Chem. Commun.*, 2012, **48**, 11301.
891   M. Endo, R. Miyazaki, T. Emura, K. Hidaka and H. Sugiyama, *J. Am. Chem. Soc.*, 2012, **134**, 2852.
892   C. Zhang, W. Wu, X. Li, C. Tian, H. Qian, G. Wang, W. Jiang and C. Mao, *Angew. Chem. Int. Ed.*, 2012, **51**, 7999.
893   H. Lee, A. K. R. Lytton-Jean, Y. Chen, K. T. Love, A. I. Park, E. D. Karagiannis, A. Sehgal, W. Querbes, C. S. Zurenko, M. Jayaraman, C. G. Peng, K. Charisse, A. Borodovsky, M. Manoharan, J. S. Donahoe, J. Truelove, M. Nahrendorf, R. Langer and D. G. Anderson, *Nature Nanotech.*, 2012, **7**, 389.
894   J. Song, J.-M. Arbona, Z. Zhang, L. Liu, E. Xie, J. Elezgaray, J.-P. Aime, K. V. Gothelf, F. Besenbacher and M. Dong, *J. Am. Chem. Soc.*, 2012, **134**, 9844.
895   M. Endo, K. Hidaka and H. Sugiyama, *Org. Biomol. Chem.*, 2011, **9**, 2075.

896   Y. Yang, M. Endo, K. Hidaka and H. Sugiyama, *J. Am. Chem. Soc.*, 2012, **134**, 20645.
897   M. Langecker, V. Arnaut, T. G. Martin, J. List, S. Renner, M. Mayer, H. Dietz and F. C. Simmel, *Science*, 2012, **338**, 932.
898   Y. Ke, L. L. Ong, W. M. Shih and P. Yin, *Science*, 2012, **338**, 1177.
899   J.-H. Lee, G.-H. Kim and J.-M. Nam, *J. Am. Chem. Soc.*, 2012, **134**, 5456.
900   T. Wang, D. Schiffels, S. Martinez Cuesta, D. Kuchnir Fygenson and N. C. Seeman, *J. Am. Chem. Soc.*, 2011, **134**, 1606.
901   Y. Ke, N. V. Voigt, K. V. Gothelf and W. M. Shih, *J. Am. Chem. Soc.*, 2011, **134**, 1770.
902   A. A. Greschner, V. Toader and H. F. Sleiman, *J. Am. Chem. Soc.*, 2012, **134**, 14382.
903   D. A. Rusling, I. S. Nandhakumar, T. Brown and K. R. Fox, *Chem. Commun.*, 2012, **48**, 9592.
904   W. Tang, D. Wang, Y. Xu, N. Li and F. Liu, *Chem. Commun.*, 2012, **48**, 6678.
905   C. Tian, C. Zhang, X. Li, Y. Li, G. Wang and C. Mao, *J. Am. Chem. Soc.*, 2012, **134**, 20273.
906   A. Kuzyk, R. Schreiber, Z. Fan, G. Pardatscher, E.-M. Roller, A. Hogele, F. C. Simmel, A. O. Govorov and T. Liedl, *Nature*, 2012, **483**, 311.
907   W. Chen and G. B. Schuster, *J. Am. Chem. Soc.*, 2011, **134**, 840.
908   H. Qian, J. Yu, P. Wang, Q.-F. Dong and C. Mao, *Chem. Commun.*, 2012, **48**, 12216.
909   T. Zhou, P. Chen, L. Niu, J. Jin, D. Liang, Z. Li, Z. Yang and D. Liu, *Angew. Chem. Int. Ed.*, 2012, **51**, 11271.
910   Z. Xiao, C. Ji, J. Shi, E. M. Pridgen, J. Frieder, J. Wu and O. C. Farokhzad, *Angew. Chem. Int. Ed.*, 2012, **51**, 11853.
911   Q. Jiang, C. Song, J. Nangreave, X. Liu, L. Lin, D. Qiu, Z.-G. Wang, G. Zou, X. Liang, H. Yan and B. Ding, *J. Am. Chem. Soc.*, 2012, **134**, 13396.
912   J. M. Yun, K. N. Kim, J. Y. Kim, D. O. Shin, W. J. Lee, S. H. Lee, M. Lieberman and S. O. Kim, *Angew. Chem. Int. Ed.*, 2012, **51**, 912.
913   R. Wang, C. Nuckolls and S. J. Wind, *Angew. Chem. Int. Ed.*, 2012, **51**, 11325.
914   K. Qu, C. Zhao, J. Ren and X. Qu, *Mol. BioSys.*, 2012, **8**, 779.
915   Y. Huang, S. Zhao, Y.-M. Liu, J. Chen, Z.-F. Chen, M. Shi and H. Liang, *Chem. Commun.*, 2012, **48**, 9400.
916   L. Y. Zhang, C. X. Guo, Z. Cui, J. Guo, Z. Dong and C. M. Li, *Chem. Eur. J.*, 2012, **18**, 15693.
917   Z. Krpetic, I. Singh, W. Su, L. Guerrini, K. Faulds, G. A. Burley and D. Graham, *J. Am. Chem. Soc.*, 2012, **134**, 8356.
918   B. Liu, L. Han and S. Che, *Angew. Chem. Int. Ed.*, 2012, **51**, 923.
919   C. K. McLaughlin, G. D. Hamblin, K. D. Hänni, J. W. Conway, M. K. Nayak, K. M. M. Carneiro, H. S. Bazzi and H. F. Sleiman, *J. Am. Chem. Soc.*, 2012, **134**, 4280.
920   J. Fu, M. Liu, Y. Liu, N. W. Woodbury and H. Yan, *J. Am. Chem. Soc.*, 2012, **134**, 5516.
921   C. Zhang, C. Tian, F. Guo, Z. Liu, W. Jiang and C. Mao, *Angew. Chem. Int. Ed.*, 2012, **51**, 3382.
922   E. Nakata, F. F. Liew, C. Uwatoko, S. Kiyonaka, Y. Mori, Y. Katsuda, M. Endo, H. Sugiyama and T. Morii, *Angew. Chem. Int. Ed.*, 2012, **51**, 2421.
923   N. D. Derr, B. S. Goodman, R. Jungmann, A. E. Leschziner, W. M. Shih and S. L. Reck-Peterson, *Science*, 2012, **338**, 662.
924   J. Timper, K. Gutsmiedl, C. Wirges, J. Broda, M. Noyong, J. Mayer, T. Carell and U. Simon, *Angew. Chem. Int. Ed.*, 2012, **51**, 7586.

925   R. Wei, T. G. Martin, U. Rant and H. Dietz, *Angew. Chem. Int. Ed.*, 2012, **51**, 4864.
926   A. Gietl, P. Holzmeister, D. Grohmann and P. Tinnefeld, *Nucl. Acids Res.*, 2012, **40**, e110.
927   G. P. Acuna, F. M. Moller, P. Holzmeister, S. Beater, B. Lalkens and P. Tinnefeld, *Science*, 2012, **338**, 506.
928   T. Bellini, G. Zanchetta, T. P. Fraccia, R. Cerbino, E. Tsai, G. P. Smith, M. J. Moran, D. M. Walba and N. A. Clark, *Proc. Natl. Acad. Sci. USA*, 2012, **109**, 1110.
929   J. B. Lee, S. Peng, D. Yang, Y. H. Roh, H. Funabashi, N. Park, E. J. Rice, L. Chen, R. Long, M. Wu and D. Luo, *Nature Nanotech.*, 2012, **7**, 816.
930   O. J. N. Bertrand, D. K. Fygenson and O. A. Saleh, *Proc. Natl. Acad. Sci. USA*, 2012, **109**, 17342.
931   H. Nishioka, X. Liang, T. Kato and H. Asanuma, *Angew. Chem. Int. Ed.*, 2012, **51**, 1165.
932   S. E. Muser and P. J. Paukstelis, *J. Am. Chem. Soc.*, 2012, **134**, 12557.
933   J. M. Thomas, H.-Z. Yu and D. Sen, *J. Am. Chem. Soc.*, 2012, **134**, 13738.
934   T. Li, D. Ackermann, A. M. Hall and M. Famulok, *J. Am. Chem. Soc.*, 2012, **134**, 3508.
935   S. Shoshani, R. Piran, Y. Arava and E. Keinan, *Angew. Chem. Int. Ed.*, 2012, **51**, 2883.
936   X. Chen, *J. Am. Chem. Soc.*, 2011, **134**, 263.
937   W. Hong, Y. Du, T. Wang, J. Liu, Y. Liu, J. Wang and E. Wang, *Chem. Eur. J.*, 2012, **18**, 14939.
938   J. Guo, T. Wang and R. Yang, *Mol. BioSys.*, 2012, **8**, 2347.
939   W. Y. Xie, W. T. Huang, N. B. Li and H. Q. Luo, *Chem. Commun.*, 2012, **48**, 82.
940   M. Zhang and B.-C. Ye, *Chem. Commun.*, 2012, **48**, 3647.
941   E. M. Cornett, E. A. Campbell, G. Gulenay, E. Peterson, N. Bhaskar and D. M. Kolpashchikov, *Angew. Chem. Int. Ed.*, 2012, **51**, 9075.
942   B.-C. Yin, B.-C. Ye, H. Wang, Z. Zhu and W. Tan, *Chem. Commun.*, 2012, **48**, 1248.
943   A. Prokup, J. Hemphill and A. Deiters, *J. Am. Chem. Soc.*, 2012, **134**, 3810.
944   M. You, Y. Chen, X. Zhang, H. Liu, R. Wang, K. Wang, K. R. Williams and W. Tan, *Angew. Chem. Int. Ed.*, 2012, **51**, 2457.
945   Z.-G. Wang, J. Elbaz and I. Willner, *Angew. Chem. Int. Ed.*, 2012, **51**, 4322.
946   S. M. Douglas, I. Bachelet and G. M. Church, *Science*, 2012, **335**, 831.
947   S. F. J. Wickham, J. Bath, Y. Katsuda, M. Endo, K. Hidaka, H. Sugiyama and A. J. Turberfield, *Nature Nanotech.*, 2012, **7**, 169.
948   K. Eom, H. Jung, G. Lee, J. Park, K. Nam, S. W. Lee, D. S. Yoon, J. Yang and T. Kwon, *Chem. Commun.*, 2012, **48**, 955.
949   T. Song and H. Liang, *J. Am. Chem. Soc.*, 2012, **134**, 10803.
950   C. Zhou, Z. Yang and D. Liu, *J. Am. Chem. Soc.*, 2012, **134**, 1416.
951   C. Liu, E. Kim, B. Demple and N. C. Seeman, *Biochemistry*, 2012, **51**, 937.
952   W. Wang, K. Chen, D. Qu, W. Chi, W. Xiong, Y. Huang, J. Wen, S. Feng and B. Zhang, *Tetrahedron Letters*, 2012, **53**, 6747.
953   C. Ligeour, A. Meyer, J.-J. Vasseur and F. Morvan, *Eur. J. Org. Chem.*, 2012, 1851.
954   J. Schoch, M. Staudt, A. Samanta, M. Wiessler and A. Jaschke, *Bioconjugate Chem.*, 2012, **23**, 1382.
955   I. Singh, C. Freeman, A. Madder, J. S. Vyle and F. Heaney, *Org. Biomol. Chem.*, 2012, **10**, 6633.

956 N. S. Selden, M. E. Todhunter, N. Y. Jee, J. S. Liu, K. E. Broaders and Z. J. Gartner, *J. Am. Chem. Soc.*, 2011, **134**, 765.
957 J. Willibald, J. Harder, K. Sparrer, K.-K. Conzelmann and T. Carell, *J. Am. Chem. Soc.*, 2012, **134**, 12330.
958 H. Lewis, A. J. Perrett, G. A. Burley and I. C. Eperon, *Angew. Chem. Int. Ed.*, 2012, **51**, 9800.
959 M. Raouane, D. Desmaele, G. Urbinati, L. Massaad-Massade and P. Couvreur, *Bioconjugate Chem.*, 2012, **23**, 1091.
960 T. Kubo, Y. Takei, K. Mihara, K. Yanagihara and T. Seyama, *Bioconjugate Chem.*, 2012, **23**, 164.
961 N. S. Petrova, I. V. Chernikov, M. I. Meschaninova, I. S. Dovydenko, A. G. Venyaminova, M. A. Zenkova, V. V. Vlassov and E. L. Chernolovskaya, *Nucl. Acids Res.*, 2012, **40**, 2330.
962 N. Tort, J. P. Salvador, A. Avino, R. Eritja, J. Comelles, E. Martínez, J. Samitier and M. P. Marco, *Bioconjugate Chem.*, 2012, **23**, 2183.
963 S. J. Kwon, K. B. Lee, K. Solakyildirim, S. Masuko, M. Ly, F. Zhang, L. Li, J. S. Dordick and R. J. Linhardt, *Angew. Chem. Int. Ed.*, 2012, **51**, 11800.
964 R. Lucas, E. Vengut-Climent, I. Gomez-Pinto, A. Avino, R. Eritja, C. Gonzalez and J. C. Morales, *Chem. Commun.*, 2012, **48**, 2991.
965 A. Sanchez, E. Pedroso and A. Grandas, *Org. Biomol. Chem.*, 2012, **10**, 8478.
966 R. J. Heinze, S. Sekerina, I. Winkler, C. Biertumpfel, T. S. Oretskaya, E. Kubareva and P. Friedhoff, *Mol. BioSys.*, 2012, **8**, 1861.
967 K. Tanabe, T. Asada, T. Ito and S.-i. Nishimoto, *Bioconjugate Chem.*, 2012, **23**, 1909.
968 S. S. Dunn, S. Tian, S. Blake, J. Wang, A. L. Galloway, A. Murphy, P. D. Pohlhaus, J. P. Rolland, M. E. Napier and J. M. DeSimone, *J. Am. Chem. Soc.*, 2012, **134**, 7423.
969 F. Huang and Y. Shi, *Bioorg. Med. Chem. Lett.*, 2012, **22**, 4254.
970 T. Takada, Y. Kawano, M. Nakamura and K. Yamana, *Tetrahedron Letters*, 2012, **53**, 78.
971 Z. Zhao, L. Wang, Y. Liu, Z. Yang, Y.-M. He, Z. Li, Q.-H. Fan and D. Liu, *Chem. Commun.*, 2012, **48**, 9753.
972 G. Shemer, Y. Atsmon, E. Karzbrun and R. H. Bar-Ziv, *J. Am. Chem. Soc.*, 2012, **134**, 3954.
973 Q. Li, C. Cui, D. A. Higgins and J. Li, *J. Am. Chem. Soc.*, 2012, **134**, 14467.
974 S. Takahashi, K. Tsuji, T. Ueda and Y. Okahata, *J. Am. Chem. Soc.*, 2012, **134**, 6793.
975 A. Bonanni, A. Ambrosi and M. Pumera, *Chem. Eur. J.*, 2012, **18**, 1668.
976 A. Kobori, T. Yamauchi, Y. Nagae, A. Yamayoshi and A. Murakami, *Bioorg. Med. Chem.*, 2012, **20**, 5071.
977 M. Su, J. Wang and X. Tang, *Chem. Eur. J.*, 2012, **18**, 9628.
978 L. I. Markova, V. L. Malinovskii, L. D. Patsenker and R. Haner, *Org. Biomol. Chem.*, 2012, **10**, 8944.
979 L. A. Smith, A. V. Makarova, L. Samson, K. E. Thiesen, A. Dhar and T. Bessho, *Biochemistry*, 2012, **51**, 8931.
980 M. J. Belousoff, T. Shapira, A. Bashan, E. Zimmerman, H. Rozenberg, K. Arakawa, H. Kinashi and A. Yonath, *Proc. Natl. Acad. Sci. USA*, 2011, **108**, 2717.
981 J. Rabl, M. Leibundgut, S. F. Ataide, A. Haag and N. Ban, *Science*, 2011, **331**, 730.
982 J. A. Dunkle, L. Wang, M. B. Feldman, A. Pulk, V. B. Chen, G. J. Kapral, J. Noeske, J. S. Richardson, S. C. Blanchard and J. H. D. Cate, *Science*, 2011, **332**, 981.

983   S. Klinge, F. Voigts-Hoffmann, M. Leibundgut, S. Arpagaus and N. Ban, *Science*, 2011, **334**, 941.
984   H. Jin, A. C. Kelley and V. Ramakrishnan, *Proc. Natl. Acad. Sci. USA*, 2011, **108**, 15798.
985   J. Zhu, A. Korostelev, D. A. Costantino, J. P. Donohue, H. F. Noller and J. S. Kieft, *Proc. Natl. Acad. Sci. USA*, 2011, **108**, 1839.
986   A. Ben-Shem, N. Garreau de Loubresse, S. Melnikov, L. Jenner, G. Yusupova and M. Yusupov, *Science*, 2011, **334**, 1524.
987   T. M. Schmeing, R. M. Voorhees, A. C. Kelley and V. Ramakrishnan, *Nature Struct. Mol. Biol.*, 2011, **18**, 432.
988   N. Demeshkina, L. Jenner, E. Westhof, M. Yusupov and G. Yusupova, *Nature*, 2012, **484**, 256.
989   C. Neubauer, R. Gillet, A. C. Kelley and V. Ramakrishnan, *Science*, 2012, **335**, 1366.
990   M. G. Gagnon, S. V. Seetharaman, D. Bulkley and T. A. Steitz, *Science*, 2012, **335**, 1370.
991   H. Tachiwana, W. Kagawa, T. Shiga, A. Osakabe, Y. Miya, K. Saito, Y. Hayashi-Takanaka, T. Oda, M. Sato, S.-Y. Park, H. Kimura and H. Kurumizaka, *Nature*, 2011, **476**, 232.
992   K.-J. Armache, J. D. Garlick, D. Canzio, G. J. Narlikar and R. E. Kingston, *Science*, 2011, **334**, 977.
993   L. J. Reha-Krantz, C. Hariharan, U. Subuddhi, S. Xia, C. Zhao, J. Beckman, T. Christian and W. Konigsberg, *Biochemistry*, 2011, **50**, 10136.
994   K. Bebenek, L. C. Pedersen and T. A. Kunkel, *Proc. Natl. Acad. Sci. USA*, 2011, **108**, 1862.
995   M. Wang, S. Xia, G. Blaha, T. A. Steitz, W. H. Konigsberg and J. Wang, *Biochemistry*, 2011, **50**, 581.
996   W. Wang, H. W. Hellinga and L. S. Beese, *Proc. Natl. Acad. Sci. USA*, 2011, **108**, 17644.
997   P. Aller, S. Duclos, S. S. Wallace and S. Doublie, *J. Mol. Biol.*, 2011, **412**, 22.
998   D. T. Nair, R. E. Johnson, L. Prakash, S. Prakash and A. K. Aggarwal, *J. Mol. Biol.*, 2011, **406**, 18.
999   S. Xia, W. H. Konigsberg and J. Wang, *J. Am. Chem. Soc.*, 2011, **133**, 10003.
1000 A. Ummat, T. D. Silverstein, R. Jain, A. Buku, R. E. Johnson, L. Prakash, S. Prakash and A. K. Aggarwal, *J. Mol. Biol.*, 2012, **415**, 627.
1001 J. Gouge, C. Ralec, G. Henneke and M. Delarue, *J. Mol. Biol.*, 2012, **423**, 315.
1002 B. J. Klein, D. Bose, K. J. Baker, Z. M. Yusoff, X. Zhang and K. S. Murakami, *Proc. Natl. Acad. Sci. USA*, 2011, **108**, 546.
1003 A. Feklistov and S. A. Darst, *Cell*, 2011, **147**, 1257.
1004 M. L. Gleghorn, E. K. Davydova, R. Basu, L. B. Rothman-Denes and K. S. Murakami, *Proc. Natl. Acad. Sci. USA*, 2011, **108**, 3566.
1005 A. C. M. Cheung and P. Cramer, *Nature*, 2011, **471**, 249.
1006 A. C. M. Cheung, S. Sainsbury and P. Cramer, *EMBO J.*, 2011, **30**, 4755.
1007 F. W. Martinez-Rucobo, S. Sainsbury, A. C. M. Cheung and P. Cramer, *EMBO J.*, 2011, **30**, 1302.
1008 B. M. Lunde, I. Magler and A. Meinhart, *Nucl. Acids Res.*, 2012, **40**, 9815.
1009 M. N. Wojtas, M. Mogni, O. Millet, S. D. Bell and N. G. A. Abrescia, *Nucl. Acids Res.*, 2012, **40**, 9941.
1010 R. A. Gosavi, A. F. Moon, T. A. Kunkel, L. C. Pedersen and K. Bebenek, *Nucl. Acids Res.*, 2012, **40**, 7518.
1011 T. Nakamura, Y. Zhao, Y. Yamagata, Y.-j. Hua and W. Yang, *Nature*, 2012, **487**, 196.

1012   B. T. Chamberlain, V. K. Batra, W. A. Beard, A. P. Kadina, D. D. Shock, B. A. Kashemirov, C. E. McKenna, M. F. Goodman and S. H. Wilson, *ChemBioChem.*, 2012, **13**, 528.

1013   K. Das, S. E. Martinez, J. D. Bauman and E. Arnold, *Nature Struct. Mol. Biol.*, 2012, **19**, 253.

1014   B. A. Kelch, D. L. Makino, M. O'Donnell and J. Kuriyan, *Science*, 2011, **334**, 1675.

1015   K. Lammens, D. J. Bemeleit, C. Mockel, E. Clausing, A. Schele, S. Hartung, C. B. Schiller, M. Lucas, C. Angermuller, J. Soding, K. Strasser and K.-P. Hopfner, *Cell*, 2011, **145**, 54.

1016   F. Yu, Y. Tanaka, K. Yamashita, T. Suzuki, A. Nakamura, N. Hirano, T. Suzuki, M. Yao and I. Tanaka, *Proc. Natl. Acad. Sci. USA*, 2011, **108**, 19593.

1017   M. Kato, Y. Araiso, A. Noma, A. Nagao, T. Suzuki, R. Ishitani and O. Nureki, *Nucl. Acids Res.*, 2011, **39**, 1576.

1018   T. Osawa, S. Kimura, N. Terasaka, H. Inanaga, T. Suzuki and T. Numata, *Nature Struct. Mol. Biol.*, 2011, **18**, 1275.

1019   C. Chen, N. Gorlatova, Z. Kelman and O. Herzberg, *Proc. Natl. Acad. Sci. USA*, 2011, **108**, 6456.

1020   P. De Ioannes, C. R. Escalante and A. K. Aggarwal, *Nucl. Acids Res.*, 2011, **39**, 7300.

1021   T. B. K. Le, M. A. Schumacher, D. M. Lawson, R. G. Brennan and M. J. Buttner, *Nucl. Acids Res.*, 2011, **39**, 9433.

1022   A. Rubio-Cosials, J. F. Sidow, N. Jimenez-Menendez, P. Fernandez-Millan, J. Montoya, H. T. Jacobs, M. Coll, P. Bernado and M. Sola, *Nature Struct. Mol. Biol.*, 2011, **18**, 1281.

1023   J. R. Stagno, A. S. Altieri, M. Bubunenko, S. G. Tarasov, J. Li, D. L. Court, R. A. Byrd and X. Ji, *Nucl. Acids Res.*, 2011, **39**, 7803.

1024   H. B. Ngo, J. T. Kaiser and D. C. Chan, *Nature Struct. Mol. Biol.*, 2011, **18**, 1290.

1025   O. V. Tsodikov and T. Biswas, *J. Mol. Biol.*, 2011, **410**, 461.

1026   A. M. Ellisdon, L. Dimitrova, E. Hurt and M. Stewart, *Nature Struct. Mol. Biol.*, 2012, **19**, 328.

1027   Y. Zhang, Y. Feng, S. Chatterjee, S. Tuske, M. X. Ho, E. Arnold and R. H. Ebright, *Science*, 2012, **338**, 1076.

1028   N. Horstmann, J. Orans, P. Valentin-Hansen, S. A. Shelburne and R. G. Brennan, *Nucl. Acids Res.*, 2012, **40**, 11023.

1029   D. Deng, C. Yan, X. Pan, M. Mahfouz, J. Wang, J.-K. Zhu, Y. Shi and N. Yan, *Science*, 2012, **335**, 720.

1030   A. N.-S. Mak, P. Bradley, R. A. Cernadas, A. J. Bogdanove and B. L. Stoddard, *Science*, 2012, **335**, 716.

1031   L. Pradhan, C. Genis, P. Scone, E. O. Weinberg, H. Kasahara and H.-J. Nam, *Biochemistry*, 2012, **51**, 6312.

1032   A. A. E. Ali, G. Timinszky, R. Arribas-Bosacoma, M. Kozlowski, P. O. Hassa, M. Hassler, A. G. Ladurner, L. H. Pearl and A. W. Oliver, *Nature Struct. Mol. Biol.*, 2012, **19**, 685.

1033   B. A. Buck-Koehntop, R. L. Stanfield, D. C. Ekiert, M. A. Martinez-Yamout, H. J. Dyson, I. A. Wilson and P. E. Wright, *Proc. Natl. Acad. Sci. USA*, 2012, **109**, 15229.

1034   B. M. Lunde, M. Horner and A. Meinhart, *Nucl. Acids Res.*, 2011, **39**, 337.

1035   E. Sauer and O. Weichenrieder, *Proc. Natl. Acad. Sci. USA.*, 2011, **108**, 13065.

1036   D. Luo, S. C. Ding, A. Vela, A. Kohlway, B. D. Lindenbach and A. M. Pyle, *Cell*, 2011, **147**, 409.

1037 N. Baburajendran, R. Jauch, C. Y. Z. Tan, K. Narasimhan and P. R. Kolatkar, *Nucl. Acids Res.*, 2011, **39**, 8213.
1038 S. Machida, H.-Y. Chen and Y. Adam Yuan, *Nucl. Acids Res.*, 2011, **39**, 7828.
1039 X. Ye, N. Huang, Y. Liu, Z. Paroo, C. Huerta, P. Li, S. Chen, Q. Liu and H. Zhang, *Nature Struct. Mol. Biol.*, 2011, **18**, 650.
1040 E. M. Gesner, M. J. Schellenberg, E. L. Garside, M. M. George and A. M. MacMillan, *Nature Struct. Mol. Biol.*, 2011, **18**, 688.
1041 Y. Tian, D. K. Simanshu, J.-B. Ma and D. J. Patel, *Proc. Natl. Acad. Sci. USA*, 2011, **108**, 903.
1042 Y. Nam, C. Chen, R. I. Gregory, J. J. Chou and P. Sliz, *Cell*, 2011, **147**, 1080.
1043 N. T. Schirle and I. J. MacRae, *Science*, 2012, **336**, 1037.
1044 K. Nakanishi, D. E. Weinberg, D. P. Bartel and D. J. Patel, *Nature*, 2012, **486**, 368.
1045 E. Elkayam, C.-D. Kuhn, A. Tocilj, Astrid D. Haase, Emily M. Greene, Gregory J. Hannon and L. Joshua-Tor, *Cell*, 2012, **150**, 100.
1046 Z. Zeng, B. Min, J. Huang, K. Hong, Y. Yang, K. Collins and M. Lei, *Proc. Natl. Acad. Sci. USA*, 2011, **108**, 20357.
1047 Z. Zhang, B. Cheng and Y.-C. Tse-Dinh, *Proc. Natl. Acad. Sci. USA*, 2011, **108**, 6939.
1048 C.-C. Wu, T.-K. Li, L. Farh, L.-Y. Lin, T.-S. Lin, Y.-J. Yu, T.-J. Yen, C.-W. Chiang and N.-L. Chan, *Science*, 2011, **333**, 459.
1049 M. Tarry, M. Jaaskelainen, A. Paino, H. Tuominen, R. Ihalin and M. Hogbom, *J. Mol. Biol.*, 2011, **409**, 642.
1050 T. J. Wendorff, B. H. Schmidt, P. Heslop, C. A. Austin and J. M. Berger, *J. Mol. Biol.*, 2012, **424**, 109.
1051 M. J. Schellenberg, C. D. Appel, S. Adhikari, P. D. Robertson, D. A. Ramsden and R. S. Williams, *Nature Struct. Mol. Biol.*, 2012, **19**, 1363.
1052 B. H. Schmidt, N. Osheroff and J. M. Berger, *Nature Struct. Mol. Biol.*, 2012, **19**, 1147.
1053 T. C. Appleby, R. Anderson, O. Fedorova, A. M. Pyle, R. Wang, X. Liu, K. M. Brendza and J. R. Somoza, *J. Mol. Biol.*, 2011, **405**, 1139.
1054 B. Montpetit, N. D. Thomsen, K. J. Helmke, M. A. Seeliger, J. M. Berger and K. Weis, *Nature*, 2011, **472**, 238.
1055 E. Kowalinski, T. Lunardi, A. A. McCarthy, J. Louber, J. Brunel, B. Grigorov, D. Gerlier and S. Cusack, *Cell*, 2011, **147**, 423.
1056 A. L. Mallam, M. Del Campo, B. Gilman, D. J. Sidote and A. M. Lambowitz, *Nature*, 2012, **490**, 121.
1057 M. P. J. Webster, R. Jukes, V. S. Zamfir, C. W. M. Kay, C. Bagneris and T. Barrett, *Nucl. Acids Res.*, 2012, **40**, 8743.
1058 S. C. Lim, M. W. Bowler, T. F. Lai and H. Song, *Nucl. Acids Res.*, 2012, **40**, 11009.
1059 I. C. Berke and Y. Modis, *EMBO J.*, 2012, **31**, 1714.
1060 S. E. Tsutakawa, S. Classen, B. R. Chapados, A. S. Arvai, L. D. Finger, G. Guenther, C. G. Tomlinson, P. Thompson, A. H. Sarker, B. Shen, P. K. Cooper, J. A. Grasby and J. A. Tainer, *Cell*, 2011, **145**, 198.
1061 E. S. Vanamee, H. Viadiu, S.-H. Chan, A. Ummat, A. M. Hartline, S.-y. Xu and A. K. Aggarwal, *Nucl. Acids Res.*, 2011, **39**, 712.
1062 M. Firczuk, M. Wojciechowski, H. Czapinska and M. Bochtler, *Nucl. Acids Res.*, 2011, **39**, 744.
1063 D. G. Sashital, M. Jinek and J. A. Doudna, *Nature Struct. Mol. Biol.*, 2011, **18**, 680.
1064 C. Dienemann, A. Boggild, K. S. Winther, K. Gerdes and D. E. Brodersen, *J. Mol. Biol.*, 2011, **414**, 713.

1065 M. Jaciuk, E. b. Nowak, K. Skowronek, A. Ta≈Ñska and M. Nowotny, *Nature Struct. Mol. Biol.*, 2011, **18**, 191.
1066 T. R. Blower, X. Y. Pei, F. L. Short, P. C. Fineran, D. P. Humphreys, B. F. Luisi and G. P. C. Salmond, *Nature Struct. Mol. Biol.*, 2011, **18**, 185.
1067 W. Siwek, H. Czapinska, M. Bochtler, J. M. Bujnicki and K. Skowronek, *Nucl. Acids Res.*, 2012, **40**, 7563.
1068 R. Sukackaite, S. Grazulis, G. Tamulaitis and V. Siksnys, *Nucl. Acids Res.*, 2012, **40**, 7552.
1069 K.-T. Wang, B. Desmolaize, J. Nan, X.-W. Zhang, L.-F. Li, S. Douthwaite and X.-D. Su, *Nucl. Acids Res.*, 2012, **40**, 5138.
1070 J. Orans, E. A. McSweeney, R. R. Iyer, M. A. Hast, H. W. Hellinga, P. Modrich and L. S. Beese, *Cell*, 2011, **145**, 212.
1071 C. Bagneris, L. C. Briggs, R. Savva, B. Ebrahimi and T. E. Barrett, *Nucl. Acids Res.*, 2011, **39**, 5744.
1072 Y.-Y. Hsiao, C.-C. Yang, C. L. Lin, J. L. J. Lin, Y. Duh and H. S. Yuan, *Nature Chem. Biol.*, 2011, **7**, 236.
1073 J. Zhang, K. A. McCabe and C. E. Bell, *Proc. Natl. Acad. Sci. USA*, 2011, **108**, 11872.
1074 Y.-Y. Hsiao, Y. Duh, Y.-P. Chen, Y.-T. Wang and H. S. Yuan, *Nucl. Acids Res.*, 2012, **40**, 8144.
1075 K. Heil, A. C. Kneuttinger, S. Schneider, U. Lischke and T. Carell, *Chem. Eur. J.*, 2011, **17**, 9651.
1076 S. Kiontke, Y. Geisselbrecht, R. Pokorny, T. Carell, A. Batschauer and L.-O. Essen, *EMBO J.*, 2011, **30**, 4437.
1077 M. Fislage, M. Roovers, I. Tuszynska, J. M. Bujnicki, L. Droogmans and W. Versées, *Nucl. Acids Res.*, 2012, **40**, 5149.
1078 L. Klipcan, N. Moor, I. Finarov, N. Kessler, M. Sukhanova and M. G. Safro, *J. Mol. Biol.*, 2012, **415**, 527.
1079 E. Antony, E. A. Weiland, S. Korolev and T. M. Lohman, *J. Mol. Biol.*, 2012, **420**, 269.
1080 N. Coquelle, Z. Havali-Shahriari, N. Bernstein, R. Green and J. N. M. Glover, *Proc. Natl. Acad. Sci. USA*, 2011, **108**, 21022.
1081 Y. Sasson, L. Navon-Perry, D. Huppert and J. A. Hirsch, *J. Mol. Biol.*, 2011, **413**, 372.
1082 A. K. Chakravarty, P. Smith and S. Shuman, *Proc. Natl. Acad. Sci. USA*, 2011, **108**, 21034.
1083 W. Joo, G. Xu, N. S. Persky, A. Smogorzewska, D. G. Rudge, O. Buzovetsky, S. J. Elledge and N. P. Pavletich, *Science*, 2011, **333**, 312.
1084 A. K. Boal, T. L. Grove, M. I. McLaughlin, N. H. Yennawar, S. J. Booker and A. C. Rosenzweig, *Science*, 2011, **332**, 1089.
1085 J. Lin, S. Lai, R. Jia, A. Xu, L. Zhang, J. Lu and K. Ye, *Nature*, 2011, **469**, 559.
1086 J. Song, O. Rechkoblit, T. H. Bestor and D. J. Patel, *Science*, 2011, **331**, 1036.
1087 J. Song, M. Teplova, S. Ishibe-Murakami and D. J. Patel, *Science*, 2012, **335**, 709.
1088 P. Tumbale, C. D. Appel, R. Kraehenbuehl, P. D. Robertson, J. S. Williams, J. Krahn, I. Ahel and R. S. Williams, *Nature Struct. Mol. Biol.*, 2011, **18**, 1189.
1089 P. Wang, C. M. Chan, D. Christensen, C. Zhang, K. Selvadurai and R. H. Huang, *Proc. Natl. Acad. Sci. USA*, 2012, **109**, 13248.
1090 J. W. Setser, G. M. Lingaraju, C. A. Davis, L. D. Samson and C. L. Drennan, *Biochemistry*, 2011, **51**, 382.
1091 T. C. Efthymiou, B. Peel, V. Huynh and J.-P. Desaulniers, *Bioorg. Med. Chem. Lett.*, 2012, **22**, 5590.

1092 B.-K. Koo, C.-J. Park, C. F. Fernandez, N. Chim, Y. Ding, G. Chanfreau and J. Feigon, *J. Mol. Biol.*, 2011, **411**, 927.
1093 J. I. Yeh, A. S. Levine, S. Du, U. Chinte, H. Ghodke, H. Wang, H. Shi, C. L. Hsieh, J. F. Conway, B. Van Houten and V. Rapic-Otrin, *Proc. Natl. Acad. Sci. USA*, 2012, **109**, E2737.
1094 M.-F. Langelier, J. L. Planck, S. Roy and J. M. Pascal, *Science*, 2012, **336**, 728.
1095 Y. Gong, D. Zhu, J. Ding, C.-N. Dou, X. Ren, L. Gu, T. Jiang and D.-C. Wang, *Nature Struct. Mol. Biol.*, 2011, **18**, 1297.
1096 G. J. Williams, R. S. Williams, J. S. Williams, G. Moncalian, A. S. Arvai, O. Limbo, G. Guenther, S. SilDas, M. Hammel, P. Russell and J. A. Tainer, *Nature Struct. Mol. Biol.*, 2011, **18**, 423.
1097 K. E. Duderstadt, K. Chuang and J. M. Berger, *Nature*, 2011, **478**, 209.
1098 C. Tu, X. Zhou, S. G. Tarasov, J. E. Tropea, B. P. Austin, D. S. Waugh, D. L. Court and X. Ji, *Proc. Natl. Acad. Sci. USA*, 2011, **108**, 10156.
1099 J. Schmitzova, N. Rasche, O. Dybkov, K. Kramer, P. Fabrizio, H. Urlaub, R. Luhrmann and V. Pena, *EMBO J.*, 2012, **31**, 2222.
1100 P.-C. Lin and R.-M. Xu, *EMBO J.*, 2012, **31**, 1579.
1101 E. Khazina, V. Truffault, R. Buttner, S. Schmidt, M. Coles and O. Weichenrieder, *Nature Struct. Mol. Biol.*, 2011, **18**, 1006.
1102 F. Ding, C. Lu, W. Zhao, K. R. Rajashankar, D. L. Anderson, P. J. Jardine, S. Grimes and A. Ke, *Proc. Natl. Acad. Sci. USA*, 2011, **108**, 7357.
1103 K. M. Hastie, T. Liu, S. Li, L. B. King, N. Ngo, M. A. Zandonatti, V. L. Woods, J. C. de la Torre and E. O. Saphire, *Proc. Natl. Acad. Sci. USA*, 2011, **108**, 19365.
1104 C. Lu, C. T. Ranjith-Kumar, L. Hao, C. C. Kao and P. Li, *Nucl. Acids Res.*, 2011, **39**, 1565.
1105 S. Hare, G. N. Maertens and P. Cherepanov, *EMBO J.*, 2012, **31**, 3020.
1106 S. M. Dibrov, K. Ding, N. D. Brunn, M. A. Parker, B. M. Bergdahl, D. L. Wyles and T. Hermann, *Proc. Natl. Acad. Sci. USA*, 2012, **109**, 5223.
1107 C. R. Buttner, M. Chechik, M. Ortiz-Lombardía, C. Smits, I.-O. Ebong, V. Chechik, G. Jeschke, E. Dykeman, S. Benini, C. V. Robinson, J. C. Alonso and A. A. Antson, *Proc. Natl. Acad. Sci. USA*, 2012, **109**, 811.
1108 T. Maehigashi, C. Hsiao, K. Kruger Woods, T. Moulaei, N. V. Hud and L. Dean Williams, *Nucl. Acids Res.*, 2012, **40**, 3714.
1109 K. Brzezinski, A. Brzuszkiewicz, M. Dauter, M. Kubicki, M. Jaskolski and Z. Dauter, *Nucl. Acids Res.*, 2011, **39**, 6238.
1110 S. Venkadesh, P. K. Mandal and N. Gautham, *Nucleosides, Nucleotides & Nucl. Acids*, 2012, **31**, 184.
1111 N. H. Campbell, N. H. A. Karim, G. N. Parkinson, M. Gunaratnam, V. Petrucci, A. K. Todd, R. Vilar and S. Neidle, *J. Med. Chem.*, 2012, **55**, 209.
1112 D. Wei, G. N. Parkinson, A. P. Reszka and S. Neidle, *Nucl. Acids Res.*, 2012, **40**, 4691.
1113 G. R. Clark, P. D. Pytel and C. J. Squire, *Nucl. Acids Res.*, 2012, **40**, 5731.
1114 G. W. Collie, R. Promontorio, S. M. Hampel, M. Micco, S. Neidle and G. N. Parkinson, *J. Am. Chem. Soc.*, 2012, **134**, 2723.
1115 H. Niyazi, J. P. Hall, K. O'Sullivan, G. Winter, T. Sorensen, J. M. Kelly and C. J. Cardin, *Nature Chem.*, 2012, **4**, 621.
1116 H. Song, J. T. Kaiser and J. K. Barton, *Nature Chem.*, 2012, **4**, 615.
1117 J. P. Hall, K. O‚ÄôSullivan, A. Naseer, J. A. Smith, J. M. Kelly and C. J. Cardin, *Proc. Natl. Acad. Sci. USA*, 2011, **108**, 17610.
1118 M. Ferraroni, C. Bazzicalupi, A. R. Bilia and P. Gratteri, *Chem. Commun.*, 2011, **47**, 4917.

1119 S. Komeda, T. Moulaei, M. Chikuma, A. Odani, R. Kipping, N. P. Farrell and L. D. Williams, *Nucl. Acids Res.*, 2011, **39**, 325.
1120 D. R. Davies, A. D. Gelinas, C. Zhang, J. C. Rohloff, J. D. Carter, D. O'Connell, S. M. Waugh, S. K. Wolk, W. S. Mayfield, A. B. Burgin, T. E. Edwards, L. J. Stewart, L. Gold, N. Janjic and T. C. Jarvis, *Proc. Natl. Acad. Sci. USA*, 2012, **109**, 19971.
1121 T. Someya, S. Baba, M. Fujimoto, G. Kawai, T. Kumasaka and K. Nakamura, *Nucl. Acids Res.*, 2012, **40**, 1856.
1122 A. Kumar, H. Park, P. Fang, R. Parkesh, M. Guo, K. W. Nettles and M. D. Disney, *Biochemistry*, 2011, **50**, 9928.
1123 L. A. Coonrod, J. R. Lohman and J. A. Berglund, *Biochemistry*, 2012, **51**, 8330.
1124 A. Kumar, P. Fang, H. Park, M. Guo, K. W. Nettles and M. D. Disney, *ChemBioChem*, 2011, **12**, 2140.
1125 O. Pikovskaya, A. Polonskaia, D. J. Patel and A. Serganov, *Nature Chem. Biol.*, 2011, **7**, 748.
1126 C. A. Shanahan, B. L. Gaffney, R. A. Jones and S. A. Strobel, *J. Am. Chem. Soc.*, 2011, **133**, 15578.
1127 K. D. Smith, C. A. Shanahan, E. L. Moore, A. C. Simon and S. A. Strobel, *Proc. Natl. Acad. Sci. USA*, 2011, **108**, 7757.
1128 E. B. Butler, Y. Xiong, J. Wang and S. A. Strobel, *Chem. Biol.*, 2011, **18**, 293.
1129 A. D. Garst, E. B. Porter and R. T. Batey, *J. Mol. Biol.*, 2012, **423**, 17.
1130 A. Ramesh, C. A. Wakeman and W. C. Winkler, *J. Mol. Biol.*, 2011, **407**, 556.
1131 A. Kiliszek, R. Kierzek, W. J. Krzyzosiak and W. Rypniewski, *Nucl. Acids Res.*, 2011, **39**, 7308.
1132 G. W. Collie, S. Sparapani, G. N. Parkinson and S. Neidle, *J. Am. Chem. Soc.*, 2011, **133**, 2721.
1133 A. Kiliszek, R. Kierzek, W. J. Krzyzosiak and W. Rypniewski, *Nucl. Acids Res.*, 2012, **40**, 8155.
1134 K. Furuita, S. Murata, J. G. Jee, S. Ichikawa, A. Matsuda and C. Kojima, *J. Am. Chem. Soc.*, 2011, **133**, 5788.
1135 H. Kato, H. van Ingen, B.-R. Zhou, H. Feng, M. Bustin, L. E. Kay and Y. Bai, *Proc. Natl. Acad. Sci. USA*, 2011, **108**, 12283.
1136 K. Kim, B. I. Khayrutdinov, C.-K. Lee, H.-K. Cheong, S. W. Kang, H. Park, S. Lee, Y.-G. Kim, J. Jee, A. Rich, K. K. Kim and Y. H. Jeon, *Proc. Natl. Acad. Sci. USA*, 2011, **108**, 6921.
1137 S. Eustermann, H. Videler, J.-C. Yang, P. T. Cole, D. Gruszka, D. Veprintsev and D. Neuhaus, *J. Mol. Biol.*, 2011, **407**, 149.
1138 S. Wang, S. Y. Kim, K.-H. Jung, V. Ladizhansky and L. S. Brown, *J. Mol. Biol.*, 2011, **411**, 449.
1139 M. Bista, S. M. Freund and A. R. Fersht, *Proc. Natl. Acad. Sci. USA*, 2012, **109**, 15752.
1140 J. N. Scarsdale, H. D. Webb, G. D. Ginder and D. C. Williams, *Nucl. Acids Res.*, 2011, **39**, 6741.
1141 A. Bazzi, L. Zargarian, F. Chaminade, C. Boudier, H. De Rocquigny, B. Rene, Y. Mely, P. Fosse and O. Mauffret, *Nucl. Acids Res.*, 2011, **39**, 3903.
1142 N. M. Baker, S. Weigand, S. Maar-Mathias and A. Mondragon, *Nucl. Acids Res.*, 2011, **39**, 755.
1143 A. Natarajan, K. Dutta, D. B. Temel, P. A. Nair, S. Shuman and R. Ghose, *Nucl. Acids Res.*, 2012, **40**, 2076.
1144 R. B. Berlow, M. Swain, S. Dalal, J. B. Sweasy and J. P. Loria, *J. Mol. Biol.*, 2012, **419**, 171.

1145 K. Oertell, Y. Wu, V. M. Zakharova, B. A. Kashemirov, D. D. Shock, W. A. Beard, S. H. Wilson, C. E. McKenna and M. F. Goodman, *Biochemistry*, 2012, **51**, 8491.
1146 K. Jaudzems, X. Jia, H. Yagi, D. Zhulenkovs, B. Graham, G. Otting and E. Liepinsh, *J. Mol. Biol.*, 2012, **424**, 42.
1147 I. V. Sergeyev, L. A. Day, A. Goldbourt and A. E. McDermott, *J. Am. Chem. Soc.*, 2011, **133**, 20208.
1148 R. Fiala, N. Spackova, S. Foldynova-Trantirkova, J. Sponer, V. Sklenar and L. Trantirek, *J. Am. Chem. Soc.*, 2011, **133**, 13790.
1149 R. Spitzer, K. Kloiber, M. Tollinger and C. Kreutz, *ChemBioChem*, 2011, **12**, 2007.
1150 V. T. Mukundan, N. Q. Do and A. T. Phan, *Nucl. Acids Res.*, 2011, **39**, 8984.
1151 J. Zavasnik, P. Podbevsek and J. Plavec, *Biochemistry*, 2011, **50**, 4155.
1152 N. Q. Do, K. W. Lim, M. H. Teo, B. Heddi and A. T. Phan, *Nucl. Acids Res.*, 2011, **39**, 9448.
1153 N. Borbone, J. Amato, G. Oliviero, V. D'Atri, V. Gabelica, E. De Pauw, G. Piccialli and L. Mayol, *Nucl. Acids Res.*, 2011, **39**, 7848.
1154 X. Tong, W. Lan, X. Zhang, H. Wu, M. Liu and C. Cao, *Nucl. Acids Res.*, 2011, **39**, 6753.
1155 D. J. E. Yue, K. W. Lim and A. T. Phan, *J. Am. Chem. Soc.*, 2011, **133**, 11462.
1156 B. Heddi and A. T. Phan, *J. Am. Chem. Soc.*, 2011, **133**, 9824.
1157 M. Marusic, P. Sket, L. Bauer, V. Viglasky and J. Plavec, *Nucl. Acids Res.*, 2012, **40**, 6946.
1158 P. Sket, A. Virgilio, V. Esposito, A. Galeone and J. Plavec, *Nucl. Acids Res.*, 2012, **40**, 11047.
1159 N. Q. Do and A. T. Phan, *Chem. Eur. J.*, 2012, **18**, 14752.
1160 A. Virgilio, V. Esposito, G. Citarella, L. Mayol and A. Galeone, *ChemBioChem.*, 2012, **13**, 2219.
1161 M. Kang, B. Heuberger, J. C. Chaput, C. Switzer and J. Feigon, *Angew. Chem. Int. Ed.*, 2012, **51**, 7952.
1162 J. Dai, M. Carver, L. H. Hurley and D. Yang, *J. Am. Chem. Soc.*, 2011, **133**, 17673.
1163 A. L. Lieblein, M. Kramer, A. Dreuw, B. Furtig and H. Schwalbe, *Angew. Chem. Int. Ed.*, 2012, **51**, 4067.
1164 A. L. Lieblein, J. Buck, K. Schlepckow, B. Furtig and H. Schwalbe, *Angew. Chem. Int. Ed.*, 2012, **51**, 250.
1165 N. Escaja, J. Viladoms, M. Garavis, A. Villasante, E. Pedroso and C. Gonzalez, *Nucl. Acids Res.*, 2012, **40**, 11737.
1166 E. Guittet, D. Renciuk and J.-L. Leroy, *Nucl. Acids Res.*, 2012, **40**, 5162.
1167 E. F. DeRose, L. Perera, M. S. Murray, T. A. Kunkel and R. E. London, *Biochemistry*, 2012, **51**, 2407.
1168 Y.-M. Lee, H.-E. Kim, C.-J. Park, A.-R. Lee, H.-C. Ahn, S. J. Cho, K.-H. Choi, B.-S. Choi and J.-H. Lee, *J. Am. Chem. Soc.*, 2012, **134**, 5276.
1169 E. N. Nikolova, F. L. Gottardo and H. M. Al-Hashimi, *J. Am. Chem. Soc.*, 2012, **134**, 3667.
1170 E. N. Nikolova, E. Kim, A. A. Wise, P. J. O/'Brien, I. Andricioaei and H. M. Al-Hashimi, *Nature*, 2011, **470**, 498.
1171 S. Mazzini, L. Scaglioni, R. Mondelli, M. Caruso and F. R. Sirtori, *Bioorg. Med. Chem.*, 2012, **20**, 6979.
1172 M. Jourdan, A. Granzhan, R. Guillot, P. Dumy and M.-P. Teulade-Fichou, *Nucl. Acids Res.*, 2012, **40**, 5115.
1173 F. Liang, S. Lindsay and P. Zhang, *Org. Biomol. Chem.*, 2012, **10**, 8654.

1174 S. Di Micco, F. Mazue, C. Daquino, C. Spatafora, D. Delmas, N. Latruffe, C. Tringali, R. Riccio and G. Bifulco, *Org. Biomol. Chem.*, 2011, **9**, 701.
1175 K. Hyz, R. Kawfôcki, A. Misior, W. Bocian, E. b. Bednarek, J. Sitkowski and L. Kozerski, *J. Med. Chem.*, 2011, **54**, 8386.
1176 S. R. Hopton and A. S. Thompson, *Biochemistry*, 2011, **50**, 4143.
1177 O. Julien, J. R. Beadle, W. C. Magee, S. Chatterjee, K. Y. Hostetler, D. H. Evans and B. D. Sykes, *J. Am. Chem. Soc.*, 2011, **133**, 2264.
1178 S. P. Pack, H. Morimoto, K. Makino, K. Tajima and K. Kanaori, *Nucl. Acids Res.*, 2012, **40**, 1841.
1179 A. Hatano, M. Okada and G. Kawai, *Org. Biomol. Chem.*, 2012, **10**, 7327.
1180 S. R. Hopton and A. S. Thompson, *Biochemistry*, 2011, **50**, 4720.
1181 A. Dallmann, A. H. El-Sagheer, L. Dehmel, C. Mugge, C. Griesinger, N. P. Ernsting and T. Brown, *Chem. Eur. J.*, 2011, **17**, 14714.
1182 E. A. Dethoff, K. Petzold, J. Chugh, A. Casiano-Negroni and H. M. Al-Hashimi, *Nature*, 2012, **491**, 724.
1183 T. Lombes, R. Moumne, V. Larue, E. Prost, M. Catala, T. Lecourt, F. Dardel, L. Micouin and C. Tisne, *Angew. Chem. Int. Ed.*, 2012, **51**, 9530.
1184 S. Nozinovic, P. Gupta, B. Fürtig, C. Richter, S. Tüllmann, E. Duchardt-Ferner, M. C. Holthausen and H. Schwalbe, *Angew. Chem. Int. Ed.*, 2011, **50**, 5397.
1185 E. Duchardt-Ferner, J. Ferner and J. Wöhnert, *Angew. Chem. Int. Ed.*, 2011, **50**, 7927.
1186 C. Camilloni, P. Robustelli, A. D. Simone, A. Cavalli and M. Vendruscolo, *J. Am. Chem. Soc.*, 2012, **134**, 3968.
1187 M. Maiti, K. Nauwelaerts, E. Lescrinier and P. Herdewijn, *Chem. Eur. J.*, 2011, **17**, 1519.
1188 M. M. Phelan, B. T. Goult, J. C. Clayton, G. M. Hautbergue, S. A. Wilson and L.-Y. Lian, *Nucl. Acids Res.*, 2012, **40**, 3232.
1189 F. E. Loughlin, L. F. R. Gebert, H. Towbin, A. Brunschweiger, J. Hall and F. H. T. Allain, *Nature Struct. Mol. Biol.*, 2012, **19**, 84.
1190 T. Ohyama, T. Nagata, K. Tsuda, N. Kobayashi, T. Imai, H. Okano, T. Yamazaki and M. Katahira, *Nucl. Acids Res.*, 2012, **40**, 3218.
1191 A. Clery, S. Jayne, N. Benderska, C. Dominguez, S. Stamm and F. H. T. Allain, *Nature Struct. Mol. Biol.*, 2011, **18**, 443.
1192 K. Tsuda, T. Someya, K. Kuwasako, M. Takahashi, F. He, S. Unzai, M. Inoue, T. Harada, S. Watanabe, T. Terada, N. Kobayashi, M. Shirouzu, T. Kigawa, A. Tanaka, S. Sugano, P. Guntert, S. Yokoyama and Y. Muto, *Nucl. Acids Res.*, 2011, **39**, 1538.
1193 A. T. Chang and E. P. Nikonowicz, *Biochemistry*, 2012, **51**, 3662.
1194 A. P. Denmon, J. Wang and E. P. Nikonowicz, *J. Mol. Biol.*, 2011, **412**, 285.
1195 A. T. Phan, V. Kuryavyi, J. C. Darnell, A. Serganov, A. Majumdar, S. Ilin, T. Raslin, A. Polonskaia, C. Chen, D. Clain, R. B. Darnell and D. J. Patel, *Nature Struct. Mol. Biol.*, 2011, **18**, 796.
1196 Z. Ren and R. Ghose, *Biochemistry*, 2011, **50**, 1875.
1197 J. Wang and E. P. Nikonowicz, *J. Mol. Biol.*, 2011, **408**, 99.
1198 G. M. Daubner, A. Clery, S. Jayne, J. Stevenin and F. H. T. Allain, *EMBO J.*, 2012, **31**, 162.
1199 K. Lu, X. Heng, L. Garyu, S. Monti, E. L. Garcia, S. Kharytonchyk, B. Dorjsuren, G. Kulandaivel, S. Jones, A. Hiremath, S. S. Divakaruni, C. LaCotti, S. Barton, D. Tummillo, A. Hosic, K. Edme, S. Albrecht, A. Telesnitsky and M. F. Summers, *Science*, 2011, **334**, 242.
1200 A. Wacker, J. Buck, D. Mathieu, C. Richter, J. Wohnert and H. Schwalbe, *Nucl. Acids Res.*, 2011, **39**, 6802.

1201 A. L. Mallam, I. Jarmoskaite, P. Tijerina, M. Del Campo, S. Seifert, L. Guo, R. Russell and A. M. Lambowitz, *Proc. Natl. Acad. Sci. USA*, 2011, **108**, 12254.
1202 E. de Almeida Ribeiro, M. Beich-Frandsen, P. V. Konarev, W. Shang, B. Vecerek, G. Kontaxis, H. Hammerle, H. Peterlik, D. I. Svergun, U. Blasi and K. Djinovic-Carugo, *Nucl. Acids Res.*, 2012, **40**, 8072.
1203 E. Harjes, A. Kitamura, W. Zhao, M. C. Morais, P. J. Jardine, S. Grimes and H. Matsuo, *Nucl. Acids Res.*, 2012, **40**, 9953.
1204 S. D. Kennedy, R. Kierzek and D. H. Turner, *Biochemistry*, 2012, **51**, 9257.
1205 J. Starega-Roslan, J. Krol, E. Koscianska, P. Kozlowski, W. J. Szlachcic, K. Sobczak and W. J. Krzyzosiak, *Nucl. Acids Res.*, 2011, **39**, 257.
1206 G. Desjardins, E. Bonneau, N. Girard, J. Boisbouvier and P. Legault, *Nucl. Acids Res.*, 2011, **39**, 4427.
1207 S. Feng, H. Li, J. Zhao, K. Pervushin, K. Lowenhaupt, T. U. Schwartz and P. Droge, *Nature Struct. Mol. Biol.*, 2011, **18**, 169.
1208 R. Parkesh, M. Fountain and M. D. Disney, *Biochemistry*, 2011, **50**, 599.
1209 Y. Huang, X. Weng and I. M. Russu, *Biochemistry*, 2011, **50**, 1857.
1210 A. Davidson, D. W. Begley, C. Lau and G. Varani, *J. Mol. Biol.*, 2011, **410**, 984.
1211 N. Martín-Pintado, M. Yahyaee-Anzahaee, R. Campos-Olivas, A. M. Noronha, C. J. Wilds, M. J. Damha and C. Gonzalez, *Nucl. Acids Res.*, 2012, **40**, 9329.
1212 K. Mayanagi, S. Kiyonari, H. Nishida, M. Saito, D. Kohda, Y. Ishino, T. Shirai and K. Morikawa, *Proc. Natl. Acad. Sci. USA*, 2011, **108**, 1845.
1213 E. P. Morris, A. Rivera-Calzada, P. C. A. da Fonseca, O. Llorca, L. H. Pearl and L. Spagnolo, *Nucl. Acids Res.*, 2011, **39**, 5757.
1214 L. Spagnolo, J. Barbeau, N. J. Curtin, E. P. Morris and L. H. Pearl, *Nucl. Acids Res.*, 2012, **40**, 4168.
1215 A. Costa, I. Ilves, N. Tamberg, T. Petojevic, E. Nogales, M. R. Botchan and J. M. Berger, *Nature Struct. Mol. Biol.*, 2011, **18**, 471.
1216 K. Hipp, K. Galani, C. Batisse, S. Prinz and B. Bottcher, *Nucl. Acids Res.*, 2012, **40**, 3275.
1217 S. Atwell, L. Disseau, A. Z. Stasiak, A. Stasiak, A. Renodon-Corniere, M. Takahashi, J.-L. Viovy and G. Cappello, *Nucl. Acids Res.*, 2012, **40**, 11769.
1218 A. Pena, K. Gewartowski, S. Mroczek, J. Cuellar, A. Szykowska, A. Prokop, M. Czarnocki-Cieciura, J. Piwowarski, C. Tous, A. Aguilera, J. L. Carrascosa, J. M. Valpuesta and A. Dziembowski, *EMBO J.*, 2012, **31**, 1605.
1219 I. C. Berke, X. Yu, Y. Modis and E. H. Egelman, *Proc. Natl. Acad. Sci. USA*, 2012, **109**, 18437.
1220 D. J. Taylor, E. R. Podell, D. J. Taatjes and T. R. Cech, *J. Mol. Biol.*, 2011, **410**, 10.
1221 R. Melero, S. Rajagopalan, M. Lazaro, A. C. Joerger, T. Brandt, D. B. Veprintsev, G. Lasso, D. Gil, S. H. W. Scheres, J. M. Carazo, A. R. Fersht and M. Valle, *Proc. Natl. Acad. Sci. USA*, 2011, **108**, 557.
1222 J. Frauenfeld, J. Gumbart, E. O. v. d. Sluis, S. Funes, M. Gartmann, B. Beatrix, T. Mielke, O. Berninghausen, T. Becker, K. Schulten and R. Beckmann, *Nature Struct. Mol. Biol.*, 2011, **18**, 614.
1223 T. Becker, S. Franckenberg, S. Wickles, C. J. Shoemaker, A. M. Anger, J.-P. Armache, H. Sieber, C. Ungewickell, O. Berninghausen, I. Daberkow, A. Karcher, M. Thomm, K.-P. Hopfner, R. Green and R. Beckmann, *Nature*, 2012, **482**, 501.
1224 T. Becker, J.-P. Armache, A. Jarasch, A. M. Anger, E. Villa, H. Sieber, B. A. Motaal, T. Mielke, O. Berninghausen and R. Beckmann, *Nature Struct. Mol. Biol.*, 2011, **18**, 715.

1225 J. Fu, J. B. Munro, S. C. Blanchard and J. Frank, *Proc. Natl. Acad. Sci. USA*, 2011, **108**, 4817.
1226 X. Agirrezabala, H. Y. Liao, E. Schreiner, J. Fu, R. F. Ortiz-Meoz, K. Schulten, R. Green and J. Frank, *Proc. Natl. Acad. Sci. USA*, 2012, **109**, 6094.
1227 B. J. Greber, D. Boehringer, V. Godinic-Mikulcic, A. Crnkovic, M. Ibba, I. Weygand-Durasevic and N. Ban, *J. Mol. Biol.*, 2012, **418**, 145.
1228 B. J. Greber, D. Boehringer, C. Montellese and N. Ban, *Nature Struct. Mol. Biol.*, 2012, **19**, 1228.
1229 M. Muhs, H. Yamamoto, J. Ismer, H. Takaku, M. Nashimoto, T. Uchiumi, N. Nakashima, T. Mielke, P. W. Hildebrand, K. H. Nierhaus and C. M. T. Spahn, *Nucl. Acids Res.*, 2011, **39**, 5264.
1230 D. J. F. Ramrath, H. Yamamoto, K. Rother, D. Wittek, M. Pech, T. Mielke, J. Loerke, P. Scheerer, P. Ivanov, Y. Teraoka, O. Shpanchenko, K. H. Nierhaus and C. M. T. Spahn, *Nature*, 2012, **485**, 526.
1231 T. Yokoyama, T. R. Shaikh, N. Iwakura, H. Kaji, A. Kaji and R. K. Agrawal, *EMBO J.*, 2012, **31**, 1836.
1232 L. F. Estrozi, D. Boehringer, S.-o. Shan, N. Ban and C. Schaffitzel, *Nature Struct. Mol. Biol.*, 2011, **18**, 88.
1233 Q. Guo, Y. Yuan, Y. Xu, B. Feng, L. Liu, K. Chen, M. Sun, Z. Yang, J. Lei and N. Gao, *Proc. Natl. Acad. Sci. USA*, 2011, **108**, 13100.
1234 B. S. Strunk, C. R. Loucks, M. Su, H. Vashisth, S. Cheng, J. Schilling, C. L. Brooks, K. Karbstein and G. Skiniotis, *Science*, 2011, **333**, 1449.
1235 B. Bradatsch, C. Leidig, S. Granneman, M. Gnadig, D. Tollervey, B. Bottcher, R. Beckmann and E. Hurt, *Nature Struct. Mol. Biol.*, 2012, **19**, 1234.
1236 D. N. Ermolenko and H. F. Noller, *Nature Struct. Mol. Biol.*, 2011, **18**, 457.
1237 K. Toropova, P. G. Stockley and N. A. Ranson, *J. Mol. Biol.*, 2011, **408**, 408.
1238 B. Gerlach, J. A. Kleinschmidt and B. Bottcher, *J. Mol. Biol.*, 2011, **409**, 427.
1239 L. Cheng, J. Sun, K. Zhang, Z. Mou, X. Huang, G. Ji, F. Sun, J. Zhang and P. Zhu, *Proc. Natl. Acad. Sci. USA*, 2011, **108**, 1373.
1240 C. Yang, G. Ji, H. Liu, K. Zhang, G. Liu, F. Sun, P. Zhu and L. Cheng, *Proc. Natl. Acad. Sci. USA*, 2012, **109**, 6118.
1241 S. E. Bakker, R. J. Ford, A. M. Barker, J. Robottom, K. Saunders, A. R. Pearson, N. A. Ranson and P. G. Stockley, *J. Mol. Biol.*, 2012, **417**, 65.
1242 P. Ge and Z. H. Zhou, *Proc. Natl. Acad. Sci. USA*, 2011, **108**, 9637.
1243 R. Arranz, R. Coloma, F. J. Chichon, J. J. Conesa, J. L. Carrascosa, J. M. Valpuesta, J. Ortín and J. Martin-Benito, *Science*, 2012, **338**, 1634.
1244 A. Lopez-Perrote, H. Munoz-Hernandez, D. Gil and O. Llorca, *Nucl. Acids Res.*, 2012, **40**, 11086.
1245 B. Wiedenheft, G. C. Lander, K. Zhou, M. M. Jore, S. J. J. Brouns, J. van der Oost, J. A. Doudna and E. Nogales, *Nature*, 2011, **477**, 486.
1246 R. Aramayo, M. B. Sherman, K. Brownless, R. Lurz, A. L. Okorokov and E. V. Orlova, *Nucl. Acids Res.*, 2011, **39**, 8960.
1247 I. Orlov, N. Rochel, D. Moras and B. P. Klaholz, *EMBO J.*, 2012, **31**, 291.
1248 R. Melero, G. Buchwald, R. Castano, M. Raabe, D. Gil, M. Lazaro, H. Urlaub, E. Conti and O. Llorca, *Nature Struct. Mol. Biol.*, 2012, **19**, 498.
1249 J.-P. J. Sobczak, T. G. Martin, T. Gerling and H. Dietz, *Science*, 2012, **338**, 1458.
1250 I. Lubitz and A. Kotlyar, *Bioconjugate Chem.*, 2011, **22**, 2043.
1251 K. Zhang, M. Jiang and D. Chen, *Angew. Chem. Int. Ed.*, 2012, **51**, 8744.
1252 D. Leipply and D. E. Draper, *Biochemistry*, 2011, **50**, 2790.
1253 H. A. Vincent, C. A. Henderson, C. M. Stone, P. D. Cary, D. M. Gowers, F. Sobott, J. E. Taylor and A. J. Callaghan, *Nucl. Acids Res.*, 2012, **40**, 8698.

1254 H. S. Kim, M. C. J. Wilce, Y. M. K. Yoga, N. R. Pendini, M. J. Gunzburg, N. P. Cowieson, G. M. Wilson, B. R. G. Williams, M. Gorospe and J. A. Wilce, *Nucl. Acids Res.*, 2011, **39**, 1117.
1255 S. Patel, J. M. Blose, J. E. Sokoloski, L. Pollack and P. C. Bevilacqua, *Biochemistry*, 2012, **51**, 9312.
1256 F. Jiang, A. Ramanathan, M. T. Miller, G.-Q. Tang, M. Gale, S. S. Patel and J. Marcotrigiano, *Nature*, 2011, **479**, 423.
1257 T. Heidebrecht, E. Christodoulou, M. J. Chalmers, S. Jan, B. ter Riet, R. K. Grover, R. P. Joosten, D. Littler, H. van Luenen, P. R. Griffin, P. Wentworth, P. Borst and A. Perrakis, *Nucl. Acids Res.*, 2011, **39**, 5715.
1258 B. N. Chaudhuri and R. Dean, *J. Mol. Biol.*, 2011, **413**, 901.
1259 N. Rochel, F. Ciesielski, J. Godet, E. Moman, M. Roessle, C. Peluso-Iltis, M. Moulin, M. Haertlein, P. Callow, Y. Mely, D. I. Svergun and D. Moras, *Nature Struct. Mol. Biol.*, 2011, **18**, 564.
1260 D. J. Billingsley, N. Crampton, J. Kirkham, N. H. Thomson and W. A. Bonass, *Nucl. Acids Res.*, 2012, **40**, e99.
1261 D. Li, Z. Yang, B. Lv and T. Li, *Bioorg. Med. Chem. Lett.*, 2012, **22**, 833.
1262 L. S. Shlyakhtenko, A. Y. Lushnikov, A. Miyagi and Y. L. Lyubchenko, *Biochemistry*, 2012, **51**, 1500.
1263 R. Buzio, L. Repetto, F. Giacopelli, R. Ravazzolo and U. Valbusa, *Nucl. Acids Res.*, 2012, **40**, e84.
1264 H. K. Tan, D. Li, R. K. Gray, Z. Yang, M. T. T. Ng, H. Zhang, J. M. R. Tan, S. H. Hiew, J. Y. Lee and T. Li, *Org. Biomol. Chem.*, 2012, **10**, 2227.
1265 Y. Amemiya, Y. Furunaga, K. Iida, M. Tera, K. Nagasawa, K. Ikebukuro and C. Nakamura, *Chem. Commun.*, 2011, **47**, 7485.
1266 I. Mela, R. Kranaster, R. M. Henderson, S. Balasubramanian and J. M. Edwardson, *Biochemistry*, 2011, **51**, 578.
1267 Y. J. Jung, J. A. Albrecht, J.-W. Kwak and J. W. Park, *Nucl. Acids Res.*, 2012, **40**, 11728.
1268 J. Pallesen, *Biochemistry*, 2011, **50**, 6170.
1269 Z. Yang, D. Li, S. Hiew, M. T. Ng, W. Yuan, H. Su, F. Shao and T. Li, *Chem. Commun.*, 2011, **47**, 11309.
1270 D. Li, Z. Yang, G. Zhao, Y. Long, B. Lv, C. Li, S. Hiew, M. T. T. Ng, J. Guo, H. Tan, H. Zhang and T. Li, *Chem. Commun.*, 2011, **47**, 7479.
1271 Z. Yang, D. Li and T. Li, *Chem. Commun.*, 2011, **47**, 11930.
1272 Z. Wu, T. Tian, J. Yu, X. Weng, Y. Liu and X. Zhou, *Angew. Chem. Int. Ed.*, 2011, **50**, 11962.
1273 J. Liang, M. Castronovo and G. Scoles, *J. Am. Chem. Soc.*, 2011, **134**, 39.
1274 A. Loksztejn, Z. Scholl and P. E. Marszalek, *Chem. Commun.*, 2012, **48**, 11727.
1275 L. Bintu, M. Kopaczynska, C. Hodges, L. Lubkowska, M. Kashlev and C. Bustamante, *Nature Struct. Mol. Biol.*, 2011, **18**, 1394.
1276 M. Endo, K. Tatsumi, K. Terushima, Y. Katsuda, K. Hidaka, Y. Harada and H. Sugiyama, *Angew. Chem. Int. Ed.*, 2012, **51**, 8778.
1277 Y. Jiang and P. E. Marszalek, *EMBO J.*, 2011, **30**, 2881.
1278 A. Evilevitch, W. H. Roos, I. L. Ivanovska, M. Jeembaeva, B. Jonsson and G. J. L. Wuite, *J. Mol. Biol.*, 2011, **405**, 18.
1279 L. S. Shlyakhtenko, A. Y. Lushnikov, A. Miyagi, M. Li, R. S. Harris and Y. L. Lyubchenko, *Biochemistry*, 2012, **51**, 6432.
1280 S. A. Shelke and S. T. Sigurdsson, *Eur. J. Org. Chem.*, 2012, 2291.
1281 A. D. J. Freeman, R. Ward, H. El Mkami, D. M. J. Lilley and D. G. Norman, *Biochemistry*, 2011, **50**, 9963.

1282 T. J. Purcell, N. Naber, K. Franks-Skiba, A. R. Dunn, C. C. Eldred, C. L. Berger, A. Malnasi-Csizmadia, J. A. Spudich, D. M. Swank, E. Pate and R. Cooke, *J. Mol. Biol.*, 2011, **407**, 79.

1283 I. Krstic, R. Hansel, O. Romainczyk, J. W. Engels, V. Dotsch and T. F. Prisner, *Angew. Chem. Int. Ed.*, 2011, **50**, 5070.

1284 M. Azarkh, O. Okle, V. Singh, I. T. Seemann, J. S. Hartig, D. R. Dietrich and M. Drescher, *ChemBioChem*, 2011, **12**, 1992.

1285 H. Atsumi, K. Maekawa, S. Nakazawa, D. Shiomi, K. Sato, M. Kitagawa, T. Takui and K. Nakatani, *Chem. Eur. J.*, 2012, **18**, 178.

1286 S. Bhattacharjee, S. Chatterjee, J. Jiang, B. K. Sinha and R. P. Mason, *Nucl. Acids Res.*, 2012, **40**, 5477.

1287 Z. Yang, M. R. Kurpiewski, M. Ji, J. E. Townsend, P. Mehta, L. Jen-Jacobson and S. Saxena, *Proc. Natl. Acad. Sci. USA*, 2012, **109**, E993.

1288 P. R. Dohrmann, C. M. Manhart, C. D. Downey and C. S. McHenry, *J. Mol. Biol.*, 2011, **414**, 15.

1289 Y. M. K. Yoga, D. A. K. Traore, M. Sidiqi, C. Szeto, N. R. Pendini, A. Barker, P. J. Leedman, J. A. Wilce and M. C. J. Wilce, *Nucl. Acids Res.*, 2012, **40**, 5101.

1290 X. Zheng, Q. Liu, C. Jing, Y. Li, D. Li, W. Luo, Y. Wen, Y. He, Q. Huang, Y.-T. Long and C. Fan, *Angew. Chem. Int. Ed.*, 2011, **50**, 11994.

1291 J. P. Pezacki, J. A. Blake, D. C. Danielson, D. C. Kennedy, R. K. Lyn and R. Singaravelu, *Nature Chem. Biol.*, 2011, **7**, 137.

1292 Y. Chen, R. Basu, M. L. Gleghorn, K. S. Murakami and P. R. Carey, *J. Am. Chem. Soc.*, 2011, **133**, 12544.

1293 A. Campion and P. Kambhampati, *Chem. Soc. Reviews*, 1998, **27**, 241.

1294 J. L. Abell, J. M. Garren, J. D. Driskell, R. A. Tripp and Y. Zhao, *J. Am. Chem. Soc.*, 2012, **134**, 12889.

1295 E. Prado, N. Daugey, S. Plumet, L. Servant and S. Lecomte, *Chem. Commun.*, 2011, **47**, 7425.

1296 E. Papadopoulou and S. E. J. Bell, *Chem. Eur. J.*, 2012, **18**, 5394.

1297 E. Papadopoulou and S. E. J. Bell, *Angew. Chem. Int. Ed.*, 2011, **50**, 9058.

1298 E. Papadopoulou and S. E. J. Bell, *Chem. Commun.*, 2011, **47**, 10966.

1299 Z. Zhang, Y. Wen, Y. Ma, J. Luo, L. Jiang and Y. Song, *Chem. Commun.*, 2011, **47**, 7407.

1300 D. van Lierop, Z. Krpetic, L. Guerrini, I. A. Larmour, J. A. Dougan, K. Faulds and D. Graham, *Chem. Commun.*, 2012, **48**, 8192.

1301 J. Wrzesien and D. Graham, *Tetrahedron*, 2012, **68**, 1230.

1302 Y.-H. Sun, R.-M. Kong, D.-Q. Lu, X.-B. Zhang, H.-M. Meng, W. Tan, G.-L. Shen and R.-Q. Yu, *Chem. Commun.*, 2011, **47**, 3840.

1303 S. Ye, Y. Yang, J. Xiao and S. Zhang, *Chem. Commun.*, 2012, **48**, 8535.

1304 J. A. Dougan, D. MacRae, D. Graham and K. Faulds, *Chem. Commun.*, 2011, **47**, 4649.

1305 J.-Y. Xu, B. Jin, Y. Zhao, K. Wang and X.-H. Xia, *Chem. Commun.*, 2012, **48**, 3052.

1306 V. Andrushchenko, D. Tsankov, M. Krasteva, H. Wieser and P. Bour, *J. Am. Chem. Soc.*, 2011, **133**, 15055.

Quinquevalent phosphorus acids

P. Bałczewski*[a,b] and A. Bodzioch[a]

DOI: 10.1039/9781782623977-00246

1 Introduction

Continuing the review of the most interesting scientific events in the field of quinquevalent organophosphorus chemistry this chapter covers the literature concerning the area published in 2012. The review is not comprehensive but shows some of the most important achievements in the area contained in leading journals. It describes compounds containing three P–O bonds (phosphates), two P–O and one P–C bonds (phosphonates) as well as one P–O and two P–C bonds (phosphinates) in addition to the phosphoryl group P=O. Each of the main sections has been divided in the same way, covering synthesis, reactions and biological aspects.

The interest in the area of synthesis of phosphorus (V) acids and their derivatives was large and has centered on several sample subjects selected below. Various, simple, phosphorus containing reagents, such as pyrophosphates, dialkyl phosphates, cyclic trialkyl phospites and 1-acylphosphonates have been utilised as phosphorylating agents for a variety of compounds, like ketones, alcohols, amino acids and acetylenes to produce higher functionalised phosphates. Tandem hetero(thia and aza)-Michael addition of phosphoric acid derivatives to conjugated aldehydes, ketones and nitro compounds followed by cyclisation led to multi-substituted thietanes and azetidines. Novel photolabile protecting groups, such as thiochromone S,S-dioxide, containing the diazomethyl group have been synthesised and successfully applied for the protection of phosphate derivatives. Bulky phosphoric acid derivatives based on BINOL substituted in the 3,3' positions have been successfully applied as catalysts in various reactions to afford high chemical yields and excellent stereoselectivities. Finally, some new reagents were designed and synthesized as potential sensors for organophosphates. In this year, a sub-section concerning use of chiral phosphoric acids as catalysts in various chemical reactions, introduced in 2009 for the first time, has been included and revealed a continuing interest in this area. Other aspects, like reactions and biological aspects of organic phosphates, have also showed considerable development.

The interest in the area of phosphonic acids and their derivatives was greater than in the previous review period. A number of new transformations with emphasis on modifications of cross coupling reactions leading to the formation of the C–P bond, produced both cyclic and

---

[a]Polish Academy of Sciences, Centre of Molecular and Macromolecular Studies, Sienkiewicza 112, 90-363 Łódź, Poland.
E-mail: pbalczew@cbmm.lodz.pl; agabodz@cbmm.lodz.pl
[b]Jan Długosz University in Częstochowa, Armii Krajowej 13/15, 42-200 Częstochowa, Poland

acyclic phosphonates and bisphosphonates. Among them, aryl and heteroaryl substituted phosphonates, as well as their aryl/heteroarylmethylene and difluoromethylene homologs, constituted a large group of new compounds. Also, variously substituted phosphonates, containing hydroxy, alkoxy, amino, isocyanate, alkenyl, alkynyl, propargyl, allenyl, nitro, nitrile and diazo groups, mostly in a position 1, have been synthesised. A limited number of multireaction total syntheses of biologically active compounds (e.g. pentalenolactones, Tamiflu) has been replaced by syntheses of new, structurally simple, enzymatically and chemically resistant phosphonates which have been further biologically tested in various therapeutic areas. In the area of phosphinic acids and their derivatives a decreasing interest, as in previous review periods, has been observed, both in sections of synthesis, reactions and biological aspects.

1.1 Synthesis of phosphoric acids and their derivatives

Simple phosphorus containing reagents, such as pyrophosphates, dialkyl phosphates, cyclic trialkyl phospites and 1-acylphosphonates have been utilised as phosphorylating agents for a variety of compounds, such as ketones and alcohols, amino acids and acetylenes to produce organic phosphates. As the first example, 1,1,3,3-tetramethylguanidine (TMG) has been found to catalyse hydrophosphorylation of 4-oxo-enoates (**1**) by diethyl- or diphenylphosphite and thus provided a convenient access to enol phosphates (**2**) under mild conditions in good yields with excellent stereoselectivities (Scheme 1).[1]

Fenton et al. reported a method for the Lewis acid catalysed phosphorylation of alcohols with pyrophosphates. In the presence of Ti(OtBu)$_4$, both primary and secondary alcohols underwent phosphorylation with tetrabenzylpyrophosphate (**3**) in 50–97% yields (Scheme 2). Other pyrophosphates with orthogonal protecting groups were synthesised and screened to validate the generality of the approach. All pyrophosphates containing benzyl, methyl, ethyl, allyl, and o-nitrobenzyl substituents, turned out to be useful phosphorylating agents under described conditions.[2]

An effective catalytic α-phosphorylation of ketones in the presence of iodobenzene, as the recyclable catalyst, and m-chloroperbenzoic acid as

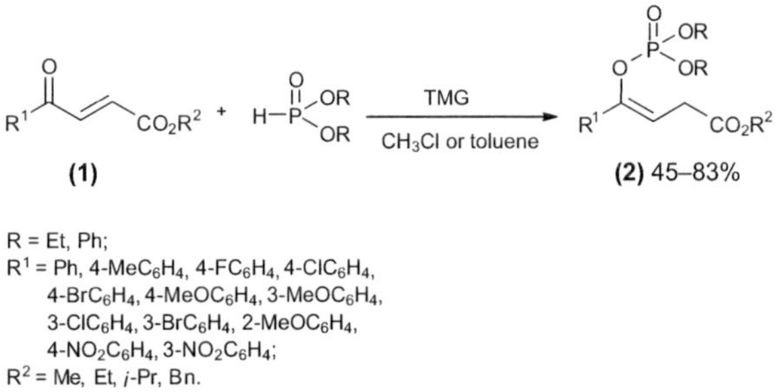

R = Et, Ph;
R^1 = Ph, 4-MeC$_6$H$_4$, 4-FC$_6$H$_4$, 4-ClC$_6$H$_4$, 4-BrC$_6$H$_4$, 4-MeOC$_6$H$_4$, 3-MeOC$_6$H$_4$, 3-ClC$_6$H$_4$, 3-BrC$_6$H$_4$, 2-MeOC$_6$H$_4$, 4-NO$_2$C$_6$H$_4$, 3-NO$_2$C$_6$H$_4$;
R^2 = Me, Et, i-Pr, Bn.

Scheme 1

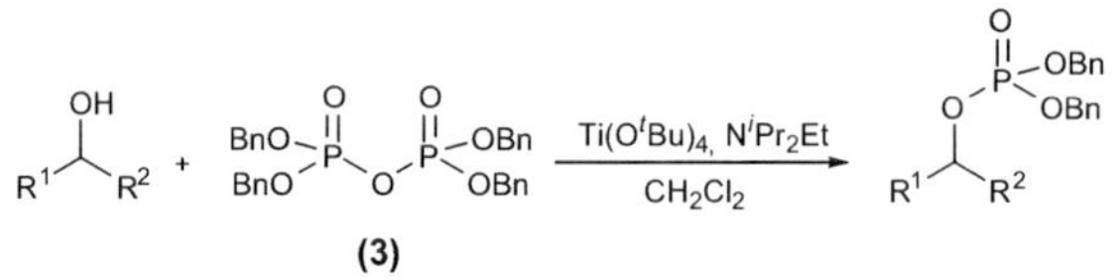

Scheme 2

R¹= Me, Cy, (CH$_2$)$_2$Ph, (CH$_2$)$_4$Ph, Ph, 2-naphthyl, CH$_2$CHC(Me)(CH$_2$)$_2$CHC(Me)$_2$;
R²= H, Me, n-C$_{10}$H$_{21}$.

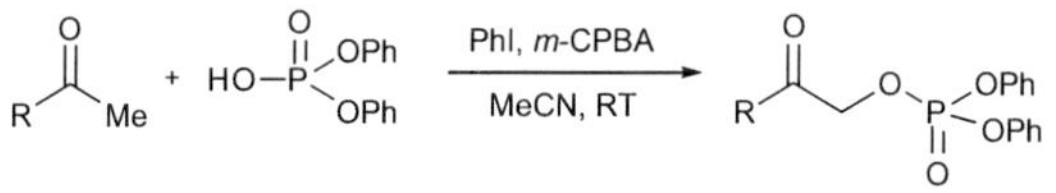

Scheme 3

R= Me, Et, Ph, 4-NO$_2$C$_6$H$_4$, 4-BrC$_6$H$_4$, 4-ClC$_6$H$_4$, 3-NO$_2$C$_6$H$_4$, 4-MeC$_6$H$_4$, 4-MeOC$_6$H$_4$, 2-thienyl.

Scheme 4

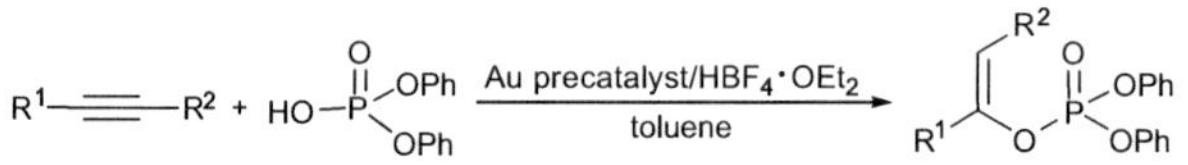

R¹= n-C$_5$H$_{11}$, n-C$_8$H$_{17}$, n-C$_6$H$_9$, t-Bu, Cl(CH$_2$)$_3$, Cl(CH$_2$)$_4$, CO$_2$Et;
R²= H, Ph.

the terminal oxidant, afforded the corresponding ketophosphates (4) in moderate to good yields (Scheme 3).[3]

Nolan et al. reported a straightforward method for the synthesis of alkenyl phosphates (6) with high selectivity, catalysed by cationic gold (I) complexes generated in situ through the reaction between the Au precatalyst and HBF$_4$·OEt$_2$. The reaction required low amounts of catalyst and gave selectively the kinetic product (6). Moreover, reactions proceeded well with a variety of substituted substrates (5) (Scheme 4).[4]

Synthesis of enol phosphates (8) has been also realised by reaction of α-aryloxyacetophenones (7) with dialkyl phosphites. Thus, in the presence of Cs$_2$CO$_3$, a variety of α-aryloxyacetophenones (7) smoothly underwent the sequential O–P bond-forming/C–O bond cleavage reaction with dialkyl phosphites at room temperature, providing the corresponding enol phosphates (8) in moderate to excellent yields (Scheme 5).[5]

Reaction of phosphorus oxytrichloride with primary alcohols and triethylamine in toluene, followed by filtration to remove triethylamine hydrochloride and treatment with steam, gave dialkyl phosphates (9) in good yields and essentially free from trialkyl phosphate contamination (Scheme 6).[6]

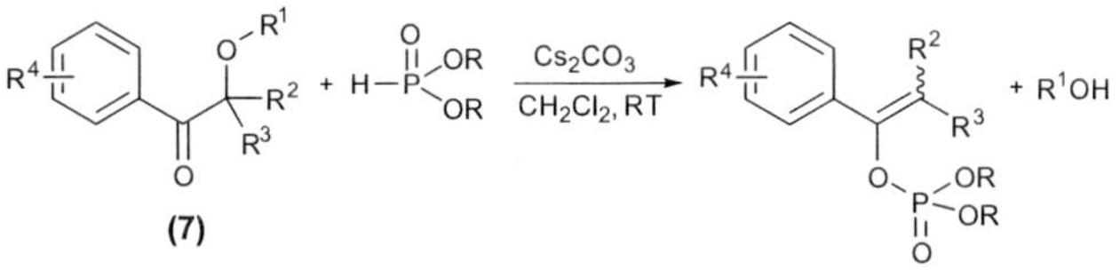

Scheme 5

R= Me, Et, i-Pr;
R^1= Ph, 4-MeC$_6$H$_4$, 4-MeOC$_6$H$_4$, 3-MeOC$_6$H$_4$,
   4-ClC$_6$H$_4$, 2-IC$_6$H$_4$, 1-naphthyl, Me;
R^2=R^3= H, Me, (CH$_2$)$_2$CN, (CH$_2$)$_2$SO$_2$Ph;
R^4= H, 4-Me, 4-MeO, 2-MeO, 4-F, 1-naphthyl.

Scheme 6

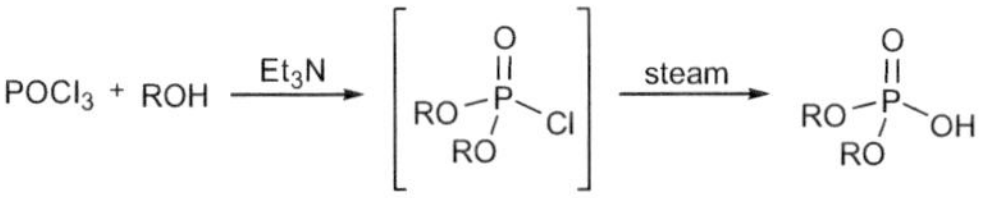

R= n-Bu, C$_5$H$_{11}$, C$_6$H$_{13}$, C$_8$H$_{17}$, C$_{10}$H$_{21}$, C$_{12}$H$_{25}$.

Scheme 7

R= Me, Et, i-Pr;
R^1= Ph, 2-BrC$_6$H$_4$, 4-ClC$_6$H$_4$, 2-Me-4-BrC$_6$H$_4$, 4-NO$_2$C$_6$H$_4$,
   2-MeO-4-BrC$_6$H$_4$, 3-MeOC$_6$H$_4$, 4-MeOC$_6$H$_4$, Me;
R^2= H, Me.

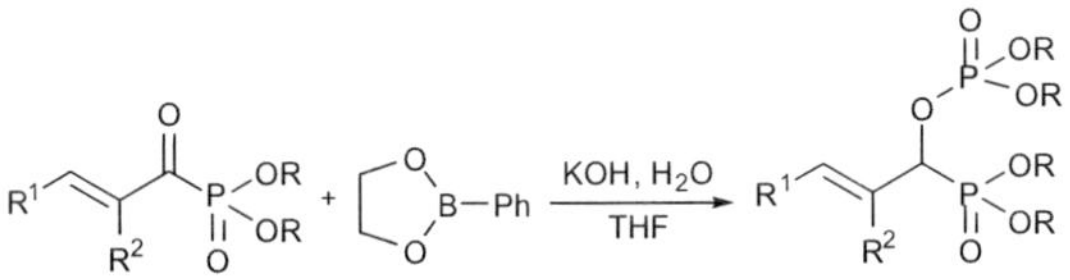

Compounds (**12**) have been prepared in moderate to good yields from dialkyl 1-acylphosphonates (**10**) in the presence of phenyldioxaborolane (**11**) and potassium hydroxide, *via* a C–P bond cleavage and a subsequent 1,2-migration of the phosphoryl group (Scheme 7).[7]

A simple, highly efficient method for the synthesis of polyanionic dendrimers (**13**) and (**14**) as potential antiviral drugs has been presented by Salamończyk. The mild conditions of both the coupling and virtually quantitative deprotection reactions, provided highly pure and water-soluble macromolecular materials in good overall yields.[8]

Petrillo and co-workers developed a highly efficient, one-pot synthesis of protected phosphoamino acids (**16**), useful building block in the synthesis of phosphopeptides, starting from inexpensive, readily available reagents. Thus, treatment of phosphorus trichloride with alcohol and an amino acid gave the corresponding phosphate intermediates (**15**), which, after oxidation with bromine-related oxidants afforded phosphoamino acids (**16**) in moderate to high yields (Scheme 8).[9]

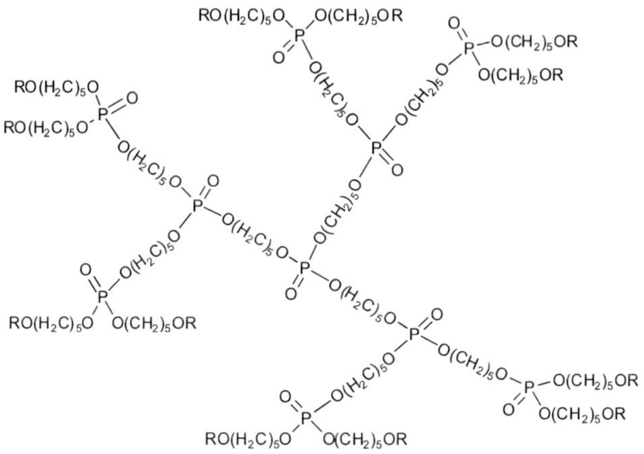

(13)

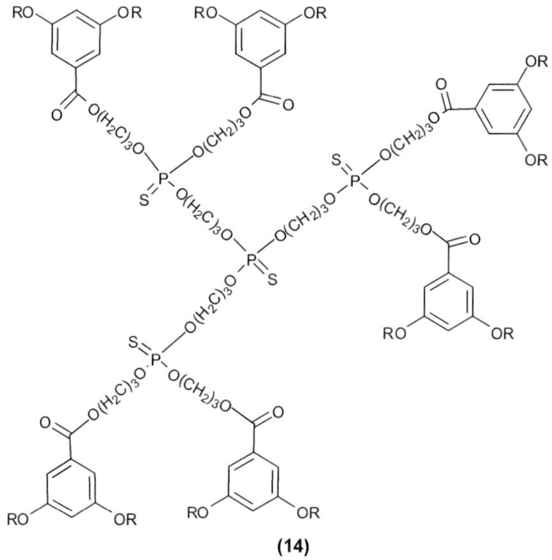

(14)

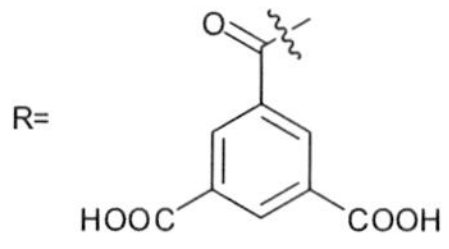

R= [structure: 3,5-dicarboxybenzoyl group with HOOC and COOH]

The synthesis of eight frame-shifted farnesyl diphosphate (FPP) analogues (**17**), bearing increased or decreased methylene units between the double bonds and/or diphosphate moieties of the isoprenoid structure, have been developed by Gibbs *et al.* Evaluation against mammalian

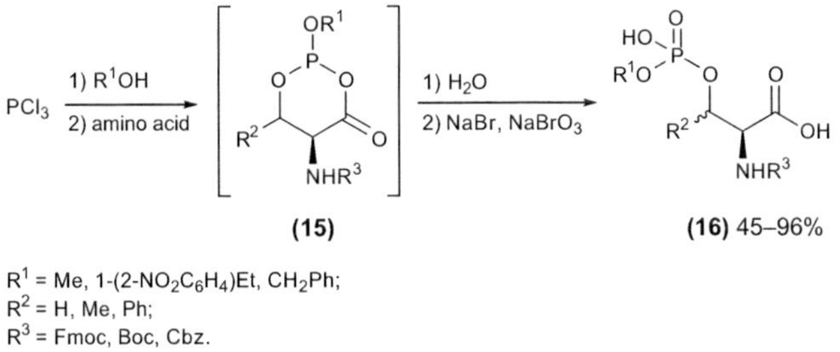

1) R¹OH
2) amino acid
PCl₃
OR¹
NHR³
1) H₂O
2) NaBr, NaBrO₃
(15)
(16) 45–96%
R¹ = Me, 1-(2-NO₂C₆H₄)Et, CH₂Ph;
R² = H, Me, Ph;
R³ = Fmoc, Boc, Cbz.

Scheme 8

FTase showed that small structural changes could lead to dramatic changes in substrate ability.[10]

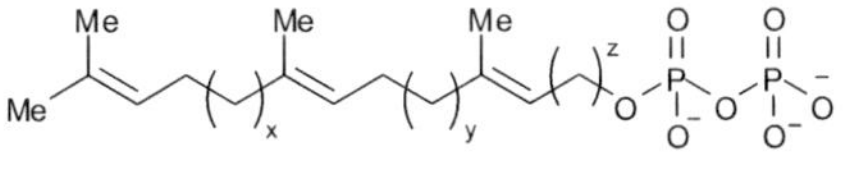

(17) x = 0, 1, 2; y = 0, 1, 2; z = 1, 2.

The synthesis of two new α,α-difluoro ester phospholipid conjugates (**18**) and (**19**) has been described. The stability of their liposomal formulations in three different aqueous buffers (pH 4.5, 7.5 and 8.5) has been also investigated. The studies confirmed that α,α-difluoro esters were much more prone to hydrolysis when positioned close to the hydrophilic head group of phospholipids than when the functionality was placed in the lipophilic part of the bilayer in liposomes.[11]

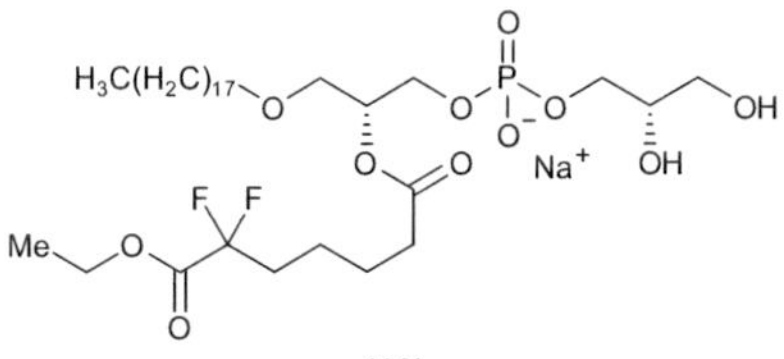

(18)

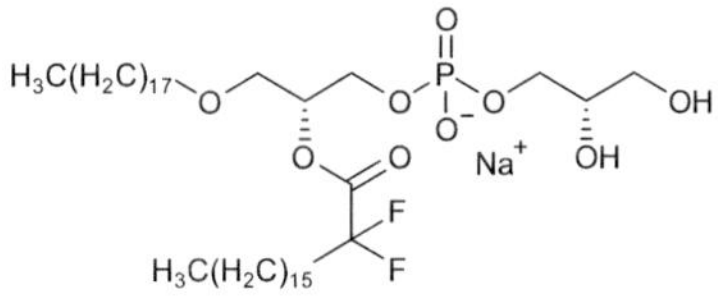

(19)

A fluorescent lipid analogue (**20**) bearing urea and phosphocholine groups has been synthesised by Sargent *et al.* Dimerisation and anion binding of phospholipid analogue (**20**) has been also investigated.[12]

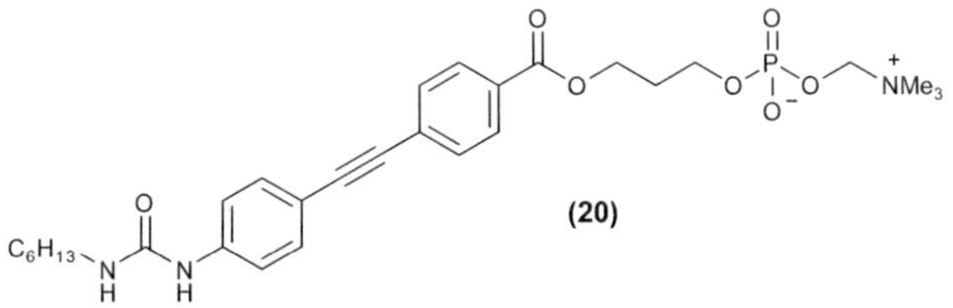

(20)

The first syntheses of a family of naturally occurring mycobacterial phosphatidylinositol (**21**) and its dimannosides (**22**)–(**24**), all possessing the naturally predominant 19:0/16:0 acylation pattern on the phosphatidyl moiety, have been reported. Complete characterisation of these compounds was achieved by a combination of NMR and MS with purity established by HPLC.[13]

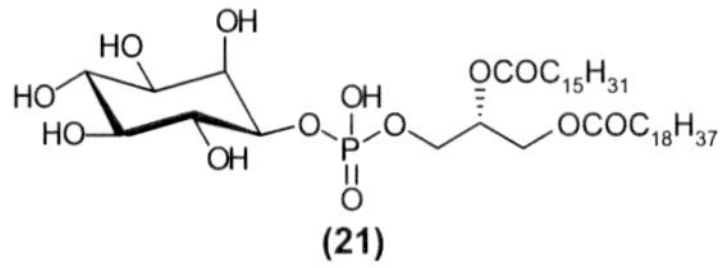

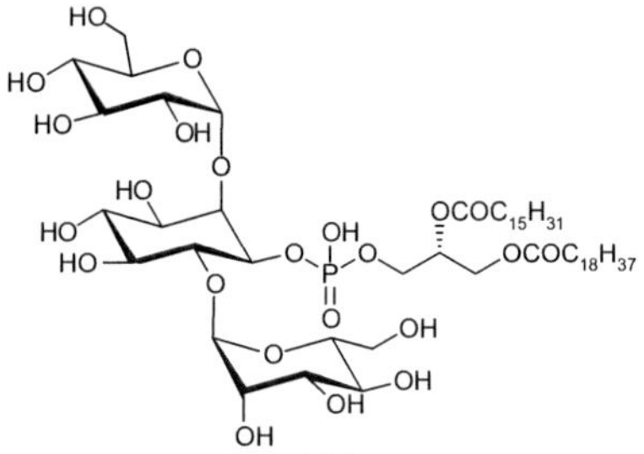

Kong and Sutton developed simple, robust and reproducible routes to 11-aminoundecyl 2-(trimethylammonio)ethyl phosphate (26). The key step of this synthesis was methyl hydrazine promoted deprotection of phosphatidylcholine (25) which gave the phosphate (26) in high yield and purity (Scheme 9).[14]

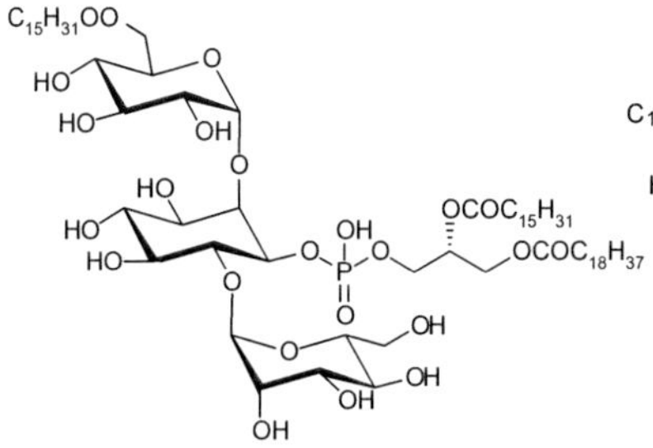

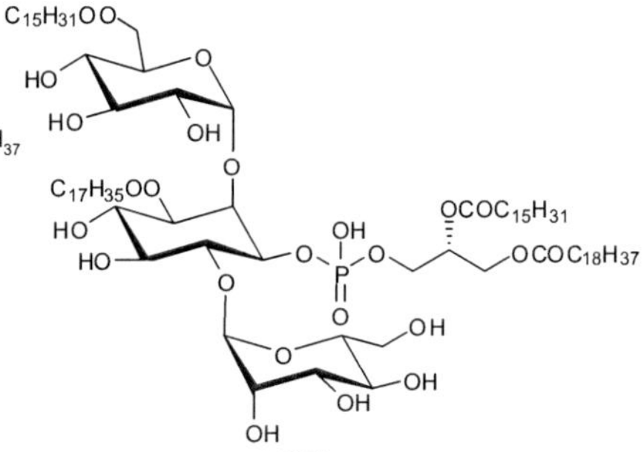

Koumbis et al. presented an efficient synthetic scheme for the preparation of a model lauryl derivative of 1-O-(ω-aminoacyl)-inositol bisphosphate (28) (12 steps from camphanyl ester (27)) in 44% overall yield (Scheme 10).[15]

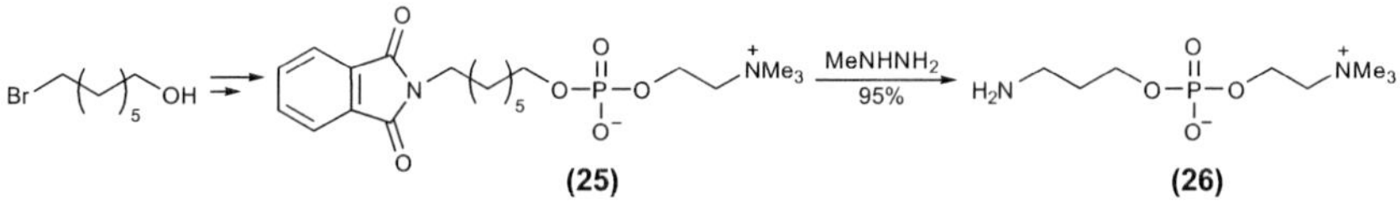

Scheme 9

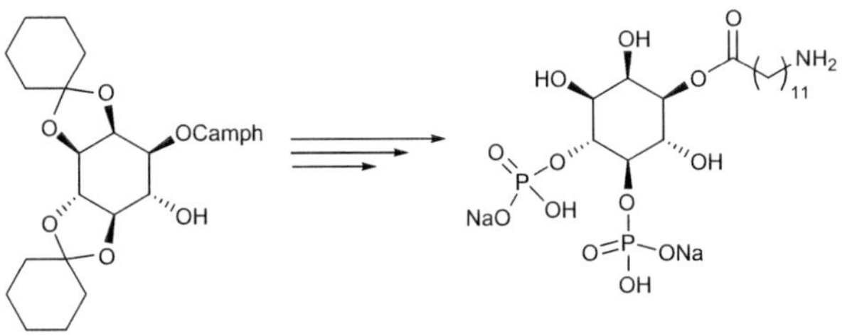

Scheme 10

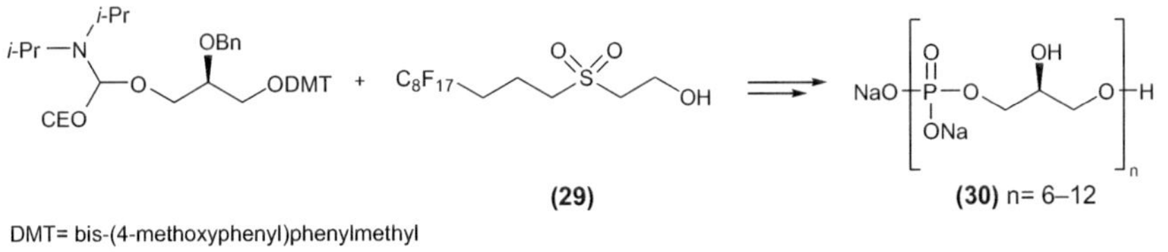

Scheme 11

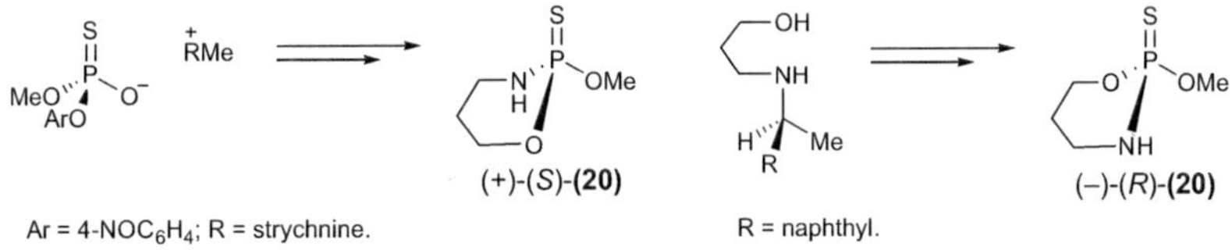

DMT= bis-(4-methoxyphenyl)phenylmethyl

Scheme 12

Ar = 4-NOC$_6$H$_4$; R = strychnine.   R = naphthyl.

An efficient fluorous synthetic strategy for the assembly of teichoic acid fragments (**30**) have been developed, based on the application of perfluorooctyl-propylsulfonylethanol (**29**) as a new fluorous phosphate protecting group. The strategy was especially useful for the assembly of multimilligram quantities of medium sized teichoic acid fragments, featuring 6–12 repeating units (Scheme 11).[16]

Mikołajczyk and co-workers developed a five-step sequence for the stereoselective synthesis of (+)-(S)- and (−)-(R)-tetramethylammonium 2-oxo-2-thio-1,3,2-oxazaphosphorinane (**31**) (Scheme 12).[17]

The D-camphor bisketal derivative (**32**) of natural mammalian lipid bis(monoacylglycero)phosphate (BMP) has been synthesised to identify the stereochemical configuration of its diglycerophosphate (DGP) backbone by ^{1}H NMR spectroscopy. As reference materials the sn-1,1' (**33**), sn-3,3' (**34**), and sn-3,1' (**35**) DGP analogues have also been prepered.[18]

A new class of zwitterionic phospholipids (iPCs) (**36**), with an inverted headgroup charge orientation relative to traditional phosphocholine (**37**) lipids have been described by Szoka and co-workers. Neutral iPC lipids with ethylated phosphate groups (**36a**) and anionic iPC lipids with non-ethylated phosphate groups (**36b**) have been obtained. The iPC lipid head

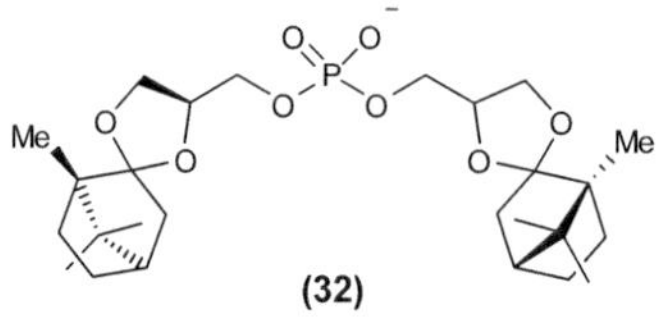

(32)

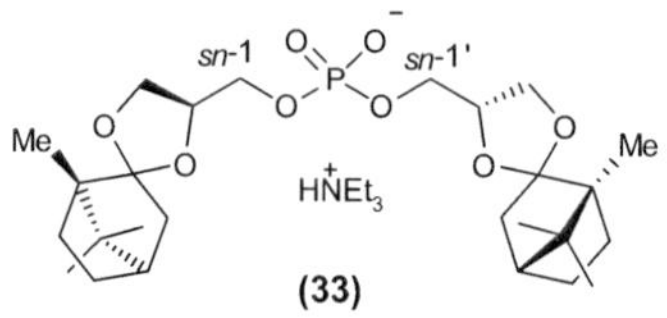

(33)

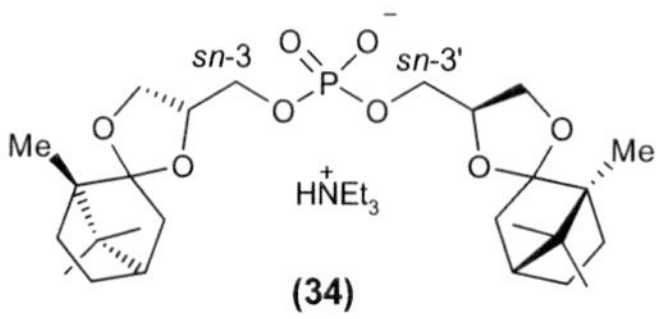

(34)

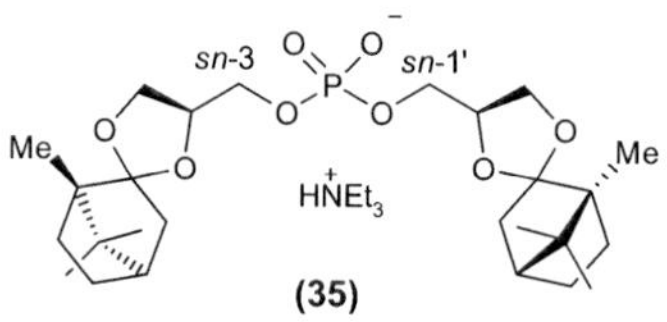

(35)

group had a quaternary amine adjacent to the bilayer interface and a phosphate that extended into the aqueous phase.[19]

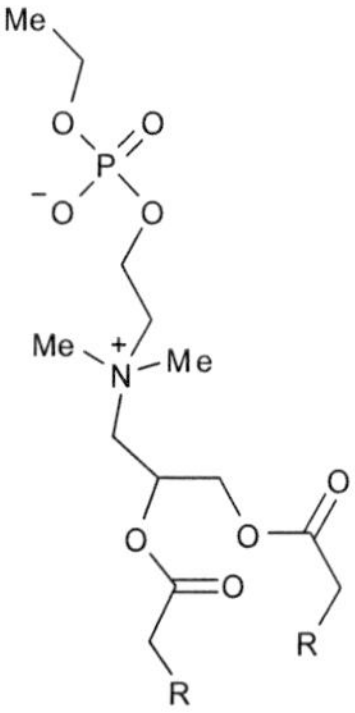

(36a)

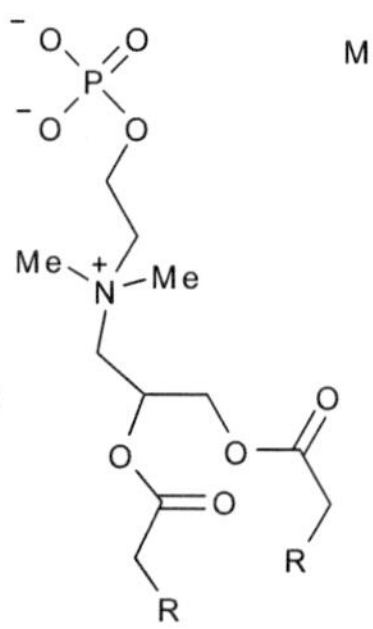

(36b)

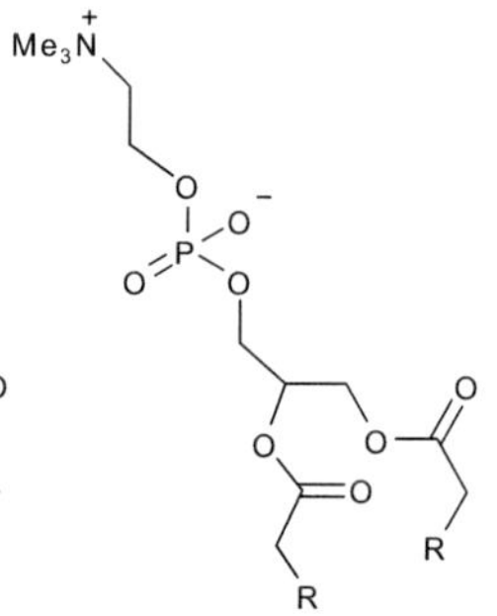

(37)

R = $C_{14}H_{29}$, $C_{16}H_{33}$, $C_{18}H_{37}$.

The successful synthesis and resolution of a chiral thiol attached to a cyclic phosphate unit (**38**), that contained a C-stereogenic center, allowed the preparation of chiral self-assembled monolayers on gold. The monolayers were used to promote the heterogeneous nucleation and growth of crystals from nonaqueous solutions of an organic molecule (the parent phencyphos) of similar structure to the compound present in the monolayer (Scheme 13).[20]

Belmant and co-workers developed two complementary synthetic pathways to phosphoantigen bromohydrin pyrophosphate (**41**). Thus, the first

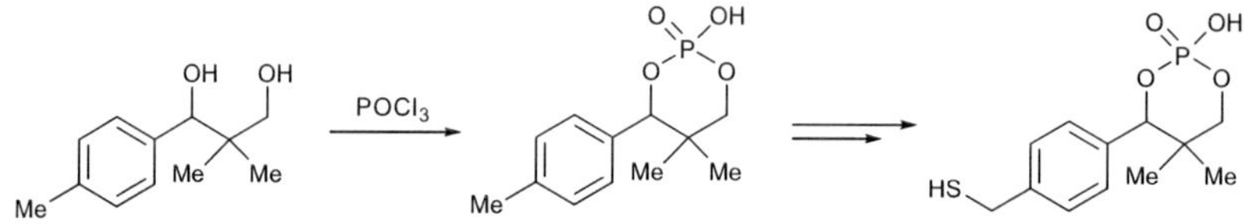

(38)

Scheme 13

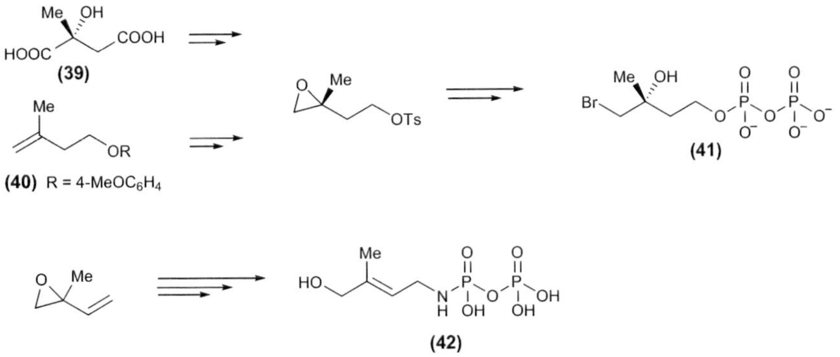

Scheme 14

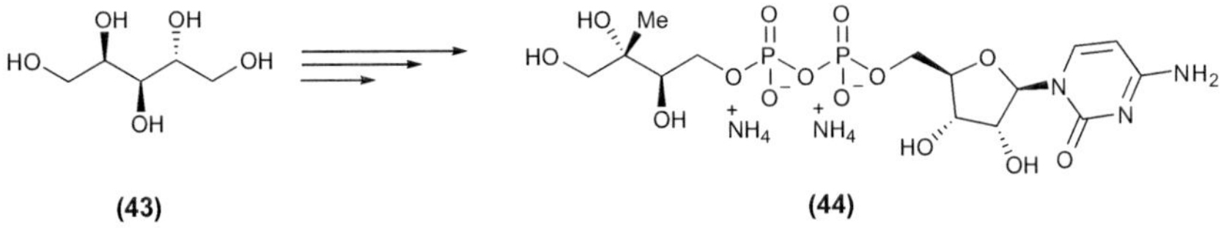

Scheme 15

route used transformation of a chiral compound (**39**) and the second one involved asymmetric synthesis starting from a prochiral building-block (**40**). The synthesis of a second-generation phosphoantigen (**42**), bearing a phosphoramidate moiety, has been also investigated (Scheme 14).[21]

A stereoselective synthesis of methylerythritol phosphate (**44**) from D-(+)-arabitol (**43**) in 7 steps and in 11.5% overall yield has been described by Lai and co-workers (Scheme 15).[22]

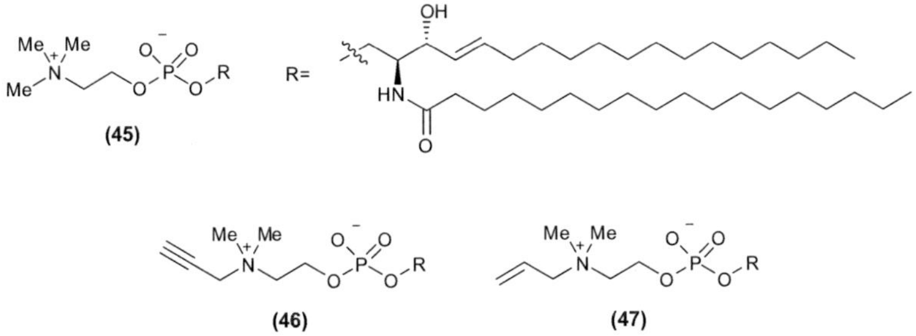

Murata and co-workers have studied the effects of chemical modification of sphingomyelin (**45**) ammonium group on formation of liquid-ordered phase. Two analogues of (**45**) with small propargyl and allyl groups, (**46**) and (**47**) respectively, on the quaternary nitrogen atom were synthesised and subjected to analysis using differential scanning calorimetry, fluorescent anisotropy, surface pressure, detergent solubilisation, and density measurements. The results demonstrated that the two analogues retained the membrane properties of (**45**), including formation of an ordered phase and the ability to interact with cholesterol. A dansyl-substituted analogue (**48**) was prepared for fluorescent measurements and showed less of a propensity to form microdomains. These findings implied the potential application of *N*-substituted (**45**) as a raft-specific molecular probe.[23]

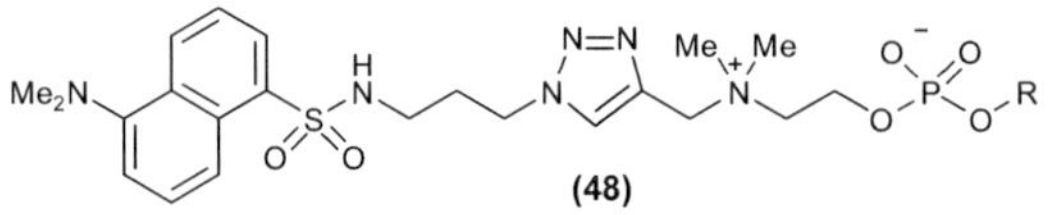

1.2 Reactions of phosphoric acids and their derivatives

A simple and convenient method for the formation of phosphorus–oxygen bonds to prepare various glycophospholipids (**50**) under microwave-assisted irradiation and in the presence of trichloroacetonitrile as an activating reagent, has been desribed. Thus, starting glycophosphates (**49**) were directly converted to the corresponding glycophospholipids (**50**) in good yields. Structurally diverse functional groups, including electron-withdrawing and electron-donating moieties, could be applied in these clean and economic microwave-assisted reactions (Scheme 16).[24]

A stereocontrolled isomerisation of α-halo-substituted propargylic phosphates (**51**) into highly functionalised 1,3-dienes, has been described by Gevorgyan *et al*. This methodology features a double 1,3-halogen and 1,3-phosphatyloxy migration relay. Depending on the choice of catalyst, synthesis of either (*Z*)- or (*E*)-1,3-dienes could be achieved selectively in high yields (up to 98%). Thus, (*E*)-dienes (**52**) have been obtained exclusively in the presence of a gold catalyst, whereas the use of a copper catalyst afforded predominantly (*Z*)-dienes (**53**) (Scheme 17).[25]

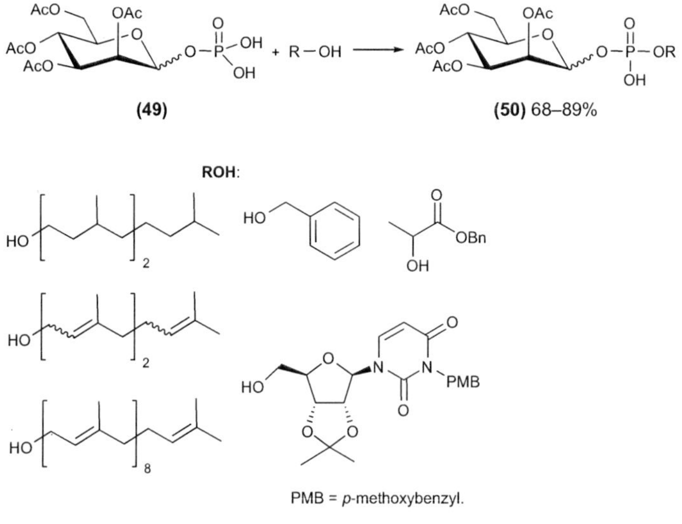

Scheme 16

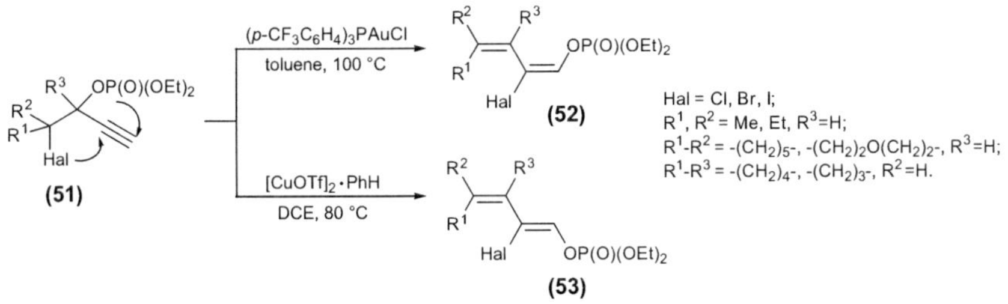

Scheme 17

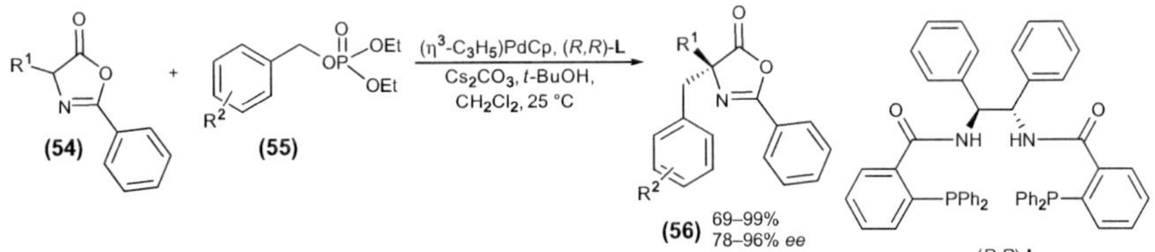

Hal = Cl, Br, I;
R^1, R^2 = Me, Et, R^3=H;
R^1-R^2 = -(CH$_2$)$_5$-, -(CH$_2$)$_2$O(CH$_2$)$_2$-, R^3=H;
R^1-R^3 = -(CH$_2$)$_4$-, -(CH$_2$)$_3$-, R^2=H.

Scheme 18

R^1 = Me, iBu, (CH$_2$)$_2$SMe, CH$_2$Ph;
R^2 = Ph; 4-OMe; 4-Ph; 4-OMOM; 3-Me,4-OMe; 4-F; 4-Me; 3,4-(Me)$_2$; 3-Cl, 4-OMe.

MOM = methoxymethyl
Cp = cyclopentadienyl

Scheme 19

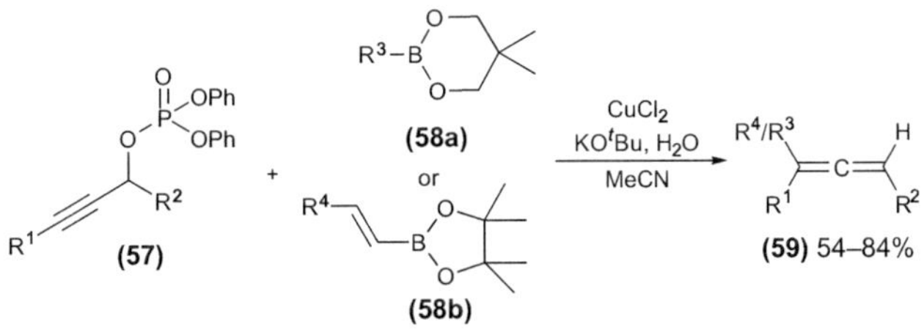

R^1 = Me, n-Bu, CH$_2$OMe, i-Bu;
R^2 = i-Pr, (CH$_2$)$_2$Ph, (CH$_2$)$_3$OTIPS;
R^3 = Ph, 2-MeC$_6$H$_4$, 4-MeOC$_6$H$_4$, 4-CF$_3$C$_6$H$_4$, 4-ClC$_6$H$_4$, 4-COMeC$_6$H$_4$, 4-CO$_2$EtC$_6$H$_4$,
R^4 = H, Ph, t-Bu, n-Hex, (CH$_2$)$_3$Cl, CH$_2$-1-indol.

TIPS = triisopropylsilyl

Trost and Czabaniuk have developed a method for palladium-catalysed asymetric benzylation of azalactones (**54**) using phosphates (**55**). A tetrasubstituted stereocenter in the product (**56**) has been generated in high enantiomeric excess (up to 96% *ee*) with a variety of electron-rich and electron-neutral elecrophiles (Scheme 18). The reported methodology represents a novel, asymmetric, carbon-carbon bond formation in an amino acid precursor.[26]

The Cu-catalysed coupling reaction between propargylic phosphates (**57**) and aryl- (**58a**) or alkenylboronates (**58b**), has been described as a versatile route to aryl- and alkenyl-conjugated allenes (**59**) (Scheme 19).

The reaction showed excellent functional group compatibility in both the boronates and the propargylic substrates. The reaction of an enantioenriched propargylic phosphate proceeded with excellent chirality transfer with 1,3-anti-stereochemistry to give axially chiral aryl- and alkenylallenes.[27]

The stereochemical courses of the copper-catalysed allyl-alkyl coupling between enantioenriched chiral allylic phosphates (**60**) and alkylboranes (**61**) have been described as switchable between 1,3-*anti* and 1,3-*syn* selectivities by the choice of solvents and achiral alkoxide bases with different steric demands (Scheme 20). Moreover, the synthetic protocols allowed the stereoselective conversion of silicon-substituted allylic phosphates into enantioenriched chiral allylsilanes with tertiary or quaternary carbon stereogenic centers. Thus, both enantiomers of the allylsilanes with high enantiomeric purities were readily available from one substrate enantiomer.[28]

Catalytic enantioselective allylic substitutions of aryl- and alkyl- substituted phosphates (**63**) with commercially available allenylboronic acid pinacol ester (**64**) resulted in the formation of tertiary or quaternary C–C bonds. Reactions were promoted by sulfonate-bearing chiral bidentate *N*-heterocyclic carbene (NHC) complexes of copper (**66**), which exhibited the unique ability to furnish chiral products arising from the S_N2' mode of addition. Allenyl-containing products (**65**) were generated in up to 95% yield, >98% S_N2' selectivity and 99:1 enantiomeric ratio (Scheme 21).[29]

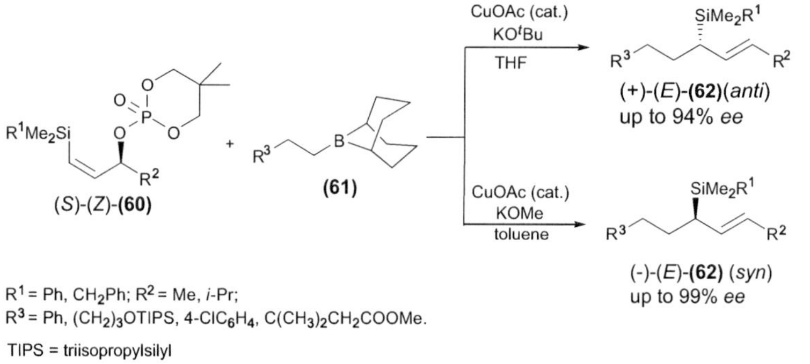

Scheme 20

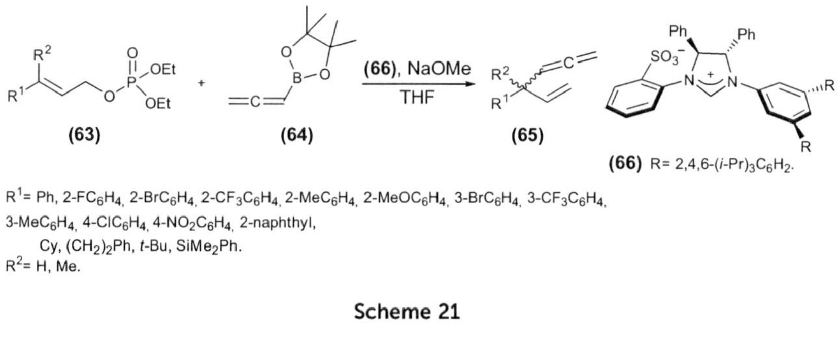

Scheme 21

The Pd-catalysed P–C cross-coupling reaction between α-amido enol phosphates (**67**) and secondary phosphine–borane complexes (**70**) or phosphine oxides (**71**) afforded under mild condition hindered tertiary α-enamido phosphine derivatives (**68**) or (**69**) with up to 99.4% *ee* and in up to 70% chemical yields (Scheme 22).[30]

A highly regioselective, Pd-catalysed allylic fluorination of phosphorothioate esters (**72**) has been reported by Lauer and Wu. The mildness of the reaction conditions have been demonstrated by its functional group tolerance (Scheme 23).[31]

Palladium-catalysed cyclisation of olefinic propargylic diethyl phosphate (**73**) afforded bicyclic trienes (**74**) (Scheme 24).[32]

A facile synthesis of chiral tetrahydrothiophenes (**76**) using phosphorothioic acids (**75**) has been reported by Wu and Robertson. Thus, treatment of the chiral acids (**75**) with NaH in DMF led to efficient ring closure and formation of the corresponding tetrahydrothiophenes (**76**) in moderate to high yields and high enantioselectivities (Scheme 25).[33]

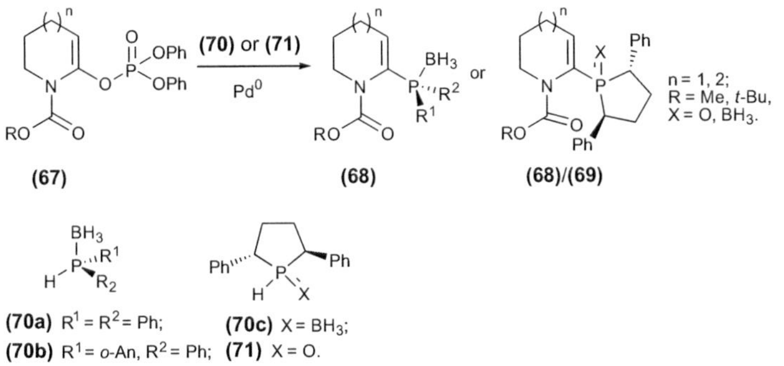

Scheme 22

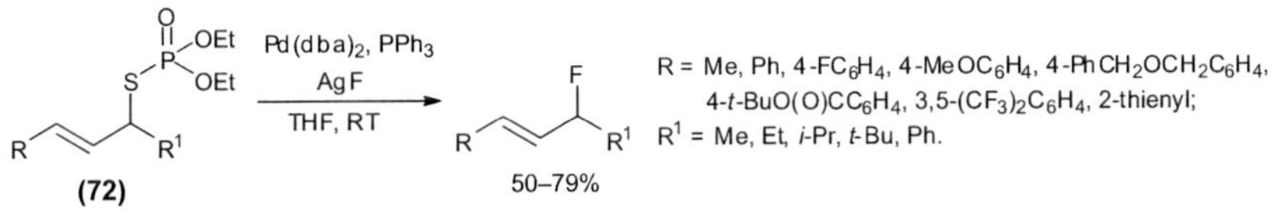

Scheme 23

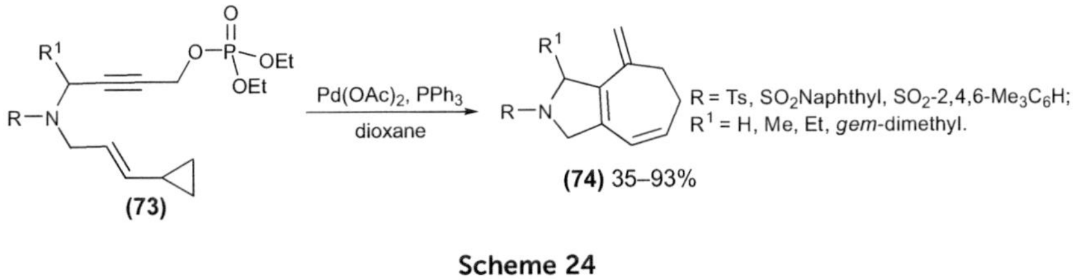

Scheme 24

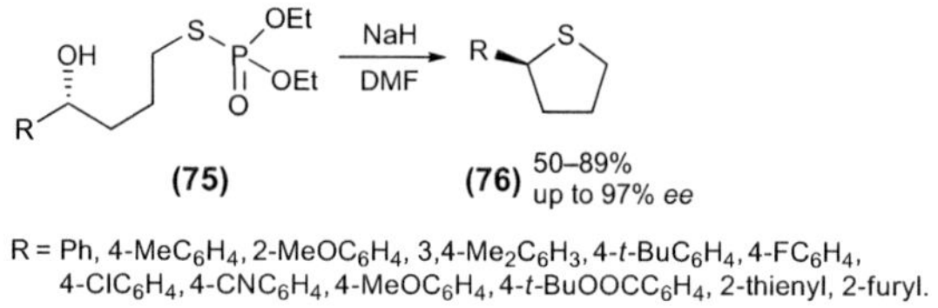

Scheme 25

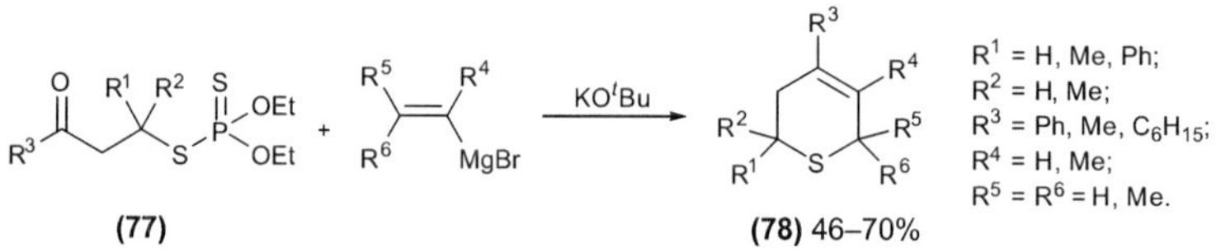

R = Ph, 4-MeC$_6$H$_4$, 2-MeOC$_6$H$_4$, 3,4-Me$_2$C$_6$H$_3$, 4-t-BuC$_6$H$_4$, 4-FC$_6$H$_4$,
4-ClC$_6$H$_4$, 4-CNC$_6$H$_4$, 4-MeOC$_6$H$_4$, 4-t-BuOOCC$_6$H$_4$, 2-thienyl, 2-furyl.

Scheme 26

Scheme 27

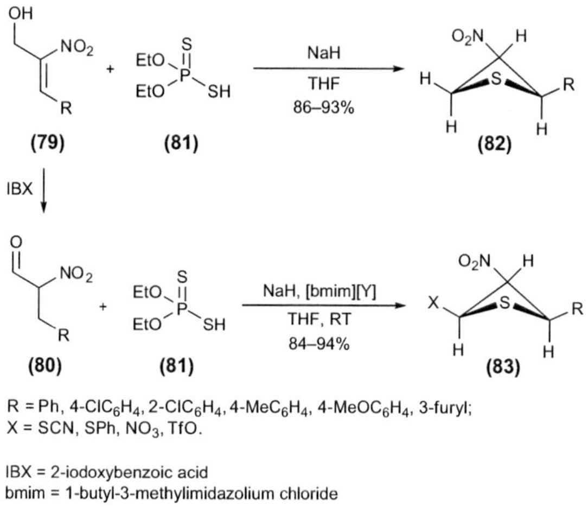

R = Ph, 4-ClC$_6$H$_4$, 2-ClC$_6$H$_4$, 4-MeC$_6$H$_4$, 4-MeOC$_6$H$_4$, 3-furyl;
X = SCN, SPh, NO$_3$, TfO.

IBX = 2-iodoxybenzoic acid
bmim = 1-butyl-3-methylimidazolium chloride

A new method for synthesising thiopyran products (**78**) from readily available keto dithiophosphates (**77**) has been also reported by Njardarson and co-workers (Scheme 26).[34]

The *thia*-Michael addition followed by an intramolecular cyclisation were used as key reactions in an efficient, one-pot, diastereoseletive synthesis of 3-nitrothietanes (**82**) and (**83**). Thus, Baylis-Hillman alcohols (**79**) and the coresponding aldehydes (**80**) were reacted with diethyl phosphorodithioate (**81**) itself or in a combination with a task-specific ionic liquid to afford the corresponding 2,3-di- and 2,3,4-trisubstituted thietanes, (**82**) and (**83**), with complete diastereoselectivity in favour of the *trans* isomers (Scheme 27).[35]

An efficient and highly enantioselective, organocatalytic *aza*-Michael addition of *N*-substituted diethyl thionophosphoramidates (**84**) to enones

(**85**) generated *aza*-Michael adducts which underwent intramolecular reductive cyclisation with (*R*)-alpine borane to afford 1,2,4-trisubstituted azetidines (**86**) in a one-pot procedure and with excellent stereocontrol (78–96% *ee*) (Scheme 28).[36]

The tandem *aza*-Claisen carbocyclisation for the synthesis of α,β-unsaturated cyclopentenimines (**88**) from *N*-phosphoryl-*N*-allyl ynamides (**87**), has been described by Hsung and co-workers (Scheme 29).[37]

The asymmetric oxidation of a variety of acyclic (**89**) and cyclic (**90**) substituted enol phosphates using commercially available Sharpless reagent (**93**), and a fructose derived chiral (**94**) as a catalyst, afforded the corresponding α-hydroxy ketones (**91**) and (**92**) in high enantioselectivity and good yields (Scheme 30). The influence of steric and electronic

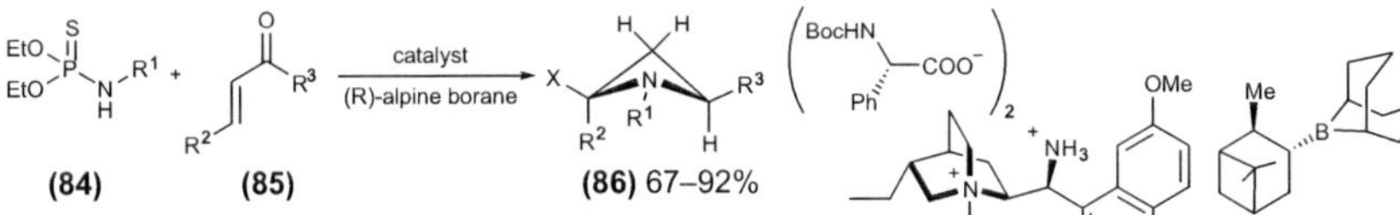

R^1 = Ts, Ph, 4-MeOC$_6$H$_4$, CH$_2$Ph, 2-MeC$_6$H$_4$, 2-BrC$_6$H$_4$;
R^2 = Ph, CH$_2$Ph,4-ClC$_6$H$_4$, Me, Et;
R^3 = Ph, CH$_2$Ph, 4-ClC$_6$H$_4$, Me, Et, *n*-Pr.

Scheme 28

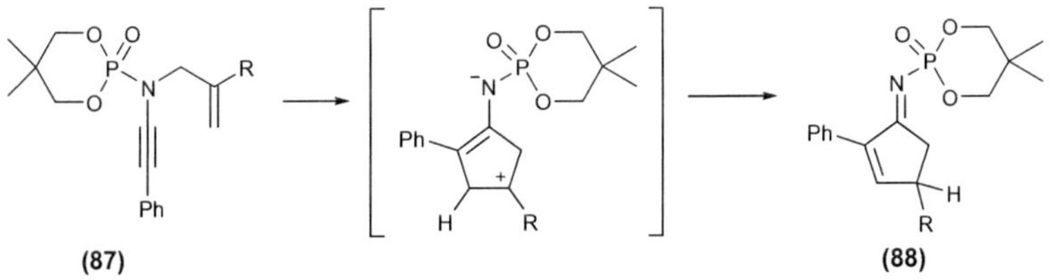

Scheme 29

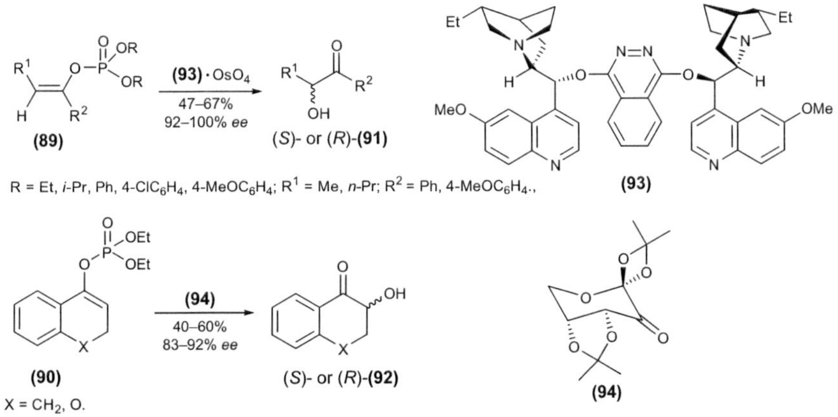

Scheme 30

factors of the substrates on a facial stereoselectivity in the reported oxidations was also studied.[38]

5-*N*-Acetyl-5-*N*,4-*O*-oxazolidinone protected α- and β-sialyl phosphates (**95**) reacted with allyltributylstannane and a variety of trimethylsilyl enol ethers to give α-sialyl *C*-glycosides (**96**) in high yield (up to 91%) and excellent selectivity (up to 100:0 α:β) (Scheme 31).[39]

A phosphate tether-mediated three-step, one-pot reaction protocol involving ring-closing metathesis (RCM), cross-metathesis (CM), and chemoselective hydrogenation has been reported by Hanson and co-workers (Scheme 32). This procedure enabled synthesis of advanced substrates in a streamlined manner.[40]

A novel, photolabile protecting group, thiochromone *S,S*-dioxide (**97**), containing the diazomethyl group has been synthesised and successfully applied for protection of phosphate derivatives (**98**). Deprotection of the phosphates (**99**) proceeded smoothly under photoirradiation using an ultrahigh-pressure mercury lamp to recover the corresponding phosphates (**98**) quantitatively (Scheme 33).[41]

Dephosphorylation reactions of mono-, di-, and triesters of 2,4-dinitrophenyl phosphate (**100**) with deferoxamine (DFO) and

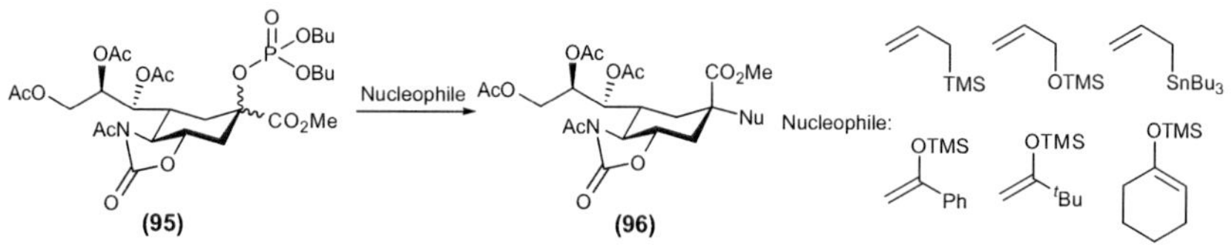

Scheme 31

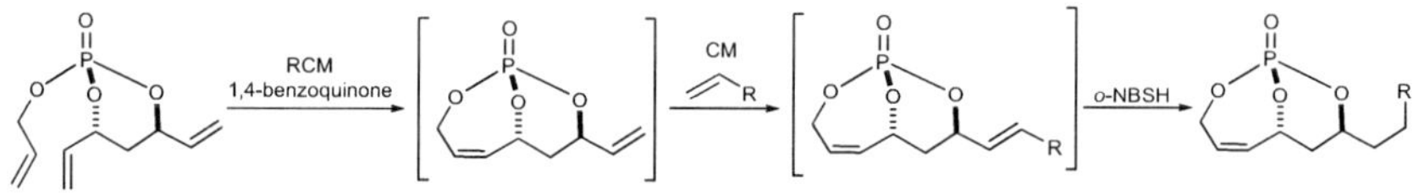

Scheme 32

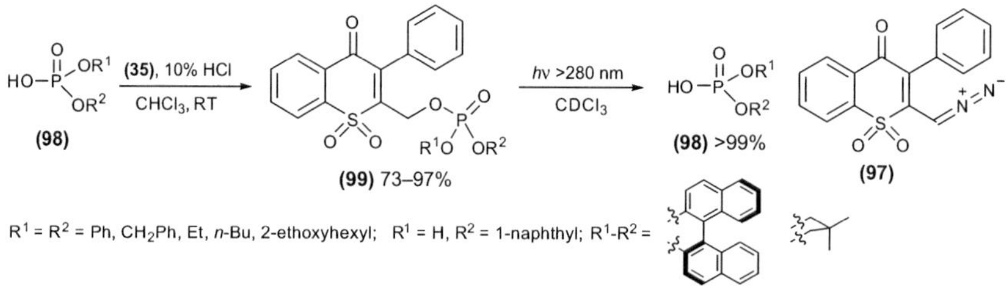

Scheme 33

benzohydroxamic acid (BHA) have been investigated by Nome and co-workers (Scheme 34). BHA showed extraordinary reactivity toward the triester diethyl 2,4-dinitrophenyl phosphate (DEDNPP) and the diester ethyl 2,4-dinitrophenyl phosphate (EDNPP) reacted very slowly with the monoester 2,4-dinitrophenyl phosphate (DNPP). DFO, showed correspondingly high nucleophilic activity toward both triester DEDNPP and diester EDNPP, which suggested a potential use for DFO in cases of acute poisoning with phosphorus pesticides.[42]

Feng and co-workers have used the structurally homologous dinuclear Zn(II) complexes (**101**), to examine the effect of introducing ligand-based hydrogen bond donor groups (NH$_2$ for (**101b**) and NHCOCH$_3$ for (**101c**)) to the selectivity and binding affinity of synthetic receptors for the pyrophosphate (**102**) (Scheme 35). Authors have demonstrated that introducing hydrogen bond donors to a receptor was an effective approach to improve both its selectivity and binding affinity for pyrophosphate in water.[43]

1.3 Phosphoric acids as catalysts

Hong and co-workers have described a kinetic resolution of α-allenic alcohols (**103**) through a chiral silver phosphate-catalysed cycloisomerisation leading to both enantiomerically enriched α-allenic alcohols (**103**) and 2,5-dihydrofurans (**104**) (Scheme 36). The reaction has been realised with high enantiostereoselectivity and tolerance of a variety of functional groups.[44]

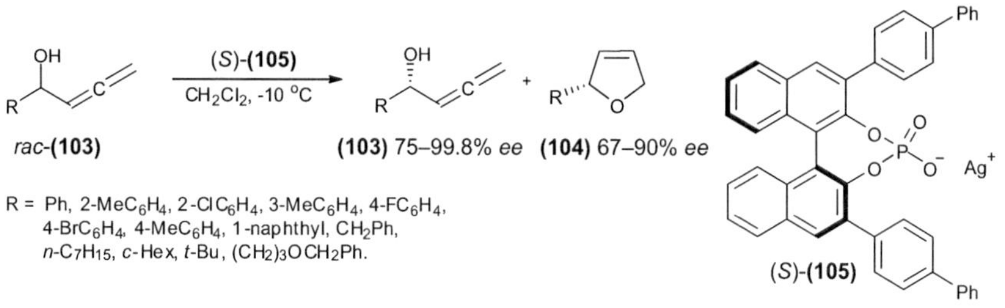
OH
(S)-(105)
CH2Cl2, -10 °C
rac-(103)
(103) 75–99.8% ee
(104) 67–90% ee
R = Ph, 2-MeC6H4, 2-ClC6H4, 3-MeC6H4, 4-FC6H4, 4-BrC6H4, 4-MeC6H4, 1-naphthyl, CH2Ph, n-C7H15, c-Hex, t-Bu, (CH2)3OCH2Ph.
Ph
Ag+
(S)-(105)

Scheme 36

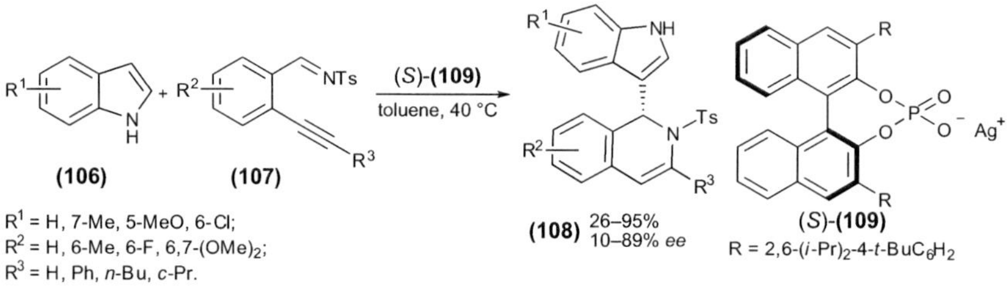
NTs
(S)-(109)
toluene, 40 °C
(106)
(107)
R1 = H, 7-Me, 5-MeO, 6-Cl;
R2 = H, 6-Me, 6-F, 6,7-(OMe)2;
R3 = H, Ph, n-Bu, c-Pr.
(108) 26–95% 10–89% ee
Ag+
(S)-(109)
R = 2,6-(i-Pr)2-4-t-BuC6H2

Scheme 37

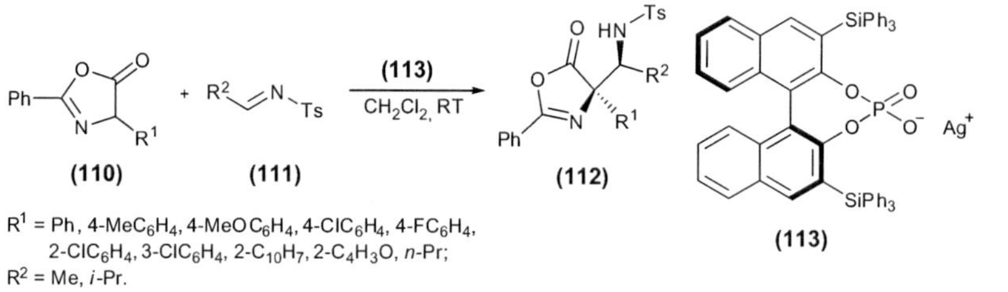
(113)
CH2Cl2, RT
(110)
(111)
(112)
R1 = Ph, 4-MeC6H4, 4-MeOC6H4, 4-ClC6H4, 4-FC6H4, 2-ClC6H4, 3-ClC6H4, 2-C10H7, 2-C4H3O, n-Pr;
R2 = Me, i-Pr.
SiPh3
Ag+
(113)

Scheme 38

The reaction of indoles (**106**) and *ortho*-alkynylaryl aldimines (**107**) catalysed by a silver binol-derived phosphate (**109**) has been realised to afford a series of enantioenriched 1,2-dihydroisoquinolines (**108**) (Scheme 37).[45]

A chiral phosphate anion combined with a silver cation has been demonstrated as a powerful ion pair catalyst (**113**) for the *aza*-Mannich addition of oxazolones (**110**) to *N*-tosyl aldimines (**111**). A series of quaternary α,β-diamino acid derivatives (**112**) has been obtained in high yields (up to 95%), and with excellent diastereo- (up to 25 : 1 *dr*) and enantioselectivities (up to 99% *ee*) (Scheme 38).[46]

The hybrid of chiral *N*-triflyl phosphoramide catalyst (**116a**) and phosphine gold complex (**116b**) has been applied in the first cascade, intramolecular hydrosiloxylation/asymmetric Diels–Alder reaction. A variety of polycyclic compounds (**115a**) and (**115b**) bearing multistereogenic

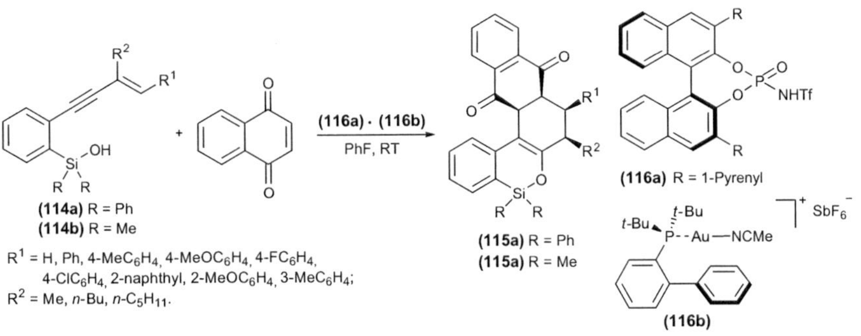

Scheme 39

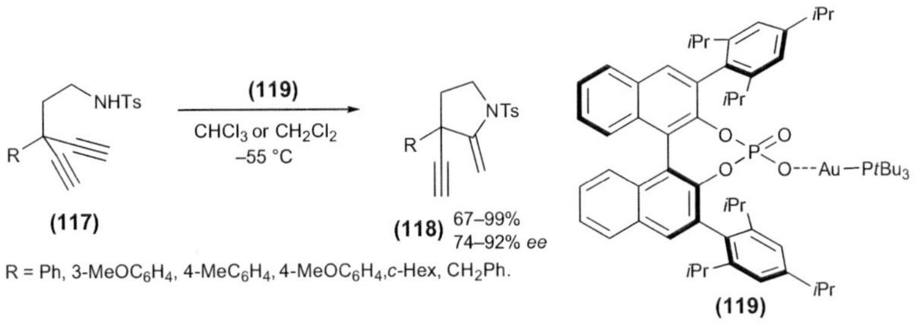

Scheme 40

centers has been obtained by this way in high yields (70–99%) and excellent enantioselectivities (87–96% ee) from phenyl (**114a**) and methyl (**114b**) substituted dialkylsilanols (Scheme 39).[47]

The efficient, enantioselective cycloisomerisation of 1,4-diynamides (**117**) to pyrrolidines (**118**), as the first example of a highly stereoselective desymmetrisation of terminal alkynes by a gold catalyst with the chiral binol phosphate counteranion (**119**), has been described by Czekelius and co-workers. The best yields have been obtained in chlorinated solvents at low temperatures (Scheme 40).[48] On the other hand, a complex of the chiral phosphate (**119**) with an achiral, gold ligand has been used as highly efficient catalyst for the asymmetric, transfer hydrogenation of quinolines (**120**) (Scheme 41).[49]

Beller and co-workers have demonstrated the enantioselective, reductive hydroamination of alkynes with primary amines in the presence of molecular hydrogen as the reducing agent. The key to success was use of a three-component catalytic system consisting of an active gold (I) complex (**122**), the Knölke's iron complex (**123**) and a chiral Brønsted acid (**119**). Excellent enantioselectivities and high yields were observed for a variety of alkynes and amines (Scheme 42).[50]

Among various chiral phosphoric acids, the derivative with bulky 2,4,6-triisopropyl phenyl groups in the 3,3′-positions of BINOL (**119**) has been also successfully used in the synthesis of enantioenriched pyrroloindoline derivatives (**126**). Thus, the reaction of readily available tryptamines

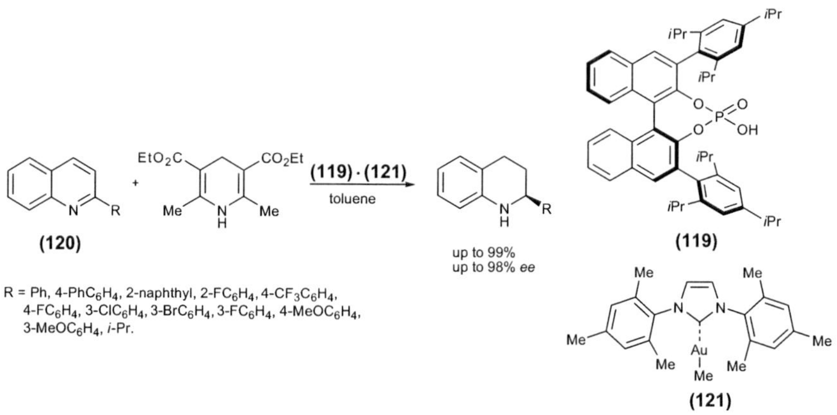

EtO2C
CO2Et
(119)·(121)
toluene
(120)
up to 99%
up to 98% ee
(119)
(121)
R = Ph, 4-PhC6H4, 2-naphthyl, 2-FC6H4, 4-CF3C6H4,
4-FC6H4, 3-ClC6H4, 3-BrC6H4, 3-FC6H4, 4-MeOC6H4,
3-MeOC6H4, i-Pr.

Scheme 41

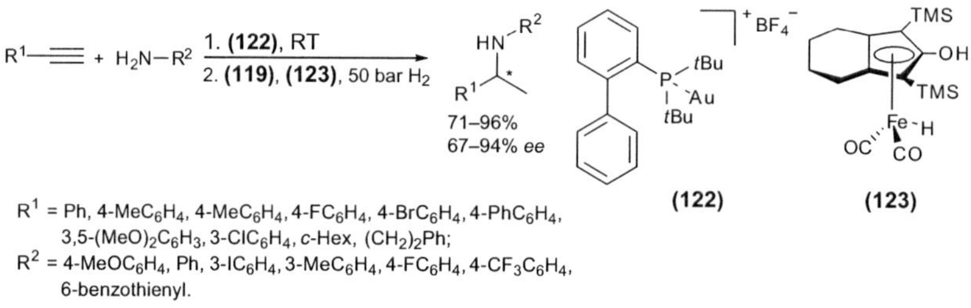

1. (122), RT
2. (119), (123), 50 bar H2
71–96%
67–94% ee
(122)
(123)
R1 = Ph, 4-MeC6H4, 4-MeC6H4, 4-FC6H4, 4-BrC6H4, 4-PhC6H4,
3,5-(MeO)2C6H3, 3-ClC6H4, c-Hex, (CH2)2Ph;
R2 = 4-MeOC6H4, Ph, 3-IC6H4, 3-MeC6H4, 4-FC6H4, 4-CF3C6H4,
6-benzothienyl.

Scheme 42

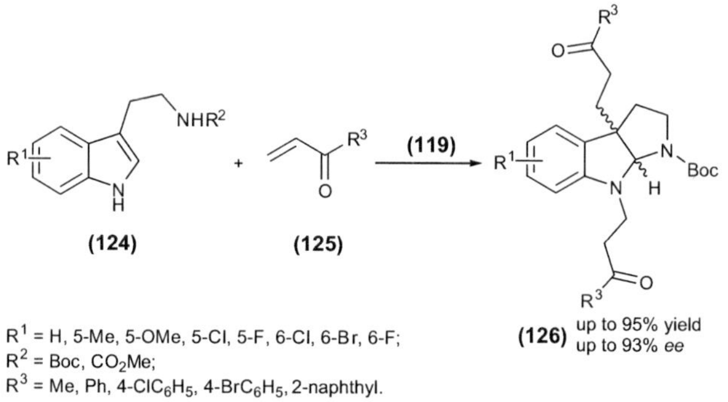

(119)
(124)
(125)
(126)
up to 95% yield
up to 93% ee
R1 = H, 5-Me, 5-OMe, 5-Cl, 5-F, 6-Cl, 6-Br, 6-F;
R2 = Boc, CO2Me;
R3 = Me, Ph, 4-ClC6H5, 4-BrC6H5, 2-naphthyl.

Scheme 43

(**124**) and enones (**125**) afforded corresponding pyrroloindolines in good yields and enantioselectivity (Scheme 43).[51,52]

The intermolecular α-amidoalkylation of the hydroxylactam (**127**) in the presence of Brønsted acids (**119**) resulted in the facile enantioselective synthesis of isoindoloisoquinolines (**128**) (Scheme 44).[53]

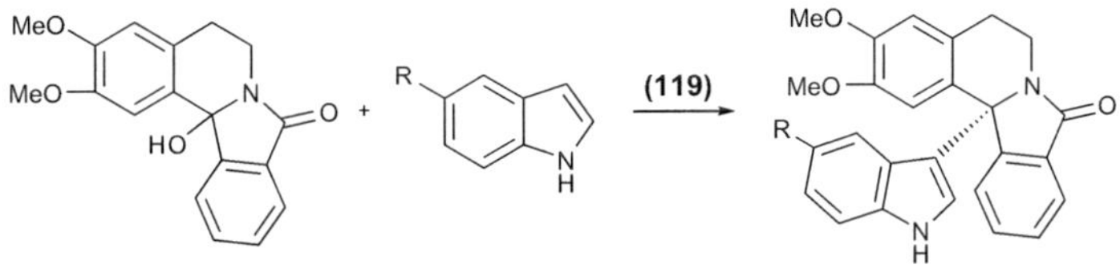

Scheme 44

(127) R = NO₂, Br, MeO.

(128) up to 95% ee

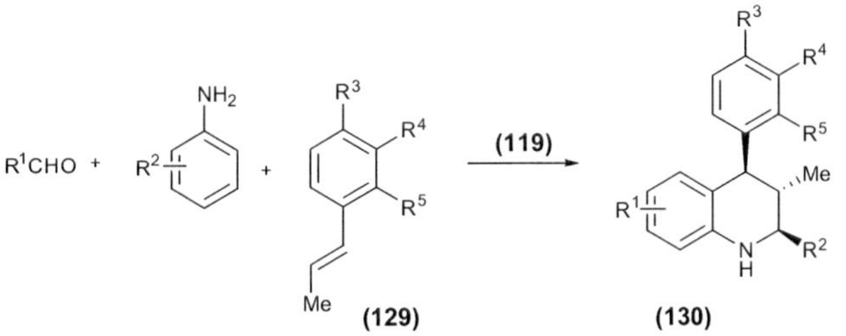

Scheme 45

R¹ = Ph, 4-BrC₆H₄, 4-FC₆H₄, 4-CNC₆H₄, 4-NO₂C₆H₄,
4-PhC₆H₄, 4-i-PrC₆H₄, 3-FC₆H₄, 3-ClC₆H₄,
3-NO₂C₆H₄, 2-FC₆H₄, 2-BrC₆H₄, 1-naphthyl, c-Hex;
R² = H, 4-OMe, 4-Me, 4-Cl, 4-Br, 4-CF₃, 4-NO₂;
R³ = H, OH, OMe;
R⁴ = H, OMe;
R⁵ = H, OH.

Highly enantioselective, three-component, inverse electron-demand *aza*-Diels-Alder reaction of aldehydes, anilines, and isoeugenol derivatives (**129**) catalysed by a phosphoric acid (**119**) has been reported by Masson and co-workers. A wide variety of 2,3,4-trisubstituted tetrahydroquinolines (**130**) containing an aryl group at the 4-position were obtained in a one-pot process with good to high yields and excellent enantioselectivities (up to >99% ee) (Scheme 45).[54]

The BINOL derivative (**119**) has been also successfully used as a chiral catalyst in synthesis of α-brominated enecarbamates (**131**),[55] indene derivatives possessing a stereogenic spirocyclic carbon center (**132**),[56] spiroketals (**133**),[57] hexahydropyrrolo[3,2-c]quinolines (**134**),[58] pyrrolo[2,1-a]isoquinolines (**135**),[59] and adducts of the Robinson-type reaction (**136**).[60]

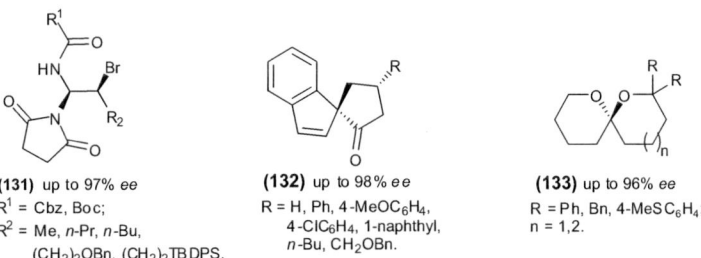

(131) up to 97% ee
R¹ = Cbz, Boc;
R² = Me, n-Pr, n-Bu,
(CH₂)₂OBn, (CH₂)₂TBDPS.

(132) up to 98% ee
R = H, Ph, 4-MeOC₆H₄,
4-ClC₆H₄, 1-naphthyl,
n-Bu, CH₂OBn.

(133) up to 96% ee
R = Ph, Bn, 4-MeSC₆H₄;
n = 1,2.

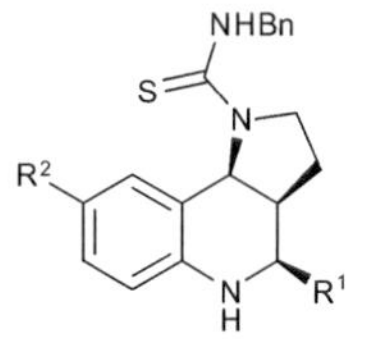

(134) up to 98% ee
R¹ = Et, i-Pr, n-Hex, 4-BrC$_6$H$_4$, 4-CNC$_6$H$_4$, 3-MeOC$_6$H$_4$, 2-NO$_2$C$_6$H$_4$, 2-furyl;
R² = H, n-Bu, CO$_2$Et, MeO, Br, CF$_3$.

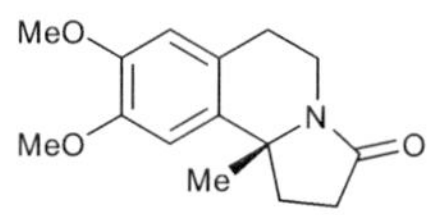

(135) 76% ee

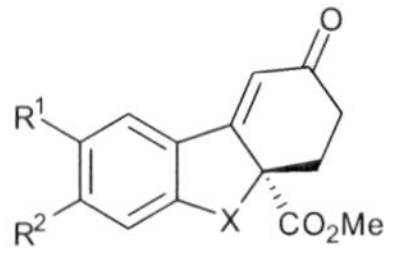

(136) up to 99% ee
X = CH$_2$, O;
R¹ = H, Cl, Br, Me; R² = H, Cl, Br.

Moreover, Mandai et al. have presented an efficient method for the kinetic resolution of secondary alcohols in the presence of the chiral catalyst (**119**), DABCO and acetyl chloride (Scheme 46).[61]

The chiral phosphoric acid (**119**) has also catalysed the deuterium transfer from benzothiazoline (**138**) as a deuterium donor to ketimines (**137**) in high yields (up to 99%) and with excellent enantioselectivities (up to 98% ee) (Scheme 47).[62]

The use of a C–8 BINOL derivative (**142**) as a chiral anionic phase-transfer catalyst in a nonpolar solvent allowed the enantioselective fluorination of enamides (**139**) using Selectfluor (**140**) as the fluorinating reagent (Scheme 48). The authors demonstrated that a wide range of stable and synthetically versatile α-(fluoro)benzoylimines (**141**) could be readily accessed with high enantioselectivity (82–99% ee).[63]

A three-component Povarov reaction involving substituted benzaldehydes and anilines as well as 2-hydroxystyrenes in the presence of the chiral phosphoric acid (**144**) provided structurally diverse cis-disubstituted tetrahydroquinolines (**143**) in high stereoselectivities of up to >99:1 dr and 97% ee (Scheme 49).[64]

The first successful enantioselective intermolecular bromoesterification of carboxylic acids has been realised by the use a 9-phenanthryl substituted chiral phosphoric acid (**145**) as a catalyst (Scheme 50).[65]

Gong and co-workers have developed a relay reaction consisting of a catalytic Friedländer condensation and a transfer hydrogenation by using

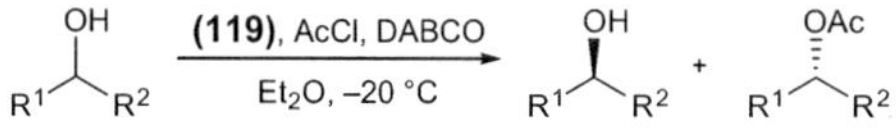

R¹ = Ph, 1-naphthyl, 2-naphthyl, Ph—≡—;
R² = Me, ≡
DABCO = 1,4-diazabicyclo[2.2.2]octane

Scheme 46

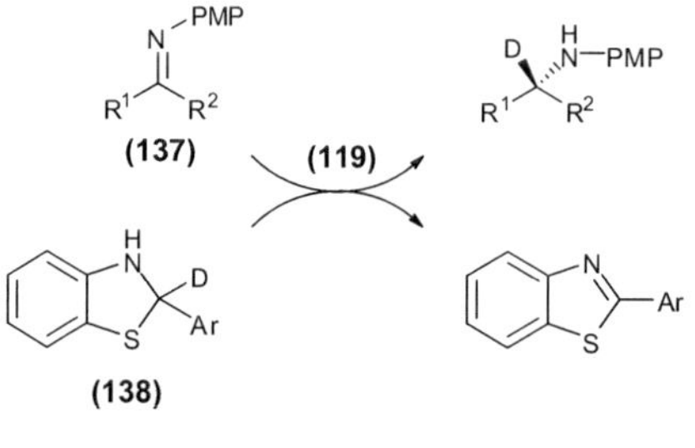

R¹ = Ph, 4-MeC$_6$H$_4$, 4-MeOC$_6$H$_4$, 4-NO$_2$C$_6$H$_4$, 4-FC$_6$H$_4$, 2-FC$_6$H$_4$, 3-FC$_6$H$_4$, 2-naphthyl, 2-thienyl;
R² = Me, CO$_2$Me;
Ar = Ph, 2-naphthyl.
PMP = p-methoxyphenyl.

Scheme 47

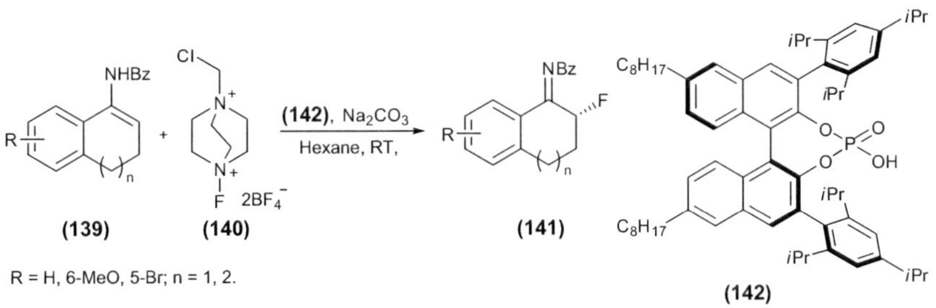

Scheme 48

R = H, 6-MeO, 5-Br; n = 1, 2.

Scheme 49

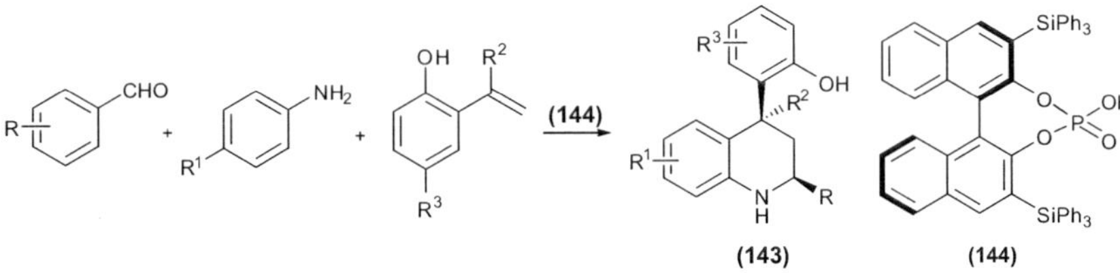

R = H, 4-NO$_2$, 3-NO$_2$, 2-NO$_2$, 4-CN, 4-Cl, 4-Br, 4-CF$_3$, 3,4-Cl$_2$, 3,4-F$_2$, 3-Me, 3-MeO; R^1 = MeO, EtO, PhO, Me, F; R^2 = H, Me, Et; R^3 = H, Me, MeO.

Scheme 50

up to 70% ee

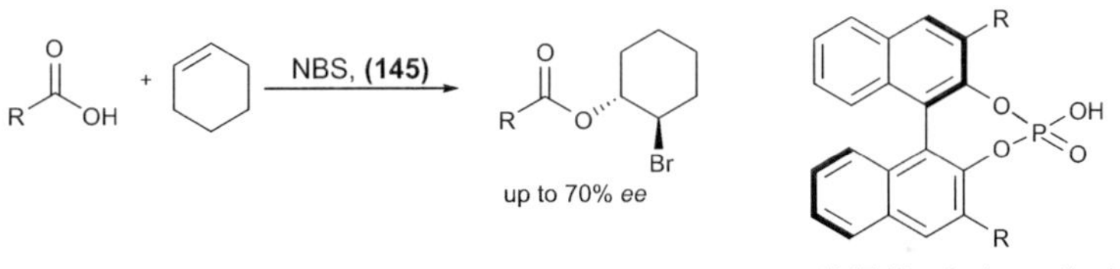

(145) R = 9-phenanthryl

R = Ph, 4-MeC$_6$H$_4$, 4-NO$_2$C$_6$H$_4$, 4-CNC$_6$H$_4$, 2-HOC$_6$H$_4$, 2-furyl, 2-thienyl, 1-naphthyl, CH$_2$Ph, CH$_2$-3-MeOC$_6$H$_4$, CH$_2$-2-MeOC$_6$H$_4$, CH$_2$-4-MeOC$_6$H$_4$, CH$_2$-2-NO$_2$C$_6$H$_4$, CH$_2$-4-FC$_6$H$_4$.

a combination of an achiral Lewis acid and a chiral Brønsted acid (**145**). This protocol provided access to highly substituted tetrahydroquinoline derivatives (**146**) with concomitant generation of multiple stereogenic centers in excellent levels of stereochemical control (Scheme 51). In this cascade reaction, the Friedländer condensation was catalysed either by Lewis acid or the chiral phosphoric acid (**145**) while the asymmetric transfer hydrogenation was promoted solely by the chiral Brønsted acid (**145**).[66]

The first catalytic ring-expansion reaction of vinyl oxetane (**147**) in the presence of chiral Brønsted sulfonamide (**149**) provided 3,6-dihydro-2*H*-pyran (**148**) with 90% ee (Scheme 52).[67]

On the other hand, a catalyst possessing double axial chirality (**151**), has been used in the enantioselective tandem oxyfluorination of

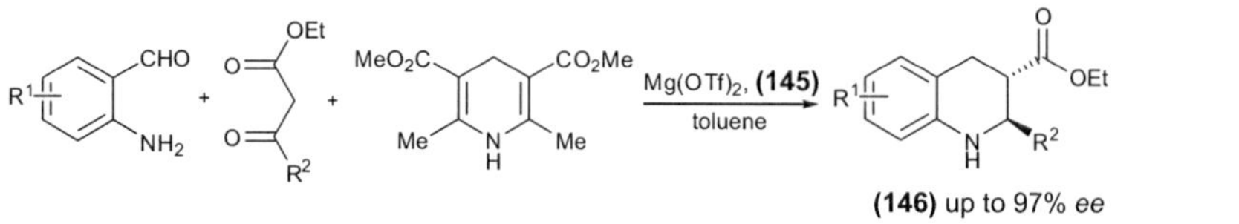

Scheme 51

R¹ = H, 5-AcO, 5-MeO, 5-F, 5-Cl, 5-Br, 4-Cl; R² = Me, n-Pr, Ph, 4-MeC$_6$H$_4$, 4-FC$_6$H$_4$, 4-ClC$_6$H$_4$, 4-NO$_2$C$_6$H$_4$, 2-naphthyl.

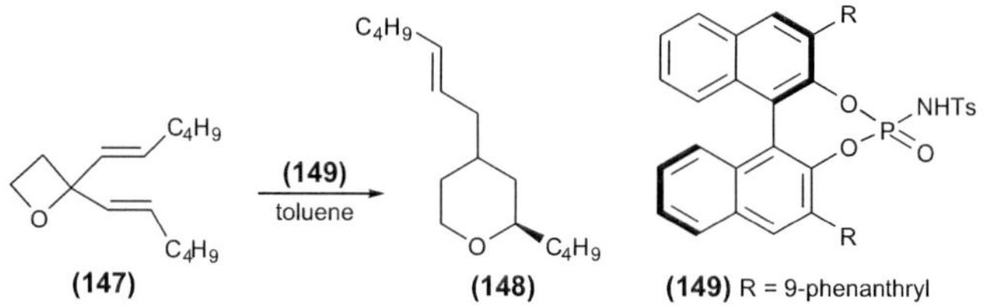

Scheme 52

Scheme 53

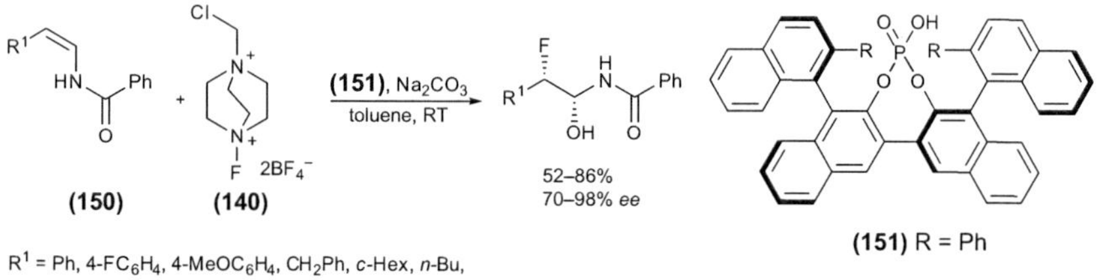

R¹ = Ph, 4-FC$_6$H$_4$, 4-MeOC$_6$H$_4$, CH$_2$Ph, c-Hex, n-Bu, Cl(CH$_2$)$_3$, CN(CH$_2$)$_3$, TBSO(CH$_2$)$_3$.

TBS = tert-butyldimethylsilyl.

enamides (**150**), taking advantage of the ability of a chiral phosphoric acid catalyst to control both fluorination, through chiral anion phase-transfer strategy, as well as addition to the resulting imine, most likely through a hydrogen-bonding mechanism (Scheme 53).[68]

The confined chiral Brønsted acid (**152**) has been shown to catalyse asymmetric oxidations of sulphides to sulphoxides with hydrogen peroxide (Scheme 54). The generality and high enantioselectivity of the described method could be compared even to the best metal-based systems and suggested utility in other asymmetric oxidations.[69]

Peng and co-workers have described the first example of the phosphoric acid (**155**) catalysed alkylation of cyclohexanone with tertiary alcohols (**153**) to afford derivatives (**154**) in high yields (up to 98%) and high enantioselectivities (up to 97% ee) (Scheme 55).[70]

The SPINOL-based phosphoric acid (**157**) has been found to be a general, highly enantioselective catalyst for asymmetric allylboration of aldehydes with pinacol allylboronates (**156**). Excellent enantioselectivities have been obtained for different types of aldehydes including aromatic

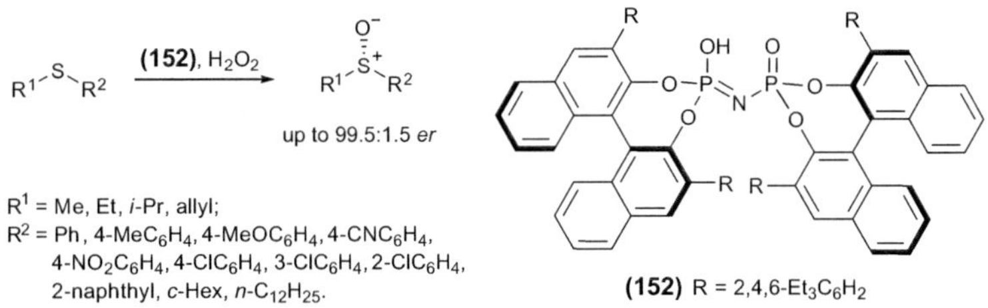

Scheme 54

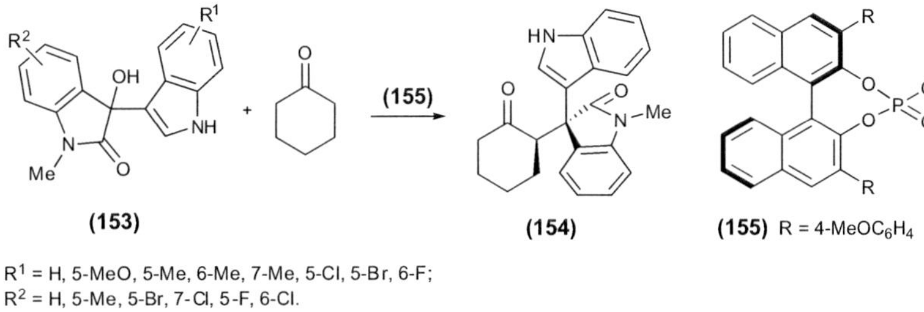

R^1 = Me, Et, *i*-Pr, allyl;
R^2 = Ph, 4-MeC$_6$H$_4$, 4-MeOC$_6$H$_4$, 4-CNC$_6$H$_4$, 4-NO$_2$C$_6$H$_4$, 4-ClC$_6$H$_4$, 3-ClC$_6$H$_4$, 2-ClC$_6$H$_4$, 2-naphthyl, *c*-Hex, *n*-C$_{12}$H$_{25}$.

(152) R = 2,4,6-Et$_3$C$_6$H$_2$

up to 99.5:1.5 *er*

Scheme 55

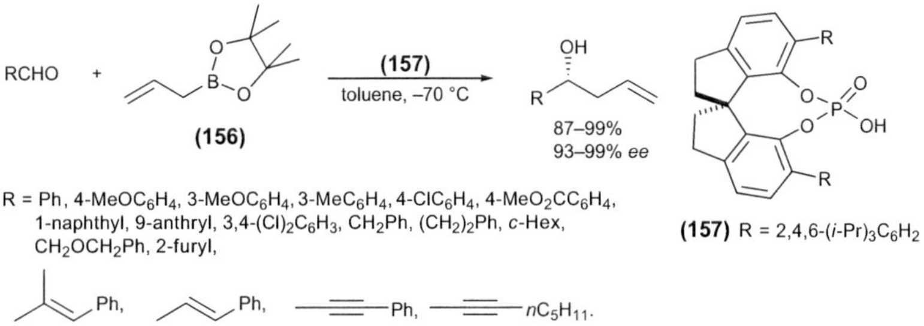

R^1 = H, 5-MeO, 5-Me, 6-Me, 7-Me, 5-Cl, 5-Br, 6-F;
R^2 = H, 5-Me, 5-Br, 7-Cl, 5-F, 6-Cl.

(155) R = 4-MeOC$_6$H$_4$

Scheme 56

R = Ph, 4-MeOC$_6$H$_4$, 3-MeOC$_6$H$_4$, 3-MeC$_6$H$_4$, 4-ClC$_6$H$_4$, 4-MeO$_2$CC$_6$H$_4$, 1-naphthyl, 9-anthryl, 3,4-(Cl)$_2$C$_6$H$_3$, CH$_2$Ph, (CH$_2$)$_2$Ph, *c*-Hex, CH$_2$OCH$_2$Ph, 2-furyl,

⋋Ph, ⋋Ph, ≡—Ph, ≡—*n*C$_5$H$_{11}$.

87–99%
93–99% ee

(157) R = 2,4,6-(*i*-Pr)$_3$C$_6$H$_2$

aldehydes, α,β-unsaturated aldehydes, propargylic aldehydes, and aliphatic aldehydes (Scheme 56).[71]

Wang et al. have developed an efficient SPINOL-phosphoric acid (**160**) catalysed asymmetric Pictet–Spengler reaction of α-naphthylmethyl tryptamines (**158**) with a wide range of aliphatic and aromatic aldehydes, affording a series of highly enantioenriched tetrahydro-β-carboline derivatives (**159**) (Scheme 57).[72]

The first peri-, regio- and enantioselective Diels–Alder (DA) and hetero-Diels–Alder (HDA) reactions of substituted cyclopentadienes (**161**) and acyclic 1,3-dienes (**162**), catalysed by a unique binary-acid (**163**)/InBr$_3$, have been described by Luo and co-workers (Scheme 58). The authors disclosed a dramatic remote *ortho*-fluoro substituent effect, on stereocontrol likely through π-interaction between the fluorinated benzene ring of (**163**) and 1,3-dienes (**162**).[73]

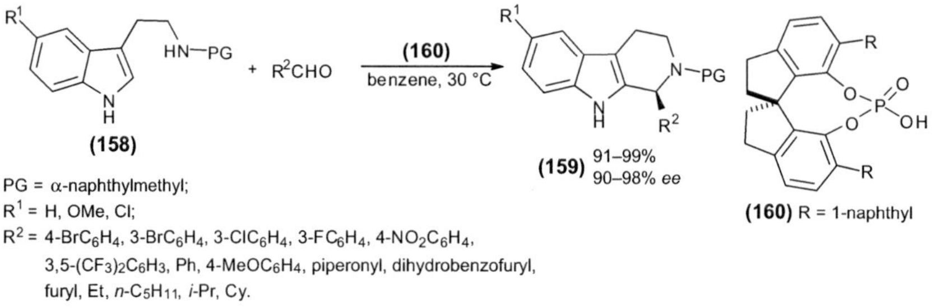
(160)
benzene, 30 °C
(158)
(159)
91–99%
90–98% ee
PG = α-naphthylmethyl;
R1 = H, OMe, Cl;
R2 = 4-BrC6H4, 3-BrC6H4, 3-ClC6H4, 3-FC6H4, 4-NO2C6H4, 3,5-(CF3)2C6H3, Ph, 4-MeOC6H4, piperonyl, dihydrobenzofuryl, furyl, Et, n-C5H11, i-Pr, Cy.
(160) R = 1-naphthyl

Scheme 57

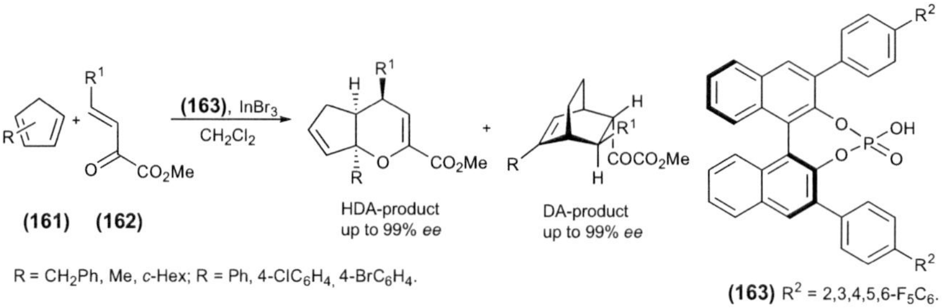
(163), InBr3
CH2Cl2
(161) (162)
HDA-product
up to 99% ee
DA-product
up to 99% ee
R = CH2Ph, Me, c-Hex; R = Ph, 4-ClC6H4, 4-BrC6H4.
(163) R2 = 2,3,4,5,6-F5C6.

Scheme 58

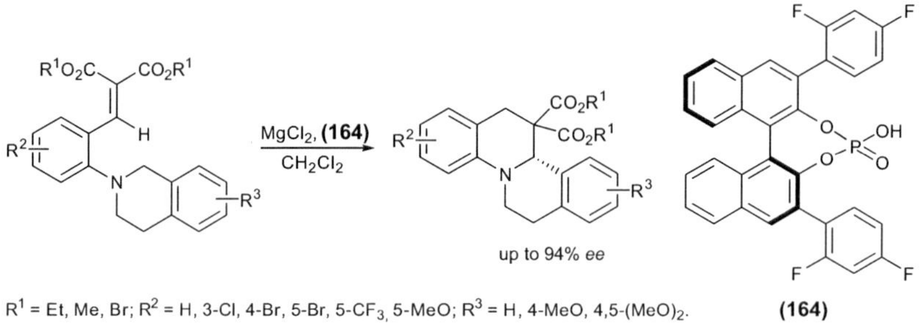
MgCl2, (164)
CH2Cl2
up to 94% ee
R1 = Et, Me, Br; R2 = H, 3-Cl, 4-Br, 5-Br, 5-CF3, 5-MeO; R3 = H, 4-MeO, 4,5-(MeO)2.
(164)

Scheme 59

A highly effective, asymmetric binary acid catalyst MgX_2/(**164**) for tert-aminocyclisation reactions has been also developed by Luo (Scheme 59). The identified catalytic system exhibited extremely high activity and good enantioselectivity for a broad range of substrates. The chiral binary acid complex selectively recognised and activated one helical conformation, required for hydrogen transfer, explaining the observed stereoselectivity.[74]

An enantioselective route to the biologically important spiro[pyrrolidin-3,2′-oxindole] scaffold (**166**) with multiple, contiguous stereogenic centers including one or two quaternary chiral centers in excellent stereoselectivities (up to > 99 : 1 *dr* and 98% *ee*) has been realised using the chiral phosphoric acid (**167**). This transformation represented the first example of a catalytic, asymmetric 1,3-dipolar cycloaddition reaction involving azo-methine ylides generated *in situ* from carbonyl compounds (**165**) in the presence of molecular sieves (Scheme 60).[75]

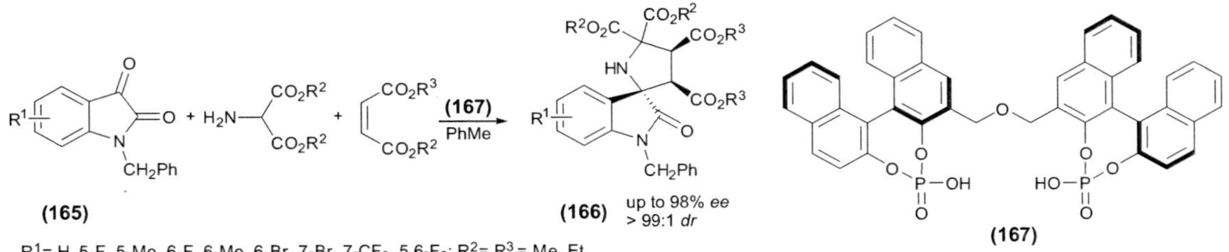

Scheme 60

R¹ = H, 5-F, 5-Me, 6-F, 6-Me, 6-Br, 7-Br, 7-CF₃, 5,6-F₂; R² = R³ = Me, Et.

The ruthenium catalyst generated *in situ* from $H_2Ru(CO)(PPh_3)_3$, (S)-SEGPHOS and the TADDOL-derived phosphoric acid (**169**) promoted butadiene hydrohydroxyalkylation to form enantiomerically enriched products (**168**) (Scheme 61). Match/mismatch effects between the chiral ligand and the chiral TADDOL-phosphate counterion have been described. For the first time, single-crystal X-ray diffraction data for a ruthenium complex modified by a chiral phosphate counterion has been reported.[76]

Asymmetric hydrogenation of *N*-alkyl ketimines (**170**) catalysed by the chiral, cationic η^6-arene-(*N*-monosulfonylated diamine) Ru(II) complex (**171**) bearing an achiral phosphate anion together with corresponding phosphoric acid as the additive has been investigated (Scheme 62). It was found that the achiral counteranion was critically important for the high enantioselectivity, which might provide a suitable platform for extending the application of this Ru(diamine) catalytic system in the asymmetric hydrogenation of other challenging substrates.[77]

A double-axially chiral bisphosphorylimide (**173**) has been demonstrated to be an efficient and highly sterically hindered Brønsted acid in asymmetric three-component Mannich reactions. Optically active *syn*-β-amino ketones were obtained in high yields (up to 99%) with excellent enantioselectivity (up to 99% *ee*) and diastereoselectivity (99:1) (Scheme 63).[78]

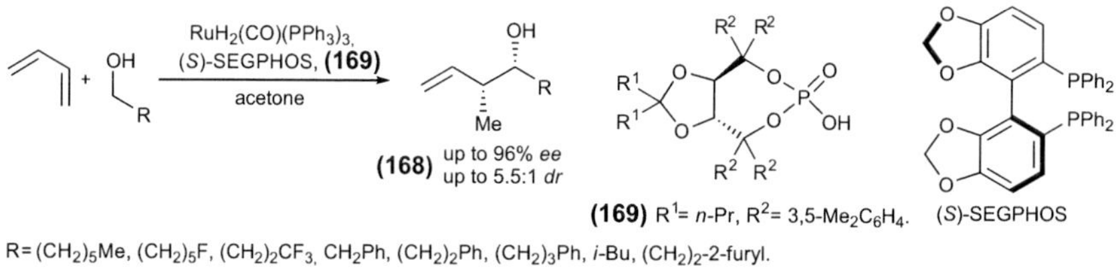

Scheme 61

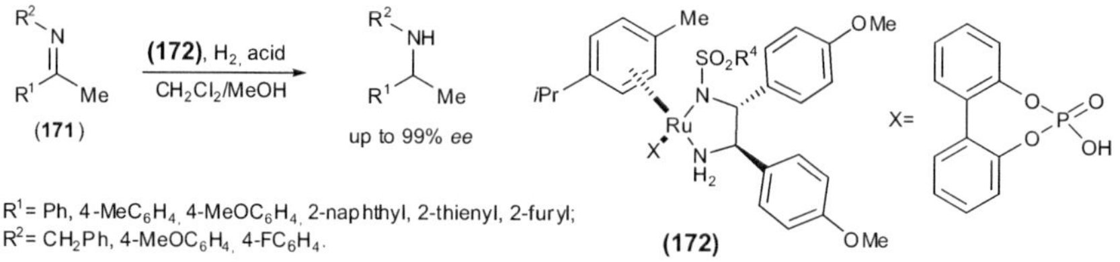

Scheme 62

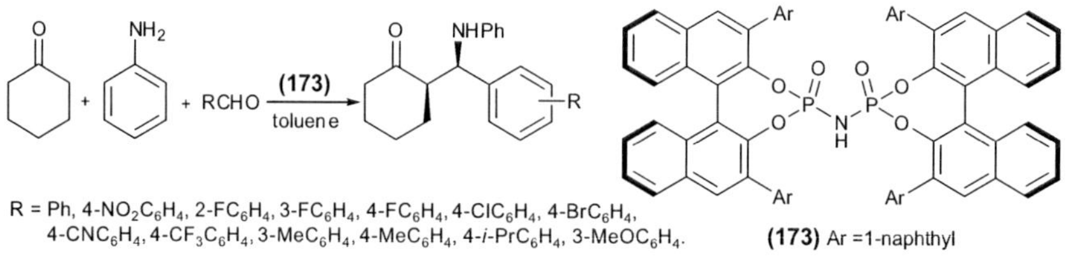

Scheme 63

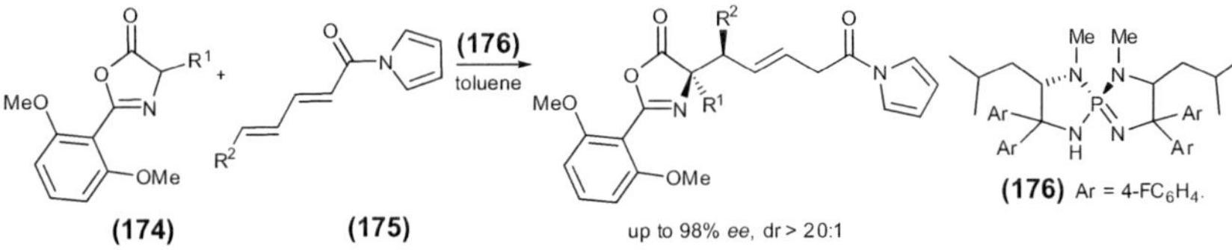

R^1 = CH_2Ph, CH_2-4-$MeOC_6H_4$, *i*-Bu, $(CH_2)_2SMe$;
R^3 = $(CH_2)_2CH_3$, $(CH_2)_6CH_3$, $(CH_2)_2Ph$, *i*-Bu, Me, $(CH_2)_3CH{=}CH_2$, CH_2OCH_2Ph, $CH_2OC(O)Ph$.

Scheme 64

The 1,6-addition of azalactones (**174**) to δ-monosubstituted dienyl *N*-acylpyrroles (**175**) has been realised with essentially complete control of regio-, diastereo- and enantioselectivities in the presence of chiral *P*-spiro triaminoimino-phosphorane (**176**) as a strong organic base catalyst (Scheme 64).[79]

1.4 Selected biological aspects

Synthesis of five new metabolically stabilised 2-MeO-lysophosphatidic acids (LPA) analogues (**178**) possessing different fatty acid residues has been performed by phosphorylation of the corresponding 1-*O*-acyl-2-MeO-glycerols (**177**) (Scheme 65). Compounds (**177**) were subjected to biological characterisation as autotaxin inhibitors using a fluorescence resonance energy transfer based synthetic autotaxin (ATX) substrate FS-3. Among tested compounds 1-*O*-oleoyl-2-MeO-LPA (**178e**) showed ATX inhibitory activity similar to that of unmodified 1-*O*-oleoyl-LPA, and was found to be resistant toward alkaline phosphatase as opposed to unmodified 1-*O*-oleoyl-LPA.[80]

A water-soluble, stable peptide-phospholipid conjugate (**179**) that possesses the necessary physical properties to enable more detailed study

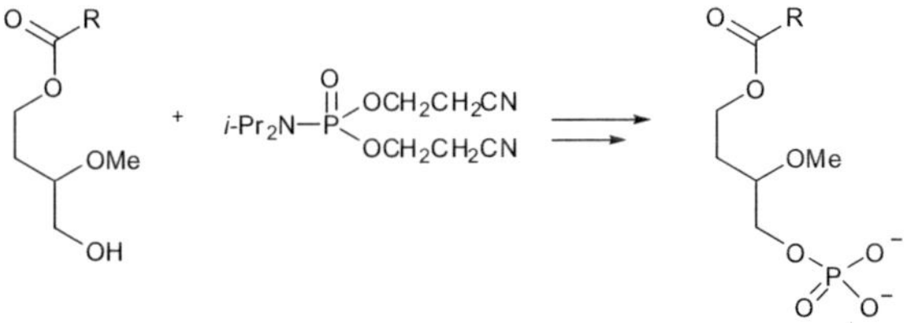

a: lauroyl;
b: myristoyl;
c: palmitoyl;
d: stearoyl;
e: oleoyl.

(177a)–(177e) **(178a)–(178e)**

Scheme 65

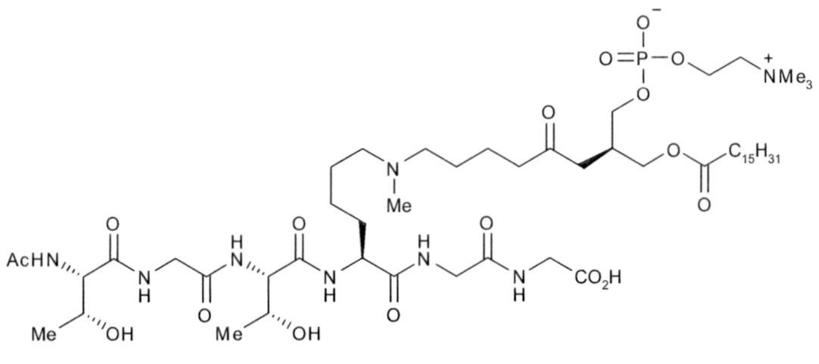

(179)

of the roles of oxidised phospholipid (OxPL) in metabolic disease has been synthesised by VanNieuwenhze and co-workers. Experimental data showed that the compound (**179**) bound to macrophage scavenger receptors was able to compete with the binding of oxidised low-density lipoprotein (OxLDL) with high affinity.[81]

A lysophosphatidylglycerol (**180**), isolated from the Korean sponge *Spirastrella abata*, showed significant cytotoxicity and Na^+/K^+-ATPase inhibition. The structure of this compound has been determined by chemical and spectroscopic methods.[82]

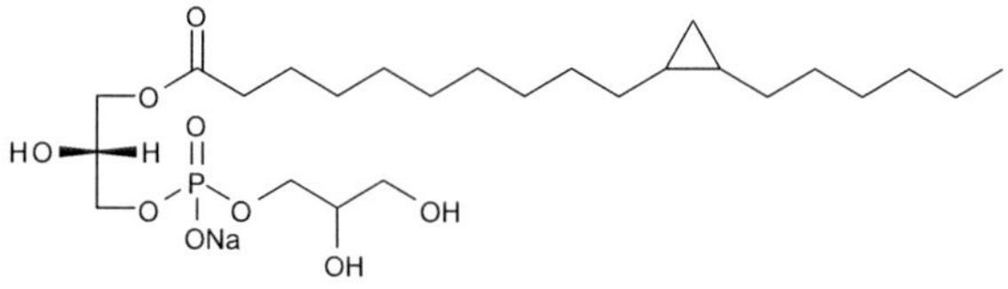

(**180**)

Two desferrioxamine B-ciprofloxacin conjugates, designed to release the antibiotic after esterase or phosphatase-mediated hydrolysis, have been synthesised. The potential esterase-sensitive conjugate (**181**) displayed moderate to good antibacterial activities against selected ferrioxamine-utilising bacteria, while, the potential phophatase-sensitive conjugate (**182**) was inactive against the same panel of organisms tested. These properties appeared to be related to the activating efficiency of the linker by the enzyme and to the outer membrane protein recognition of the chemically modified siderophore used in the conjugate.[83]

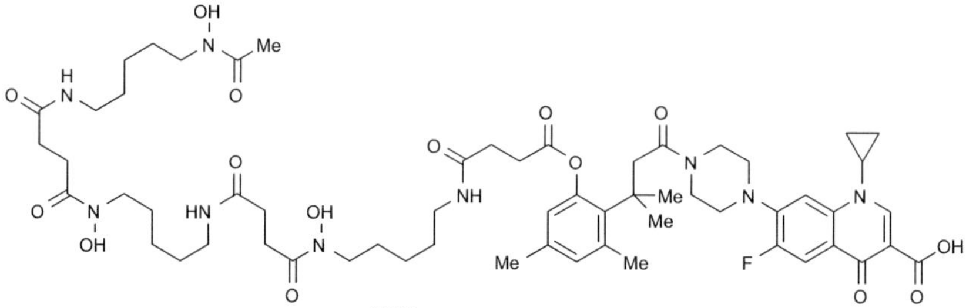

(**181**)

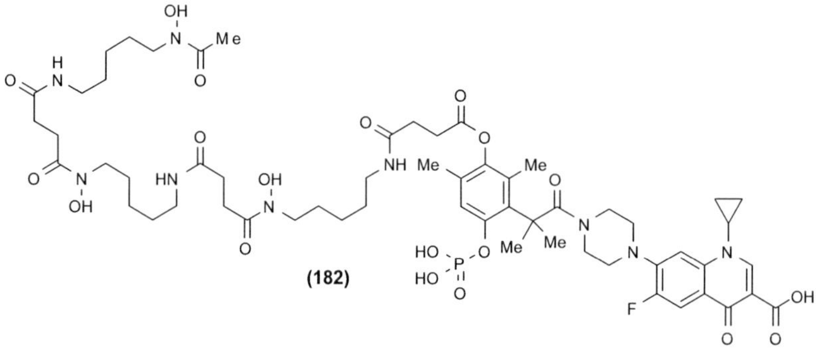

(**182**)

(E)-4-Hydroxy-3-methyl-2-butenyl pyrophosphate (**183**) (HMBPP) is a highly potent innate immunogen that stimulates human γδT cells expressing the Vγ2Vδ2 T cell antigen receptor. To determine if a glycoside conjugate could retain potency, the 4-β-glucoside of HMBPP (**184**) has been synthesised and evaluated. The glycoside HMBPP conjugate (**184**) stimulated human γδT cells with an EC_{50}, and was 2400 times less potent than HMBPP (**183**) itself. Thus, HMBPP glycosylated at the 4-OH position stimulated γδT cells as long as the pyrophosphate moiety was present.[84]

The synthesis of six novel phospholipids (**185**) and (**186**), including two γ-tocopheryl succinate (TOS) phospholipid (**185a,b**) conjugates has been described by Clausen and co-workers The enzyme activity of secretory phospholipase A_2 towards various succinate-phospholipid conjugates has been also investigated. The studies revealed that the TOS conjugates were poor substrates for the enzyme whereas the phospholipids with alkyl and phenyl succinate moieties were hydrolysed by the enzyme to a high extent.[85]

(**185a**) R= Me
(**186b**) R= H

(**186a**) R = C_8H_{17}
(**186b**) R = Ph
(**186c**) R = 2,6-$Me_2C_6H_3$
(**186d**) R = 3,4,5-$Me_3C_6H_2$

Synthesis of a new series of O-3 and O-4 phosphate prodrugs of N-acetyl-(D)-glucosamine (**187**) and (**188**), respectively, bearing a 4-methoxy phenyl group and different amino acid esters on the phosphate moiety has been reported by McGuigan *et al.* Among the compounds, the (*L*)-proline amino acid-containing prodrugs proved to be the most active of the series, and well processed in chondrocytes *in vitro*. Data on human cartilage supported the notion that these novel O-3 and O-4 regioisomers might represent novel promising leads for drug discovery for osteoarthritis.[86]

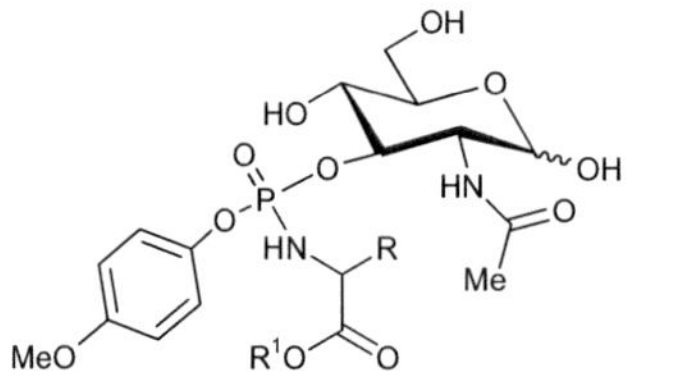

(**187**) O-3
Aminoacid: (L)-Ala, (L)-Val, (L)-Pro, (L)-Ser; R^1=Et, CH$_2$Ph;

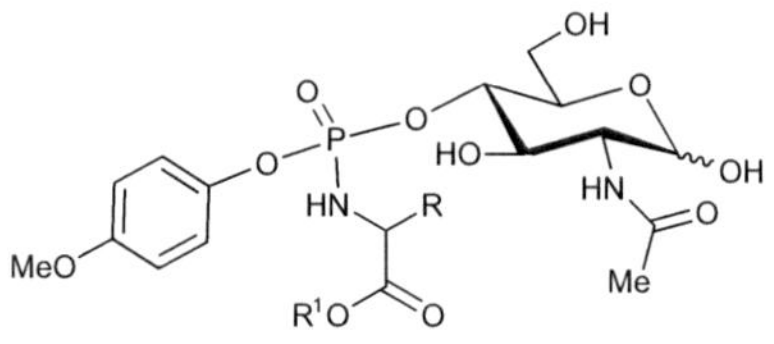

(**188**) O-4
Aminoacid: (L)-Pro; R^1=Et, *n*-Bu, *i*-Pr, *c*-Hex, 2-Bu, CH$_2$Ph

A phosphonooxymethyl prodrug (**189**) of azaindole-based HIV-1 attachment inhibitor (**190**) has been prepared and profiled in a variety of pre-clinical *in vitro* and *in vivo* models designed to assess its ability to deliver the parent drug following oral administration. The data showed that (**189**) had excellent potential to significantly reduce dissolution rate-limited absorption following oral dosing in humans. Clinical studies in normal healthy subjects confirmed the potential of (**189**), and provided guidance for further efforts to obtain an effective HIV-1 attachment inhibitor.[87]

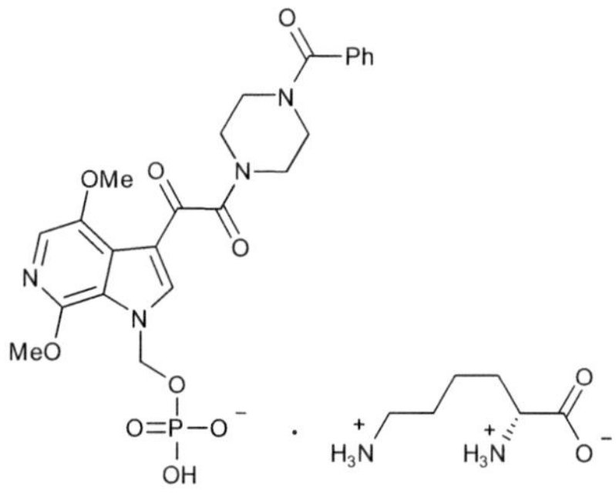

(**189**)

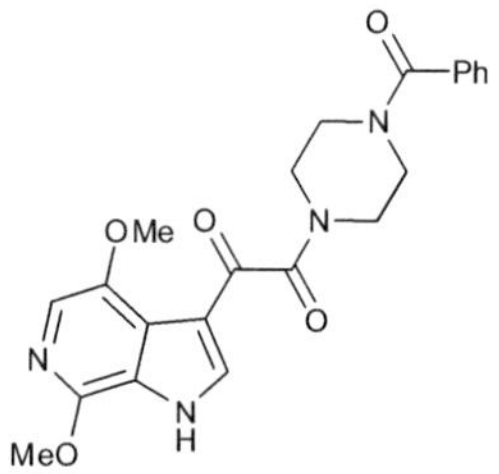

(**190**)

A systematic investigation of a glycosyltransferase (GT) enzyme inhibition by peptidoglycan precursor analogues with variations in the carbohydrate, lipid, pyrophosphate, and peptide moieties showed that saturated C$_{16}$ phosphoglycerate linked to (**191**) and to the disaccharide dipeptide (**192**) were good inhibitors of the GT activity *in vitro* and were able to induce growth defects or lysis of bacterial cells. These molecules were promising leads for the design of new antibacterial GT inhibitors.[88]

(191) R = OH
(192) R = (L)-Ala-γ-(D)-Glu-(L)-Lys

Synthesis and evaluation of tetrahedral intermediate mimic (193), inhibitors of 3-deoxy-D-manno-octulosonate 8-phosphate synthase (KDO8P) has been described by Parker and co-workers. The intermediate (193) was found to inhibit the metal-dependent KDO8P synthase from *Neisseria meningitides* and the metal-dependent KDO8P synthase from *Acidithiobacillus ferrooxidans* with inhibition constants in the low micromolar range. Additionally, monophosphorylated inhibitors (194) and (195) were synthesised to determine the relative importance of the two phosphate groups of this bisphosphate analogue for enzyme inhibition. The removal of either of these two phosphate groups gave less potent inhibitors for both enzymes.[89]

The pseudopentasaccharide (201) repeating unit of the lipoteichoic acid of *Streptococcus pneumonia* has been efficiently prepared from precursors (196)–(200) (Scheme 66). This molecule, lacking a lipid moiety did not stimulate a pro-inflammatory response in human monocytes (hMNCs).[90]

Van der Hoorn *et al.* have detected and confirmed selective inhibition of carboxylesterase (CXE12) by paraoxon (202) and profenofos (203), and showed selective inhibition of prolyloligopeptidase (POPL) by (202), and methylesterase-2 (MES2) by (202) and (203). These observations could be used for the design of novel probes and selective inhibitors and might help to assess the physiological effects of agrochemicals on crop plants.[91]

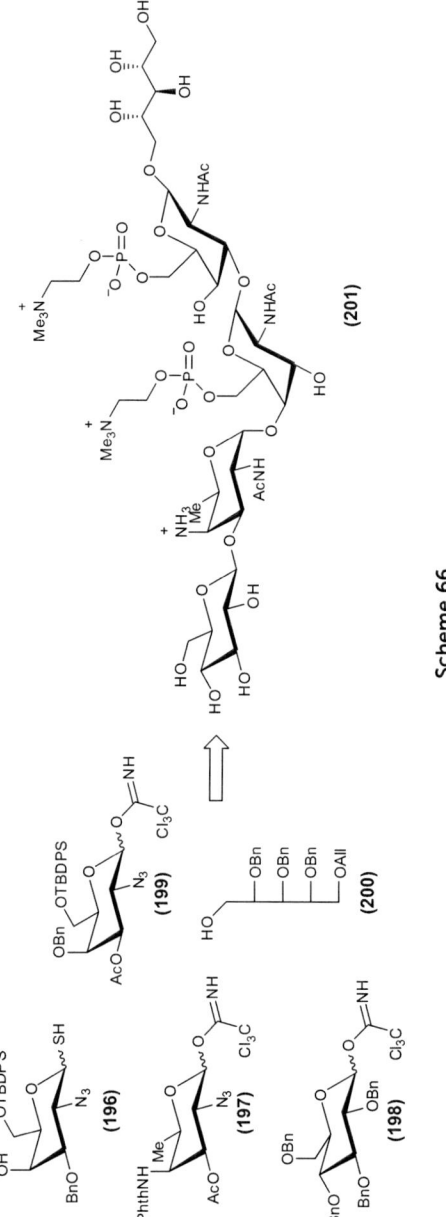

Scheme 66

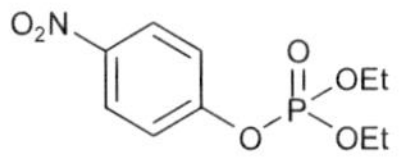

(202)

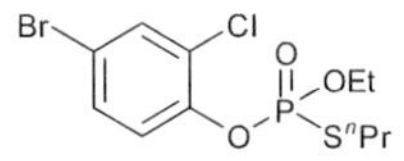

(203)

1.5 Miscellaneous

The spontaneous hydrolysis of a series of triaryl and dialkyl aryl phosphate esters (**204**) (Scheme 67), previously studied experimentally, has been examined theoretically using two different hybrid density functional methods, B3LYP and M06, and the Gaussian 09 program. The B3LYP/6−31+G(d) methodology combined excellent accuracy with minor computational cost, and the calculations showed excellent quantitative agreement with experiment, which was best in the presence of three discrete water molecules.[92]

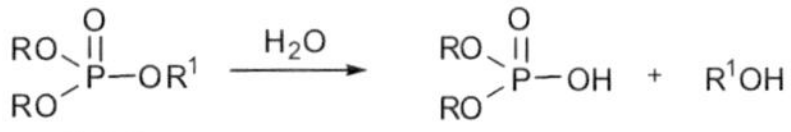

(204)

R= Ph, 4-Cl-C$_6$H$_4$, 4-NO$_2$C$_6$H$_4$, 2-pirydyl, Me, CH$_2$CF$_3$;
R^1= 4-NO$_2$C$_6$H$_4$, 2-pirydyl.

Scheme 67

The use of a combination of chemical approaches and 2D NMR spectroscopy enabled the elucidation of the structure of the O-polysaccharide (**205**) containing D-glyceramide 2-phosphate isolated from the *Providencia alcalifaciens* O22. The polisaccharide (**205**) was found to contain a novel non-carbohydrate substituent, D-glyceramide 2-phosphate, which enlarged the list of known components of bacterial polysaccharides. The phosphate group and the free amino group of the 2-acetamido-4-amino-2,4,6-trideoxy-D-galactose residue confered a zwitterionic character to the O-polysaccharide, which was believed to promote the adaptation of bacteria to environmental conditions at different pH.[93]

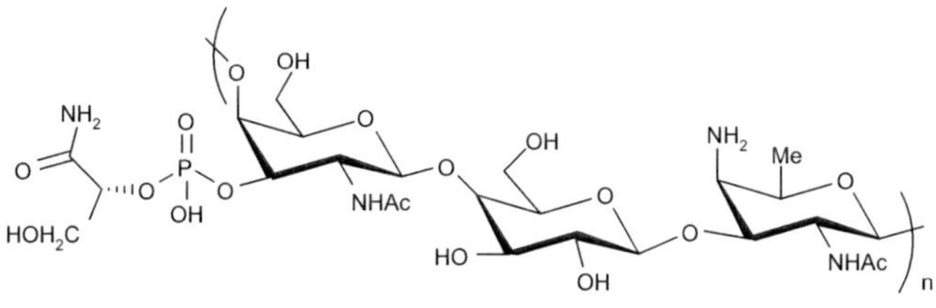

(205)

The biophysical study of phosphodistearylglycerol (**206**) fragment of glycosylphosphatidylinositols, revealed the unprecedented crystalline two-dimensional structure of monolayers. These monolayers were characterised by two commensurate lattices: the oblique lattice of the alkyl chains and the molecular lattice formed owing to highly ordered head groups. This structure was reminiscent of sub-gel phase structures observed in lipid dispersions after partial dehydration of the head groups during long incubation periods at low temperature.[94]

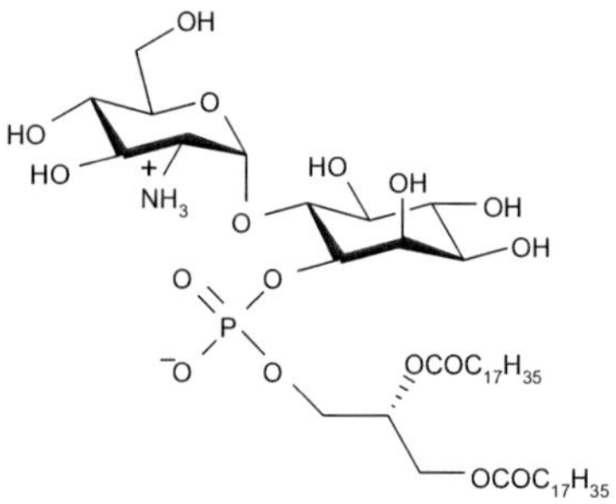

(206)

An N,N-carbonyl-bridged dipyrrinone oxime (**207**) has been synthesised and studied as a potential sensor for organophosphates. Thus molecular sensor (**207**) underwent a drastic colorimetric response upon formation of the adduct (**208**) (Scheme 68).[95]

The N-(Rhodamine B)-deoxylactam-5-amino-1-pentanol (**209**) was designed as the chromo-fluorogenic sensor for detection of a nerve agent simulant *via* analyte triggered tandem phosphorylation and intramolecular opening of the deoxylactam. The successful detection of diethyl chlorophosphate suggested the utility of rhodamine-deoxylactams (**209**) as the chromo-fluorogenic signal reporting platform for design of sensors targeting reactive chemical species *via* various chemistries (Scheme 69).[96]

Chiral organolithiums (**212**) have been prepared from the tributylstannylmethanol (**210**) for testing their configurational stability. Thus, (**210**) has been converted into corresponding diisopropyl S-tributylstannylmethyl thiophosphate derivatives (**211**) transmetallation in THF at $-78\,^\circ$C with MeLi added dropwise every 3 s. A thermodynamically-driven migration of the phosphoryl group from the sulfur to the carbon atom led to the formation of mercaptomethylphosphonates (**213**) (Scheme 70). Temperature dependent studies of the rearrangement of thiophosphates (**211**) to mercaptomethylphosphonates (**213**) revealed that chiral organolithiums (**212**) were configurationally-labile down to $-95\,^\circ$C on the time scale of the thiophosphate-α-mercaptophosphonate rearrangement.[97]

Quaternary onium halides (**214**) were efficiently converted into the corresponding quaternary onium salts (**215**) possessing various anions [NO_3^-, BF_4^-, PF_6^-, $CF_3SO_3^-$, $CH_3SO_3^-$, ClO_4^-, p-$CH_3C_6H_4SO_3^-$, $CF_3CO_2^-$, 2,4-$(NO_2)_2C_6H_3O^-$] by treating the onium halides (**214**) with trimethyl phosphate under neat conditions in the presence of an equivalent amount of conjugate acid of the desired anion (Scheme 71).

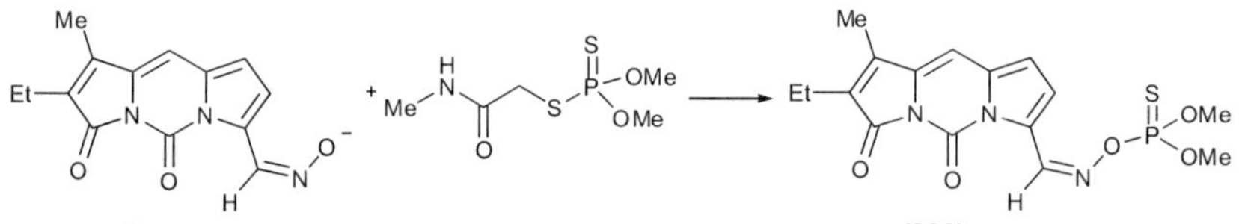

(**207**) - deep red   (**208**) - yellow

Scheme 68

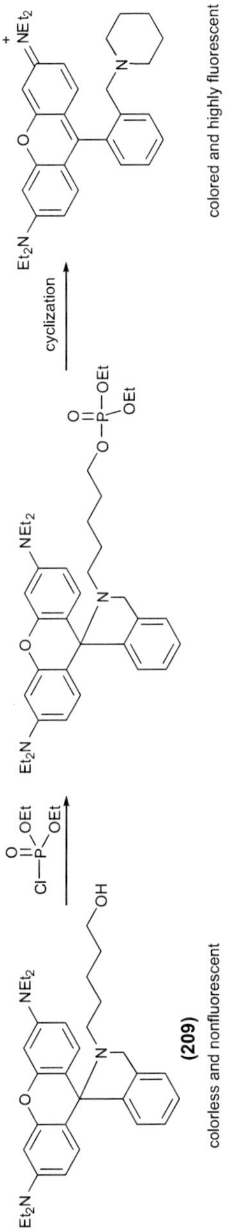

Scheme 69

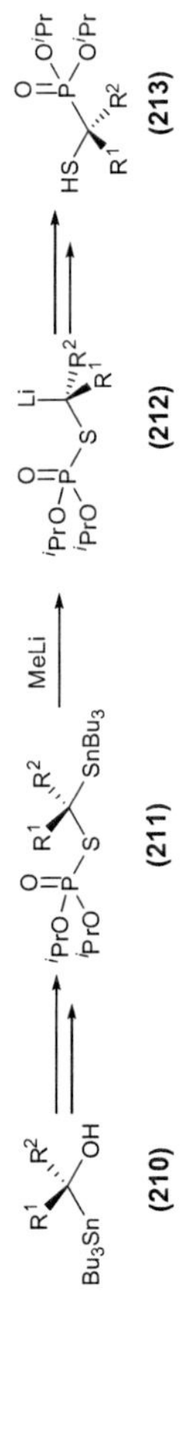

Scheme 70

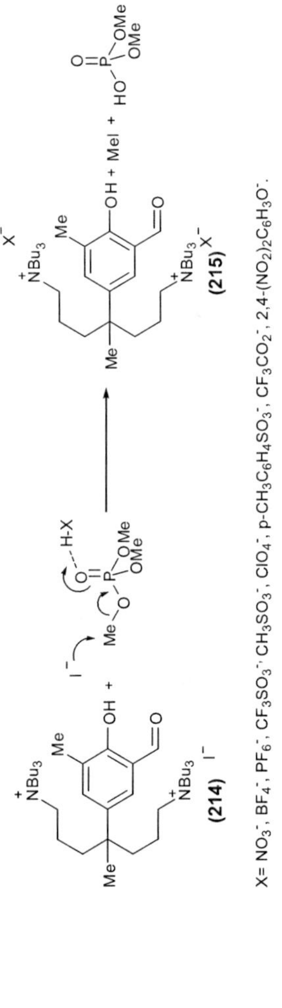

Scheme 71

X = NO$_3^-$, BF$_4^-$, PF$_6^-$, CF$_3$SO$_3^-$, CH$_3$SO$_3^-$, ClO$_4^-$, p-CH$_3$C$_6$H$_4$SO$_3^-$, CF$_3$CO$_2^-$, 2,4-(NO$_2$)$_2$C$_6$H$_3$O$^-$.

This protocol could be applied to large-scale syntheses, allowing economical preparation of key intermediates for a highly active catalysts for CO_2/epoxide co-polymerisations.[98]

2 Phosphonic acids and their derivatives

2.1 Synthesis of phosphonic acids and their derivatives

A copper-mediated oxidative cross-coupling reaction of terminal alkynes with readily available α-silyldifluoromethylphosphonates (**216**) has been developed by Qing and co-workers. Thus, a series of synthetically useful α,α-difluoropropargylphosphonates (**217**) has been obtained in moderate yields and with excellent functional group compatibility (Scheme 72).[99]

A copper-catalysed cross-coupling of iodobenzoates (**218**) with bromozinc-difluorophosphonate, generated from diethyl bromodifluoromethylphosphonate (**219**) and zinc in dioxane, has been also reported. This method provided a useful and facile access to aryldifluorophosphonates (**220**) of interest in life science (Scheme 73).[100]

A silver-catalysed dehydrogenative cross-coupling reaction of substituted furans, thiophene, thioazole, and pyrrole (**221**) with dialkyl phosphites (**222**) has been developed to give corresponding heteroarylphosphonates (**223**) in good to excellent yields and good regioselectivities. An unprecedented coupling of various substituted pyridines with dialkyl phosphites (**222**) using $AgNO_3$ as a catalyst and $K_2S_2O_8$ as an oxidant, followed by reduction with $Na_2S_2O_3$, has been also realised to afford desired pyridine phosphonates (**224**) in satisfactory yields (Scheme 74).[101]

The direct oxidative C-phosphorylation of indoles with dialkyl phosphites in the presence of silver (I) acetate has been demonstrated by Wan *et al.* A wide variety of indoles and dialkyl phosphites participated in the reaction to afford corresponding indolylphosphonates (**225**) in up to 71% yields (Scheme 75).[102]

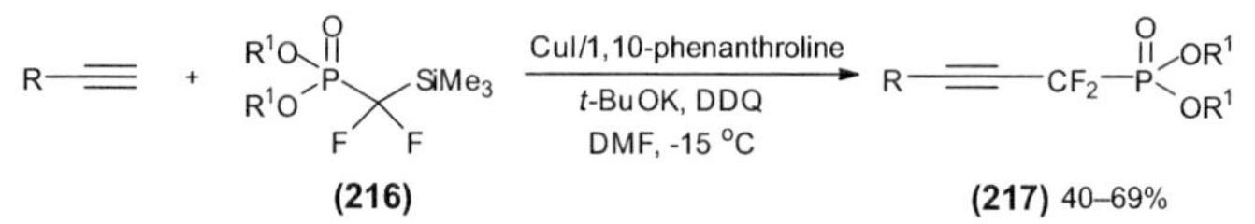

R = Ph, 4-MeOC6H4, 4-BrC6H4, 4-CF3C6H4, 3-NO2C6H4, 4-t-BuC6H4, 3-ClC6H4, 4-CNC6H4, 4-CO2EtC6H4, 1-naphthyl, 3-thienyl, n-Bu;
R1 = Et, i-Pr, n-Bu.

Scheme 72

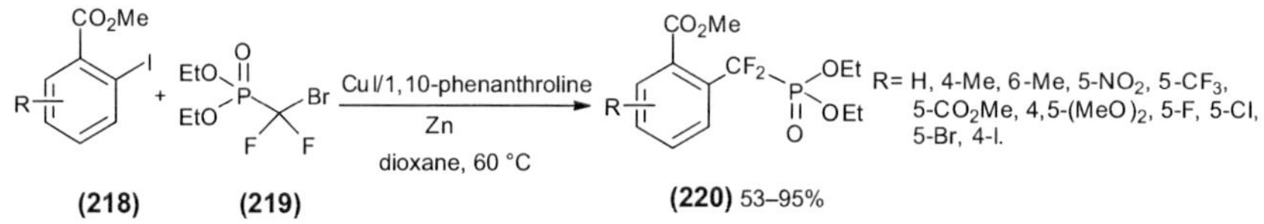

Scheme 73

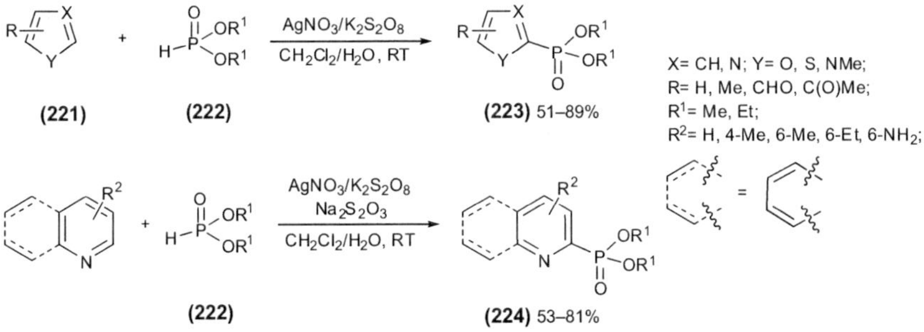

AgNO3/K2S2O8
CH2Cl2/H2O, RT
(221)
(222)
(223) 51–89%
X= CH, N; Y= O, S, NMe;
R= H, Me, CHO, C(O)Me;
R1= Me, Et;
R2= H, 4-Me, 6-Me, 6-Et, 6-NH2;
AgNO3/K2S2O8
Na2S2O3
CH2Cl2/H2O, RT
(222)
(224) 53–81%

Scheme 74

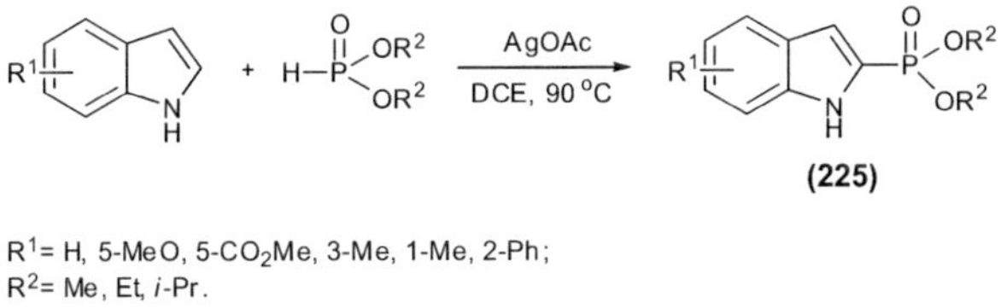

AgOAc
DCE, 90 °C
(225)
R1= H, 5-MeO, 5-CO2Me, 3-Me, 1-Me, 2-Ph;
R2= Me, Et, i-Pr.

Scheme 75

The asymmetric allylic substitution reaction of Morita-Baylis-Hillman carbonates (**226**) with diphenyl phosphite in the presence of chiral multifunctional thiourea-phosphine catalyst (**228**) provided allylic phosphites (**227**) in high yields and with excellent enantioselectivities (Scheme 76).[103]

The rare earth metal amido complexes (**230**) have been found to be an excellent catalysts for the Pudovik hydrophosphonylation of wide range of aldehydes and unactivated ketones. The catalysed reactions between diethyl phosphite and aldehydes in the presence of the complexes (**230**) afforded the α-hydroxy aryl phosphonates (**229**) ($R^1 = H$) in high yields (up to 99%) at room temperature in short reaction times of 5 to 10 min. The reaction of ketones with diethyl phosphite has been also realised providing corresponding α-hydroxy phosphonates (**229**) ($R \neq H$) in moderate to high yields (50–99%) (Scheme 77).[104] Moreover, Carpentier and co-workers have shown for the first time that the selective Pudovik addition of diethylphosphite to aldehydes or ketones to yield α-hydroxy phosphonates (**229**) could be also effectively promoted by complexes of the large alkaline earth metals (**231**), Ca, Sr, and Ba.[105]

The direct oxidative coupling of tertiary amines and dialkyl- or diphenyl-substituted phosphonates has been developed by Prabhu and Wan. Thus, DDQ (2,3-dichloro-5,6-dicyano-*p*-benzoquinone)[106,107] or MoO_3-catalysed[108] C-phosphorylation of *N*-aryl tetrahydroisoquinolines (**232**) gave the corresponding α-aminophosphonates (**233**) in moderate to excellent yields (Scheme 78).

A general approach to C-P bond formation through the cross-coupling of aryl halides with dimethyl phosphite by using [$NiCl_2$(dppp)] as a catalyst has been described (Scheme 79). This catalyst displayed a broad applicability that was capable of catalysing the cross-coupling of aryl

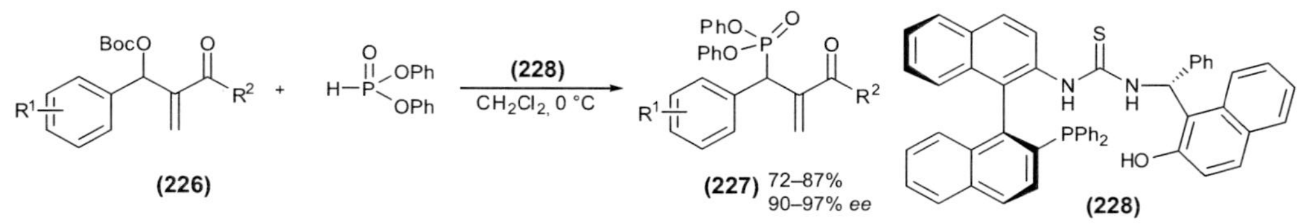

Scheme 76

(226) + H-P(=O)(OPh)(OPh) → (227) 72–87%, 90–97% ee

Conditions: (228), CH$_2$Cl$_2$, 0 °C

R^1 = H, 4-NO$_2$, 3-NO$_2$, 4-CF$_3$, 4-CN, 4-Br, 4-MeSO$_2$, 4-Cl, 3-Cl, 2-Cl, 4-Me, 3,4-Cl$_2$;
R^2 = Me, Et.

Scheme 77

R-C(=O)-R^1 + H-P(=O)(OEt)(OEt) → (229) 50–99%

Conditions: (230) or (231)

(230) Ln = Y, Nd, Sm, Dy, Yb
(231) M = Ca, Sr, Ba; [M{N(SiMe$_3$)$_2$}$_2$(THF)$_2$]

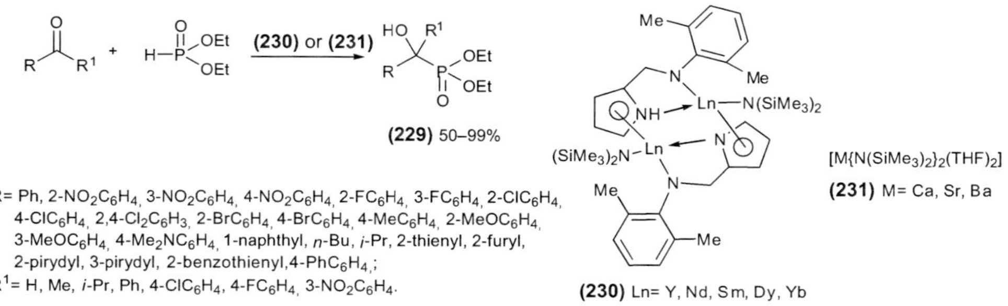

R = Ph, 2-NO$_2$C$_6$H$_4$, 3-NO$_2$C$_6$H$_4$, 4-NO$_2$C$_6$H$_4$, 2-FC$_6$H$_4$, 3-FC$_6$H$_4$, 2-ClC$_6$H$_4$, 4-ClC$_6$H$_4$, 2,4-Cl$_2$C$_6$H$_3$, 2-BrC$_6$H$_4$, 4-BrC$_6$H$_4$, 4-MeC$_6$H$_4$, 2-MeOC$_6$H$_4$, 3-MeOC$_6$H$_4$, 4-Me$_2$NC$_6$H$_4$, 1-naphthyl, n-Bu, i-Pr, 2-thienyl, 2-furyl, 2-pirydyl, 3-pirydyl, 2-benzothienyl, 4-PhC$_6$H$_4$, ;
R^1 = H, Me, i-Pr, Ph, 4-ClC$_6$H$_4$, 4-FC$_6$H$_4$, 3-NO$_2$C$_6$H$_4$.

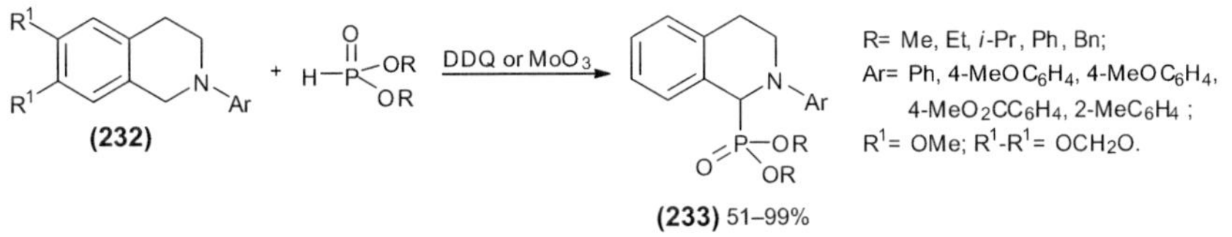

Scheme 78

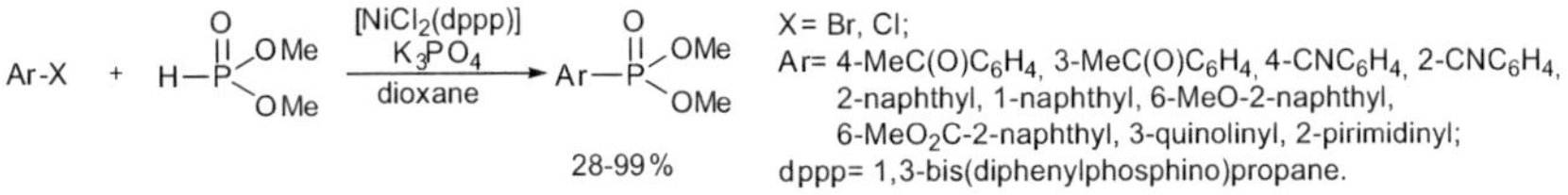

Scheme 79

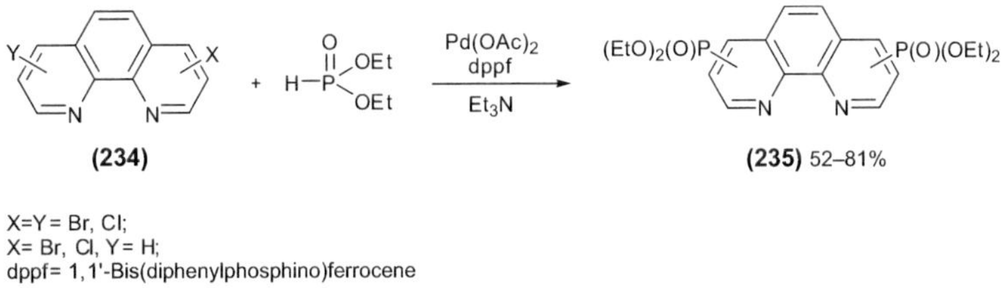

Scheme 80

bromides and unreactive aryl chlorides, with phosphorus substrates. Moreover, the reaction proceeded at a much lower temperature (100–120 °C) relative to the classic Arbuzov reaction (ca. 160–220 °C), without the need of external reductants and supporting ligands, and tolerated a range of labile groups, such as ether, ester, ketone or cyano groups.[109]

The synthesis of diethyl phosphonate derivatives of 1,10-phenanthroline (235) has been achieved by coupling of mono- and dihalo-1,10-phenanthrolines (234) with diethyl phosphite in the presence of Pd(OAc)$_2$/dppf catalytic system (Scheme 80).[110]

A convenient one-pot, three-component approach leading to synthesis of a series of novel heterocyclic α-aminophosphonates (237) has been developed by Zhao and co-workers. Using diisopropyl or diphenyl phosphites along with 2-hydroxyacetophenones (236a,b) and primary amines a wide range of the corresponding heterocyclic α-aminophosphonates (237) have been synthesised in relatively high yields (Scheme 81).[111]

The choline chloride based ionic liquid has been applied as an efficient and eco-friendly catalyst for the one pot, three-component synthesis of α-aminophosphonates (238) under solvent-free conditions. The reaction was completed in short reaction times (40–120 min) and gave corresponding products in good to excellent yields (70–98%) (Scheme 82).[112] Ordóñez et al. have described three-component synthesis of α-aminophosphonates (238) using phenylboronic acids as catalysts under solvent-free conditions. A wide range of carbonyl compounds have been compatible with this reaction, producing tertiary α-aminophosphonates (238) in 62–93% yields in short reaction times (15–60 min) (Scheme 82).[113]

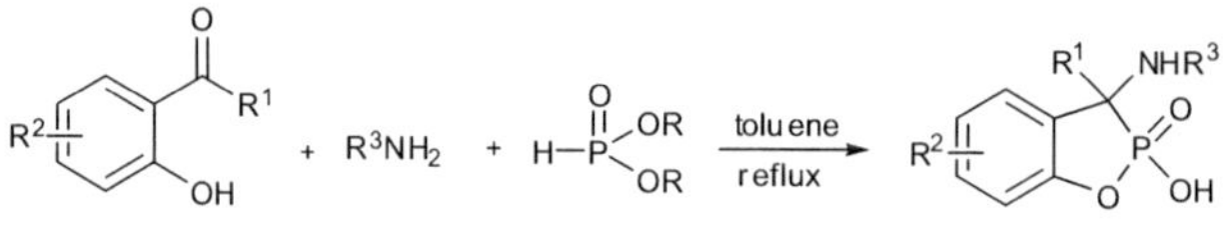

(236a) R¹= Me
(236b) R¹= H

(237) 65–90%

R²= H, 6-OMe, 5-NO₂, 3-NO₂, 5-Br, 5-I, 3,5-Br₂, 3,5-I₂;
R³= H, Me, Et, n-Pr, n-Bu, n-Hex, n-Oct, Ph, CH₂CH₂NH₂;
R= i-Pr, Ph.

Scheme 81

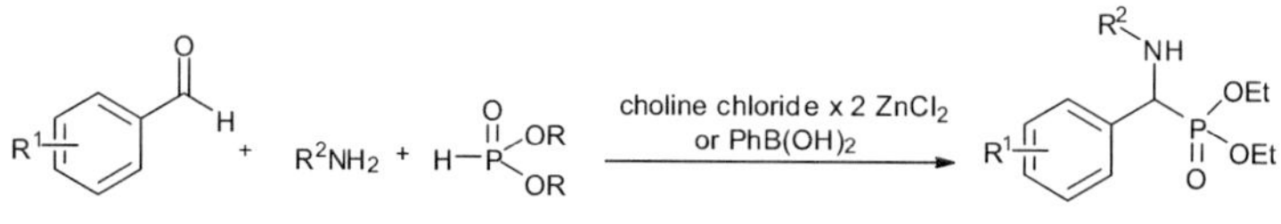

(238)

R= Me, Et;
R¹= H, 3-Br, 4-NO₂, 3-Cl, 4-MeO, 2-MeO, 3,4-(MeO)₂,
4-OH, 4-Br, 2-Cl, i-Pr, s-Bu, t-Bu;
R²= Ph, PhCH₂, 4-FC₆H₄, 4-ClC₆H₄, 4-BrC₆H₄, 4-FC₆H₄.

Scheme 82

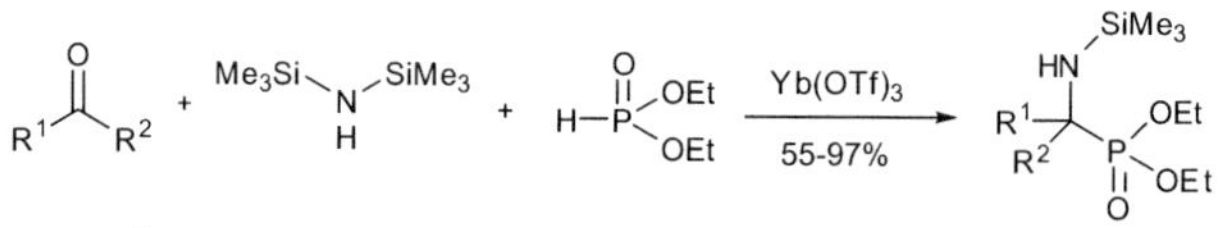

(239a) R¹= 4-MeOC₆H₄, 4-ClC₆H₄, 4-NO₂C₆H₄, i-Pr, n-Pr, R²=H;
(239b) R¹= Me, R²= C₁₀H₂₁; R¹=Me, R²=Ph; R¹-R²=-CH₂(CH₂)₃CH₂-.

Scheme 83

N-silylated α-substituted α-aminophosphonates (**239a**) and α,α-disubstituted α-aminophosphonates (**239b**) have been obtained from aldehydes or ketones, hexamethyldisilazane and diethylphosphite in the presence of ytterbium(III) triflate under mild conditions (Scheme 83).[114]

One-pot reaction of arylaldehydes with malononitrile and triethyl phosphite has been carried out in the presence of ethylenediamine diacetate (EDDA) as a catalyst. Thus, the biologically interesting phosphonates (**240**) have been obtained in good yields, under mild reaction conditions (Scheme 84).[115]

Ordóñez et al. have developed a practical and efficient strategy to obtain N-substituted dimethyl 3-oxoisoindolin-1-yl phosphonates (**241**) in a one-pot reaction of 2-formylbenzoic acid, primary amines and dimethyl phosphite. This reaction proceeded in a short time either with conventional heating or microwave irradiation, under catalyst-free conditions, in good yield (69–82%) (Scheme 85).[116]

The nucleophilic addition of diethyl phosphite or trimethyl phosphite to imidazole-derived imines (**242**) provided imidazole-2-yl-(amino)methylphosphonates (**243a**) and imidazole-2-yl-(amino)methylphosphonic acids (**243b**).

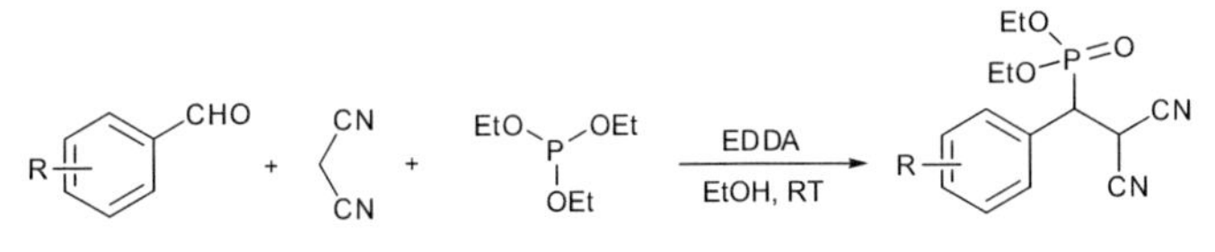

(240) 75–90%

R= 3-Me, 3-MeO, 4-Me, 4-MeO, 4-F, 4-Br, 4-Cl.

Scheme 84

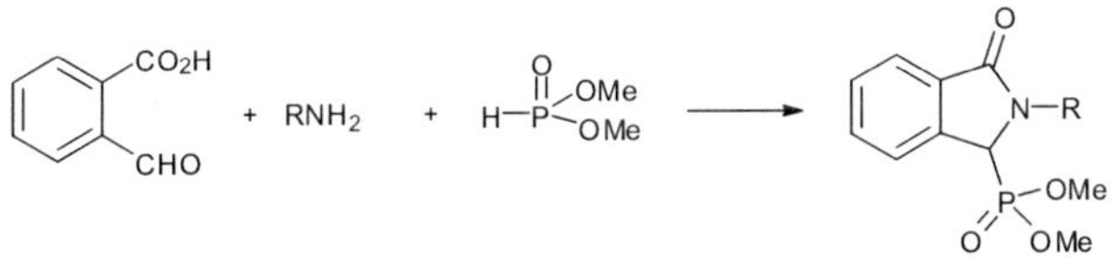

(241)

R = Ph, 4-$MeOC_6H_4$, 3,4-$(MeO)_2C_6H_4$, 3-$CF_3C_6H_4$, 2-pyridyl, CH_2Ph, *t*-Bu, (*S*)-CH(Me)Ph.

Scheme 85

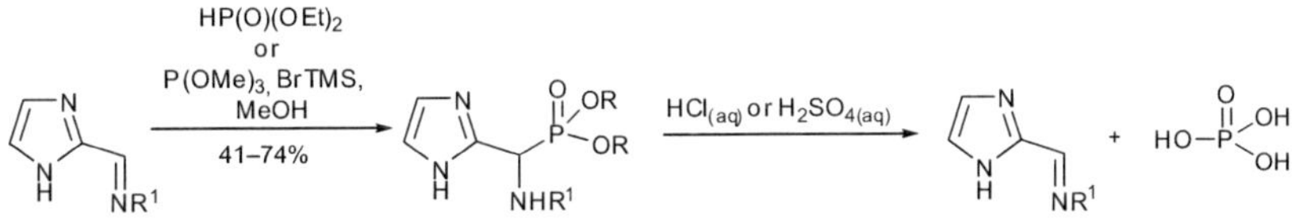

(242) **(243a)** R = Et, **(243b)** R = H. **(242)**

R^1= *n*-Bu, CH_2Ph, 2-pyridylCH_2, 3-pyridylCH_2, imidazol-1-yl-$(CH_2)_3$.

Scheme 86

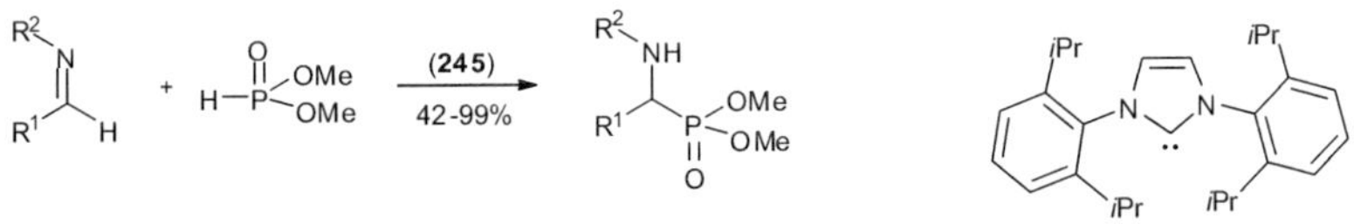

(244) **(245)**

R^1 = Ph, 4-MeC_6H_4, 4-$MeOC_6H_4$, 4-ClC_6H_4, 2-ClC_6H_4, 2,4-$Cl_2C_6H_4$, 3-BrC_6H_4, 3-$NO_2C_6H_4$, 2-furyl, 2-naphthyl, *n*-Pr, Cy;
R^2 = Ts, Boc, CH_2Ph, 4-nitrosulphonyl.

Scheme 87

It has been also discovered that heating of products (**243**) with aqueous HCl or H_2SO_4 led to their decomposition resulting in a cleavage of the C–P bond, elimination of the phosphorous-containing fragment and formation of the corresponding imine (**242**) and phosphoric acid (Scheme 86).[117]

Aldimines underwent efficient hydrophosphonylation reactions with dimethyl phosphite in the presence of the nucleophilic heterocyclic carbene (**245**) as organocatalyst to give the corresponding (α-aminoalkyl)-phosphonates (**244**) in moderate to excellent yields (Scheme 87).[118]

Bis[(dimethoxyphosphoryl)methyl)]amino terminated poly(ethyleneglycols) (PEGs) (246) and (247) have been prepared from commercially available PEG by means of adaptations of the Moedritzer–Irani and Kabachnik–Fields procedures, respectively (Scheme 88).[119]

Stable derivatives of dialkyl 2-(dialkoxyphosphoryl)succinates (250) and (252) have been prepared in water, in good yields using multicomponent reactions involving activated acetylenic compounds (249), trimethyl phosphite or triphenyl phosphate, 1-(6-hydroxy-2-isopropenyl-1-benzofuran-yl)-1-ethanone (248) or 4-hydroxycoumarin (251) (Scheme 89).[120]

A convenient and efficient synthesis of 4-halo-2,5-dihydro-1,2-oxaphosphole 2-oxides (254) through CuX_2-mediated direct halocyclisation of diethyl 1,2-allenylphosphonates (253) has been described by Ma and Xin. Further Suzuki cross-coupling of the resultant chloro derivatives (254) with phenylboronic acid, has been also established (Scheme 90).[121]

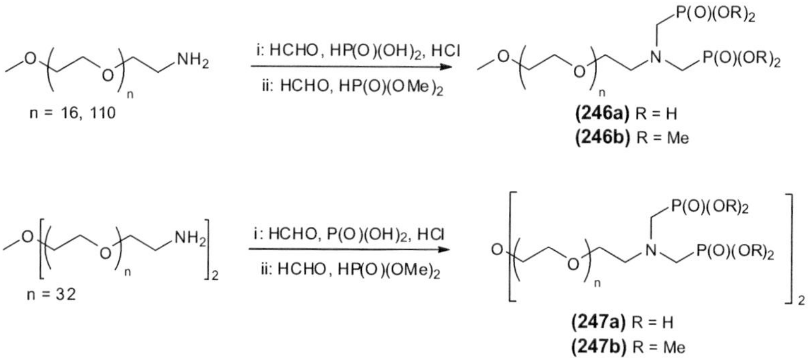

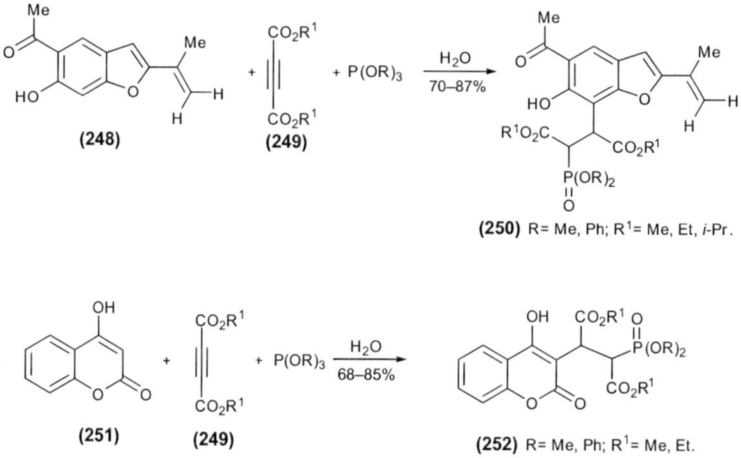

α-Amino allenephosphonates (258) have been easily prepared in two steps from protected amines (255), 3-bromopropargyl alcohols (257) and chlorophosphites. In the first step, a copper (II) catalysed coupling reaction of unprotected 1-bromopropargyl alcohols (256) and amines (256)

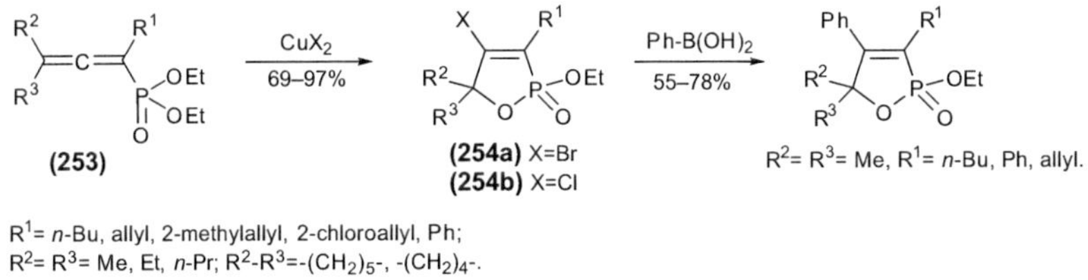

Scheme 90

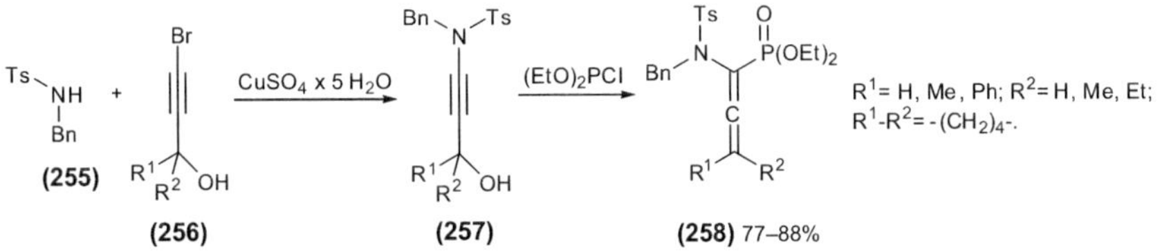

Scheme 91

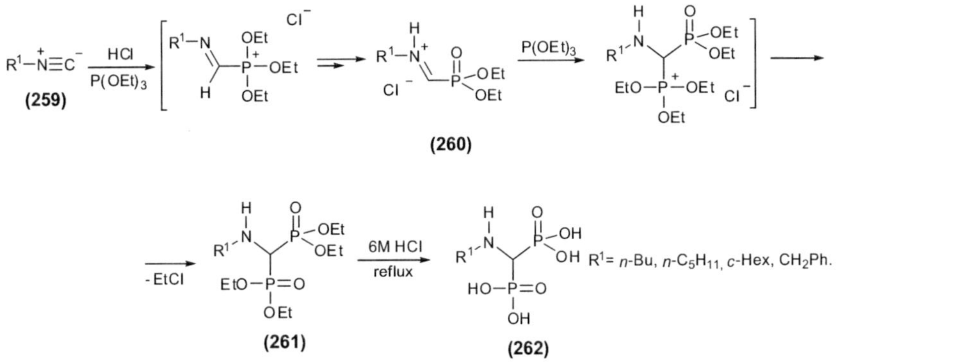

Scheme 92

afforded the corresponding ynamides (**257**) which were then transformed directly to allenephosphonates (**258**) through a [2,3]-sigmatropic rearrangement of propargyl phosphites. This efficient method led to the formation of a series of α-amino allenephosphonates (**259**) with diverse substituents on the amine, the phosphonate, and the allene moieties (Scheme 91).[122]

The reaction of isonitriles (**259**) with triethyl phosphites in the presence of hydrogen chloride gave tetraalkyl *N*-substituted aminomethylenebisphosphonates (**261**) *via* *N*-methylideneaminium (isonitrilium) salts (**260**). Hydrolysis or dealkylation of these tetraalkyl esters (**261**) gave the corresponding *N*-substituted bisphosphonic acids (**262**) in 77–98% yields (Scheme 92).[123]

Addition of trialkyl phosphites to (*S*)-*N*-*tert*-butanesulphinyl imine (**263**) afforded under mild conditions the corresponding *N*-*tert*-butanesulphinyl α-aminophosphonates (**264**) in moderate to high yields and diastereoselectivity (Scheme 93).[124]

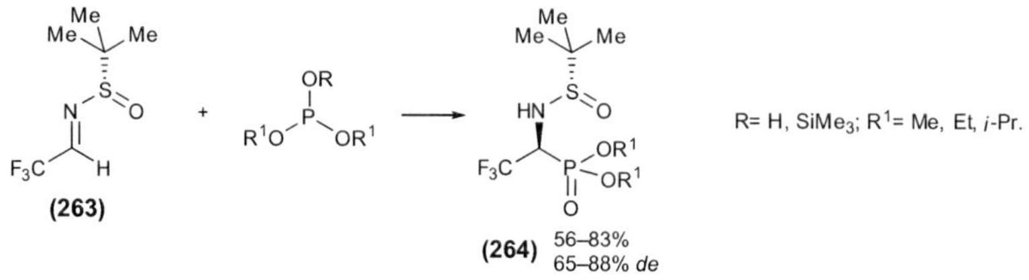

Scheme 93

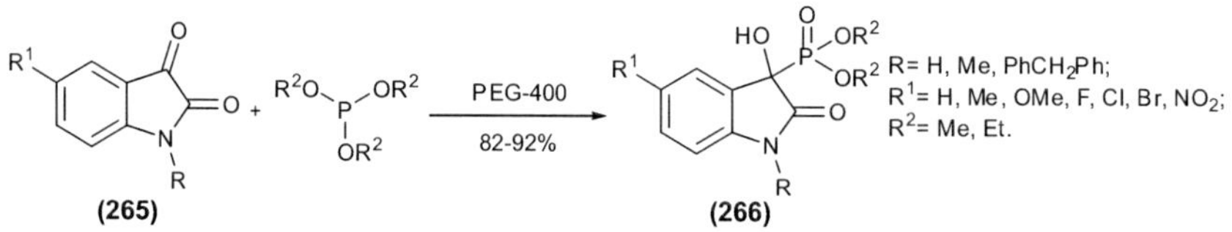

Scheme 94

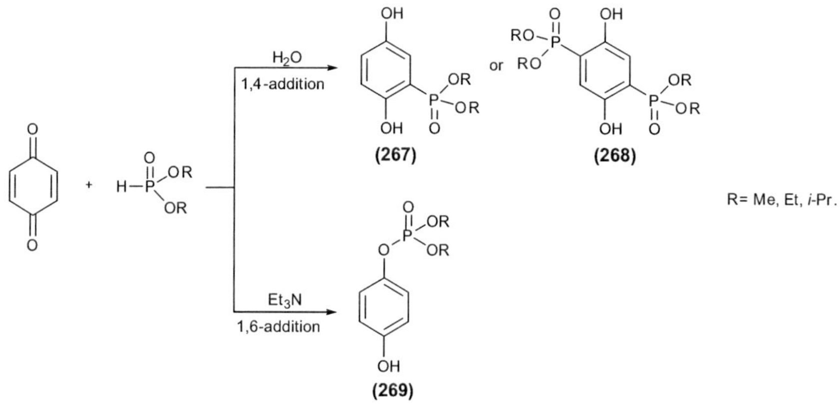

Scheme 95

A simple, efficient, cost-effective, and environmentally-friendly method has been developed for the synthesis of α^1-oxindole-α-hydroxyphosphonate derivatives (**266**) by a one-pot reaction of trialkyl phosphites with isatins (**265**) under catalyst-free-conditions using inexpensive, non-toxic polyethylene glycol (PEG-400) (Scheme 94).[125]

The reaction of dialkyl phosphite with *p*-quinones proceeded through the 1,4-addition pathway in the presence of water and gave the corresponding C-phosphoryl hydroquinone derivatives (**267**). Oxidative, double 1,4-addition of dialkyl phosphite to *p*-quinones has been also achieved in polar solvents, such as DMSO or DMF, affording a facile access to bis-substituted hydroquinones (**268**). On the other hand, O-phosphoryl hydroquinone derivatives (**269**) have been obtained from dialkyl phosphite and *p*-quinones in the presence of triethylamine through a selective 1,6-addition reaction (Scheme 95).[126]

The synthesis of four types of fosmidomycin analogues (**270**)–(**273**) for the inhibition of 1-deoxy-D-xylulose-5-phosphate reductoisomerase has been described by Virieux et al. In these structures, the hydroxamic acid moiety was replaced by various phosphinic acid based chelators linked to different functional groups capable of forming five- or six-membered chelating rings.[127]

(**270**) R= H,Me,Ph;X=OH,NH$_2$.

(**271**) R= Et,CH$_2$Ph,Ph,4–FC$_6$H$_4$.

(**272**) Ar= 2–AcOC$_6$H$_4$, 2-AcHNC$_6$H$_4$.

(**273**) Het= 2–pyridyl, 2–thiazolyl, 2–thienyl, 2–pyrimidinyl, 2–(4–trifluoromethyl) pyridyl

An efficient synthesis of sodium alendronate (**276**), an osteoporosis drug, has been developed through counterattack reaction involving treatment of the respective acyl phosphonate (**275**), obtained from the carbocylic acid (**274**), with P(OMe)$_3$ and TMSBr followed by deprotection of the bisphosphonate ester with aqueous HCl (Scheme 96).[128]

The microwave-assisted synthesis of polysubstituted pyrimidines (**279a**) and 1,3,5-triazines (**279b**) containing one or two phosphonic acid groups has been described, starting from the readily available chloro substituted derivatives (**277**). The Michaelis-Arbuzov reaction afforded the corresponding heteroaryl phosphonic esers (**278**) in good to excellent yields (72–93%), which when treated with bromotrimethylsilane, gave the free phosphonic acids (**279**) (Scheme 97).[129]

A convenient method for the synthesis of novel 1-isothiocyano-alkyl-phosphonate diaryl ester derivatives (**280**), building blocks in organic chemistry and anticancer agents, has been described by Oleksyszyn and co-workers. The synthesis was based on dithiocarbamates obtained *in situ* from carbon disulfide and aminoalkylphosphonates under basic conditions, and their desulfurisation using hydrogen peroxide (Scheme 98). The synthesised compounds demonstrated high antiproliferative activity against several cancer cell lines *in vitro*, and also showed some activity as serine protease inhibitors.[130]

New ligands (**291**) and (**292**) with a chelating pocket comprising phosphonate, phosphonic acid and mixed phosphonate/carboxylate units, have been constructed based on a pyridine-ethynylnaphthalene platform. Selective hydrolysis provided pockets with four to six anionic carboxylate or phosphate functions suitable for lanthanide complexation. The long emission lifetimes in the ms range and the high quantum yields displayed by the emergent lanthanide complexes of these ligands made them excellent candidates for bioconjugation and use in time-resolved fluoroimmuno assays.[131]

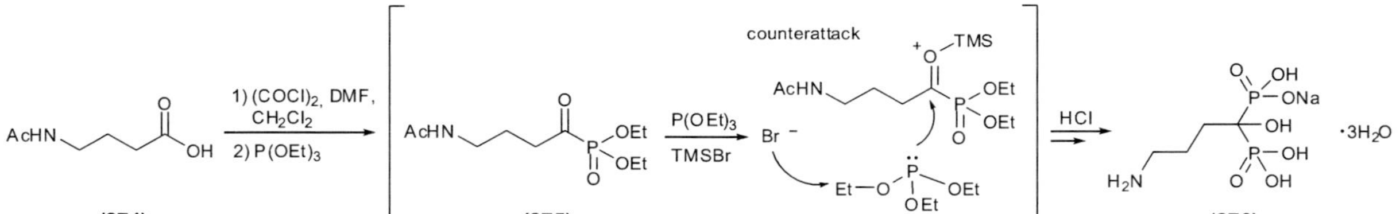

(274) **(275)** **(276)**

Scheme 96

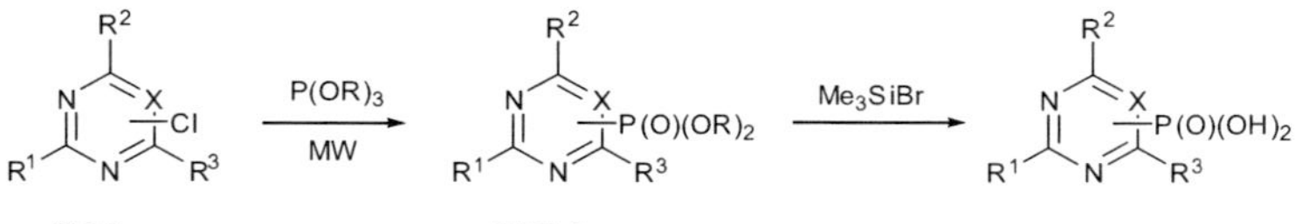

(277) **(278a)** X = CH **(278b)** X = N **(279a)** X = CH **(279b)** X = N

R= *i*-Pr, Et; R^1= F, NH_2; R^2= NH_2, Cl; R^3= NH_2, OH, Cl.

Scheme 97

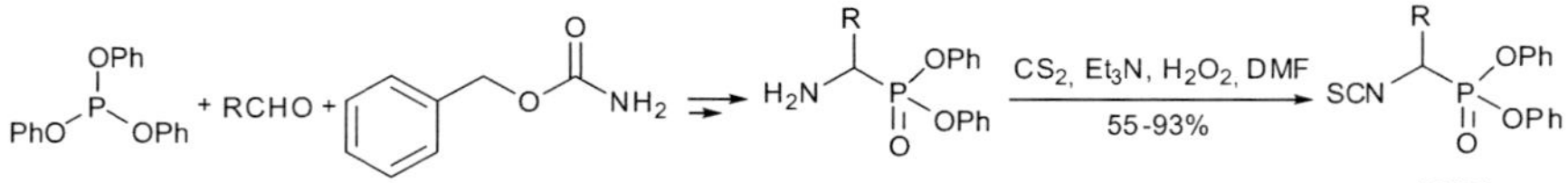

(290)

R= Me, *n*-Bu, *n*-C_5H_{11}, *n*-C_9H_{19}, $CH(CH_3)CH_2CH_3$, $(CH_2)_2Ph$, Ph, 4-$NO_2C_6H_4$, 4-ClC_6H_4, 3,4-$(MeO)_2C_6H_3$.

Scheme 98

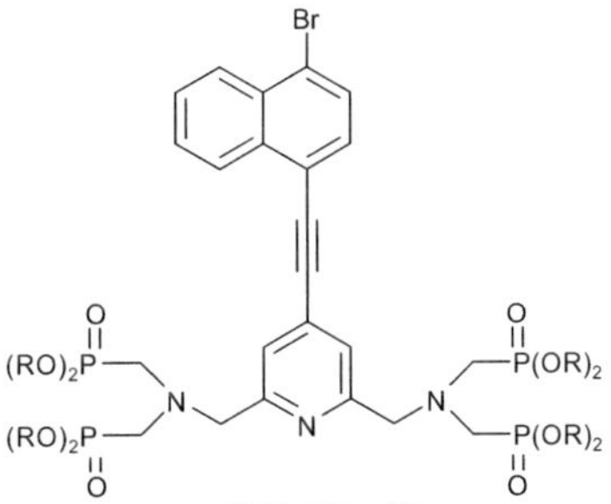

(291a) R = Et
(291b) R = H

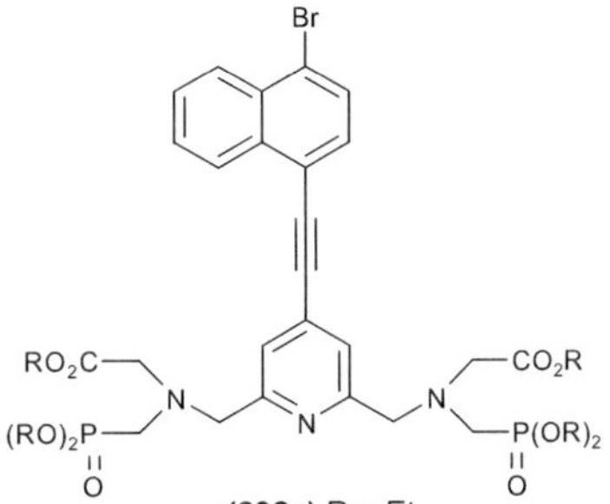

(292a) R = Et
(292b) R = H

2.2 Reactions of phosphonic acids and their derivatives

A novel approach to cyclic phosphonates (296) has been developed *via* a quinine thiourea catalysed, enatioselective Michael addition of quaternary α-nitroalkylphosphonates (293) to enones (294). Selected nitrophosphonates (295) have been conveniently transformed to the phosphonates (296) and (297) *via in situ* reduction-intramolecular cyclisation or Baeyer-Villiger oxidation followed by *in situ* reduction-intramolecular cyclisation, respectively (Scheme 99).[132]

Michael addition of α-substituted α-nitroalkylphosphonates (293) to various nitroolefins (299) have been shown to produce α,γ-diaminophosphonic esters (300) in the presence of a quinine-derived thiourea catalyst (298), with high diastereo- and enantioselectivity (Scheme 100).[133]

Enantioselective synthesis of tertiary α-hydroxyalkylphosphonates (301) has been realised by addition reaction of α-ketophosphonates (302) and trimethylsilyl cyanide (TMSCN) catalysed by carbohydrate/ thiourea organocatalysts (303) (Scheme 101).[134]

Highly enantioselective aldol reactions of acetylphosphonates (304), in which the acetylphosphonate was directly used as an enolate precursor for the first time, and activated carbonyl compounds (305) have been realised with cinchona alkaloid derived catalysts (307) or (308). Through an *in situ* methanolysis of the phosphonate group of the original aldol

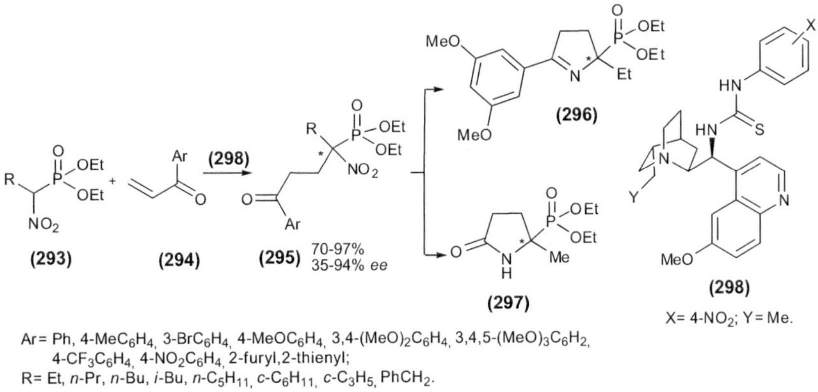

Ar= Ph, 4-MeC$_6$H$_4$, 3-BrC$_6$H$_4$, 4-MeOC$_6$H$_4$, 3,4-(MeO)$_2$C$_6$H$_4$, 3,4,5-(MeO)$_3$C$_6$H$_2$, 4-CF$_3$C$_6$H$_4$, 4-NO$_2$C$_6$H$_4$, 2-furyl,2-thienyl;
R= Et, *n*-Pr, *n*-Bu, *i*-Bu, *n*-C$_5$H$_{11}$, *c*-C$_6$H$_{11}$, *c*-C$_3$H$_5$, PhCH$_2$.

X= 4-NO$_2$; Y = Me.

Scheme 99

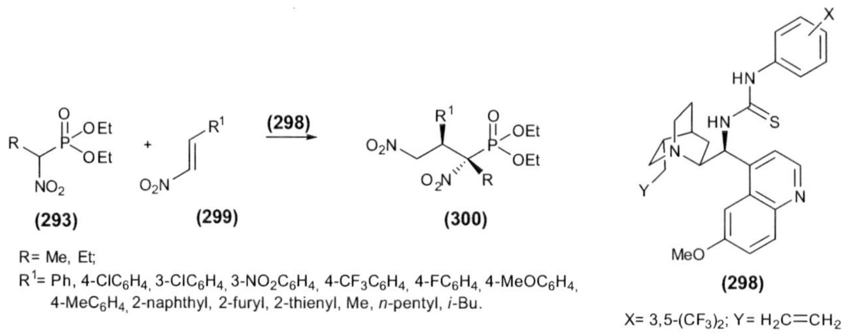

Scheme 100

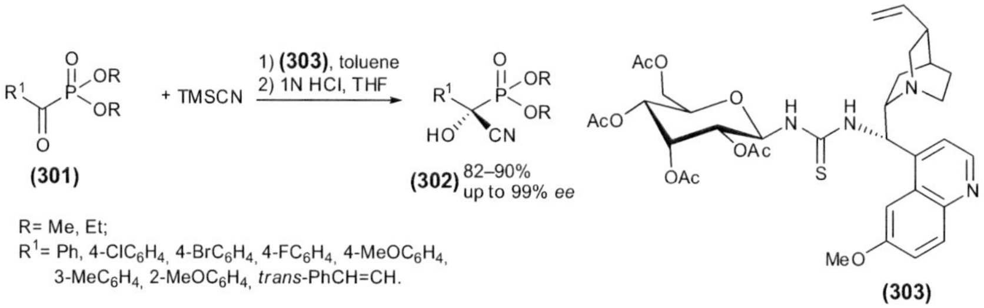

R= Me, Et;
R¹= Ph, 4-ClC$_6$H$_4$, 3-ClC$_6$H$_4$, 3-NO$_2$C$_6$H$_4$, 4-CF$_3$C$_6$H$_4$, 4-FC$_6$H$_4$, 4-MeOC$_6$H$_4$, 4-MeC$_6$H$_4$, 2-naphthyl, 2-furyl, 2-thienyl, Me, n-pentyl, i-Bu.

(298) X= 3,5-(CF$_3$)$_2$; Y= H$_2$C=CH$_2$

Scheme 101

R= Me, Et;
R¹= Ph, 4-ClC$_6$H$_4$, 4-BrC$_6$H$_4$, 4-FC$_6$H$_4$, 4-MeOC$_6$H$_4$, 3-MeC$_6$H$_4$, 2-MeOC$_6$H$_4$, trans-PhCH=CH.

products obtained in this reaction, the corresponding esters (**306**) were obtained (Scheme 102). It was found that the presence of the *N*-trityl protecting group was necessary for achieving high enantioselectivity in this reaction.[135]

2-Hydrazinoethylphosphonic (**309**) acid has been prepared by the Michael addition of hydrazine hydrate to diethyl ethenylphosphonate, followed by hydrolysis with hydrochloric acid. The product has been then converted in moderate to good yield into five- and six-membered heterocycles, (**310**) and (**311**), respectively (Scheme 103).[136]

Direct esterification of a labile alkoxyamine (**312**) has been achieved by the Mitsunobu reaction or a nucleophilic substitution. Thus, corresponding phosphonic ester derivatives (**313**) and (**314**) have been obtained under smooth conditions (Scheme 104).[137]

An efficient method involving the use of chiral phosphoric acids (**119**) as catalysts in the asymmetric Mannich reaction of dialkyl diazomethylphosphonate (**315**) and *N*-carbamoyl imines (**316**) has been developed. The reaction proceeded smoothly and produced the corresponding β-amino-α-diazophosphonates (**317**) with up to 97% yields and >99% *ee* (Scheme 105).[138]

(Halodiazomethyl)phosphonates (**319**), generated by a one-pot procedure from diethyl diazomethylphosphonate (**318**), have been applied in the intermolecular Rh-catalysed cyclopropanation to afford the corresponding halocyclopropylphosphonates (**320**) in moderate to high yields and high diastereomeric ratios (Scheme 106).[139]

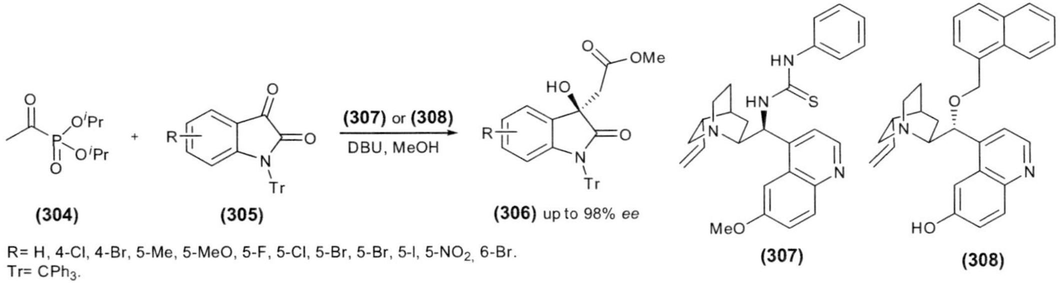

(304)
(305)
(307) or (308)
DBU, MeOH
(306) up to 98% ee
(307)
(308)
R= H, 4-Cl, 4-Br, 5-Me, 5-MeO, 5-F, 5-Cl, 5-Br, 5-Br, 5-I, 5-NO2, 6-Br.
Tr= CPh3.

Scheme 102

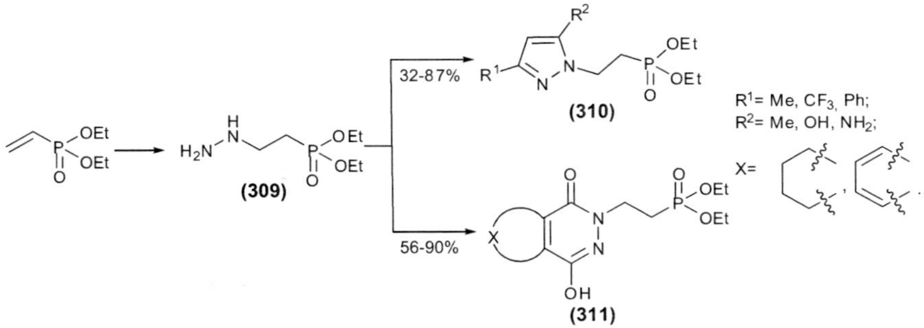

(309)
32-87%
56-90%
(310)
(311)
R1= Me, CF3, Ph;
R2= Me, OH, NH2;
X=

Scheme 103

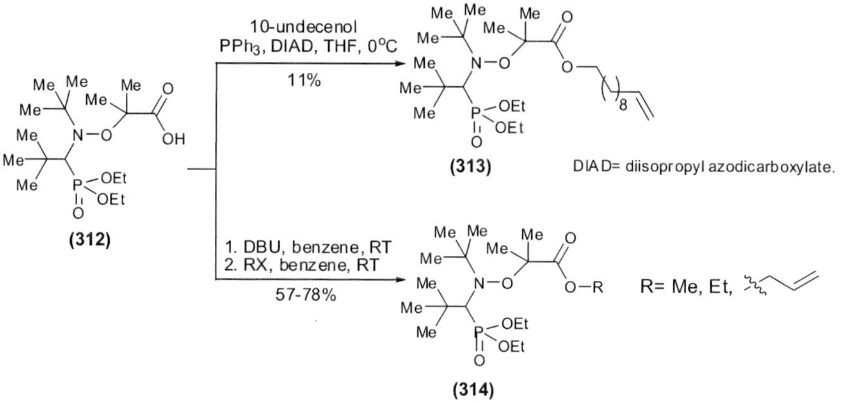

Scheme 104

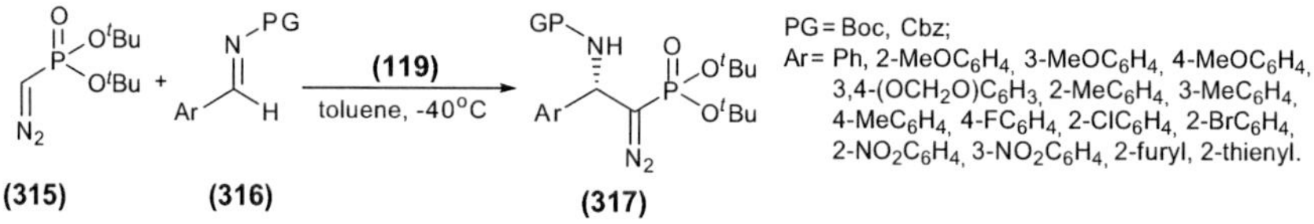

Scheme 105

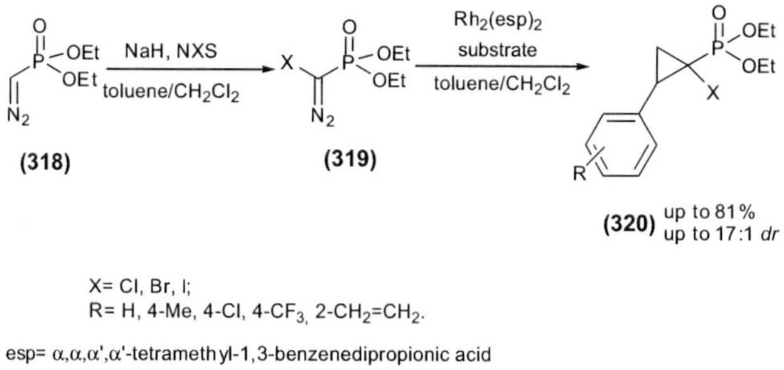

Scheme 106

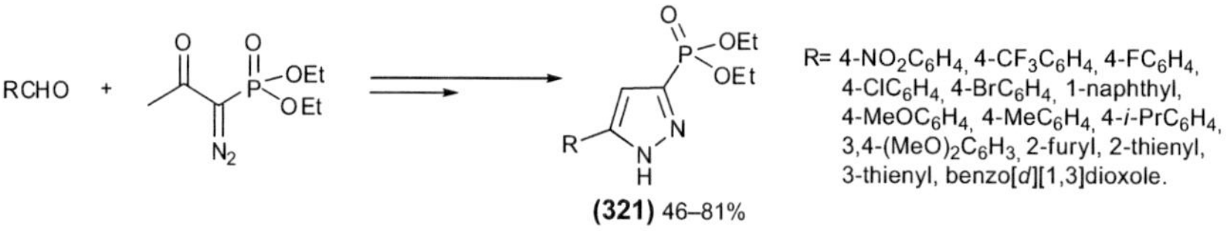

X= Cl, Br, I;
R= H, 4-Me, 4-Cl, 4-CF$_3$, 2-CH$_2$=CH$_2$.

esp= α,α,α',α'-tetramethyl-1,3-benzenedipropionic acid

Scheme 107

A one-pot, two-step method for the conversion of commercially available aldehydes to 3-phosphorylpyrazoles (**321**) has been developed by Namboothiri and co-workers (Scheme 107).[140]

Tverdomed *et al.* have described a new synthetic approach towards the syntheses of phosphorylated polysubstituted 4-aminoquinolines (**324**) through a heterocyclisation reaction. Various CF_2-containing alkynylphosphonates and alkynediylbis(phosphonates) (**323**) underwent heterocyclisation in good to excellent yields with 2-amino-benzonitriles (**322**) mediated by K_2CO_3 (Scheme 108).[141]

A metal-catalysed three-component coupling reaction of α-diazophosphonates (**325**), anilines, and aldehydes using a chiral rhodium catalyst, afforded a series of α-amino-β-hydroxyphosphonates (**326**) in good to high yields and with good to high enantioselectivities (Scheme 109).[142]

Highly substituted four- and five-membered heterocycles (**328**) have been prepared starting from phosphonates (**327**) by using a one-pot method involving base induced cyclisation and the Horner-Wadsworth-Emmons olefination reaction (Scheme 110).[143]

A convenient and general protocol for the asymmetric hydrogenation of a variety of alkyl- and aryl-substituted β-ketophosphonates (**329**) in the presence of ruthenium catalyst (**330**) has been developed by Zhang and co-workers. Good to excellent enantioselectivities (up to 99.9% *ee*) and excellent diastereoselectivities (96 : 4) have been obtained (Scheme 111).[144]

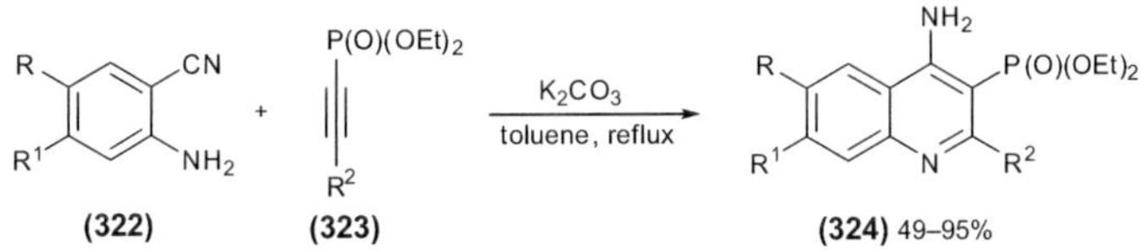

R= H, NO_2, MeO; R^1= H, MeO; R^2= CF_3, C_2F_5, CF_2Cl, CF_2Br, CF_2H, $P(O)(OEt)_2$.

Scheme 108

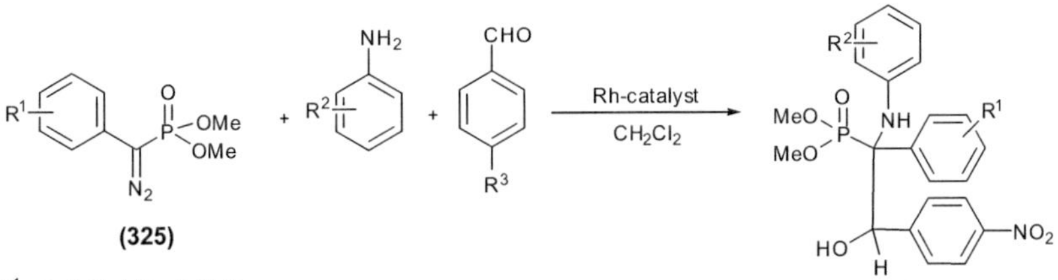

R^1= H, 4-Cl, 4-Me, 4-MeO;
R^2= 2-Me, 2-Br, 2-Cl, 4-Cl, 3-Cl, 2-Me, 3-MeO, 2-F;
R^3= NO_2, CO_2Me.

Scheme 109

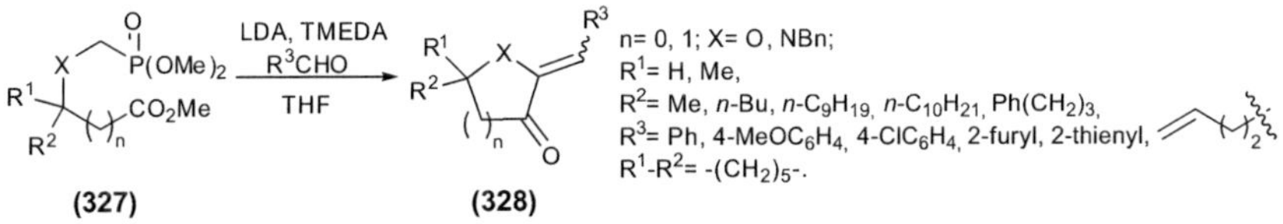

n= 0, 1; X= O, NBn;
R^1= H, Me,
R^2= Me, *n*-Bu, *n*-C_9H_{19}, *n*-$C_{10}H_{21}$, $Ph(CH_2)_3$,
R^3= Ph, 4-$MeOC_6H_4$, 4-ClC_6H_4, 2-furyl, 2-thienyl,
R^1-R^2= -$(CH_2)_5$-.

Scheme 110

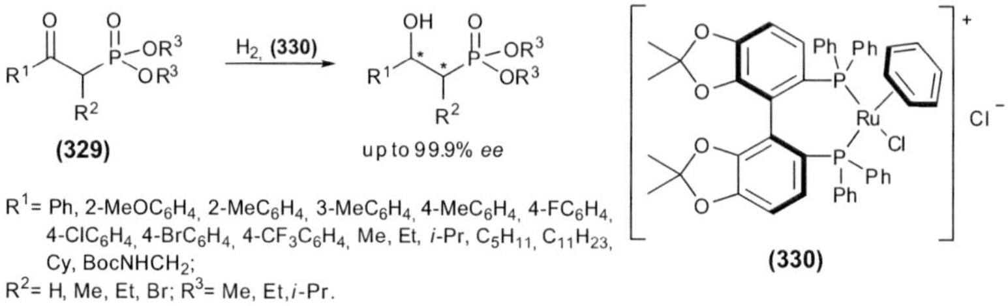
H2, (330)
up to 99.9% ee
R1= Ph, 2-MeOC6H4, 2-MeC6H4, 3-MeC6H4, 4-MeC6H4, 4-FC6H4, 4-ClC6H4, 4-BrC6H4, 4-CF3C6H4, Me, Et, i-Pr, C5H11, C11H23, Cy, BocNHCH2;
R2= H, Me, Et, Br; R3= Me, Et, i-Pr.
(329)
(330)

Scheme 111

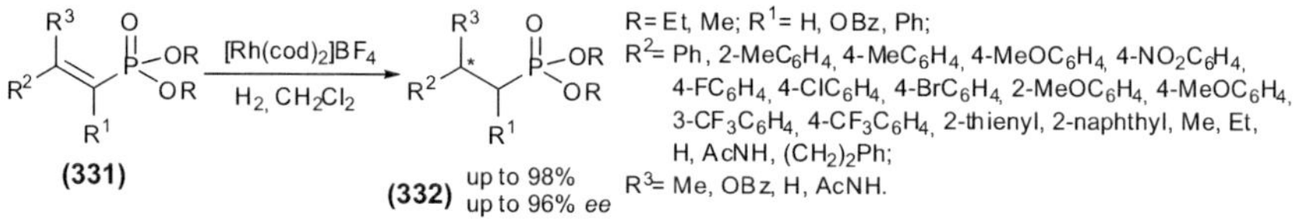
[Rh(cod)2]BF4
H2, CH2Cl2
(331)
(332)
up to 98%
up to 96% ee
R= Et, Me; R1= H, OBz, Ph;
R2= Ph, 2-MeC6H4, 4-MeC6H4, 4-MeOC6H4, 4-NO2C6H4, 4-FC6H4, 4-ClC6H4, 4-BrC6H4, 2-MeOC6H4, 4-MeOC6H4, 3-CF3C6H4, 4-CF3C6H4, 2-thienyl, 2-naphthyl, Me, Et, H, AcNH, (CH2)2Ph;
R3= Me, OBz, H, AcNH.

Scheme 112

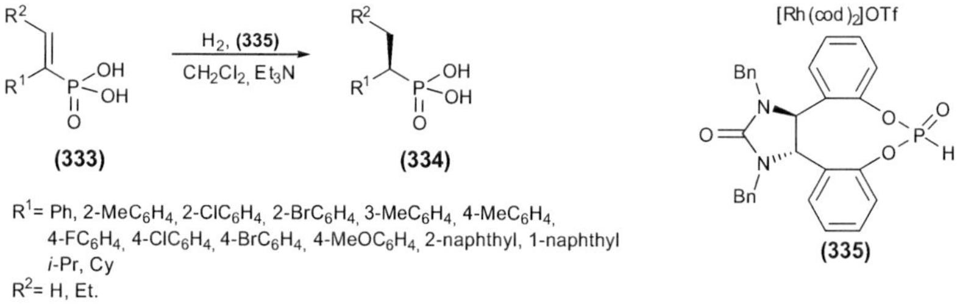
H2, (335)
CH2Cl2, Et3N
(333)
(334)
[Rh(cod)2]OTf
(335)
R1= Ph, 2-MeC6H4, 2-ClC6H4, 2-BrC6H4, 3-MeC6H4, 4-MeC6H4, 4-FC6H4, 4-ClC6H4, 4-BrC6H4, 4-MeOC6H4, 2-naphthyl, 1-naphthyl i-Pr, Cy
R2= H, Et.

Scheme 113

The use of a rhodium catalyst has been examined in the hydrogenation of functionalised α,β-unsaturated phosphonates (**331**). The reaction produced the corresponding chiral phosphonates (**332**) in good yields with high enantioselectivities (Scheme 112). This catalytic system is potentially practical for the synthesis of optically active β-substituted alkylphosphonates.[145]

A chiral rhodium complex (**335**) demonstrated unprecedented enantioselectivities (98–99% *ee*) in asymmetric hydrogenations of a wide variety of α-aryl-/alkyl-substituted ethenylphosphonic acids (**333**), providing a facile approach to corresponding enantiopure phosphonic acids (**334**) (Scheme 113).[146]

The synthesis of diethyl cyclopenta[*c*]furanyl phosphonates (**337**) has been realised by molybdenum-mediated cyclisation of 3-allyloxy-1-propynylphosphonates (**336**) (Scheme 114).[147]

A selective metalation of ω-sulphonyl substituted phosphonates (**338**) resulted in sufficiently stable carbanions that underwent the chemoselective Julia-Kocienski condensation with various aldehydes to provide (*E*)-allylic phosphonates (**339**) in good yields (up to 85%) and selectivities (up to 95 : 5). Synthesis of *trans*-dienes, trienes, and tetraenes (**340**) based on the Horner-Wadsworth-Emmons olefination of the resulting allylic phosphonates (**339**), has been also described (Scheme 115).[148]

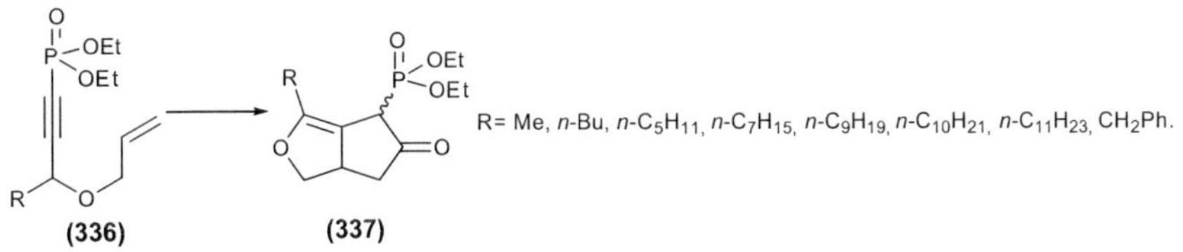

Scheme 114

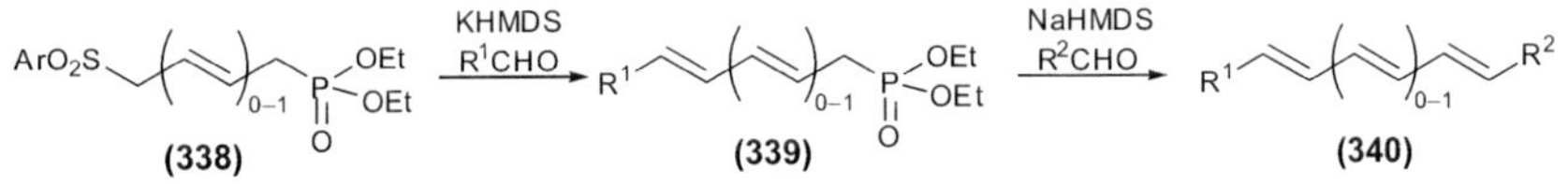

R= Me, n-Bu, n-C$_5$H$_{11}$, n-C$_7$H$_{15}$, n-C$_9$H$_{19}$, n-C$_{10}$H$_{21}$, n-C$_{11}$H$_{23}$, CH$_2$Ph.

Scheme 115

Ar= 1-phenyl-1H-tetrazole, 2-benzo[d]thiazole;
R^1= Ph(CH$_2$)$_2$, Et, i-Pr, t-Bu, n-C$_5$H$_{11}$, c-C$_6$H$_{11}$, c-C$_5$H$_9$, (E)-Ph(CH)$_2$;
R^2= t-Bu, c-C$_6$H$_{11}$.

Spilling and Roy have reported a short and very efficient syntheses of both diastereomers of the naturally occurring oxylipids (**342a**) and (**342b**) using a combination of alkene cross metathesis, palladium (0) cyclisation followed by the oxidative cleavage of phosphonates (**341a**) and (**341b**) (Scheme 116).[149]

A cross metathesis (CM) reaction involving γ,δ-unsaturated β-ketophosphonate (**343**) catalysed by a ruthenium catalyst, has been also described by Cossy and co-workers. Depending on the olefinic partners and conditions, a CM/1,4-addition sequence afforded functionalised tetrahydrofurans (**344**), tetrahydropyrans (**345**), and pyrrolidines (**346**) (Scheme 117).[150]

A novel asymmetric [4 + 2] cycloaddition of β,γ-unsaturated α-ketophosphonates (**347**) with allenic esters (**348**), afforded the corresponding phosphonate-substituted functionalised pyran (**349a**) or dihydropyran (**349b**) derivatives in good yields and enantioselectivities which were important structural subunits in many natural products, drugs and biologically active molecules (Scheme 118).[151]

A protocol for the two-carbon homologation of esters (**350**) to α,β-unsaturated esters (**352**) has been described by Jamison and Webb. Application of multireactor allowed to carry out three-operations procedure involving the ester (**350**) reduction, phosphonate (**351**) deprotonation, and the Horner-Wadsworth-Emmons olefination (Scheme 119).[152]

Yamamoto et al. have successfully developed the first copper-catalysed enantioselective Claisen rearrangement of enolphosphonates (**353**). A number of desired α-ketophosphonates (**354**) have been obtained in excellent yields and stereoselectivities (Scheme 120).[153]

A remarkable, synthetic pathway for conversion of readily available phospha-Wittig-Horner reagent (**355**) into highly functionalised phospholes (**356**) and 1,2-oxaphospholes (**357**) has been described by Ott and co-workers (Scheme 121).[154]

Diethyl α-hydroxy-benzylphosphonate (**358**) underwent nucleophilic substitution with primary amines under solventless MW conditions to afford the corresponding α-aminophosphonates (**359**) (Scheme 122).

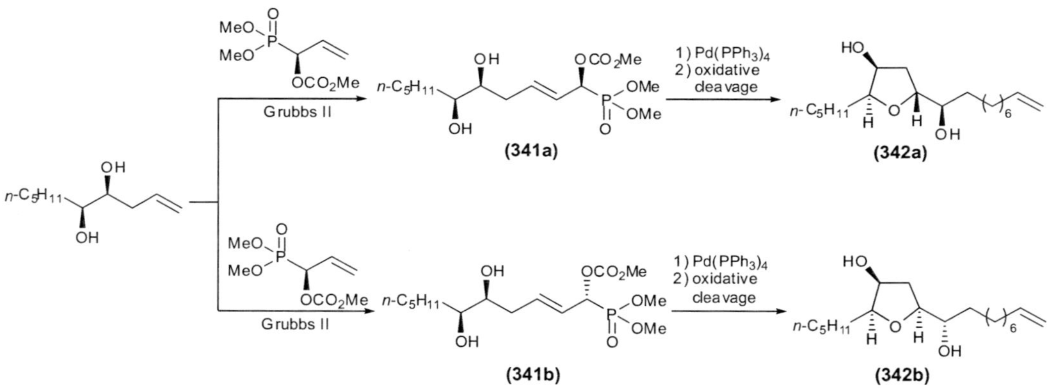

Scheme 116

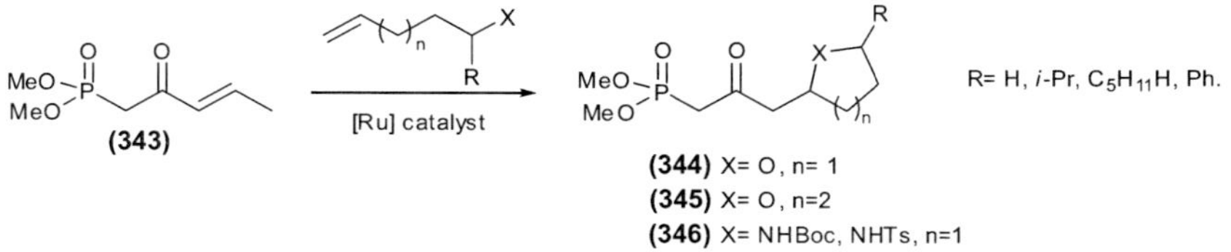

Scheme 117

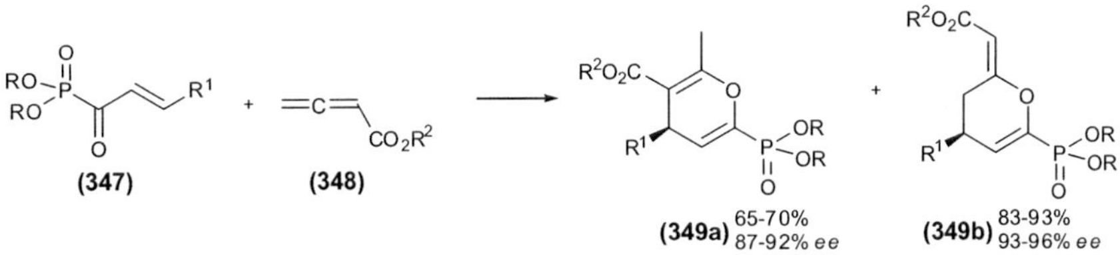

(347)
(348)
(349a) 65-70% 87-92% ee
(349b) 83-93% 93-96% ee

R= *i*-Pr, Me, *t*-Bu;
R^1= Me, *n*-Pr, Ph, 4-ClC_6H_4, 4-BrC_6H_4, 4-MeC_6H_4, 4-MeOC_6H_4, 3-MeC_6H_4, 3-MeOC_6H_4, 2-furyl, 1-naphthyl;
R^2= Et, Bn.

Scheme 118

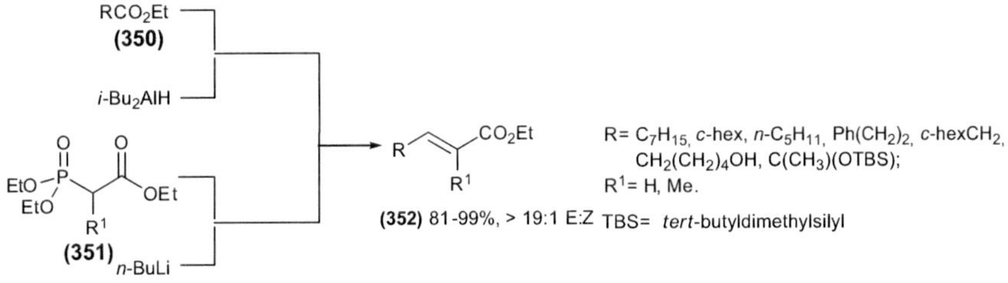

RCO_2Et
(350)
i-Bu_2AlH
(351)
n-BuLi
(352) 81-99%, > 19:1 E:Z
R= C_7H_{15}, *c*-hex, *n*-C_5H_{11}, $Ph(CH_2)_2$, *c*-hexCH_2, $CH_2(CH_2)_4OH$, $C(CH_3)(OTBS)$;
R^1= H, Me.
TBS= *tert*-butyldimethylsilyl

Scheme 119

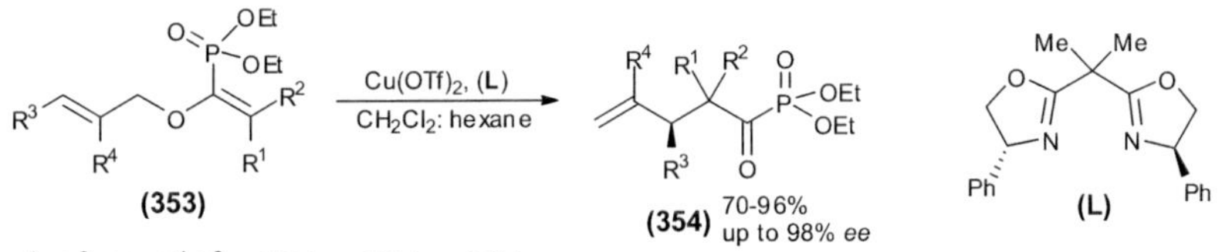

$Cu(OTf)_2$, (L)
CH_2Cl_2: hexane
(353)
(354) 70-96% up to 98% ee
(L)

R^1= R^2= Me; R^1-R^2= -$(CH_2)_3$-, -$(CH_2)_4$-, -$(CH_2)_5$-;
R^3= Ph, 1-naphthyl, 2-naphthyl, CH_2Ph, Me, *n*-hex, 4-BrC_6H_4, *trans*-PhCH=CH;
R^4= H, Me, Br.

Scheme 120

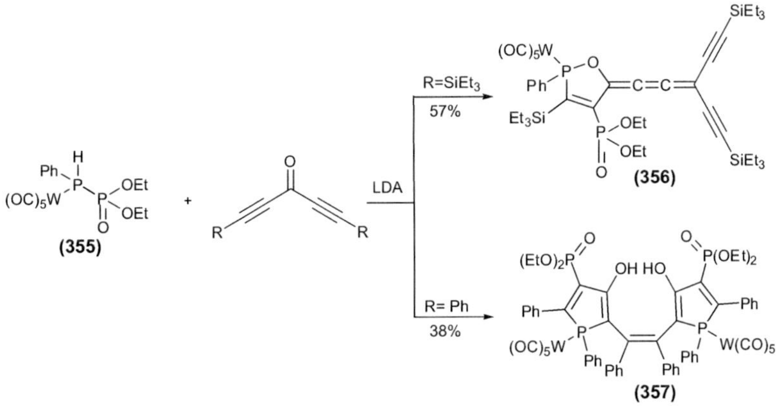

(355)
LDA
R=$SiEt_3$
57%
(356)
R= Ph
38%
(357)

Scheme 121

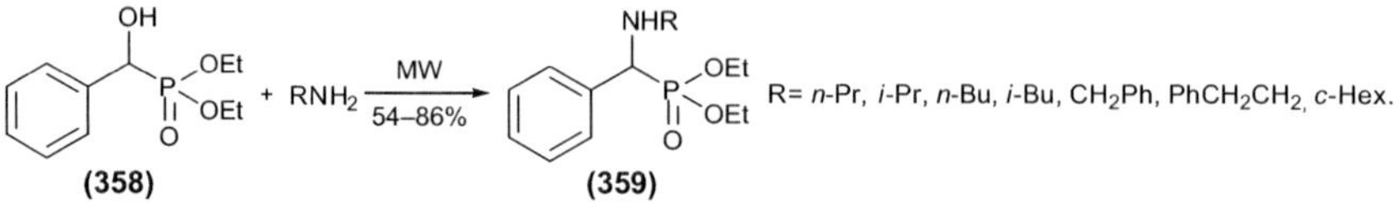

Scheme 122

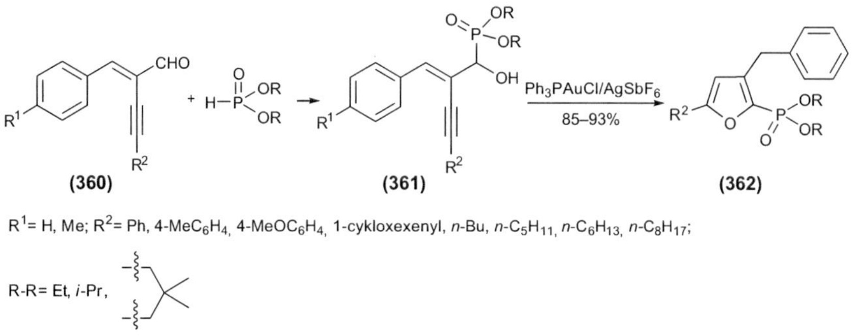

Scheme 123

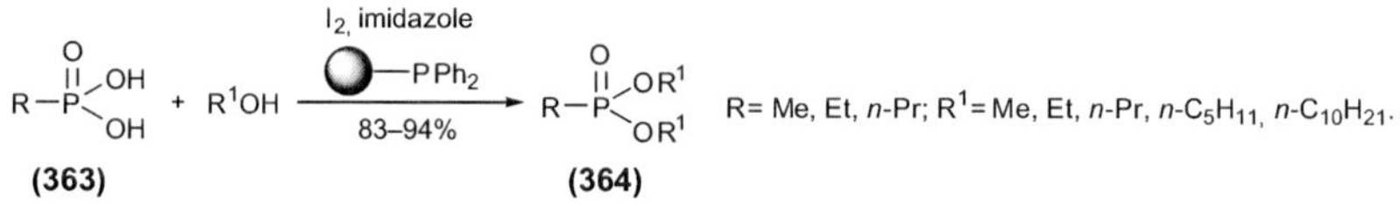

R^1 = H, Me; R^2 = Ph, 4-MeC$_6$H$_4$, 4-MeOC$_6$H$_4$, 1-cykloxenyl, n-Bu, n-C$_5$H$_{11}$, n-C$_6$H$_{13}$, n-C$_8$H$_{17}$;

R-R = Et, i-Pr,

Scheme 124

The substitution was enhanced by the neighbouring effect of the adjacent P=O group.[155]

The reaction of 1-hydroxy substituted phosphonates (**361**), obtained by addition of the corresponding dialkyl phosphites to 2-alkynylcinnamaldehydes (**360**), underwent cycloisomerisation in the presence of AgOTf or Ph$_3$PAuCl-AgSbF$_6$ catalyst to afford 2-furylphosphonates (**362**) in good to excellent yields (Scheme 123).[156]

An efficient and convenient method for the synthesis of a variety of O,O'-dialkyl alkylphosphonates (**364**) from corresponding alkylphosphonic acids (**363**) using iodine, imidazole and polymer-bound triphenylphosphine has been developed by Dubeyl and co-workers (Scheme 124).[157]

Kusukawa et al. reported the formation of water-soluble hetero-capsules (**367**) resulting from the ionic interactions between positively-charged flexible aniline hydrochloride (**365**) and the negatively charged phosphonate (**366**) having rigid homooxacalix[3]arene units (Scheme 125). The formation of the molecular capsules was studied by NOESY, DOSY NMR spectroscopy and ESI-Mass spectrometry. The water-solubility of the capsules was improved by the introduction of mono- or triethylene glycol substituents in the homooxacalix [3]arene-based phosphonate units (**366**).[158]

The reaction of 1-chloroacetylenephosphonates (**368**) with 1H-1,2,4-triazole-3-thiols (**369**) in anhydrous acetonitrile afforded fused heterocycles, 6-(dialkoxyphosphoryl)-3H-thiazolo[3,2-b][1,2,4]triazol-7-ylium chlorides (**370**) with high regioselectivity (Scheme 126).[159]

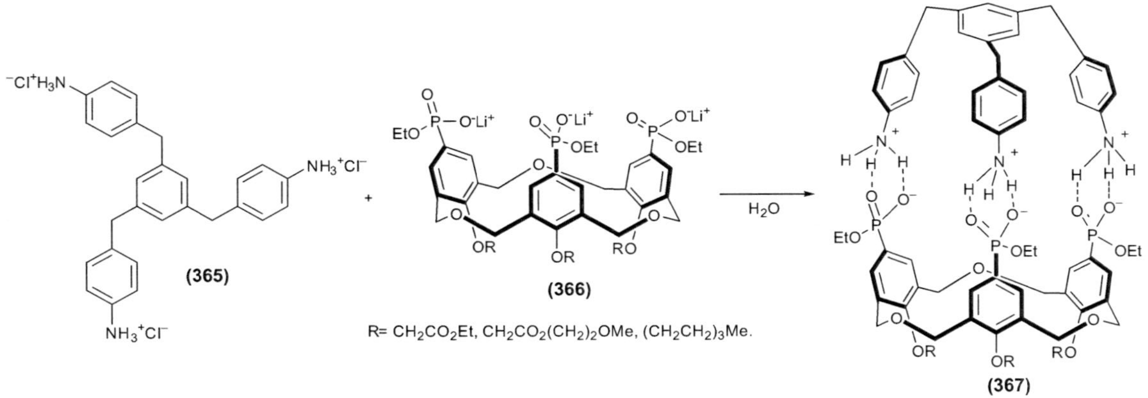
(365)
(366)
(367)
H2O
R= CH2CO2Et, CH2CO2(CH2)2OMe, (CH2CH2)3Me.

Scheme 125

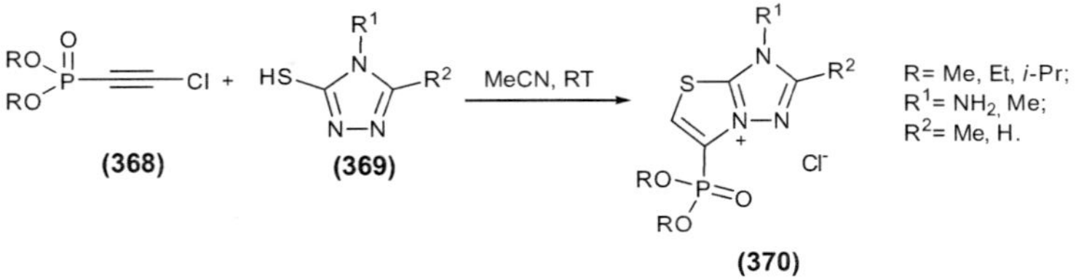
(368)
(369)
(370)
MeCN, RT
R= Me, Et, i-Pr;
R1= NH2, Me;
R2= Me, H.

Scheme 126

Two, exocyclic (**371a**) and endocyclic (**371b**), olefin tautomers of chromone phosphonates (**371**) have been obtained at different temperatures (exocyclic olefin −17 °C; endocyclic olefin −30 °C) by utilising an *N*-heterocyclic carbene catalyst (**372**) in the intramolecular reactions between the alkynylphosphonates and formyl group (Scheme 127). The exocyclic olefins could isomerise to the endocyclic derivatives completely when treated with a thiazolium salt precatalyst and potassium carbonate at 30 °C.[160]

The copper-catalysed aerobic oxidative trifluoromethylation of readily accessible dialkyl phosphites has been described for the first time, providing an alternative method for the facile synthesis of a series of biologically important CF_3-phosphonates (Scheme 128).[161]

N,N-Diethylamides of phenylphosphonic acid (**373**) have been subjected to a double modification of its phenyl substituent through the directed *ortho*-metalation followed by dearomatisation of the aryl substituent under Birch reduction conditions (Scheme 129).[162]

The catalytic direct aldol condensation of α-hydroxy trialkyl phosphonoacetates (**374**) to aldehydes to afford α-hydroxy β-phosphoryloxy esters (**375**) has been developed by Johnson and co-workers. The reaction worked well for a variety of alkyl, alkenyl, aryl, and heteroaryl aldehydes to afford the desired products in good to excellent yields but in low to moderate diastereoselectivities (Scheme 130).[163]

Severin and co-workers have demonstrated that borophosphonate cages of the general formula [*t*-BuPO$_3$BR]$_4$ (**376**) could be obtained in good to excellent yields by condensation of *tert*-butylphosphonic acid with boronic acids (Scheme 131).[164]

Treatment of α,α-dialkyl substituted phosphonoesters (**377**) with LiAlH$_4$ caused the one-pot reduction of the ester moiety followed by dephosphorylation to yield the corresponding primary alcohols (**378**) bearing a controllable β secondary carbon center (Scheme 132).[165]

The diethoxyphosphoryl group has been validated as an effective agent to achieve negative charge migration in Type II Anion Relay Chemistry (ARC). The process involved a [1,4]-phosphorus-Brook rearrangement that proceeded *via* a phosphacyclic intermediate (**379**). In the absence of an exogenous electrophile, the anion derived *via* phosphorus migration underwent internal displacement of the phosphoryl group to produce a diastereomeric mixture of cyclopropanes (**380**) (Scheme 133).[166]

An efficient access to (*Z*)-β-fluoroallyl alcohols (**382**), based on the two carbon homologation of aromatic aldehydes by the Horner–Wadsworth–Emmons reaction with 2-(diethoxyphosphoryl)-2-fluoro-ethanethioic acid *S*-ethyl ester (**381**), followed by reduction with sodium borohydride, has been devepoped by Rolando and co-workers (Scheme 134).[167]

A short route to enantiopure γ-butenolides (**383**) (up to 99% *ee*) has been developed from readily available starting materials. The strategy involved a sequential organocatalytic α-aminoxylation followed by the *cis*-Wittig olefination of aldehydes (Scheme 135).[168]

A novel three-component coupling reaction between trimethylsilylmethylphosphonates (**384**), acyl fluoride (**385**) and aldehydes has been developed by Matsuda and co-workers. A sequential nucleophilic addition of 1-lithio-trimethylsilylmethylphosphonates to (**385**) followed

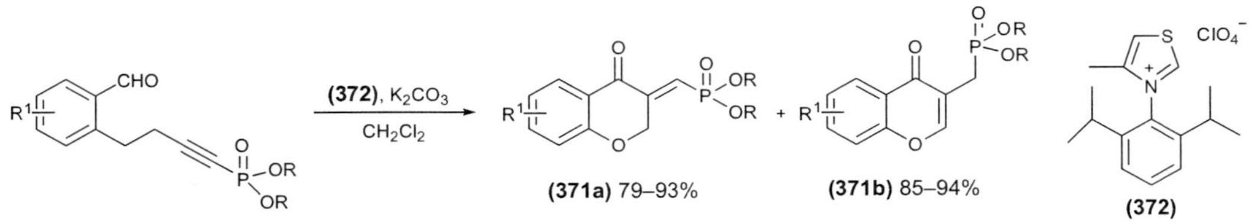

R= Et, *n*-Bu, *i*-Pr; R^1= H, 4-NO_2, 4-F, 6-Br-4-*t*-Bu, 6-Me, 4,6-(*i*-Bu)$_2$.

Scheme 127

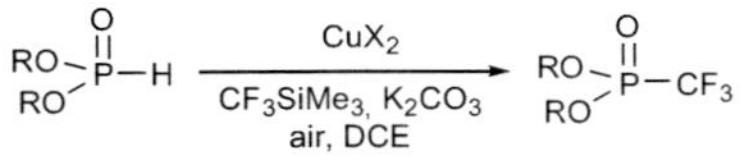

X= OH, OEt;
R= Et, *n*-Pr, *n*-Bu, *i*-Bu, $(CH_2)_5Me$, $(CH_2)_4Cl$, *c*-Hex.

Scheme 128

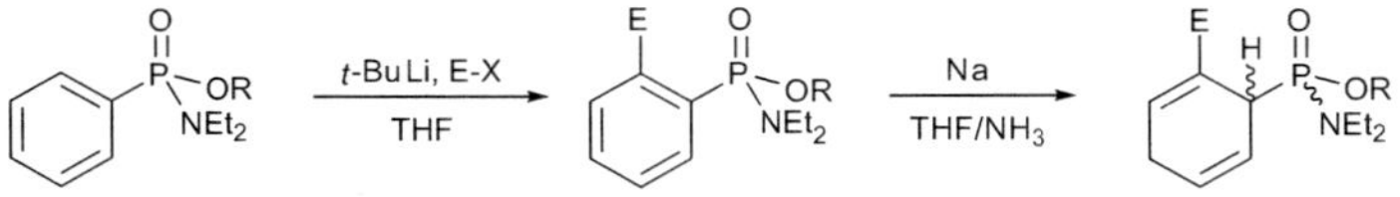

(373)

up to 99%

up to 89%,
up to 79:21 *dr*

R= Et, menthyl;
E= Me, $SiMe_3$, -CH(Ph)O-, -$C(CH_3)_2$O-, Et, *i*-Pr.

Scheme 129

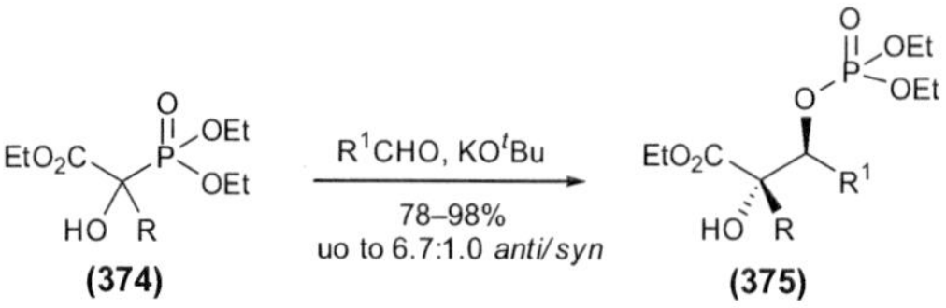

R= Me, CH_2Ph, CH_2CHCH_2, CH_2CHCH;
R^1= Ph, 4-ClC_6H_4, 2-FC_6H_4, 4-FC_6H_4, 2-$NO_2C_6H_4$, 3-$NO_2C_6H_4$, 4-$NO_2C_6H_4$, 4-$CF_3C_6H_4$, 4-CNC_6H_4, 4-MeC_6H_4, 4-$MeOC_6H_4$, 2-thienyl, (*E*)-CHCHPh, CH_2CH_2Ph.

Scheme 130

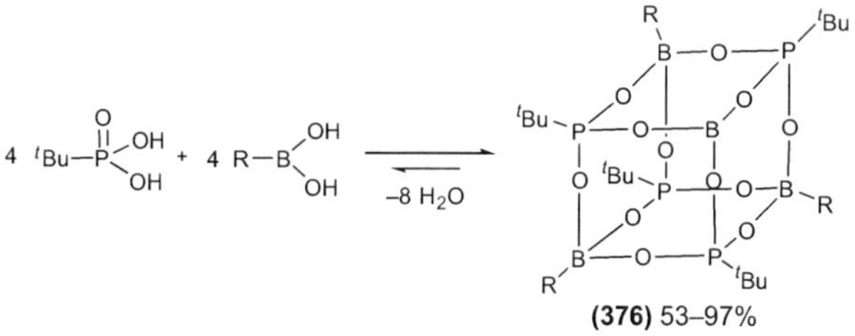

R= Me, *n*-Pr, *n*-Bu, Cy, Ph, *p*-Tol, 4-$CHOC_6H_4$, 3,5-$(CHO)_2C_6H_3$.

Scheme 131

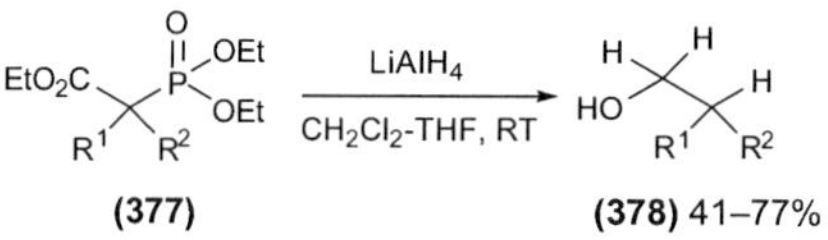

R^1= R^2= 4-pentenyl, *c*-Hex, *c*-C_5H_{10};
R^1= *n*-Pr, Et, 2-naphthyl;
R^2= 3-butenyl, CH_2Ph, Me, *c*-Hex.

Scheme 132

by the Horner-Wadsworth-Emmons reaction with aldehydes of the 1-lithio-β-ketophosphonate (**387**) (generated *in situ* by the fluoride anion desilylation of the α-silyl-β-ketophosphonate (**386**)), proceeded smoothly in a one-pot operation. Various *E*- and *Z*-enones (**379**) were obtained in high yields and with high stereoselectivities (Scheme 136).[169]

A short and practical synthesis of oseltamivir phosphate (Tamiflu) (**391**), an effective neuraminidase inhibitor, has been accomplished in 11 steps in 2.9% overall yield, from inexpensive and abundant D-mannose (**389**). The key step of the synthesis include formation of the cyclohexene ring (**390**) by an intramolecular Horner–Wadsworth–Emmons reaction (Scheme 137).[170]

The selective conversion of serine or threonine units of di- and tripeptides (**392**) into substituted, unsaturated aminoacids (**394**) has been reported by Hernández *et al.* Thus, α-aminoacids esters (**392**) underwent a scission-phosphorylation process to give phosphonate derivatives (**393**). The Horner-Wadsworth-Emmons reaction of the latter with aldehydes or ketones afforded the final products (**394**) with excellent *Z*-stereoselectivity (Z:E > 98:2) (Scheme 138).[171]

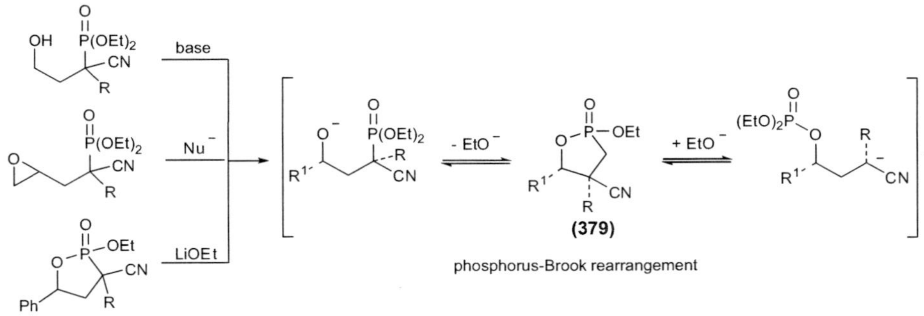

R= H, Me, CH_2Ph;
R^1= Nu or Ph.

Scheme 133

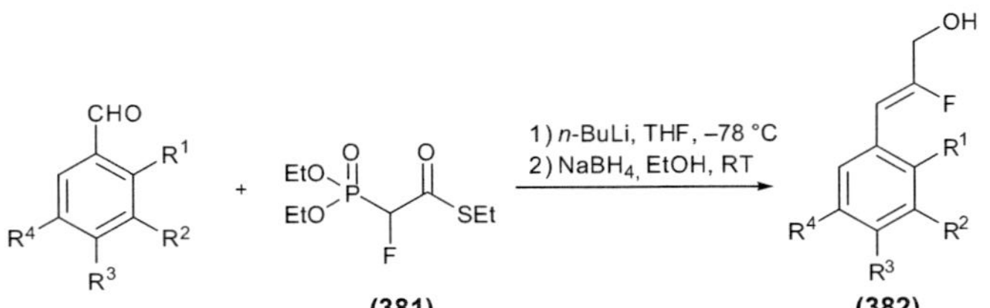

R^1= R^2= R^3= R^4= H;
R^1=MeO; R^2= GluO; R^3= Br, R^4= Me.
GluO= 2,3,4,6-tetra-O-acetyl-β-D-glucopyranosyl.

Scheme 134

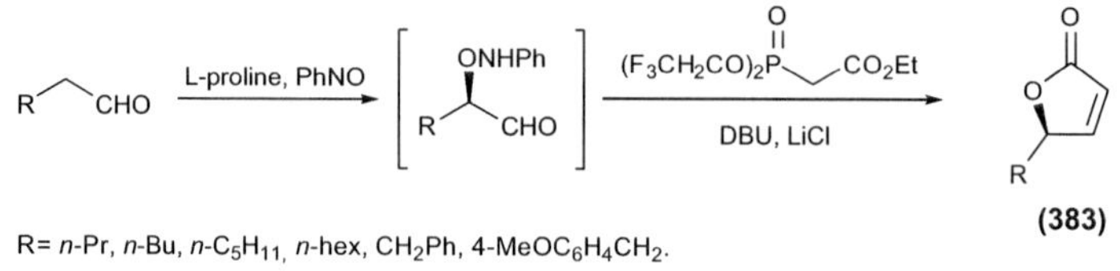
L-proline, PhNO
ONHPh
R
CHO
$(F_3CH_2CO)_2P$
CO_2Et
DBU, LiCl
(383)
R= n-Pr, n-Bu, n-C_5H_{11}, n-hex, CH_2Ph, 4-$MeOC_6H_4CH_2$.

Scheme 135

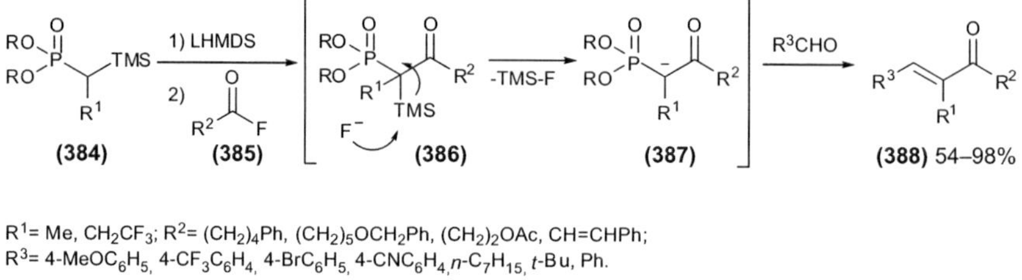
1) LHMDS
2)
-TMS-F
R^3CHO
(384)
(385)
(386)
(387)
(388) 54–98%
R^1= Me, CH_2CF_3; R^2= $(CH_2)_4Ph$, $(CH_2)_5OCH_2Ph$, $(CH_2)_2OAc$, CH=CHPh;
R^3= 4-$MeOC_6H_5$, 4-$CF_3C_6H_4$, 4-BrC_6H_5, 4-CNC_6H_4, n-C_7H_{15}, t-Bu, Ph.

Scheme 136

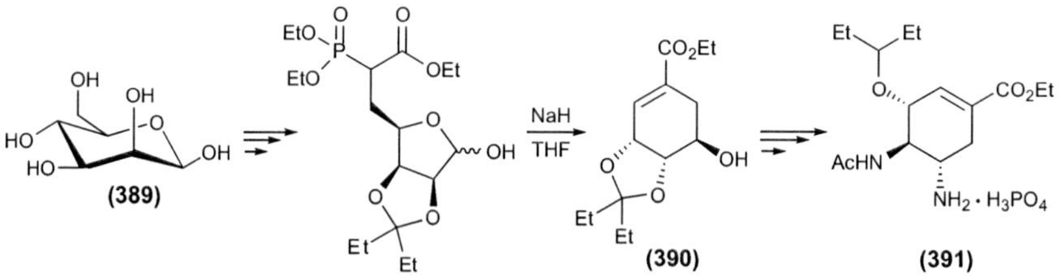
NaH
THF
(389)
(390)
(391)
$NH_2 \cdot H_3PO_4$

Scheme 137

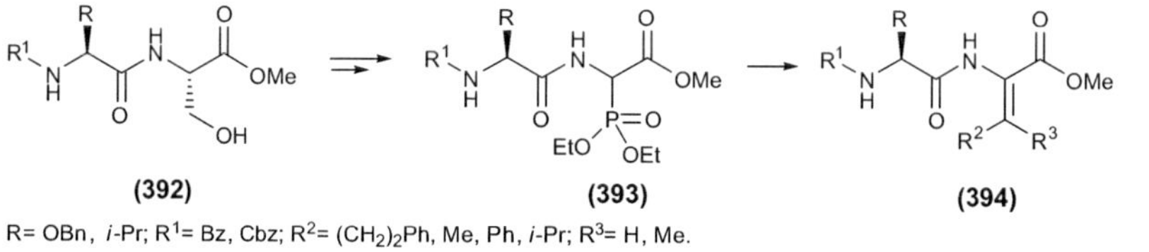
(392)
(393)
(394)
R= OBn, i-Pr; R^1= Bz, Cbz; R^2= $(CH_2)_2Ph$, Me, Ph, i-Pr; R^3= H, Me.

Scheme 138

The synthesis of two α-(diethoxyphosphoryl)-α,β-unsaturated γ-lactones (**396**) with 98% *ee* has been developed starting form (*S*)-3-aryl-3-hydroxy-2,2-dimethylpropanals (**395**) *via* a direct, asymmetric, organocatalytic aldol reaction. The resulting lactones (**396**) were then transformed into the corresponding α-methylene-γ-lactones (**397**) in a two-step process involving Michael addition followed by Horner-Wadsworth-Emmons olefination without loss of enantiomeric purity (Scheme 139).[172]

Jørgensen *et al.* have reported a new, asymmetric, catalytic strategies for the synthesis of biologically important α-methylene-δ-lactones (**401**)

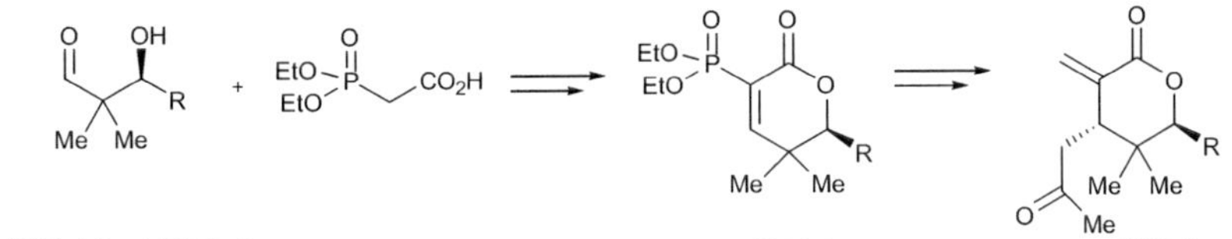

(**395a**) R= 4-NO$_2$C$_6$H$_4$
(**395b**) R= 4-NCC$_6$H$_4$

(**396a**)
(**396b**)

(**397a**)
(**397b**)

Scheme 139

and δ-lactams (**402**). The elaborated protocols utilised pyrrolidinium Michael addition of trimethyl phosphonoacetate (**398**) to α,β-unsaturated aldehydes (**399**) as the key step. Enantiomerically enriched Michael adducts (**400**) were next employed in two different reaction pathways. Transformation into α-methylene-δ-lactones (**401**) was realised by a sequence of reactions involving chemoselective reduction of the aldehyde, followed by a trifluoroacetic acid-mediated cyclisation and Horner–Wadsworth–Emmons olefination with formaldehyde. On the other hand, reductive amination of the Michael adducts (**400**) and cyclisation followed by the Horner–Wadsworth–Emmons olefination with formaldehyde gave α-methylene-δ-lactams (**402**) (Scheme 140).[173]

A series of new α-methylidene-δ-lactones with 3,4-dihydrocoumarin skeletons (**406**) have been synthesised in a three step reaction sequence. The Friedel–Crafts alkylation of phenols or naphthols using ethyl 3-methoxy-2-diethoxyphosphorylacrylate (**403**) in the presence of trifluoromethanesulphonic acid gave 3-diethoxyphosphorylchromen-2-ones (**404**) which were employed as Michael acceptors in the reaction with Grignard reagents. The obtained adducts (**405**) were finally used as Horner–Wadsworth–Emmons reagents for the olefination with formaldehyde (Scheme 141). All obtained 3-methylidenechroman-2-ones (**406**) were tested against two human leukemia cell lines as well as breast cancer and colon cancer adenocarcinomas.[174]

A novel strategy for the stereoselective synthesis of pentalenolactones has been successfully developed. One of the critical steps in this strategy was a TMS-promoted, intramolecular Michael/olefination reaction of the ketophosphonate (**407**) for the stereoselective synthesis of α-methylidene-δ-pentyrolactone (**408**). The implementation of this novel strategy allowed for the first reported stereoselective total synthesis of the methyl ester of pentalenolactone A (**409**) (Scheme 142).[175]

Details for the synthesis of a resorcylic acid lactone (**413**) incorporating a *trans*-enone and an amide in the macrocyclic ring have been desrcibed. The key step of this synthesis included intermolecular olefination reaction of the aldehyde (**410**) with phosphonate (**411**) which led to isolation of the (*E*)-enone (**412**). The complete removal of MOM and Boc groups from (**412**) followed by lactamisation gave the desired lactam (**413**) (Scheme 143).[176]

A convergent and flexible, synthetic approach to stereochemically defined spiroketals (**417**), has been reported by Azman and co-workers. Thus, the Horner–Wadsworth–Emmons olefination of the β-ketophosphonate (**414**) and the aldehyde (**415**), followed by a selective reduction of the

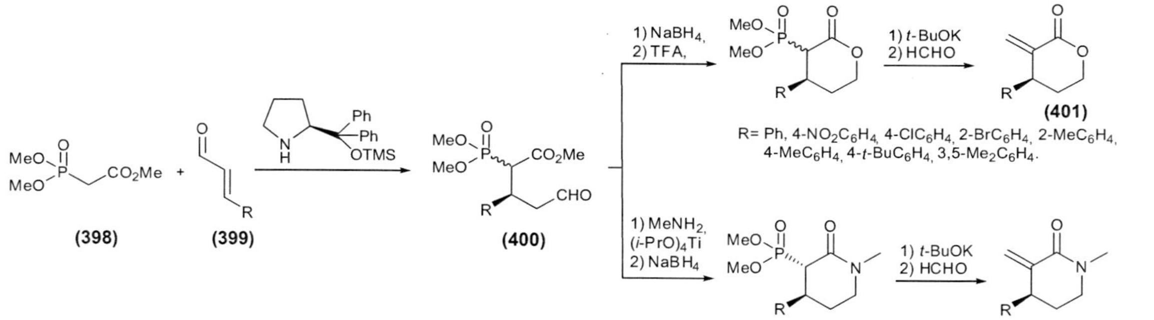

(398) (399) (400) (401) (402)

R= Ph, 4-NO$_2$C$_6$H$_4$, 4-ClC$_6$H$_4$, 2-BrC$_6$H$_4$, 2-MeC$_6$H$_4$, 4-MeC$_6$H$_4$, 4-t-BuC$_6$H$_4$, 3,5-Me$_2$C$_6$H$_4$.

R= Ph, 4-NO$_2$C$_6$H$_4$, 4-ClC$_6$H$_4$, 2-MeC$_6$H$_4$, 3,5-Me$_2$C$_6$H$_4$.

Scheme 140

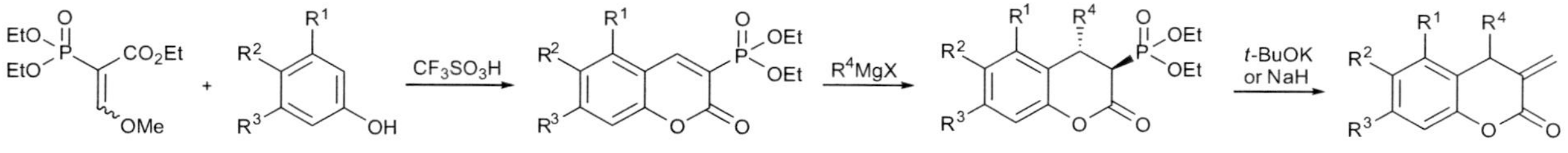

(403) (404) (405) (406)

R^1= H, OMe; R^2= H; R^3= H, OMe; R^1-R^2= CHCHCHCH, CHC(OH)CHCH; R^2-R^3= CH$_2$OCH$_2$;
R^4= Me, n-Bu, i-Pr, vinyl.

Scheme 141

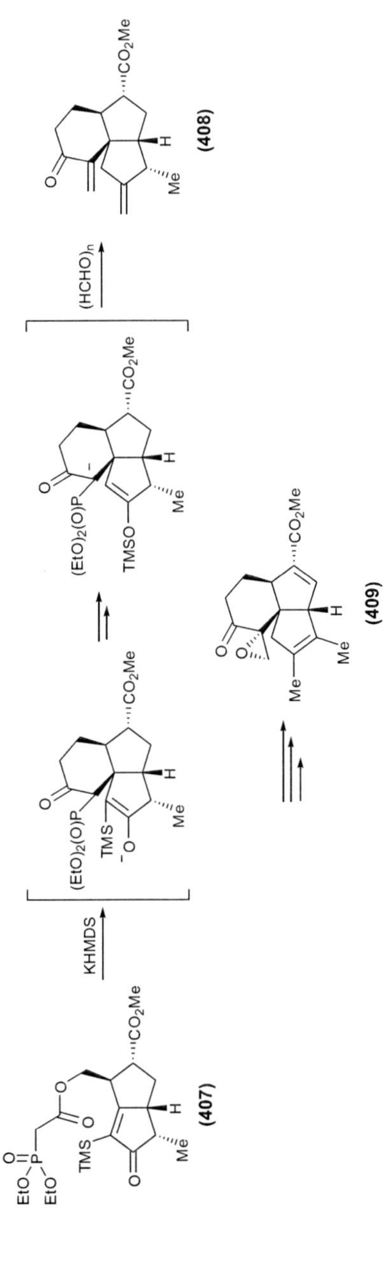

Scheme 142

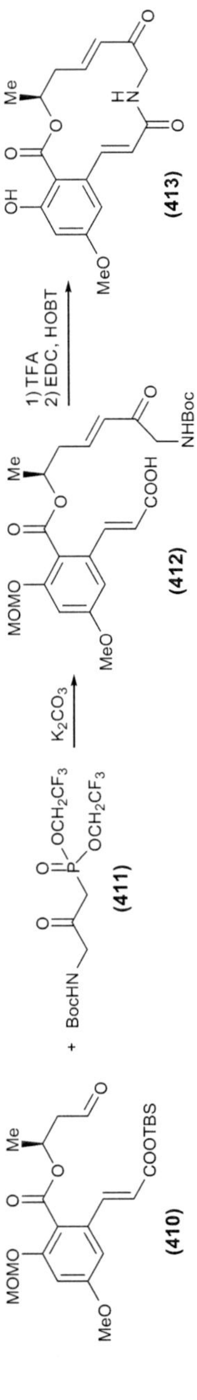

Scheme 143

EDC= 1-ethyl-3-(3-dimethylaminopropyl)carbodiimide,
HOBT= hydroxybenzotriazole.

α,β-unsaturated ketone, provided the saturated ketone (**416**). Deprotection of triethylsilyl ether and then treatment of the intermediate with a catalytic amount of PPTS converted the dihydroxyketone into the desired spiroketal (**417**) in 86% yield as a single diastereomer (Scheme 144).[177]

A significant effort toward the model study of the jatrophane skeleton (**421**) has been made by Mohan and co-workers. To synthesise an important synthon (**420**), Horner–Emmons–Wadsworth olefination using the phosphonate (**418**) and the bis-MOM aldehyde (**419**) was attempted (Scheme 145). Unfortunately, the two fragments did not couple under various conditions.[178]

An efficient synthesis of the fused morpholine oxadiazoline core structure (**429**) has been accomplished in an effort to identify optimum gamma secretase modulator (GSM) leads in treatment of Alzheimer's disease. A highly diastereoselective synthesis of the potent GSM (**429**) started from the bromooxime (**422**) which underwent [3+2] dipolar cycloaddition with the ketimine (**423**) to provide the oxadiazoline lactone (**424**). The lactone (**424**) was then converted in three steps to the bromide (**426**), which treated with the aldehyde (**427**) gave the vinyl-bromide (**428**), the key product in synthesis of the seven-member ring product (**429**) (Scheme 146).[179]

Al-Horani and Desai have reported two complementary, glycine donor-based strategies for synthesis of highly substituted, electron rich 1,2,3,4-tetrahydroisoquinoline-3-carboxylic acid esters (**432**) using the Boc-α-phosphonoglycine trimethyl ester (**430**) and the Cbz-α-phosphonoglycine trimethyl ester (**431**). A high yielding approach for the desired esters (**432**) has been best achieved using the ester (**431**) in three mild and efficient steps including Horner–Wittig–Emmons olefination with aromatic aldehydes (Scheme 147).[180]

Brynaert and co-workers have successfully exploited the [4+2] cycloaddition of 1-diethoxyphosphoryl-1,3-butadiene (**434**) and N-substituted maleimide derivatives (**435**) to produce a small library of phosphorylated bicyclic cycloadducts (**436**), particularly well adapted for coordination with metal cations (Scheme 148). A short arm (five to six atoms) terminated by different functional groups and attached to the imide nitrogen atom has been designed to achieve a good coordinating pocket. Metal coordination properties of the ligands equipped with functionalised N-lateral chains were proven by an ESI-HRMS study. The stoichiometry of one selected Eu^{III} complex with a diphosphorylated ligand has been also determined by photoluminescence spectroscopy in the emission mode.[181]

A series of novel chiral-bridged atropisomeric monophosphine ligands (**437**) have been found to be highly effective in the Pd-catalysed asymmetric Suzuki-Miyaura cross-coupling reaction between diethyl 1-bromo-2-naphthylphosphonate and aryl boronic acids. Corresponding products (**438**) were obtained in moderate to high yield (65–98%) and with high enantioselectivity (78–97% ee) (Scheme 149).[182]

Meyer and co-workers discovered a photoconversion reaction of the ruthenium bisphosphonic complex (**439**) by sensitised formation of $^1O_2$. In this reaction, the bisphosphonic functional groups were converted

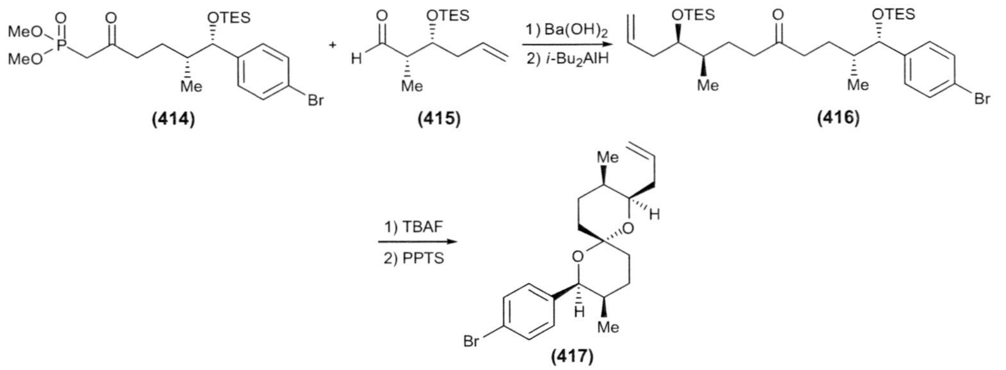

Scheme 144

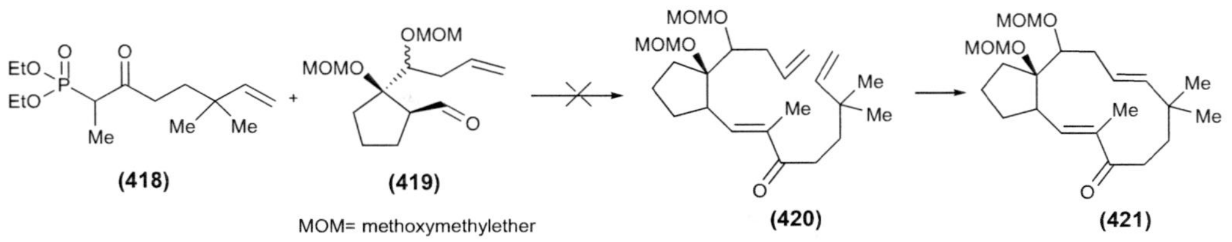

Scheme 145

MOM= methoxymethylether

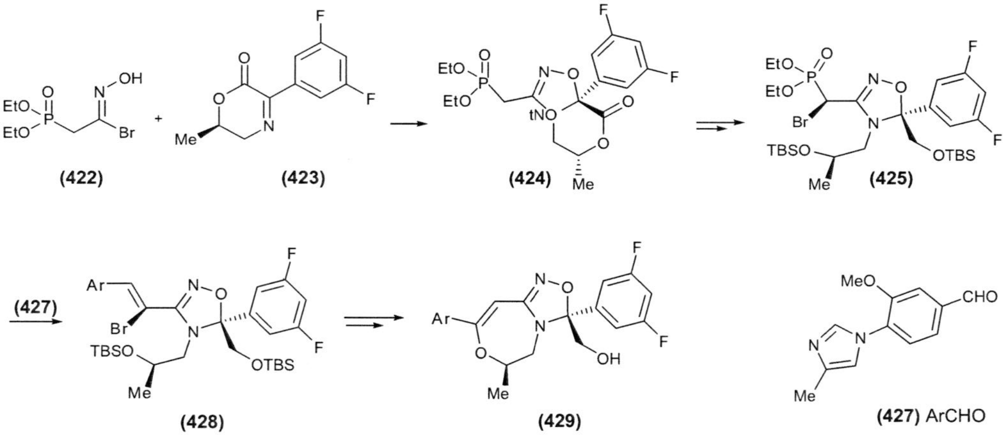

Scheme 146

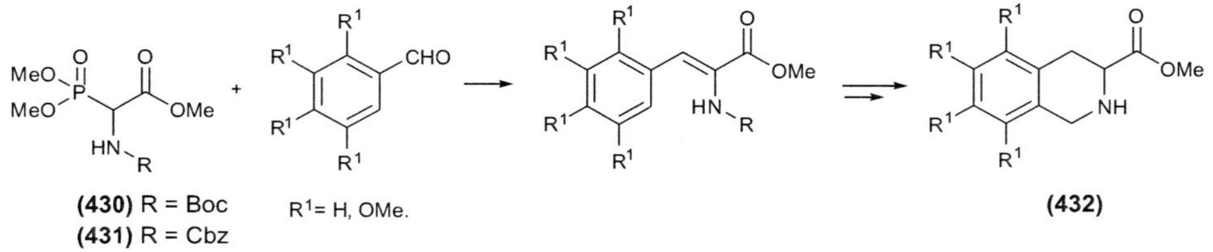

Scheme 147

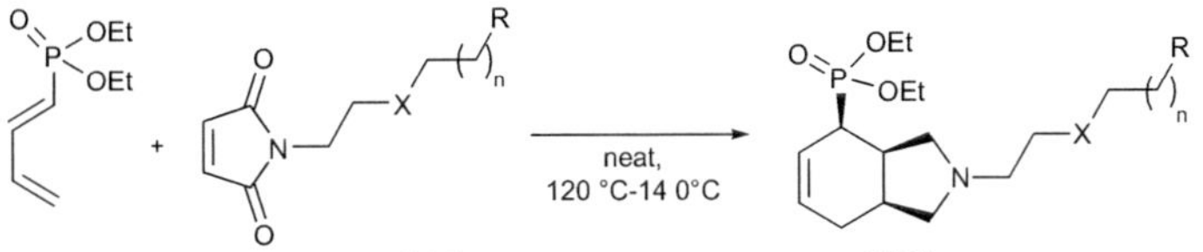

Scheme 148

X= CH$_2$, O; n= 0, 1, 2;
R= Me, Cl, CN, NHBoc, CCH, OH, Mal.

Mal= N-(methoxycarbonyl)maleimide

Scheme 149

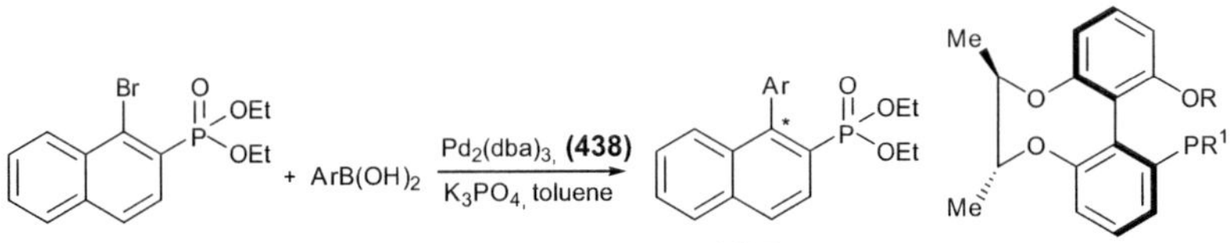

Ar = naphthyl, 4-MeO-naphthyl, 4-Me$_2$N-naphthyl, 2-PhC$_6$H$_4$, 2-MeOC$_6$H$_4$, 2-EtOC$_6$H$_4$, 2-n-PrOC$_6$H$_4$, 2-i-PrOC$_6$H$_4$, 2-ClC$_6$H$_4$, 2,3-Me$_2$C$_6$H$_3$.

R= Me, Et, i-Pr, CH$_2$Ph;
R^1= Ph, 4-MeC$_6$H$_4$, 3,5-Me$_2$C$_6$H$_3$, 3,5-i-BuC$_6$H$_3$.

Scheme 150

into carboxylic groups and orthophosphoric acid. The reaction occured by sensitised formation of $^1O_2$ by the lowest metal-to-ligand charge transfer excited state(s) followed by the $^1O_2$ oxidation of the bisphosphonic substituents (Scheme 150).[183]

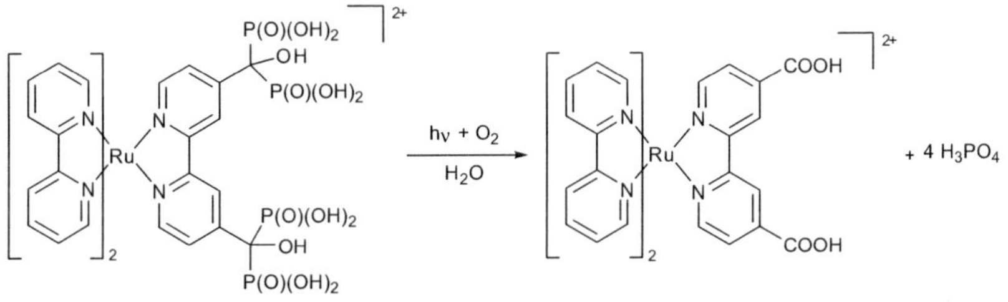

A simple and efficient method for the synthesis of alcohols from the corresponding carboxylic acids has been described by Sureshbabu and co-workers. Activation of aryl and alkyl carboxylic acids with 1-propanephosphonic acid cyclic anhydride (**440**) and subsequent reduction in the presence of NaBH$_4$ yielded the alcohol in good yields (83–93%) and with good purity. The reduction of N^α-protected amino acids (**441**) has been also successfully carried out to obtain corresponding alcohols (**442**) (Scheme 151).[184]

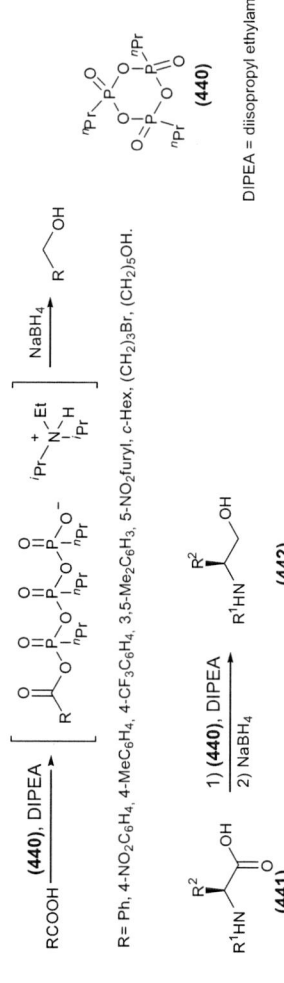

Scheme 151

Propanephosphonic anhydride (**440**) has been demonstrated to be an efficient activator for the chemoselective acetalisation and thioacetalisation of aldehydes in the presence of ketones. Thus, cyclic and acyclic acetals (**443**) and thioacetals (**444**) were obtained in excellent yields (89–96%) at room temperature in the presence of a catalytic amount of (**440**) (Scheme 152).[185]

The sequential activities of PhnY, an α-ketoglutarate (αKG) /Fe(II)-dependent dioxygenase, and PhnZ, a Fe(II)-dependent enzyme of the histidine-aspartate motif hydrolase family, cleaved the carbon-phosphorus bond of the organophosphonate natural product, 2-aminoethylphosphonic acid (**445**). PhnY added a hydroxyl group to the α-carbon, yielding 2-amino-1-hydroxyethylphosphonic acid (**446**), which was oxidatively converted by PhnZ to phosphoric acid and glycine (Scheme 153). The PhnZ reaction represented a new enzyme mechanism for metabolic cleavage of a carbon-phosphorus bond.[186]

A selective protection and deprotection of *ortho*-functionalised arylphosphonates (**447**) have been reported by Marchand-Brynaert and co-workers. The key step of the developed synthesis consisted of the introduction of a phosphoryl group by an *ortho*-metallation reaction from protected *ortho*-anisidine. Chemoselective deprotections have been investigated and mono-, bis-, and ter-deprotected aromatic derivatives have been obtained. A practical route to phosphorylated benzoxazoles (**448**) has been thus discovered (Scheme 154).[187]

2.3 Selected biological aspects

Sieńczyk and co-workers have presented the synthesis and the measurement of the inhibitory activity of novel peptidyl derivatives of α-aminoalkylphosphonate diaryl esters (**449**) as human neutrophil elastase (HNE) inhibitors. Their selectivity against other serine proteases, including porcine pancreatic elastase, chymotrypsin, and trypsin, has also been demonstrated. The compound (**450**) displayed the best balance of high inhibitory activity against HNE and selectivity for HNE over porcine

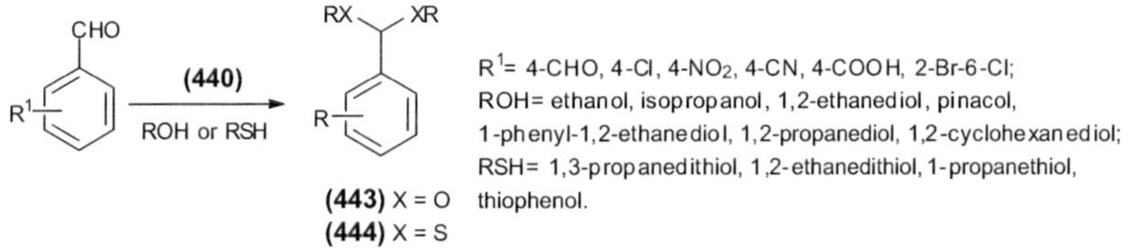

Scheme 152

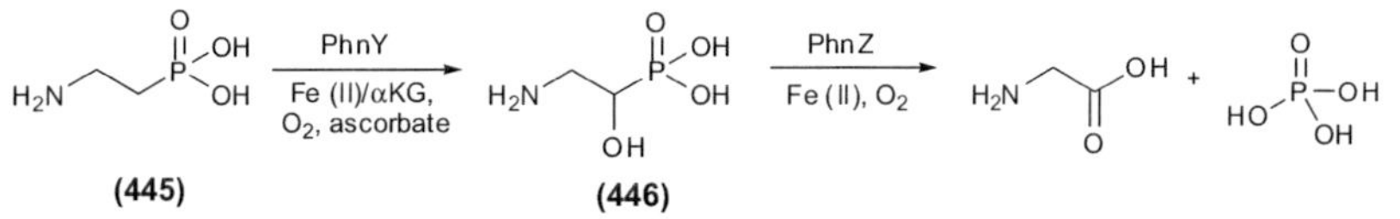

Scheme 153

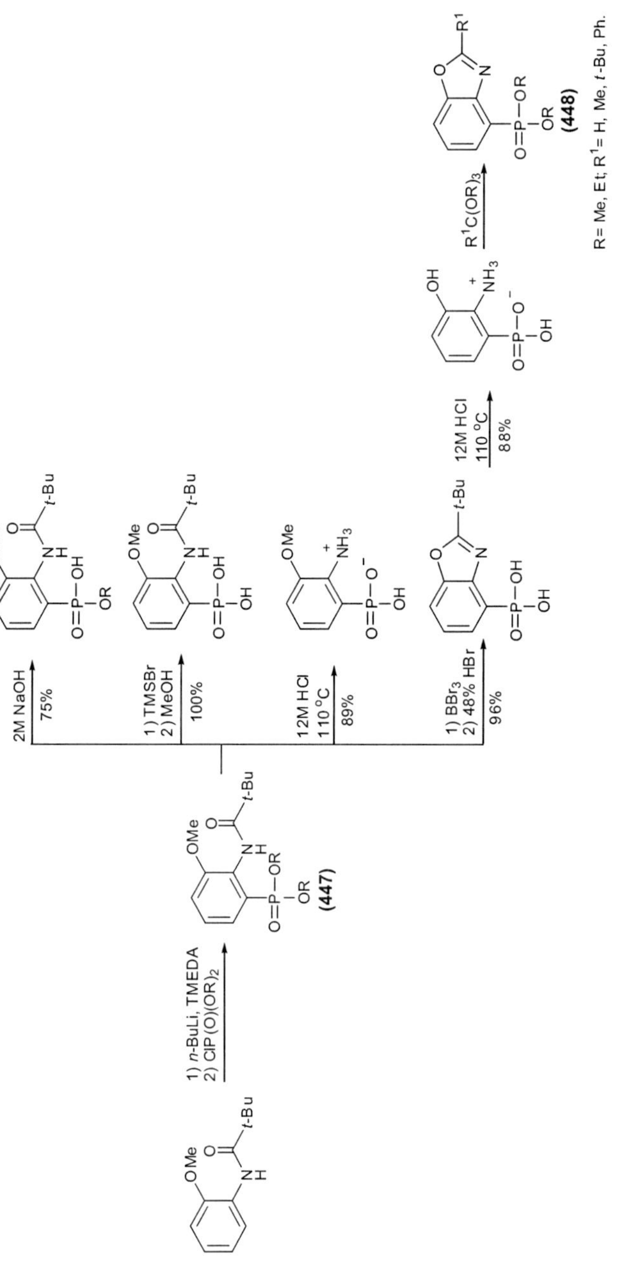
1) n-BuLi, TMEDA
2) ClP(O)(OR)2
(447)
2M NaOH
75%
1) TMSBr
2) MeOH
100%
12M HCl
110 °C
89%
1) BBr3
2) 48% HBr
96%
12M HCl
110 °C
88%
R1C(OR)3
(448)
R = Me, Et; R1 = H, Me, t-Bu, Ph.

Scheme 154

pancreatic elastase, chymotrypsin, and trypsin. Additionally, all obtained inhibitors displayed high stability in either phosphate buffered saline (PBS) or human plasma.[188]

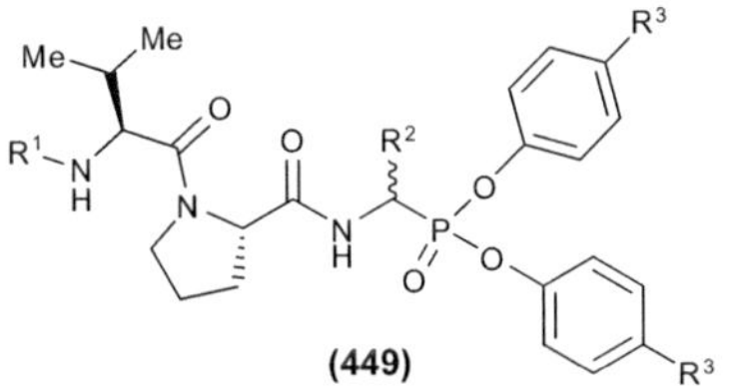

(449)

R=Boc-Gly, Cbz-Gly, Boc-D-Ala, Cbz-L-Ala, Cbz-Aib, Cbz-L-Val, Cbz-L-Leu, Boc-L-Leu, Cbz-L-Met, C(O)Ot-Bu; R^2=Me, Et, n-Pr, i-Pr, n-Bu, 2-methylpropyl, 1-methylethyl; R^3=H, 4-Et, 4-t-B

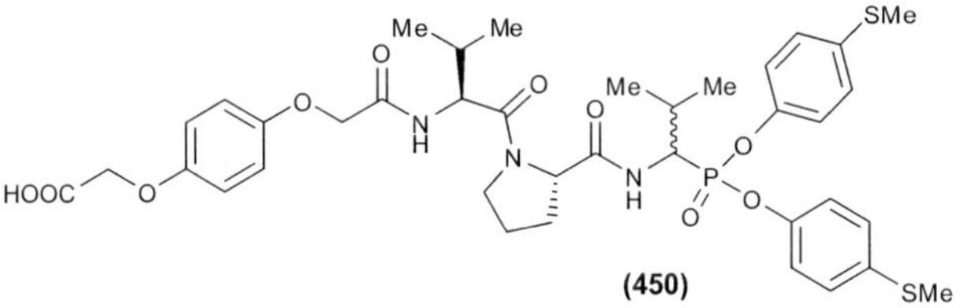

(450)

The Sieńczyk group has also presented the design and synthesis of novel α-aminophosphonic analogues of glutamine (**451**), as well as their peptidyl derivatives (**452**) and (**453**). The inhibitory effects of these compounds towards the newly discovered SplB serine protease from *S. aureusare* have been characterised. Moreover, the authors investigated the influence of aromatic ester substituents on inhibitory potency towards SplB. Introduction of electron-donating groups at the *para* position of the phenyl ester ring such as ethyl (**453d**) or *t*-butyl (**453c**) resulted in decreased inhibitory properties. Despite this fact, the highest potency towards the SplB protease has been observed for the derivative (**453b**).[189]

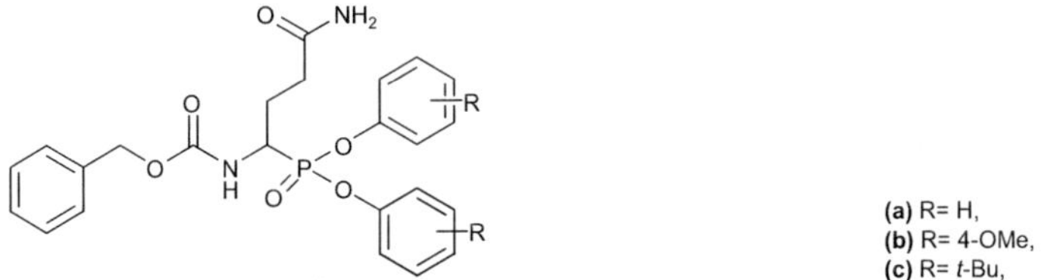

(451a–d)

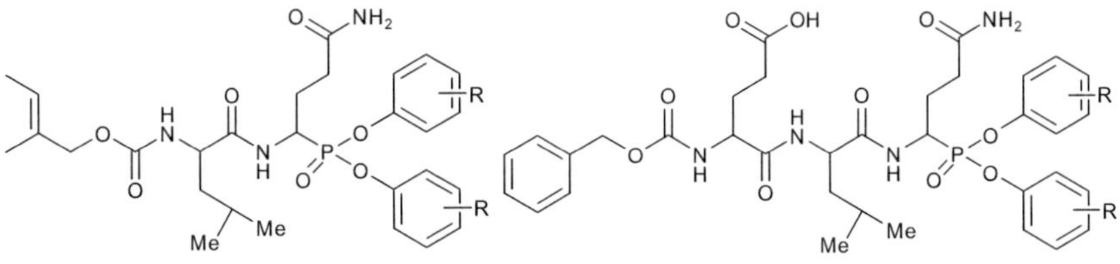

(a) R= H,
(b) R= 4-OMe,
(c) R= t-Bu,
(d) R= Et.

(452a–d)

(453a–d)

Kratz et al. designed two low-molecular-weight and water-soluble prodrugs (**454**) and (**444**) which incorporated a bisphosphonate group as a bone targeting ligand, doxorubicin as the anticancer agent, and either an acid-sensitive hydrazone bond (**454**) or a cathepsin B cleavable Val-Ala bond (**444**) for ensuring effective release of doxorubicin at the site of action. Sufficient stability in human plasma as well as high affinity for hydroxyapatite and native bone was demonstrated, making the novel prodrugs (**454**) and (**455**) suitable candidates for *in vivo* evaluation in a bone metastases model.[190]

Members of a novel series of benzothiophene derived phosphonate esters (**456**)–(**458**) have been shown to be potent antagonists of the transient receptor potential melastatin 8 (TRPM8) channel. Structure-activity relationships were developed by modification of the core structure and subsequently by variation of the aromatic substituents and the phosphonate ester. Among various phosphonate esters, 3,4-difluoro substituted derivative (**456**) has been identified as a highly selective and robust TRPM8 antagonist *in vitro* that inhibited icilin-induced behaviors in a rat WDS model.[191]

Oxoquinoline acyclonucleoside phosphonate analogues (**459**) with different substituents at the C6 or C7 positions of the oxoquinoline nucleus and an N1-bonded phosphonate group have been investigated as a new class of specific inhibitors of human immunodeficiency virus type 1. The fluoro-substituted derivatives (**459f**) and (**459g**) presented excellent EC50 values of 0.4 ± 0.2 μM (**459f**) and 0.2 ± 0.005 μM (**459g**) and selectivity index values (SI) of 6240 and 14675, respectively. These results also indicated that oxoquinolinephosphonate (**459g**) was a notably promising compound for the development of a new, more potent and selective anti-HIV-1 drug and consequently represented a significant advance in the field of anti-HIV-1 oxoquinoline derivatives.[192]

(a) R= H
(b) R= 6-Cl,
(c) R= 7-Cl,
(d) R= 6-Br,
(e) R= 7-Br,
(f) R= 6-F,
(g) R=7-F,
(h) R=6-Me,
(i) R=7-Me,
(j) R=6-NO_2,
(k) R=7-NO_2.

(**459**)

A convenient synthesis of inositol phosphonate analogues (**461**)–(**463**) using *myo*-inositol (**460**) as the starting material has been demonstrated (Scheme 155). Two phosphonate analogues (**461**) and (**462**) exhibited good cytotoxic activity against non-small cell lung cancer (NSCLC) cell line A549.[193]

Three glucosamine-6-phosphate (GlcN6P) analogues (**465**)–(**567**) have been prepared to investigate their effect on the *glmS* riboswitch, a ribozyme found in numerous Gram-positive bacteria that responds to the cellular concentrations of (GlcN6P). The Pudovik reaction has been applied to synthesise the α-hydroxyphosphonates (**464**), key substrates in the synthesis of analogues (**465**) and (**466**) (Scheme 156). Among the three analogues (**465**)–(**467**), the carba-sugar (**467**) induced moderate self-cleavage of the glmS riboswitch from *B. subtilis*. The displacement of the phosphate in GlcN6P by α-hydroxyphosphonate led to a massive loss of activation. These results provided valuable information for further elucidation of the structure activity relationships and drug design for glmS riboswitch antibiotics.[194]

α-Aryl-substituted β-oxa isosteres of fosmidomycin (**468**) with a reverse orientation of the hydroxamic acid group have been synthesised (Scheme 157). Evaluation of their inhibitory activity against recombinant 1-deoxy-D-xylulose 5-phosphate reductoisomerase (IspC) of *Plasmodium falciparum* and for their *in vitro* antiplasmodial activity against chloroquine-sensitive and resistant strains of *P. falciparum* have been also performed. The most active derivative inhibited IspC protein of *P. falciparum* (PfIspC) with an IC_{50} value of 12 nM and showed potent

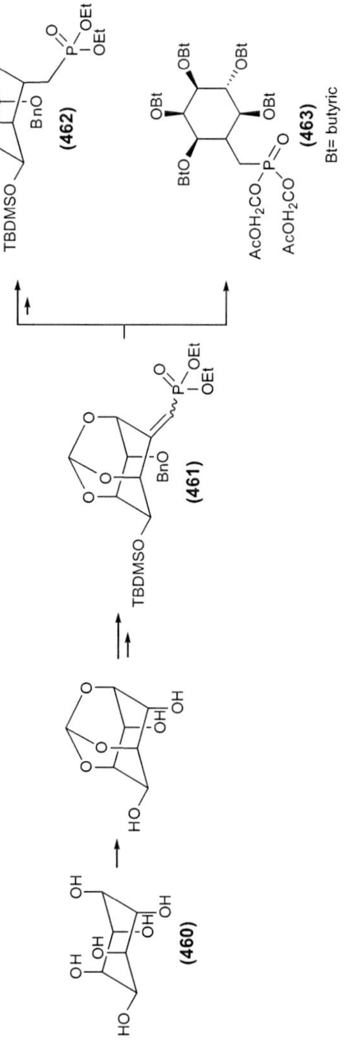

Scheme 155

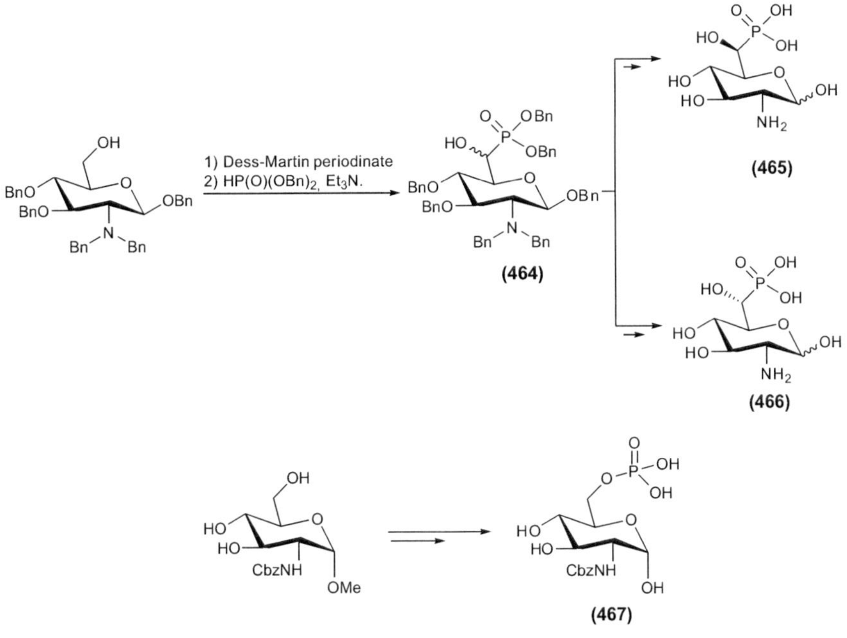

1) Dess-Martin periodinate
2) HP(O)(OBn)2, Et3N.
(464)
(465)
(466)
(467)

Scheme 156

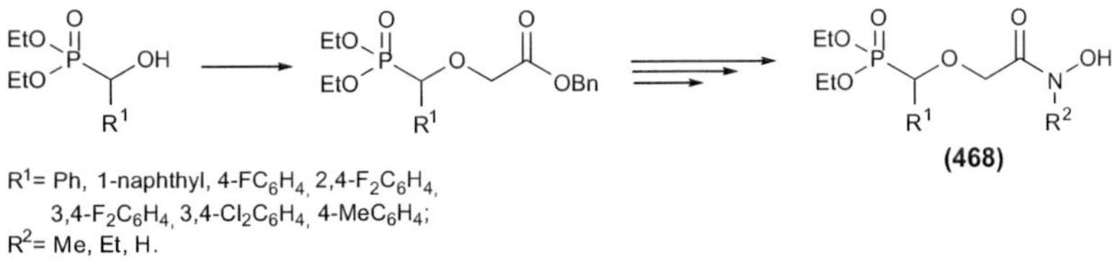

(468)
R1= Ph, 1-naphthyl, 4-FC6H4, 2,4-F2C6H4, 3,4-F2C6H4, 3,4-Cl2C6H4, 4-MeC6H4;
R2= Me, Et, H.

Scheme 157

in vitro antiplasmodial activity. In addition, lipophilic ester prodrugs demonstrated improved *P. falciparum* growth inhibition *in vitro*.[195]

Fluorinated derivatives of 2-(phosphonomethyl)pentanedioic acid (**469**) have been designed and synthesised (Scheme 158) to explore whether this fluorine-substituent was tolerated in the pentanedioic acid moiety that was common to almost all prostate specific membrane antigen (PSMA) targeting small molecule inhibitors. The binding affinities of the racemic and individual stereoisomers of 2-fluoro-4-(phosphonomethyl)pentanedioic acid (**469**) have been determined and showed that the introduction of fluorine was well tolerated. The radiosynthesis of the analogous 2-[^{18}F]fluoro-4-(phosphonomethyl)pentanedioic acid (**470**) has been developed (Scheme 159) and evaluated *in vivo* with the PSMA positive LNCaP human prostate cancer cell. The biological results demonstrated specific binding of the tracer to PSMA positive tumors in mice. These results warranted the further evaluation of this class of compounds as radiolabeled tracers for the detection and staging of prostate cancer.[196]

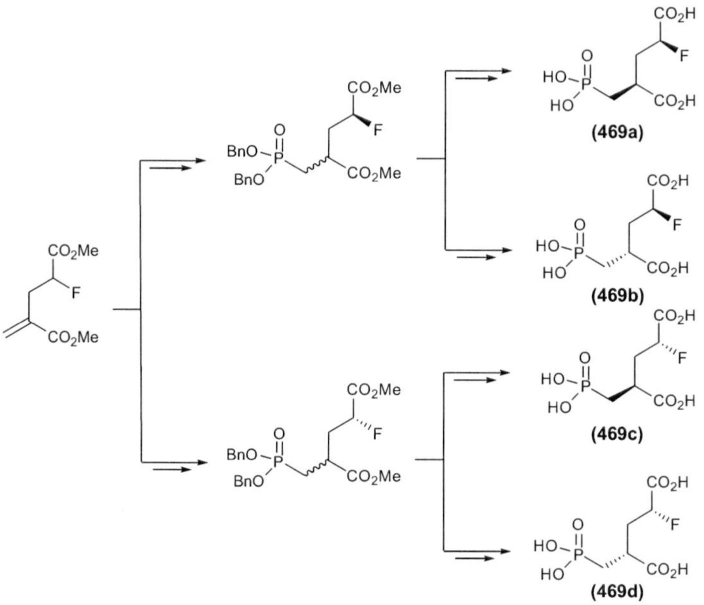

Scheme 158

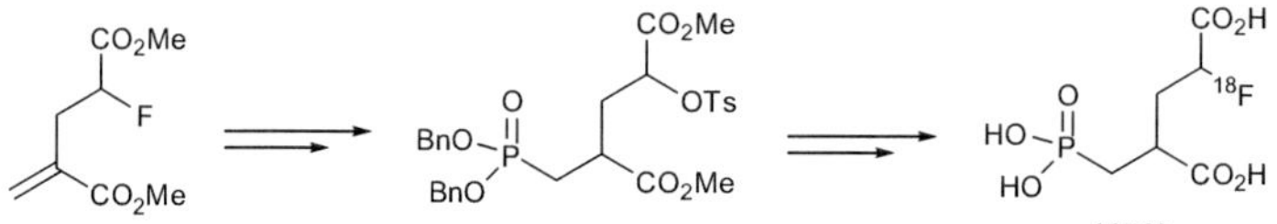

Scheme 159

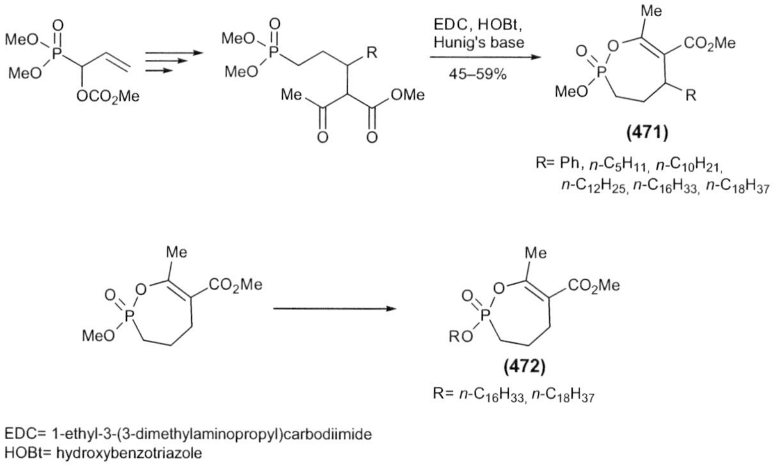

Scheme 160

EDC= 1-ethyl-3-(3-dimethylaminopropyl)carbodiimide
HOBt= hydroxybenzotriazole

A new series of customisable diastereomeric *cis*-and *trans*-monocyclic enol-phosphonate analogues to Cyclophostin (**471**) and Cyclipostins (**472**) has been synthesised (Scheme 160). Their potencies and mechanisms of inhibition toward six representative lipolytic enzymes belonging to distinct lipase families have been also examined. The best inhibitors displayed a *cis* conformation (H and OMe) and exhibited higher inhibitory activities than the lipase inhibitor Orlistat toward the same enzymes. These results have revealed that chemical group at the γ-carbon of the phosphonate ring strongly impacted on the inhibitory efficiency, leading to a significant improvement in selectivity toward one target lipase over another. The powerful and selective inhibition of microbial (fungal and mycobacterial) lipases suggested that these seven-membered monocyclic enol-phosphonates should provide useful leads for the development of novel and highly selective antimicrobial agents.[197]

A new series of nitrogen-containing bisphosphonates (**473**) and (**474**) has been designed that interacted with a larger portion of the geranyl pyrophosphate (GPP) subpocket, as compared to the current therapeutic drugs, and rigidified the C-terminal basic (^{364}KRRK^{367}) tail of human farnesyl pyrophosphate synthase (hFPPS) in the closed conformation in the absence of isopentenyl pyrophosphate (IPP). An analogue of this series has been used to demonstrate inhibition of the intended biological target, resulting in apoptosis and down-regulation of extracellular-signal regulated kinase-ERK phosphorylation in human Multiple Myeloma cell lines.[198]

Y= CH$_2$, NH; X= CH, N;
R= H, F; R^1= Ph, 6-MeOC$_6$H$_4$, 5-MeOC$_6$H$_4$, 4-MeOC$_6$H$_4$,
4-*i*-PrC$_6$H$_4$, 4-*i*-PrC$_6$H$_4$, 6-MeC$_6$H$_4$, 5-MeC$_6$H$_4$, 4-MeC$_6$H$_4$,
3,5-(CF$_3$)$_2$C$_6$H$_4$, 5-CF$_3$C$_6$H$_4$, 4-CF$_3$C$_6$H$_4$, 4-FC$_6$H$_4$, 4-ClC$_6$H$_4$,
4-PhOC$_6$H$_4$, 3-thienyl, 2-thienyl, 2-naphthyl, 3-pirydyl.

(**473**)   (**474**)

Carbamoylphosphonates (**475**) have been identified as inhibitors of matrix metalloproteinases (MMPs) and as orally-active, bioavailable, and safe antimetastatic agents. In this article, the authors focused on the direct antitumor activity of the (**475**). It has been discovered that (**475**) also inhibited carbonic anhydrases (CAs), especially the IX and XII isoforms identified as cancer promoting factors. Thus, (**475**) could be regarded as novel non-toxic drug candidates for tumor microenvironment targeted chemotherapy acting by two synergistic mechanisms, namely, inhibiting CAs and MMPs, simultaneously. It has been also demonstrated that the ionised acid (**475**) was unable to cross the cell membrane and thus limited to interact with the extracellular domains of isozymes CAIX and CAXII. Finally, applying (**475**) against cancer cells in hypoxic conditions resulted in a dose-dependent release of lactate dehydrogenase, confirming the direct interaction of the (**475**) with the cancer related isozymes CAIX and XII and thereby promoting cellular damage.[199]

R=*i*-Pr, *c*-C$_5$H$_9$, *c*-Hex, *c*-C$_7$H$_{15}$, (CH$_2$)$_2$Ph, CH$_2$-*c*-Hex, (CH$_2$)$_4$NHC(O)P(O)(OH)$_2$.

(**475**)

Oseltamivir phosphonic acid (tamiphosphor, **476**), its monoethyl ester (**476a**), guanidino-tamiphosphor (**477**), and its monoethyl ester (**477a**) have been found to be potent inhibitors of influenza neuraminidases. They inhibited the replication of influenza viruses, including the oseltamivir-resistant H275Y strain, at low nanomolar to picomolar levels, and significantly protected mice from infection with lethal doses of influenza viruses when orally administered with 1 mg/kg or higher doses. Moreover, these compounds were stable in simulated gastric fluid, liver microsomes, and human blood and were largely free from binding to plasma proteins. The pharmacokinetic properties of these inhibitors were thoroughly studied in dogs, rats, and mice. The absolute oral bioavailability of these compounds was lower than 12%. No conversion of monoester (**477a**) to phosphonic acid (**477**) was observed in rats after intravenous administration, but partial conversion of (**477a**) was observed with oral administration. An advanced formulation may be investigated to develop these new anti-influenza agents for better therapeutic use.[200]

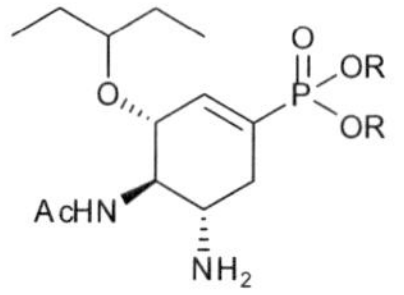

(**476**) R= H
(**476a**) R= Et

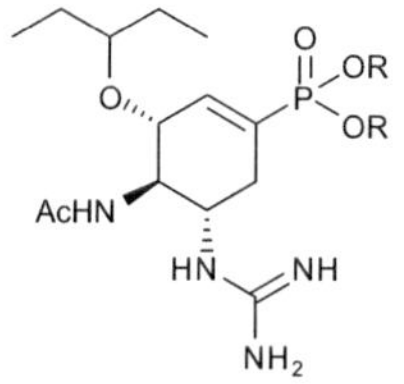

(**477**) R= H
(**477a**) R= Et

Fosmidomycin derivatives (**478**)–(**482**) in which the hydroxamic acid group was replaced by several bidentate chelators as potential hydroxamic alternatives, have been prepared and tested as inhibitors of the metalloenzyme 1-deoxyxylulose 5-phosphate reductoisomerase (DXR) from *Escherichia coli*. Even, if divalent Mg^{2+}, regarded a hard ion, prefered hard ligands with oxygen being the most preferred coordinating

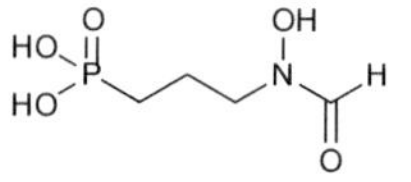

Fosmidomycine

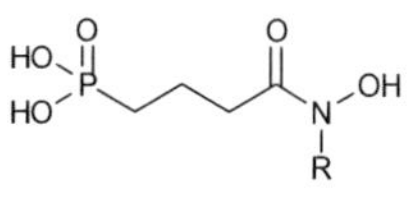

(**478**) R= H, Me

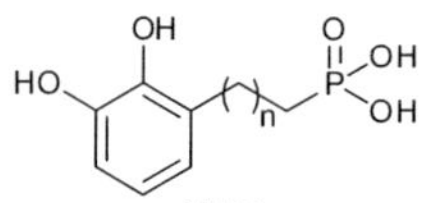

(**479**) n= 1, 2

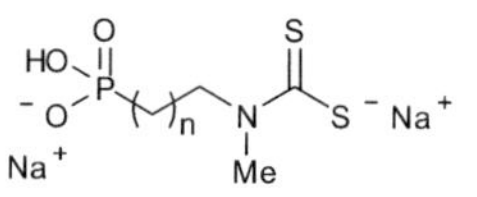

(**480**) n= 1, 2

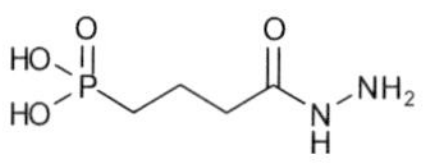

(**481**)

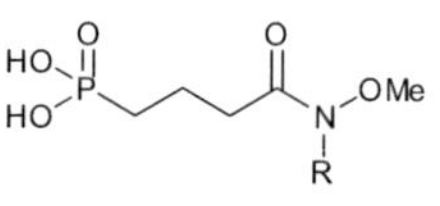

(**482**) R= H, Me

atom followed by nitrogen, the replacement of the hydroxamic group by hydrazine, catechol and *O*-methylated hydroxamate moieties resulted in an increase of the IC_{50} value. These results showed that the attempts to replace the hydroxamate group by other chelating groups were ineffective and illustrated the predominant role of the hydroxamate functional group as the most effective metal-binding group in DXR inhibitors.[201]

A novel series of diphenyl 1-(arylamino)-1-(pyridine-3-yl)ethylphosphonates (**483**) has been obtained in high yields by reaction of 3-acetylpyridine with aromatic amines and triphenylphosphite in the presence of lithium perchlorate as a catalyst (Scheme 161). The compounds (**483**) showed high antimicrobial activities against *Escherichia coli* as a Gram-negative bacterium, *Bacillus subtilis* and *Staphylococcus aureus* as Gram-positive bacteria and *Candida albicans* and *Saccharomyces cerevisiae* as fungi, at low concentrations (10–100 μg/mL). Moreover, the synthesised compounds showed significant cytotoxicity and anticancer activities against liver carcinoma cell line (HepG2) and human breast adenocarcinoma cell line (MCF7). The lethal dose of the synthesised compounds was also determined and indicated that most compounds were safe to use.[202]

Three, novel 1-aminophosphonic acid diesters (**485**), bearing anthracene moieties have been synthesised and characterised by Troev and co-workers (Scheme 162). The aminophosphonates and their precursors were evaluated for *in vitro* antitumor activity on a panel of seven human epithelial cancer cell lines. Compounds (**485c**), showed optimal anti proliferative activity to human tumor cells from colon carcinoma and from malignant tumors of the breast and urinary bladder. Moreover, the aminophosphonate (**485c**) exhibited higher activity against all tested cancer cell lines, than its synthetic precursor Schiff base (**484b**).[203]

Young and co-workers have described the synthesis of radiolabelled bisphosphonates conjugates (**486**), (**487**) and (**488**). All compounds have

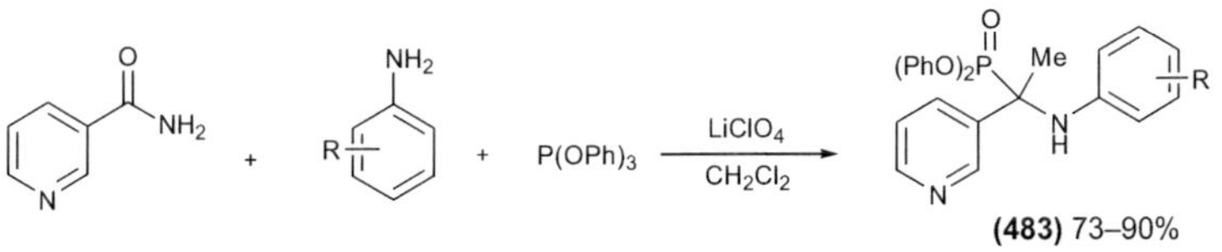

(483) 73–90%

R= H, 4-Cl, 4-Me, 4-OH.

Scheme 161

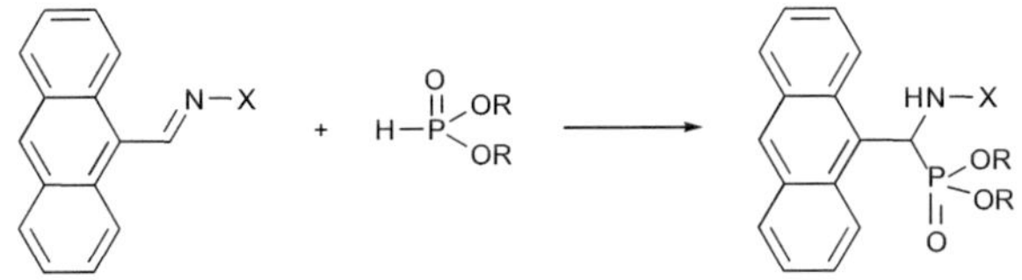

(484a) X= 4-MeC$_6$H$_4$
(484b) X= CH$_2$-2-furyl

(485a) X= 4-MeC$_6$H$_4$, R= Me
(485b) X= 4-MeC$_6$H$_4$, R= Et
(485c) X= CH$_2$-2-furyl, R= Et

Scheme 162

been evaluated *in vivo* in rats for uptake of the conjugate into bone and subsequent release of the EP4 agonists over time. While the conjugate (**486**) was taken up (9.0% of initial dose) but not released over two weeks, conjugates (**487**) and (**488**) were absorbed at 9.4% and 5.9% uptake of the initial dose and slowly released with half-lives of approximately 2 weeks and 5 days respectively. These conjugates were well tolerated and offer potential for sustained release and dual synergistic activity through their selective bone targeting and local release of the complimentary active components.[204]

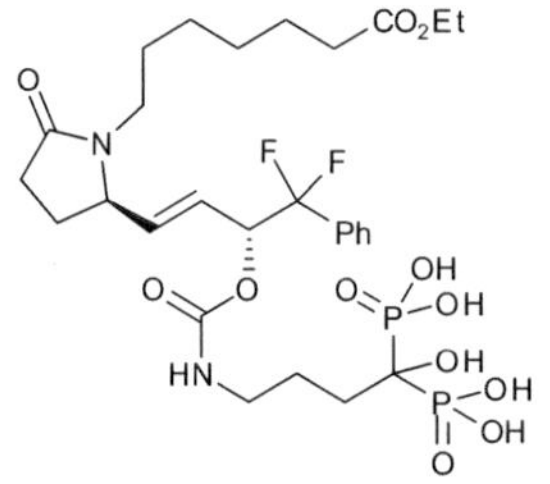

(**486**)

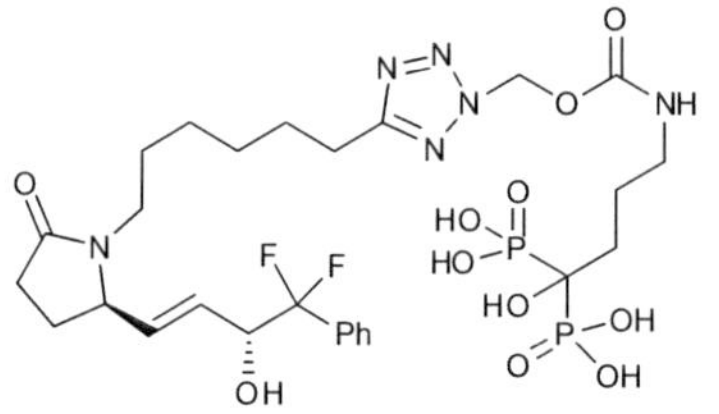

(**487**)

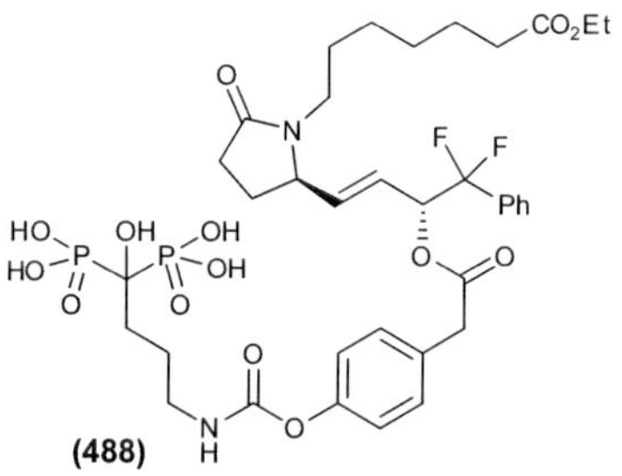

(**488**)

Kaiser *et al.* have presented the development of a *para*-nitrophenol phosphonate activity-based probe (**489**) with structural similarities to the potent agrochemical paraoxon. The authors demonstrated that these probes labels distinguished serine hydrolases with the carboxylesterase

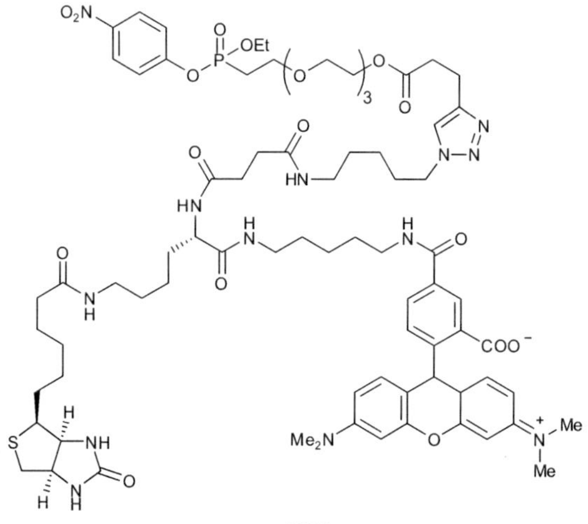

(**489**)

CXE12 as the predominant target in *Arabidopsis thaliana*. The designed probe featured a distinct labeling pattern and therefore represented a promising chemical tool to investigate physiological roles of selected serine hydrolases such as CXE12 in plant biology.[205]

2.4 Miscellaneous

A mechanistic investigation of methylphosphonate synthase (MPnS), a non-heme iron-dependent oxygenase, that converted 2-hydroxyethylphosphonate (**490**) to methylphosphonate (**491**), has been performed by van der Donk and co-workers. Experimental data suggested that the pro-*S* hydrogen was removed first and showed that the pro-*R* hydrogen was transferred from C2 of 2-hydroxyethylphosphonate (**490**) to the methyl group of methylphosphonate (**491**) (Scheme 163). Kinetic studies revealed that neither hydrogen transfer was rate limiting under saturating substrate conditions. A mechanism for these reaction have been proposed that was consistent with the available data.[206]

Two radical clock analogues of (*R*)-2-hydroxypropylphosphonic acid, cyclopropyl- and methylenecyclopropyl-containing compounds, (**492**) and (**493**) have been synthesised and employed to study the mechanism of (*S*)-2-hydroxypropylphosphonic acid epoxidase (HppE) catalysis. Enzymatic assays indicated that the (*S*)- and (*R*)-isomers of the cyclopropyl-containing analogues (**492**) were efficiently converted by HppE to epoxide and ketone products, respectively. On the other hand, the ultrafast methylenecyclopropyl-containing probe (**493**) inactivated HppE, consistent with a rapid radical-triggered ring-opening process that leads to enzyme inactivation (Scheme 164). These results provided, for the first time, experimental evidence for the involvement of a C2-centered radical

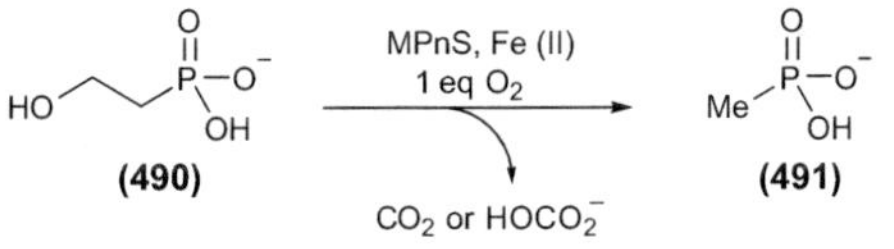

Scheme 163

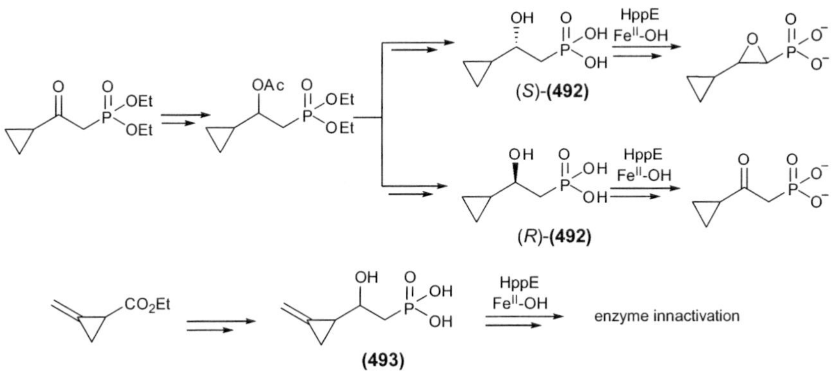

Scheme 164

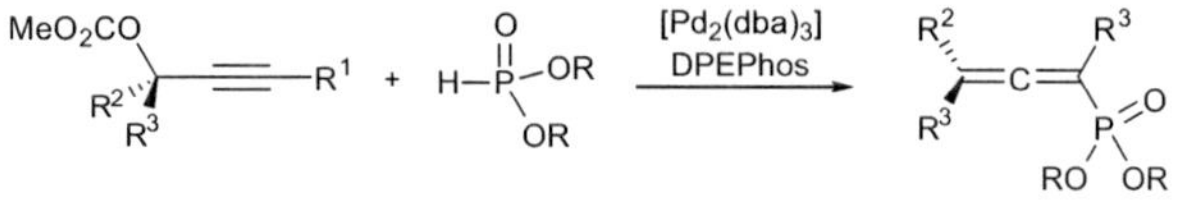

Scheme 165

intermediate with a lifetime on the order of nanoseconds in the HppE-catalysed oxidation of (R)-HPP.[207]

The mechanism and sources of selectivity in the palladium-catalysed propargylic substitution reaction involving phosphorus nucleophiles, and yielding predominantly allenyl-phosphonates (**494**), have been studied in detail by using density functional theory (DFT) calculations (Scheme 165). Multiple alternative pathways for each step of the reaction have been examined and their relative free energies evaluated, thereby resulting in a plausible catalytic cycle. It was found that the C-P bond formation took place by means of a three-center reductive elimination, which was the rate-limiting step of the reaction. In addition, it was found that these transition states changed when using a different phosphorous nucleophile, which added to the complexity of the system.[208]

Landry et al. have developed a new class of biologically suitable MRI contrast agents that capitalised on the strong binding affinity of the imidodi(methanediphosphate) (NDP$_2$) ligand and the large internal surface area of mesoporous silica. In this study, the authors immobilised a unique phosphonate-containing ligand onto mesoporous silica particles with a range of pore diameters, pore volumes, and surface areas, and Gd (III) ions were then chelated to the particle (**495**). The per-Gd relaxivities were significantly higher than those of commercially available contrast agents, and the per-particle relaxivities were among the highest values measured in the literature. As acid-prepared mesoporous silica (APMS) was nontoxic, nonimmunogenic, and excreted over time, these are exciting new materials for future *in vivo* studies.[209]

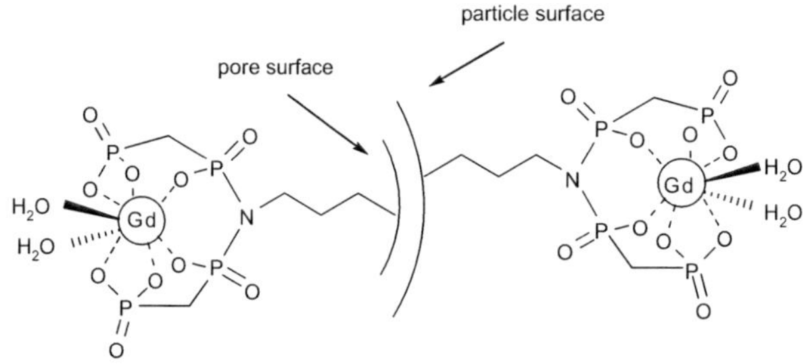

A sample pre-treatment was evaluated to enable the production of intact cationic species of synthetic polymers holding a phosphonate group using MALDI-mass spectrometry. Thus, the polymer (**496**) obtained by nitroxide-mediated polymerisation was stirred for a few hours in trifluoroacetic acid to induce the substitution of a *tert*-butyl group on the nitrogen of the nitroxide end-group by a hydrogen atom (Scheme 166).

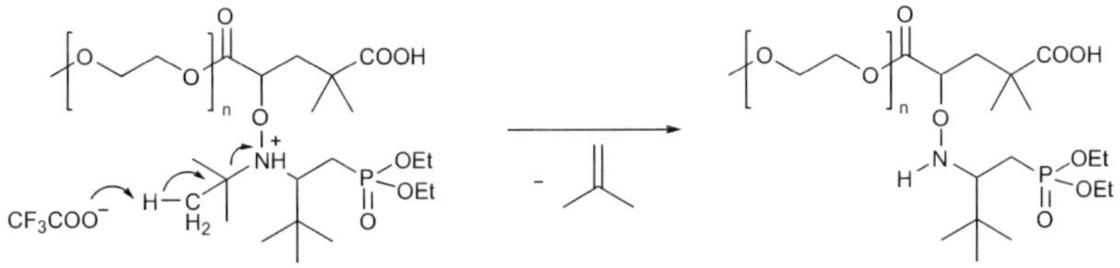

Scheme 166

The substitution reaction has been found to increase the dissociation energy of the fragile C–ON bond to a sufficient extent to prevent this bond being spontaneously cleaved during MALDI analysis.[210]

3 Phosphinic acids and their derivatives

3.1 Synthesis of phosphinic acids and their derivatives

New enantiopure crown ethers containing either a proton-ionisable di-arylphosphinic acid unit [(S,S)-497] or an ethyl diarylphosphinate moiety [(S,S)-498] have been synthesised by Huszthy and co-workers. Electronic circular dichroism studies on the complexation of these crown ethers with the enantiomers of α-(1-naphthyl)ethylammonium perchlorate and with α-(2-naphthyl)ethylammonium perchlorate have been also carried out.[211]

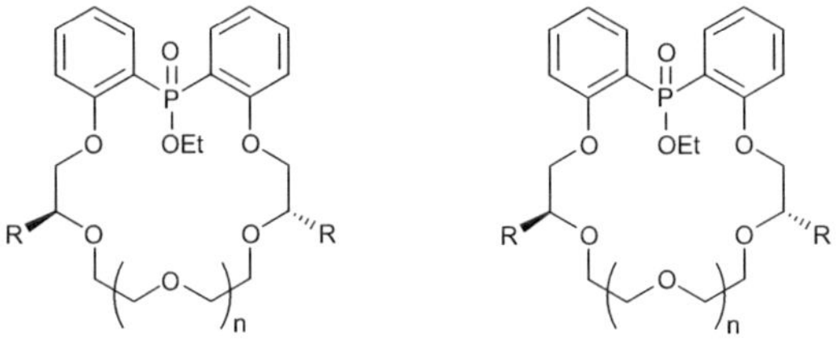

(a) R= Me, n= 1;
(b) R= Me, n= 2;
(c) R= Oct, n= 1;
(d) R= Oct, n= 2.

Synthesis of 1-amino-2-vinylcyclopropane-1-phosphinates (503) have been achieved by conversion of diethyl 1-amino-2-vinylcyclopropanephosphonates (499) and accomplished by using either nucleophilic or electrophilic carbon reagents. Thus, hydrolysis of the phosphonate diethyl ester (499) to the mono acid (500) followed by oxalyl chloride treatment provided the phosphonomonochloridate (501), which was treated with nucleophilic organometallic agents to afford phosphinate ethyl esters (503a). Alternatively, the chloridate (501) was reduced to the phosphonous ethyl ester (502), and then alkylated with various electrophilic alkylating agents to obtain phosphinate ethyl esters (503b) (Scheme 167).[212]

3.2 Reaction of phosphinic acids and their derivatives

A novel and simple method for the synthesis of α-amino-α'-hydroxyalkylphosphinic acids (506) has been developed by Kaboudin and co-workers. Treatment of α-hydroxyalkylphosphinic acids (504) with diimines (505) in the presence of trimethylsilyl chloride gave a mixture of two diastereomeric forms of α-amino-α'-hydroxyalkylphosphinic acids (506), which were easily separated by difference in solubility in organic solvents (Scheme 168).[213]

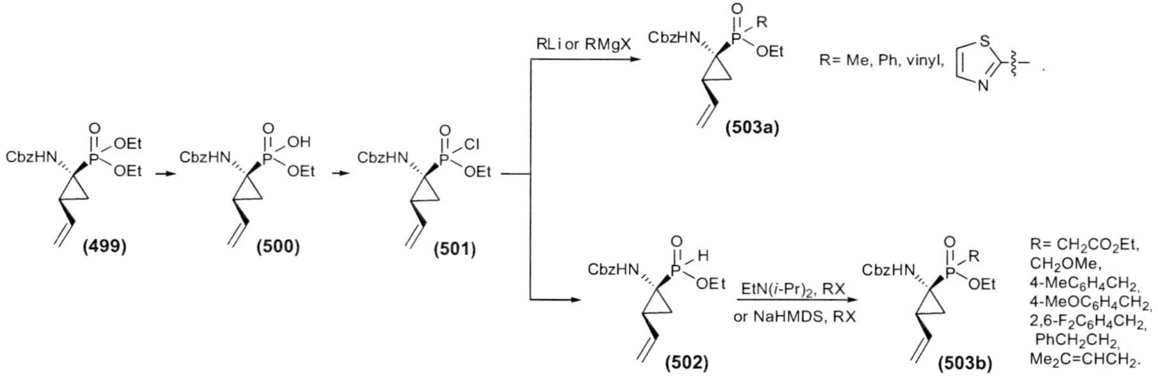
CbzHN
OEt
OEt
(499)
OH
(500)
Cl
(501)
RLi or RMgX
R
(503a)
R= Me, Ph, vinyl,
S
N
H
(502)
EtN(i-Pr)2, RX
or NaHMDS, RX
(503b)
R= CH2CO2Et,
CH2OMe,
4-MeC6H4CH2,
4-MeOC6H4CH2,
2,6-F2C6H4CH2,
PhCH2CH2,
Me2C=CHCH2.

Scheme 167

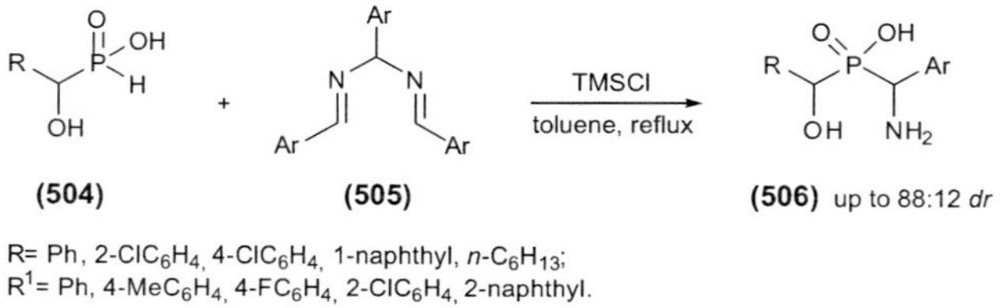
R
OH
H
(504)
Ar
N
(505)
TMSCl
toluene, reflux
NH2
(506) up to 88:12 dr
R= Ph, 2-ClC6H4, 4-ClC6H4, 1-naphthyl, n-C6H13;
R1= Ph, 4-MeC6H4, 4-FC6H4, 2-ClC6H4, 2-naphthyl.

Scheme 168

Chiral non-racemic N-(diphenylphosphinoyl)-protected propargylic amines (**509**) have been prepared by addition of terminal alkynes (**507**) to N-(diphenylphosphinoyl)aldimines (**508**) in the presence of dimethylzinc and 3,3′-dibromo-BINOL (**510**) as a catalyst. The reaction worked with a variety of (hetero)aromatic aldimines and with different alkynes, providing products in good yields and enantiomeric excesses (up to 96%) (Scheme 169).[214]

Matsunaga et al. have developed an asymmetric Mannich-type synthesis of imines (**508**) and 3-isothiocyanato oxindoles (**511**), catalysed by a strontium/Schiff base (**513**) complex, which provided a concise access to enantiomerically enriched spiro[imidazolidine-4,3′-oxindole] compounds (**512**) (Scheme 170).[215]

A highly diastereo- and enantioselective zinc-catalysed Mannich reaction of imines (**508**) with 5H-oxazol-4-ones (**514**), which leads to the first catalytic asymmetric synthesis of syn-α-alkyl norstatine derivatives (**515**) has been also described (Scheme 171).[216]

A highly enantioselective [3 + 2] cyclisation between diphenylphosphinoyl imines (**516**) and allenoate (**517**) by employing the chiral phosphine (**518**) as catalyst has been developed by Lu and co-workers (Scheme 172).[217]

A general and efficient procedure for converting 1,1-diethoxyalkylphosphinates (**519**) into phosphonates (**520**) or phosphonamides (**521**) has been described with moderate to high yields and good purities in a one-pot reaction. H-Phosphinates (**519**) reacted stereospecifically with bromine and subsequently coupled with nucleophiles to form the corresponding optically active products (**520**) or (**521**) with retention of configuration at the phosphorus center (Scheme 173).[218]

A new dimension in the chemistry of (hydroxymethyl)phosphinates (**522**) for organophosphorus synthesis, either through functionalisation preserving the methylene carbon or through oxidative cleavage to unmask the H-phosphinate moiety have been reported by Montchamp et al. Thus, conversion of (hydroxymethyl)phosphinates (**522**) into H-phosphinates (**523**) has been achieved using the Corey-Kim oxidation. Other reactions preserving the methylene carbon have been also reported (Scheme 174).[219]

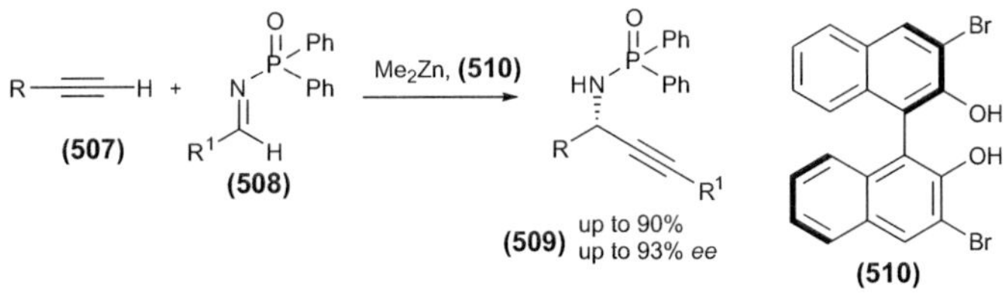

R= Ph, 4-MeC$_6$H$_4$, 4-FC$_6$H$_4$, 4-ClC$_6$H$_4$, 3,5-(MeO)$_2$C$_6$H$_3$, 4-MeOC$_6$H$_4$, 3-thienyl, n-Bu, c-Pr.
R^1= Ph, 4-MeC$_6$H$_4$, 4-MeOC$_6$H$_4$, 4-FC$_6$H$_4$, 4-ClC$_6$H$_4$, 3-MeC$_6$H$_4$, 2-MeC$_6$H$_4$, 2-MeOC$_6$H$_4$, 2-naphthyl, 2-furanyl, 2-thienyl, PhCH$_2$CH$_2$.

Scheme 169

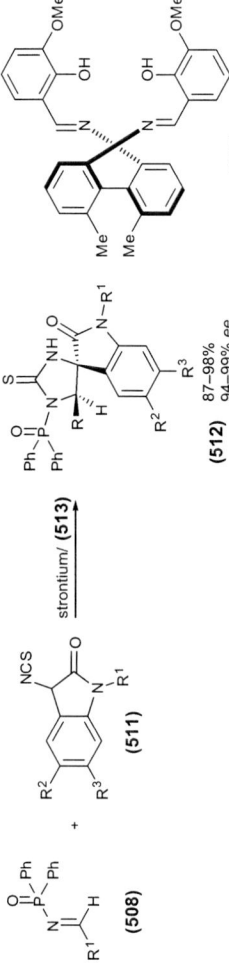

Scheme 170

R= Ph, 4-MeC$_6$H$_4$, 3-MeC$_6$H$_4$, 4-MeOC$_6$H$_4$, 4-FC$_6$H$_4$, 3-FC$_6$H$_4$, 4-ClC$_6$H$_4$, 3-ClC$_6$H$_4$, 4-BrC$_6$H$_4$, 3-thienyl; R^1= Me, CH$_2$CH=CH; R^2= H, Me; R^3= H, Cl.

Scheme 171

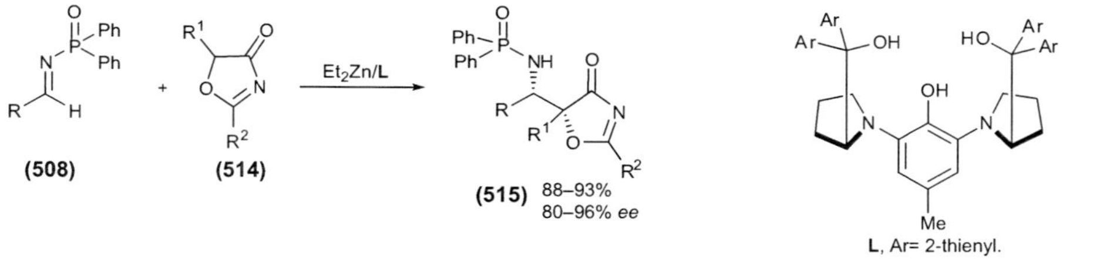

R= Ph, 4-MeC$_6$H$_4$, 3-MeC$_6$H$_4$, 4-MeOC$_6$H$_4$, 3-MeOC$_6$H$_4$, 4-FC$_6$H$_4$, 4-ClC$_6$H$_4$, 3-ClC$_6$H$_4$, 4-BrC$_6$H$_4$, 2-naphthyl, 3-furyl, c-Pr; R^1= Me, Et, n-Pr, n-Bu, CH$_2$Ph; R^2= Ph, 4-ClC$_6$H$_4$.

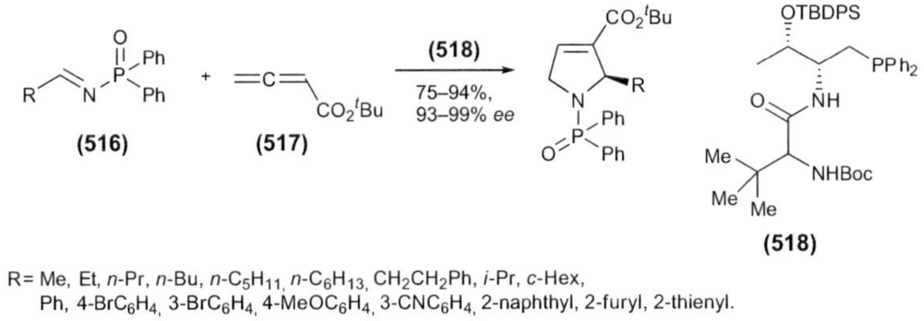

Scheme 172

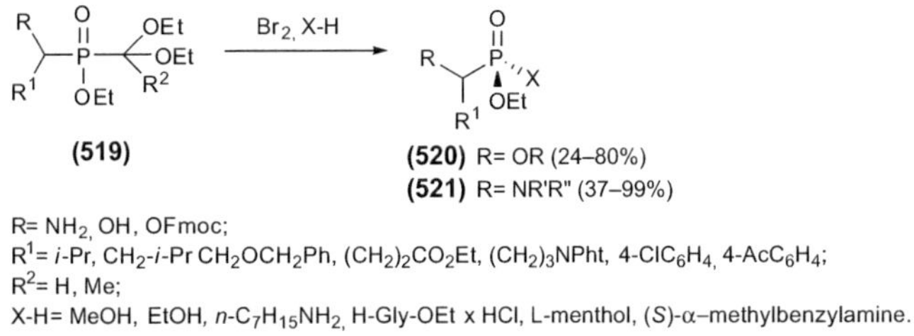

R= Me, Et, n-Pr, n-Bu, n-C$_5$H$_{11}$, n-C$_6$H$_{13}$, CH$_2$CH$_2$Ph, i-Pr, c-Hex, Ph, 4-BrC$_6$H$_4$, 3-BrC$_6$H$_4$, 4-MeOC$_6$H$_4$, 3-CNC$_6$H$_4$, 2-naphthyl, 2-furyl, 2-thienyl.

Scheme 173

(520) R= OR (24–80%)
(521) R= NR'R" (37–99%)

R= NH$_2$, OH, OFmoc;
R^1= i-Pr, CH$_2$-i-Pr CH$_2$OCH$_2$Ph, (CH$_2$)$_2$CO$_2$Et, (CH$_2$)$_3$NPht, 4-ClC$_6$H$_4$, 4-AcC$_6$H$_4$;
R^2= H, Me;
X-H= MeOH, EtOH, n-C$_7$H$_{15}$NH$_2$, H-Gly-OEt x HCl, L-menthol, (S)-α–methylbenzylamine.

Directed *ortho*-lithiation of diphenylphosphinic acid and subsequent electrophilic trapping provided mono *ortho*-functionalised derivatives (**524a,b**) including enantiopure γ-aminophosphinic acids (**524c**) in moderate yields (Scheme 175).[220]

Treating of Woollins' reagent (**525**) with an equimolar amount of sodium 2-bromoalkanolates, prepared *in situ* by treating of bromoalkanol with NaH, gave five- or six-membered phosphorus-selenium heterocycles, (**526**) and (**527**). Thus, Woollins' reagent (**525**) reacted with 1,4-diphenylbuta-1,3-diene to give a five-membered 2,3,5-triphenyl-4-styryl-1,2,5-selenadiphospholane 2,5-diselenide (**526**). On the other hand, reaction of Woollins' reagent (**525**) with an equimolar amount of 2,3-dibenzyl-1,3-butadiene in refluxing toluene led to 6-membered 4,5-dibenzyl-2-phenyl-3,6-dihydro-2H-1,2-selenaphosphinine 2-selenide (**527**) (Scheme 176).[221]

3.3 Selected biological aspects

A new series of cyclic oxaphosphinates (**528**) has been designed, synthesised and screened for their anti-proliferative activity against a large panel of NCI cancer cell lines. Due to easy synthesis and low EC$_{50}$ value (500 nM against the C6 rat glioma cell line), the compound (**528a**) was selected for further biological study. The specific biological effect of (**528a**) on the glioblastoma phylogenetic cluster from the NCI was dependent on its stereochemistry.[222]

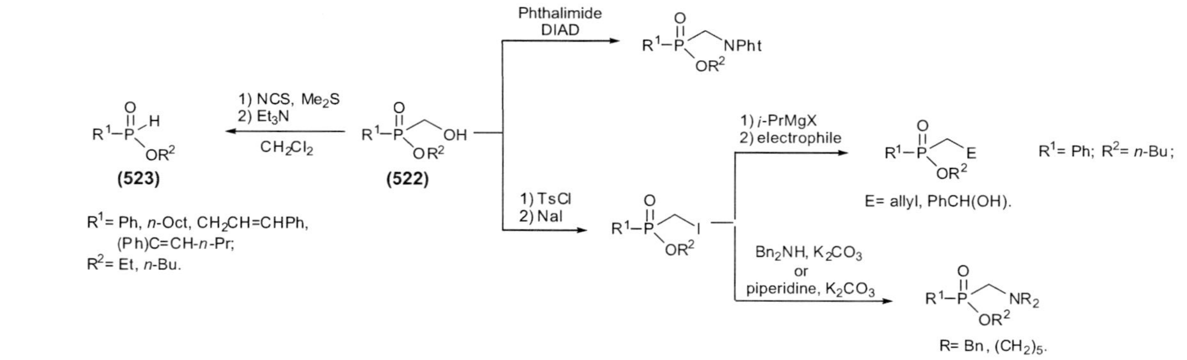

Scheme 174

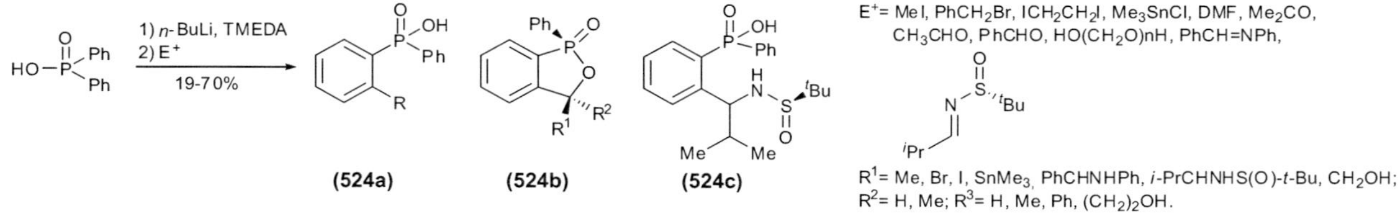

Scheme 175

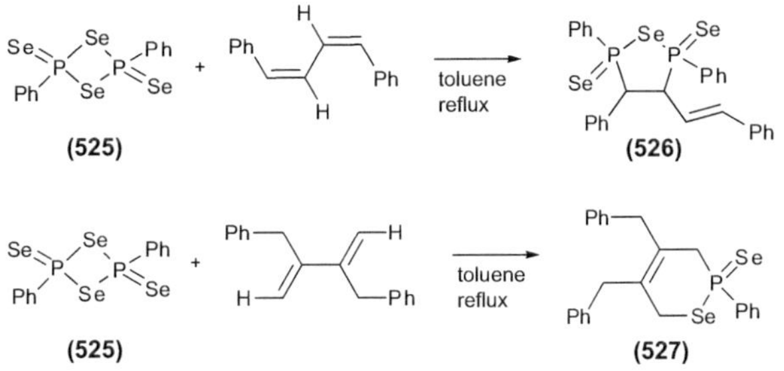

Scheme 176

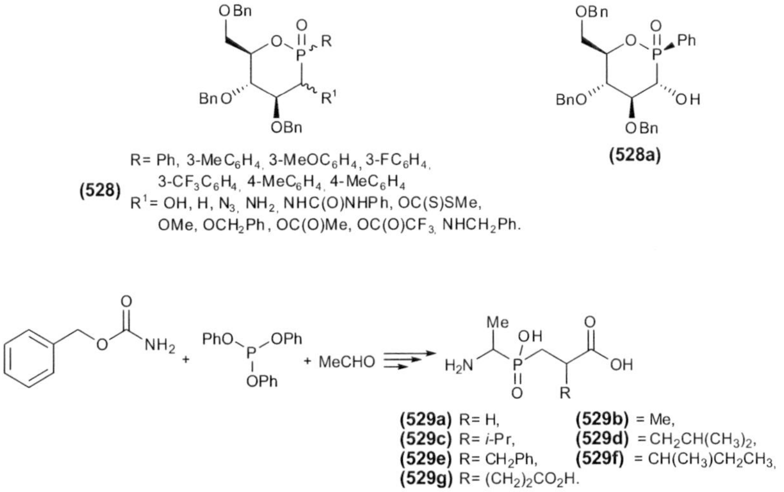

Scheme 177

The synthesis and characterisation of seven phosphonate dipeptide analogues (529) of D-ala-D-ala with various substituents have been reported. The 11 step synthesis was employed for preparation of the phosphonates (529) in overall yields 16.8–21.2% (Scheme 177). The activity evaluation indicated that six of these phosphonate analogues (529a–529f) inhibited VanX, a Zn(II)-dependent D-ala-D-ala dipeptidase, with IC_{50} of 0.48–8.21 mM. These data revealed a structure-activity relationship which showed that the large substituent group on β-carbon resulted in low binding affinity of the phosphonate analogue (529) to VanX. This information will be helpful to guide the design and synthesis of the tightly-binding inhibitors for VanX.[223]

3.4 Miscellaneous

A kinetic study of the nucleophilic substitution reaction of substituted phenyl diphenylphosphinates (530a) and diphenylphosphinothioates (530b) with alkali-metal ethoxides (MOEt; M = Li, Na, K) in anhydrous ethanol has been reported (Scheme 178). Experimental studies revealed the wide-ranging and important effects of structural change P=O to P=S bond on reactivity, mechanisms (concerted versus stepwise), and on

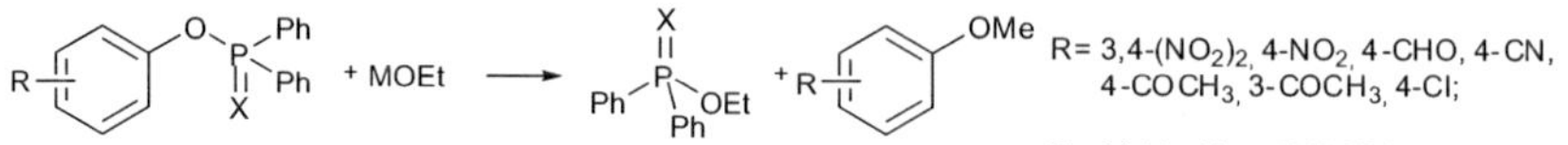

Scheme 178

selectivity towards alkali-metal ions. The P=O compounds (**530a**) were approximately 80-times more reactive than the P=S compounds (**530b**) toward the dissociated EtO⁻ (regardless of the electronic nature of phenyl substituent) but were up to 3.1×10^{-3}-fold more reactive toward ion-paired LiOEt.[224]

References

1. X.-Y. Zhu, J.-R. Chen, L.-Q. Lu and W.-J. Xiao, *Tetrahedron*, 2012, **68**, 6032.
2. O. S. Fenton, E. E. Allen, K. P. Pedretty, S. D. Till, J. E. Todaro and B. R. Sculimbrene, *Tetrahedron*, 2012, **68**, 9023.
3. Y. Pu, L. Gao, H. Liu and J. Yan, *Synthesis*, 2012, **44**, 99.
4. P. Nun, J. D. Egbert, M.-J. Oliva-Madrid and S. P. Nolan, *Chem. Eur. J*, 2012, **18**, 1064.
5. R.-J. Song, Y.-Y. Liu, J.-C. Wu, Y.-X. Xie, G.-B. Deng, X.-H. Yang, Y Liu and J.-H. Li, *Synthesis*, 2012, **44**, 1119.
6. R. A. Aitken, C. J. Collett and S. T. E. Mesher, *Synthesis*, 2012, **44**, 2515.
7. Y.-W. Sun, P.-L. Zhu, Q. Xu and M. Shi, *Tetrahedron*, 2012, **68**, 9924.
8. G. M. Salamończyk, *Tetrahedron*, 2012, **68**, 10209.
9. D. E. Petrillo, D. R. Mowrey, S. P. Allwein and Roger P. Bakale, *Org. Lett.*, 2012, **14**, 1206.
10. A. T. Placzek, J. L. Hougland and R. A. Gibbs, *Org. Lett.*, 2012, **14**, 4038.
11. P. J. Pedersen, T. L. Andresen and M. H. Clausen, *Eur. J. Org. Chem.*, 2012, 6656.
12. D. M. Jessen, A. N. Wercholuk, B Xiong, A. L. Sargent and W. E. Allen, *J. Org. Chem.*, 2012, **77**, 6615.
13. G. M. Rankin, B. J. Compton, K. A. Johnston, C. M. Hayman, G. F. Painter and D. S. Larsen, *J. Org. Chem.*, 2012, **77**, 6743.
14. F. Kong, G. Ribeiro Morais, R. A. Falconer and C. W. Sutton, *Tetrahedron Lett.*, 2012, **53**, 546.
15. A.-A. C. Varvogli, K. C. Fylaktakidou, T. Farmaki, J. G. Stefanakis and A. E. Koumbis, *Eur. J. Org. Chem.*, 2012, 5855.
16. W. F. J. Hogendorf, L. N. Lameijer, T. J. M. Beenakker, H. S. Overkleeft, D. V. Filippov, J. D. C. Codée and G. A. Van der Marel, *Org. Lett.*, 2012, **14**, 848.
17. M. Mikołajczyk, J. Łuczak, L. Sieroń and M. W. Wieczorek, *Tetrahedron*, 2012, **68**, 126.
18. H.-H. Tan, A. Makino, K. Sudesh, P. Greimel and T. Kobayashi, *Angew. Chem. Int. Ed.*, 2012, **51**, 533.
19. E. K. Perttu, A. G. Kohli and F. C. Szoka, Jr, *J. Am. Chem. Soc.*, 2012, **134**, 4485.
20. Á. Bejarano-Villafuerte, M. W. van der Meijden, M. Lingenfelder, K. Wurst, R. M. Kellogg and D. B. Amabilino, *Chem. Eur. J.*, 2012, **18**, 15984.
21. D. Brégeon, L. Ferron, A. Chrétien, F. Guillen, V. Zgonnik, M. Rivaud, M.-R. Mazières, G. Coquerel, C. Belmant and J.-C. Plaquevent, *Bioorg. Med. Chem. Lett.*, 2012, **22**, 5807.

22 S. I. Odejinmi, R. G. Rascon, W. Chen and K. Lai, *Tetrahedron*, 2012, **68**, 8937.
23 S. A. Goretta, M. Kinoshita, S. Mori, H. Tsuchikawa, N. Matsumori and M. Murata, *Bioorg. Med. Chem.*, 2012, **20**, 4012.
24 H.-W. Shih, K.-T. Chen and W.-C. Cheng, *Tetrahedron Lett.*, 2012, **53**, 243.
25 R. K. Shiroodi, A. S. Dudnik and V. Gevorgyan, *J. Am. Chem. Soc.*, 2012, **134**, 6928.
26 B. M. Trost and L. C. Czabaniuk, *J. Am. Chem. Soc.*, 2012, **134**, 5778.
27 M. Yang, N. Yokokawa, H. Ohmiya and M. Sawamura, *Org. Lett.*, 2012, **14**, 816.
28 K. Nagao, U. Yokobori, Y. Makida, H. Ohmiya and M. Sawamura, *J. Am. Chem. Soc.*, 2012, **134**, 8982.
29 B. Jung and A. H. Hoveyda, *J. Am. Chem. Soc.*, 2012, **134**, 1490.
30 M. Cieslikiewicz, A. Bouet, S. Jugé, M. Toffano, J. Bayardon, C. West, K. Lewinski and I. Gillaizeau, *Eur. J. Org. Chem.*, 2012, 1101.
31 A. M. Lauer and J. Wu, *Org. Lett.*, 2012, **14**, 5138.
32 H. M. Oh, S. H. Sim, S. I. Lee, J. Kim and Y. K. Chung, *Synlett*, 2012, **23**, 2657.
33 F. J. Robertson and J. Wu, *J. Am. Chem. Soc.*, 2012, **134**, 2775.
34 F. Li, D. Calabrese, M. Brichacek, I. Lin and J. T. Njardarson, *Angew. Chem. Int. Ed.*, 2012, **51**, 1938.
35 A. Rai and L. D. S. Yadav, *Tetrahedron*, 2012, **68**, 2459.
36 R. Kapoor, R. Chawla, S. Singh and L. D. S. Yadav, *Synlett*, 2012, **23**, 1321.
37 K. A. DeKorver, X.-N. Wang, M. C. Walton and R. P. Hsung, *Org. Lett.*, 2012, **14**, 1768.
38 E. Krawczyk, G. Mielniczak, K. Owsianik and J. Łuczak, *Tetrahedron Asymmetry*, 2012, **23**, 1480.
39 A. Noel, B. Delpech and D. Crich, *Org. Lett.*, 2012, **14**, 1342.
40 P. K. M. Venukadasula, R. Chegondi, G. M. Suryn and P. R. Hanson, *Org. Lett.*, 2012, **14**, 2634.
41 Y. Zhang, H. Tanimoto, Y. Nishiyama, T. Morimoto and K. Kakiuchi, *Synlett*, 2012, **23**, 367.
42 M. Medeiros, E. S. Orth, A. M. Manfredi, P. Pavez, G. A. Micke, A. J. Kirby and F. Nome, *J. Org. Chem.*, 2012, **77**, 10907.
43 F. Huang, C. Cheng and G. Feng, *J. Org. Chem.*, 2012, **77**, 11405.
44 Y. Wang, K. Zheng and R. Hong, *J. Am. Chem. Soc.*, 2012, **134**, 4096.
45 J.-W. Zhang, Z. Xu, Q. Gu, X.-X. Shi, X.-B. Leng and S.-L. You, *Tetrahedron*, 2012, **68**, 5263.
46 S.-H. Shi, F.-P. Huang, P. Zhu, Z.-W. Dong and X.-P. Hui, *Org. Lett.*, 2012, **14**, 2010.
47 Z.-Y. Han, D.-F. Chen, Y.-Y. Wang, R. Guo, P.-S. Wang, C. Wang and L.-Z. Gong, *J. Am. Chem. Soc.*, 2012, **134**, 6532.
48 A. K. Mourad, J. Leutzow and C. Czekelius, *Angew. Chem. Int. Ed.*, 2012, **51**, 11149.
49 X.-F. Tu and L.-Z. Gong, *Angew. Chem. Int. Ed.*, 2012, **51**, 11346.
50 S. Fleischer, S. Werkmeister, S. Zhou, K. Junge and M. Beller, *Chem. Eur. J.*, 2012, **18**, 9005.
51 Q. Cai, C. Liu, X.-W. Liang and S.-L. You, *Org. Lett.*, 2012, **14**, 4588.
52 Z. Zhang and J. C. Antilla, *Angew. Chem. Int. Ed.*, 2012, **51**, 11778.
53 E. Aranzamendi, N. Sotomayor and E. Lete, *J. Org. Chem.*, 2012, **77**, 2986.
54 L. He, M. Bekkaye, P. Retailleau and G. Masson, *Org. Lett.*, 2012, **14**, 3158.
55 A. Alix, C. Lalli, P. Retailleau and G. Masson, *J. Am. Chem. Soc.*, 2012, **134**, 10389.
56 Z. Chai and T. J. Rainey, *J. Am. Chem. Soc.*, 2012, **134**, 3615.

57 Z. Sun, G. A. Winschel, A. Borovika and P. Nagorny, *J. Am. Chem. Soc.*, 2012, **134**, 8074.
58 G. Dagousset, P. Retailleau, G. Masson and J. Zhu, *Chem. Eur. J.*, 2012, **18**, 5869.
59 A. Gómez-SanJuan, N. Sotomayor and E. Lete, *Tetrahedron Lett.*, 2012, **53**, 2157.
60 M. Yamanaka, M. Hoshino, T. Katoh, K. Mori and T. Akiyama, *Eur. J. Org. Chem.*, 2012, 4508.
61 H. Mandai, K. Murota, K. Mitsudo and S. Suga, *Org. Lett.*, 2012, **14**, 3486.
62 T. Sakamoto, K. Mori and T. Akiyama, *Org. Lett.*, 2012, **14**, 3312.
63 R. J. Phipps, K. Hiramatsu and F. D. Toste, *J. Am. Chem. Soc.*, 2012, **134**, 8376.
64 F. Shi, G.-J. Xing, Z.-L. Tao, S.-W. Luo, S.-J. Tu and L.-Z. Gong, *J. Org. Chem.*, 2012, **77**, 6970.
65 G.-x. Li, Qi.-q. Fu, X.-m. Zhang, J. Jiang and Z. Tang, *Tetrahedron Asymm.*, 2012, **23**, 245.
66 L. Ren, T. Lei, J.-X. Ye and L.-Z. Gong, *Angew. Chem. Int. Ed.*, 2012, **51**, 771.
67 B. Guo, G. Schwarzwalder and J. T. Njardarson, *Angew. Chem. Int. Ed.*, 2012, **51**, 5675.
68 T. Honjo, R. J. Phipps, V. Rauniyar and F. D. Toste, *Angew. Chem. Int. Ed.*, 2012, **51**, 9684.
69 S. Liao, I. Čorić, Q. Wang and B. List, *J. Am. Chem. Soc.*, 2012, **134**, 10765.
70 L. Song, Q.-X. Guo, X.-C. Li, J. Tian and Y.-G. Peng, *Angew. Chem. Int. Ed.*, 2012, **51**, 1899.
71 C.-H. Xing, Y.-X. Liao, Y. Zhang, D. Sabarova, M. Bassous and Q.-S. Hu, *Eur. J. Org. Chem.*, 2012, 1115.
72 D. Huang, F. Xu, X. Lin and Y. Wang, *Chem. Eur. J.*, 2012, **18**, 3148.
73 J. Lv, L. Zhang, S. Hu, J.-P. Cheng and S. Luo, *Chem. Eur. J.*, 2012, **18**, 799.
74 L. Chen, L. Zhang, J. Lv, J.-P. Cheng and S. Luo, *Chem. Eur. J.*, 2012, **18**, 8891.
75 F. Shi, Z.-L. Tao, S.-W. Luo, S.-J. Tu and L.-Z. Gong, *Chem. Eur. J.*, 2012, **18**, 6885.
76 E. L. McInturff, E. Yamaguchi and M. J. Krische, *J. Am. Chem. Soc.*, 2012, **134**, 20628.
77 F. Chen, Z. Ding, Y. He, J. Qin, T. Wang and Q.-H. Fan, *Tetrahedron*, 2012, **68**, 5248.
78 Y.-Y. Chen, Y.-J. Jiang, Y.-S. Fan, D. Sha, Q. Wang, G. Zhang, L. Zheng and S. Zhang, *Tetrahedron:Asymmetry*, 2012, **23**, 904.
79 D. Uraguchi, K. Yoshioka, Y. Ueki and T. Ooi, *J. Am. Chem. Soc.*, 2012, **134**, 19370.
80 E. Gendaszewska-Darmach, E. Laska, P. Rytczak and A. Okruszek, *Bioorg. Med. Chem. Lett.*, 2012, **22**, 2698.
81 W. W. Turner, K. Hartvigsen, A. Boullier, E. N. Montano, J. L. Witztum and M. S. VanNieuwenhze, *J. Med. Chem.*, 2012, **55**, 8178.
82 K. H. Jang, Y. Lee, C. J. Sim, K.-B. Oh and J. Shin, *Bioorg. Med. Chem. Lett.*, 2012, **22**, 1078.
83 C. Ji and M. J. Miller, *Bioorg. Med. Chem.*, 2012, **20**, 3828.
84 J.-L. Giner, H. Wang and C. T. Morita, *Bioorg. Med. Chem. Lett.*, 2012, **22**, 811.
85 P. J. Pedersen, H. M.-F. Viart, F. Melander, T. L. Andresen, R. Madsen and M. H. Clausen, *Bioorg. Med. Chem.*, 2012, **20**, 3972.
86 M. Serpi, R. Bibbo, S. Rat, H. Roberts, C. Hughes, B. Caterson, M. J. Alcaraz, A. T. Gibert, C. R. A. Verson and C. McGuigan, *J. Med. Chem.*, 2012, **55**, 4629.

87 J. F. Kadow, Y. Ueda, N. A. Meanwell, T. P. Connolly, T. Wang, C.-P. Chen, K.-S. Yeung, J. Zhu, J. A. Bender, Z. Yang, D. Parker, P.-F. Lin, R. J. Colonno, M. Mathew, D. Morgan, M. Zheng, C. Chien and D. Grasela, *J. Med. Chem.*, 2012, **55**, 2048.
88 S. Dumbre, A. Derouaux, E. Lescrinier, A. Piette, B. Joris, M. Terrak and P. Herdewijn, *J. Am. Chem. Soc.*, 2012, **134**, 9343.
89 A. N. Harrison, S. Reichau and E. J. Parker, *Bioorg. Med. Chem. Lett.*, 2012, **22**, 907.
90 C. M. Pedersen, I. Figueroa-Perez, A. J. Ulmer, U. Zähringer and R. R. Schmidt, *Tetrahedron*, 2012, **68**, 1052.
91 F. Kaschani, S. Nickel, B. Pandey, B. F. Cravatt, M. Kaiser and R. A. L. van der Hoorn, *Bioorg. Med. Chem.*, 2012, **20**, 597.
92 J. R. Mora, A. J. Kirby and F. Nome, *J. Org. Chem.*, 2012, **77**, 7061.
93 O. G. Ovchinnikova, N. A. Kocharova, M. Bialczak-Kokot, A. S. Shashkov, A. Rozalski and Y. A. Knirel, *Eur. J. Org. Chem.*, 2012, 3500.
94 C. Stefaniu, I. Vilotijevic, M. Santer, D. V. Silva, G. Brezesinski and P. H. Seeberger, *Angew. Chem. Int. Ed.*, 2012, **51**, 12874.
95 I. Walton, M. Davis, L. Munro, V. J. Catalano, P. J. Cragg, M. T. Huggins and K. J. Wallace, *Org. Lett.*, 2012, **14**, 2686.
96 Z. Wu, X. Wu, Y. Yang, T. Wen and S. Han, *Bioorg. Med. Chem. Lett.*, 2012, **22**, 6358.
97 A. Wieczorek and F. Hammerschmidt, *J. Org. Chem.*, 2012, **77**, 10021.
98 J. Y. Jeon, J. K. Varghese, J. H. Park, S.-H. Lee and B. Y. Lee, *Eur. J. Org. Chem.*, 2012, 3566.
99 X. Jiang, L. Chu and F.-L. Qing, *Org. Lett.*, 2012, **14**, 2870.
100 Z. Feng, F. Chen and X. Zhang, *Org. Lett.*, 2012, **14**, 1938.
101 C.-B. Xiang, Y.-J. Bian, X.-R. Mao and Z.-Z. Huan, *J. Org. Chem.*, 2012, **77**, 7706.
102 H. Wang, X. Li and Fan Wu, B. Wan, *Synthesis*, 2012, **44**, 941.
103 H.-P. Deng and M. Shi, *Eur. J. Org. Chem.*, 2012, 183.
104 S. Zhou, Z. Wu, J. Rong, S. Wang, G. Yang, X. Zhu and L. Zhang, *Chem. Eur. J.*, 2012, **18**, 2653.
105 B. Liu, J.-F. Carpentier and Y. Sarazin, *Chem. Eur. J.*, 2012, **18**, 13259.
106 K. Alagiri, P. Devadig and K. R. Prabhu, *Chem. Eur. J.*, 2012, **18**, 5160.
107 H. Wang, X. Li, F. Wu and B. Wan, *Tetrahedron Lett.*, 2012, **53**, 681.
108 K. Alagiri, P. Devadig and K. R. Prabhu, *Tetrahedron Lett.*, 2012, **53**, 1456.
109 Y.-L. Zhao, G.-J. Wu, Y. Li, L.-X. Gao and F.-S. Han, *Chem. Eur. J.*, 2012, **18**, 9622.
110 A. Mitrofanov, A. B. Lemeune, C. Stern, R. Guilard, N. Gulyukina and I. Beletskaya, *Synthesis*, 2012, **44**, 3805.
111 Z. Qu, X. Chen, J. Yuan, Y. Bai, T. Chen, L. Qu, F. Wang, X. Li and Y. Zhao, *Tetrahedron*, 2012, **68**, 3156.
112 S. T. Disale, S. R. Kale, S. S. Kahandal, T. G. Srinivasan and R. V. Jayaram, *Tetrahedron Lett.*, 2012, **53**, 2277.
113 G. D. Tibhe, M. Bedolla-Medrano, C. Cativiela and M. Ordóñez, *Synlett*, 2012, **23**, 1931.
114 Y. Heo, D. Hyan Cho, M. Kumar Mishra and D. O. Jang, *Tetrahedron Lett.*, 2012, **53**, 3897.
115 S. R. Kolla and Y. R. Lee, *Tetrahedron*, 2012, **68**, 226.
116 M. Ordóñez, G. D. Tibhe, A. Zamudio-Medina and J. L. Viveros-Ceballos, *Synthesis*, 2012, **44**, 569.
117 B. Boduszek, T. K. Olszewski, W. Goldeman, K. Grzegolec and P. Blazejewska, *Tetrahedron*, 2012, **68**, 1223.

118   Z.-H. Cai, G.-F. Du, B. Dai and L. He, *Synthesis*, 2012, **44**, 694.
119   C.-O. Turrin, A. Hameau and A.-M. Caminade, *Synthesis*, 2012, **44**, 1628.
120   F. Rostami-Charati and Z. Hossaini, *Synlett*, 2012, **23**, 2397.
121   N. Xin and S. Ma, *Eur. J. Org. Chem.*, 2012, 3806.
122   F. Gomes, A. Fadel and N. Rabasso, *J. Org. Chem.*, 2012, **77**, 5439.
123   W. Goldeman, A. Kluczyński and M. Soroka, *Tetrahedron Lett.*, 2012, **53**, 5290.
124   G.-V. Röschenthaler, V. P. Kukhar, I. B. Kulik, M. Y. Belik, A. E. Sorochinsky, E. B. Rusanov and V. A. Soloshonok, *Tetrahedron Lett.*, 2012, **53**, 539.
125   L. Nagarapu, R. Mallepalli, U. N. Kumar, P. Venkateswarlu, R. Bantu and L. Yeramanchi, *Tetrahedron Lett.*, 2012, **53**, 1699.
126   B. Xiong, R. Shen, M. Goto, S.-F. Yin and L.-B. Han, *Chem. Eur. J.*, 2012, **18**, 16902.
127   S. Montel, C. Midrier, J.-N. Volle, R. Braun, K. Haaf, L. Willms, J.-L. Pirat and D. Virieux, *Eur. J. Org. Chem.*, 2012, 3237.
128   M. Seki, *Synthesis*, 2012, **44**, 1556.
129   P. Jansa, O. Hradil, O. Baszczyński, M. Dračínský, B. Klepetářová, A. Holý, J. Balzarini and Z. Janeba, *Tetrahedron*, 2012, **68**, 865.
130   M. Psurskia, M. Piguła, J. Ciekot, Ł. Winiarski, J. Wietrzyk and J. Oleksyszyn, *Tetrahedron Lett.*, 2012, **53**, 5845.
131   R. Ziessel, A. Steffen and M. Starck, *Tetrahedron Lett.*, 2012, **53**, 3713.
132   K. Bera and I. N. N. Namboothiri, *Org. Lett.*, 2012, **14**, 980.
133   C. Bhushan Tripathi, S. Kayal and S. Mukherjee, *Org. Lett.*, 2012, **14**, 3296.
134   S. Kong, W. Fan, G. Wu and Z. Miao, *Angew. Chem. Int. Ed.*, 2012, **51**, 8864.
135   J. Guang, Q. Guo and J. C.-G. Zhao, *Org. Lett.*, 2012, **14**, 3174.
136   D. Cal, *Tetrahedron Lett.*, 2012, **53**, 3774.
137   P. Brémond, K. Kabytaev and S. R. A. Marque, *Tetrahedron Lett.*, 2012, **53**, 4543.
138   H. Zhang, X. Wen, L. Gan and Y. Peng, *Org. Lett.*, 2012, **14**, 2126.
139   C. Schnaars and T. Hansen, *Org. Lett.*, 2012, **14**, 2794.
140   R. Kumar, D. Verma, S. M. Mobin and I. N. N. Namboothiri, *Org. Lett.*, 2012, **14**, 4070.
141   B. Duda, S. N. Tverdomed, B. I. Ionin and G.-V. Röschenthaler, *Eur. J. Org. Chem.*, 2012, 3684.
142   C.-Y. Zhou, J.-C. Wang, J. Wei, Z.-J. Xu, Z. Guo, K.-H. Low and C.-M. Che, *Angew. Chem. Int. Ed.*, 2012, **51**, 11376.
143   T. Maegawa, K. Otake, K. Hirosawa, A. Goto and H. Fujioka, *Org. Lett.*, 2012, **14**, 4798.
144   X. Tao, W. Li, X. Ma, X. Li, W. Fan, L. Zhu, X. Xie and Z. Zhang, *J. Org. Chem.*, 2012, **77**, 8401.
145   T. Konno, K. Shimizu, K. Ogata and S. Fukuzawa, *J. Org. Chem.*, 2012, **77**, 3318; Z. Duan, L. Wang, X. Song, X. Hu and Z. Zheng, *Tetrahedron: Asymmetry*, 2012, **23**, 508.
146   K. Dong, Z. Wang and K. Ding, *J. Am. Chem. Soc.*, 2012, **134**, 12474.
147   D. Moradov, A. Rubinstein, D. Gelman and M. Srebnik, *Synthesis*, 2012, **44**, 1258.
148   N. R. Cichowicz and P. Nagorny, *Org. Lett.*, 2012, **14**, 1058.
149   S. Roy and Christopher D. Spilling, *Org. Lett.*, 2012, **14**, 2230.
150   T. Cochet, D. Roche, V. Bellosta and J. Cossy, *Eur. J. Org. Chem.*, 2012, 801.
151   C.-K. Pei, Y. Jiang, Y. Wei and M. Shi, *Angew. Chem. Int. Ed.*, 2012, **51**, 11328.
152   D. Webb and T. F. Jamison, *Org. Lett.*, 2012, **14**, 2465.
153   J. Tan, C. l-H. Cheon and H. Yamamoto, *Angew. Chem. Int. Ed.*, 2012, **51**, 8264.

154 A. I. Arkhypchuk, M.-P. Santoni and S. Ott, *Angew. Chem. Int. Ed.*, 2012, **51**, 7776.
155 N. Z. Kiss, A. Kaszás, L. Drahos, Z. Mucsi and G. Keglevich, *Tetrahedron Lett.*, 2012, **53**, 207.
156 R. Kotikalapudi and K. C. K. Swamy, *Tetrahedron Lett.*, 2012, **53**, 3831.
157 A. K. Purohit, D. Pardasani, V. Tak, A. Kumar, R. Jain and D. K. Dubeyl, *Tetrahedron Lett.*, 2012, **53**, 3795.
158 T. Kusukawa, C. Katano and C. Kim, *Tetrahedron*, 2012, **68**, 1492.
159 E. B. Erkhitueva, A. V. Dogadina, A. V. Khramchikhin and B. I. Ionin, *Tetrahedron Lett.*, 2012, **53**, 4304.
160 Z. Wang, Z. Yu, Y. Wang and D. Shi, *Synthesis*, 2012, **44**, 1559.
161 L. Chu and F.-L. Qing, *Synthesis*, 2012, **44**, 1521.
162 M. Stankevič and J. Bazan, *J. Org. Chem.*, 2012, **77**, 8244.
163 M. T. Corbett, D. Uraguchi, T. Ooi and J. S. Johnson, *Angew. Chem. Int. Ed.*, 2012, **51**, 4685.
164 J. Tönnemann, R. Scopelliti, Konstantin O. Zhurov, L. Menin, S. Dehnen and K. Severin, *Chem. Eur. J.*, 2012, **18**, 9939.
165 J.-L. Zhu, J.-S. Bau and Y.-C. Shih, *Synlett*, 2012, **23**, 863.
166 A. Sokolsky and A. B. Smith III, *Org. Lett.*, 2012, **14**, 4470.
167 M. Kajjouta, R. Zemmouri, S. Eddarir and C. Rolando, *Tetrahedron*, 2012, **68**, 3225.
168 D. A. Devalankar, P. V. Chouthaiwale and A. Sudalai, *Tetrahedron: Asymm.*, **23**, 240.
169 T. Umezawa, T. Seino and F. Matsuda, *Org. Lett.*, 2012, **14**, 4206.
170 N. Chuanopparat, N. Kongkathip and B. Kongkathip, *Tetrahedron*, 2012, **68**, 6803.
171 C. J. Saavedra, A. Boto and R. Hernández, *Org. Lett.*, 2012, **14**, 3788.
172 Ł. Albrecht, D. Deredas, J. Wojciechowski, W. M. Wolf and H. Krawczyk, *Synthesis*, 2012, **44**, 247.
173 A. Albrecht, F. Morana, A. Fraile and K. A. Jørgensen, *Chem. Eur. J*, 2012, **18**, 10348.
174 J. Modranka, A. Albrecht, R. Jakubowski, H. Krawczyk, M. Różalski, U. Krajewska, A. Janecka, Anna Wyrębska, Barbara Różalska and T. Janecki, *Bioorg. Med. Chem.*, 2012, **20**, 5017.
175 Q. Liu, G. Yue, N. Wu, G. Lin, Y. Li, J. Quan, C.-c. Li, G. Wang and Z. Yang, *Angew. Chem. Int. Ed.*, 2012, **51**, 12072.
176 C. Napolitano, V. R. Palwai, L. A. Eriksson and P. V. Murphy, *Tetrahedron*, 2012, **68**, 5533.
177 M. T. Crimmins and A. M. Azman, *Synlett*, 2012, **23**, 1489.
178 P. Mohan, K. Koushik and M. J. Fuertes, *Tetrahedron Lett.*, 2012, **53**, 2730.
179 X. Huang, D. Pissarnitski, H. Li, T. Asberom, H. Josien, X. Zhu, M. Vicarel, Z. Zhao, M. Rajagopalan, A. Palani, R. Aslanian, Z. Zhu, W. Greenlee and A. Buevich, *Tetrahedron Lett.*, 2012, **53**, 6451.
180 R. A. Al-Horani and U. R. Desai, *Tetrahedron*, 2012, **68**, 2027.
181 E. Villemin, M.-F. Herent and J. Marchand-Brynaert, *Eur. J. Org. Chem.*, 2012, 6165.
182 S. Wang, J. Li, T. Miao, W. Wu, Q. Li, Y. Zhuang, Z. Zhou and L. Qiu, *Org. Lett.*, 2012, **14**, 1966.
183 K. Hanson, D. L. Ashford, J. J. Concepcion, R. A. Binstead, S. Habibi, H. Luo, C. R. K. Glasson, J. L. Templeton and T. J. Meyer, *J. Am. Chem. Soc.*, 2012, **134**, 16975.
184 G. Nagendra, C. Madhu, T. M. Vishwanatha and V. V. Sureshbabu, *Tetrahedron Lett.*, 2012, **53**, 5059.

185 J. K. Augustine, A. Bombrun, W. H. B. Sauer and P. Vijaykumar, *Tetrahedron Lett.*, 2012, **53**, 5030.
186 F. R. McSorley, P. B. Wyatt, A. Martinez, E. F. DeLong, B. Hove-Jensen and D. L. Zechel, *J. Am. Chem. Soc.*, 2012, **134**, 8364.
187 E. Lagadic, Y. Garcia and J. Marchand-Brynaert, *Synthesis*, 2012, **44**, 93.
188 Ł. Winiarski, J. Oleksyszyn and M. Sieńczyk, *J. Med. Chem.*, 2012, **55**, 6541.
189 E. Burchacka, M. Walczak, M. Sieńczyk, G. Dubin, M. Zdżalik, J. Potempa and J. Oleksyszyn, *Bioorg. Med. Chem. Lett.*, 2012, **22**, 5574.
190 K. Hochdörffer, K. Abu Ajaj, C. Schaäfer-Obodozie and F. Kratz, *J. Med. Chem.*, 2012, **55**, 7502.
191 J. M. Matthews, N. Qin, R. W. Colburn, S. L. Dax, M. Hawkins, J. J. McNally, L. Reany, M. A. Youngman, J. Baker, T. Hutchinson, Yi Liu, M. L. Lubinh, M. Neeper, M. R. Brandt, D. J. Stone and C. M. Flores, *Bioorg. Med. Chem. Lett.*, 2012, **22**, 2922.
192 L. V. Faro, J. M. de Almeida, C. C. Cirne-Santos, V. A. Giongo, L. R. Castello-Branco, I. de, B. Oliveira, J. E. F. Barbosa, A. C. Cunha, V. F. Ferreira, M. C. de Souzaa, I. C. N. P. Paixão and M. C. B. V. de Souza, *Bioorg. Med. Chem. Lett.*, 2012, **22**, 5055.
193 F. Song, J. Zhang, Q. Cui, T. Wang, W. Chen, L. Li and Z. Xi, *Tetrahedron Lett.*, 2012, **53**, 1102.
194 G.-N. Wang, P. S. Lau, Y. Li and X.-S. Ye, *Tetrahedron*, 2012, **68**, 9405.
195 K. Brücher, B. Illarionov, J. Held, S. Tschan, A. Kunfermann, M. K. Pein, A. Bacher, T. Gräwert, L. Maes, B. Mordmüller, M. Fischer and T. Kurz, *J. Med. Chem.*, 2012, **55**, 6566.
196 K. Graham, R. Lesche, A. V. Gromov, N. Böhnke, M. Schäfer, J. Hassfeld, L. Dinkelborg and G. Kettschau, *J. Med. Chem.*, 2012, **55**, 9510.
197 V. Point, R. K. Malla, S. Diomande, B. P. Martin, V. Delorme, F. Carriere, S. Canaan, N. P. Rath, C. D. Spilling and J.-F. Cavalier, *J. Med. Chem.*, 2012, **55**, 10204.
198 Y.-S. Lin, J. Park, J. W. De Schutter, X. F. Huang, A. M. Berghuis, M. Sebag and Y. S. Tsantrizos, *J. Med. Chem.*, 2012, **55**, 3201.
199 R. Reich, A. Hoffman, A. Veerendhar, A. Maresca, A. Innocenti, C. T. Supuran and E. Breuer, *J. Med. Chem.*, 2012, **55**, 7875.
200 T.-J. R. Cheng, S. Weinheimer, E. B. Tarbet, J.-T. Jan, Y.-S. E. Cheng, J.-J. Shie, C.-L. Chen, C.-A. Chen, W.-C. Hsieh, P.-W. Huang, W.-H. Lin, S.-Y. Wang, J.-M. Fang, O. Y.-P. Hu and C.-H. Wong, *J. Med. Chem.*, 2012, **55**, 8657.
201 C. Zinglé, L. Kuntz, D. Tritsch, C. Grosdemange-Billiard and M. Rohmer, *Bioorg. Med. Chem. Lett.*, 2012, **22**, 6563.
202 M. F. Abdel-Megeeda, B. E. Badr, M. M. Azaam and G. A. El-Hiti, *Bioorg. Med. Chem.*, 2012, **20**, 2252.
203 I. Kraicheva, I. Tsacheva, E. Vodenicharova, E. Tashev, T. Tosheva, A. Kril, M. Topashka-Ancheva, I. Iliev, Ts. Gerasimova and K. Troev, *Bioorg. Med. Chem.*, 2012, **20**, 117.
204 S. Arns, R. Gibe, A. Moreau, M. Monzur Morshed and R. N. Young, *Bioorg. Med. Chem.*, 2012, **20**, 2131.
205 S. Nickel, F. Kaschani, T. Colby, R. A. L. van der Hoorn and M. Kaiser, *Bioorg. Med. Chem.*, 2012, **20**, 601.
206 H. A. Cooke, S. C. Peck, B. S. Evans and W. A. van der Donk, *J. Am. Chem. Soc.*, 2012, **134**, 15660.
207 H. Huang, W.-c. Chang, P.-Ji Pai, A. Romo, S. O. Mansoorabadi, D. H. Russell and H.-w. Liu, *J. Am. Chem. Soc.*, 2012, **134**, 16171.

208 J. O. C. Jiménez-Halla, M. Kalek, J. Stawinski and F. Himo, *Chem. Eur. J.*, 2012, **18**, 12424.
209 A. K. Duncan, P. J. Klemm, K. N. Raymond and C. C. Landry, *J. Am. Chem. Soc.*, 2012, **134**, 8046.
210 C. Barrére, C. Chendo, T. N. T. Phan, V. Monnier, T. Trimaille, S. Humbel, S. Viel, D. Gigmes and L. Charles, *Chem. Eur. J*, 2012, **18**, 7916.
211 G. Székely, B. Csordás, V. Farkas, J. Kupai, P. Pogány, Z Sánta, Z. Szakács, T. Tóth, M. Hollósi, J. Nyitrai and P. Huszthy, *Eur. J. Org. Chem.*, 2012, 3396.
212 H.-J. Pyun, M. O. Clarke, A. Cho, A. Casarez, M. Ji, M. Fardis, R. Pastor, X. C. Sheng and C. U. Kim, *Tetrahedron Lett.*, 2012, **53**, 2360.
213 B. Kaboudin, H. Haghighat, S. Alaie and T. Yokomatsu, *Synlett*, 2012, **23**, 1965.
214 G. Blay, E. Ceballos, A. Monle on and J. R. Pedro, *Tetrahedron*, 2012, **68**, 2128.
215 S. Kato, T. Yoshino, M. Shibasaki, M. Kanai and S. Matsunaga, *Angew. Chem. Int. Ed.*, 2012, **51**, 7007.
216 D. Zhao, L. Wang, D. Yang, Y. Zhang and R. Wang, *Angew. Chem. Int. Ed.*, 2012, **51**, 7523.
217 X. Han, F. Zhong, Y. Wang and Y. Lu, *Angew. Chem. Int. Ed.*, 2012, **51**, 767.
218 Q. Yao and C. Yuan, *J. Org. Chem.*, 2012, **77**, 10985.
219 O. Berger, L. Gavara and J.-L. Montchamp, *Org. Lett.*, 2012, **14**, 3404.
220 V. Y. Rodríguez, M. A. del Águila, M. J. Iglesias and F. L. Ortiz, *Tetrahedron*, 2012, **68**, 7355.
221 G. Hua, A. L. Fuller, A. M. Z. Slawin and J. D. Woollins, *Synlett.*, 2012, **23**, 2453.
222 L. Clarion, C. Jacquard, O. S. Catherine, S. Loiseau, D. Filippini, M.-H. Hirlemann, J.-N. Volle, D. Virieux, M. Lecouvey, J.-L. Pirat and N. Bakalara, *J. Med. Chem.*, 2012, **55**, 2196.
223 C. Jia, K.-W. Yang, C.-C. Liu, L. Feng, J.-M. Xiao, L.-S. Zhou and Y.-L. Zhang, *Bioorg. Med. Chem. Lett.*, 2012, **22**, 482.
224 I.-H. Um, Y.-H. Shin, J.-E. Park, J.-S. Kang and E. Buncel, *Chem. Eur. J.*, 2012, **18**, 961.

Pentacoordinated and hexacoordinated compounds

Romana Pajkert and Gerd-Volker Röschenthaler*

DOI: 10.1039/9781782623977-00348

1 Introduction

This chapter covers recent developments in the title area published during 2012. Surprisingly, the past year has seen a significant decrease in the number of papers including the synthesis of novel pentacoordinated compounds. In this respect, there has been a review highlighting methods for the asymmetric synthesis of *P*-chiral pentacoordinated spirophosphoranes.[1] The unexpected formation of 1,2-λ^5 oxaphospholenes upon phosphine-mediated cycloisomerization of alkynyl hemiketals[2] as well as the preparation of novel pentacoordinated phosphoranes in the reaction of aromatic hydrazides and phosphoryl reagent bearing at least two leaving groups bound to the phosphorus atom[3] have also been reported in a recent year.

A lot more emphasis has been placed on understanding chemical and biochemical mechanisms in which pentacoordinated species are involved. Among them, the most significant examples include: semipinacol rearrangements of diols,[4,5] reduction of phosphine oxides to the corresponding phosphines,[6] formation of bis-phosphonium ylides,[7] hydrolysis of phosphate triesters,[8] cleavage and isomerization of uridine 3′-alkylphosphates in the presence of a zinc complex[9] and spontaneous reactivation reactions of the sarin-phosphorylated wild-type enzyme of human butyrylcholinesterase.[10] The field of theoretical research relating to pentacoordinated phosphoranes has been mostly aimed at the theoretical explanations of a well-known mechanism of the Wittig-reaction, based on the dynamic processes observed in the structure of 1,2-oxaphosphethanes.[11,12] Finally, the field of the chemistry of hexacoordinated compounds has also been targeted, with some novel synthetic pathways, such as oxidative addition to phosphorus(III) halides[14] and the thermolysis of hexafluorophosphates.[15] The well-established topic including the synthesis of hexacoordinated phosphorus compounds bearing transannular N–P bond[13] has also been examined as well as the formation of hypervalent betaine derivatives[16] and the chemistry of *meso*-triarylcorroles.[17]

2 Pentacoordinated phosphorus compounds

2.1 Synthesis

The synthetic routes to pentacoordinated phosphorus compounds have continued to generate considerable interest, however little new works has

School of Engineering and Science, Jacobs University Bremen gGmbH, P.O. Box 750 561, D-28725 Bremen, Germany. E-mail: g.roeschenthaler@jacobs-university.de

appeared in the last year. For example, a review has been published on methods for the asymmetric synthesis of *P*-chiral pentacoordinated spirophosphoranes.[1] Other investigations included the unexpected formation of a diastereomeric pair of 1,2-λ^5 oxaphospholenes (**1**) which was observed during phosphine-mediated cyclo-isomerization of alkynyl hemiketal (**2**).[2] Usually this cycloisomerization gives cyclic keto enol ethers, however in this case the expected product (**3**) was formed in only 14% yield. The alternate pathway to the major product (**1**) could be explained by the addition of phosphine to the alkynoate to generate zwitterion (**4**). Upon a proton transfer alkoxide (**5**) was formed which then attacked the carbonyl carbon (instead of α-carbon to the carbonyl) resulting in the formation of spirocycle (**1**) (Scheme 1). The species on the left and right show how the diastereoisomers of (**1**) (α and β) arised by bond rotation in the acyclic precursors (**4**) and (**5**).

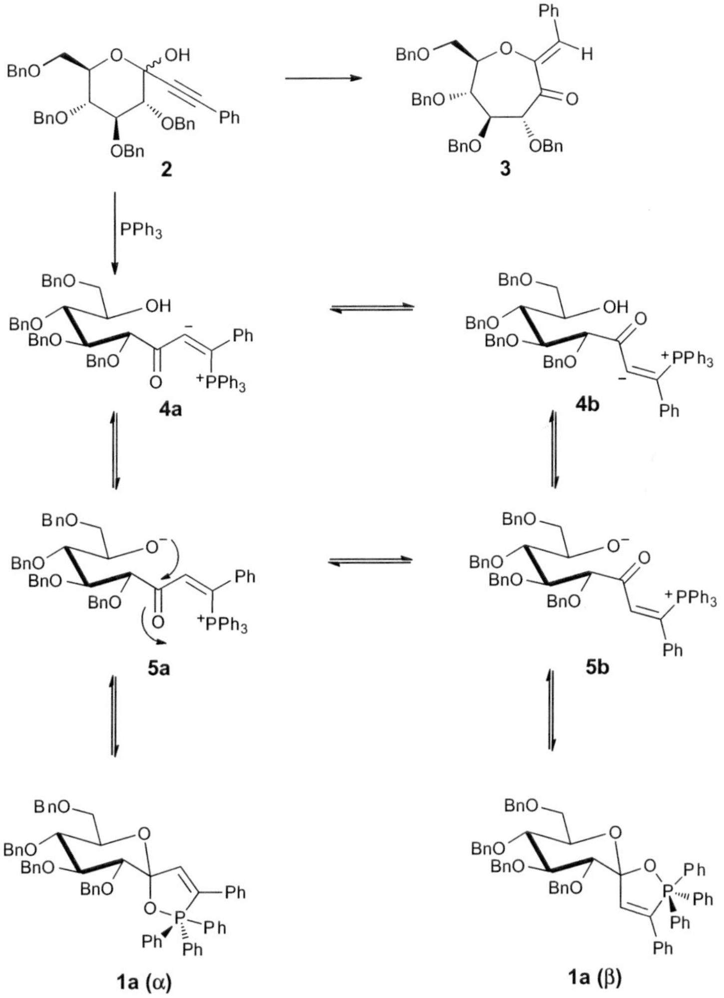

Scheme 1

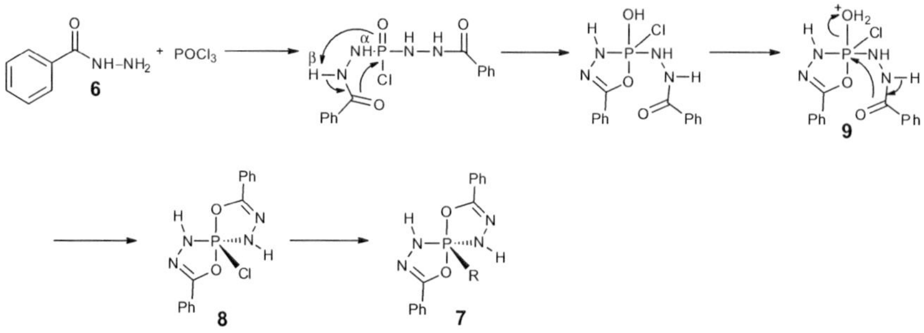

Scheme 2

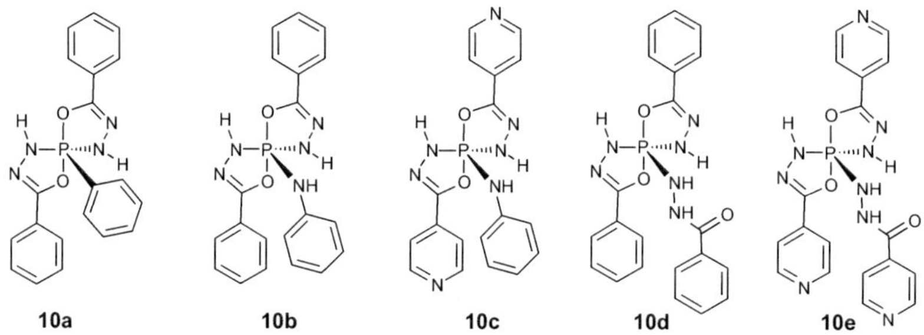

Fig. 1

Noteworthy, the formation of a six-membered ring is favoured over a seven-membered ring and the lack of rigidity in the substrate gave rise to oxaphospholenes (**1**).

The preparation of spiro-phosphoranes was achieved by a dehydration-cyclization rearrangement in the reaction of a hydrazide (**6**) and a phosphoryl reagent bearing at least two leaving groups (Cl) bound to the phosphorus atom ($POCl_3$ or $PhPOCl_2$).[3] Thus bicyclophosphoranes (**7**) were obtained upon treatment of benzhydrazide with $POCl_3$ in refluxing acetonitrile to give the intermediate (**8**). Compound (**8**) was then reacted with the appropriate amine (morpholine and *tert*-butylamine) to furnish phosphoranes (**7**). The proposed mechanism involved the cyclization of the initially phosphorylated hydrazide accompanied by the elimination of the β-amidic proton, prior to the dehydration of phosphorane (**9**) (Scheme 2).

In a similar manner, reaction of a mixture of triethylamine and benhydrazide or 4-pyridinecarboxylic acid hydrazide with $PhPOCl_2$, $PhNHPOCl_2$ and $POCl_3$, gave spirophosphoranes (**10a–e**) (Fig. 1). All of these structural rearrangements were supported by X-Ray analysis as well as NMR and IR experiments.

2.2 Intermediates and transition states

As in previous years, the participation of pentacoordinated species in various chemical and biochemical processes has been emphasized by numerous examples. For instance, the 6-azabicyclo[3.2.1]octane ring system (**11**), prevalent in a range of biologically active molecules, was prepared through a novel semipinacol rearrangement of *cis*-fused β-lactam diols (**12**) utilizing *in situ* generated cyclic pentacoordinate phosphorane intermediate (**13**).[4] The rearrangement proceeded with exclusive *N*-acyl group migration of a β-lactam ring and resulted in carbonyl functionalities at the 7- and bridging 8-position of the bicyclic product. Moreover this is the first example where cyclic phosphoranes have been used to promote rearrangement involving carbon-carbon bond migration (Scheme 3). A phosphorane-mediated pinacol-like rearrangement has been also applied in the stereoselective synthesis of the *trans*-hydrindane core (**14**) of the marine natural product dictyoxetane (**15**), to establish the requisite *trans* ring junction.[5] Treatment of the diol (**16**) with dichlorotriphenylphosphine, generated from triphenylphosphine and hexachloroethane, gave cyclic phosphorane (**17**), which after heating gave the requisite *trans*-hydrindane (**14**) *via* a formal 1,2-hydride shift and the expulsion of $Ph_3P=O$. In this case, the use of Ph_3PCl_2 in the pinacol-like rearrangement notably avoided the need to selectively functionalise the more hindered tertiary alcohol in (**16**) (Scheme 4).

Quantum mechanical calculations have provided new mechanistic insights on the reduction of phosphine oxides to the corresponding phosphines with opposite stereoselectivities.[6] It was shown that phosphine oxides reacted *via* conventional phosphorane intermediate (**18**). The free energy surface for the reaction of $Me_3P=O$ with Si_2Cl_6 is presented in Scheme 5. The sequence of transformation, beginning with formation of the $Me_3P=O\cdots Si_2Cl_6$ adduct (**19**), led to the phosphine (**20**) with inversion of configuration. The rate limiting step was the cleavage of the Si-Si bond in the initial adduct (**TS2**) which generated the

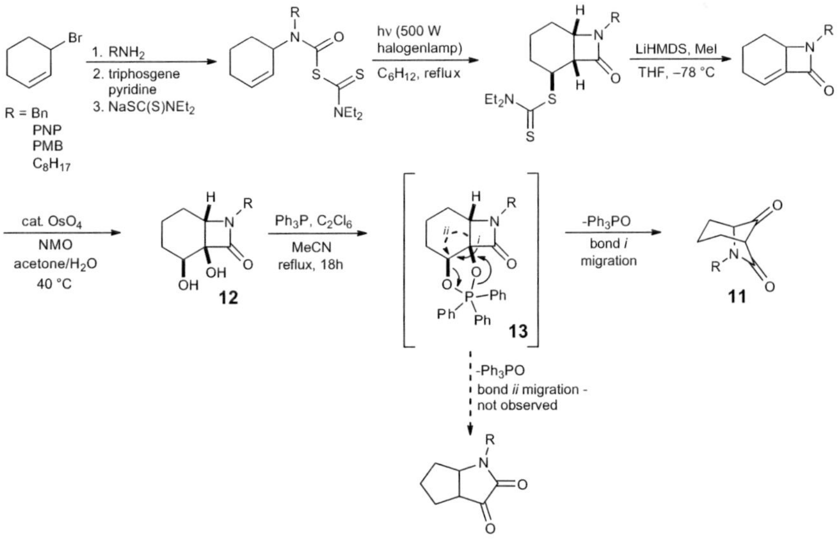

Scheme 3

phosphonium salt (**21**). Subsequent backside addition of $SiCl_3^-$ (**TS3**) provided phosphorane (**18**) which lost $OSiCl_3^-$ from the opposite face (**TS4**) to give (**22**). The last step led to the liberation of the phosphine (**20**) (Scheme 5).

The preparation and characterization of acyclic diphosphonium bis-ylides (**23**) from the corresponding diphosphonium precursors (**24**) have been systematically investigated with respect to the steric demand at the phosphorus centres and the substitution pattern of the phenylene bridge.[7] In the *o*-phenylene series it was shown, that instead of the corresponding bis-ylide (**25**), bis-phosphonium ylides (**23**) were produced. The formation of (**23**) was attributed to the construction of the cyclic five-membered-ring intermediate (**26**) which resulted from the attack of the mono-ylide intermediate at the adjacent phosphonium centre. The cyclic ylidophosphorane (**26**), which existed as a 95:5 mixture of two stereoisomers corresponding to different occupations of the axial and equatorial positions of the trigonal-bipyramidal configuration, underwent subsequent rearrangement involving phenylene-P(σ^5) bond cleavage to give carbodiphosphorane (**27**) which after protonation gave the bis-ylide (**23**) (Scheme 6).

Detailed evidence on the mechanism of spontaneous hydrolysis of selected phosphate triesters with common leaving groups and varying

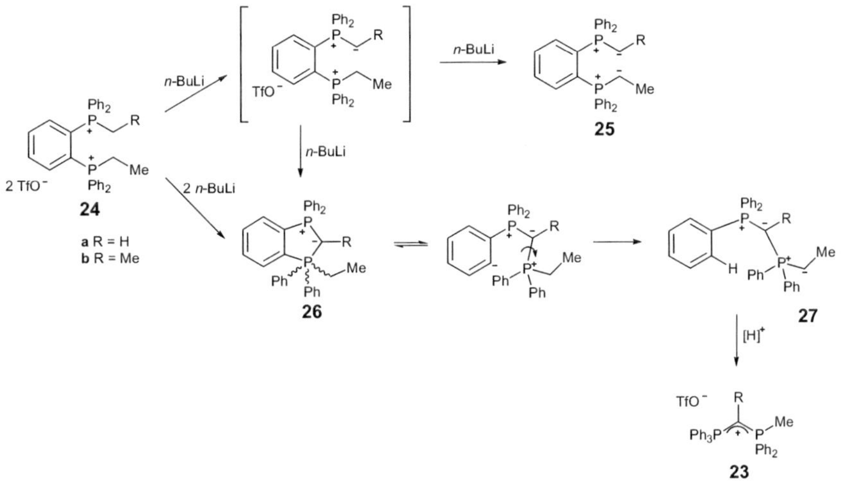

Scheme 6

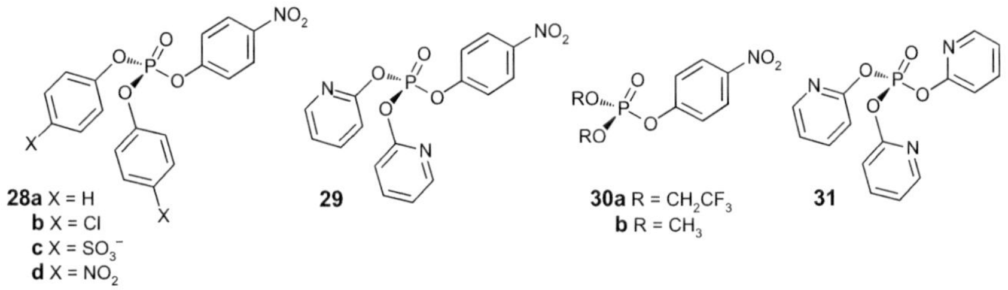

Fig. 2

non-leaving groups (Fig. 2) has been examined theoretically using DFT calculations.[8] The calculations have shown excellent quantitative agreement with previous experimental studies, which was best in the present of three discrete water molecules. The results supported a two-step mechanism involving a pentacovalent phosphorus intermediate (**32**), with a lifetime of tenths of a millisecond. The rate-determining formation of this intermediate involved general base catalysis, defined by concerted proton transfers in a six-membered activated complex (**TS1**), which involved two hydrogen bonded water molecules supporting a well-developed $H_2O\cdots P$ bond. The third water molecule, hydrogen-bonded to P=O is subsequently involved in product formation through (**TS2**) (Scheme 7).

The cleavage and isomerization of uridine 3′-alkylphosphates (**33**) has been studied in the presence of a dinuclear Zn^{2+} complex (**34**) (Fig. 3).[9] The kinetic results suggested that the dominant background reactions of the substrates (**33**) at pH 6.5 assumed base-catalysed cleavage via a di-anionic phosphorane intermediate that is sensitive to the pK_a of the leaving group, and pH independent isomerization that is virtually insensitive to the substituents (Scheme 8).

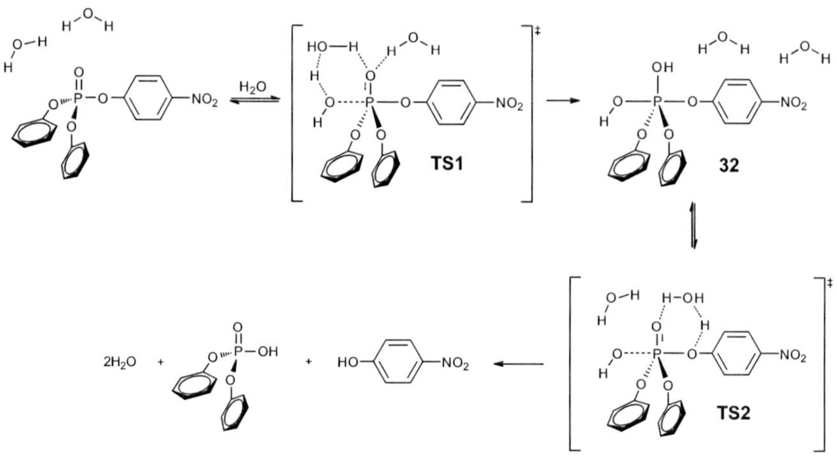

Scheme 7

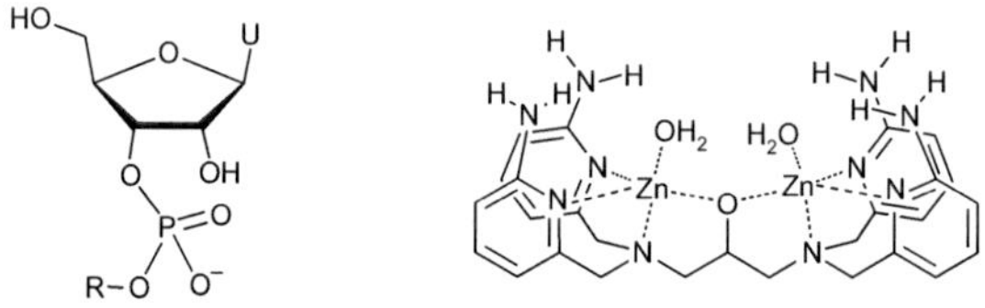

33a R = isopropyl
 b R = methyl
 c R = ethoxyethyl
 d R = 2,2-difluoroethyl
 e R = 2,2-dichloroethyl
 f R = 2,2,2-trichloroethyl
 g R = phenyl

Fig. 3

isomerization

cleavage

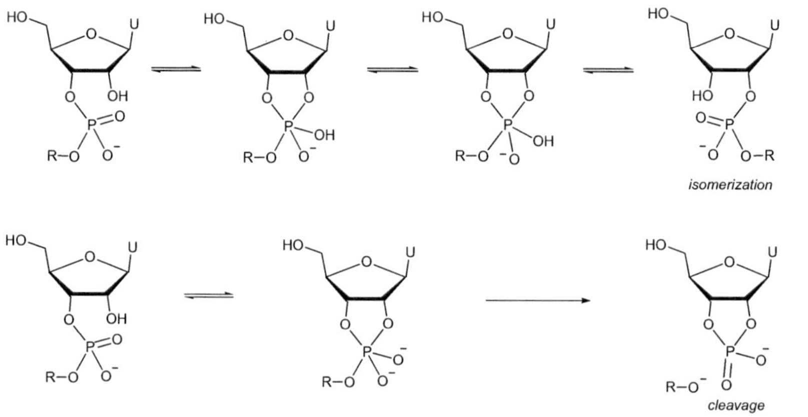

Scheme 8

Further investigations have shown that the cleavage mechanism involved the reversible formation of a phosphorane intermediate that was rate limiting, as well as general acid catalyzed loss of the leaving group (with the potential role of Zn bound water as a general acid catalyst). On the other hand, the strong interactions between the Zn-complex and the phosphorane dianion could stabilize the pentacoordinated intermediate to such an extent that it is able to pseudorotate and consequently lead to Zn-promoted isomerization. This pseudorotation seems to be the rate-limiting step of the isomerization reaction however it occurs significantly slower than the cleavage, in contrast to the analogous triesters. This might result from the fact that the initial phosphorane can be generated with both oxyanions substituents in equatorial positions but after pseudorotation it must adopt an axial position, leading to a higher energy intermediate and thus giving rise to a more substantial barrier for the process. Moreover, the change in the geometry between the oxyanions might compromise the electrostatic stabilization that complex (**34**) provides (Scheme 9).

Spontaneous reactivation of the sarin-phosphorylated wild-type enzyme and G117H mutant of human butyrylcholinesterase (BChE) – the most promising bioscavenger for organophosphorus warfare nerve agents – have been investigated using quantum mechanical calculations.[10] It was shown that the reaction pathway consists of two steps. The first one included nucleophilic attack of a water molecule, in WAT532, on the phosphorus atom to produce the pentacoordinated intermediate (**35**) while in the second step, involving the decomposition of (**35**), results in the departure of the oxygen of the S198 side chain and the generation of a

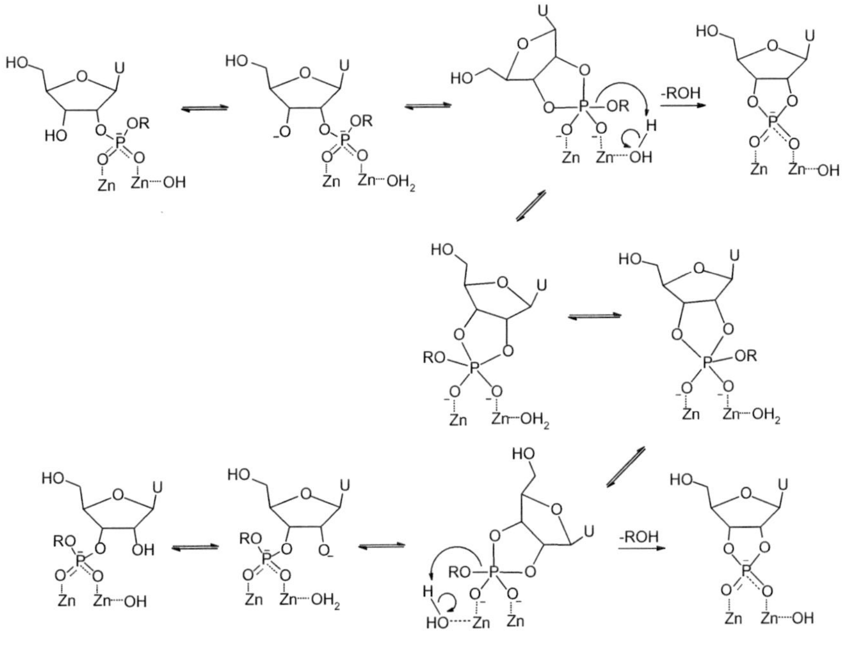

Scheme 9

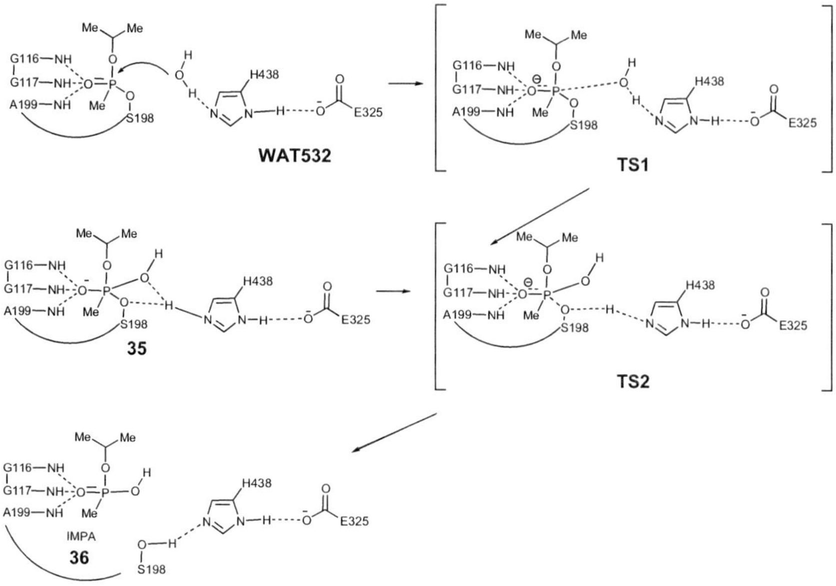

Scheme 10

nontoxic product, *i.e.*, *O*-isopropyl methylphosphonic acid (IMPA, **36**) (Scheme 10).

2.3 Structural investigations

Investigations of the Wittig reaction using a non-stabilized ylide (ethylidene triphenylphosphorane) (**37**) and various hindered aldehydes (**38**) under the same reaction conditions were performed in order to understand the factors which influence the stereochemistry of this process.[11] It was experimentally shown, that the prevailing outcome toward the *Z* olefin is influenced only by changes in steric factors or different overcrowding in the pentacoordinated oxaphosphetane intermediates. The general reaction mechanism with nonstabilized ylides is presented in Scheme 11.

As already known, the first step is the cycloaddition of the carbonyl group to the ylide to furnish pentacoordinate oxaphosphetane intermediates with trigonal bypiramidal (TBP) structures ***A-cis*** and ***A-trans***, which contained the oxygen atom in apical position (according to the general propensity of electronegative substituents to prefer the apical positions, whereas bulky ligands prefer the equatorial positions in the TBP). To obtain the final alkenes, a complete pseudorotation of these intermediates must occur, giving rise to two new pentacoordinated species ***B-cis*** and ***B-trans*** bearing the oxygen atom in equatorial position and the P–C bond in apical position. Since species **B** are generally less stable than species **A**, due to the equatorial position of oxygen, the irreversible decomposition of intermediates **B** to alkene and triphenylphosphine oxide is more likely to occur. Moreover, isomer ***B-cis*** is more unstable because of the greater steric hindrance and should decompose to the *Z* olefin much faster, *i.e.* $k_1 \gg k_2$, and hence increases the quantity of the

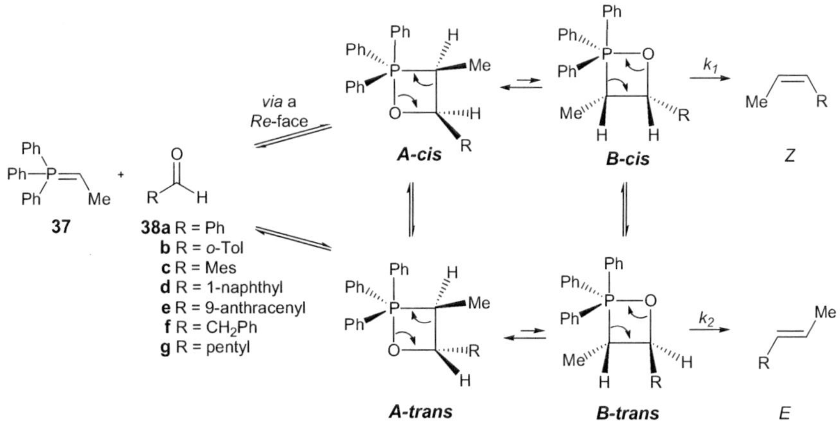

Scheme 11

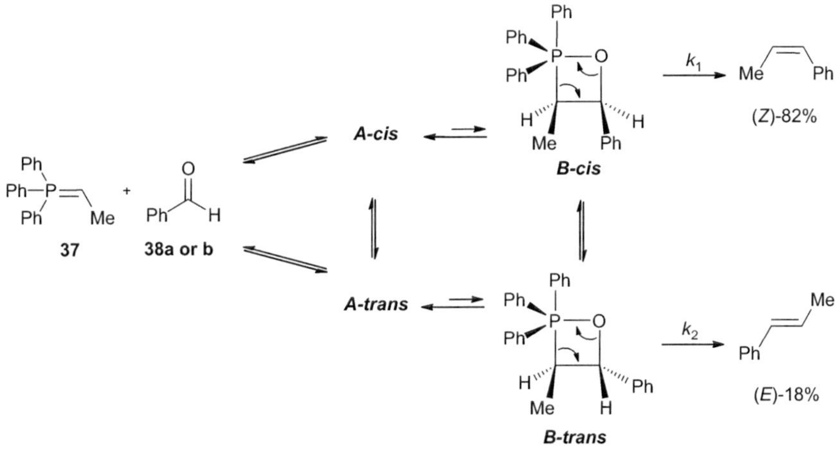

Scheme 12

Z alkene (Scheme 11). For instance, in the case of benzaldehyde (**38a**) and o-tolyl aldehyde (**38b**), the corresponding **B-cis** form is more unstable than **B-trans** ($k_1 \gg k_2$), giving rise to the predominant formation of a Z isomer in a 18:82 and 16:84 E/Z ratio, respectively (Scheme 12).

In fact, the E/Z ratio strongly depends on the steric hindrance of an aldehyde (**38**) and thus on the structure of pentacoordinate oxaphosphetanes **A** and **B**. Thus with more hindered mesityl aldehyde (**38c**), the overcrowding on the pentacoordinated intermediates became very high, both for *cis* and *trans* intermediates ($k_1 \sim k_2$), resulting in an increase of the E/Z ratio (48:52) of the products (Scheme 13).

Other theoretical insights into the second step of the Wittig reaction have also been reported.[12] These studies included isomerization and thermal decomposition of isolable spiro-1,2-oxaphosphetanes (OPAs). Key features of these investigations were: (1) the synthesis of enantiopure derivatives, (2) the identification of three mechanisms of OPA

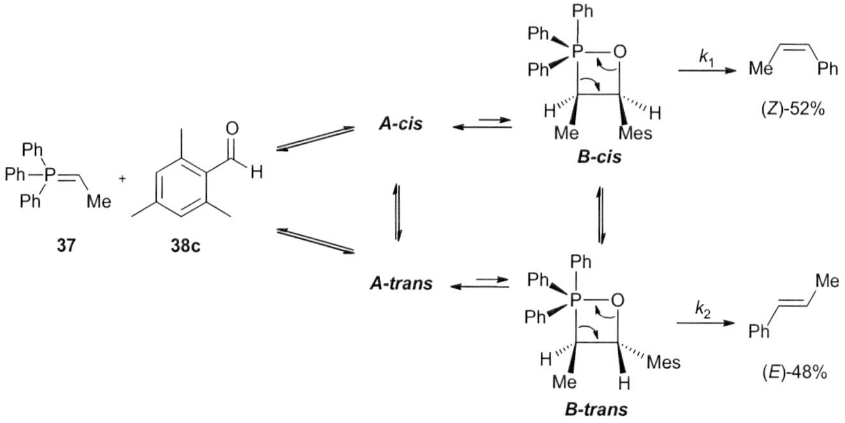

Scheme 13

stereomutation (M_{B2}, M_{B3} and a new one M_{B4}), and (3) description of olefination as a single step reaction. Spiro-oxaphosphetanes (**39–43**), stabilized by the presence of the *o*-benzamide moiety were prepared according to the known procedures (Scheme 14).

Thermal decomposition of OPAs (**39–43**) afforded alkenes (**44–48**) and a heterocycle (**49**). However upon thermolysis of OPAs (**40–43**), partial isomerization to give intermediates (**40′–43′**) bearing the opposite configuration at the phosphorus atom, occurred. In the decomposition of enantiopure phosphorane (**43**), this stereomutation led to partially racemized phosphine oxide (**49**), at *ca.* 20 times lower rate than for (**42**), whereas the stereospecific formation of olefins (**47**) and (**48**) resulted from the participation of three possible OPAs with square pyramid (SP) TBP geometry (Scheme 15). Using DFT calculations the mechanism of the formation of olefins from OPAs (**39, 42**) and (**43**) was investigated. For pentacoordinated oxaphosphetane (**39**), the potential energy surface (PES) consists of seven stationary points: energy minima for OPAs (**39**) and (**39″**), alkene (**44**), and benzoazaphosphole (**49**). The fragmentation had transition states ([**39 · A**]‡ and ([**39″ · B**]‡) and for the stereomutation ([**39 · B**]‡) (Fig. 4). Permutational isomers (**39**) and (**39″**) have distorted TBP geometries however (**39**) is 6 kcal/mol more stable than (**39″**). Isomerization of (**39**) into (**39″**) proceeds through [**39 · B**]‡ state which exhibit a TBP-like geometry with the O1, N and C_8 atoms in the equatorial plane and C_3 and C_{13} in apical positions. Moreover this process can be described as a 120° cyclic rotation of one apical and two equatorial ligands (M_{B2} mechanism) and is the first example of the interconversion of OPAs by this mechanism. The decomposition of spirophosphorane (**39**) in a single, highly exothermic, irreversible step proceeds through [**39 · A**]‡ showing TBP geometry with C_3 and C_8 in apical positions indicating that the ring opening involves P pseudorotation, without the participation of an anti-apicophilic P-O$_{eq}$ intermediate. Similarly, the decomposition of (**39″**) proceeds *via* [**39 · B**]‡ to furnish (**44**) and (**49**) and is disfavored by 7.1 kcal/mol compared to (**39**).

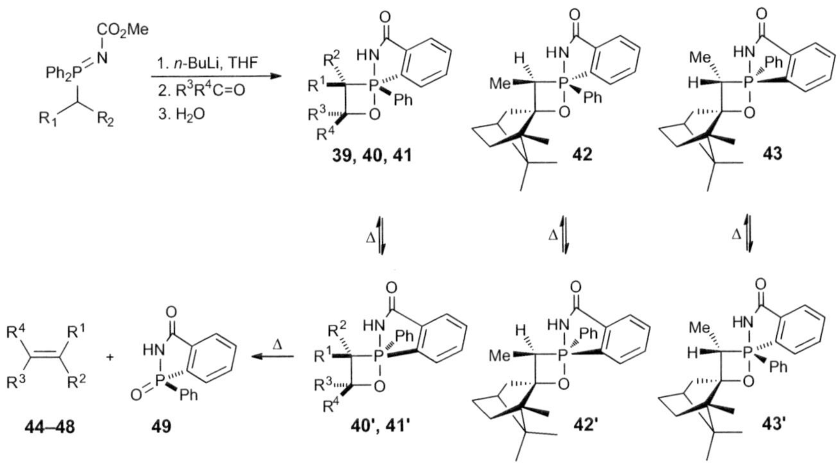

comp.	R^1	R^2	R^3	R^4	yield (%)
39	Me	Me	Ph	Ph	85
40(40')	Me	H	(CH$_2$)$_5$		82 (2.6)[a]
41(41')	Me	Me	Me	Ph	89 (32.7)[a]
42[b](42')	Me	H	 (camphyl)		44 (2.0)[a]
43[b](43')	H	Me	 (camphyl)		22 (7.3)[a]
44	Me	Me	Ph	Ph	100
45	Me	H	(CH$_2$)$_5$		100
46	Me	Me	Me	Ph	100
47[b]	Me	H	 (camphyl)		100
48[b]	H	Me	 (camphyl)		100

[a] Conversion of permutational isomers **40'–43'** was observed by ^{31}P NMR.
[b] *ent*-**42/43** and *ent*-**47/48** were prepared from D-(+)-camph or (>97% *ee*).

Scheme 14

In the case of OPAs (**43**), the PES for its decomposition includes OPAs at minima (**43**), (**43'**), and (**43"**) and products (**48**) and (**49**). The motion of the P substituents of (**43**) may follow two courses: two consecutive 60° clockwise rotations of the N, C8, and C13 ligands (M$_{B2}$ mechanism) that convert (**43**) into (**43"**) *via* (**TS**) [**43 · B**]‡ having a TBP geometry (Fig. 5). Most importantly, two successive 120° anticlockwise rotations produce the same isomerization through the new (**TS**) [**43 · C**]‡ having a TBP configuration with O$_1$ and C$_{13}$ in apical positions. This process contains four Berry pseudorotations (BPRs) and represents a new mechanism of

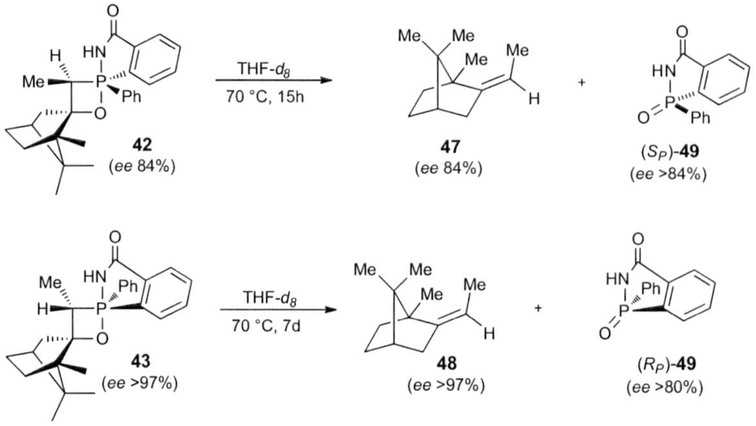

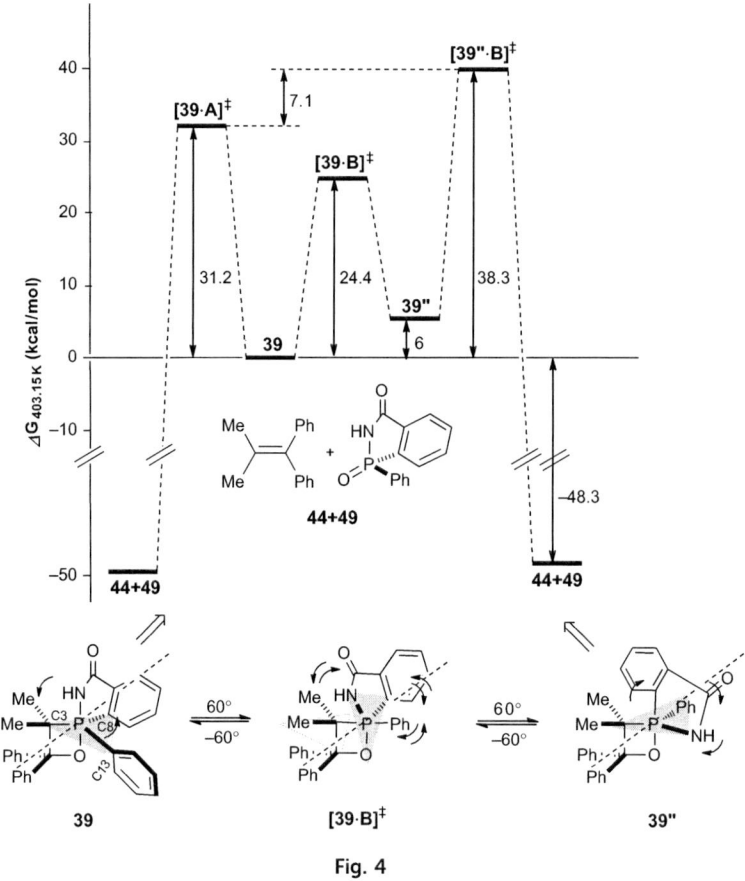

Scheme 15

Fig. 4

stereomutation identified as M_{B4}. The barrier is only 1 kcal/ mol higher than that of the M_{B2} mechanism, which makes M_{B4} a reasonable alternative for the (**43**) ⇌ (**43″**) interconversion. Interchange of the C8 and N ligands in (**43″**) through a 180° rotation leads to (**43′**), the isomer of (**43**)

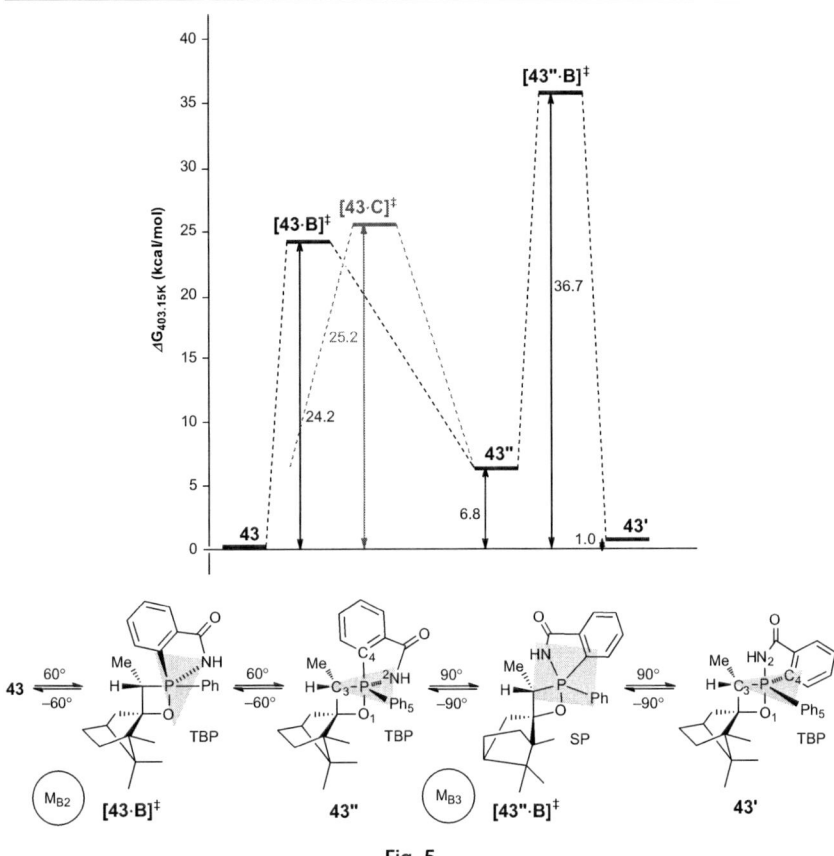

Fig. 5

experimentally observed *via* P stereomutation. The (**TSs**) for this process, [**43″ · B**]‡ and [**43″ · C**]‡, arise from clockwise and anticlockwise rotations, respectively. Both TSs have SP geometry and are located halfway through three consecutive Berry pseudorotations, as expected for an M_{B3} mechanism. Stereomutation through [**43″ · B**]‡ is highly favored because its barrier (30 kcal/mol) is 7.1 kcal/mol lower than that of [**43″ · C**]‡. The barrier for olefination of (**43**) is 0.5 and 6.8 kcal/mol lower than those for decomposition of (**43′**) and the (**43**) ⇌ (**43′**) inter-conversion, respectively. The unfavorable formation of (**43′**) may be accelerated by the presence of acidic species in the reaction medium. The detection of (**43′**) only after 57.5 h of heating supports this assumption. Importantly, The decomposition of OPA (**42**) follows the same pathway but is destabilized by 5.1 kcal/mol with respect to (**43**), and the barrier for cycloreversion (27.3 kcal mol^{-1}) is 1.7 kcal mol^{-1} lower than for (**43**). The destabilization of (**42**) arises from the interaction of the C_3-Me protons with the camphor moiety.

3 Hexacoordinated compounds

As in previous year, the chemistry of hexacoordinated compounds has not been explored a great deal and was generally limited to the synthesis and properties of novel polycyclic and acyclic derivatives. Once again, well-established topic such as the preparation of hexacoordinated compounds (50) formed in the reaction of alkylenechlorophosphites (51) with hydroxylated bisazomethines (52) has continued to attract attention.[13] In order to study the limits of applicability of this transformation, hydroxylated alkylene- and butylenechlorophosphites were reacted with diimine (52) derived from the 2-hydroxy-5-chloroacetophenone to furnish polycyclic salt structures (53) bearing transannular N–P bond. These compounds underwent further dehydrochlorination upon treatment with triethylamine to give phosphorates (50) with a neutral structure (Scheme 16).

Two independent papers were published discussing the preparation and structure of carbene-stabilized complexes of phosphorus(V) fluorides. In the first paper, cyclic and acyclic 2,2-(difluoroalkylamines) (54a,b) were for the first time used as a carbene precursors in oxidative addition to substituted and unsubstituted phosphorus halides (fluorides or chlorides) to form adducts (55a–d).[14] These octahedral and hydrolytically stable complexes were obtained in quantitative yields and were fully characterized in the solid state by single-crystal X-Ray diffraction (Scheme 17).

In the second paper, the non-hindered Arduengo carbene PF_5 adduct (56) was prepared by thermolysis of 1,3-dimethyl-1H-imidazolium hexafluorophosphate (57) under reduced pressure.[15] This observation represented the first example of a direct bond rearrangement of the type [C–H] + [P–F] → [C–P] + H–F. Thermolysis of 1,2,3-trimethyl-1H-imidazolidinium hexafluorophosphate (**58**) gave zwitterionic [(1,3-dimethyl-1H-imidazolium-2-yl)methyl]pentafluorophosphate (**59**), although in a lower yield it supported the generality of this transformation (Scheme 18).

A study on the formation of P(VI)-derivatives (**60**) from P(IV)-containing betaines, such as 6-bromo-2-hydroxy-4-tributylphosphonium naphtyl-1-ate (**61**) with bis(phenylenedioxy)chlorophosphoranes (**62**) has been performed.[16] These high-yielding reactions resulted in the betaine structures (**60**), which are of considerable interest as reagents in the Wittig reaction, organocatalysts or as reagents for creating C–C and C–X bonds, where X = O, N, S (Scheme 19).

A series of novel hexacoordinated P^V-*meso*-triarylcorroles (**63**) was prepared by refluxing the corresponding free-base corroles (**64**) with $POCl_3$ in pyridine (Scheme 20).[17] These compounds, bearing two axial hydroxyl groups, were isolated and characterized using various spectroscopic techniques. Detailed ^1H and ^{31}P NMR analysis in $CDCl_3$ however indicated that P^V-*meso*-triarylcorroles (**63**) were prone to axial-ligand dissociation to form pentacoordinated P^V-*meso*-triarylcorroles in non-coordinating solvents (toluene, CH_2Cl_2, $CHCl_3$). This suggested that the axial hydroxyl groups were weakly bound to the central phosphorus ion and could readily dissociate in solution. However, in the presence of strongly coordinating solvents such as MeOH, THF and DMSO, the P^V-*meso*-triarylcorroles preferred the hexacoordinated geometry in which the corresponding solvent molecules acted as axial ligand. To confirm this hypothesis, the titration of compound (**63a**) (in $CDCl_3$) with

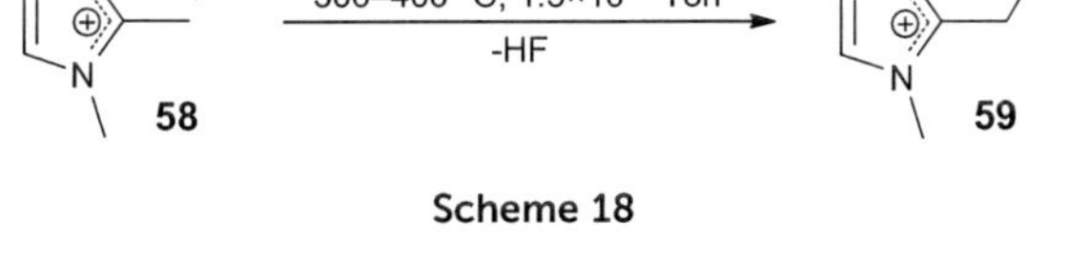

Scheme 18

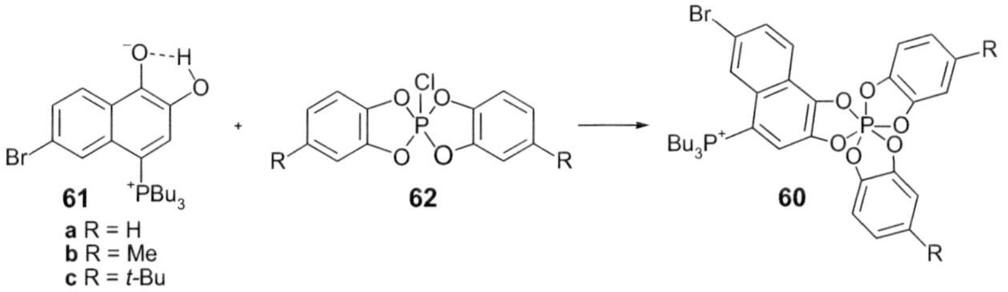

Scheme 19

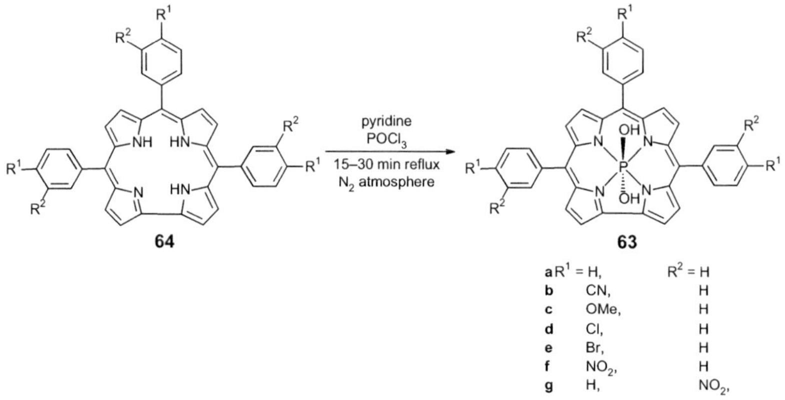

Scheme 20

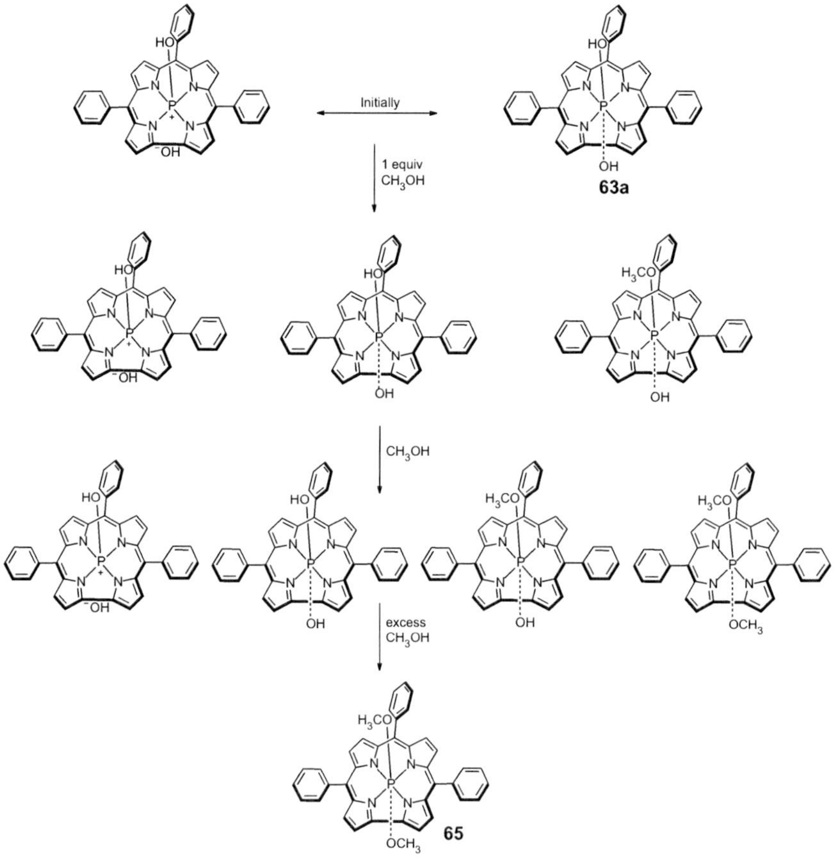

a R¹ = H, R² = H
b CN, H
c OMe, H
d Cl, H
e Br, H
f NO₂, H
g H, NO₂,

Scheme 21

increasing amounts of methanol (in CDCl₃) was carried out. All possible intermediate complexes during the NMR titration are presented in Scheme 21.

In the presence of an access of MeOH only compound (**65**) was observed, indicating hexacoordinated environment of phosphorus with two OCH_3 groups as axial ligands. This result was also clearly evident in the gradual disappearance of the signal corresponding to the axial OH protons and appearance of the doublet corresponding to the axial OCH_3 protons. Moreover, the hexacoordinated geometry of two selected complexes was additionally confirmed by X-Ray crystal analysis and their absorption spectra in coordinating solvents indicated that P^V-*meso*-triarylcorroles (**63**) were brightly fluorescent with high quantum yields what could be very useful for various molecular device applications.

References

1. O. I. Kolodiazhnyi, *Tetrahedron: Assymetry*, 2012, **23**, 1.
2. J. Saha, C. Lorenc, B. Surana and M. W. Peczuch, *J. Org. Chem.*, 2012, **77**, 3846.
3. K. Gholivand, H. R. Mahzouni, F. Molaei and A. A. Kalateh, *Tetrahedron Lett.*, 2012, **53**, 5944.
4. R. S. Grainger, M. Betou, L. Male, M. B. Pitak and S. J. Coles, *Org. Lett.*, 2012, **14**, 2234.
5. B. Defaut, T. B. Parsons, N. Spencer, L. Male, B. M. Kariuki and R. S. Grainger, *Org. Biomol. Chem.*, 2012, **10**, 4926.
6. E. H. Krenske, *J. Org. Chem.*, 2012, **77**, 1.
7. C. Maaliki, M. Abdalilah, C. Barthes, C. Duhayon, Y. Canac and R. Chauvin, *Eur. J. Inorg. Chem.*, 2012, 4057.
8. J. R. Mora, A. J. Kirby and F. Nome, *J. Org. Chem.*, 2012, **77**, 7061.
9. H. Korhonen, S. Mikkola and N. H. Williams, *Chem. Eur. J.*, 2012, **18**, 659.
10. Y. Yao, J. Liu and Ch.-G. Zhan, *Biochemistry*, 2012, **51**, 8980.
11. G. Baccolini, C. Delpivo and G. Micheletti, *Phosphorus, Sulfur and Silicon*, 2012, **187**, 1291.
12. J. G. López, A. M. Ramallal, J. González, L. Roces, S. Garcia-Ganda, M. J. Iglesias, P. Oña-Burgos and F. L. Ortis, *J. Am. Chem. Soc.*, 2012, **134**, 19504.
13. L. K. Kibardina, S. A. Terent'eva, A. R. Burilov and M. A. Pudovik, *Russ. J. Gen. Chem.*, 2012, **82**, 1309.
14. T. Böttcher, O. Shyshkov, M. Bremer, B. S. Bassil and G.-V. Röschenthaler, *Organometallics*, 2012, **31**, 1278.
15. Ch. Tian, W. Nie, M. V. Borzov and P. Su, *Organometallics*, 2012, **31**, 1751.
16. N. R. Khasiyatullina, V. F. Mironov and O. I. Gnezdilov, *Russ. J. Gen. Chem.*, 2012, **82**, 939.
17. A. Ghosh and M. Ravikanth, *Chem. Eur. J.*, 2012, **18**, 6386.

Phosphazenes

Frederick F. Stewart

DOI: 10.1039/9781782623977-00366

1 Introduction

The unique chemistry of phosphazenes continues to be developed. A phosphazene, which is defined as a P(V) and N structure with these atoms attached to each other by double bonds, exhibits a unique two-step general synthesis in which the P=N skeleton is intially formed, followed by nucleophilic substitution to provide final form. In general, a pure inorganic phosphazene is assembled, typically with halide substitution at P. The lability of the P–X bond allows substitution with a nearly unlimited variety of nucleophiles. Final substitution can yield stable structures amenable to both characterization and function. This article discusses some of the more recent developments in phosphazene chemistry and seeks to provide the reader a picture of the current state of the art.

2 Non-ionic phosphazenes for drug delivery

Phosphazenes are known for their biocompatible character that has spawned much recent research.[1] Further, their chemical nature and adaptability allows them to be developed as water swellable and water soluble compounds suited for applications such as drug delivery. Polymer hydrogels are of particular interest. A review article recognized the value of several phosphazenes, inlcuding those substituted with amino acids and polyethylene glycol variants.[2] Advantages of these molecules include lower critical solution temperatures (LCST) at temperatures of biological interest, and the fact the degradation products are largely non-toxic. Formation of hydrogels is dependent on the pendant group speciation on the phosphazene. For example, hydrogels have been formed by balancing the polymer solubility by using mixtures of hydrophobic and hydrophilic pendant groups (1).[3] Aminated 550 Dalton polyethylene glycol (AMPEG550) serves as the hydrophilic component while the amino acid ester serves as the biocompatible hydrophobic portion. The utility of this scheme is shown in this example where the phosphazene was complexed with cobalt ferrite nanoparticles and paclitaxel forming a potential drug therapy. This phosphazene also showed effectiveness in anti-tumor activity when combined with the drug silibinin where the phosphazene exhibited slow degradation allowing for steady release of the drug.[4]

The ability to insert functional groups into phosphazenes facilitates numerous chemical attachments, such as drug conjugates. Camptothecin, an anti-tumor drug, was attached to a phosphazene hydrogel (2) through a hydrolyzable linkage that could be used to deliver the drug to the desired tumor site.[5] Functionality also can allow for control mechanisms, such as

Biological and Chemical Processing Department, Idaho National Laboratory, P.O. Box 1625, Idaho Falls, ID, USA. E-mail: Frederick.Stewart@inl.gov

pH, for drug delivery. Diisopropylamino pendant groups (3) sucessfully demonstrated pH sensitivity, while limiting deleterious effects of the cellular environment.[6]

Glycosylation of polyphosphazene, which adds hydrophilicity, was demonstrated through a photochemically induced "thio-yne" click reaction with poly[bis-propargylphosphazene], which itself was synthesized through reaction of poly[bis-chlorophosphazene] with propargylamine, Scheme 1.[7] Due to the highly hindered nature of the glucofuranose ring, only approximately 55% of the propargyl groups undergo the addition chemistry with 8% of those attaching only one ring yielding an alkene. Removal of the acetate protecting groups proceeded readily with sodium methoxide in methanol. To probe the steric effect of the incoming group in the addition reaction, 1-pentanethiol was found to add quantitatively suggesting greater conversion with less sterically bulky thiols (Scheme 2). An additional phosphazene was formed by a click reaction with 60% propargylamine and 40% non-reactive 1-butylamine. By reducing the alkyne density on the backbone, greater conversion to the alkane was acheived with virtually no alkene observed. Similar chemistry has also been deonstrated using a vinyl moiety in place of the propargyl group.[8]

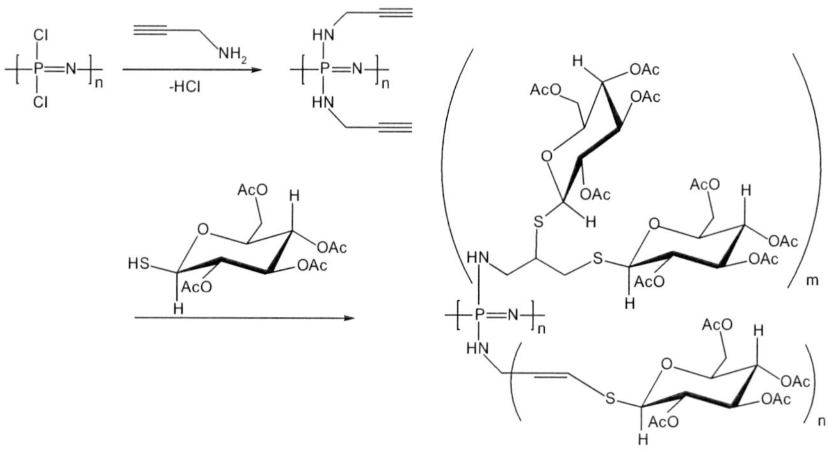

Scheme 1

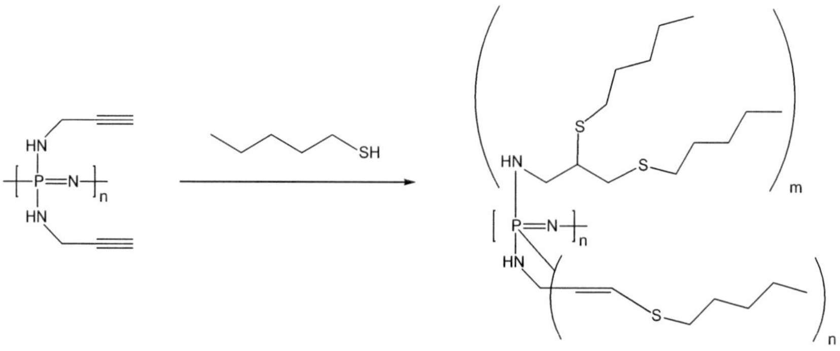

Scheme 2

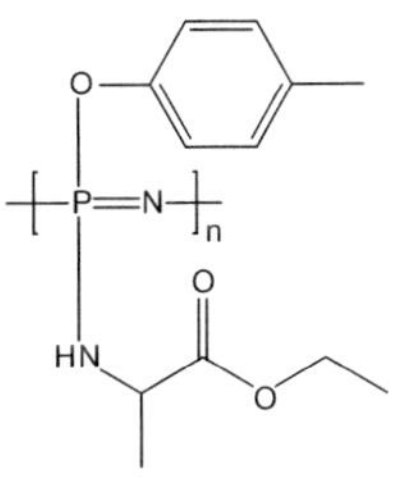

Hypericin is a drug with demonstrated interest in its use as an antiviral, anti-tumor, and anti-bacterial agent that has also demonstrated some use as a photosensitizer in photodynamic therapy.[9] It has poor water solubility which limits its use. However, being a functionalized molecule, a pathway was found to attach it to a linear phosphazene, Scheme 3. Loadings of the hypericin ranged from 5 to 20 %. Clearly, the significant steric bulk of the drug will limit is loading on the polymer. To stabilize the polymer, 1000 and 2000 molecular weight polypropylene oxide (PPO)/polyethylene oxide (PEO) chains were used to substitute the remaining chlorines. The shorter PPO/PEO substituents allowed higher loadings of the hypericin.

A route to bioerodible phosphazene materials was shown using phenolic esters.[10] The chemistry of the attachment of ethylsalicylate was explored as a function of the base used in the substitution. Use of sodium hydride gave the expected bis-substitution, Scheme 4. Replacement of

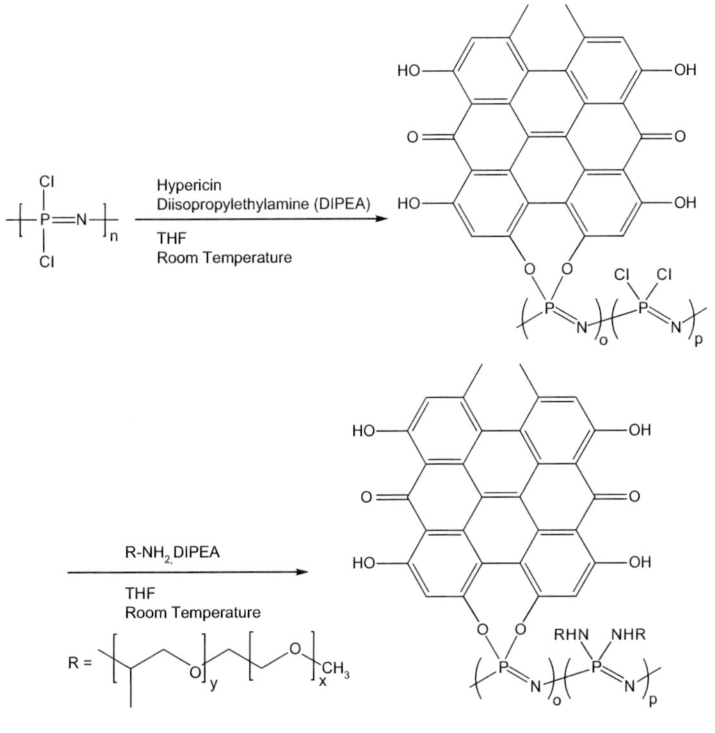

Scheme 3

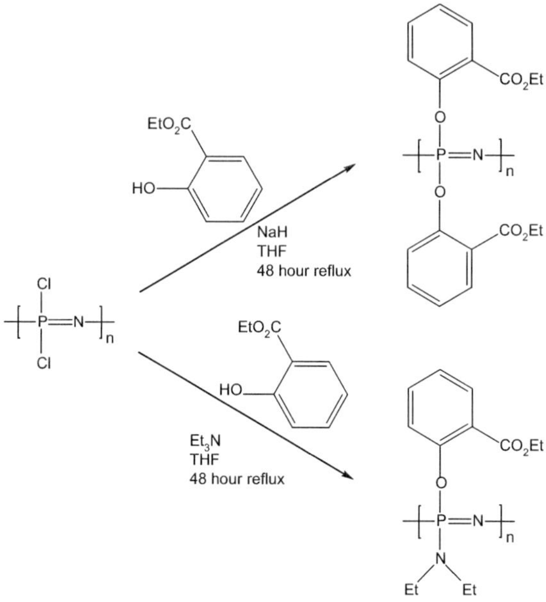

Scheme 4

sodium hydride with triethylamine yields the unexpected amine adduct. The authors proposed a mechanism for the amine attachment in which the amine initially attaches as the ammonium chloride salt, followed by loss of ethylchloride.

A cyclotriphosphazene variant on the PEG/amino acid substitution scheme was shown to give materials that exhibited self-assembly behavior.[11] Specifically, a cyclotriphosphazene with amino acid and PEG750 substitution gave micelles in aqueous solution that were found to perform as stable drug encapsulants. Use of these structures to encapsulate the drug docetaxel did not result in largely improved activity compared to other drugs currently in clinical use; however significantly lower cellular toxicity was observed.

3 Phosphazenes for tissue engineering applications

The materials science and engineering aspects of phosphazenes are driven by their biocompatible nature. The goal is to develop phosphazenes with specific material properties that can be used in the place of natural tissues. Towards this goal, differing chemistries are employed to develop these materials. For example, to form a material for use in tendon augmentation, electrospun poly(ε-caprolactone) (PCL) fibers were functionalized with (4) through dip coating.[12,13] The phosphazene coating was found to increase cell ahesion and, over the longer term, cell penetration into the matrix.

An intriguing route to materials with "built-in" anti-oxidant capability was demonstrated through the addition of ferrulic acid.[14] The strategy for attaching the acid included initial protection of the acidic moeity,

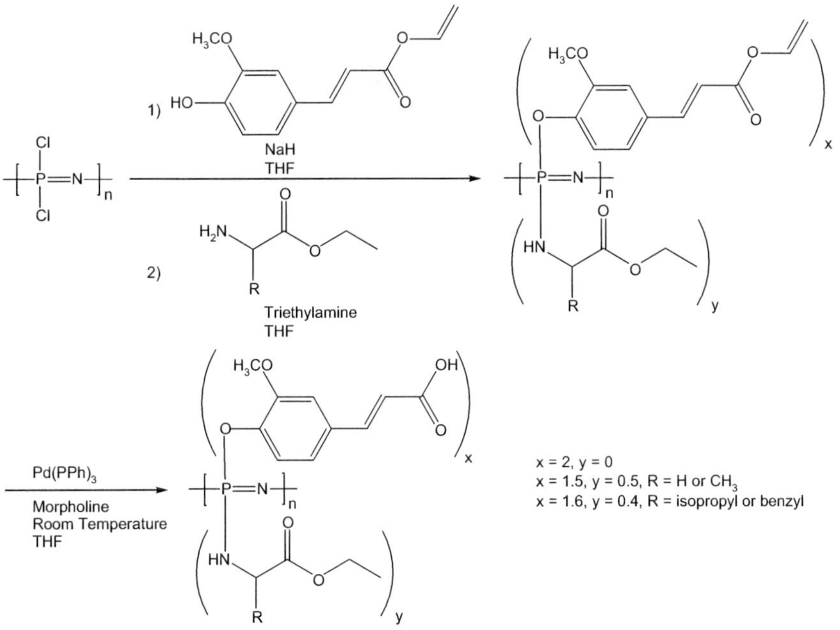

Scheme 5

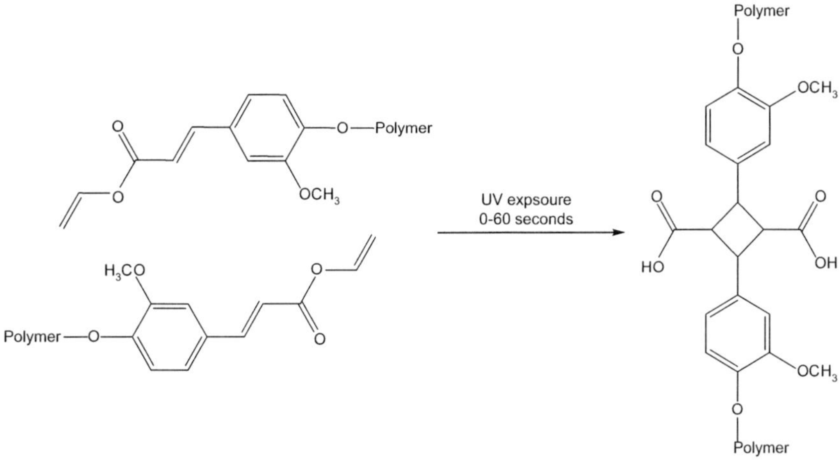

x = 2, y = 0
x = 1.5, y = 0.5, R = H or CH₃
x = 1.6, y = 0.4, R = isopropyl or benzyl

Scheme 6

followed by attachment to the phosphazene and deprotection, Scheme 5. Ultra-violet induced cross-linking studies revealed that a significant portion of the ferrulic acid moieties, up to 63%, reacted through a 2 + 2 cycloaddition reaction, Scheme 6. Ferrulic acid substituted phosphazene was found to mineralize with calcium on exposure to simulated body fluid.[15] The polymers were found to selectively mineralize monocalcium phosphate monohydrate; however the degree to which they complexed was a factor of the hydrophilicity of the polymer. Increased hydrophilicity gave higher degrees of mineralization. Additionally, no interference from sodium chloride was observed in the process.

4 Phosphazenes as immunoadjuvants

Significant work has been reported on the development of immunoadjuvants based on polyphosphazene, which has been recently reviewed.[16] A mechanistic study of the activity of poly[bis-(sodium carboxylatoethylphenoxy)phosphazene] (PCEP) as an immunoadjuvant.[17] This study revealed that PCEP stimulates antigen-specific immune responses. Localized increases in production of cytokines, chemokines, and innate immune receptors were seen at and near the site of injection, suggesting enhanced responses induced by the phosphazene. In a targeted study to enhance the effectiveness of live attenuated measles virus immunization, PCEP was found to increase TH1/TH2-type immune response.[18] A similar response was seen in a study where PCEP was employed with a vaccine that targets *Actinobacillus pleuropneumoniae*, which is a resporatory virus that is typically found in pigs.[19] Coacervation of PCEP to form microparticle vaccines was demonstrated in a study targeting the respiratory syncytial virus that is common to humans.[20]

5 Thermal stability and decomposition

As seen in the biomaterial discussion, one of the principal drivers for research in phosphazene chemistry is controllable hydrolytic decomposition *in-vivo*. However, in non-biological applications, the understanding of the controlled decomposition and thermal stability of phosphazenes is critical. To emphasize this importance, several recent papers have appeared that report on specific stability studies.

Cellulose fibers were spun using a wet spinning method that allowed for the inclusion of hexaphenoxycyclotriphosphazene (HPTP) as an inbedded flame retardant.[21] Fibers containing HPTP were studied by Themogravimetric Analysis (TGA), which revealed that the initial decomposition point increased by 20 °C, as compared to the untreated fiber. Repeated washing cycles indicated that the HPTP exhibited a low leaching rate.

Highly fluorinated cyclotriphosphazenes have been studied as high performance lubricants in applications such as hard disk drives using the Heat Assisted Magnetic Recording method.[22] In this method, areas of disks are heated by lasers to temperatures > 400 °C, which necessitates a high degree of thermal stability in materials, such as lubricants, associated with the device. Decomposition of (5) was studied by TGA where it was determined the phosphazene was stable to 318 °C. Laser treatments of the lubricant on surfaces revealed that the decomposition seen in the phosphazene material occurred in the organic pendant groups.

Copolymers of phosphazene and polyarylamides (6) were prepared by a condensation process.[23] The presence of the phosphazene core stabilised the polymers as seen in higher char yield upon thermal decomposition. Interestingly, the glass transition temperature (Tg) decreases with increasing phosphazene content.

A halogen-free epoxy resin (7) was formed through the sequential attachment of phenolic groups, followed by bisphenol A and epichlorohydrin.[24] Curing of the epoxy resin was studied using several hardeners: dicyandiamide, 4,4′-diaminophenylmethane, and novolak resin. For the cured phosphazenes, Tg values were high at > 150 °C and exhibited high char yields by TGA. Further, the material gave high LOI (loss on ignition) values indicating high fire resistance. Similar observations were made for a structurally related material (**8**).[25] Another example supported the conclusions of these two works using a phosphazene trimer with geminally pendant ethylenediamine.[26]

6 Elastomeric polyphosphazenes

Polyphosphazenes tend to be elastomeric materials and the end properties of the polymers are dependent on the pendant group substitution.[27] Although the knowledge on the polymerization and basic chemistry of the polyphosphazene backbone is well-established, interesting novelties continue to emerge. For example, quaternization of backbone nitrogens has been reported.[28] This report, which indicates that an organophosphazene can be quaternized with a cation delivered with a non-coordinating anion, is further support for the basicity and reactivity of backbone nitrogens. This reactivity was initially indicated with lithium ion, but now has been shown using a sterically more bulky methyl cation. Phosphazenes included in this study were substituted with pendant groups such as phenyl, n-butyl, and methoxyethoxyethoxy, testifying to the ubiquitousness of backbone nitrogen basicity.

A report on the synthesis of halogenated phosphazenes has indicated a facile pathway to these materials.[29] A straightforward synthesis was conducted that substituted 100% of the backbone with 2,2,2-trichloroethoxy groups. Hetero-polymers with two pendant groups also were reported using phenoxy, 2,2,2-trifluoroethoxy, or methoxyethoxyethoxy groups as the second substituent. The physical properties and characterization data are typical for this type of polymer.

Functionalization of phosphazenes with phosphonic acid derivatives was reported in the 1990's; however the synthesis was plagued by unwanted side reactions, such as cross-linking between adjacent phosphonic groups. An improved procedure was employed to synthesize phosphazene heteropolymers (**9–11**) containing 2,2,2-trifluoroethoxy groups, as well as a phosphonated homopolymer (**12**), Scheme 7.[30] However, it was noted that synthesis of (**12**) was performed at slightly lower yield that what was seen for (**9–11**) suggesting that some unwanted side reactions did occur. In an additional report, two new routes to phosphonated phosphazenes was demonstrated.[31] Two methods that were developed included a prior-side-group assembly, Scheme 8, where the phosphonate is attached to the pendant group prior to attachment to the phosphazene, and a post-side-group method, Scheme 9, where the phosphonate is attached to a pendant group located on the polymer backbone. In prior-side-group method, the phosphoric acid moiety was attached to the pendant group prior to attachment to the phosphazene. Loadings ranged from 25% to 60%. For the post-side-group strategy, a protected amine was attached to the backbone, followed by deprotection and Michael addition of diethylvinylphosphonate. Conversions were limited in this strategy with 16% as the highest phosphoric acid loading obtained. It should also be noted that the Michael addition did yield a small amount of doubly phosphonated groups.

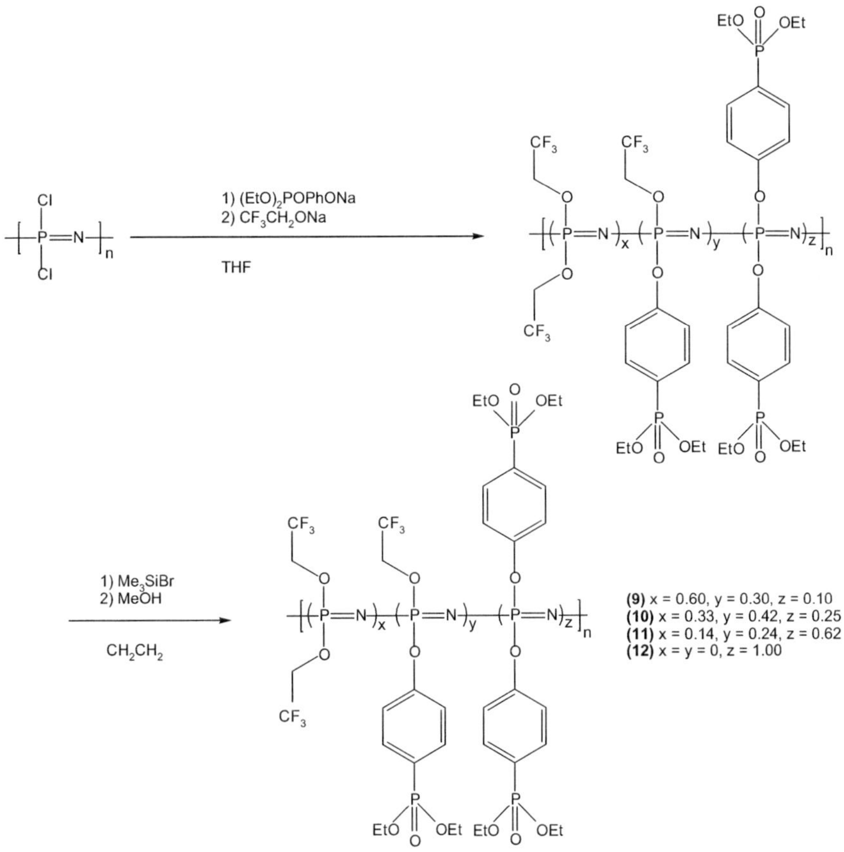

Scheme 7

The common ring opening polymerization route to linear polyphosphazenes is limited by some difficulty in forming copolymers. However, it is posible to functionalise end groups by a phosphoraminime condensation route.[32]

Propargylamine reacts with bromo-(bis-trifluoroethoxy)-N-trimethylsilylphosphoranimine (**13**) to give the expected condensation product, that can then be grown to a longer chain length by N-trimethylsilyl-trichlorophosphoranime, Scheme 10. Substitution with 2,2,2-trifluoroethoxy groups produces a stable polymer that can undergo grafting to an organic polyacrylate precursor, that upon subsequent polymerization, gives the co-polymer. A variant on this strategy was performed using both linear (**14**) and cyclic (**15**) brominated precursors.[33] Using the bromo functionality, polystyrene, poly(tert-butylacrylate), and poly(N-isopropylacrylamide) fragments were successfully grafted using an atom transfer radical polymerization technique.

Substitution of poly[bis-chlorophosphazene] typically occurs with little regiospecificity and the use of multiple pendant groups tends to result in random distribution of pendant groups. Block systems typically cannot be created using this method. N-silylphosphoranimines, as shown for the

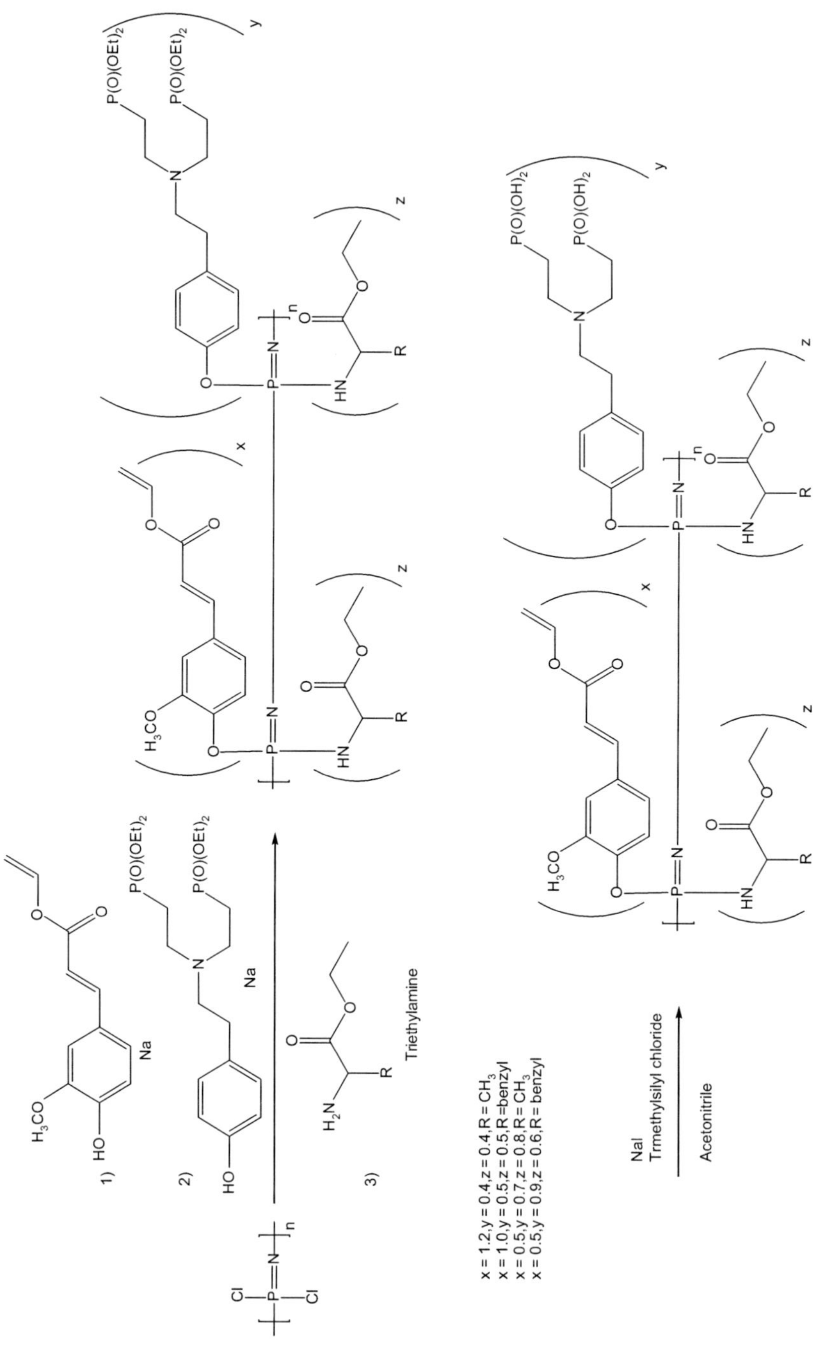

1)
2)
3)
Na
Na
Triethylamine
x = 1.2,y = 0.4,z = 0.4,R = CH3
x = 1.0,y = 0.5,z = 0.5,R =benzyl
x = 0.5,y = 0.7,z = 0.8,R = CH3
x = 0.5,y = 0.9,z = 0.6,R = benzyl
NaI
Trmethylsilyl chloride
Acetonitrile

Scheme 8

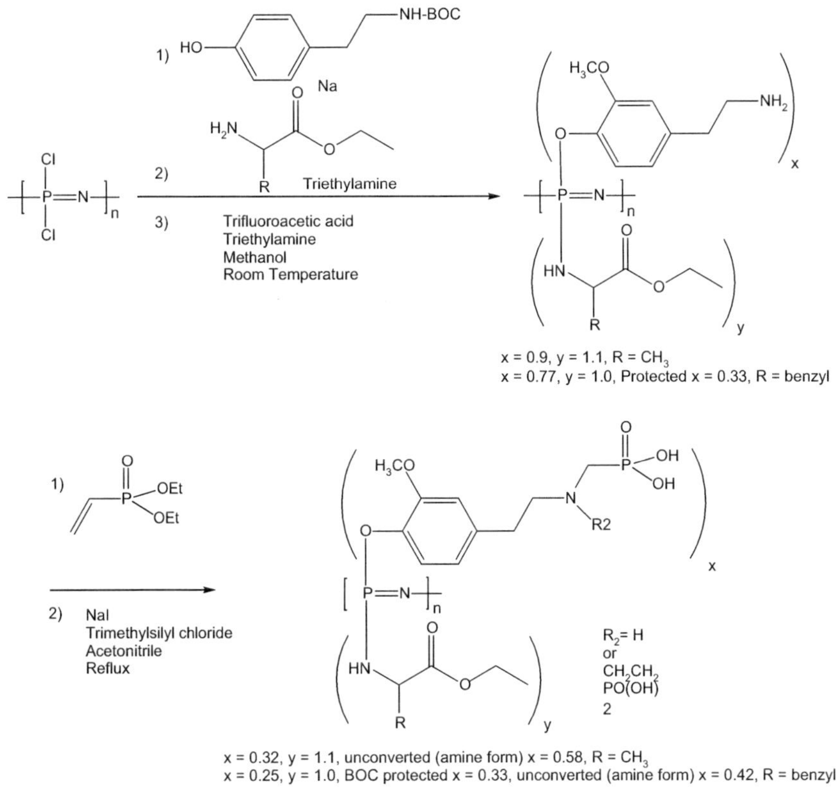

Scheme 9

phosphonation example above, can be manipulated to yield the increased order seen in block copolymers. This method was demonstrated for methoxy and phenoxy substituted phosphazenes plus polymers with 2,2′-phenylenedioxy groups, Scheme 11.[34] A theoretical study of the phosphoranimine condensation process was provided.

Substituent exchange in linear polymers is a rarely observed phenomenon that was studied through the reaction of various aromatic and fluorinated phosphazenes with excess nucleophile.[35] For example, poly[bis-2,2,2-trifluoroethoxyphosphazene] (PTFEP) was allowed to react with excess sodium 2,2,2-trifluoroethoxide over a period of up to 58 days where long term instability is observed. The molecular weight decreased, presumably due to basic hydrolysis of the backbone. Further, additional peaks in the P-31 NMR spectrum initially were observed that transformed into broad envelopes suggesting increased polydispersity. Additionally, 2,2,3,3,4,4,5,5-octafluoropentanol was found to exchange with PTFEP to the degree of approximately 61% after one days reaction. Molecular weight decrease was also observed over a longer term experiment. In a reaction that yielded little observable degradation, poly[bis-2,2,2-trichloroethoxyphosphazene] was converted to PTFEP by exposure to excess sodium 2,2,2-trifluoroethoxide. Curiously, polymers with phenoxy substituents were not observed, presumably due to the increased steric bulk imposed by these groups. Similar

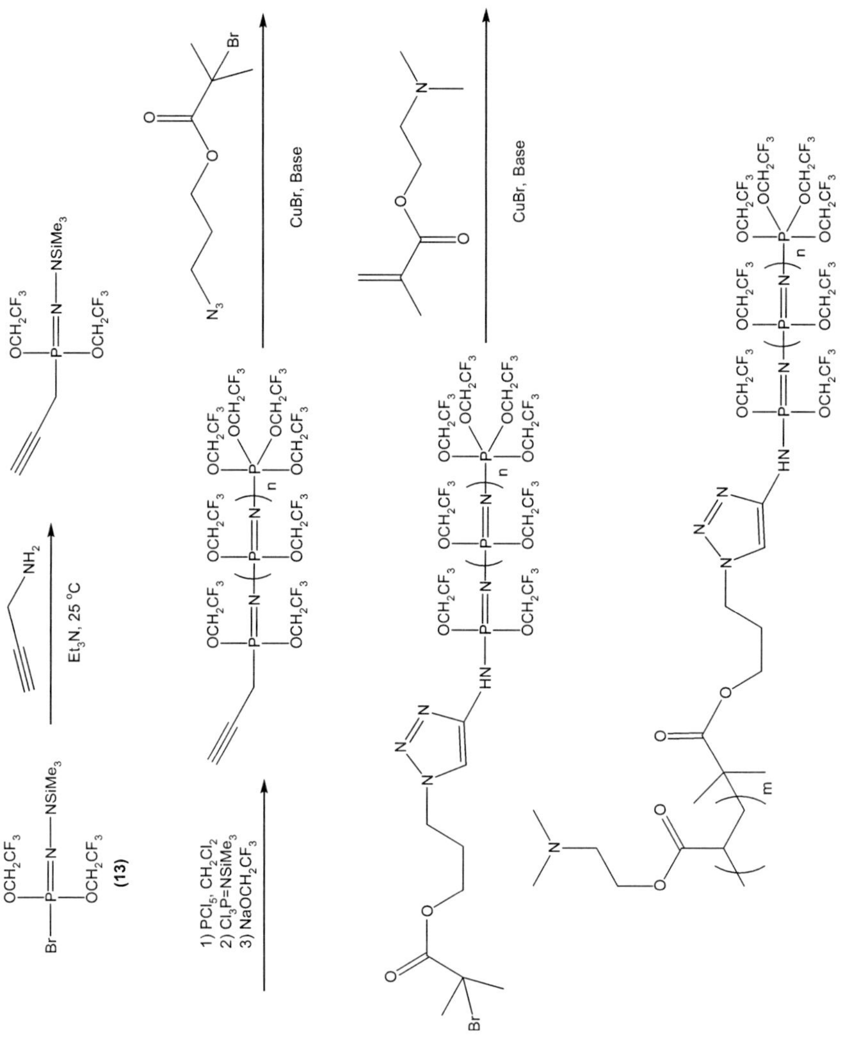
(13)
Et3N, 25 °C
1) PCl5, CH2Cl2
2) Cl3P=NSiMe3
3) NaOCH2CF3
CuBr, Base
CuBr, Base

Scheme 10

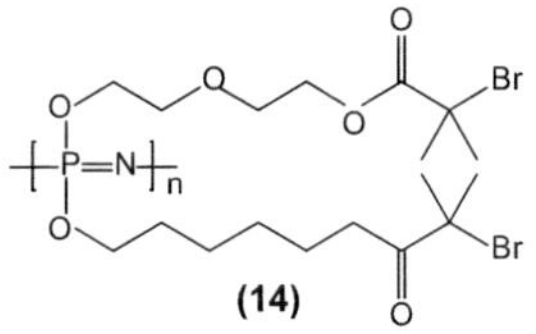

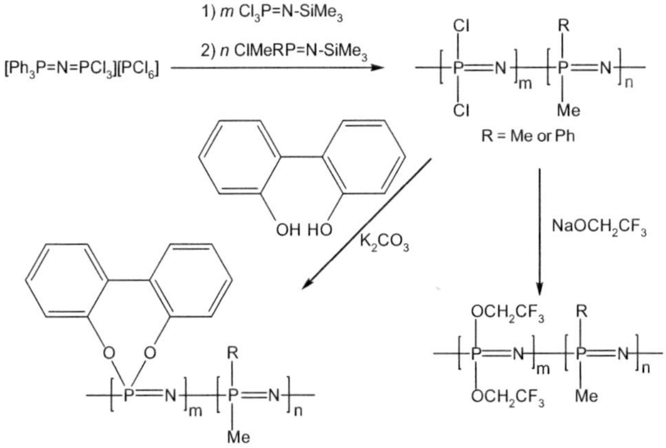

Scheme 11

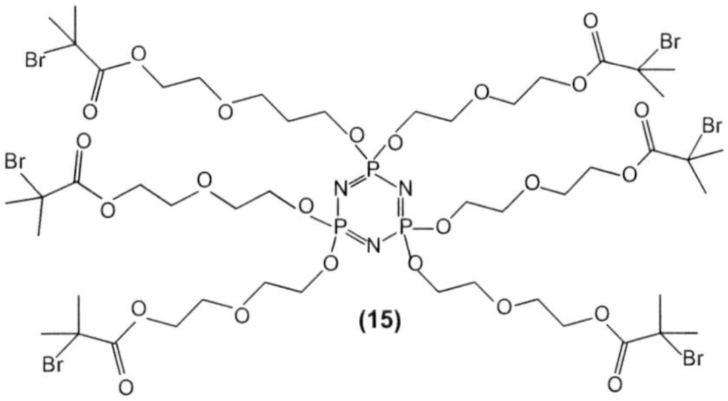

experiments performed on cyclic trimeric and tetrameric phosphazenes revealed that aryloxide subsituents could be displaced by 2,2,2-trifluoroethoxide and the displacement rate increased with the addition of electron withdrawing groups on the aryloxy ring.[36]

Heterpolymeric phosphazenes with 2,2,2-trifluorethanol and 4-hydroxyaniline were used as stabilizers in polymer/liquid crystal polymer (LCP) blends.[37] Often, LCP materials are not compatible with other organic polymers. The phosphazene added with multi-walled carbon nanotubes reduced the LCP domain size in the blends. A 4-bromophenoxy analog (in place of the 4-hydroxyaniline) also showed promise.[38]

In a purely application driven paper, PTFEP, which is known for it excellent membrane forming properties was characterized as a membrane for the removal of thiophene from gasoline.[39] PTFEP in this work was synthesized using the common ring opening polymerization procedure for poly[bis-chlorophosphazene], followed by substitution with sodium 2,2,2-trifluoroethoxide in THF solvent. Solution cast membranes were formed and characterized for transmembrane flux and selectivity using pervaporation as the separation technique. An inverse relationship between selectivity and flux was observed.

In a probe of structure, a series of poly[alkoxyphosphazenes] were studied for their electrooptical properties, polarity, and chain rigidity.[40,41] Alkyl chain lengths used in this work include 3,4,6,and 7 carbons.

Surprisingly, this paper concluded that these materials are not flexible chain polymers. Thermodynamic equilibrium in these polymers is observed to a rigidity, characterized by a Kuhn segment, which increases from 47 Å to 64 Å as the chain length increased from 3 to 7 carbons. The kinetic rigidity in these polymers is also asserted with a low frequency dispersion of dielectric permittivity.

7 Cyclotriphosphazenes

The phosphorus content in phosphazenes provides materials with good fire resistant behavior. For example, hexa-(4-nitrophenoxy)-cyclotriphosphazene (**16**) can be blended with poly[ethylene terephthalate] (PET) to create a composite with enhanced properties.[42] Thermal stability was assessed using TGA and it was found that the phosphazene increased char yields when either air or nitrogen was used as the purge gas. Blends of the phosphazene and PET were formed through melting the organic polymer and mixing the phosphazene into the melt.

In these relatively simple phenoxy-substituted cyclotriphosphazenes, conformation plays a role in their structure. An x-ray crystallographic study of hexa-(4-fluorophenoxy)cyclotriphosphazene (**17**) yielded four polymorphs.[43] The polymorphs, three of which have not been previously reported, were identified through variable temperature x-ray crystallographic analysis and differential scanning calorimetry (DSC). Crystals of the α form were found to melt at 128.4 °C and then re-crystallize on cooling to 53.4 °C. A β polymorph was found to have an endotherm at 98.4 °C and then a melt at 128.4 °C, which is the same as the α polymorph. This suggested a transition from the β to the α form at 98.4 °C. Further studied revealed that the various forms all undergo transitions from one form to another.

Phenoxy-substituted cyclotriphosphazenes can be functionalized to allow grafting of organic polymeric structures. Sequential substitution with phenol and 4-methoxyphenol, followed by reaction with BBr$_3$ yields phenolic structure (**18**) that then can be converted to the cyanurate (**19**),

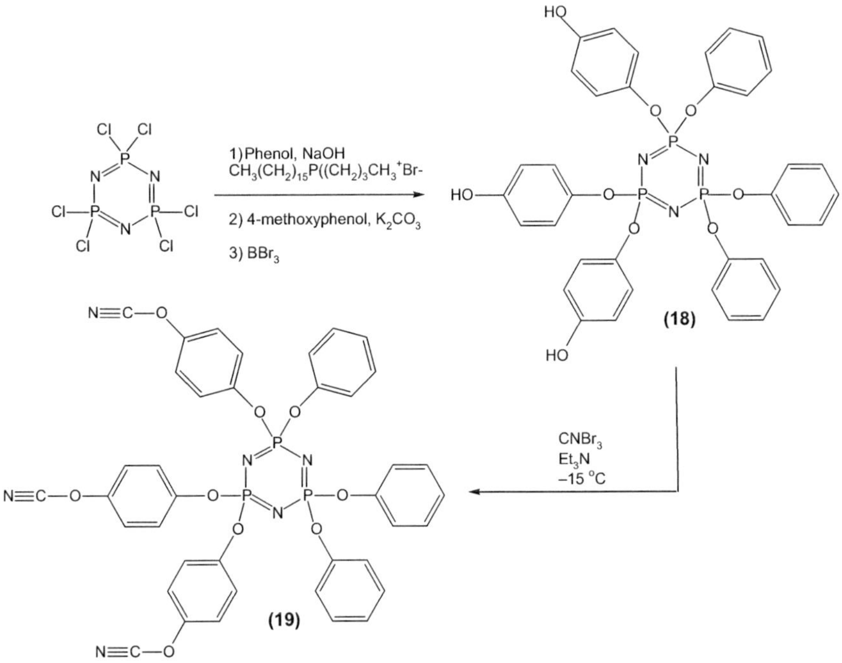

Scheme 12

Scheme 12. Polycyanurates were formed through thermal polymerization of (**19**) with 2,2-bis(4-cyanatophenyl)propane with curing onset temperature of around 100 °C. Several ratios of the two components were investigated. Thermal analysis showed increased mass loss with increasing organic content, which is a function of the dilution of the phosphazene. Further, the Tgs followed the same trend for the cured polymeric materials.

Larger phenoxy substituents were demonstrated through a functionalized stilbene structure (**20**).[44] In this synthesis, the functionalized stilbene pendant group is assembled and then attached the phosphazene ring in THF using sodium hydride as the base. The resulting structures were characterized using multinuclear NMR spectroscopy, matrix assisted laser desorption/ionization time of flight mass spectrometry and elemental analysis. These larger pendant groups were attached; although steric considerations always plays a role in this type of chemistry. Adamantane groups also embody a degree of structural bulk. Reaction of the hexachlorocyclotriphosphazene with thiol and amine adamantyl derivatives gave controllable substitution, Scheme 13.[45] Reaction with 2 equivalents of adamantyl thiol gave (**21**) with geminal substitution. Increasing the reaction to 4 equivalents also gave the geminal tetra-substituted trimer (**22**). Using adamantylamine, the mono-substituted derivative (**23**) was formed. Substitution of chlorines on octa-chlorocyclotetraphosphazene (**24**) gave mono-substituted and non-geminal di-substituted products, Scheme 14.

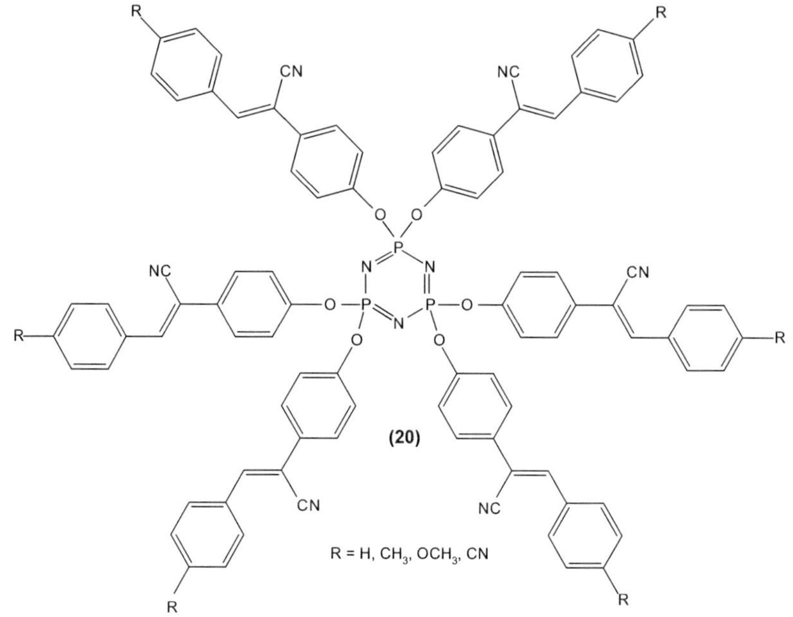

(20)

R = H, CH$_3$, OCH$_3$, CN

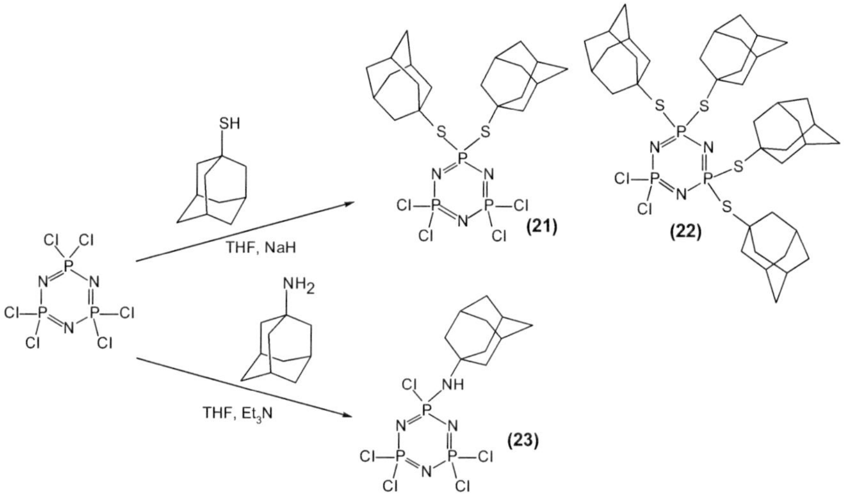

Scheme 13

Polyether substituted phosphazenes have been known for three decades as conductive for lithium ion. A new method for forming electrolytes involved the radical polymerization of polyether substituted cyclotriphosphazenes.[46] In this scheme, a protected etherial chain is attached to the cyclotriphosphazene ring, followed by deprotection to yield a free hydroxyl group, Scheme 15. Reaction of the hydroxyl with methacryloyl chloride yield a reactive trimer (25). Cross-linking of (25) was performed *in-situ* with appropriate lithium salts upon heating in the presence of tert-amyl peroxide as an initiator. This chemistry yielded a highly conductive (10^{-4} S/cm) gel. Another strategy employed bis-hydroxy-terminated

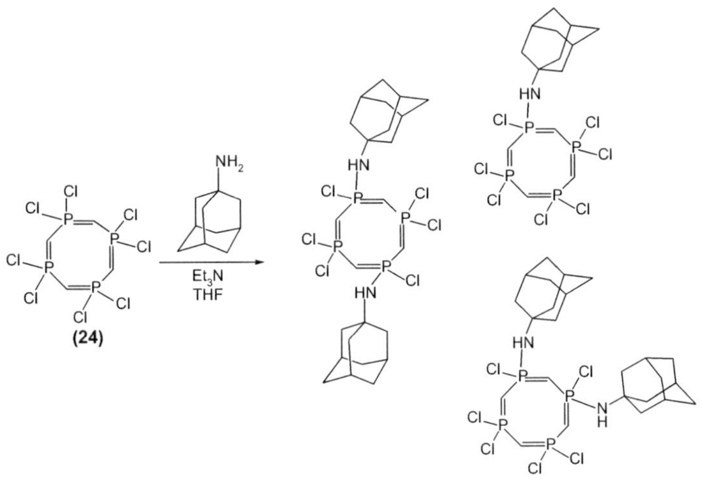

Scheme 14

etherial chains in the absence of protecting groups. With both reactive sites available for attachment, cross-linked gels were created that acted as plasticizers to increase lithium ion conductivity.

Other star-shaped molecules have been formed with naphthoxy groups.[47] In this example, sequential addition of 2-naphthol and hydroquinone gave cyclophosphazene (**26**) as a pendant group, which was then attached to the phosphazene cyclic trimer (**27**) and the tetramer (**28**). Additionally, a dimer was formed using bisphenol as a linkage between the two ring systems. This was accomplished by initial reaction of hexachlorocyclotriphosphazene with bisphenol, followed by replacement of the remaining chlorine atoms with structure (**26**) to yield dicyclophosphazene (**29**).

Grafting of poly(phenylene oxide) to a cyclic trimer is possible using a reactive phenolic phosphazene, Scheme 16.[48] 4-Methoxyphenol, deprotonated using NaOH, was rigorously dried and reacted with hexachlorocyclotriphosphazene, followed by potassium phenoxide, prepared from phenol and KOH. Demethylation was performed using HCl and pyridine at 200 °C. A series of steps were then conducted to graft the poly(phenylene oxide). The resulting materials were characterized chemically and for electrical properties and flame retardancy.

Mechanistic studies were conducted to characterize the reaction of ethynyl lithium reagents with hexafluorocyclotriphosphazene.[49] In this study, monosubstituted trimers were formed with lithiophenylacetylene or p-propenylphenyl lithium and then reacted with a second equivalent of the reagent. In mixing the reagents, a regiochemical result was observed as a function of the order of addition. For example, initial reaction with lithiophenylacetylene followed by p-propenylphenyl lithium yields largerly geminal disubstitution. However, reversing the reagents gave nongeminal product with a slight preference for the cis isomer. Use of lithiophenylacetylene for both additions resulted in a similar preference for nongeminal substitution.

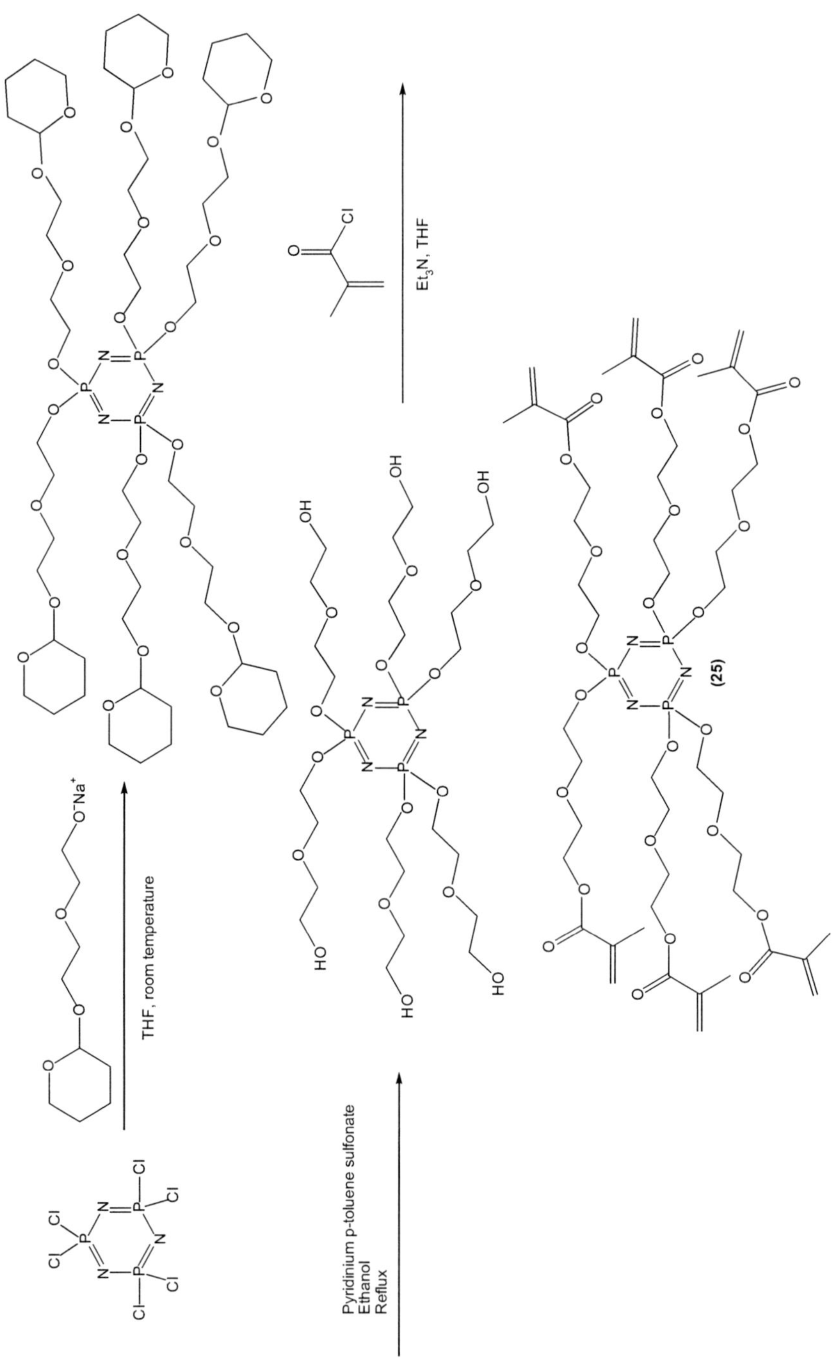
THF, room temperature
Et$_3$N, THF
Pyridinium p-toluene sulfonate
Ethanol
Reflux
(25)

Scheme 15

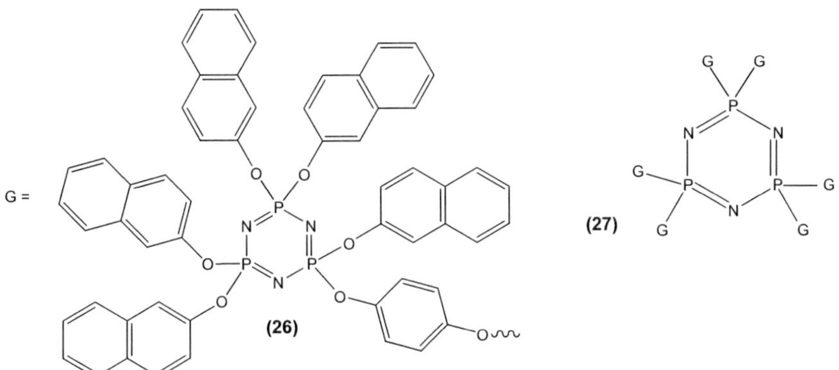

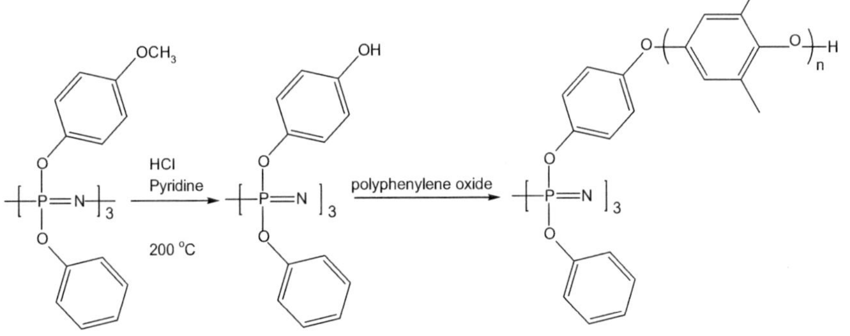

Scheme 16

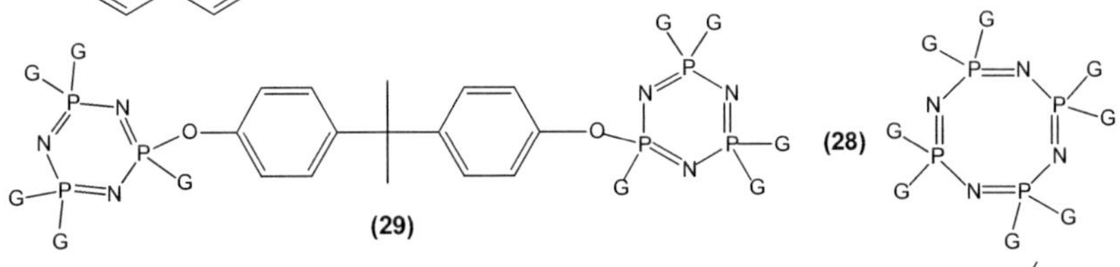

8 Phosphazenes in ring systems

Although cyclotriphosphazenes are themselves ring systems, far more elegant and complex ring systems can be formed by taking advantage of the ability for bidentate attachment of pendant groups. Many examples have been discussed in the literature that show both geminal and non-geminal chemistries. The resulting ring systems can be large or small, depending on the nature of the incoming pendant group. A recent report shows a facile attachment that resulted in an 18 membered ring, Scheme 17.[50] The products included alkyl substituted derivatives (**30, 31**) and halogenated species (**32, 33**). Other spirocyclic compounds were obtained through reaction of phosphazenes with a bis-anilino species, Scheme 18.[51] Geminal substitution was observed when the phosphazene had only geminal availability due to the attachment of two phenylenedioxy groups, resulting in the formation spirocyclic compound (**34**). Hexachlorocyclotriphosphazene, with availability for both geminal and non-geminal chemistries, yielded the ansa compound (**35**). An additional

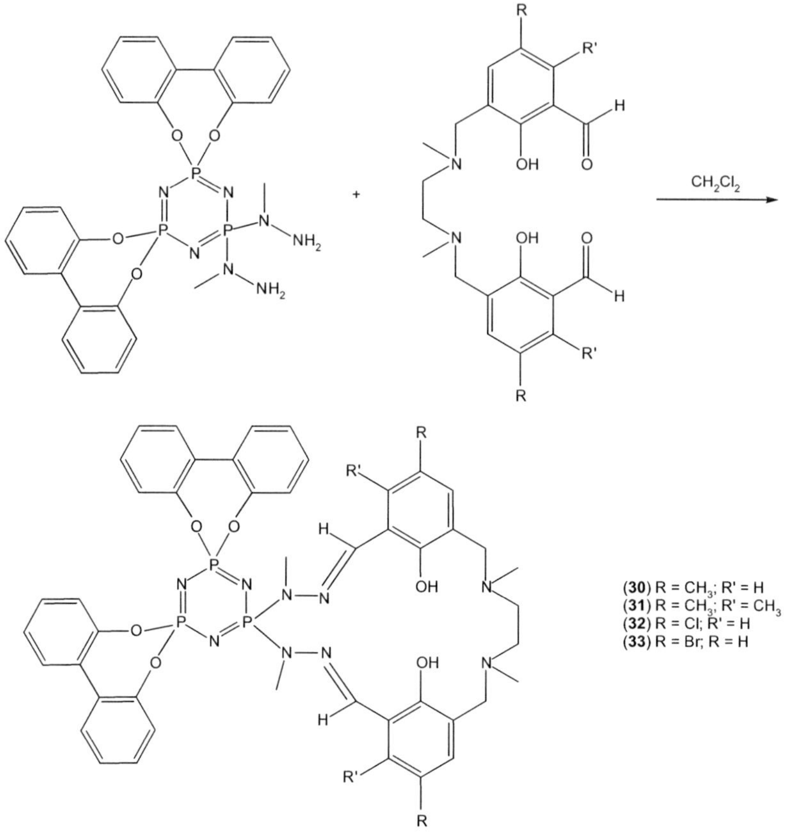

Scheme 17

(30) R = CH$_3$; R' = H
(31) R = CH$_3$; R' = CH$_3$
(32) R = Cl; R' = H
(33) R = Br; R = H

observation was the formation of what the authors termed a "bino" compound (**36**), which also can be considered a cross-link between two phosphazene ring, Scheme 19. Phenylenedioxy phosphazenes also served as a basis for oxime and oxime derivatives.[52] The synthesis began with the attachment of two phenylenedioxy groups, which due to their configuration, attach geminally. This dichloride is then reacted with 4-hydroxybenzophenone yielding (**37**) (Scheme 20). Conversion of the ketone to an oxime was accomplished using hydroxylamine hydrochloride. Functionalization of the oxime subsequently was performed with a variety of organic halides to create various derivatives.

In another strategy, amine-containing pendant groups were attached through an aldehyde, Scheme 21.[53] One phenylenedioxy is attached followed by four 4-hydroxybenzaldehyde groups yielding (**38**). The aldehyde moieties do not interact with P-Cl substitution chemistry resulting in good yields of cleanly substituted product. Sodium borohydride is then used to reduce the aldehydes to alcohols, which are brominated. Reaction with various cyclic amines was accomplished with loss of HBr. The goal of these compounds was to create new compounds that could cleave DNA under hydrolytic conditions.

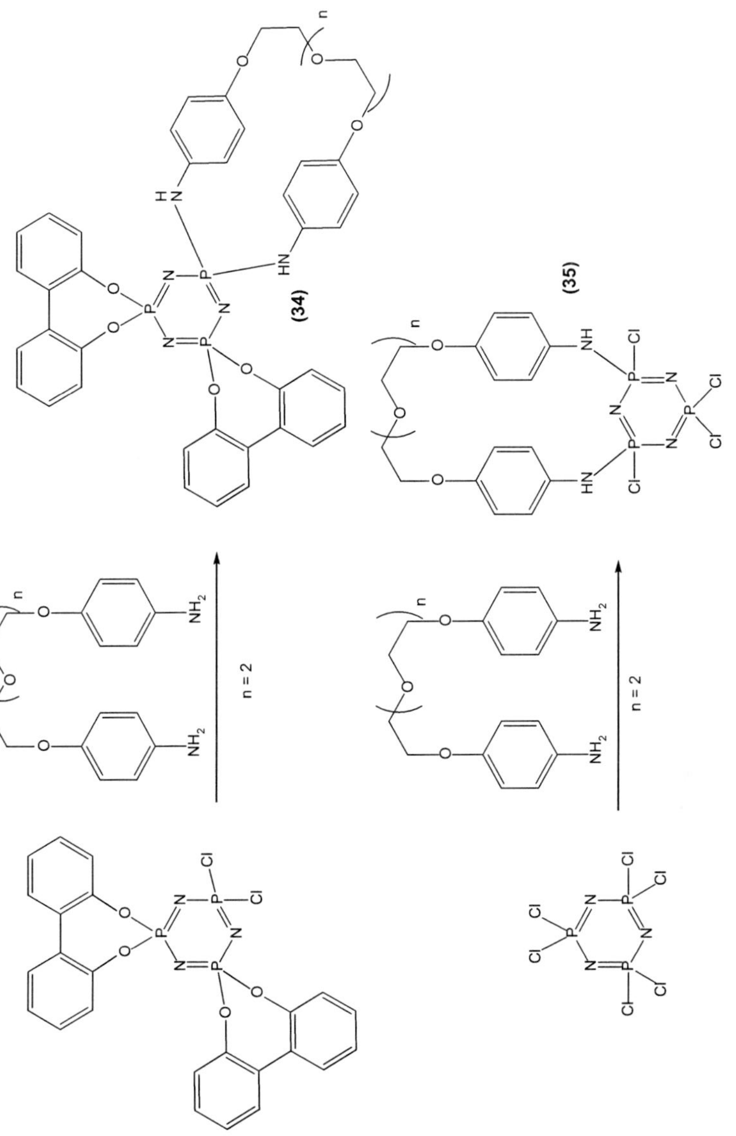

Scheme 18

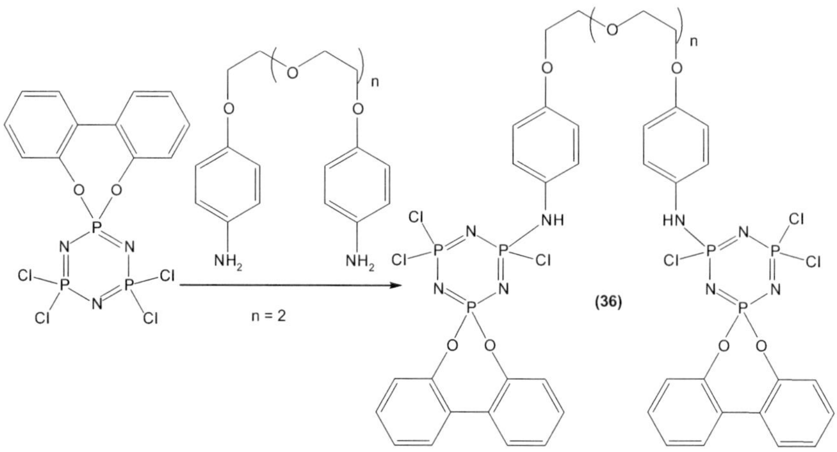

Scheme 19

Tris-o-phenylenedioxycyclotriphosphazene (**39**) can be converted into finite structures. Ordering provides channels into which molecules may intercalate. A recent example showed how molecular rotors may be assembled by insertion of a functionalized dodecaborane into a channel comprised of (**39**).[54] A rotation barrier of 1.2–9 kcal/mole, as measured by dielectric spectroscopy, suggested differing degrees of insertion, which, in turn, gave differing amounts of steric interaction. Molecular rotors are of interest as potential ultrafast organic dielectrics for use in electronics, energy generation, and for sensors. An additional application for phosphazenes with o-phenylenedioxy pendant groups includes their attachment onto linear polymers, Scheme 22.[55] The polymer is nitrated using a mixture of nitric and sulfuric acids. Reduction using a borohydride results in the aminated species, which was studied as a support for immobilization of a Baeyer-Villager monooxygenase, phenylacetone monooxygenase. This was obtained from *Thermobifida fusca*, and glucose-6-phosphate dehydrogenase, which is a NADPH recycling enzyme. The most effective loading of the amine groups on the polymer was 0.5 amines per repeat unit.

Assembly of ring systems into trimeric phosphazenes is often dictated by steric effects of the pendant groups. Generally, it is thought that the larger the pendant group, the more difficult it is to assemble structures. In a surprising result, porphyrins have been placed at all six positions about the ring of phosphazene (**40**) in 85 % yield.[56] The same chemistry was demonstrated for the tetrameric phosphazene (**41**), as well. Metal ion complexation was studied in these new materials using the chlorides of Cu(II) and Ni (II). The fully aminated variant of cyclotriphosphazene (**40**) was reacted with either Cu or Ni to give the porphyrin complexes in 90% and 85% yield, respectively. Similar chemistry was afforded by phosphazene (**40**) (x = NH; y = S); although to maintain an oxidation state of 2 +, a chloride was retained in each case. This chemistry was demonstrated for the tetrameric phosphazene with similar results.

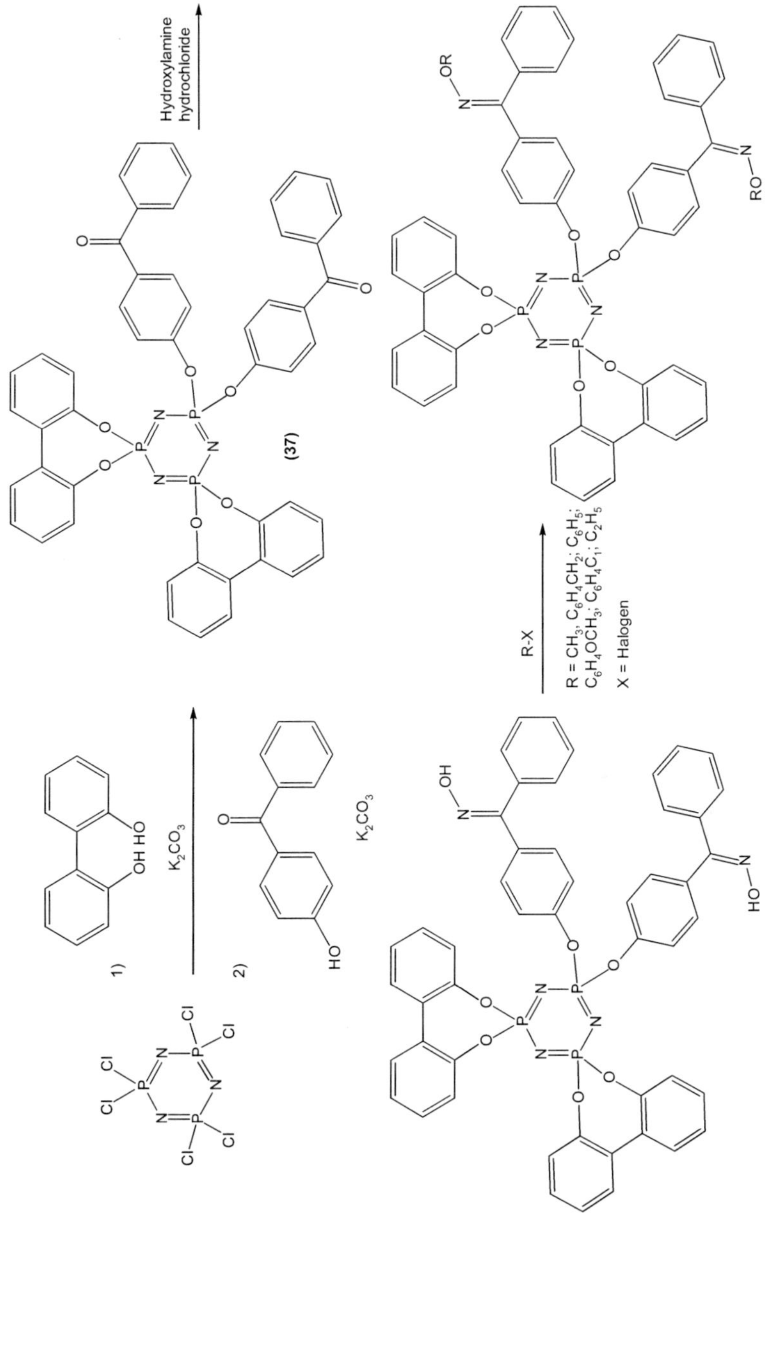
1)
K2CO3
2)
K2CO3
(37)
Hydroxylamine
hydrochloride
R-X
R = CH3, C6H4CH2; C6H5;
C6H4OCH3; C6H4C1; C2H5
X = Halogen

Scheme 20

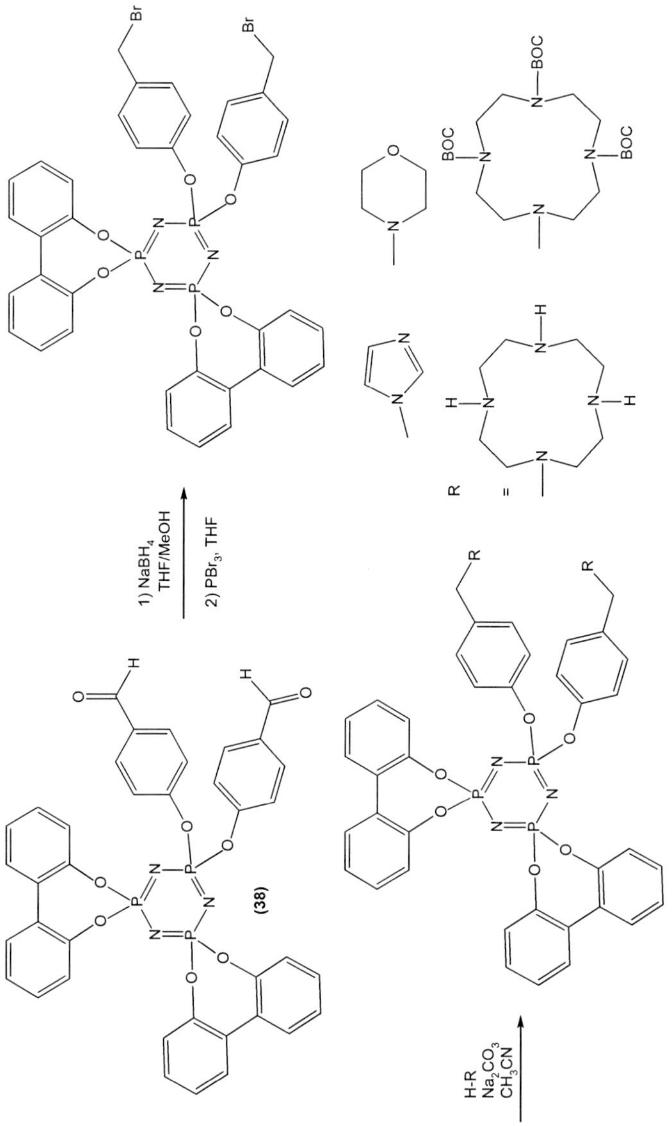

Scheme 21

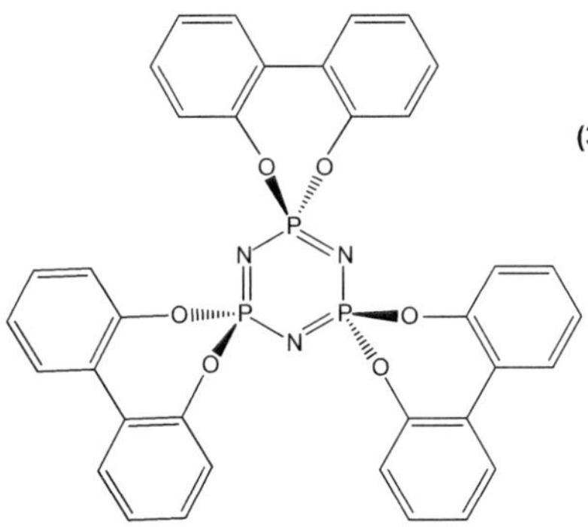

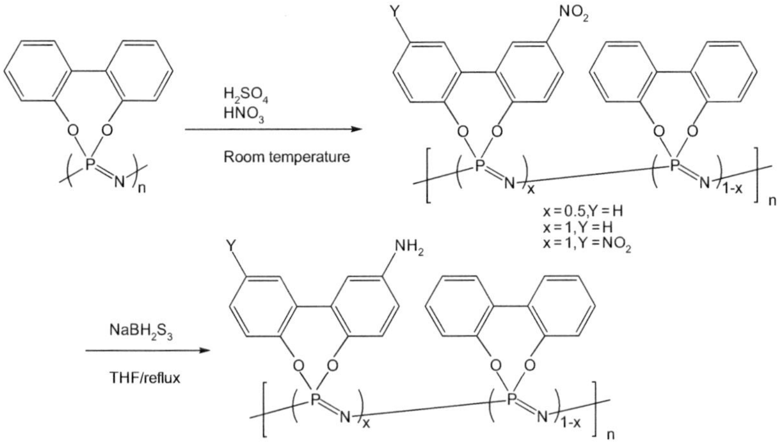

Scheme 22

Relatively complex ring systems were the result of the attachment of a series of tetradentate pendant groups to octachlorocyclotetraphosphazene.[57] With the attachment, three multiple ring systems are formed, Scheme 23. Curiously, and perhaps quite expectedly, the length of the R group in the pendant groups had an effect on speciation. The most sterically accessible conformation is probably (**42**), and its formation was seen for all R groups (R = $(CH_2)_2$; $(CH_2)_3$; $(CH_2)_4$; $(CH_2)_6$). The "ansa-spiro-ansa" configuration (**43**) was only seen for the two shorter R groups; while the bicyclo configuration (**44**) was only seem for R = $(CH_2)_4$. To stabilize the compounds, morpholine and pyrrolidine were used to replace chlorine. Support for the proposed structures were provided by x-ray crystallography.

Regio-directing groups have the ability to control the direction of subsequent nucleophilic addition. An example of this has been shown with spirocyclic derivative (**45**), Scheme 24.[58] A comparison between two diols are seen where the shorter three-carbon unit give both ansa and spiro conformations. In contrast, the longer chain gives only ansa products, which can be attributed to the chain length that discourages spirocycle formation. Another consideration is that the ansa products are produced on the face of the phosphazene ring that is opposite the

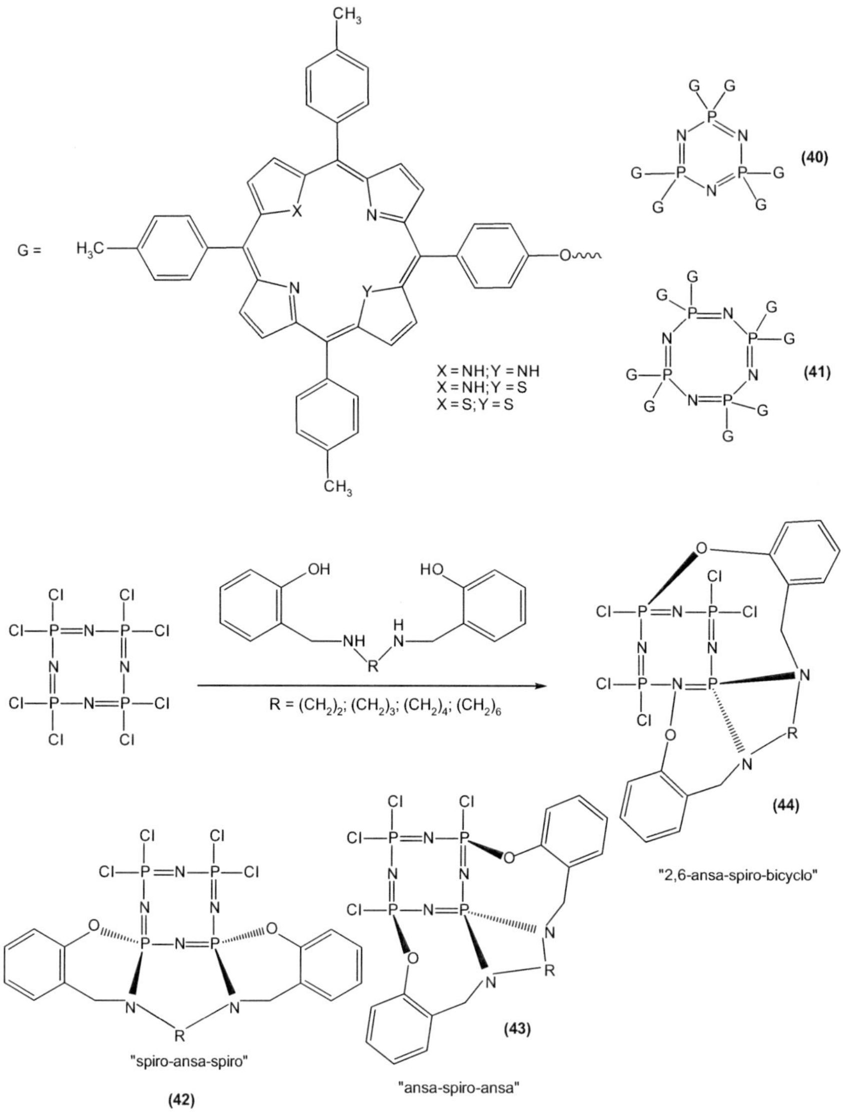

Scheme 23

N-methyl group. Another report shows that a variety of products can be formed from a spirocyclic phosphazene trimer with either ethanolamine or ethylene glycol as a substituent.[59] Ethanolamine reaction with 1,3-propanediol (**48**) gives compounds (**52**) and (**54**) as the only observed products. Replacement of the incoming nucleophile with 2,2,3,3-tetrafluorobutane-1,4-diol (**50**) yields (**54**) and (**55**). The effect of the initial spirocycle was studied by replacement of ethanolamine with ethylene glycol. Reaction of 1,3-propanediol with spirocyclic phosphazene (**49**) gave products (**53**), (**54**), and (**55**). Finally, 2,2,3,3-tetrafluorobutane-1,4-diol (**52**) was found to yield the same products, (**54**) and (**55**), as seen with the ethanolamine spirocycle suggesting that the nature of the incoming

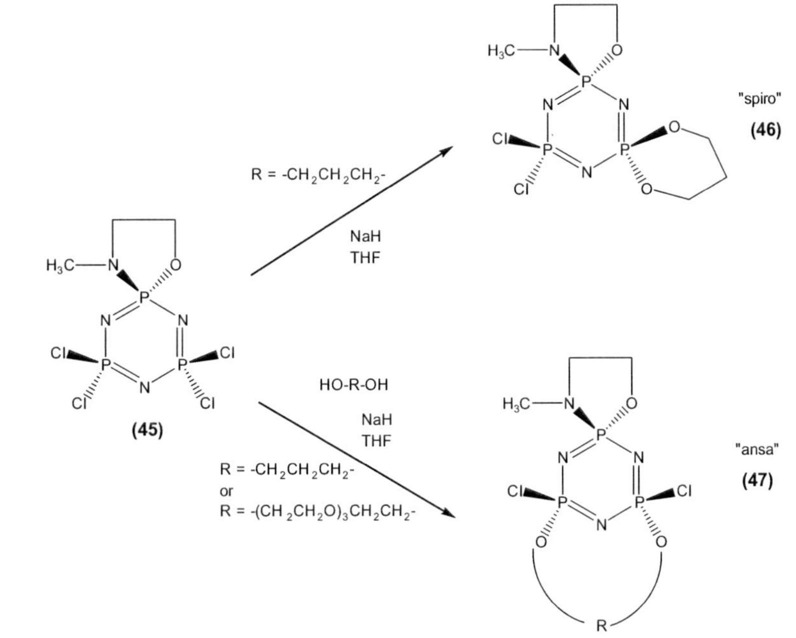

Scheme 24

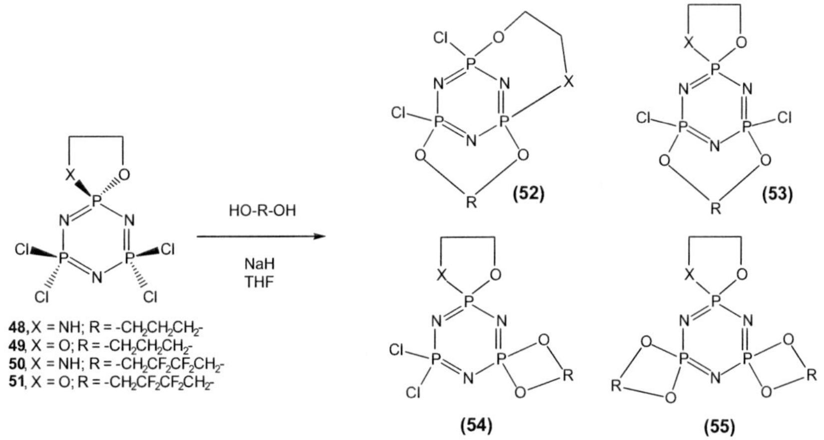

Scheme 25

nucleophile was more significant than the spirocyclic component. A discussion of the mechanism of both the attachment chemistry and the rearrangement to ansa product (52) and the role of phosphazene ring conformation is discussed (Scheme 25).

Additional N/O donor pendant groups were shown to form spirocyclic compounds.[60] In this work, six membered 1,3,2 oxazaphosphorine rings (56) were formed, Scheme 26. Stabilization of these new complexes was provided by replacement of the chlorines with either pyrroline,

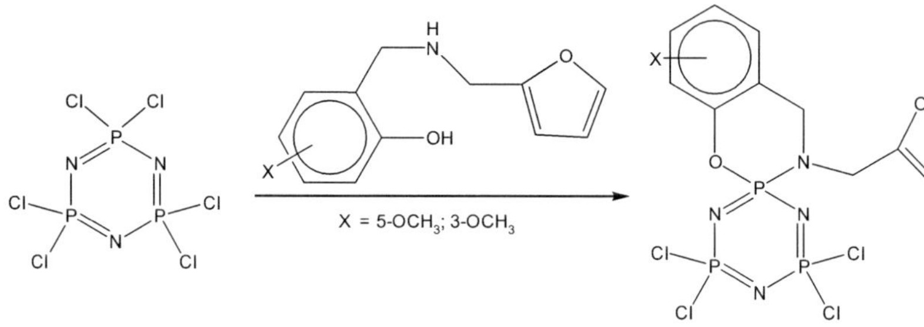

Scheme 26

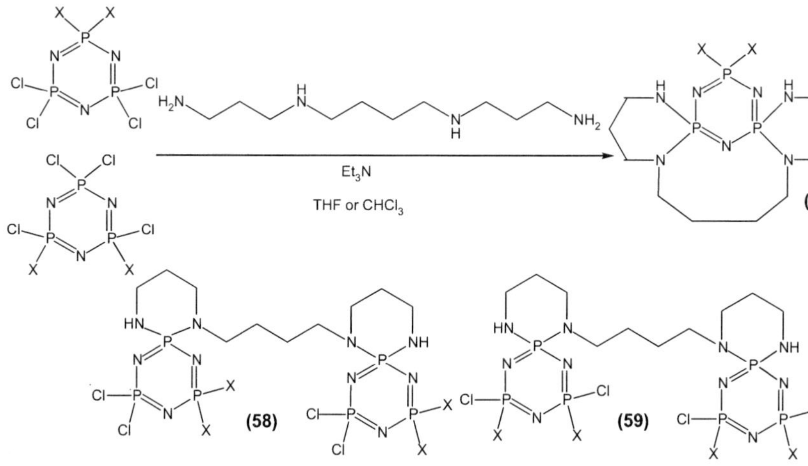

Scheme 27

morpholine, or 1,4-dioxa-8-azaspiro[4,5] decane. Biological testing of these new compounds revealed little antimicrobial activity; however they were found to change the mobility of pBR322 plasmid DNA.

Dispiro-bino and dispiro-ansa compounds were formed from the attachment of spermine, Scheme 27.[61] Initially, geminal and nongeminal disubstituted cyclotriphosphazenes were formed with NHPh, SPh, or linear diols ([$OCH_2(CF_2)_2CH_2O$]$_{0.5}$ and [$OCH_2C(CH_3)_2CH_2O$]$_{0.5}$), denoted as X in Scheme 27. Subsequent reaction with spermine was found to yield three new general structure types (**57–59**). Intramolecular binding of spermine gave the multi-ring system (**57**) using the geminal cyclotriphosphazene. Intermolecular attachment was seen for both starting configurations, yielding diphosphazenes (**58**) and (**59**). Characterization of these new compounds revealed cytotoxicity against two cancer cell lines suggesting that these structures are worthy of further study.

In contrast to the ring chemistries demonstrated with long multidentate organic pendant groups, a report of the intermolecular coupling of phosphazene rings facilitated by a hindered secondary amines revealed interesting new structures, Scheme 28.[62] For example, use of various monosubstituted butyl amines as pendant groups gave differing

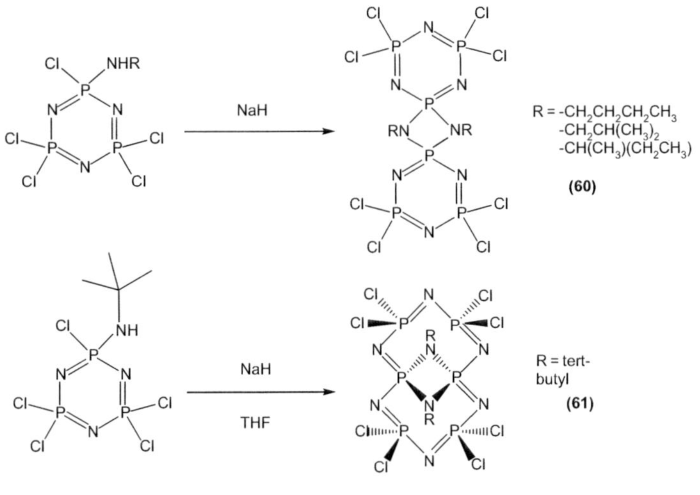

Scheme 28

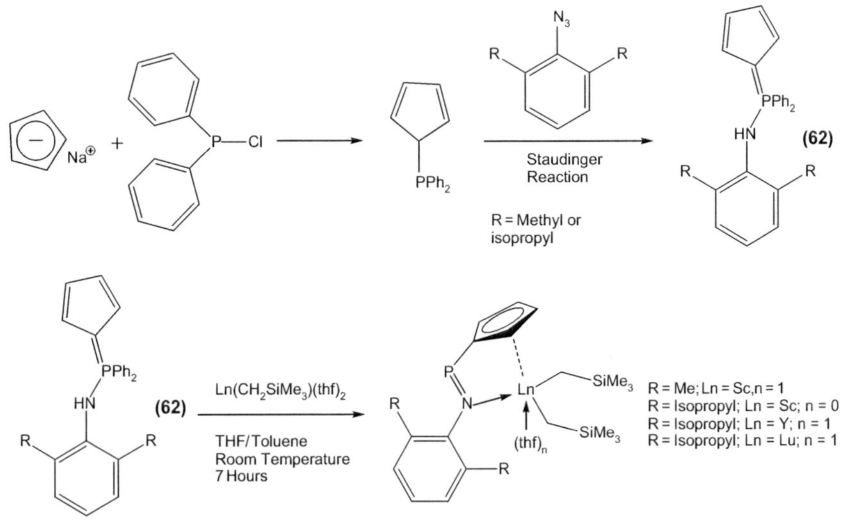

Scheme 29

chemistries. n-Butyl, sec-butyl, and isobutyl amines gave a coupling reaction in which a four-membered ring bridges two cyclotriphosphazene rings (**60**); however the rings stay intact. For the instance where a highly hindered tert-butylamine was employed, a very different compound was formed in which the phosphazene ring was expanded (**61**).

9 Inorganic complexes

There are numerous ways to create complexes of phosphazenes with metal centers to create materials with catalytic activity. A mono-phosphazene can be synthesized using the Staudinger reaction to form the P-N linkage (**62**), Scheme 29.[63] Complexes with the rare earth elements,

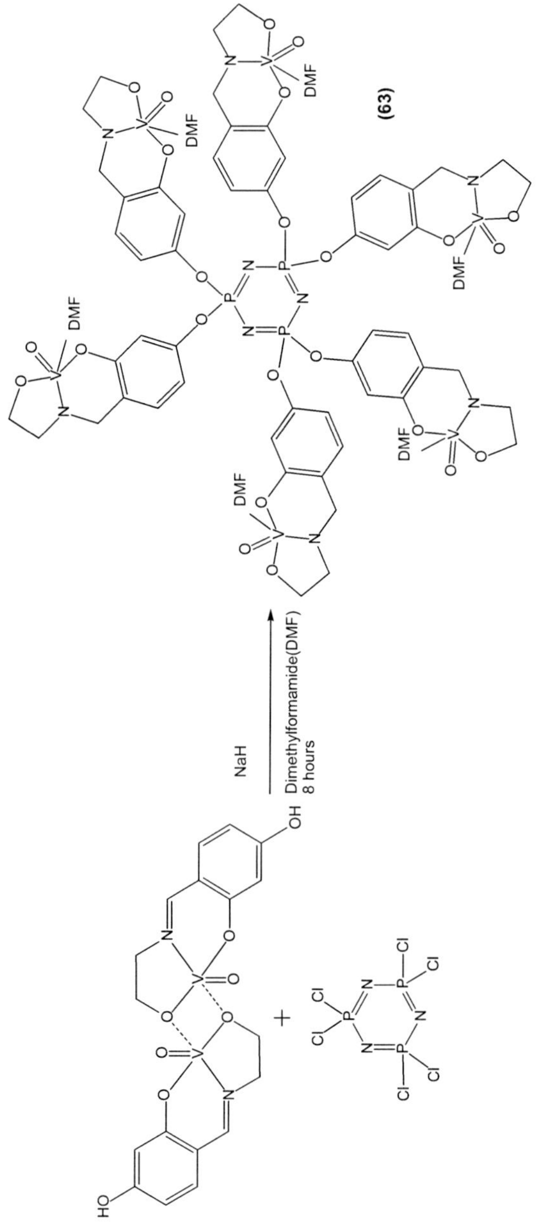

Scheme 30

Sc, Y, and Lu, were formed at room temperature with suitable precursors. The new complexes were found to exhibit some activity as ethylene polymerization catalysts; however their function was dictated by the steric hindrance of the complex, the electronic environment of the metal center, co-catalysts used, temperature, pressure, and reaction time. Another catalyst was prepared by attachment of a functionalized organo-vanadium complexes to a cyclotriphosphazene core, Scheme 30.[64] In this attachment reaction, products identified included the hexa-substituted trimer (**63**), penta-substituted, tetra-substituted, and tri-substituted compounds. It should be noted that the tri-substituted one had one substituent on each P, which is reasonable considering the steric bulk of the incoming nucleophile (V complex). A mixture of the four products was found to be an effective catalyst for the regioselective oxidation of phenols and naphthols to quinones using tert-butylhydroperoxide as the oxidizing agent.

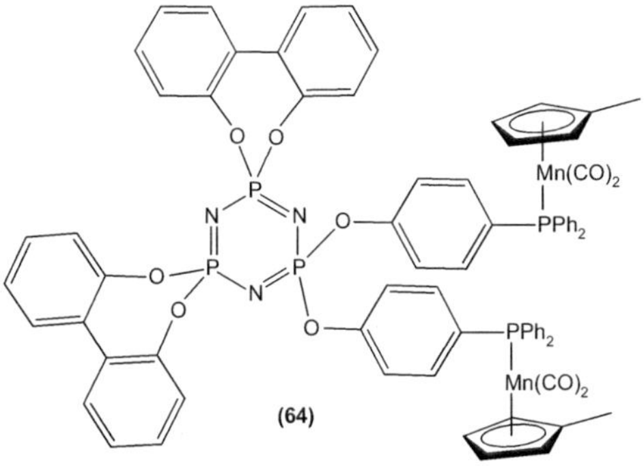

Functionalization of pendant groups enables complexation of phosphazene with a cyclopentadienyl manganese center yielding bis complex (**64**).[65] Pyrrolysis gave single crystals of $Mn_2P_2O_7$ that could be formed in either the nano- or microscale. The pyrrolysis was performed in the absence of solvents and appears to create a layered graphitic host structure for the complex. Iron cyclopentadienyl structures have been attached to poly(alkyl/arylphosphazenes) through a lithiated intermediate, Scheme 31.[66] Stable materials were formed with a degree of ferrocene

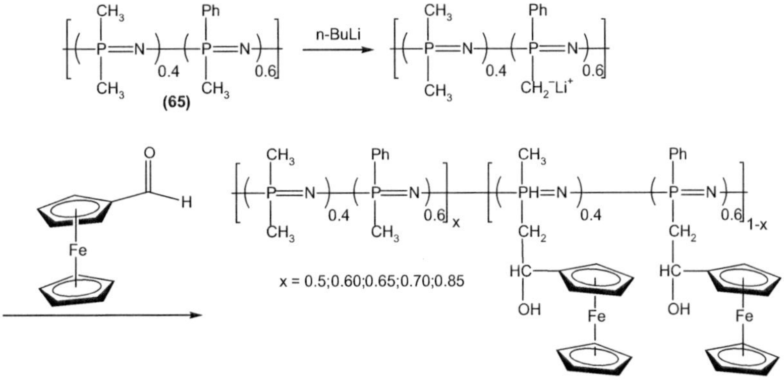

Scheme 31

(66)

substitution varying between 50% and 85%. A study of this material by DSC revealed a steady increase in Tg with increase in ferrocene content. For example, the Tg for polymer (**65**) was 2 °C. For the 85% ferrocene loaded polymer it was 92 °C. These new complexes were characterized electrochemically where reversible redox behavior of Fe was observed. Another report displayed an alternate pathway to ferrocenyl phosphazenes resulting in a hexa-substituted star-type compound (**66**) using an acrylate linkage.[67]

Another interesting Fe complex was created by the coordination of tridentate nitrogen donor ligands attached to either cyclotriphosphazenes or linear polyphosphazenes.[68] These new complexes were targeted as possible spin-crossover materials that are of interest for quantum computers and large data storage devices. An x-ray crystal structure report suggested that the cyclotriphosphazene variant involved two complexed ring systems (**67**). Further work showed that the same type of ligand could be attached to linear polymer (**68**) as well.

Silver coordination compounds of oxadiazole (**69**) were formed with both the nitrate and triflate salts of the metal.[69] The phosphazene was formed by assembly of the ligand followed by attachment to bis(2,2-phenylenedioxy)-bis-chlorocyclotriphosphazene.

Carbaphosphazenes, with carbon in the place of two ring phosphorus atoms, represent a far less studied inorganic ring system, as compared to cyclotriphosphazenes. An example of their chemistry led to the development of ligands in silver and copper complexes.[70] Using a diphenyldichlorocarbaphosphazene (**70**) as a starting material, imidazolium ions were attached that formed complexes with silver, Scheme 32. Additionally, complexation to copper was demonstrated. Another carbaphosphazene was constructed with a multi-dentate ligand for heterometallic complexation.[71]

(67)

(68)

(69)

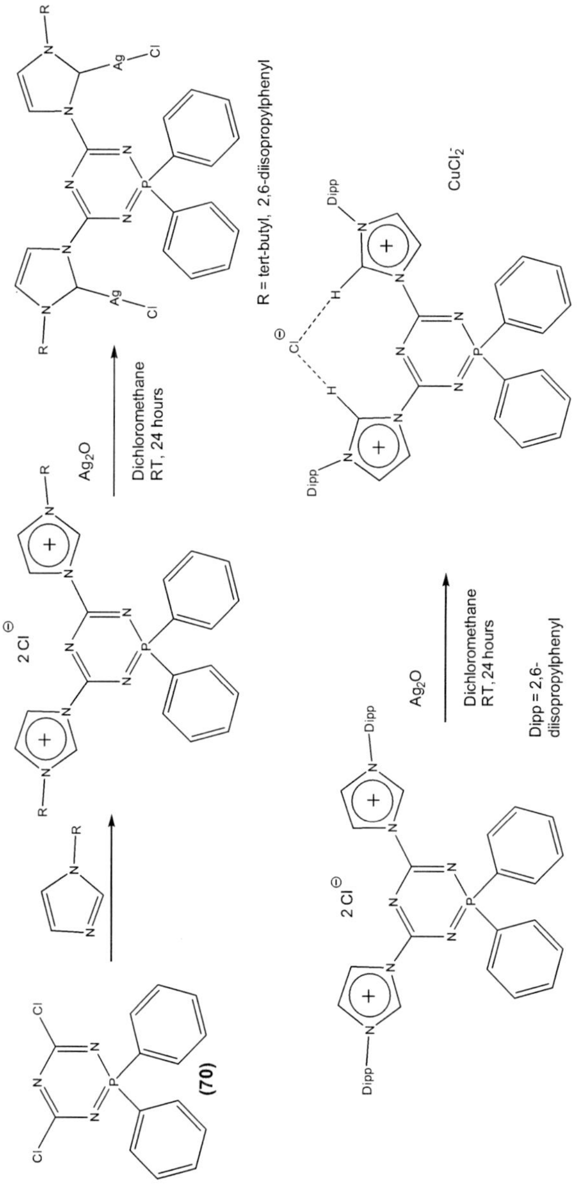
(70)
Ag$_2$O
Dichloromethane
RT, 24 hours
R = tert-butyl, 2,6-diisopropylphenyl
Dipp = 2,6-diisopropylphenyl
$CuCl_2^-$

Scheme 32

Carbaphosphazene (71) was shown to form complexes by sequential treatment with Cu and Ca to yield complexes 1) (71)CuCa, and 2) (71)Cu$_2$Ca$_2$. Sequential treatment of (71) with Cu(NO$_3$)$_2$ and Dy(NO$_3$)$_3$ yielded the (71)CuDy complex. Similar chemistry was demonstrated with the phosphazene analog (72).

Fluxionality was observed in a zinc-phosphazene coordination compound.[72] In this work, a pyridyloxyphosphazene was found to exhibit two

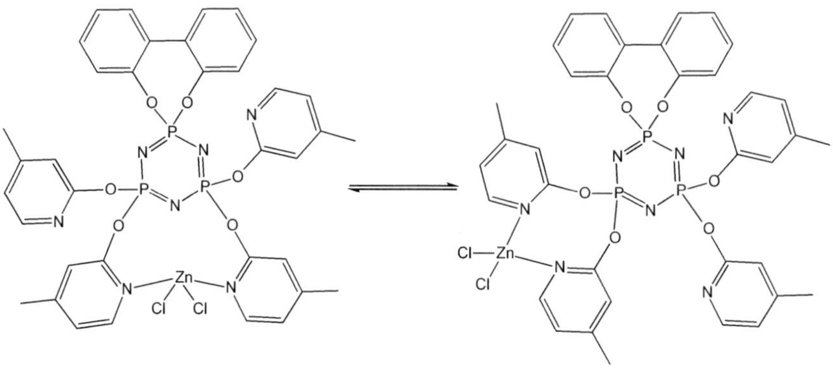

Scheme 33

modes of coordination with zinc, Scheme 33, where the complexation occurred at both geminal and non-geminal pyridyloxy pendant groups. Differentiation between the two complexes was made through single crystal X-ray structure determinations. In another report, flexible structures were reported using a novel pyridylamino-substituted tetrameric phosphazene ring (**73**).[73] Coordination with Cd(II) resulted in compounds that exhibited a strong blue emission.

A palladium complex can be formed with a dual purpose phosphazene trimer.[74] First, groups that ligate the metal were attached, followed by functionlization to yield a condensable silicate precursor. Condensation by the sol-gel method resulted in a recyclable Pd catalyst. Synthesis of this novel bifunctional ligand was conducted in four steps. First, the biphenyl linkage was attached, followed by the phosphine-containing aryl groups. The biphenyl groups is then manipulated to yield pentaphosphine (**74**). Other organometallic complexes were formed in a SiO_2 matrix using sol-gel chemistry without direct bonding between the silicate precursors and the phosphazene.[75] Examples include copper containing structures (**75**), with the metal coordinated through a pyridyl nitrogen, and (**76**), where copper binding is accomplished through a cyano group. Other examples included W, Ti, Ag, and Ru in similar complexes that wereformed and then mixed with a sol-gel precursor to create an immobilized solid materials that upon pyrolysis at 800 °C gave nanostructured metal oxides.

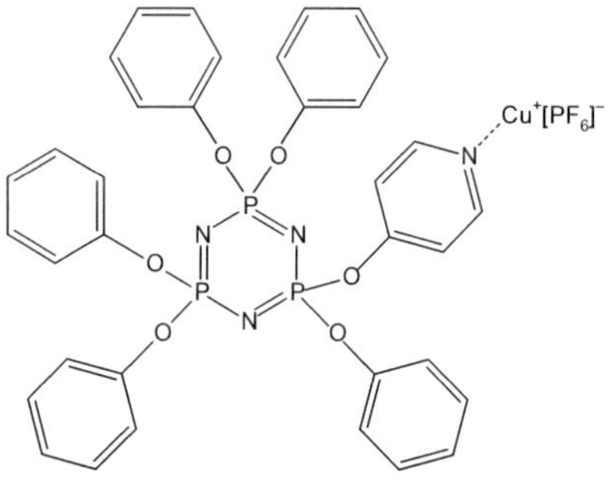

(**75**)

10 Nanomaterials

The study of cyclomatrix 4,4′-sulfonyldiphenol structures has continued with a variety of new applications. In this strategy, a bifunctional pendant group is used to create a highly cross-linked network of cyclotriphosphazene rings (77), which when condensed can form nanostructures. Further, pyrolysis also can yield nanomaterials in a variety of conformations. Using a templating strategy, phosphazene (77) can be formed as nanotubes that could be converted into Au, Pd, or Ag-containing nanocomposites.[76] The resulting nanocomposites were characterized using X-ray powder diffraction, thermal analysis, and various microscopic techniques. Thermal stability was measured in these complexes to be as high as 460 °C. Other applications for nanotubes formed from phosphazene (77) include a polypropylene flame retardant,[77] an additive to a polyethylene oxide based solid polymer electrolyte,[78] and as a support for Pt-Co catalysts in a direct methanol fuel cell.[79] Another report showed that free hydroxyls (from the 4,4′sulfonyldiphenol) can react with bis-isocyanates to cross-link the ananotubes into a network.[80] The result of this chemistry was polyurethanes with enhanced tensile properties.

By changing the preparation conditions, (77) can be formed into nanospheres that can be further manipulated into heteroatom containing porous carbon nanospheres.[81] This transformation was accomplished by soaking nanospheres of (77) in aqueous NaOH for 2 hours to allow the base to inpregnate into the structures, drying at 110 °C, and then heating at a 5 °C/min ramp rate to 750 °C in a nitrogen atmosphere. BET analysis revealed a surface area of 1140 m^2/g and a total pore volume of of 0.9 m^3/g. Also, a bimodal pore distribution was observed (3–5 nm and 0.6–0.8 nm diameters).

Benzidine, used in place of 4,4′-sulfonyldiphenol, yielded similar cyclomatrix structures (78) that formed into microspheres.[82] It should be noted that both mono and bidentate attachment to the phosphazene rings was noted. Due to the abundance of aromaticity and the availability of amines, luminescence was observed without the need for further treatments. Luminescence in the this structure was used to detect several nitro-aromatics, including 2,4,6-trinitrotoluene (TNT) and picric acid.

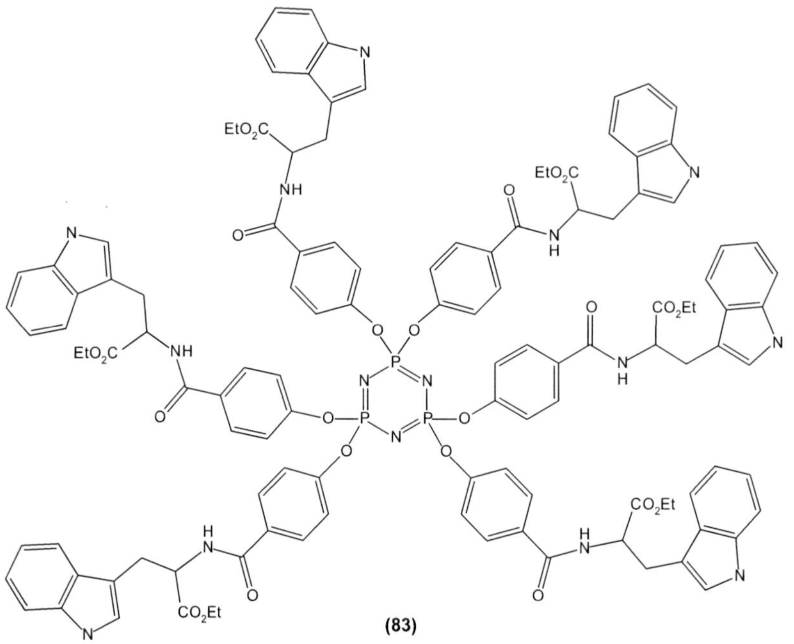

(83)

Other fluorescent monodisperse microspheres were formed using phloroglucinol (benzene-1,3,5-triol) (79) as the cross-linker.[83] The size of the spheres ranged from 600 to 1100 nm, and the size was controlled the concentration of the hexachlorocyclotriphosphazene precursor during synthesis.

Attachment of three phenylenedioxy groups yields the discrete trimer with geminal pendant group attachments (80).[84] Nanoparticles have been formed from (80) by physically mixing it with AuCl(PPh$_3$), followed by pyrolysis at 800 °C. The particle morphology was found to be dependent

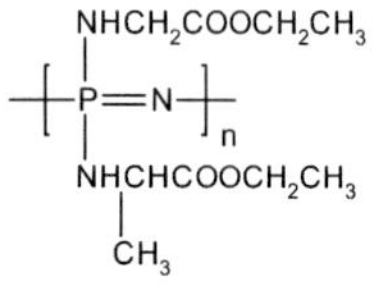

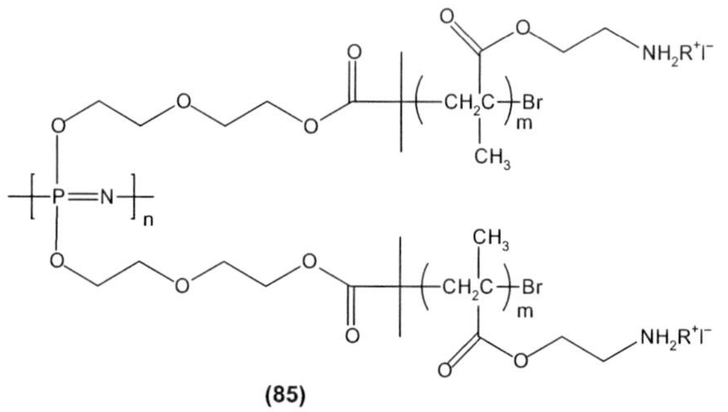

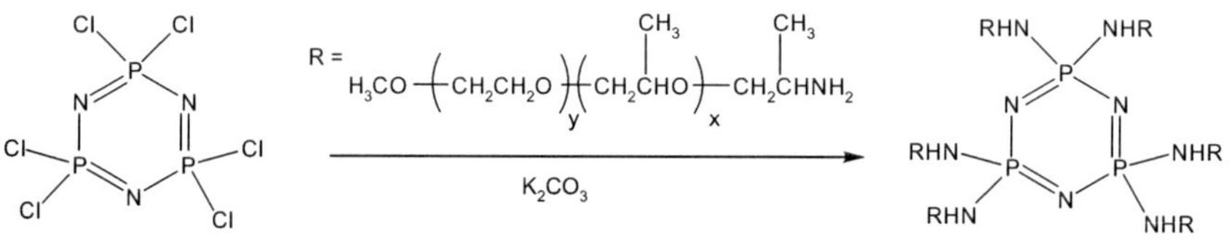

Scheme 34

on the pyrrolysis conditions and sizes as small at 3.5 nm were reported. Introduction of fluorine was performed by using a functionalized biphenyl pendant group that yielded (81).[85] Particles formed in this work ranged from 0.57 µm to 4.33 µm in diameter. Characterization techniques reported for these particles included scanning electron microscopy, FT-IR, energy dispersive X-ray spectroscopy, NMR imaging, and X-ray diffraction. Nanoparticles also have been formed from cyclotriphosphazenes containing amino acid fragments (82).[86] The procedure for formation of the particles focused on a solution method that yielded largely monodisperse particle distributions with diameters of ∼300 nm. A density functional theory (DFT) study supported this work and proposed that the most stable conformation of acid (82) had the pendant groups above and below the plane defined by the cyclotriphosphazene ring. A related work extended the length of the amino acid fragment attachments with tryptophan ethyl ester as a pendant group, yielding indole (83).[87] Cyclotriphosphazene (83) was observed to self-assemble into nanoparticles of approximately 400 nm in diameter. DFT studies supported the association of the pendant groups above and below the plane of the ring, which the authors termed a "basket" conformation.

Nanofibers containing polyphosphazene have been formed through electrospinning.[88] The polymer used in this study was one of potential biomedical interest, poly[(alanino ethyl ester)$_{0.67}$ (glycino ethyl ester)$_{0.33}$] (84). In this paper the effects of the various processing parameters were examined. For example, polymer concentration in the casting solution was found to be critical where at lower concentration, such as 5 weight percent (wt. %), numerous polymer beads were located on the fibers. Increasing the concentration to 15 wt. % eliminated bead formation. Also 25 °C was found to be an optimum temperature for smooth fiber formation. Temperatures both higher and lower gave irregular formations. Electrospinning also was successfully performed with a phosphazene-methacrylate copolymer (85).[89] Structural information was also reported for the cyclotriphosphazene variant of (85). Antimicrobial activity was noted for these compounds.

(86)   (87)   (88)

Composites can be formed between phosphazenes and other dissimilar materials. Nanohybrids of a functionalized phosphazene and montmorillonite clay.[90] The phosphazene was a cyclic trimer substituted with a poly[ethylene-propylene] chain, Scheme 34. The pendant group was prepared and substitution was facilitated by K_2CO_3 as the base. Prior to the phosphazene intercalation, the montmorillonite was treated with base to activate sites for bonding to occur with the phosphazene.

The phosphazene was found to expand the clay and liberate silicate platelets. The phosphazene-montmorillonite composite was dispersed into polyurethane, which resulted in enhanced thermal and photochemical stability, presumably due to the presence of the platelets.

11 Computational methods

Computational methods provide insight into the structure and conformation of complex three dimensional systems, such as phosphazenes, and can be of particular value if paired with empirical data. For example, a study reported the use of the Amsterdam Density Functional package to probe adducts (B, Al, Ga, In, Tl) with the ring nitrogen atoms of a cyclotriphosphazene.[91] Methods explored included Hartree-Fock, B3LYP*, B3LYP, PW91, OLYP, BP, LDA. Studies suggested that the Al adduct (86) was the most stable and that the LDA (local density approximation) method yielded data that most closely matched the X-ray crystallographic data. Further, calculations revealed that over 50% of the HOMO electron density is located in the P_z orbital of nitrogen. These data are significant since this adduct has been proposed to play a role in $AlCl_3$ catalyzed ring-opening polymerization.

The COMPASS force field was applied to a series of biomedically relevant polyphosphazenes, including poly[bis-chlorophosphazene], poly[glycine ethyl ester phosphazene], poly[bis(carboxylatophenoxy)phosphazene], and poly[di(imidazole)-phosphazene] (87).[92] The computations were found to predict polymer densities and Tgs with a high degree of accuracy. Furthermore, solubility parameters were calculated and they suggested that the electrostatic interactions with the pendant groups were a significant factor with poly[bis(carboxylatophenoxy)phosphazene] having the highest solubility parameter due to its hydrogen bonding capacity with the carboxylic acid.

DFT was employed to interpret the infrared spectrum of hexa(2,2,2-trifluoroethoxy)cyclotriphosphazene (88). Agreement between calculation and experiment was considered to be excellent and this allowed for credible assignments of vibrational data. The tetrameric version of phosphazene (**88**) also was examined with similar result. Application of DFT to a larger dendrimeric structure (**89**) revealed a concave structure and a slighly non-planar phosphazene ring.[93] The computational data was used to interpret the IR data for the dendrimeric phosphazene. The method was also successfully applied to a related structure (**90**) with similar results.[94]

(91)

A series of Co complexes with hexa(2-pyridyloxy)cyclotriphosphazene (**91**) in the most stable high spin states were studied by DFT using hybrid B3LYP basis set.[95] Electron density was studied using the Quantum Theory of Atoms in Molecules (QTAIM) method. This is an example of a system with two Co atoms, one tetrahedral and bound to one pyridyl nitrogen, while the other octahedral with bonding at one of the ring and four of the pyridyl nitrogens. Atomic charge determinations on nitrogen revealed that the phosphazene ring nitrogens are approximately twice as basic as the pyridyl nitrogen atoms. Furthermore, the bonds from metal center to ring nitrogen are shorter and stronger.

Acknowledgement

This review was supported by the U.S. Department of Energy, Office of Nuclear Energy, under DOE Idaho Operations Office Contract DE-AC07-05ID14517.

References

1. H. R. Allcock and N. L. Morozowich, *Polymer Chem.*, 2012, **3**, 578.
2. Y. L. Li, J. Rodrigues and H. Tomas, *Chem. Soc. Rev.*, 2012, **41**, 2193.
3. J. I. Kim, B. S. Lee, C. Chun, J. K. Cho, S. Y. Kim and S. C. Song, *Biomaterials*, 2012, **33**, 2251.
4. J. K. Cho, J. W. Park and S. C. Song, *J. Pharmaceut. Sci.*, 2012, **101**, 2382.
5. J. K. Cho, C. Chun, H. J. Kuh and S. C. Song, *Eur. J. Pharmaceut. Biopharm.*, 2012, **81**, 582.
6. L. Y. Qiu, C. Zheng and Q. H. Zhao, *Molecul. Pharmceut.*, 2012, **9**, 1109.
7. N. Ren, X. J. Huang, X. Huang, Y. C. Qian, C. Wang and Z. K. Xu, *J. Polym. Sci. Part A Polym. Chem.*, 2012, **50**, 3149.
8. Y. C. Qian, X. J. Huang, C. Chen, N. Ren, X. Huang and Z. K. Xu, *J. Polym. Sci. Part A Polym. Chem.*, 2012, **50**, 5170.

9   I. Teasdale, M. Waser, S. Wilfert, H. Falk and O. Bruggemann, *Monat. Chem.*, 2012, **143**, 355.
10 A. M. Amin, L. Wang, H. J. Yu, W. A. Amer, J. M. Gao, J. Huo, Y. L. Tai and L. Zhang, *J. Inorg. Organomet. Polym. Mat.*, 2012, **22**, 196.
11 Y. J. Jun, V. B. Jadhav, J. H. Min, J. X. Cui, S. W. Chae, J. M. Choi, I. S. Kim, S. J. Choi, H. J. Lee and Y. S. Sohn, *Int. J. Pharmaceut.*, 2012, **422**, 374.
12 M. S. Peach, R. James, U. S. Toti, M. Deng, N. L. Morozowich, H. R. Allcock, C. T. Laurencin and S. G. Kumbar, *Biomed. Mat.*, 2012, 7.
13 M. S. Peach, S. G. Kumbar, R. James, U. S. Toti, D. Balasubramaniam, M. Deng, B. Ulery, A. D. Mazzocca, M. B. McCarthy, N. L. Morozowich, H. R. Allcock and C. T. Laurencin, *J. Biomed. Nantech.*, 2012, **8**, 107.
14 N. L. Morozowich, J. L. Nichol, R. J. Mondschein and H. R. Allcock, *Polym. Chem.*, 2012, **3**, 778.
15 N. L. Morozowich, J. L. Nichol and H. R. Allcock, *Chem. Mat.*, 2012, **24**, 3500.
16 A. K. Andrianov and G. Mutwiri, *Vaccine*, 2012, **30**, 4355.
17 S. Awate, H. L. Wilson, K. Lai, L. A. Babiuk and G. Mutwiri, *Molecular Immun.*, 2012, **51**, 292.
18 L. M. Lobanova, N. F. Eng, M. Satkunarajah, G. K. Mutwiri, J. M. Rini and A. N. Zakhartchouk, *Vaccine*, 2012, **30**, 3061.
19 A. Dar, K. Lai, D. Dent, A. Potter, V. Gerdts, L. A. Babiuk and G. K. Mutwiri, *Vet. Immun. Immunopath.*, 2012, **146**, 289.
20 S. Garlapati, R. Garg, R. Brownlie, L. Latimer, E. Simko, R. E. W. Hancock, L. A. Babiuk, V. Gerdts, A. Potter and S. V. Littel-van den Hurk, *Vaccine*, 2012, **30**, 5206.
21 X. Wang, Q. S. Li, Y. B. Di and G. Z. Xing, *Fibers Polym.*, 2012, **13**, 718.
22 R. Ji, Y. S. Ma and J. W. H. Tsai, *IEEE Trans. Magnetics*, 2012, **48**, 4475.
23 Z. P. Zhao, Q. Guo, X. Li, J. L. Sun and Z. J. Nie, *Express Polym. Lett.*, 2012, **6**, 308.
24 Y. W. Bai, X. D. Wang and D. Z. Wu, *Ind. Eng. Chem. Res.*, 2012, **51**, 15064.
25 J. Liu, J. Y. Tang, X. D. Wang and D. Z. Wu, *RSC Adv.*, 2012, **2**, 5789.
26 J. Sun, X. D. Wang and D. Z. Wu, *ACS Appl. Mat. Interfaces*, 2012, **4**, 4047.
27 H. R. Allcock, *Soft Matter*, 2012, **8**, 7521.
28 C. Chen, A. R. Hess, A. R. Jones, X. Liu, G. D. Barber, T. E. Mallouk and H. R. Allcock, *Macromolecules*, 2012, **45**, 1182.
29 C. Chen, X. Liu, Z. C. Tian and H. R. Allcock, *Macromolecules*, 2012, **45**, 9085.
30 F. Hacivelioglu, E. Okutan, S. U. Celik, S. Yesilot, A. Bozkurt and A. Kilic, *Polymer*, 2012, **53**, 3659.
31 N. L. Morozowich, T. Modzelewski and H. R. Allcock, *Macromolecules*, 2012, **45**, 7684.
32 Z. C. Tian, X. Liu, C. Chen and H. R. Allcock, *Macromolecules*, 2012, **45**, 2502.
33 X. Liu, Z. C. Tian, C. Chen and H. R. Allcock, *Macromolecules*, 2012, **45**, 1417.
34 S. S. Suarez, D. P. Soto, G. A. Carriedo, A. P. Soto and A. Staubitz, *Organometallics*, 2012, **31**, 2571.
35 X. Liu, J. P. Breon, C. Chen and H. R. Allcock, *Molecules*, 2012, **45**, 9100.
36 X. Liu, J. P. Breon, C. Chen and H. R. Allcock, *Dalton Trans*, 2012, **41**, 2100.
37 G. C. Nayak, S. Sahoo, S. Das, G. Karthikeyan, C. K. Das, A. K. Saxena and A. Ranjan, *J. Appl. Polym. Sci.*, 2012, **124**, 629.
38 G. Hatui, S. Sahoo, C. K. Das, A. K. Saxena, T. Basu and C. Y. Yue, *Mat. Design*, 2012, **42**, 184.
39 Z. J. Yang, W. Y. Zhang, J. D. Li and J. X. Chen, *Sep. Purif. Tech.*, 2012, **93**, 15.
40 N. Yevlampieva, D. Tur, A. Kovshik and E. Rjumtsev, *J. Inorg. Organomet. Polym. Mat*, 2012, **22**, 1156.

41 N. P. Yevlampieva, D. R. Tur, A. S. Gubarev and E. I. Ryumtsev, *Polym. Sci. A*, 2012, **54**, 349.
42 X. Zhang, Y. Zhong and Z. P. Mao, *Polym. Degrad. Stabil.*, 2012, **97**, 1504.
43 H. Wahl, D. A. Haynes and T. le Roex, *Crystal Growth Design*, 2012, **12**, 4031.
44 R. Bao, M. Pan, Y. Zhou, J. J. Qiu, H. Q. Tang and C. M. Liu, *Synth. Comm.*, 2012, **42**, 1661.
45 I. Un, H. Ibisoglu, A. Kilic, S. S. Un and F. Yuksel, *Inorg. Chimica Acta*, 2012, **387**, 226.
46 D. He, S. Y. Cho, D. W. Kim, C. Lee and Y. Kang, *Macromolecules*, 2012, **45**, 7931.
47 B. Cosut and S. Yesilot, *Polyhedron*, 2012, **35**, 101.
48 Y. Tada, N. Moriya, M. Kanazawa, K. Asanuma, A. Suzuki and S. Koyama, *Polym. Chem.*, 2012, **3**, 2815.
49 M. Bahadur and C. W. Allen, *Inorg. Chem.*, 2012, **51**, 5465.
50 P. Thilagar, P. Sudhakar, P. C. A. Swamy and S. Mukherjee, *Inorg. Chimica Acta*, 2012, **390**, 163.
51 D. Erdener and M. Yildiz, *Asian J. Chem.*, 2012, **24**, 1530.
52 F. Ozen, E. Cil and M. Arslan, *J. Chem. Soc. Pakistan*, 2012, **34**, 690.
53 X. Yang, R. Y. Zou, R. Li, J. L. Yang, Y. Ye and Y. F. Zhao, *Phos. Silicon Sulfur*, 2012, **187**, 722.
54 L. Kobr, K. Zhao, Y. Q. Shen, A. Comotti, S. Bracco, R. K. Shoemaker, P. Sozzani, N. A. Clark, J. C. Price, C. T. Rogers and J. Michl, *J. Am. Chem. Soc.*, 2012, **134**, 10122.
55 A. Cuetos, A. Rioz-Martinez, M. L. Valenzuela, I. Lavandera, G. de Gonzalo, G. A. Carriedo and V. Gotor, *J. Molecul. Cat. B Enzymatic*, 2012, **74**, 178.
56 Y. Pareek and M. Ravikanth, *Chem. Eur. J.*, 2012, **18**, 8835.
57 G. Elmas, A. Okumus, Z. Kilic, T. Hokelek, L. Acik, H. Dal, N. Ramazanoglu and L. Y. Koc, *Inorg. Chem.*, 2012, **51**, 12841.
58 E. T. Ecik, S. Besli, G. Y. Ciftci, D. B. Davies, A. Kilic and F. Yuksel, *Daltron Trans.*, 2012, **41**, 6715.
59 S. Besli, S. J. Coles, L. S. Coles, D. B. Davies and A. Kilic, *Polyhedron*, 2012, **43**, 176.
60 N. Asmafiliz, Z. Kilic, Z. Hayvali, L. Acik, T. Hokelek, H. Dal and Y. Oner, *Spectrochim. A Part A Macromol. Biomolec. Spect.*, 2012, **86**, 214.
61 T. Yildirim, K. Bilgin, G. Y. Ciftci, E. T. Ecik, E. Senkuytu, Y. Uludag, L. Tomak and A. Kilic, *Eur. J. Med. Chem.*, 2012, **52**, 213.
62 S. Besli, F. Yuksel, D. B. Davies and A. Kilic, *Inorg. Chem.*, 2012, **51**, 6434.
63 Z. B. Jian, A. R. Petrov, N. K. Hangaly, S. H. Li, W. F. Rong, Z. H. Mou, K. A. Rufanov, K. Harms, J. Sundermeyer and D. M. Cui, *Organometallics*, 2012, **31**, 4267.
64 P. K. Khatri and S. L. Jain, *Cat. Lett.*, 2012, **142**, 1020.
65 C. Diaz, M. L. Valenzuela, V. Lavayen and C. O'Dwyer, *Inorg. Chem.*, 2012, **51**, 6228.
66 D. C. Kraiter, P. Wisian-Neilson, C. P. Zhang and A. L. Crumbliss, *Macromolecules*, 2012, **45**, 3658.
67 A. G. Xiao, Z. J. Li, S. B. Zhou, Q. Y. Zheng, Y. M. Shen, Z. G. Chen, W. Q. Zheng and A. P. Hao, *Polym. Plastics Tech. Eng.*, 2012, **51**, 521.
68 R. J. Davidson, E. W. Ainscough, A. M. Brodie, G. B. Jameson, M. R. Waterland, H. R. Allcock, M. D. Hindenlang, B. Moubaraki, K. S. Murray, K. C. Gordon, R. Horvath and G. N. L. Jameson, *Inorg. Chem.*, 2012, **51**, 8307.
69 X. W. Wu, X. Y. Wang, Q. L. Li, J. P. Ma and Y. B. Dong, *J. Coord. Chem.*, 2012, **65**, 3299.

70 R. Senkuttuvan, V. Ramakrishna, K. Bakthavachalam and N. D. Reddy, *J. Organomet. Chem.*, 2013, **723**, 72.
71 V. Chandrasekhar, T. Senapati, A. Dey, S. Das, M. Kalisz and R. Clerac, *Inorg. Chem.*, 2012, **51**, 2031.
72 E. W. Ainscough, A. M. Brodie, P. J. B. Edwards, G. B. Jameson, C. A. Otter and S. Kirk, *Inorg.Chem.*, 2012, **51**, 10884.
73 X. J. Li, F. L. Jiang, L. Chen, M. Y. Wu, Q. H. Chen, Y. Bu and M. C. Hong, *Dalton Trans.*, 2012, **41**, 14038.
74 V. I. de Paula, C. A. Sato and R. Buffon, *J. Braz. Chem. Soc.*, 2012, **23**, 258.
75 C. Diaz, M. L. Valenzuela, D. Carrillo, J. Riquelme and R. Diaz, *J. Inorg. Organomet Polym Mat.*, 2012, **22**, 1101.
76 J. W. Fu, M. H. Wang, C. Zhang, Q. Xu, X. B. Huang and X. Z. Tang, *J. Mat. Sci.*, 2012, **47**, 1985.
77 K. Y. Chen, X. B. Huang, X. Z. Tang and L. Zhu, *J. Macromol. Sci. Part B Phys.*, 2012, **51**, 269.
78 J. W. Zhang, X. B. Huang, H. Wei, J. W. Fu, Y. W. Huang and X. Z. Tang, *J. Solid State Electrochem.*, 2012, **16**, 101.
79 J. P. Qian, W. Wei, X. B. Huang, Y. M. Tao, K. Y. Chen and X. Z. Tang, *J. Power Sources*, 2012, **210**, 345.
80 W. Liu, Y. L. Zheng, J. Li, L. Liu, X. B. Huang, J. W. Zhang, X. Q. Kang and X. Z. Tang, *Polym. Adv. Tech.*, 2012, **23**, 1.
81 J. W. Fu, M. H. Wang, C. Zhang, P. Zhang and Q. Xu, *Mat. Lett.*, 2012, **81**, 215.
82 W. Wei, X. B. Huang, K. Y. Chen, Y. M. Tao and X. Z. Tang, *RSC Adv.*, 2012, **2**, 3765.
83 T. J. Pan, X. B. Huang, H. Wei, W. Wei and X. Z. Tang, *Macromol. Chem. Phys.*, 2012, **213**, 1590.
84 C. D. Valenzuela, G. A. Carriedo, M. L. Valenzuela, L. Zuniga and C. O'Dwyer, *J. Inorg. Organomet. Polym. Mat.*, 2012, **22**, 447.
85 Y. Wang, J. X. Mu, L. Li, L. L. Shi, W. H. Zhang and Z. H. Jiang, *High Perf. Polym.*, 2012, **24**, 229.
86 X. L. Li, Z. G. Li, Y. Q. Jing, B. C. Bing and B. Li, *J. Colloid Interface Sci.*, 2012, **375**, 41.
87 X. L. Li, B. Li, Z. G. Li and S. Y. Zhang, *RSC Adv.*, 2012, **2**, 5997.
88 Y. J. Lin, Q. H. Deng and R. G. Jin, *J. Wuhan Univ. Tech. Mat. Sci. Ed.*, 2012, **27**, 207.
89 X. Liu, H. Zhang, Z. C. Tian, A. Sen and H. R. Allcock, *Polym. Chem.*, 2012, **3**, 2082.
90 T. K. Huang, Y. C. Wang, K. H. Hsieh and J. J. Lin, *Polymer*, 2012, **53**, 4060.
91 A. Akbari, H. Ghatezadeh, B. Golzadeh and S. Arshadi, *E-J. Chem.*, 2012, **9**, 2097.
92 J. L. Kroger and J. R. Fried, *J. Inorg Organomet. Polym. Mat.*, 2012, **22**, 973.
93 V. L. Furer, A. E. Vandyukov, S. Fuchs, J. P. Majoral, A. M. Caminade and V. I. Kovalenko, *Spectrochim Acta Part A Molec. Biomolec. Struct.*, 2012, **91**, 97.
94 V. L. Furer, A. E. Vandyukov, S. Fuchs, J. P. Majoral, A. M. Caminade and V. I. Kovalenko, *J. Molec. Struct.*, 2012, **1026**, 17.
95 M. Gall and M. Breza, *Polyhedron*, 2012, **31**, 570.

Physical methods

Robert N. Slinn

DOI: 10.1039/9781782623977-00413

1 Introduction

This year's coverage of the literature is in a similar format to that in Volume 42. Because of the large number of publications, it has once again been necessary to be very selective in certain areas (especially in organometallics), choice of publication, and class of compound covered. The physical methods used for examining *nucleotides and nucleic acids*- NMR spectroscopy, X-ray crystallography (XRD), electron microscopy, atomic force microscopy (AFM) and surface plasmon resonance (SPR)- are covered within that particular chapter. For each class of compound, the relevant physical methods employed are described in the sections specified. Section 2 continues again with theoretical and computational chemistry methods, whereas studies relating to specific physical methods are covered in their appropriate sections. In those cases where more than one physical method or analytical technique is used, particularly in the characterization of a compound, then the principal technique reported is normally referenced first, followed then by the other methods. As before, the compounds discussed in each subsection are covered in the order of increasing coordination number of phosphorus, where this is appropriate. Within their formulae, the letter R normally represents hydrogen, alkyl or aryl, while X represents an electronegative substituent, Ch represents a chalcogenide (oxygen, sulphur, and selenium), and Y and Z are used for groups of a more varied nature.

2 Theoretical and computational chemistry methods

This section again covers *ab initio*, density functional theory (DFT), semi-empirical and empirical calculations, molecular mechanics and molecular dynamics methods. Other interesting theoretical and computational chemistry techniques reported include quantitative structure-retention relationships (QSRR), and quantitative structure-property relationships (QSPR) for the prediction of various physicochemical properties. Again, all areas of theoretical and computational chemistry have continued to expand rapidly. These methods have been used to predict, support and also validate the majority of the observed experimental data.

The rotational barriers of the P–N bonds in three model *P*-substituted aminophosphanes ($R^1R^2PNX_2$), $R^1=R^2=Ph$ or Cl, and $R^1=Ph$, $R^2=Cl$, have been studied using DFT calculations with B3LYP functional and 6-311+G* and DZP2 basis sets to establish the electronic effect.[1]

Visiting Researcher, Department of Chemistry, University of Liverpool, Crown Street, Liverpool, L69 7ZD. E-mail: r_n_slinn@yahoo.co.uk

The calculations were performed allowing for the solvent (toluene) effect using the *Polarizable Continuum Model* (PCM) to simulate the experimental conditions. The anomeric effect was established as an electronic effect that drives the P–N rotational barriers and characterized by the highest chlorine substitution on the phosphorus atom using *Natural Bonding Orbital* (NBO) analysis. The bond energy was calculated using a modified version of the bond energy change. It was shown that the *N* lone pair has a *gauche* conformation with respect to the lone pair of the *P* atom, via interaction of the *N* lone pair with the *P–Cl* σ^* orbital, or, in the extreme case, it could form a *N=P* double bond and, as a consequence, the rotational energy increases. This was confirmed by second-order energy analysis where the main interactions are the *N* lone pair with the *P–C* σ^* orbital. There are other interactions that could establish the characteristics of the *N–P* bond. The stabilization criterion was established by the normalized value of the bond energy. The values show that the *N=P* double bond is the most stable and that there are different energy values for the *N–P* single bond. Also, the energy of the molecular orbitals could 'intercross' during the rotational barriers, changing the stability and reactivity of the molecules, which could be taken as another criterion of conformation stabilization. The molecular structures of 1,2-diphosphinoethane conformers have been studied in the gas phase.[2] *Ab initio* calculations at HF/6-31G* and MP2/6-311G* basis sets were performed and twelve conformers were located on the potential energy surface. Their geometries, total electronic energy, HOMO, LUMO, HOMO-LUMO gaps, their 1st, 2nd, and 3rd ionization potentials, and harmonic frequency were all investigated.

The static dipole (*hyper*)polarizabilities (α, β, and γ) of *p*-nitrophenylphosphine (PNPhP), a hypothetical *Push-Pull* model, have been studied by *ab initio* (HF, MP2, MP4) and DFT (B3LYP, CAM-B3LYP, WB97X-D) calculations with 6-31+G(d,p) and 6-311++G(3d,3p) basis sets and similar calculations were carried out on the *p*-nitroaniline (PNA) molecule for comparison.[3] These properties were evaluated within the *Finite Field* methodology based in the *Kurtz* equations, where the effects of replacing the -NH$_2$ donor group in PNA by the -PH$_2$ group in PNPhP were examined. The results of α, β and γ properties at different levels show that PNPhP is much more (hyper)polarizable than PNA. Electron correlation effects, at MP2, MP4, DFT/B3LYP, CAM-B3LYP and WB97X-D levels, show that α_{ave} increases between 7 to 16% with respect to the HF values. Likewise, the largest responses are found with the most extended basis sets, such as 6-311++G(3d,3p) where the polarizability increases between 4 to 9% with respect to the 6-31+G(d,p) set. The tendencies for β and γ nonlinear optical (NLO) values follow the electron density delocalization of the molecular system, which is more extended in PNPhP than in PNA. Electron correlation effects increase β and γ properties in the order of two or three times with respect to HF values. The impact of dynamic and solvent effects on the first hyperpolarizability of the PNA molecule were also taken into account. The results reveal that PNPhP is an interesting hypothetical molecule which can be useful as a building block for the design and creation of new materials with enhanced NLO applications.

The spatial structures of molecular clusters for modelling a solvate shell around P-containing methyl- and butyl-derivatives of phosphine and betaine molecules dissolved in different solvents (acetone, toluene, and formamide) have been calculated using different variants of DFT: UB3LYP, PBE, and OLYP, with 6-31G(d,p) and 6-31G++(d,p) basis sets.[4] The ^{31}P nuclear magnetic shielding constants (σ) for the structures were calculated using *gauge-including atomic orbitals* (GIAO) in the UB3LYP/6-31G(d,p) and 6-31G++(d,p) methods. Modelling of the molecular clusters was performed using the supermolecular model, the molecular mechanics method, and the combination of quantum mechanics and molecular mechanics (QM/MM) methods. The *own N-layered integrated molecular orbital method* (ONIOM) was applied to modelling and calculating the isotropic ^{31}P nucleus magnetic shielding of clusters of trimethylphosphine and trimethylbetaine molecules in acetone using combinations of UB3LYP/6-31G(d,p) (higher level) and (UHF)/6-31G(d,p) (lower level) methods, and also by using the semi-empirical AM1 method. The applicability of the ONIOM approach and different ways of modelling to the calculation of the ^{31}P nuclear magnetic shielding constants were studied and the results obtained by the DFT, ONIOM and MM methods were compared. Regarding ONIOM, the results showed that, with appropriate partitioning, this approach furnishes shielding which represents close approximations to the corresponding UB3LYP-GIAO values for the entire molecule. Such a method may be a highly efficient technique for accurate shielding calculations in biological molecules. In contrast to large biological molecules, application of the ONIOM method to solutions does not require breaking of chemical bonds since, within the ONIOM approach, the resulting 'free valences' are saturated through the addition of terminal hydrogen atoms. Computational studies on phosphine (and diphosphine) ligands include, respectively, a DFT analysis of phosphine ligand dissociation versus hemilability in a Grubbs-type precatalyst containing a bidentate ligand during alkene metathesis,[5] and a QM examination of the donor-acceptor capacities of two diphosphine ligands. The ligands were 1,2-bis(diphenylphosphino)ethane (dppe) and *cis*-1,2-*bis*(diphenylphosphino)ethene (cis-dppen),[6] in [Fe(NH$_3$)$_4${bpy/dppe/cis-dppen}]$^{2+}$ and [Fe(bpy)$_2${(NH$_3$)$_2$/dppe/cis-dppen}]$^{2+}$ where bpy is 2,2′-bipyridine (the π-acceptor molecule).

(1)   (2)

The use of dynamic ^{1}H NMR spectroscopy along with *ab initio* and DFT theoretical methods to investigate restricted rotation in the phosphorus ylides (**1**) and (**2**) has been reported. In the case of the ylide (**1**),[7] restricted rotation around the C=C bond was examined using dynamic ^{1}H NMR at variable temperatures, and theoretical calculations at both HF/6-31G(d,p) and DFT/B3LYP/6-31G(d,p) levels of theory. The results showed that the activation parameters for the interchangeable process between the two (*E*) and (*Z*) isomers ($\Delta G^{\neq}$ = 69.29 and 67.42 kJ mol^{-1}), respectively, were in good agreement with the experimental dynamic ^{1}H NMR data ($\Delta G^{\neq}$ = 68.71 kJ mol^{-1}). In the case of ylide (2),[8] restricted rotation around the C–C bond and partial C=C bond was also examined by dynamic ^{1}H NMR, at variable temperatures, along with calculations at both the HF/6-31G(d,p) and DFT/B3LYP/6-31G(d,p) levels of theory for the two (*E*) and (*Z*) isomers. Activation and kinetic parameters, including $\Delta H^{\neq}$, $\Delta G^{\neq}$, $\Delta S^{\neq}$, and E_a, were obtained in accord with the dynamic ^{1}H NMR data for three rotational processes, and theoretical studies based upon rotation around the same bonds investigated using the *ab initio* and DFT methods. The theoretical activation and kinetic parameters, $\Delta H^{\neq}$, $\Delta G^{\neq}$, $\Delta S^{\neq}$, and E_a, were calculated at 298 K and experimental temperatures for five rotational processes. These results, experimental and theoretical, taken together, showed that the rotational energy barrier around the C=C bond was fairly high, but observation of the two rotational isomers was impossible at high temperatures (and appeared as a single isomer). In this case, rotation around the C=C bond was too fast on the NMR time scale. When the temperature was relatively low, the rate of rotation was sufficiently slow and, thus, observation of the two *Z*- and *E*-isomers was then possible. Calculations at the HF/6-31G(d,p) level of theory gave results in agreement with the experimental data on rotation around the C=C bond, whereas the DFT/B3LYP/6-31G(d,p) results gave reasonable data for restricted rotations around the C–C and C–N bonds. *Ab initio* and DFT, with

NMR, IR and XRD data, have been used for studying the conformations of monoylidic diester triphenylphosphonium ylides.[9] In the ylidic triphenylphosphonium carboxylic esters, the ester oxygen can be oriented towards (*syn*) or away (*anti*) from phosphorus, and, apart from small ylidic ester groups such as methyl, the *anti* conformer is dominant. For suitable crystals, the conformations can be established by XRD but, in general, HF and DFT techniques, along with NMR and IR spectroscopy, are useful methods. Bulky ylidic or non-ylidic groups strongly favour the *anti* conformer and, even with small carboxylic groups such as ethoxy, the *anti* conformers are preferred in solution and dominant in the crystal. The balance of attractive interactions between anionoid oxygen and cationoid phosphorus and nonbonding interactions controls the conformations, as indicated by HF and DFT, and NMR, IR, and X-ray observations. A space-restricted wave function (SRW) theoretical method for the analysis of various types of intramolecular interactions and its applicability to the P–O bond in phosphine oxides (R_3PO; R=H, Me, and F), has been reported.[10] An interesting character of this bond has been studied extensively, focusing on the negative hyperconjugation of the O lone pair (n_O) with the R_3P group. The nature of the bond was re-investigated in terms of a change in total energy to produce evidence for the validity of this method. The electronic states without the interaction involving three n_O orbitals (R_3P^+-O^-) produced by the method were used as reference states in the assessment of the effects of the n_O-R_3P interaction. The result confirmed that this interaction plays an essential role in the nature of the bond and occurs between the n_O orbitals and the P–R antibonding orbitals, in agreement with previous studies. A molecular orbital (MO)-pair analysis technique showed that the n_O-R_3P interaction is comprised of negative hyperconjugation and Pauli repulsion. Considering a reference state where the P–O bond is completely broken ($R_3P^{2+}\cdots O^{2-}$) at an interacting distance, P–O bond formation is attributed to one σ bond plus two 0.5 π bonds, equivalent to three 'banana' bonds highly polarized to the O atom. In consequence, the SRW method improved explanations of the nature of the P–O bond.

(3)   (4)   (5)   (6)

The hydrogen-bond acceptor properties of a strain phosphine molecule (**3**), and its deprotonated derivatives (**4**), (**5**), and (**6**) have been examined at the MP2/6-311++G(d,p) level of theory.[11] In addition, complexation of these with metallic atoms, *i.e*, the anion (**4**) with Li^+, dianion (**5**) with Be^{2+}, and the trianion (**6**) with B^{3+} (in order to obtain neutral complexes) has been considered, and the effects of such complexation in their hydrogen-bonding characteristics were analyzed. The results obtained for the complexes were compared with the analogous ones of trimethylphosphine oxide. It was shown that the strain decreases the complexation energy with metallic atoms as well as the thermodynamic and hydrogen-bond acceptor ability of the tetrahedric derivatives. Conformational analysis of *bis*(2-phenylalkyl)phosphine selenides has been performed using the dipole moment method and different quantum-chemical calculation methods,[12] and it was found that *bis*(2-phenylpropyl)phosphine selenide exists as a mixture of several conformers, the most energetically-favourable being characterized by gauche (non-eclipsed) orientation of the P=Se and C_{sp3}–C_{sp3} bonds. Similarly, the conformational analyses of 1,3,2-dioxaphospholan-2-yl 2,2,2-trifluoroacetate (**7**) and 4,5-benzo-1,3,2-dioxaphosphol-2-yl 2,2,2-trifluoroacetate (**8**) were carried out using the methods of dipole moments and (here) DFT/B3LYP/6-31G* quantum-chemical calculations.[13] Both compounds (**7**) and (**8**) were found to exist as an *axial* conformer with preferably syn arrangement of the carbonyl group and the lone electron pair.

The first full study on the structure-property relationships of 2,2′-bis-(benzo[b]phosphole), (**9**), and 2,2′-benzo[b]phosphole-benzo[b]heterole hybrid π-systems (**10**), (**11**), (**12**), (**13**), and (**14**) has been reported.[14] The structure-property relationships of all the benzo[b]phosphole derivatives were investigated by using X-ray crystallography, UV-visible absorption and fluorescence spectroscopy, cyclic voltammetry, fluorescence lifetime measurements, and DFT/B3LYP/6-31G* calculations. The steady-state UV-visible absorption and fluorescence spectroscopy, fluorescence lifetime measurements, and DFT theoretical calculations of the non-fused and acetylene-fused benzo[b]phosphole-benzo[b]heterole π-systems showed that their emissive excited states consist of two different conformers in rapid equilibrium. A novel *bis*(azidophenyl)phosphole sulfide (**15**) was developed to yield phosphole-containing π-conjugated systems by a simple route using 'click chemistry'.[15] This was explored for the reaction of the two azido moieties with phenyl-, pyridyl-, and thienyl-acetylenes, to give *bis*(aryltriazolyl)-extended π-systems with the phosphole sulfide (**16a**), or phosphole (**16b**) group as the central ring. These conjugated frameworks exhibited intriguing photophysical and electrochemical properties that vary with the nature of the aromatic end-group

(**16a**) X = S (**16b**) X = lonepair

(Ar = Ph, 2-pyridyl, or 2-thienyl)

TDDFT calculations on the extended π-systems showed some variation in the shape of the HOMOs, which was found to have an effect on the extent of charge transfer, depending on the aromatic end-group. Some fine-tuning of the emission maxima was observed, showing a decrease in conjugation in the order: thienyl < phenyl < pyridyl. These results show that variations in the distal ends of such π-systems have a subtle but significant effect on photophysical properties.

The stereochemistry of unsaturated phosphonic acids dichlorides (**17**), (**18**), (**19**), and (**20**) has been studied by ^{31}P NMR spectroscopy for

^{31}P chemical shift values (δ_{31P}) with GIAO-B3LYP/DZP theoretical calculations.[16] Thus, (**17**) exists in solution as a mixture of *E*- and *Z*-isomers with the prevailing *s-cis*-orientation of the pyrazole ring and the C=C bond, and (**18**), (**19**), and (**20**) exist as individual *Z*-isomers in the form of a mixture of three conformers: *gauche*, *s-cis*, and *s-trans*, with predominance of the *gauche*-form; (**20**) is also characterized by the prevailing *s-cis*-orientation of the N=PCl$_3$ fragment with respect to the C=N bond. DFT has been used to investigate the Mo-catalyzed intramolecular *Pauson-Khand* reaction of 3-allyloxy-1-propynylphosphonates, with all intermediates and transition states optimized at the B3LYP/6-31G(d,p) level (and the LANL2DZ(f) basis set for Mo),[17] and, along with low-temperature ^{31}P NMR spectroscopy, DFT has also been used in studying the mechanism of the *Stec* reaction between phosphoroselenanilidate or phosphonoanilidate and CS$_2$.[18] DFT computations, performed at the B3LYP level with the 6-311++G (d,P) standard basis set, have been used to characterize the ^{17}O and ^{2}H electric field gradient (EFG) in various bisphosphonate derivatives.[19] The calculated EFG tensors were used to determine the ^{17}O and ^{2}H nuclear quadrupole coupling constant (χ) and asymmetry parameter (η). For a better understanding of the bonding and electronic structure of bisphosphonates, the isotropic and anisotropic NMR chemical shieldings were calculated for the ^{13}C, ^{17}O and ^{31}P nuclei using the GIAO method for the optimized structure of intermediate bisphosphonates at B3LYP level of theory using 6-311++G (d, p) basis set. The results showed that various substituents have a strong effect on the nuclear quadrupole resonance (NQR) parameters (χ, η) of ^{17}O in contrast with the ^{2}H NQR parameters. The NMR and NQR parameters were studied in order to find the correlation between the electronic structure and activity of the bisphosphonates. Also, the effect of substitutions on the bisphosphonates polarity was investigated and the molecular polarity determined via the DFT-calculated dipole moment vectors. The results showed that substitution of Br on the ring would increase the activity of the bisphosphonates. The interaction, and binding enthalpies of chemical nerve agent 'simulants' to a silica surface have been examined using DFT/B3LYP calculations, and compared to experimental IR data for the compounds dimethyl methylphosphonate (DMMP), di*iso*propyl methylphosphonate (DIMP), di*iso*propyl fluorophosphate (DFP), and *Sarin*.[20] Comparison with the experimental IR shifts indicated that the theoretically-modelled adsorption sites are similar to those found by experiment. The calculated binding enthalpies agreed fairly well with experimental data for *Sarin*, $\Delta H_{ads,443K} = -22.0$ (-18.8), and DIMP, $\Delta H_{ads,463K} = -26.9$ (-29.3), but not as well for DMMP, $\Delta H_{ads,463K} = -19.7$ (-26.1), and DFP, $\Delta H_{ads,423K} = -20.4$ calc. (-27.5 expt.) kcal/mol. The mechanism of phosphodiester hydrolysis in the DNA analogue *bis*(4-nitrophenyl) phosphate (BNPP), catalysed by a mononuclear Zn(II) complex, was also studied using DFT,[21] and the multidimensional free energy surface for phosphoester hydrolysis in the monomethyl phosphate dianion by using hybrid quantum mechanics and MM simulations.[22]

A DFT/B3LYP and MP2/6-311G** study of the possible conformations in the most stable form of *Cyclophosphamide* was performed, and its IR and Raman spectra assigned using DFT.[23] The *axial* structure was calculated to be 5–6 kcal/mol lower in energy than the *equatorial* form due to a very weak anomeric effect and weak conjugation. The FT-IR and Raman spectra of solid hexachlorocyclotriphosphazene have been recorded and conformational energies calculated using MP2 and DFT (B3LYP and B3PW91) methods with a variety of basis sets up to 6-311 + G(d).[24] On the basis of D_{3h} symmetry, the simulated vibrational spectra from the MP2 and DFT methods were in excellent agreement with those obtained experimentally. Also, frontier molecular orbitals and electronic transitions were predicted by steady state and time-dependent DFT(B3LYP)/PCM calculations, respectively, each using the 6-311 + G(d,p) optimized structural parameters. The predicted wavelengths were in excellent agreement with experimental values with CH_2Cl_2 as solvent. The ^{14}N and ^{31}P NMR chemical shifts were predicted by B3LYP/6-311 + G(2d,p) calculations using the GIAO technique with the solvent effect modelled using the PCM method. The computed structural parameters of the planar hexachlorocyclotriphosphazene (D_{3h}) agreed with experimental values from both the XRD and electron diffraction data, with slight distortions seen due to lattice defects in the solid phase. The experimental/computational results favoured a slightly distorted D_{3h} symmetry in the gas, solution and solid phases. A computational study of new *geminal bis*(supermesityl)-substituted phosphorus compounds has been carried out and the novel compounds fully characterized by multinuclear NMR, IR and Raman spectroscopy, mass spectrometry, and single-crystal XRD.[25] DFT calculations at both the B3LYP and M06-2X level showed that the strain energies of the *geminally*-substituted compounds are extremely high (180 to 250 kJ/mol), and most of the strain is stored as *boat* distortions to the phenyl rings of the supermesityl (Mes*) substituents.

A DFT study has been carried out on $P^{III}=C-P^V$ type diphosphaproprenes that are coordinated to transition metals, as in (**21a–c**).[26] DFT calculations at the BP86/6-311G*(d,p) level allowed energetic and geometric analysis of possible coordination modes of the diphosphaproprenes to $W(CO)_5$, $PtCl_2(CO)$, and $PdCl_2(CO)$ to be evaluated, revealing a general preference for coordination to tungsten through the double-bonded phosphorus atom as in (**21a**), whereas coordination to Pt occurs through either the *P* atom (**21a**) or the *Ch* atom (Ch = O, S) in (**21b**), without any apparent energetic preference, and the least favourable coordination is through the P=C double bond. For the Pd complexes, no preference for either coordination mode was seen when varying the diphosphapropene or substituents on the *P* atom. An exception was observed for R=H or Ph and Ch=O, for which the coordination is through the P=C bond, (**21c**). Also, the first 2-aminophosphasilene (**22**), having a hydride ligand on the three-coordinate *Si* atom, has been synthesized and structurally characterized.[27] X-ray single-crystal analysis of (**22**) and DFT/B3LYP/6-311+G(d) level calculations on the model compound (**23**) indicate that the amino group on the *Si* atom is involved in significant N-Si-P π-conjugation. The various possible molecular conformations of sodium dimyristoylphosphatidylglycerol (DMPG) have been examined using a DFT-based method,[28] and high-level post-Hartree-Fock (HF) methods have been used to investigate the possibility of α-effects occurring during methyl transfers in esters such as in dimethyl *P*-methylphosphonate (DMMP).[29]

(**22**)   (**23**)

(Ar = 2,6-iPr$_2$C$_6$H$_3$, Ar' = 2,6-Mes$_2$C$_6$H$_3$)

Several theoretical studies have been performed on chemical nerve agents and simulants. DFT/B3LYP/6-31+G(d,p) studies have been carried out to explore mechanisms for the atmospheric degradation of *iso*propyl methyl methylphosphonate (IMMP),[30] also for the mechanism of aminolysis of *O, S*-dimethyl methylphosphonothiolate by both DFT (using MO62X method) and by *ab initio* MP2 with 6-311+G(d,p) basis sets.[31] The micro-hydration of the agent *Sarin* ((*RS*)-*O*-*iso*propyl methylphosphonofluoridate) has been investigated using DFT/B3LYP and *ab initio* MP2 calculations,[32] and a molecular dynamics simulation study has been performed on the sorption and diffusion of *Sarin* in hydrated polyelectrolyte membranes (PEMs).[33] The behaviour was found to be similar to that of its common simulant, DMMP. Finally, a DFT and *ab initio* theoretical and experimental study has been carried out on the role of the P–F bond in the fluoride-promoted aqueous hydrolysis of the agent *VX* (*O*-ethyl *S*-2-(di*iso*propylamino)ethyl methylphosphonothioate).[34]

A major use of theoretical and computational methods is the prediction or validation of observed experimental data. New applications involving quantitative structure-property relationships (QSPR) and quantitative structure-retention relationships (QSRR) for the prediction of various physicochemical properties have also been reported. A series of alkyltributylphosphonium chloride ionic liquids has been synthesized and their experimental density and viscosity data interpreted using QSPRs and group contribution methods.[35] A QSRR study has also been performed on a series of new phosphoramidic acid derivatives.[36] Their retention factors (k'') were predicted using a model obtained from a set of previously-synthesized phosphoramidic acid derivatives.

The experimental retention factors were determined by reversed-phase high-performance liquid chromatography HPLC). Stepwise multiple linear regression (MLR) and partial least square (PLS) methodology were used to investigate the correlation between the retention factor and a number of molecular descriptors for these compounds. The MLR and PLS equations were found to be useful for the estimation and comparison of retention factors (k'') for new synthesized phosphoramidic acid derivatives using the selected molecular descriptors. More details are given in the section on HPLC.

3 Nuclear magnetic resonance spectroscopy

3.1 Analytical applications

Multinuclear (^{1}H, ^{13}C, and ^{31}P) NMR spectroscopy, along with IR, UV-visible spectroscopy, mass spectrometry, X-ray crystallography, chromatographic purity, and elemental analysis, completes the suite of methods used for the full characterization of novel organic compounds.

3.2 Applications including chemical shifts, shielding effects and spin–spin coupling

^{31}P NMR is normally accompanied by ^{1}H, ^{13}C, and X NMR spectroscopy; thus applications mentioned may cross-refer to multinuclear NMR unless specifically stated. Characterizations include ^{13}C and ^{1}H NMR analysis, but studies relating specifically to these nuclei have been somewhat limited during 2012; on account of this, they are grouped in this subsection. For ^{31}P NMR, positive chemical shifts (δ_{31P}) are expressed downfield of the external reference of 85% phosphoric acid and are normally given without the appellation (ppm) unless stated.

Spectroscopic and other experimental data are now often predicted and/or validated by using theoretical and computational chemistry methods. For NMR spectroscopy in particular, those methods described earlier are cross-referenced.[4,7–9,16,19,25] Thus, the ^{31}P nuclear magnetic shielding constants (σ) of derivatives of betaine and phosphine molecules were calculated in different solvents using combined QM/MM methods,[4] and dynamic variable-temperature ^{1}H NMR spectroscopy, along with *ab initio* and DFT methods, was used to investigate restricted rotation in the phosphorus ylides (**1**),[7] and (**2**).[8] *Ab initio* and DFT theoretical methods, along with NMR, IR and XRD experimental data, were used to establish the conformations of the monoylidic diester triphenylphosphonium ylides.[9] Also the stereochemistry of the unsaturated phosphonic acids dichlorides (**17**), (**18**), (**19**), and (**20**) was determined using experimental ^{31}P NMR measurements and GIAO-B3LYP/DZP-calculated chemical shift values, (δ_{31P}).[16] A GIAO-B3LYP/6-311++G(d,P) study and the NMR and NQR parameters of bisphosphonate derivatives were also reported.[19] Novel *geminal bis*(supermesityl)-substituted phosphorus compounds were fully characterized by NMR, IR and Raman spectroscopy, MS, and XRD, and DFT calculations carried out at B3LYP and M06-2X levels to examine strain in the molecules.[25]

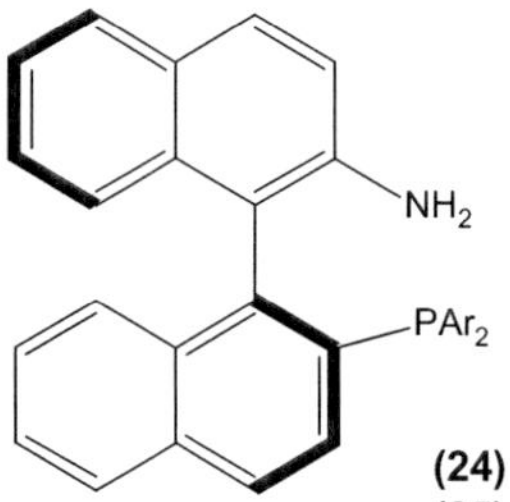

(24) Ar = Ph
(25) Ar = 3,5-F(C_6H_3)
(26) Ar = 4-F(C_6H_4)
(27) Ar = 4-MeO(C_6H_4)

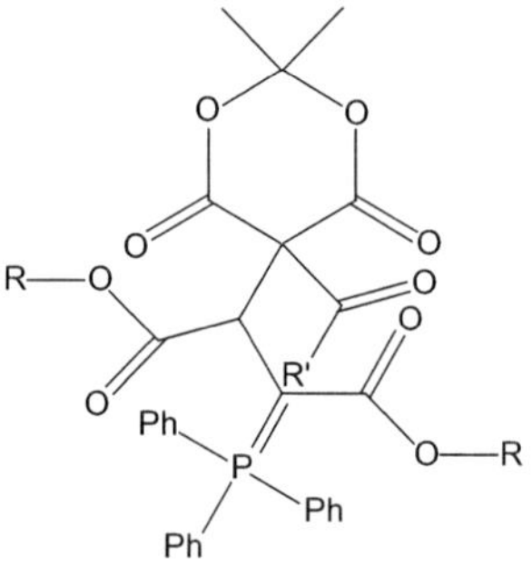

(24)–(27)

(28)

New BINAP-based aryl-substituted aminophosphines (24), (25), (26), and (27) have been synthesized {where BINAP = [2,2′-bis(diphenylphosphino)-1,1′-binaphthyl]}, and their ^{31}P NMR chemical shifts (δ_{31P} = −11.3, −7.3, −13.2, and −14.2 ppm, respectively) obtained experimentally.[37] The values of δ_{31P} varied, due primarily to the resonance effects for *para*-substitution, whereas only the field (induction) effect is important in *meta*-substitution. A computational analysis was performed to linearly quantitate contributions to the shifts from both the resonance and field effects of the substituents. This correlation may be useful for designing and preparing other related aminophosphines with varying ligand properties. A series of *tris*-aryl phosphanes, designed as 'residual' enantiomers or diastereoisomers with substituents differing in size and electronic properties on the aryl rings, have been characterized and the configurational stability of the residual phosphanes evaluated by dynamic HPLC on a chiral stationary phase and/or by using dynamic ^{1}H and ^{31}P NMR spectroscopy.[38] Luminescent Au^I-Cu^I triphosphane clusters, containing extended linear aryl-acetylenes, have been characterized using X-ray crystallography and NMR spectroscopy,[39] and new *N,N-bis*(diphenylphosphino)naphthylamine chalcogenides similarly characterized.[40] New stable ylide derivatives of dimethyl (5-acetyl-2,2-dimethyl-4,6-dioxo-1,3-dioxane-5-yl)-3-(triphenyl-λ^5-phosphanylidene) succinate (28) have been synthesized, and a dynamic NMR study of the rotamers carried out and compared with previous reports.[41]

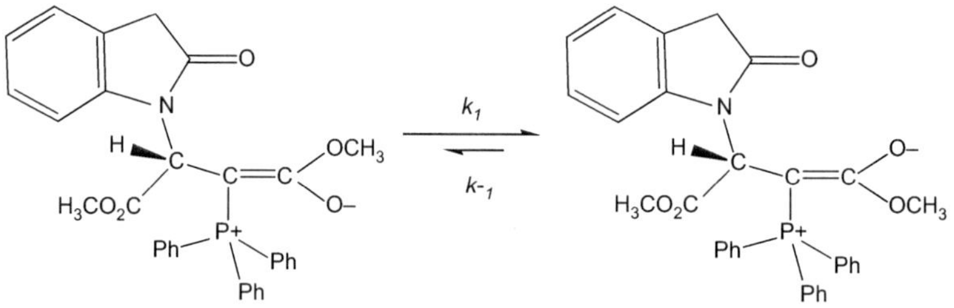

E-(29)   Z-(29)

For the first time, a kinetic investigation has been carried out on the interchange process for the equilibrium between the Z- and E-isomers of a stable phosphorus ylide involving a 2-indolinone (29), by using

^{1}H NMR spectroscopy.[42] The results show that the interchange process of the rotational isomers follows first-order kinetics with respect to the forward (k_1) and reverse (k_{-1}) reactions. Activation parameters (k_1, k_{-1}, k_{total}, E_{a1}, E_{a-1}, $E_{a(total)}$, $\Delta S^{\neq}$, $\Delta H^{\neq}$ and $\Delta G^{\neq}$), and thermodynamic parameters (K_e, ΔH^0, ΔG^0 and ΔS^0) were also obtained experimentally. The synthesis and full characterization of the previously unknown, highly-labile, amino-(azido)phosphenium salt [(Me$_3$Si)$_2$N=P–N$_3$][GaCl$_4$] (30), has been reported.[43]

The low temperature reactions of the precursor (Me$_3$Si)$_2$N–PCl$_2$ (31) with GaCl$_3$ followed by Me$_3$SiN$_3$ were studied using low-temperature ^{31}P NMR spectroscopy for (30), (δ_{31P}=367 ppm), and for (31), (32), (33) and (34), and also by using low-temperature XRD. The [(Me$_3$Si)$_2$N=P–N$_3$]$^+$ ion in (30) can be regarded as the first-known phosphapentacenium ion. It has a planar molecular skeleton and is only stable at temperatures below −40 °C; above, it does not cyclize to a tetraazaphosphole but, instead, slowly decomposes by a *Staudinger* reaction (−N$_2$) to oligomeric PN compounds. Multinuclear (^{1}H, ^{13}C, ^{31}P) NMR spectroscopy has been used along with other techniques in the characterization of novel, anellated 4H-1,4,2-diazaphospholes,[44] *spiro*oxaphosphirane complexes,[45] and new, biologically-active compounds including 1,3,2-dioxaphosphinane derivatives (35),[46] antifungal compounds (36),[47] and α-hydroxyphosphonates prepared under solvent-free conditions, using a simple method,[48] and via a rapid, 'green' chemistry route.[49]

(35) R = O-aryl or NH-aryl; X = O or S

(36)

A series of 2-phenyl-4-aryl-1,3,2-dioxaphosphonates derived from xylofuranose derivatives have been synthesized from diacetone-D-glucose in order to carry out conformational and configurational studies.[50] The absolute stereochemistry at the P atom was assigned by the ^{1}H NMR chemical shifts (δ_{1H}) for each diastereoisomeric pair, and confirmed by XRD studies. In the case of two cyclic phosphonates, spontaneous conversion into two more stable diastereoisomeric phosphonates occurred either during column chromatography or in CDCl$_3$ solution. This was attributed to cleavage of the C–O bond, within the P-heterocyclic ring, which is induced by the anomeric effect ($n_O \rightarrow \pi^*_{P=O}$). The results confirm that the use of a xylofuranose carbohydrate template facilitates the conformational and configurational study of six-membered ring phosphonates, and the study provides further evidence that the anomeric effect plays a key role in both the conformational behaviour and isomerization process of these cyclic phosphonates. A study has been carried out on a series of aminobisphosphonates, H$_2$N(CH$_2$)$_n$C(OH)[P(O)(OH)$_2$]$_2$, (n = 2–5, 7–11 and 15), and their aqueous solubility, pH and pK$_a$-values, thermal stability, IR absorptions, and NMR spectra data obtained by both solution (^{1}H, ^{13}C, ^{31}P)- and solid–state (^{13}C, ^{15}N, ^{31}P–CP/MAS)–NMR.[51] The ^{1}H–NMR spectra were complicated due to the prochiral R–CH$_2$–C(OH)P$_2$ fragment but all other spectra were typical for aminobisphosphonates. A novel aryldithiofluorophosphonic acid and its PPh$_4^+$ salt (37)[52] have been characterized using IR, ^{1}H, ^{13}C, ^{19}F, and ^{31}P NMR spectroscopy, ESI-mass spectrometry, and single-crystal XRD, and similarly the new class of O,O-diphenyl-α-[(5-fluorouracil-1-yl)-acetic]amino-aryl phosphonates (38) characterized.[53]

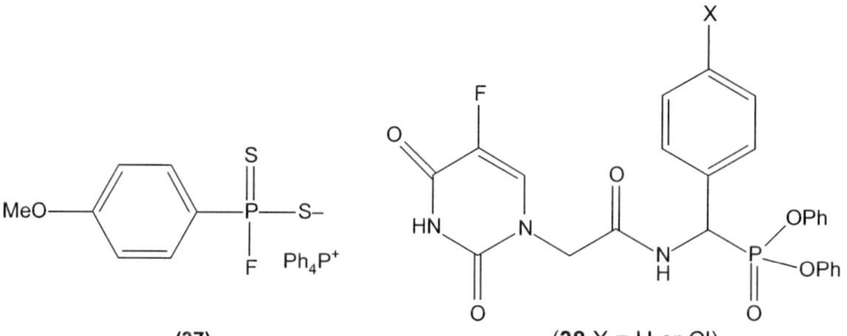

(37)

(38 X = H or Cl)

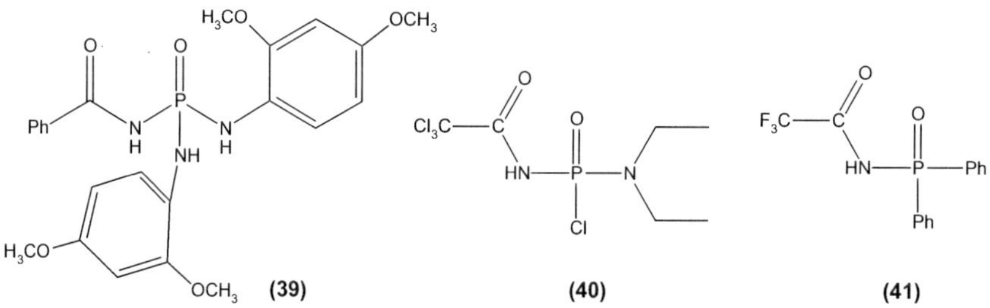

(39)

(40)

(41)

A bifunctional 2-aminoethyl dihydrogen phosphate (AEPH$_2$) has been characterized using ^{1}H and ^{31}P NMR titrations in pH range 1–12 to determine the zwitterionic properties at different pH regions in H$_2$O and D$_2$O, and also the pH range where AEPH$_2$ exists as a zwitterion.[54] It was shown that the phosphate group has two deprotonation points, around pH 1 and 6, while the amino group is deprotonated at pH 11. The zwitterionic form of AEPH$_2$ (NH$_3^+$-CH$_2$-CH$_2$-OPO$_3$H$^-$) exists as the main ion between pH 1 and 6 in water, and in the solid state. The new phosphoramidates (39), (40), and (41) have been characterized by IR, and ^{1}H, ^{13}C and ^{31}P NMR spectroscopy,[55] and have δ_{31P} values at -2.77, -18.07, and 23.32 ppm, respectively. The compounds contain the RC(O)XP(O) moiety (X = NH and O) and substitution of different groups on the P atom leads to various shielding effects. Comparing the δ_{31P} values, δ_{31P} in (40) appears upfield relative to δ_{31P} in (39), and results indicate that the halogen atoms are good electron donors to the P atom and cause shielding. Attaching two phenyl groups to the P atom gives a more downfield shift, with δ_{31P} at 23.32 in (41) rather than at -18.07 ppm in (40). Another series of six new phosphoramidates, (42), (43), (44), (45), (46), and (47), has been similarly characterized,[56] with δ_{31P} values at -1.13, 15.11, -23.00, -21.65, 18.07, and 1.39 ppm, respectively. It can be seen, among the compounds (42), (43), (46), and (47), that the P atom is the most shielded in (42) due to the attachment of aromatic groups. In (46) and (47) containing aliphatic amido moieties, the P atoms are the most deshielded. Interestingly, the P atoms of (44) and (45) are the most upfielded atoms, again indicating that the Br atom acts as a strong electron donor to the P atom via a resonance interaction. The ^{1}H and ^{13}C NMR spectra of (46) each reveal three separate sets of peaks for the aliphatic, non-equivalent CH$_2$ protons of the three five-membered rings, which is explained by different spatial orientations of the aliphatic rings relative to each other. In the ^{13}C NMR spectrum of (46), $^3J_{(P,C)} = 2.7$ Hz and $^2J_{(P,C)} = 5.2$ and 4.8 Hz, and, in compound (43), coupling of the *ipso* carbon with the phosphorus atom was $^3J_{(P,C)} = 3.4$ Hz

Multinuclear (^{1}H, ^{13}C, ^{31}P) NMR- and UV-visible spectroscopy studies have been carried out to rationalize the variation in the air-stability of the P-heterocyclic biradicals (48)–(55).[57] The NMR spectra and photoabsorption parameters of diradical (48) were discussed on the basis of previous results and DFT calculations on model compounds, suggesting

(48) R¹ = Et, R² = Me
(49) R¹ = Buᵗ, R² = Me
(50) R¹ = R² = Et
(51) R¹ = R² = Me
(52) R¹ = Buᵗ, R² = CH₂Ph
(53) R¹ = Buᵗ, R² = CH₂OMe
(54) R¹ = Buᵗ, R² = CH₂CCH
(55) R¹ = Buᵗ, R² = CH₂CCPh

Mes* = 2,4,6-(Buᵗ)₃C₆H₂

(48)–(55)

particular structural characteristics of the biradical skeleton and aromatic substituent effects on the sp^2-C atoms in the 4-membered P_2C_2 ring. Introduction of the CH_2OMe substituent to the biradical gave more-stabilized, 1,3-diphosphacyclobutane-2,4-diyl derivatives, as in diradical (53). In comparison with the much less stable biradicals, (54) and (55), that have propargyl substituents, the relatively-higher LUMO energies suggest reluctant oxidation of the P-heterocyclic skeleton. A comparative study has also been carried out on the rotational dynamics of the PF_6^- anions in the crystalline and liquid states of the 'room temperature' ionic liquid (RTIL) 1-butyl-3-methylimidazolium hexafluorophosphate ([C_4mim]PF_6), using ^{31}P NMR spectroscopy line shape analyses and spin-lattice relaxation time measurements.[58] The PF_6^- anion was found to perform isotropic rotation in all three polymorphic crystals phases (α, β, and γ) as well as in the liquid state, with a characteristic time scale ranging from a few ps to a few hundred ps over a temperature range of 180–280 K. The rotational correlation time (τ_c) for the PF_6^- rotation follows the sequence of γ-phase < α-phase ~ liquid < β-phase. However, in the liquid state, the δ-CH_3 rotation in the [C_4mim]$^+$ cation, as well as its global rotational reorientation, are characterized by time scales that are slower compared to that for the PF_6^- anion rotation. The first crystalline phosphorus oxonitride imide (i.e., $P_8O_8N_6(NH)_3$) has been synthesized and characterized by using powder XRD and solid–state NMR spectroscopic techniques.[59] Information on the hydrogen atoms was obtained by 1D ^1H MAS, 2D homo- and heteronuclear (along with ^{31}P) NMR correlation spectroscopy, and a ^1H spin-diffusion experiment with a hard-pulse sequence designed for selective excitation of a single peak. Two hydrogen sites with a multiplicity ratio of 2:1 were identified and the empirical formula thus determined. The protons were assigned to particular *Wyckoff* positions. Some new, biologically-active, spermine derivatives of *cyclo*triphosphazene have been characterized using single-crystal XRD, IR, and ^1H and ^{31}P NMR spectroscopy in solution.[60]

4 Electron paramagnetic (spin) resonance spectroscopy

Uses of EPR (ESR) spectroscopy reported during 2012 include the characterizations of the triarylphosphine radical cations (56) and (57),[61] and a stable tetraaryldiphosphine radical cation (**58a**) and dication (**58b**).[62] Salts containing the radical cations (**56**) and (**57**) were isolated and characterized by EPR and UV-visible absorption spectroscopy, and single-crystal XRD. It was found that, whereas structure (**56**) exhibits a relaxed pyramidal geometry, radical cation (**57**) becomes fully planar. EPR studies and SOMO calculations at the UB3LYP/6-31G(d) level showed that introduction of bulky aryl groups leads to enhanced p character of the singly occupied molecular orbital (SOMO) with the radicals becoming less pyramidal or fully flattened. Salts containing the tetraaryldiphosphine radical cation (**58a**) and dication (**58b**) were similarly characterized. Whereas the radical cation (**58a**) has a relaxed pyramidal geometry, the dication (**58b**) prefers a planar, olefin-like geometry with a two-electron π bond. The alteration in geometry of the tetraaryldiphosphine on oxidation is rationalized by the nature of the bonding. The EPR spectrum showed that the spin density of (**58a**) is mainly localized on the P atoms, supported by SOMO calculations at the UB3LYP/6-31G(d) level. EPR spectroscopy has also been used to measure the N and P hyperfine coupling constants (a_N and $a_{P\beta}$) of β-phosphorylated nitroxides in order to probe cybotactic effects of solvents.[63] It was shown that the cybotactic

(**56**)

(**57**) X = SbF$_6$; Al[OC(CF$_3$)$_3$]$_4$; Al[OCMe(CF$_3$)$_2$]$_4$

(**58a**) $\xrightarrow{-1e^-}$ (**58b**)

effect on a_N and $a_{P\beta}$ depends on the structure and conformation adopted by the groups attached to the nitroxyl moiety and cannot be predicted easily. However, $a_{P\beta}$ provides complementary data to that provided by a_N on the solvent cage and is a powerful tool to investigate the solvent cybotactic effect on reactivity. The EPR and UV-visible spectra of an iodo-Co(I) tripodal phosphine complex have also been reported.[64]

5 Vibrational (IR and Raman) spectroscopy

The use of IR and Raman spectroscopy as complementary analytical techniques (to other physical and theoretical methods) is unlimited, especially in the case of characterizations. Also, a major use of theoretical calculations is the prediction and validation of experimental (including IR and Raman) spectroscopic data, and some applications have been reported earlier.[20,23,51] Thus, in the surface binding of DMMP, DIMP, DFP, and *Sarin* to silica, a DFT comparison with the experimental IR shifts showed that the theoretically-modelled adsorption sites are similar to those found by experiment,[20] and in the case of *Cyclophosphamide*, the IR and Raman spectra were assigned from DFT calculations,[23] also a homologous series of aminobisphosphonates were studied and characterized using IR and NMR spectroscopy.[51]

The first detailed vibrational characterization of five analogues of N-benzylamino- (boronphenyl)methylphosphonic acids has been carried out using FT-IR, FT-Raman, and surface-enhanced Raman spectroscopy (SERS), along with DFT/B3LYP theoretical calculations.[65] Analysis of the FT-IR and FT-Raman spectra showed that these compounds exist in the solid state as dimeric species, formed by H-bonding between the -B(OH)$_2$ moieties of each monomer. In addition, comparison of the wavenumbers, intensities, and broadness of bands from the FT-Raman and SERS spectra provided information regarding their adsorption onto an electrochemically-roughened silver substrate. Also, the molecular structure and vibrational assignments of N,N',N''-*tris*(2-aminoethyl)-phosphoric acid triamide (TEDAP) were confirmed by FT-IR spectroscopy with DFT/B3LYP calculations.[66] An FTIR-attenuated total reflectance (FTIR-ATR) technique has been developed to rapidly monitor tri-n-butyl phosphate (TBP) and TBP degradation product ratios required in nuclear forensics,[67] and the ozonolysis of 1-palmitoyl-2-oleoyl-sn-glycero-3-phosphocholine (POPC), itself adsorbed on salt mixtures used as models for sea-salt particles, was studied in real time by *diffuse reflection infrared Fourier transform spectrometry* (DRIFTS) at room temperature, with and without added water vapour, with the products identified by FTIR and MALDI-TOF mass spectrometry.[68] SERS has also been used for the trace analysis of organophosphorus compounds including the pesticides *Trichlorfon* and *Glyphosate* and the nerve agent o-ethyl methylphosphonothioate (EMPT) and simulator dimethyl methylphosphonate (DMMP).[69]

Doubly-deprotonated adenosine 5'-diphosphate ([ADP-2H]$^{2-}$) and adenosine 5'-triphosphate ([ATP-2H]$^{2-}$) dianions have been investigated in the gas phase using infrared multiple photon dissociation (IR-MPD) and photoelectron spectroscopy (PES).[70] Vibrational spectra obtained in the X-H stretch region (X = C, N, O), augmented by $^{13}C^{15}N$-labelling, were compared to DFT calculations at the B3LYP/TZVPP level. From this, in [ATP-2H]$^{2-}$ the two phosphate groups adjacent to the ribose ring are preferentially deprotonated. The photoelectron spectra recorded at 4.66 and 6.42 eV revealed adiabatic detachment energies (ADE) of 1.35 eV for [ADP-2H]$^{2-}$ and 3.35 eV for [ATP-2H]$^{2-}$, and the 'repulsive Coulomb barriers' (RCB) were estimated at 2.2 eV for [ADP-2H]$^{2-}$ and 1.9 eV for [ATP-2H]$^{2-}$. TDDFT calculations were used to simulate the photoelectron spectra. It was shown that photodetachment occurs primarily from lone-pair orbitals on the O atoms within the phosphate chain. More details on PES are given in the following section and on IR-MPD in the *Mass spectrometry* section.

6 Electronic spectroscopy

6.1 Absorption spectroscopy

6.1.1 UV-visible spectroscopy.
UV-visible (UV-Vis) spectroscopy is also used as a complementary technique to the other methods available (IR, NMR, XRD, and mass spectrometry) for the characterization of new compounds, along with other photophysical methods, and some applications have been mentioned earlier.[14,15,24,39,57,61,64] In particular, in 2,2′-benzo[b]phosphole-benzo[b]heterole hybrid π-systems (**10**), (**11**), (**12**), (**13**), and (**14**),[14] steady-state UV-visible absorption and fluorescence spectroscopy, fluorescence lifetime measurements, and DFT calculations showed that their emissive excited states consisted of two different conformers in rapid equilibrium. Also, in the *bis*(aryltriazolyl)-extended π-systems with the phosphole sulfide (**16a**) or phosphole (**16b**) group as central ring, these conjugated frameworks showed intriguing photophysical and electrochemical properties varying with the nature of the aromatic end-group.[15] A theoretical analysis of the UV and FT-IR/Raman spectra of hexachlorocyclotriphosphazene was carried out,[24] also the EPR and UV-visible spectra of an iodo-Co(I) tripodal phosphine complex reported.[64] Its absorption spectrum had a metal-to-ligand charge-transfer peak at 320 nm ($\varepsilon = 8790$) and a d-d band at 850 nm ($\varepsilon = 840$). Two d-d bands were also seen in the NIR region at 8650 ($\varepsilon = 450$) and 7950 cm^{-1} ($\varepsilon = 430$ M^{-1} cm^{-1}).

(**59**) Ch = S
(**60**) Ch = O

(**61**)

The novel benzofuran-fused phosphole derivatives (**59**), (**60**), and (**61**) show optical and electrochemical properties that differ from their benzothiophene analogues.[71] The optical properties of (**59**), (**60**), and (**61**) were studied in CH_2Cl_2 by UV-visible absorption and fluorescence spectroscopy. The three compounds displayed broad absorption in the visible range ($\lambda_{max} = 380$ nm), which is attributed to a π-π* transition of the conjugated backbone and confirmed by DFT/B3LYP/6-31+G (d, p) level calculations. The substitution on the P atom only weakly modified the optical transitions. Compared to their benzothiophene analogues, phospholes (**59**), (**60**), and (**61**) display slightly blue-shifted transitions, which are in accordance with a reduced HOMO-LUMO gap in the oligothiophene series compared to its oligofuran analogue. The effect of the P atom environment is more pronounced in the emission properties of these phospholes, as reported in subsection 6.1.2. Preliminary results show that compound (**60**) can be used as an emitter in 'organic light-emitting diodes' (OLEDs), illustrating the potential of these new compounds for optoelectronic applications.

6.1.2 Fluorescence and luminescence spectroscopy.

Within the general grouping of photophysical methods, fluorescence and luminescence are included with UV-visible spectroscopy where appropriate electronic transitions within a molecule are allowed. As reported above in the case of the benzofuran-fused phosphole derivatives (**59**), (**60**), and (**61**),[71] the effect of the P atom environment is more pronounced in their emission properties. A gradual decrease in the maximum emission wavelength is observed in the order (**60**)–(**59**)–(**61**), as already observed in the phosphole oligomers. The P-atom substitution also affects the fluorescence quantum yield and, notably, the compounds (**60**) and (**61**) display high quantum yields (65–73%), making them good candidates for use as emitters in light-emitting devices and, furthermore, the solid–state emission matches the emission in diluted solution. Some references that involve fluorescence spectroscopy and properties have already been mentioned earlier,[14,15,39] including the *bis*(aryltriazolyl)-extended π-systems with the phosphole sulfide (**16a**) and phosphole (**16b**) group as central ring.[15] These systems exhibited intriguing photophysical and electrochemical properties that vary with the nature of the aromatic end-group. The λ^3-phospholes (**16b**) display blue fluorescence ($\lambda_{em} = 460$–469 nm) with high quantum yields ($\Phi_F = 0.134$–0.309).

The luminescence properties of two new blue phosphorescent iridium(III) diazine PPh_3 complexes have also been reported,[72] and the chemiluminescence of phosphonate carbanions generated during an *oxy-Wittig* type reaction has been reviewed, with the phospha-1,2-dioxetane being the most likely high-energy intermediate.[73] The luminescence properties of several new transition metal phosphonates[74] and carboxydiphosphonates[75] have also been reported. Fluorescent nano CdSe '*quantum dot*' probes have been prepared and patented for the detection of nanomolar amounts of *Parathion*[76] and *Methamidophos*[77] organophosphorus pesticides. Finally, a comparison of the circularly polarized luminescence (CPL) and circular dichroism (CD) characteristics of four *axially* chiral binaphthyl-2,2'-diyl hydrogen phosphate derivatives has been carried out,[78] with more details given in the next subsection 6.1.3.

6.1.3 Circular dichroism (CD) and circularly polarized luminescence (CPL) spectroscopy.

Circularly polarized luminescence (CPL) spectroscopy is the emission analogue of circular dichroism (CD) spectroscopy. In order to understand the relationship between chiral organic fluorophores and pendant groups, four types of chiral, binaphthyl-based chromophores with different side groups, R-(62), R-(63), R-(64), and R-(65), were studied in chloroform at room temperature.[78] There was no sign of polarized luminescence or of CPL in phosphonic acid (63), possibly due to the nitrophenyl groups, while acid (62) exhibited circularly polarized luminescence. CPL was also observed in acids (64) and (65) due to the naphthalene and anthracene units, respectively. It is evident that CPL in acid (65) arises from the chirally-oriented anthracene units by efficient intramolecular chirality transfer. This led to the belief that by introducing two chirally-arranged fluorescent substituents to a chiral binaphthyl unit, very efficient CPL could be possible due to efficient photoexcited chirality transfer from the *axially*-chiral binaphthyl unit. The results illustrated how high-performance chiral binaphthyl-based CPL fluorophores with high Kuhn's anisotropy factors [g_{em}] and high quantum yields (Φ_F) can be designed.

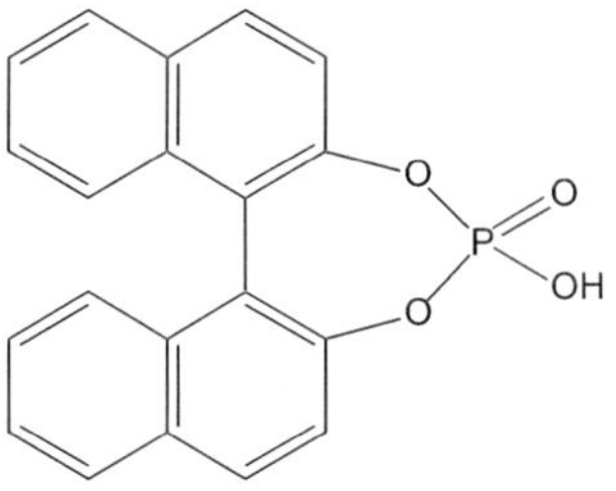

(62) 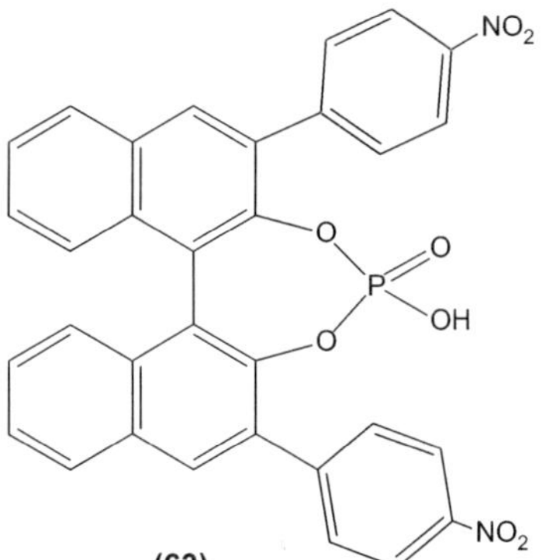   (63)

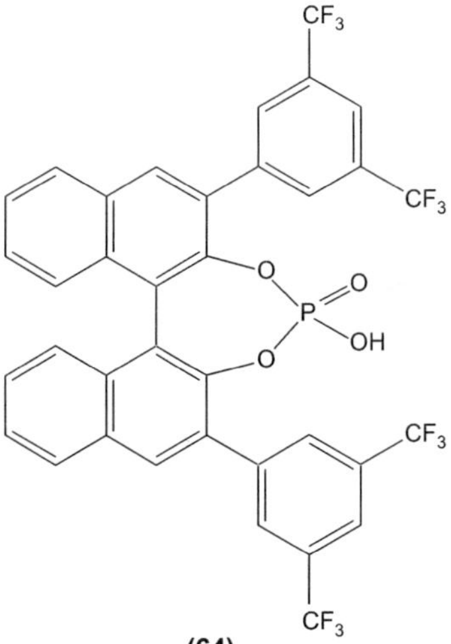

(64) 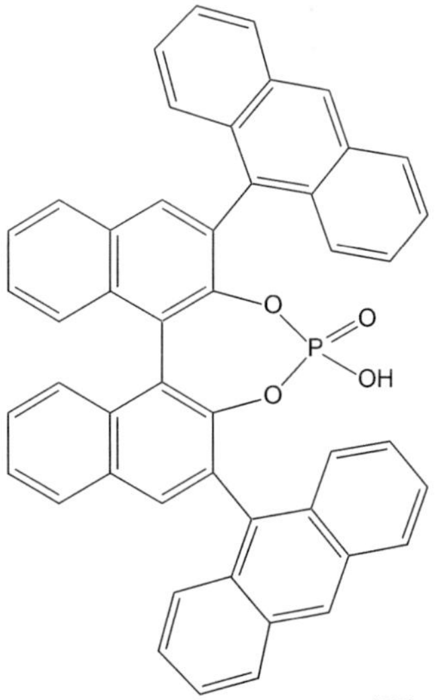   (65)

6.2 Photoelectron spectroscopy (PES)

Photoelectron spectroscopy (PES) was mentioned earlier, together with IR-MPD, in the examination of doubly-deprotonated adenosine 5′-diphosphate ([ADP-2H]$^{2-}$) and adenosine 5′-triphosphate ([ATP-2H]$^{2-}$) dianions in the gas phase.[70] The photoelectron spectra at 266 and 193 nm, along with DFT/B3LYP/TZVPP calculations, identified bands corresponding to the electronic ground states of the monoanions, as well as bands corresponding to excited states. The adiabatic detachment energies (ADE) and vertical detachment energies (VDE) of [ADP-2H]$^{2-}$ were found to be 1.35 and 1.90 eV, respectively, and the repulsive coulomb barrier (RCB) was estimated at 2.2 eV. Also, the ADE and VDE of [ADP-2H]$^{2-}$ were found to be 3.35 eV and 4.01 eV, respectively, and the RCB estimated at 1.9 eV. It was shown that the photodetachment dynamics are dominated by the lone-pair orbitals of O atoms in the phosphate chains, similar to the dynamics of $H_2P_2O_7{}^{2-}$ and $H_3P_3O_{10}{}^{2-}$ reported elsewhere.

7 X-ray diffraction (XRD) structural studies

Along with IR/Raman, UV, NMR spectroscopy and mass spectrometry, X-ray diffraction (XRD) is a complementary technique used in characterizations. Thus, some applications have been mentioned earlier,[27,39,40,43,61,74] particularly for the first-known phosphapentacenium ion [(Me$_3$Si)$_2$N=P-N$_3$]$^+$ in azide (30),[43] and for salts containing the radical cations (56) and (57).[61]

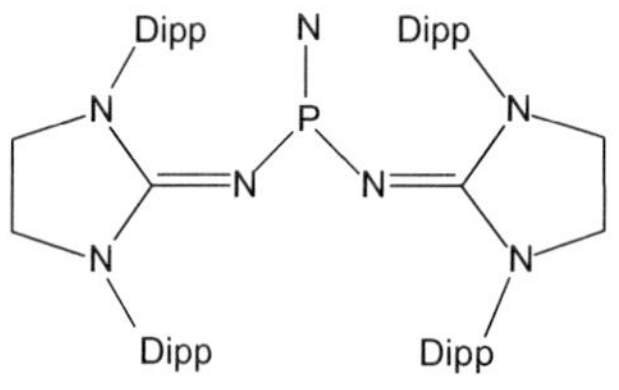

(66) Dipp = 2,6-diisopropylphenyl

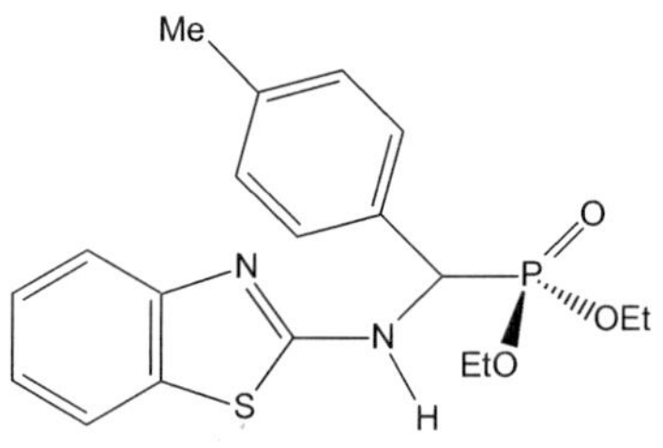

(67)

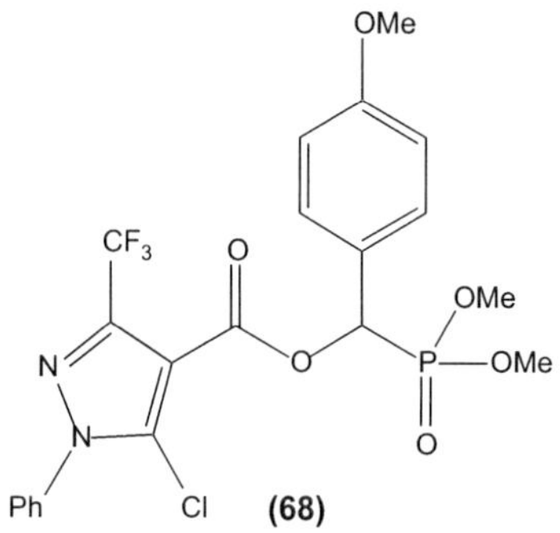

(68)

A stable, crystalline singlet bis(imidazolidin-2-iminato)phosphinonitrene (66) has been isolated and a single-crystal XRD study showed that the phosphorus atom is in a planar environment and the P–N bond length is very short at 1.457 Å.[79] The bonding between phosphorus and nitrogen is analogous to that observed in metallonitrenes. It was shown that this phosphinonitrene can be used to transfer a N atom to organic fragments, a difficult task for metallonitrenes. The X-ray crystal structure of *bis*(diphenylphosphino)ethane monoxide, the first member of the $R_2P(CH_2)_nP(O)R_2$ ligand class to be structurally characterized, has been reported,[80] also the structure of the $[Rh(PhCOCHCO(CH_2)_3CH_3)(CO)(PPh_3)]$ square-planar complex, along with conformational analysis of the PPh_3 coordinated to it.[81] Structural characterization of cDHAP, the cyclic form of dihydroxyacetone phosphate, has also been carried out, revealing *chair* and *skew* conformations of the 1,3,2-dioxaphosphorinane ring.[82]

From XRD analysis, the crystal structure of *rac*-ethyl(phenyl)phosphinic acid features O–H···O=P–OH···O=P hydrogen bonds which link molecules related by the *b*-glide plane into continuous chains along [010],[83] and the short P–O···O=P distance of 2.4931 Å indicates a strong hydrogen bond. This is slightly shorter than the average O···O interaction distance for other phosphinic acids but is equal to that for methyl(phenyl)phosphinic acid (2.4838 Å). In hydrogen 4-ammoniophenylphosphonate, existing as the zwitterion $H_3N^+C_6H_4PO_3H^-$, in the crystal the molecules are linked by O–H···O and N–H···O hydrogen-bond bridges giving a 3D network structure.[84] The strongest H-bonds are formed between adjacent PO_3H groups with O···O distances of 2.577 Å. The α-aminophosphonates are biologically important and several new compounds have been synthesized. Ten diethyl aryl(benzo[d]thiazol-2-ylamino)methyl phosphonates have been characterized and the crystal structure of phosphonate (**67**) determined.[85] In this ester there are weak intermolecular P=O···H–N hydrogen bonds, effective in the stabilization of the structure, and a pair of hydrogen bonds are seen in a centrosymmetric dimer. Fourteen new α-arylmethylphosphonates have been characterized and the structure of phosphonate (**68**) also determined.[86] Four novel metal phosphonates with a 3D molecular structure, (**69**), (**70**), (**71**) and (**72**), have also been structurally characterized.[87] Compounds (**69**), (**70**) and (**71**) are isostructural and adopt a 3D supramolecular structure. Every two $\{M(1)O_5\}$ polyhedra are interconnected by phosphonate $\{CPO_3\}$ tetrahedra to form a unit which are linked by $\{M(2)O_6\}$ octahedra to form a 1D chain by corner-sharing down the c-axis. Such 1D infinite chains are cross-linked via oxalate anions into a 2D layer, which are further connected through hydrogen bonding interactions to give a 3D

$Fe_{1.5}(H_2O)[(H_2L)(C_2O_4)_{0.5}]$

(**69**)

$Co_{1.5}(H_2O)[(H_2L)(C_2O_4)_{0.5}]$

(**70**)

$Zn_{1.5}(H_2O)[(H_2L)(C_2O_4)_{0.5}]$

(**71**)

$Cu_{1.5}[(H_2L)(C_2O_4)_{0.5}]$

(**72**)

$(H_4L = C_5H_4NCH_2CH(OH)(PO_3H_2)_2)$

supramolecular structure. The overall structure of (72) is similar to that of (69), (70) and (71) except that the {M(2)O$_6$} octahedra are replaced by {Cu(2)O$_4$} planar squares. The surface photovoltage properties of the four compounds were also studied.

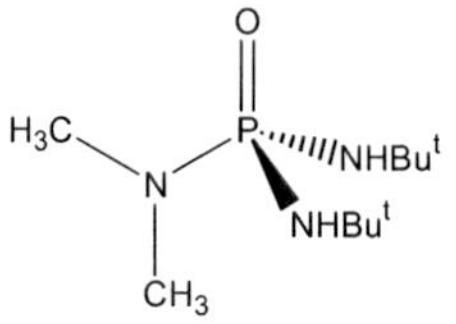

(73)

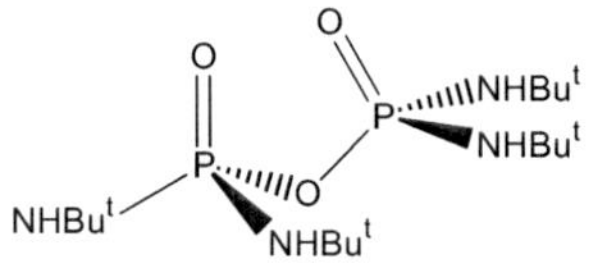

(74)

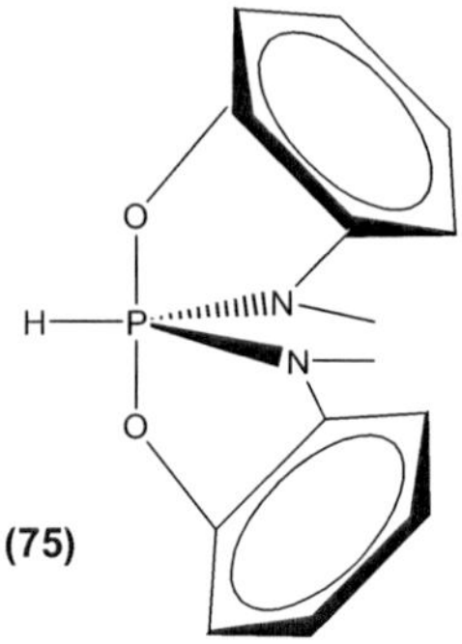

(75)

Two new phosphoramidates, (73) and (74), have also been structurally characterized using XRD.[88] They have extended structures that are mediated by P(O)···(H–N)$_2$ interactions. The asymmetric unit of (73) consists of six independent molecules which aggregate through P(O)···(H–N)$_2$ hydrogen bonds forming two independent chains parallel to the a axis. In phosphoramidate (74), the asymmetric unit contains one molecule. The P(O)···(H–N)$_2$ hydrogen bonds lead to the molecules forming extended chains parallel to the c axis. Their structures along with similar structures with (N)P(O)(NH)$_2$ and (NH)$_2$P(O)(O)–P(O)(NH)$_2$ skeletons from the *Cambridge Structural Database*, were used to compare hydrogen-bond patterns in phosphoramidates. The strengths of P(O)[···H–N]$_x$ (x = 1, 2 or 3) hydrogen bonds were also analysed from these and previous structures with (N)$_2$P(O)(NH) and P(O)(NH)$_3$ fragments. Also, the crystal structure of the hydridophosphorane (75) has been determined using XRD, along with DFT/B3LYP calculations.[89] Both the calculated and experimental structure display a slightly-distorted trigonal bipyramidal geometry about the phosphorus centre, with *axial* oxygen atoms and *equatorial* nitrogen atoms. The planar geometry of the nitrogen atoms, along with the shortened P–N bond lengths and molecular orbital calculations, are evidence of N → P π-bonding. Also, a comparison of the structure of phosphorane (75) with previous metallophosphorane structures featuring [P(OC$_6$H$_4$N(CH$_3$))$_2$]- as a ligand gives further support to the importance of M → P π-bonding in the case of metallophosphoranes.

8 Electrochemical methods

8.1 Dipole Moments

As mentioned earlier, by using the dipole moment method and different quantum-chemical methods, the conformational analyses of *bis*(2-phenylalkyl)phosphine selenides,[12] and of 1,3,2-dioxaphospholan-2-yl 2,2,2-trifluoroacetate (7) and 4,5-benzo-1,3,2-dioxaphosphol-2-yl 2,2,2-trifluoroacetate (8)[13] were carried out. The experimental dipole moments

(μ_{exp}) were determined using the *Debye* method based on dielectric constants recorded on dilute solution in non-polar solvents, and the calculated dipole moments (μ_{calc}) were obtained from quantum-chemical and vector-addition methods, and compared with the theoretical dipole moments (μ_{theor}). Thus, it was shown that *bis*(2-phenylpropyl)phosphine selenide exists as a mixture of several conformers, the most energetically-favourable characterized by gauche orientation of the P=Se and C_{sp3}–C_{sp3} bonds,[12] and also that both compounds (**7**) and (**8**) exist as the *axial* conformer with preferably syn arrangement of the carbonyl group and the lone electron pair.[13]

8.2 Cyclic Voltammetry

As also mentioned earlier,[15] the phosphole sulfide (**16a**) and free phosphole (**16b**) exhibit intriguing photophysical and electrochemical properties that vary with the nature of the aromatic (phenyl, 2-pyridyl or 2-thienyl) end-group. From the cyclic voltammetry (CV) results, the π-systems of (**16a**) and (**16b**) were shown to be readily reduced. Thus, the phosphole sulfides (**16a**) can be reduced between -1.91 and -2.04 V, consistent with the low-lying LUMO of the phosphole-based π-systems, and no anodic waves are detected below $+1.0$ V, whereas the phospholes (**16b**) exhibit slightly more negative cathodic waves between -2.25 and -2.35 V, in line with the DFT data. The redox properties of phospholes (**59**), (**60**) and (**61**) were also studied by cyclic voltammetry.[71] All the compounds display amphoteric redox character with oxidation and reduction waves at relatively low potentials, and substitution on the *P* atom was shown to have only a weak impact on their redox properties. Compared to their thiophene analogues, they display lower oxidation and reduction potentials with a global increase of the electrochemical gap. This is in line with the observed general trend that exchanging *S* for *O* in conjugated five-membered oligomers results in changes such as increasing the HOMO-LUMO gap and lowering the oxidation potential. Also mentioned above is a series of *tris*-aryl 'blade' phosphane oxides, existing as residual enantiomers or diastereoisomers with substituents on the aryl rings differing in size and electronic properties, that have been characterized.[38] In a related extended study,[90] their electronic properties, together with those of the corresponding 'blade bromides', were evaluated on the basis of oxidation and reduction potentials and determined by cyclic voltammetry. The configurational stability of the residual *tris*-aryl phosphane oxides was found to be little influenced by the electronic properties of the substituents on the aromatic rings constituting the blades, whereas the steric effects played a major role. Cyclic voltammetry has also been used to determine the redox properties of new heterobidentate phosphane-olefin ligands based on the dibenzophosphepine scaffold.[91]

8.3 Electrochemical sensors and biosensors

Two gas sensors have been patented for the detection of organophosphorus compounds, nerve agents and simulators, based on carbon nanotube sensor layers with palladium[92] and gold[93] electrodes. The simulator dimethyl methylphosphonate (DMMP) has also been detected by a fluorosiloxane-coated quartz crystal microbalance (QCM) sensor,[94] and organophosphorus volatiles by a surface acoustic wave (SAW) sensor array coated with selective polymers deposited by laser induced forward transfer (LIFT) using a XeCl laser.[95]

9 Acidities, basicities and thermochemistry

The thermodynamic and hydrogen-bond basicities of tetrahedral phosphine oxide derivatives were mentioned earlier,[11] where it was shown that ring strain decreases the complexation energy with metallic atoms and also the thermodynamic and hydrogen-bond acceptor ability of the derivatives. The pK_a values of a series of aminobisphosphonates have been determined along with other physicochemical properties in a detailed study,[51] where it was found that lengthening of the methylene chain in $H_2N(CH_2)_nC(OH)[P(O)(OH)_2]_2$ between the bisphosphonate and amino groups decreases the value of the first protonation constant (pK_a) for the NH_2 group but increases the values of the other protonation constants for the phosphonate groups.

The basicities of several phosphanes and diphosphanes in acetonitrile have been determined in order to rationalize their basicity trends compared with amines of similar structure.[96] In the case of the phosphanes studied, pK_a values ranged from 4 to 16 with a significantly weaker change in basicity observed as compared to the amines when the alkyl groups were replaced by aryl groups, mainly caused by resonance between the aryl group and the lone pair of the basicity centre, which is strong in the case of the *N* atom and weak in the case of the *P* atom. For the diphosphanes, their basicity led to the conclusion that there is no intramolecular hydrogen bond in protonated diphosphanes with an alkyl backbone, but that there may be a weak intramolecular hydrogen bond in protonated diphosphanes with an aromatic backbone. The thermal decomposition behaviours of amino trimethylene phosphonic acid (ATMP) and 1-hydroxyethylidene-1,1-diphosphonic acid (HEDP) have also been studied.[97] A comparison of the experimental results from the thermal decomposition using TGA-FTIR and pyrolysis GC-MS with modelling of the formation reactions showed the usefulness of the latter method in predicting the possible decomposition products. There are more details in subsection 11.1.

10 Mass spectrometry techniques

10.1 Mass spectrometry (MS)

Mass spectrometry is included also in section 11 when used as a stand-alone detector for identifying the eluents of gas and liquid chromatography column separations (as in GC-MS and LC-MS). As with the other methods (IR, UV-visible, NMR spectroscopy and XRD), mass spectrometry is a complementary technique used for the characterization of organic compounds and various applications have been mentioned.[25,52,53,68,70]

R = Ph or Ph-d$_5$
X = H or D,
Y = H or D

cis-(76)　　　　　　　　　　　　trans-(76)

Electron ionization mass spectrometry (EI-MS) and DFT calculations have been used to study the fragmentation processes and differentiate the diastereoisomers of diethyl 5-phenyl-(2-thioxoimidazolidine-4-yl)phosphonates (76).[98] The loss of a diethoxyphosphoryl group and the elimination of diethyl phosphonate were found to be competitive fragmentation processes and used to differentiate both stereoisomers. Selective deuterated analogues and product- and precursor-ion mass spectra allowed the elucidation of the fragmentation mechanisms. The structures of the transition states and product ions were optimized using DFT, and free energy calculations confirmed the observed differences in the formation and relative intensities of specific fragment ions. The negative-ion mass spectrometry of N-benzyloxycarbonyl and N-ethoxycarbonyl 1-aminoarylmethylphosphonic acid monoesters (Me, Et and Ph) has been investigated under electrospray ionization (ESI) conditions.[99] Their fragmentation pathways were proposed and supported by collisional activated dissociation (CAD) product-ion spectrometry. All of the deprotonated molecules preferentially eliminate a molecule of benzyl alcohol or ethanol to yield isocyanato-alkylphosphonate anions which further generate phosphonate ions by the loss of CO, phenol alcohol, CO plus arenes. The isocyanato-sulfonate anions can then further cyclize to generate 3-aryl-2-phenoxy-1,2,4-oxaphosphazolidine-2,5-dione amide anions, which can undergo rearrangements by the loss of CO_2, metaphosphoric acid phenyl ester, or CO_2 plus metaphosphoric acid phenyl ester, respectively, to give rise to nitrogen-containing anions. The compounds show different fragmentation in the negative- and positive-ion modes under ESI conditions. An EI-MS study has been carried out on the CWC-related O(S)-alkyl N,N-dimethylamino alkyl-phosphonates (alkyl-phosphonothiolates), and their GC *retention indices* and MS data and fragmentation patterns given.[100] The mass spectrometric studies revealed that their fragmentations were dominated by alkene elimination, McLafferty rearrangement, α-cleavage, and amine elimination, and were confirmed by using MS/MS experiments, fragment ions of deuterated analogues, and DFT calculations. More details are given in the subsection on GC-MS.

The formation and dissociation of phosphorylated peptide radical cations $[_pM]^+$ has been studied.[101] Phosphoserine- and phosphothreonine-containing peptide radical cations were formed by low-energy collision-induced dissociation (CID) of the ternary metal-ligand phosphorylated peptide $[Cu^{II}(terpy)_pM]^{2+}$ and $[Co^{III}(salen)_pM]^+$ complexes. Subsequent CID of the phosphorated peptide radical cations $[_pM]^+$ revealed fascinating gas-phase radical chemistry yielding charge-directed *b*- and *y*-type product ions, radical-driven product ions via cleavage of peptide backbones and side chains, and different degrees of formation of $[M-H_3PO_4]^+$ species from phosphate ester bond cleavage. CID spectra of the $_pM^+$ species and their non-phosphorylated analogues featured fragment ions of similar sequence suggesting that the phosphoryl group did not play a significant role in the fragmentation of the peptide backbone or side chain. The extent of neutral H_3PO_4 loss was influenced by the peptide sequence and initial sites of the charge and radical. A DFT study of the neutral loss of H_3PO_4 from a model, *N*-acetylphosphorylserine methylamide, revealed several factors governing the elimination of neutral phosphoryl groups via charge- and radical-induced mechanisms.

10.2 Infrared multiple photon dissociation (IR-MPD) spectroscopy

This is a technique used in mass spectrometry to fragment molecules in the gas phase for structural analysis of the parent molecule. It allows for the measurement of vibrational spectra of species that can only be prepared in the gas phase. Such species include molecular ions but also neutral species like metal clusters that can be gently ionized after interaction with a mid-IR laser source for mass spectrometric detection. As mentioned, IR-MPD- and photoelectron spectroscopy were used to examine the molecular and electronic structure of $[ADP-2H]^{2-}$ and $[ATP-2H]^{2-}$ dianions in the gas phase.[70] The IR-MPD spectra of $[ATP-2H]^{2-}$ dianions were recorded in the O–H and N–H stretching region of the spectrum (2600–4000 cm^{-1}) and DFT calculations used to assign the vibrational modes. The spectra and DFT calculations confirm that both the phosphate groups on $[ADP-2H]^{2-}$ are deprotonated and that $\alpha\beta$-deprotonated $[ATP-2H]^{2-}$ is the dominant deprotonation route.

11 Chromatography and related separation techniques

11.1 Gas chromatography (GC) and gas chromatography-mass spectrometry (GC-MS)

As reported, the thermal decomposition behaviour of amino trimethylene phosphonic acid (ATMP) and 1-hydroxyethylidene-1,1-diphosphonic acid (HEDP) have been studied and a comparison of the experimental results from thermal decomposition by TGA-FTIR and pyrolysis GC-MS, together with modelling of the formation reactions, showed the usefulness of the latter method in predicting the possible decomposition products.[97] Thus, pyrolysis GC-MS was used to determine the gaseous decomposition products of ATMP and HEDP at temperatures corresponding to the main decomposition steps detected by TGA-FTIR spectroscopy and, from a comparison of the experimental results with theoretical modelling, it was established that the decomposition process should follow the formation mechanism, *i.e.* the thermal decomposition can be understood as the reverse reaction of phosphonic acids.

$$\text{Me}_2\text{N}-\overset{\overset{\text{O}}{\|}}{\underset{R^2}{P}}-O(S)R^1$$

(77)

Also as mentioned above, an EI-MS study was carried out on freshly-prepared $O(S)$-alkyl N,N-dimethylamino alkylphosphonates (alkylphosphonothiolates) (77) following their GC-MS separation, and their GC retention indices and MS data and fragmentation routes reported.[100]

New GC and GC-MS methods have been reported for simultaneous determinations of organophosphorus flame retardants in textiles, including a GC method combined with microwave-assisted extraction,[102] and a GC-MS method following ultrasonic extraction,[103] also a phosphate-based flame retardant in textiles by GC-MS,[104] and in styrene-based polymers from waste electrical equipment by both GC (NP detector) and GC-MS.[105] The gas chromatographic retentions of alkyl phosphates on ionic liquid stationary phases have been studied,[106] as well as a comparative study on the determination of di-n-butyl phosphate in spent nuclear solvents by both gas- (GC) and ion chromatography (IC) methods.[107]

11.2 High performance liquid chromatography (HPLC) and mass spectrometry (LC-MS)

The determination of the retention factors (k'') of new phosphoramidic acid derivatives by reversed-phase HPLC was mentioned earlier in a QSSR study,[36] also the configurational stability of 'residual' phosphanes were evaluated by both dynamic HPLC on a chiral phase and dynamic ^{1}H and ^{31}P NMR spectroscopy.[38] The determination of di-n-butyl phosphate (DBP) in spent nuclear solvents was also compared by using both gas- (GC) and ion chromatography (IC) methods,[107] and it was shown that ion chromatography was much faster than gas chromatography since it involves minimum steps and no derivatization of DBP.

The effects of the stationary phase and solvent on the racemization of 2,2'-bis-phosphino- and 2,2'-bis(phosphinyl)-1,1'-biphenyls (BIPHEP) and the racemization kinetics of the 3,3'-disubstitued BIPHEP ligands have been determined by chiral dynamic HPLC (DHPLC) using a novel three-column continuous flow technique.[108] Also, phosphatidylcholine has been determined by ultra performance liquid chromatography (UPLC) with MS/MS detection.[109]

11.3 Electrokinetic capillary chromatography (EKC)

The applicability of phosphonium-based ionic liquids, as pseudo-stationary phases in electrokinetic capillary chromatography (EKC), with UV-detection, has been elucidated for the separation of model analytes including neutral benzene and benzene-derivatives.[110] It was found that the electroosmotic flow was strongly influenced by the type of ionic liquid, and the strongest impact on the flow was achieved with tetraalkylammoniumphosphonium ioic liquids with one long alkyl chain (C_{14} or C_{16}). EKC was performed using a capillary electrophoresis system.

12 Kinetics

A kinetic investigation (mentioned earlier) was carried out by ^{1}H NMR spectroscopy on the equilibrium between the Z- and E-isomers of a stable phosphorus ylide involving a 2-indolinone (**29**), which revealed that the interchange process for the rotational isomers follows first-order kinetics with respect to the forward (k_1) and reverse (k_{-1}) reactions.[42] The activation parameters and thermodynamic parameters were also obtained experimentally.

Third-order rate constants have been determined for the alkaline hydrolysis of four series of alkylphenylphosphonium- and alkylphenylbenzyl-phosphonium salts in 50–70% aqueous THF and 70% aqueous methanol at various temperatures.[111] The thermodynamic activation parameters were calculated for the reactions of each substrate and the effects of varying the ratio of alkyl to phenyl groups were compared, as well as the effects of changes in the nature of the alkyl group. It was shown that solvation, as revealed by trends in entropy of activation, plays a largely counter-balancing role with respect to the enthalpy and energy of activation. The role of the isokinetic effect was also discussed. In aqueous THF, the influence of solvation effects on the hydrolyses of phosphonium salts changes as the mole fraction of water changes, and in aqueous methanol the trends in the thermodynamic activation parameters actually reverse. Kinetic studies have also been carried out to help understand the mechanism for the synthesis of stable phosphorus ylides in the presence of different NH-acids such as imidazole, 2-methylimidazole or 4-methylimidazole used as a protic/nucleophilic reagent.[112]

The kinetics of the decomposition of 4-nitrophenyl diethylphosphonate and 4-nitrophenyl diethyl phosphate by $H_2O_2/B(OH)_3/HO^-$ with variations in pH and initial concentrations of H_2O_2 and $B(OH)_3$ have been studied in water at room temperature.[113] It was found that additions of boric acid either weakly accelerate the reaction or retard it, depending on the ratio of the initial concentrations of peroxide and boric acid and also the nature of the substrate. The peroxoborate anions, $(HO)_3BOOH^-$ and $(HO)_2B(OOH)_2^-$, and the $(HO)_3BOOB(OH)_3^{2-}$ anions generated do not affect the decomposition rate and, in both compounds, linear relationships were found between the observed rate constants (k) and concentrations of peroxide anion. A kinetic study has also been carried out on the hydrolysis of mono-4-methyl-2-nitroaniline phosphate in 0.1 to 7.0 M HCl at 50 °C,[114] and it was found that the phosphate ester is hydrolyzed exclusively via conjugate acid species. Based on these results, a bimolecular hydrolysis with P–N bond fission of the conjugate acid species has been proposed and an $S_N2(P)$ mechanism suggested. The enhanced kinetics of hydrolysis of *bis*(4-nitrophenyl) phosphate diester, catalysed by a lanthanum(III) tetraaza-macrocycle complex, has also been studied.[115]

References

1. M. Gueizado-Rodriguez and G. Ramirez-Galicia, *Comput. Theor. Chem.*, 2012, **992**, 48.
2. A. A. Ahmed, *Rasayan J. Chem.*, 2012, **5**, 131.
3. J. Urdaneta, Y. Bermudez, R. Bracho, R. Moreno and H. Soscun, *J. Comput. Methods Sci. Eng.*, 2012, **12**, 407.
4. R. M. Aminova, E. R. Baisupova and A. V. Aganov, *Appl. Magn. Reson.*, 2011, **40**, 147.
5. M. Jordaan and H. C. M. Vosloo, *Industrial Applications of Molecular Simulations*, 2012, 171.
6. L.-S. Sbirna, A. Oubraham, I. Dabuleanu and F. Ciolan, *Analele Universitatii din Craiova, Ser. Chimie.*, 2011, **40**, 34 (*Chem. Abs.*, 2013, **158**, 272365).
7. S. M. Habibi-Khorassani, M. T. Maghsoodlou, A. Ebrahimi, F. Ghodsi and R. Doostmohammadi, *Orient. J. Chem*, 2012, **28**, 1345.
8. S. M. Habibi-Khorassani, M. T. Maghsoodlou, A. Ebrahimi, M. Mohammadi, M. Shahraki and E. Aghdaei, *J. Phys. Org. Chem.*, 2012, **25**, 1328.
9. F. Castaneda, P. Silva, C. Acuna, M. Teresa Garland and C. A. Bunton, *J. Mol. Struct.*, 2013, **1034**, 51.
10. K. Yamada and N. Koga, *J. Comput. Chem.*, 2013, **34**, 149.
11. C. Trujillo, G. Sanchez-Sanz, I. Alkorta and J. Elguero, *Comput. Theor. Chem*, 2012, **994**, 81.
12. Y. A. Vereshchagina, D. V. Chachkov, A. Z. Alimova, E. A. Ishmaeva and S. F. Malysheva, *Russ. J. Org. Chem.*, 2012, **48**, 1320.
13. E. A. Ishmaeva, Y. A. Vereshchagina, D. V. Chachkov and A. Z. Alimova, *Russ. J. Gen. Chem.*, 2012, **82**, 1777.
14. Y. Hayashi, Y. Matano, K. Suda, Y. Kimura, Y. Nakao and H. Imahori, *Chem. Eur. J.*, 2012, **18**, 15972.
15. W. Weymiens, F. Hartl, M. Lutz, J. C. Slootweg, A. W. Ehlers, J. R. Mulder and K. Lammertsma, *Eur. J. Org. Chem.*, 2012, **34**, 6711.
16. K. A. Chernyshev, L. I. Larina, E. A. Chirkina, V. G. Rozinov and L. B. Krivdin, *Russ. J. Org. Chem.*, 2012, **48**, 676.
17. Q. Meng and M. Li, *J. Mol. Model.*, 2012, **18**, 3489.
18. M. D. C. Michelini, N. Russo, S. Alcaro and L. A. Wozniak, *Tetrahedron*, 2012, **68**, 5554.
19. H. Aghabozorg, B. Sohrabi, S. Mashkouri and H. R. Aghabozorg, *J. Mol. Model.*, 2012, **18**, 929.
20. D. E. Taylor, K. Runge, M. G. Cory, D. S. Burns, J. L. Vasey, J. D. Hearn, K. Griffith and M. V. Henley, *J. Phys. Chem. C*, 2013, **117**, 2699.
21. X. Zhang, H. Gao, H. Xu, J. Xu, H. Chao and C. Zhao, *J. Mol. Catal. A: Chem.*, 2013, **368**, 53.
22. W. Li, T. Rudack, K. Gerwert, F. Grater and J. Schlitter, *J. Chem. Theory Comput.*, 2012, **8**, 3596.
23. H. M. Badawi and W. Furner, *Z. Naturforsch., B: Chem. Sci.*, 2012, **67**, 1305.
24. W. M. Zoghaib, J. Husband, U. A. Soliman, I. A. Shaaban and T. A. Mohamed, *Spectrochim. Acta, Part A*, 2013, **105**, 446.
25. C. G. E. Fleming, A. M. Z. Slawin, K. S. Athukorala Arachchige, R. Randall, M. Buehl and P. Kilian, *Dalton Trans.*, 2013, **42**, 1437.
26. R. Septelean, P. M. Petrar and G. Nemes, *Studia Universitatis Babes-Bolyai, Chemia*, 2011, **56**, 131 (*Chem. Abs.*, 2013, **158**, 273120).
27. H. Cui, J. Zhang and C. Cui, *Organometallics*, 2013, **32**, 1.
28. D. Mishra, S. Pal and S. Krishnamurty, *Mol. Simul.*, 2011, **37**, 953.
29. D. Afzal and K. R. Fountain, *Can. J. Chem.*, 2011, **89**, 1343.

30 S.-X. Hu, J.-G. Yu and E. Y. Zeng, *Int. J. Quantum Chem.*, 2013, **113**, 1128.
31 D. Mandal, K. Sen and A. K. Das, *J. Phys. Chem. A*, 2012, **116**, 8382.
32 T. M. Alam, C. J. Pearce and J. E. Jenkins, *Comput. Theor. Chem.*, 2012, **995**, 24.
33 M.-T. Lee, A. Vishnyakov, G. Y. Gor and A. V. Neimark, *J. Phys. Chem. B*, 2013, **117**, 365.
34 D. Marciano, I. Columbus, S. Elias, M. Goldvaser, O. Shoshanim, N. Ashkenazi and Y. Zafrani, *J. Org. Chem.*, 2012, **77**, 10042.
35 G. Adamova, R. L. Gardas, M. Nieuwenhuyzen, A. V. Puga, L. P. N. Rebelo, A. J. Robertson and K R. Seddon, *Dalton Trans.*, 2012, **41**, 8316.
36 M. Petric, L. Crisan, M. Crisan, A. Micle, B. Maranescu and G. Ilia, *Heteroat. Chem.*, 2013, **24**, 138.
37 C. Anstiss, P. Karuso, M. Richardson and F. Liu, *Molecules*, 2013, **18**, 2788.
38 S. Rizzo, T. Benincori, V. Bonometti, R. Cirilli, P. R. Mussini, M. Pierini, T. Pilati and F. Sannicolo, *Chem. Eur. J*, 2013, **19**, 182.
39 J. R. Shakirova, E. V. Grachova, A. A. Melekhova, D. V. Krupenya, V. V. Gurzhiy, A. J. Karttunen, I. O. Koshevoy, A. S. Melnikov and S. P. Tunik, *Eur. J. Inorg. Chem.*, 2012, **25**, 4048.
40 H. T. Al-Masri, *Synth. React. Inorg. Met.-Org. Nano-Met. Chem.*, 2013, **43**, 102.
41 A. Habibi, K. Eskandari and A. Alizadeh, *Phosphorus, Sulfur, Silicon Relat. Elem.*, 2012, **187**, 1109.
42 S. M. Habibi-Khorassani, M. T. Maghsoodlou, E. Aghdaei and M. Shahraki, *Prog. React. Kinet. Mech.*, 2012, **37**, 301.
43 C. Hering, A. Schulz and A. Villinger, *Angew. Chem., Int. Ed.*, 2012, **51**, 6241.
44 W. Betzl, C. Hettstedt and K. Karaghiosoff, *New J. Chem.*, 2013, **37**, 481.
45 R. Streubel, E. Schneider and G. Schnakenburg, *Organometallics*, 2012, **31**, 4707.
46 G. S. Reddy, S. H. J. Prakash, K. U. M. Rao, E. Dadapeer, B. S. Krishna and C. S. Reddy, *Int. J. Pharm. Bio. Sci.*, 2012, **3**, 343.
47 V. K. Pandey, K. Chaturvedi, R. Chandra, O. P. Pandey and S. K. Sengupta, *Phosphorus, Sulfur, Silicon Relat. Elem.*, 2012, **187**, 1401.
48 G. S. Reddy, C. S. Sundar, S. S. Prasad, E. Dadapeer, C. N. Raju and C. S. Reddy, *Der Pharma Chemica.*, 2012, **4**, 2208.
49 K. S. Kumar, C. B. Reddy, M. V. N. Reddy, C. R. Rani and C. S. Reddy, *Org. Commun.*, 2012, **5**, 50.
50 L. Quintero, H. Hopfl, M. Sosa-Rivadeneyra, A. Ortiz, R. L. Meza-Leon, S. Cruz-Gregorio and F. Sartillo-Piscil, *Tetrahedron Lett.*, 2012, **53**, 4727.
51 A.-L. Alanne, H. Hyvonen, M. Lahtinen, M. Ylisirnio, P. Turhanen, E. Kolehmainen, S. Peraniemi and J. Vepsalainen, *Molecules*, 2012, **17**, 10928.
52 E. G. Saglam, O. Celik, H. Yilmaz and N. Acar, *Phosphorus, Sulfur, Silicon Relat. Elem.*, 2012, **187**, 1339.
53 Y. Li, A.-B. Wang, J.-S. Wang, H.-B. Ma and B. Wang, *Huaxue Shiji*, 2012, **34**, 902 (*Chem. Abs.*, 2013, **158**, 272933).
54 A. T. Myller, J. J. Karhe, M. Haukka and T. T. Pakkanen, *J. Mol. Struct.*, 2013, **1033**, 171.
55 Z. Shariatinia, M. Sohrabi, M. Yousefi, T. Koval and M. Dusek, *Main Group Chem.*, 2012, **11**, 125.
56 Z. Shariatinia, M. Sohrabi, M. Yousefi, T. Koval and M. Dusek, *Heteroat. Chem.*, 2012, **23**, 478.
57 S. Ito, M. Kikuchi, J. Miura, N. Morita and M. Yoshifuji, *J. Phys. Org. Chem.*, 2012, **25**, 733.
58 T. Endo, H. Murata, M. Imanari, N. Mizushima, H. Seki, S. Sen and K. Nishikawa, *J. Phys. Chem. B*, 2013, **117**, 326.
59 S. J. Sedlmaier, V. R. Celinski, J. Schmedt auf der Gunne and W. Schnick, *Chem. Eur. J.*, 2012, **18**, 4358.

60 G. Y. Ciftci, E. T. Ecik, T. Yildirim, K. Bilgin, E. Senkuytu, F. Yuksel, Y. Uludag and A. Kilic, *Tetrahedron*, 2013, **69**, 1454.
61 X. Pan, X. Chen, T. Li, Y. Li and X. Wang, *J. Am. Chem. Soc.*, 2013, **135**, 3414.
62 X. Pan, Y. Su, X. Chen, Y. Zhao, Y. Li, J. Zuo and X. Wang, *J. Am. Chem. Soc.*, 2013, **135**, 5561.
63 G. Audran, P. Bremond, S. R. A. Marque and G. Obame, *ChemPhysChem*, 2012, **13**, 3542.
64 M. J. Rose, D. E. Bellone, A. J. Di Bilio and H. B. Gray, *Dalton Trans.*, 2012, **41**, 11788.
65 N. Piergies, E. Proniewicz, A. Kudelski, A. Rydzewska, Y. Kim, M. Andrzejak and L. M. Proniewicz, *J. Phys. Chem. A*, 2012, **116**, 10004.
66 O. Unsalan, B. Szolnoki, A. Toldy and G. Marosi, *Spectrochim. Acta, Part A*, 2012, **98**, 110.
67 A. R. Gillens and B. A. Powell, *J. Radioanal. Nucl. Chem.*, 2013, **296**, 859.
68 C. W. Dilbeck and B. J. Finlayson-Pitts, *Phys.Chem. Chem. Phys.*, 2013, **15**, 1990.
69 J. C. S. Costa, R. A. Ando, A. C. Sant'Anab and P. Corio, *Phys.Chem. Chem. Phys.*, 2012, **14**, 15645.
70 F. Schinle, P. E. Crider, M. Vonderach, P. Weis, O. Hampe and M. M. Kappes, *Phys.Chem. Chem. Phys.*, 2013, **15**, 6640.
71 H. Chen, W. Delaunay, J. Li, Z. Wang, P.-A. Bouit, D. Tondelier, B. Geffroy, F. Mathey, Z. Duan, R. Reau and M. Hissler, *Org. Lett.*, 2013, **15**, 330.
72 G.-P. Ge, C.-Y. Li and H.-Q. Guo, *Fuguang Xuebao*, 2012, **33**, 591 (*Chem. Abs.*, 2013, **158**, 158759).
73 J. Motoyoshiya, *Yuki Gosei Kagaku Kyokaishi*, 2012, **70**, 1018 (*Chem. Abs.*, 2013, **158**, 36983).
74 R. Fu, S. Hu and X. Wu, *CrystEngComm*, 2013, **15**, 802.
75 S.-F. Tang, X.-B. Pan, X.-X. Lv and X.-B. Zhao, *J. Solid State Chem.*, 2013, **197**, 139.
76 J. Ma, Q. Xiao, S. Huang, J. Cui, W. Su and F. Luo, *Faming Zhuanli Shenqing CN* 102,849,690, 2011 (*Chem. Abs.*, 2013, **158**, 203920)
77 Q. Xiao, S. Huang, J. Ma, J. Cui, W. Su and J. Chen, *Faming Zhuanli Shenqing CN* 102,849,691, 2011 (*Chem. Abs.*, 2013, **158**, 203921)
78 T. Amako, T. Kimoto, N. Tajima, M. Fujiki and Y. Imai, *Tetrahedron*, 2013, **69**, 2753.
79 F. Dielmann, O. Back, M. Henry-Ellinger, P. Jerabek, G. Frenking and G. Bertrand, *Science*, 2012, **337**, 1526.
80 A. R. Head, D. S. Amenta, J. W. Gilje and G. P. A. Yap, *Global J. Inorg. Chem*, 2011, **2**, 1.
81 N. F. Stuurman, A. Muller and J. Conradie, *Inorg. Chim. Acta*, 2013, **395**, 237.
82 K. Slepokura, *Carbohydr. Res.*, 2013, **368**, 96.
83 R. A. Burrow and R. M. Siqueira da Silva, *Acta Crystallogr., Sect. E: Struct. Rep.*, 2012, **68**, 3488.
84 K. Thiele, C. Wagner and K. Merzweiler, *Acta Crystallogr., Sect. E: Struct. Rep.*, 2012, **68**, 263.
85 M. Lashkari, N. Hazeri, M. T. Maghsoodlou, S. M. Habibi-Khorassani, N. A. Torbati, A. Hosseinian, S. Garcia-Granda and L. Torre-Fernandez, *Heteroat. Chem.*, 2013, **24**, 58.
86 Y. Liang, H.-W. He, H.-F. He and Z. Yang, *Youji Huaxue*, 2012, **32**, 1513 (*Chem. Abs.*, 2013, **158**, 158492).
87 C. Li, C.-Q. Jiao, Z.-G. Sun, K. Chen, C.-L. Wang, Y.-Y. Zhu, J. Zhu, Y. Zhao, M.-J. Zheng, S.-H. Sun, W. Chu and H. Tian, *CrystEngComm*, 2012, **14**, 5479.
88 M. Pourayoubi, A. Tarahhomi, F. Karimi Ahmadabad, K. Fejfarova, A. van der Lee and M. Dusek, *Acta Crystallogr., Sect. C: Cryst. Struct. Commun*, 2012, **68**, 164.

89   C. D. Montgomery, R. T. Boere, A. Dawn and B. J. Jelier, *J. Chem. Crystallogr.*, 2013, **43**, 127.
90   T. Benincori, V. Bonometti, R. Cirilli, P. R. Mussini, A. Marchesi, M. Pierini, T. Pilati, S. Rizzo and F. Sannicolo, *Chem. Eur. J*, 2013, **19**, 165.
91   V. Lyaskovskyy, R. J. A. van Dijk-Moes, S. Burck, W. I. Dzik, M. Lutz, A. W. Ehlers, J. C. Slootweg, B. de Bruin and K. Lammertsma, *Organometallics*, 2013, **32**, 363.
92   U.-Y. Lee, Y.-J. Kim, H.-H. Choi, S.-G. Choi and H.-J. Yoon, *Repub. Korean Kongkae Taeho Kongbo KR 2012* 76,852 (*Chem. Abs.*, 2012, **157**, 281297)
93   B.-I. Seo, S.-G. Choi, J.-O. Lee, Y.-S. Jung and J.-H. Kwak, *Repub. Korean Kongkae Taeho Kongbo KR 2012* 102,902 (*Chem. Abs.*, 2012, **157**, 565241)
94   W. Li, X.-Y. Wen, S.-M. Li, X. Wang, J.-Z. Wang and H. Tang, *Adv. Mater. Res*, 2012, **542-3**, 959.
95   F. Di Pietrantonioa, M. Benetti, D. Cannata, E. Verona, A. Palla-Papavlu, V. Dinca, M. Dinescu, T. Mattle and T. Lippert, *Sens. Actuators, B*, 2012, **174**, 158.
96   K. Haav, J. Saame, A. Kuett and I. Leito, *Eur. J. Org. Chem.*, 2012, **11**, 2167.
97   T. Hoffmann, P. Friedel, C. Harnisch, L. Haeussler and D. Pospiech, *J. Anal. Appl. Pyrolysis*, 2012, **96**, 43.
98   E. Drabik, G. Krasinski, M. Cypryk, R. Błaszczyk, T. Gajda and M. Sochacki, *J. Am. Soc. Mass Spectrom.*, 2013, **24**, 388.
99   X.-G. Lv and J. Xu, *Phosphorus, Sulfur, Silicon Relat. Elem.*, 2012, **187**, 1151.
100 H. Saeidian, D. Ashrafi, M. Sarabadani, M. T. Naseri and M. Babri, *Int. J. Mass Spectrom.*, 2012, **319**, 9.
101 R. P. W. Kong, Q. Quan, Q. Hao, C.-K. Lai, C.-K. Siu and I. K. Chu, *J. Am. Soc. Mass Spectrom.*, 2012, **23**, 2094.
102 C.-Y. Wang, T.-T. Xie, L.-L. Xiao, W.-Y. Zhang, B.-H. Jin, C.-M. Liu and L.-X. Li, *Fenxi Shiyanshi*, 2011, **30**, 38 (*Chem. Abs.*, 2012, **157**, 398818).
103 C.-Y. Wang, Y.-Q. Gong, T.-T. Xie, W.-Y. Zhang, Y.-L. Shen, E.-S. Zhang, C.-M. Liu and L.-X. Li, *Fenxi Kexue Xuebao*, 2012, **28**, 652 (*Chem. Abs.*, 2013, **158**, 232436).
104 G. Wu, S.-H. Zhao, J.-J. Wu, F.-L. Guo, L.-J. Wang, T. Liu and M.-Y. Zhang, *Faming Zhuanli Shenqing CN* 102,662,020, 2012 (*Chem. Abs.*, 2012, **157**, 467414)
105 T. Roth, R. Urpi Bertran, M. Poehlein, M. Wolf and R. van Eldik, *J. Chromatogr. A*, 2012, **1262**, 188.
106 B. M. Weber and J. J. Harynuk, *J. Chromatogr. A*, 2013, **1271**, 170.
107 P. Velavendan, S. Ganesh, N. K. Pandey, U. Kamachi Mudali and R. Natarajan, *Desalin. Water Treat.*, 2012, **49**, 123.
108 F. Maier and O. Trapp, *Angew. Chem., Int. Ed.*, 2012, **51**, 2985.
109 X. Huang, Q.-H. Sheng, Z.-G. Zhang and Y.-F. Sun, *Shipin Gongye Keji*, 2012, **33**, 96 (*Chem. Abs.*, 2013, **158**, 65043).
110 S. K. Wiedmer, A. W. T. King and M.-L. Riekkola, *J. Chromatogr. A*, 2012, **1253**, 171.
111 J. G. Dawber, R. G. Skerratt, J. C. Tebby and A. A. C. Waite, *Phosphorus, Sulfur, Silicon Relat. Elem.*, 2012, **187**, 1261.
112 M. Zakarianezhad, S. M. Habibi-Khorassani, M. T. Maghsoodlou and B. Makiabadi, *Orient. J. Chem*, 2012, **28**, 1259.
113 Y. S. Sadovskii, T. N. Solomoichenko, T. M. Prokopeva, Z. P. Piskunova, N. G. Razumova, B. V. Panchenko and A. F. Popov, *Theor. Exp. Chem*, 2012, **48**, 163.
114 B. Bairagi and S. A. Bhoite, *J. Chem. Pharm. Res.*, 2012, **4**, 2312.
115 S. Cai, F. Feng, L. Zou, C. Huang and J.-Q. Xie, *Prog. React. Kinet. Mech.*, 2012, **37**, 398.